Dendrites

Kazuo Emoto • Rachel Wong • Eric Huang •
Casper Hoogenraad
Editors

Dendrites

Development and Disease

Springer

Editors
Kazuo Emoto
Graduate School of Science
The University of Tokyo
Tokyo, Japan

Rachel Wong
Department of Biological Structure
University of Washington
Seattle, Washington
USA

Eric Huang
Department of Pathology
University of California San Francisco
San Francisco, California
USA

Casper Hoogenraad
Faculty of Science
Utrecht University
Utrecht, The Netherlands

ISBN 978-4-431-56048-7 ISBN 978-4-431-56050-0 (eBook)
DOI 10.1007/978-4-431-56050-0

Library of Congress Control Number: 2016942328

Printed on acid-free paper

This Springer imprint is published by Springer Nature
The registered company is Springer Japan KK

Contents

Part I
Introduction

Chapter 1
Introduction

Kazuo Emoto, Rachel Wong, Eric Huang, and Casper Hoogenraad

Abstract Dendrites are cellular structures essential for the integration of neuronal information. The immense variety but stereotypic architecture of dendritic arbors across the nervous system, have long suggested that dendritic structure and function are organized to meet the processing needs unique to each circuit. Technological advances have greatly pushed the frontiers in research on dendrites at macroscopic and microscopic levels. As such, there has been increasing efforts and knowledge gained in elucidating the structural, functional and molecular mechanisms that regulate the development of dendritic arbors, and that maintain their form and function throughout life. Moreover, as we seek to repair the damaged nervous system, it is clear that a better understanding of how dendrites are perturbed in neurodegenerative disease is needed. In this book, we introduce the basic biology of dendrites, and discuss current knowledge of the mechanisms that control cellular, molecular and functional aspects of dendritic development and maintenance in health and in disease.

Dendrites are the essential substrate for receiving and integrating neuronal information. These elegant but complex structures have been studied for more than a century since their visualization using cellular staining methods devised in the late nineteenth century by Camillo Golgi (Fig. 1.1a; Golgi 1873). Santiago Ramon y Cajal's (1899–1904) initial and insightful definition of the role of dendrites in neuronal processing was borne out by electrophysiological investigations that became feasible in the mid-1900s. Since then, an entire field has evolved – the study of dendritic structure

K. Emoto (✉)
Department of Biological Sciences, Graduate School of Science, The University of Tokyo, Tokyo, Japan
e-mail: emoto@bs.s.u-tokyo.ac.jp

R. Wong
Department of Biological Structure, University of Washington, Seattle, WA, USA

E. Huang
Department of Pathology, University of California San Francisco, San Francisco, CA, USA

C. Hoogenraad
Cell Biology, Faculty of Science, Utrecht University, Utrecht, The Netherlands

K. Emoto et al. (eds.), *Dendrites*, DOI 10.1007/978-4-431-56050-0_1

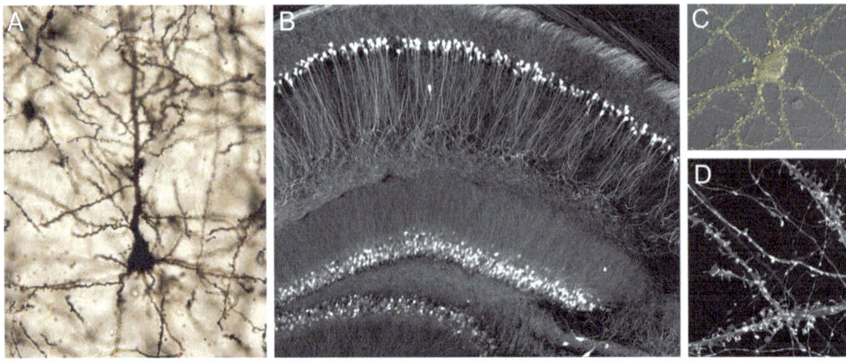

Fig. 1.1 (**a**) Cortical neuron labeled by the Golgi method (provided by E. Huang). (**b**) Dendritic arbors of adult hippocampal neurons in transgenic mice expressing yellow fluorescent protein. Confocal microscopy reconstruction from a brain slice (provided by H. Okawa, K. Oda, and R. Wong). (**c**) An image of a hippocampal neuron in culture immunostained with antibodies against two postsynaptic scaffolding proteins, PSD-95 (*green*) and Homer (*red*) (provided by S. Okabe). Fluorescence images were overlaid with a DIC (differential interference contrast) image. (**d**) Dendritic spines of hippocampal neurons in culture expressing myristoylated alanine-rich protein kinase C substrate (1–40) tagged with GFP (Marcks-GFP) to highlight the plasma membrane (provided by C. Hoogenraad). *Source*: A cover picture of the journal (Microscopy vol 62, no 1, 2013)

and function is now one of the leading areas of investigation in basic and translational neuroscience research. In this book, we wish to bring together current knowledge of how dendrites gain their unique and necessary cellular architecture and connectivity during development, as well as highlight how a failure to attain or maintain these attributes result in devastating disease. It is our hope that the reader will find new comprehension and appreciation for the many strategies that neurons engage to assemble and maintain their dendritic structure and function, and thus the workings of the nervous system, throughout life.

Dendritic arbors are highly patterned across the nervous system, but vary tremendously in their size and fine architecture, each designed to best serve specific computations within their networks. Despite the immense diversity in architecture, dendritic arbors of neurons from each region of the nervous system are stereotypic in their branching patterns and synaptic organization. Such stereotypic patterns make dendrites highly amenable for studies across animals, as well as in tissue explants and in dissociated cell culture in which major features of dendrites are preserved (Fig. 1.1b–d). Both in vivo and in vitro studies have greatly illuminated the cytoskeletal organization and cell biological processes that dendrites depend on in order to perform properly. Also, much is now known about the cellular interactions and molecules that pattern individual dendritic arbors as well as regulate their spatial arrangements across their populations. Importantly, in vivo and in vitro approaches have in parallel greatly increased our understanding of the structure and function of synapses onto dendrites, as well as the computations that result from these inputs. Proper development of dendrites is thus an essential step in the establishment of neuronal connectivity and function. Equally important is the maintenance of dendrites and their connectivity, a process that is seriously disrupted in neurological disease.

By comparing dendritic morphologies across ages, Cajal was able to describe the successive stages in the development of dendrites and their connectivity. Over the past several decades, many molecular mechanisms that control dendrite development and maintenance have been revealed, largely due to technical advances in cell and molecular biology. High-resolution imaging of living neurons expressing various fluorescent protein-tagged reporters, together with calcium imaging and electrophysiological recording, has greatly facilitated simultaneous real-time monitoring of morphological and functional changes in dendrites in response to physiological stimuli. Methods to introduce or manipulate genes of interest in specific neuronal cell types or entire networks of cells are also improving continuously. These methods are also being applied to elucidate the changes in dendrites in many neurological diseases, either as a consequence of developmental defects or due to the failure in maintaining the circuits that have formed. The chapters in this book aim to relay how such diverse approaches have been applied to generate a comprehensive understanding of the cellular and molecular mechanisms responsible for shaping the form and function of dendrites.

The authors of this book describe recent work that we have grouped into five major themes, which collectively cover a wide array of topics that we hope will provide the reader with a broad, but sufficiently detailed, view of dendrite development and disruption in disease. The chapters will highlight work across species including flies, frogs, and rodents. In Chaps. 2, 3, 4, 5, 6, and 7, we focus on several major aspects of the "cell biology" of dendrites, including their cytoskeletal organization, the trafficking of proteins, and the signals that convey input at dendrites to the cell nucleus to generate longer-term regulation of neuronal function (Fig. 1.2, Cell biology). The next theme (Chaps. 8, 9, 10, and 11) directs the reader's attention to the stereotypic architectural arrangements of dendritic arbors, which are designed to execute specific functions of the respective cell types. In particular, the authors underscore the basic rules that govern dendrite branching patterns and review how specific branching patterns relate to function (Fig. 1.2, Patterning). Then, in Chaps. 12, 13, and 14, we lead the reader to current findings unraveling the cellular and molecular control of dendrite morphology. Chapters 15, 16, 17, 18, and 19 discuss many aspects of the development and maturation of synaptic connections onto dendrites (Fig. 1.2, Connectivity). This block of chapters emphasizes a variety of molecular, imaging, and electrophysiological techniques that are used to unravel the mechanisms that govern synapse assembly on dendrites. Finally, in Chaps. 20, 21, and 22, we consider dendritic organization in common neurological diseases: autism, amyotrophic lateral sclerosis, frontotemporal dementia, and glaucoma. The discussions in these disease models hopefully will lead to more discoveries in the future of the critical role of dendritic growth and maintenance in neurodevelopmental and neurodegenerative diseases.

We thank the authors for their dedication and excellent contributions to the book, without which this project would not have been possible. We hope that the book will provide novel information to students and established investigators alike,

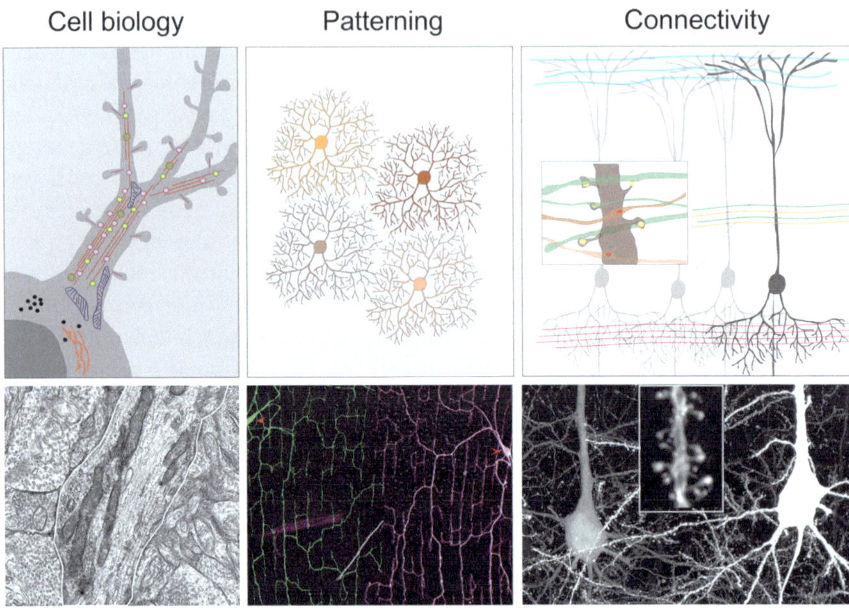

Fig. 1.2 Schematics and an example showing the major organizational plans of dendrites covered in the chapters of this book: cell biological organization and processes, patterning of dendritic arbors, and global and local arrangements of synaptic connectivity. *Cell biology*: The cytoskeleton and trafficking of proteins to and away from the cell body are essential for dendrite maintenance and function. Image: Transmission electron microscopy showing ultrastructure of a mouse retinal ganglion cell dendrite (outlined; provided by A. Bleckert and R. Wong). *Patterning*: Dendrites of individual neurons do not cross each other (self-avoidance), and in some systems, dendritic arbors of cell-like types tile. Image: *Drosophila* sensory neurons labeled genetically, showing "self-avoidance" and tiling (provided by K. Emoto). *Connectivity*: Dendritic arbors are contacted by many types of inputs (colorized differently), which are distributed in stereotypic patterns across the cell and also at the level of individual synapses. Inset shows excitatory connections in green/yellow on spiny protrusions on the dendrites and inhibitory connections in red on dendritic "shafts," smooth portions of dendrites. Image: Dendrites of fluorescent protein expressing cortical neurons from a transgenic line; inset shows spines (provided by H. Okawa, K. Oda, and R. Wong)

as well as to clinicians interested in the basic pathology of dendrite degeneration. We also hope that the many concepts and ideas presented in the book will contribute toward escalating research on this crucial compartment of the neuron.

References

Cajal SR (1899–1904) Textura del systema nervioso del hombre y de los vertebrados. Moya, Madrid

Golgi C (1873) Sulla struttura della sostanza grigia del cervello. Gaz Med Lomb 6:244–246

Part II
Basic Biology of Dendrites

Chapter 2
Cytoskeletal Organization: Actin

Amr Abou Elezz, Rimante Minkeviciene, and Pirta Hotulainen

Abstract The regulation of cytoskeletal organization is key to the establishment of proper neuronal morphology. The neuronal cytoskeleton comprises three different elements: microtubules, intermediate filaments, and actin filaments. Actin cytoskeleton dynamics play a key role in the establishment and maintenance of both dendritic arborization and spines, as well as in spine shape changes associated with synaptic plasticity. Not surprisingly, abnormalities in dendritic arborization and dendritic spine morphology arising from the faulty regulation of the cytoskeleton have been linked to several neurological diseases. In this review, we describe the basic mechanisms regulating the actin cytoskeleton in spinogenesis and discuss examples of mutations in actin regulators associated with neurological disorders.

Keywords Actin cytoskeleton • Dendrites • Dendritic spines • Actin-binding proteins • Rho GTPases • Neurological diseases

Neurons are the excitable cells responsible for processing and transmitting information in the nervous system. The nature of a neuron as a computational unit is partly the result of a complex specialized morphology. A neuron typically extends a long thin axon to transmit information to target cells and several shorter dendrites that receive input from other cells through specialized connections known as synapses. Input received through synapses is integrated in dendrites and the cell body and may give rise to an output in the form of an action potential that is transmitted downstream through the axon. This largely unidirectional flow of information – first postulated by Ramon y Cajal (1999) – is made possible through neuronal polarity. The majority of excitatory synapses in the central nervous system exist on small bulbous structures on dendrites known as dendritic spines. Dendritic spines are found in varying sizes and shapes (Matus 2000) but mostly have a spherical head and a thin narrow neck (Harris and Stevens 1989; Noguchi et al. 2005). Dendritic spines hold the postsynaptic machinery (Sheng and Hoogenraad 2007) and provide an electrical and biochemical compartment

---

A. Abou Elezz • R. Minkeviciene • P. Hotulainen (✉)
Neuroscience Center, University of Helsinki, Viikinkaari 4, 00014 Helsinki, Finland
e-mail: pirta.hotulainen@helsinki.fi

K. Emoto et al. (eds.), *Dendrites*, DOI 10.1007/978-4-431-56050-0_2

(Nimchinsky et al. 2002; Newpher and Ehlers 2009). Spines offer means to regulate synaptic signaling on a synapse-by-synapse basis. It is believed that information in the brain can be stored through modulating the strength of individual synapses (Kasai et al. 2010) and changes in dendritic spine morphology are associated with synaptic function (Yuste and Bonhoeffer 2001).

The complex morphology of different neuronal cell types emphasizes the importance of the cytoskeleton (Fig. 2.1). The regulation of cytoskeletal organization is key to the establishment and maintenance of neuronal polarity (Tahirovic and Bradke 2009; Stiess and Bradke 2011). The neuronal cytoskeleton comprises three different elements: microtubules, intermediate filaments, and actin filaments. During early neuronal development, the actin cytoskeleton is the major structural component of growth cones. Growth cones are protrusive tips of extending dendrites or axons. The actin cytoskeleton provides the force to push the plasma membrane of growth cones forward. In more mature dendrites, filamentous actin is mainly located in dendritic spines where it plays a crucial role in spinogenesis and changes in spine morphology (Hotulainen and Hoogenraad 2010). Furthermore, the actin cytoskeleton is also a common substrate for several plastic processes in the brain (Luo 2002). Not surprisingly, abnormalities in dendritic and dendritic spine morphology arising from faulty regulation of the cytoskeleton have been linked to several neuropsychiatric diseases (Penzes et al. 2011; Kulkarni and Firestein 2012).

The different structural and biochemical properties of microtubules, intermediate filaments, and actin filaments make each of them suitable for fulfilling specific functions in a neuron. Actin filaments are filamentous polymers of actin.

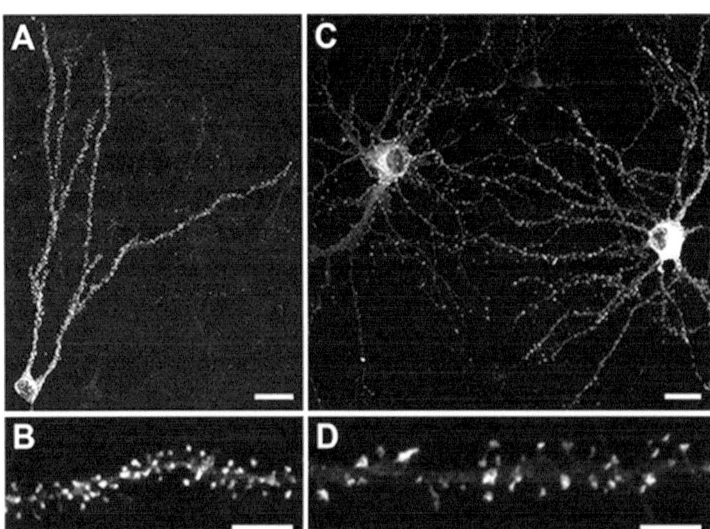

Fig. 2.1 GFP-actin-transfected neurons. (**a**, **b**) Granule cell of the dentate gyrus (P16 in vivo). (**c**, **d**) Hippocampal neurons in culture (15 days in vitro). *Bars*: *top* 20 μm, *bottom* 5 μm

Filamentous actin (F-actin) is highly dynamic and under the tight control of a number of regulatory elements and cascades that function in a temporally and spatially controlled manner. In this chapter we will look at the structure and the regulatory mechanisms governing the actin filament component of the cytoskeleton in mature dendrites, as well as neurological diseases associated with dysregulation of the actin cytoskeleton.

2.1 The Actin Cytoskeleton in the Development and Maintenance of Dendrites

2.1.1 Introduction to Actin

The monomeric form of actin is a globular 43 kDa protein capable of binding an adenosine triphosphate (ATP) nucleotide. Actin monomers (G-actin) polymerize to form polarized, right-handed, double-helical filamentous actin (F-actin) (Amann and Pollard 2000). The ends of the polar actin filaments are named the "barbed" and "pointed" ends. In general, in a cellular context there is a net addition of monomers at the barbed end and net dissociation at the pointed end, leading to filament elongation occurring mainly at the barbed end. The structural and biochemical properties of actin filaments enable them to function in providing structural support, in generating force for cell shape changes and motility, and in acting as tracks for a number of motor proteins. The constant incorporation and dissociation of monomers at filament ends results in a "treadmilling" effect. This dynamic nature of actin filaments makes it possible to quickly remodel the actin cytoskeleton in response to cellular cues. In addition, actin filaments can be organized to build structures of different architectures and mechanical properties such as parallel bundles or branched networks (Blanchoin et al. 2014). The structure and dynamics of the actin cytoskeleton are under tight control of regulatory machineries that operate in a temporally and spatially controlled manner (Michelot and Drubin 2011).

The de novo formation of an actin filament from free G-actin monomers is an energetically unfavorable process that is not likely to spontaneously happen in a cellular environment (Goley and Welch 2006). As a result, the de novo assembly of actin filaments requires the activity of an actin nucleator. Several classes of actin nucleators have been identified, of which formins and the actin-related protein 2/3 (Arp2/3) complex are the best-studied. The identity of the actin nucleator strongly influences the structure and properties of the nascent filament; Arp2/3-mediated nucleation leads to the formation of branched filaments or networks of filaments, whereas formin-mediated nucleation gives rise to straight unbranched filaments that can form bundles (Chesarone et al. 2010; Campellone and Welch 2010). Actin nucleators are differentially distributed in the cell, and their activity is triggered and can be modulated by the action of several regulatory proteins. This enables the formation of actin filaments of various structures and biochemical properties in a tightly controlled manner (Michelot and Drubin 2011; Bovellan et al. 2014).

Following the formation of a stable nucleus, ATP-bound G-actin monomers are incorporated into the nascent actin filament in a rapid and spontaneous process. The rate of polymerization is partly a function of G-actin concentration. The simultaneous polymerization and depolymerization of actin monomers gives rise to a state of equilibrium at which there is no net change in filament length. The concentration of G-actin at which this equilibrium is reached is known as the *critical concentration* (Karp 2007). The critical concentration is higher at the pointed ends than at the barbed ends, partly explaining why polymerization mainly takes place at the barbed end (Wegner and Isenberg 1983). At equilibrium, the net polymerization at the barbed end and the net depolymerization at the pointed end give rise to the "treadmilling effect." ATP-bound G-actin monomers have a higher affinity for the barbed ends of filaments compared to adenosine diphosphate (ADP)-bound monomers. However, after incorporation into an actin filament, ATP is hydrolyzed to ADP, leading to an abundance of ADP-bound monomers toward the pointed end. The phosphorylation state of the actin-bound nucleotide affects the efficiency of interactions with regulatory actin-binding proteins (ABPs) (Fechheimer and Zigmond 1993).

The regulation of the structure and dynamics of the actin cytoskeleton occurs through the concerted action of a number of ABPs. Actin dynamics can be regulated through controlling the de facto concentration of actin monomers available for polymerization. Thymosin β_4, for example, inhibits polymerization through binding to G-actin (Sanders et al. 1992). Conversely, profilins enhance polymerization through binding monomers and facilitating their incorporation onto the barbed ends of actin filaments (Pring et al. 1992). ABPs also regulate actin by directly binding to actin filaments; polymerization can be inhibited by the action of one of several capping proteins that bind to the barbed ends (Cooper and Schafer 2000; Menna et al. 2011); proteins of the actin depolymerizing factor (ADF)/cofilin family sever actin filaments and enhance depolymerization at the pointed ends (Lappalainen and Drubin 1997; Hotulainen et al. 2005; Andrianantoandro and Pollard 2006; Jansen et al. 2014); gelsolin also severs actin filaments and caps the barbed ends of the severed fragments (Sun et al. 1999); tropomodulins cap pointed ends and inhibit both polymerization and depolymerization (Yamashiro et al. 2012).

The action of ABPs is orchestrated and regulated by several molecular pathways and signaling cascades. Rho family GTPases act as messengers that can trigger actin regulatory pathways in response to intra- or extracellular signals (Lee and Dominguez 2010). The Rho family consists of many different Rho GTPases. The most extensively studied Rho GTPases are RhoA, Rac1, and Cdc42. Rho GTPase activity is regulated by its binding to either guanosine triphosphate (GTP) (active state) or guanosine diphosphate (GDP) (inactive state). Exchange between GDP and GTP is regulated by guanine nucleotide exchange factors (GEFs) and GTPase activating factors (GAPs). GEFs activate and GAPs inactivate Rho GTPases. Rho GTPase activation leads to the activation or inactivation of specific effectors.

Effectors of RhoA, Rac1, and Cdc42 are often ABPs or proteins regulating the activity or localization of ABPs.

In addition to Rho GTPase-mediated actin regulation, proteins belonging to the tropomyosin family interact with actin filaments and recruit specific ABPs and inhibit the activity of others in an isoform-specific manner (Vindin and Gunning 2013). Furthermore, several ABPs are modular and can have multiple effects on actin structure or dynamics or both; besides severing filaments and capping barbed ends, gelsolin contributes to the regulation of the G-actin pool by binding free G-actin (Khaitlina et al. 2004); cofilin binds to and inhibits nucleotide exchange on monomeric ADP-actin (Paavilainen et al. 2004); tropomodulin-3 also binds monomeric G-actin (Fischer et al. 2006). Thus, the multiplicity of regulatory elements and pathways enables precise differential spatial and temporal regulation of actin structure and dynamics within the cell. Vertebrates have about 150 ABPs, all of which probably contribute to the regulation of the neuronal actin cytoskeleton.

2.1.2 The Dendritic Cytoskeleton: Actin

The dendritic cytoskeleton plays an important role in the development and maintenance of dendrites (Fig. 2.1). Actin filaments are enriched in dendritic spines, whereas the shaft is dominated by microtubules (Matus et al. 1982; Peters et al. 1991; Scott and Luo 2001; Squire et al. 2012). Dendritic shaft actin is relatively difficult to detect except for small areas of local enrichment of F-actin on the dendritic shaft membrane. These actin patches are associated with dendritic microtubules and constitute points of future filopodia outgrowth (Korobova and Svitkina 2010; Willig and Barrantes 2014) (Fig. 2.1b, d). Dendritic spines have a high concentration of F-actin, up to 18 times that of the shafts (Capani et al. 2001; Honkura et al. 2008) (see Fig. 2.1d). Electron micrographs showed that spine heads predominantly contain branched actin networks, while necks contain networks of branched as well as straight filaments (Korobova and Svitkina 2010). Honkura et al. (2008) detected three populations of F-actin in dendritic spines that showed different dynamic properties and are necessarily differently regulated. Of these they found the pool of actin close to the tip of the spine capable of modulating its treadmilling rate as to rapidly exert an expansive force that causes the enlargement of the spine head in response to specific patterns of synaptic activity (Honkura et al. 2008).

Initiation of dendrite outgrowth is believed to be preceded by a local destabilization of cortical actin that leads to the formation of a filopodia-like extension which is then invaded by microtubules (Georges et al. 2008). This is similar to the model proposed for axon outgrowth (Yu et al. 1994). Indeed, live imaging of developing hippocampal pyramidal neurons revealed that filopodia are constantly extended and retracted from dendritic shafts but are occasionally sustained and develop into a dendritic branch (Dailey and Smith 1996). Rho family GTPases play a crucial role in orchestrating the modifications in the actin cytoskeleton necessary for dendrite morphogenesis (Threadgill et al. 1997; Nakayama et al. 2000; Georges et al. 2008; Stankiewicz and Linseman 2014).

As the neuron develops, thin motile dendritic filopodia are replaced by more stable spines (Dailey and Smith 1996; Schachtele et al. 2011). A better understanding of the process of filopodia initiation is beginning to materialize owing to recent studies focusing on the mechanisms involved. BIN/Amphiphysin/Rvs (BAR) domain superfamily proteins have been shown to be important for proper neuronal morphogenesis (Qualmann et al. 2011). These proteins can either generate positive membrane curvature to facilitate the formation of plasma membrane invaginations (e.g., BAR, N-BAR, and F-BAR domain proteins) or induce negative membrane curvature to promote the formation of plasma membrane protrusions (I-BAR and IF-BAR domain proteins) (Suetsugu et al. 2006; Mattila et al. 2007; Guerrier et al. 2009). Data collected recently by ourselves and others strongly suggest that I-BAR/IF-BAR domain proteins play a major role in the initiation of dendritic filopodia (Carlson et al. 2011; Saarikangas et al. 2015) (Fig. 2.2). MIM (missing in metastasis/MTSS1) is an actin and membrane-binding protein that comprises an actin-monomer-binding WASP homology 2 (WH2) domain and an I-BAR domain (Mattila et al. 2003). The I-BAR domain binds to membranes with an abundance in phosphatidylinositol (4,5)-bisphosphate ($PI(4,5)P_2$) and induces the formation of a tubular structure (Mattila et al. 2007). Slit-Robo GAP 3 (srGAP3), also known as mental disorder-associated GAP protein (MEGAP) or WAVE-associated GAP (WRP), consists of an N-terminal IF-BAR domain, a central Rac1 GAP domain, and a C-terminal Src-homology 3 (SH3) domain (Wong et al. 2001; Endris et al. 2002; Soderling et al. 2002; Guerrier et al. 2009). Similar to MTSS1/MIM, the IF-BAR domain of WRP/srGAP3 binds specifically to a number of lipids which include phosphatidic acid, $PI(4,5)P_2$, and phosphatidylinositol (3,4,5)-triphosphate ($PI(3,4,5)P_3$) (Carlson et al. 2011). We have observed that $PI(4,5)P_2$ recruits MIM to dendritic membranes where it clusters and forms buds that give rise to filopodia. Arp2/3 complex-based actin polymerization follows MIM clustering. Actin polymerization is required to start the elongation of newly formed filopodia (Saarikangas et al. 2015). However, the overexpression of MIM resulted in an increased density of dendritic protrusions but inhibited spine head formation. Vice versa, knockdown of MIM reduced spine density and increased spine head size (Saarikangas et al. 2015). Similarly, binding of the srGAP3 IF-BAR domain to the plasma membrane induces filopodia formation in neurons (Carlson et al. 2011; Coutinho-Budd et al. 2012). Knockdown of srGAP3 reduced spine density, especially during early development, when the rate of filopodia initiation is at its maximum (Carlson et al. 2011).

The subsequent elongation of filopodia is driven by actin polymerization. The Rho GTPase Rif was shown to induce the elongation of filopodia through activating the formin mDia2 (Hotulainen et al. 2009). Formin nucleation leads to the rapid polymerization of straight filaments (Fig. 2.2). It is possible that formins other than mDia2 are also involved in filopodia elongation. Ena/VASP has also been suggested as an alternative elongation factor (Yang and Svitkina 2011).

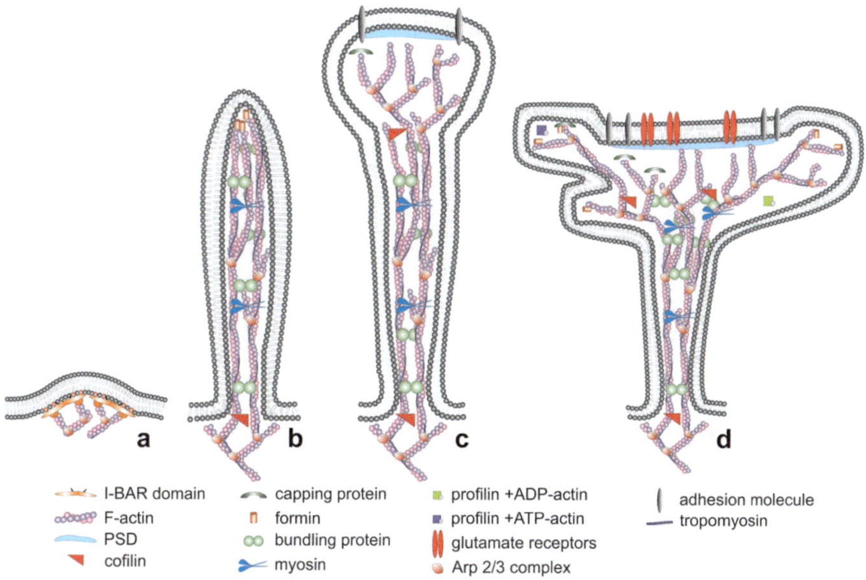

Fig. 2.2 Working model of molecular mechanisms underlying spine initiation, elongation, and stabilization. (**a**) *Initiation*: I-BAR or IF-BAR domain-containing proteins facilitate membrane curving inducing the initiation of filopodia. Arp2/3 complex-based actin polymerization is required for the initial elongation of filopodia. (**b**) *Elongation*: Formins elongate the actin filaments at barbed ends resulting in straight actin filaments and elongation of filopodia. (**c**) Simultaneous with synapse formation (postsynaptic density (PSD)), the spine head starts to grow through Arp2/3 complex-based actin polymerization resulting in a branched actin filament network. Cofilins and capping proteins control the proper length of actin filaments and prevent the formation of protrusions from spine heads. (**d**) *Growth and stabilization of the spine head*: After stable synapse formation, the spine head and PSD grow. PSD contains the adhesion molecules and glutamate receptors. Arp2/3 complex-based nucleation occurs close to the PSD, but the filaments' barbed ends are found in protrusions next to the PSD. At the barbed ends, formins and Ena/VASP facilitate filament elongation. Profilins help formins by changing the ADP to ATP in actin molecules and by delivering ATP-actin monomers to the barbed ends. Cofilins and capping proteins control the proper length of the filaments. Actin-bundling proteins cross-link actin filaments together forming a stable "core" for dendritic spines

Upon making contact with an axon, the highly dynamic filopodium stabilizes to form a spine. Drebrin was shown to accumulate in filopodia (Mizui et al. 2005), and this accumulation is necessary for the recruitment of the scaffold protein PSD-95 (Takahashi et al. 2003) which is followed by the recruitment of glutamate receptors. The subsequent formation of the spine head requires a shift from formin-mediated to Arp2/3 complex-mediated nucleation of actin filaments (Hotulainen et al. 2009) (Fig. 2.2). Beli et al. (2008) showed that WAVE binds to and inhibits mDia2 and simultaneously activates the Arp2/3 complex, which inhibits the elongation of filopodia and induces membrane ruffles in fibroblasts. A similar mechanism could drive the switch from filopodia elongation to spine head formation and enlargement. Alternatively, the transient invasion of dynamic microtubules into a spine can lead to Arp2/3 complex activation and morphological changes (Jaworski et al. 2009). Importantly, although Arp2/3 complex has been shown to be necessary for dendritic spine head formation and maintenance by us and others (Wegner et al. 2008; Hotulainen et al. 2009; Kim et al. 2013), it seems that various formins and Ena/VASP help in actin filament elongation in dendritic spine heads. This idea came from a recent study using super-resolution photoactivated localization microscopy (PALM) (Chazeau et al. 2014). Using high-resolution localization of different actin regulators, the authors showed that the Arp2/3 complex and its activators localized in the vicinity of the PSD. However, actin filament elongation occurred at the tips of fingerlike protrusions (Chazeau et al. 2014). They further showed that Ena/VASP and formin-like protein-2 localized to the tips of these protrusion (Fig. 2.2). In addition, capping proteins Eps8 and capping protein have been shown to play important roles in dendritic spine morphogenesis (Fan et al. 2011; Menna et al. 2013). Filament-severing proteins gelsolin (Star et al. 2002) and cofilin (Hotulainen et al. 2009) also have fundamental roles in maintaining the dynamic actin cytoskeleton. The processes of inhibiting polymerization, filament severing, and enhancing depolymerization are especially important in dendritic spines where the average length of the filaments is relatively short (200 nm) (Frost et al. 2010). Furthermore, myosin IIb enhances actin dynamics through its motor activity, possibly through breaking nonaligned actin filaments (Koskinen et al. 2014).

As the spine maturates, F-actin accumulates in the spine head and is stabilized (Koskinen et al. 2014). We have shown that stabilization depends on the cross-linking activity of myosin IIb but it is probable that stabilization is carried out by the action of various actin cross-linkers. Actin cross-linking seems to form a stable core at the base of the spine head. Other actin cross-linkers with positive roles in actin stabilization in dendritic spines are CaMKII (Okamoto et al. 2007) and drebrin (Ivanov et al. 2009; Mikati et al. 2013; Worth et al. 2013). At least drebrin partially stabilizes and protects actin filaments through cofilin inhibition (Shirao and Gonzalez-Billault 2013; Grintsevich and Reisler 2014). Furthermore, the actin cross-linking protein girdin could play a stabilizing role in dendritic spines (Enomoto et al. 2006, 2009; Nakai et al. 2014).

2.2 The Actin Cytoskeleton in Neurological Diseases

Disease-specific disruptions in dendritic spine shape, size, or number accompany a large number of brain disorders, suggesting that dendritic spines may serve as a common substrate for many psychiatric and neurological diseases, particularly those that involve deficits in information processing (Penzes et al. 2011). Actin cytoskeleton dynamics play a key role in the establishment and maintenance of both dendritic arborizations and spines, as well as in spine shape changes in synaptic plasticity. Thus, it is not surprising that actin regulating proteins are linked to many neurological diseases. In the following section, we describe examples of mutations in actin regulators associated with neurological diseases.

2.2.1 Rho GTPase Signaling in Neurological Diseases

Mutations in Rho family GTPase regulators in particular are frequently associated with neurological diseases (Newey et al. 2005; DeGeer and Lamarche-Vane 2013; Ba et al. 2013). Through the control of the actin cytoskeleton, Rho GTPases and their regulators and effectors are involved in the regulation of neuronal morphology, including migration, polarity, axon growth and guidance, dendrite elaboration and plasticity, and synapse formation. All these steps are essential for the maturation of neurons and for the correct establishment of a neuronal network.

Nonspecific X-linked mental retardation (MRX) is characterized solely by cognitive impairment (Chelly and Mandel 2001). As such, the genes associated with this disease are considered to be especially important for normal brain function rather than the homeostasis of the whole body. Among the genes associated with MRX, four encode proteins linked to Rho GTPase-dependent signaling pathways: oligophrenin-1 (*OPHN1*), a RhoGAP protein (Billuart et al. 1998); αPIX/Cool2 (*ARHGEF6*), a Rac/Cdc42 GEF (Kutsche et al. 2000); collybistin (*ARHGEF9*), a Cdc42 GEF (Reid et al. 1999; Xiang et al. 2006; Marco et al. 2009); and PAK3 (*p21 (CDKN1A)-activating kinase 3*), a Rac/Cdc42 downstream effector (Allen et al. 1998).

Oligophrenin-1 is expressed in the developing spinal cord and later in brain areas that are highly plastic. At the cellular level, oligophrenin-1 is expressed in both glial and neuronal cells where it co-localizes with actin, notably at the tips of growing neurites (Fauchereau et al. 2003). Oligophrenin-1 inhibits RhoA (Billuart et al. 1998).

αPIX has been localized to dendritic spines and is a Rac1- and Cdc42-specific GEF (Meyer 2014). Mutant mice deficient for αPIX display abnormal dendritic morphology, defects in synapse formation, and altered long-term potentiation as well as learning and behavioral deficits. These abnormalities are accompanied by a reduction in active Rac1 and Cdc42 (Ramakers et al. 2012). A mutation in the gene coding for collybistin is linked to MRX with severe mental retardation and clinical symptoms of hyperekplexia and epilepsy. Collybistin is a Cdc42 GEF (Reid et al. 1999). In addition, collybistin can modulate GABAergic neurotransmission through binding the scaffolding protein gephyrin (Marco et al. 2009; Ba et al. 2013).

The Rac1 and Cdc42 downstream signaling effector p21-activated kinase (PAK) activates LIM-kinase which phosphorylates – and thereby inactivates – cofilin-1. The inhibition of PAK results in an increased growth of new, unstable spines, occurring in clusters, and an impairment of the maturation and stabilization of existing spines (Dubos et al. 2012). *PAK3* mutations are sufficient to cause severe mental retardation, also consistent with a potential causal role of PAK defects in cognitive deficits in Alzheimer's disease (Zhao et al. 2006). PAK deficit leads to cofilin-1 activation. At least in Alzheimer's disease, cofilin-1 activation leads to the release of drebrin from actin filaments (Zhao et al. 2006). Altogether, this leads to unstable actin filaments and actin filament depolymerization, which leads to a smaller spine head size and less stable dendritic spines (Fig. 2.3a).

SrGAP3 is ubiquitously expressed in the developing nervous system (Bacon et al. 2009) and interacts with a number of proteins implicated in various aspects of nervous system development (Wong et al. 2001; Soderling et al. 2002; Endris et al. 2011; Bacon et al. 2011). The srGAP3-encoding gene was reported to be deleted in a patient with a severe form of mental retardation, the 3p syndrome (Endris et al. 2002; Shuib et al. 2009). Furthermore, an *srGAP3* copy number variant (CNV) has been identified in schizophrenia (Addington and Rapoport 2009; Wilson et al. 2011). In line with these discoveries, srGAP3 knockout mice suffered lethal hydrocephalus or "schizophrenia-related" behavior (Carlson et al. 2011; Kim et al. 2012; Waltereit et al. 2012). srGAP3 binds WAVE-1 and downregulates Rac1 signaling (Endris et al. 2002; Guerrier et al. 2009; Soderling et al. 2002). Knocking out srGAP3 decreases the number of dendritic filopodia in early development, and thus, it is plausible that srGAP3 deficiency leads to decreased dendritic spine density also in human patients (Carlson et al. 2011) (Fig. 2.3b).

Disrupted-in-schizophrenia 1 (*DISC1*) was originally discovered in a Scottish family in which a chromosomal translocation disrupts the structure of the gene (Millar et al. 2000). This translocation is significantly linked to a clinical phenotype that includes schizophrenia and affective disorders (St Clair et al. 1990; Blackwood et al. 2001). DISC1 plays a major role in early brain development by regulating a number of essential neurodevelopmental events, including cell proliferation, neurite outgrowth, synapse formation, and neuronal migration (Ishizuka et al. 2006; Brandon et al. 2009; Soares et al. 2011; Narayan et al. 2013; Thomson et al. 2013). DISC1 has numerous binding partners, making it a scaffold protein with impacts on many diverse brain functions. DISC1 is enriched in the postsynaptic density where it anchors kalirin-7 in a protein complex, among other proteins, and limits and controls kalirin-7 access to Rac1 in response to NMDA receptor activation (Hayashi-Takagi et al. 2010). As kalirin-7 is a Rac1 GEF, the lack of DISC1 results in a higher, unregulated Rac1 activity as its GEF is continuously "free." Interestingly, inhibition of the Rac1 effector PAK rescued the DISC1 knockdown phenotype, thus providing at least an idea for further drug development (Hayashi-Takagi et al. 2014).

In addition to Rac1 regulation, DISC1 interacts directly with the actin-binding protein girdin (Enomoto et al. 2006, 2009; Steinecke et al. 2014). DISC1 also co-localizes with F-actin itself within the tip of the leading process in migrating interneurons (Steinecke et al. 2014). Girdin cross-links actin filaments (Enomoto et al. 2006, 2009), and its activity is regulated by Akt (protein kinase B)-driven phosphorylation (Enomoto et al. 2005, 2009). Akt was also reported to interact with DISC1 (Steinecke et al. 2014). It was recently found that phosphorylation-deficient girdin knock-in mice, in which S1416 was replaced with alanine, exhibited

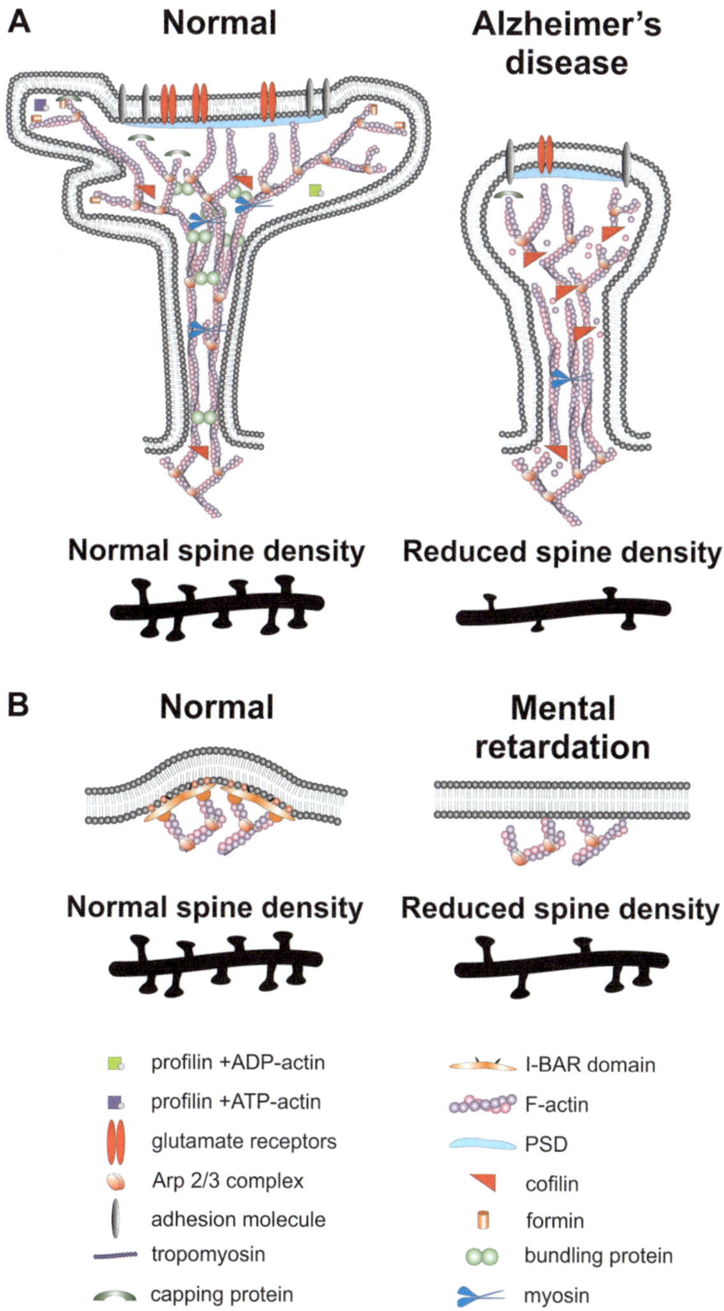

Fig. 2.3 Working models of molecular mechanisms underlying Alzheimer's disease and mental retardation. (**a**) *Healthy condition*: Actin-bundling protein drebrin stabilizes actin filaments, and the activity of cofilins is strictly regulated by phosphorylation. Spines with active synapses are stable. *Alzheimer's disease*: Decreased expression of PAK3 decreases the activity of LIM-kinase which normally inactivates cofilins by phosphorylation. Increased cofilin activity leads to the release of drebrin from actin filaments. Altogether, this leads to unstable actin filaments and actin

shrinkage of spines, deficits in hippocampal long-term potentiation, and memory impairment (Nakai et al. 2014).

2.2.2 Actin-Binding Proteins in Neurological Diseases

A mutation in the *diaphanous homolog 3* (*DIAPH3*) gene is associated with autism spectrum disorders (ASD) (Vorstman et al. 2011) as well as with progressive hearing loss (Starr et al. 2004; Schoen et al. 2010). This gene encodes the actin polymerization factor formin DRF3 (called mDia2 in mouse). During filopodia elongation, mDia2/DRF3 polymerizes straight actin filaments at the barbed ends. In addition to the regulation of spine morphogenesis, mDia2/DFR3 has a significant function in neurite extension (Dent et al. 2007).

Amyotrophic lateral sclerosis (ALS) is a late-onset neurodegenerative disorder resulting from motor neuron death. Mutations in the *PFN1* gene encoding profilin-1 have recently been shown to be a cause of familial amyotrophic lateral sclerosis (Wu et al. 2012; Ingre et al. 2013). Expression of mutated profilin-1 induced ubiquitinated, insoluble aggregates in cells that in many cases contained the ALS-associated protein TDP-43. Furthermore, primary motor neurons expressing mutant profilin-1 displayed smaller growth cones with a reduced F/G-actin ratio. These observations show that cytoskeletal alterations are closely linked to ALS pathogenesis (Wu et al. 2012).

2.2.3 Actin Mutations in Neurological Diseases

Baraitser-Winter syndrome is a rare but well-defined developmental disorder recognized by brain malformation, intellectual disability, seizures, and hearing loss (Riviere et al. 2012). There are currently six known de novo actin mutations in γ- and β-actin encoding genes found in 15 studied Baraitser-Winter syndrome patients (Riviere et al. 2012). Eight of the 15 individuals studied carried an R196H mutation in β-actin. This mutation increases actin filament stability (Riviere et al. 2012). As the dendritic actin cytoskeleton is known to be highly dynamic (Star et al. 2002; Honkura et al. 2008; Hotulainen et al. 2009; Frost et al. 2010; Koskinen et al. 2014),

Fig. 2.3 (continued) filament depolymerization which leads to a smaller spine head size and less stable dendritic spines. Increased spine turnover results in decreased spine density. (**b**) *Healthy condition*: SrGAP3 facilitates filopodia initiation, regulating dendritic spine density. *Mental retardation*: The srGAP3-encoding gene is deleted in a patient with a severe form of mental retardation. As the srGAP3 knockout has been shown to decrease the number of dendritic filopodia in mouse models, it is plausible that srGAP3 deficiency leads to decreased dendritic spine density also in human patients

it can be speculated that overly stable actin filaments can interfere with fast morphological changes during development and plasticity. However, further studies are needed to test these ideas.

2.3 Final Words

Large-scale exome sequencing and genome-wide association studies have identified many mutations that impact dendritic actin regulation in association with synapse dysfunction. It is very likely that many more are yet to be identified. Rather than being caused by a few highly penetrant mutations, neuropsychiatric disorders appear to be associated with subtle defects in numerous genes, each conferring a small increase in disease risk (Fromer et al. 2014). Although thousands of genes are affected, genes involved in F-actin remodeling seem to be particularly affected in neurological diseases (Fromer et al. 2014). Future challenges will be to clarify how these disease mutations affect protein function as well as neuron morphology and functionality. Different mutations in the same gene may have opposite effects depending on their location. In parallel, it is important to study the basic mechanisms underlying neuronal development in order to have a comprehensive understanding of synaptic plasticity, brain function, and neurological diseases. Only when we know how dendrite development is affected in neurological diseases, and the underlying molecular mechanisms, will it be possible to design new treatments. Rho GTPase signaling is especially an attractive target for drug development as the activity of its effectors, especially kinases, can be manipulated by small molecules.

References

Addington AM, Rapoport JL (2009) The genetics of childhood-onset schizophrenia: when madness strikes the prepubescent. Curr Psychiatry Rep 11(2):156–161

Allen KM, Gleeson JG, Bagrodia S, Partington MW, MacMillan JC, Cerione RA, Mulley JC, Walsh CA (1998) PAK3 mutation in nonsyndromic X-linked mental retardation. Nat Genet 20 (1):25–30. doi:10.1038/1675

Amann KJ, Pollard TD (2000) Cellular regulation of actin network assembly. Curr Biol: CB 10 (20):R728–R730

Andrianantoandro E, Pollard TD (2006) Mechanism of actin filament turnover by severing and nucleation at different concentrations of ADF/cofilin. Mol Cell 24(1):13–23. doi:10.1016/j.molcel.2006.08.006

Ba W, van der Raadt J, Nadif Kasri N (2013) Rho GTPase signaling at the synapse: implications for intellectual disability. Exp Cell Res 319(15):2368–2374. doi:10.1016/j.yexcr.2013.05.033

Bacon C, Endris V, Rappold G (2009) Dynamic expression of the Slit-Robo GTPase activating protein genes during development of the murine nervous system. J Comp Neurol 513(2): 224–236. doi:10.1002/cne.21955

Bacon C, Endris V, Andermatt I, Niederkofler V, Waltereit R, Bartsch D, Stoeckli ET, Rappold G (2011) Evidence for a role of srGAP3 in the positioning of commissural axons within the ventrolateral funiculus of the mouse spinal cord. PLoS One 6(5):e19887. doi:10.1371/journal.pone.0019887

Beli P, Mascheroni D, Xu D, Innocenti M (2008) WAVE and Arp2/3 jointly inhibit filopodium formation by entering into a complex with mDia2. Nat Cell Biol 10(7):849–857. doi:10.1038/ncb1745

Billuart P, Bienvenu T, Ronce N, des Portes V, Vinet MC, Zemni R, Carrie A, Beldjord C, Kahn A, Moraine C, Chelly J (1998) Oligophrenin 1 encodes a rho-GAP protein involved in X-linked mental retardation. Pathol Biol 46(9):678

Blackwood DH, Fordyce A, Walker MT, St Clair DM, Porteous DJ, Muir WJ (2001) Schizophrenia and affective disorders—cosegregation with a translocation at chromosome 1q42 that directly disrupts brain-expressed genes: clinical and P300 findings in a family. Am J Hum Genet 69(2):428–433. doi:10.1086/321969

Blanchoin L, Boujemaa-Paterski R, Sykes C, Plastino J (2014) Actin dynamics, architecture, and mechanics in cell motility. Physiol Rev 94(1):235–263. doi:10.1152/physrev.00018.2013

Bovellan M, Romeo Y, Biro M, Boden A, Chugh P, Yonis A, Vaghela M, Fritzsche M, Moulding D, Thorogate R, Jegou A, Thrasher AJ, Romet-Lemonne G, Roux PP, Paluch EK, Charras G (2014) Cellular control of cortical actin nucleation. Curr Biol: CB 24(14): 1628–1635. doi:10.1016/j.cub.2014.05.069

Brandon NJ, Millar JK, Korth C, Sive H, Singh KK, Sawa A (2009) Understanding the role of DISC1 in psychiatric disease and during normal development. J Neurosci 29(41): 12768–12775. doi:10.1523/jneurosci.3355-09.2009

Cajal SRy (1999) Texture of the nervous system of man and the vertebrates (trans: Pasik P, Pasik T), vol I. Springer, New York

Campellone KG, Welch MD (2010) A nucleator arms race: cellular control of actin assembly. Nat Rev Mol Cell Biol 11(4):237–251. doi:10.1038/nrm2867

Capani F, Martone ME, Deerinck TJ, Ellisman MH (2001) Selective localization of high concentrations of F-actin in subpopulations of dendritic spines in rat central nervous system: a three-dimensional electron microscopic study. J Comp Neurol 435(2):156–170

Carlson BR, Lloyd KE, Kruszewski A, Kim IH, Rodriguiz RM, Heindel C, Faytell M, Dudek SM, Wetsel WC, Soderling SH (2011) WRP/srGAP3 facilitates the initiation of spine development by an inverse F-BAR domain, and its loss impairs long-term memory. J Neurosci 31(7): 2447–2460. doi:10.1523/jneurosci.4433-10.2011

Chazeau A, Mehidi A, Nair D, Gautier JJ, Leduc C, Chamma I, Kage F, Kechkar A, Thoumine O, Rottner K, Choquet D, Gautreau A, Sibarita JB, Giannone G (2014) Nanoscale segregation of actin nucleation and elongation factors determines dendritic spine protrusion. EMBO J. doi:10.15252/embj.201488837

Chelly J, Mandel JL (2001) Monogenic causes of X-linked mental retardation. Nat Rev Genet 2 (9):669–680. doi:10.1038/35088558

Chesarone MA, DuPage AG, Goode BL (2010) Unleashing formins to remodel the actin and microtubule cytoskeletons. Nat Rev Mol Cell Biol 11(1):62–74. doi:10.1038/nrm2816

Cooper JA, Schafer DA (2000) Control of actin assembly and disassembly at filament ends. Curr Opin Cell Biol 12(1):97–103

Coutinho-Budd J, Ghukasyan V, Zylka MJ, Polleux F (2012) The F-BAR domains from srGAP1, srGAP2 and srGAP3 regulate membrane deformation differently. J Cell Sci 125(Pt 14): 3390–3401. doi:10.1242/jcs.098962

Dailey ME, Smith SJ (1996) The dynamics of dendritic structure in developing hippocampal slices. J Neurosci 16(9):2983–2994

DeGeer J, Lamarche-Vane N (2013) Rho GTPases in neurodegeneration diseases. Exp Cell Res 319(15):2384–2394. doi:10.1016/j.yexcr.2013.06.016

Dent EW, Kwiatkowski AV, Mebane LM, Philippar U, Barzik M, Rubinson DA, Gupton S, Van Veen JE, Furman C, Zhang J, Alberts AS, Mori S, Gertler FB (2007) Filopodia are required for cortical neurite initiation. Nat Cell Biol 9(12):1347–1359. doi:10.1038/ncb1654

Dubos A, Combeau G, Bernardinelli Y, Barnier JV, Hartley O, Gaertner H, Boda B, Muller D (2012) Alteration of synaptic network dynamics by the intellectual disability protein PAK3. J Neurosci 32(2):519–527. doi:10.1523/jneurosci.3252-11.2012

Endris V, Wogatzky B, Leimer U, Bartsch D, Zatyka M, Latif F, Maher ER, Tariverdian G, Kirsch S, Karch D, Rappold GA (2002) The novel Rho-GTPase activating gene MEGAP/srGAP3 has a putative role in severe mental retardation. Proc Natl Acad Sci U S A 99(18):11754–11759. doi:10.1073/pnas.162241099

Endris V, Haussmann L, Buss E, Bacon C, Bartsch D, Rappold G (2011) SrGAP3 interacts with lamellipodin at the cell membrane and regulates Rac-dependent cellular protrusions. J Cell Sci 124(Pt 23):3941–3955. doi:10.1242/jcs.077081

Enomoto A, Murakami H, Asai N, Morone N, Watanabe T, Kawai K, Murakumo Y, Usukura J, Kaibuchi K, Takahashi M (2005) Akt/PKB regulates actin organization and cell motility via Girdin/APE. Dev Cell 9(3):389–402. doi:10.1016/j.devcel.2005.08.001

Enomoto A, Ping J, Takahashi M (2006) Girdin, a novel actin-binding protein, and its family of proteins possess versatile functions in the Akt and Wnt signaling pathways. Ann N Y Acad Sci 1086:169–184. doi:10.1196/annals.1377.016

Enomoto A, Asai N, Namba T, Wang Y, Kato T, Tanaka M, Tatsumi H, Taya S, Tsuboi D, Kuroda K, Kaneko N, Sawamoto K, Miyamoto R, Jijiwa M, Murakumo Y, Sokabe M, Seki T, Kaibuchi K, Takahashi M (2009) Roles of disrupted-in-schizophrenia 1-interacting protein girdin in postnatal development of the dentate gyrus. Neuron 63(6):774–787. doi:10.1016/j.neuron.2009.08.015

Fan Y, Tang X, Vitriol E, Chen G, Zheng JQ (2011) Actin capping protein is required for dendritic spine development and synapse formation. J Neurosci 31(28):10228–10233. doi:10.1523/jneurosci.0115-11.2011

Fauchereau F, Herbrand U, Chafey P, Eberth A, Koulakoff A, Vinet MC, Ahmadian MR, Chelly J, Billuart P (2003) The RhoGAP activity of OPHN1, a new F-actin-binding protein, is negatively controlled by its amino-terminal domain. Mol Cell Neurosci 23(4):574–586

Fechheimer M, Zigmond SH (1993) Focusing on unpolymerized actin. J Cell Biol 123(1):1–5

Fischer RS, Yarmola EG, Weber KL, Speicher KD, Speicher DW, Bubb MR, Fowler VM (2006) Tropomodulin 3 binds to actin monomers. J Biol Chem 281(47):36454–36465. doi:10.1074/jbc.M606315200

Fromer M, Pocklington AJ, Kavanagh DH, Williams HJ, Dwyer S, Gormley P, Georgieva L, Rees E, Palta P, Ruderfer DM, Carrera N, Humphreys I, Johnson JS, Roussos P, Barker DD, Banks E, Milanova V, Grant SG, Hannon E, Rose SA, Chambert K, Mahajan M, Scolnick EM, Moran JL, Kirov G, Palotie A, McCarroll SA, Holmans P, Sklar P, Owen MJ, Purcell SM, O'Donovan MC (2014) De novo mutations in schizophrenia implicate synaptic networks. Nature 506(7487):179–184. doi:10.1038/nature12929

Frost NA, Shroff H, Kong H, Betzig E, Blanpied TA (2010) Single-molecule discrimination of discrete perisynaptic and distributed sites of actin filament assembly within dendritic spines. Neuron 67(1):86–99. doi:10.1016/j.neuron.2010.05.026

Georges PC, Hadzimichalis NM, Sweet ES, Firestein BL (2008) The Yin-Yang of dendrite morphology: unity of actin and microtubules. Mol Neurobiol 38(3):270–284. doi:10.1007/s12035-008-8046-8

Goley ED, Welch MD (2006) The ARP2/3 complex: an actin nucleator comes of age. Nat Rev Mol Cell Biol 7(10):713–726. doi:10.1038/nrm2026

Grintsevich EE, Reisler E (2014) Drebrin inhibits cofilin-induced severing of F-actin. Cytoskeleton (Hoboken, NJ) 71(8):472–483. doi:10.1002/cm.21184

Guerrier S, Coutinho-Budd J, Sassa T, Gresset A, Jordan NV, Chen K, Jin WL, Frost A, Polleux F (2009) The F-BAR domain of srGAP2 induces membrane protrusions required for neuronal migration and morphogenesis. Cell 138(5):990–1004. doi:10.1016/j.cell.2009.06.047

Harris KM, Stevens JK (1989) Dendritic spines of CA 1 pyramidal cells in the rat hippocampus: serial electron microscopy with reference to their biophysical characteristics. J Neurosci 9(8):2982–2997

Hayashi-Takagi A, Takaki M, Graziane N, Seshadri S, Murdoch H, Dunlop AJ, Makino Y, Seshadri AJ, Ishizuka K, Srivastava DP, Xie Z, Baraban JM, Houslay MD, Tomoda T, Brandon NJ, Kamiya A, Yan Z, Penzes P, Sawa A (2010) Disrupted-in-Schizophrenia 1 (DISC1) regulates spines of the glutamate synapse via Rac1. Nat Neurosci 13(3):327–332. doi:10.1038/nn.2487

Hayashi-Takagi A, Araki Y, Nakamura M, Vollrath B, Duron SG, Yan Z, Kasai H, Huganir RL, Campbell DA, Sawa A (2014) PAKs inhibitors ameliorate schizophrenia-associated dendritic spine deterioration in vitro and in vivo during late adolescence. Proc Natl Acad Sci U S A 111 (17):6461–6466. doi:10.1073/pnas.1321109111

Honkura N, Matsuzaki M, Noguchi J, Ellis-Davies GCR, Kasai H (2008) The subspine organization of actin fibers regulates the structure and plasticity of dendritic spines. Neuron 57(5): 719–729. doi:http://dx.doi.org/10.1016/j.neuron.2008.01.013

Hotulainen P, Hoogenraad CC (2010) Actin in dendritic spines: connecting dynamics to function. J Cell Biol 189(4):619–629. doi:10.1083/jcb.201003008

Hotulainen P, Paunola E, Vartiainen MK, Lappalainen P (2005) Actin-depolymerizing factor and cofilin-1 play overlapping roles in promoting rapid F-actin depolymerization in mammalian nonmuscle cells. Mol Biol Cell 16(2):649–664

Hotulainen P, Llano O, Smirnov S, Tanhuanpaa K, Faix J, Rivera C, Lappalainen P (2009) Defining mechanisms of actin polymerization and depolymerization during dendritic spine morphogenesis. J Cell Biol 185(2):323–339. doi:10.1083/jcb.200809046

Ingre C, Landers JE, Rizik N, Volk AE, Akimoto C, Birve A, Hubers A, Keagle PJ, Piotrowska K, Press R, Andersen PM, Ludolph AC (2013) Weishaupt JH (2013) A novel phosphorylation site mutation in profilin 1 revealed in a large screen of US, Nordic, and German amyotrophic lateral sclerosis/frontotemporal dementia cohorts. Neurobiol Aging 34(6):1708.e1–1708.e6. doi:10.1016/j.neurobiolaging.2012.10.009

Ishizuka K, Paek M, Kamiya A, Sawa A (2006) A review of Disrupted-In-Schizophrenia-1 (DISC1): neurodevelopment, cognition, and mental conditions. Biol Psychiatry 59(12): 1189–1197. doi:10.1016/j.biopsych.2006.03.065

Ivanov A, Esclapez M, Pellegrino C, Shirao T, Ferhat L (2009) Drebrin A regulates dendritic spine plasticity and synaptic function in mature cultured hippocampal neurons. J Cell Sci 122(Pt 4): 524–534. doi:10.1242/jcs.033464

Jansen S, Collins A, Golden L, Sokolova O, Goode BL (2014) Structure and mechanism of mouse Cyclase-associated Protein (CAP1) in regulating actin dynamics. J Biol Chem 289(44): 30732–30742. doi:10.1074/jbc.M114.601765

Jaworski J, Kapitein LC, Gouveia SM, Dortland BR, Wulf PS, Grigoriev I, Camera P, Spangler SA, Di Stefano P, Demmers J, Krugers H, Defilippi P, Akhmanova A, Hoogenraad CC (2009) Dynamic microtubules regulate dendritic spine morphology and synaptic plasticity. Neuron 61 (1):85–100. doi:10.1016/j.neuron.2008.11.013

Karp G (2007) Cell and molecular biology: concepts and experiments, 5th edn. Wiley, Hoboken,

Kasai H, Hayama T, Ishikawa M, Watanabe S, Yagishita S, Noguchi J (2010) Learning rules and persistence of dendritic spines. Eur J Neurosci 32(2):241–249. doi:10.1111/j.1460-9568.2010.07344.x

Khaitlina S, Walloscheck M, Hinssen H (2004) Calcium-induced conformational changes in the C-terminal half of gelsolin stabilize its interaction with the actin monomer. Biochemistry 43 (40):12838–12845. doi:10.1021/bi049548z

Kim IH, Carlson BR, Heindel CC, Kim H, Soderling SH (2012) Disruption of wave-associated Rac GTPase-activating protein (Wrp) leads to abnormal adult neural progenitor migration associated with hydrocephalus. J Biol Chem 287(46):39263–39274. doi:10.1074/jbc.M112.398834

Kim IH, Racz B, Wang H, Burianek L, Weinberg R, Yasuda R, Wetsel WC, Soderling SH (2013) Disruption of Arp2/3 results in asymmetric structural plasticity of dendritic spines and

progressive synaptic and behavioral abnormalities. J Neurosci 33(14):6081–6092. doi:10.1523/jneurosci.0035-13.2013
Korobova F, Svitkina T (2010) Molecular architecture of synaptic actin cytoskeleton in hippocampal neurons reveals a mechanism of dendritic spine morphogenesis. Mol Biol Cell 21(1):165–176. doi:10.1091/mbc.E09-07-0596
Koskinen M, Bertling E, Hotulainen R, Tanhuanpaa K, Hotulainen P (2014) Myosin IIb controls actin dynamics underlying the dendritic spine maturation. Mol Cell Neurosci 61:56–64. doi:10.1016/j.mcn.2014.05.008
Kulkarni VA, Firestein BL (2012) The dendritic tree and brain disorders. Mol Cell Neurosci 50(1):10–20. doi:10.1016/j.mcn.2012.03.005
Kutsche K, Yntema H, Brandt A, Jantke I, Nothwang HG, Orth U, Boavida MG, David D, Chelly J, Fryns JP, Moraine C, Ropers HH, Hamel BC, van Bokhoven H, Gal A (2000) Mutations in ARHGEF6, encoding a guanine nucleotide exchange factor for Rho GTPases, in patients with X-linked mental retardation. Nat Genet 26(2):247–250. doi:10.1038/80002
Lappalainen P, Drubin DG (1997) Cofilin promotes rapid actin filament turnover in vivo. Nature 388(6637):78–82. doi:10.1038/40418
Lee SH, Dominguez R (2010) Regulation of actin cytoskeleton dynamics in cells. Mol Cells 29(4):311–325
Luo L (2002) Actin cytoskeleton regulation in neuronal morphogenesis and structural plasticity. Annu Rev Cell Dev Biol 18:601–635. doi:10.1146/annurev.cellbio.18.031802.150501
Marco EJ, Abidi FE, Bristow J, Dean WB, Cotter P, Jeremy RJ, Schwartz CE, Sherr EH (2009) ARHGEF9 disruption in a female patient is associated with X linked mental retardation and sensory hyperarousal. BMJ Case Rep 2009. doi:10.1136/bcr.06.2009.1999
Mattila PK, Salminen M, Yamashiro T, Lappalainen P (2003) Mouse MIM, a tissue-specific regulator of cytoskeletal dynamics, interacts with ATP-actin monomers through its C-terminal WH2 domain. J Biol Chem 278(10):8452–8459. doi:10.1074/jbc.M212113200
Mattila PK, Pykalainen A, Saarikangas J, Paavilainen VO, Vihinen H, Jokitalo E, Lappalainen P (2007) Missing-in-metastasis and IRSp53 deform PI(4,5)P2-rich membranes by an inverse BAR domain-like mechanism. J Cell Biol 176(7):953–964. doi:10.1083/jcb.200609176
Matus A (2000) Actin-based plasticity in dendritic spines. Science (New York, NY) 290(5492):754–758
Matus A, Ackermann M, Pehling G, Byers HR, Fujiwara K (1982) High actin concentrations in brain dendritic spines and postsynaptic densities. Proc Natl Acad Sci U S A 79(23):7590–7594
Menna E, Fossati G, Scita G, Matteoli M (2011) From filopodia to synapses: the role of actin-capping and anti-capping proteins. Eur J Neurosci 34(10):1655–1662. doi:10.1111/j.1460-9568.2011.07897.x
Menna E, Zambetti S, Morini R, Donzelli A, Disanza A, Calvigioni D, Braida D, Nicolini C, Orlando M, Fossati G, Cristina Regondi M, Pattini L, Frassoni C, Francolini M, Scita G, Sala M, Fahnestock M, Matteoli M (2013) Eps8 controls dendritic spine density and synaptic plasticity through its actin-capping activity. EMBO J 32(12):1730–1744. doi:10.1038/emboj.2013.107
Meyer MA (2014) Highly expressed genes within hippocampal sector CA1: implications for the physiology of memory. Neurol Int 6(2):5388. doi:10.4081/ni.2014.5388
Michelot A, Drubin DG (2011) Building distinct actin filament networks in a common cytoplasm. Curr Biol: CB 21(14):R560–R569. doi:10.1016/j.cub.2011.06.019
Mikati MA, Grintsevich EE, Reisler E (2013) Drebrin-induced stabilization of actin filaments. J Biol Chem 288(27):19926–19938. doi:10.1074/jbc.M113.472647
Millar JK, Wilson-Annan JC, Anderson S, Christie S, Taylor MS, Semple CA, Devon RS, St Clair DM, Muir WJ, Blackwood DH, Porteous DJ (2000) Disruption of two novel genes by a translocation co-segregating with schizophrenia. Hum Mol Genet 9(9):1415–1423
Mizui T, Takahashi H, Sekino Y, Shirao T (2005) Overexpression of drebrin A in immature neurons induces the accumulation of F-actin and PSD-95 into dendritic filopodia, and the formation of large abnormal protrusions. Mol Cell Neurosci 30(4):630–638

Nakai T, Nagai T, Tanaka M, Itoh N, Asai N, Enomoto A, Asai M, Yamada S, Saifullah AB, Sokabe M, Takahashi M, Yamada K (2014) Girdin phosphorylation is crucial for synaptic plasticity and memory: a potential role in the interaction of BDNF/TrkB/Akt Signaling with NMDA receptor. J Neurosci 34(45):14995–15008. doi:10.1523/jneurosci.2228-14.2014

Nakayama AY, Harms MB, Luo L (2000) Small GTPases Rac and Rho in the maintenance of dendritic spines and branches in hippocampal pyramidal neurons. J Neurosci 20(14): 5329–5338

Narayan S, Nakajima K, Sawa A (2013) DISC1: a key lead in studying cortical development and associated brain disorders. Neuroscientist 19(5):451–464. doi:10.1177/1073858412470168

Newey SE, Velamoor V, Govek EE, Van Aelst L (2005) Rho GTPases, dendritic structure, and mental retardation. J Neurobiol 64(1):58–74. doi:10.1002/neu.20153

Newpher TM, Ehlers MD (2009) Spine microdomains for postsynaptic signaling and plasticity. Trends Cell Biol 19(5):218–227. doi:10.1016/j.tcb.2009.02.004

Nimchinsky EA, Sabatini BL, Svoboda K (2002) Structure and function of dendritic spines. Annu Rev Physiol 64:313–353. doi:10.1146/annurev.physiol.64.081501.160008

Noguchi J, Matsuzaki M, Ellis-Davies GC, Kasai H (2005) Spine-neck geometry determines NMDA receptor-dependent Ca2+ signaling in dendrites. Neuron 46(4):609–622. doi:10.1016/j.neuron.2005.03.015

Okamoto K, Narayanan R, Lee SH, Murata K, Hayashi Y (2007) The role of CaMKII as an F-actin-bundling protein crucial for maintenance of dendritic spine structure. Proc Natl Acad Sci U S A 104(15):6418–6423. doi:10.1073/pnas.0701656104

Paavilainen VO, Bertling E, Falck S, Lappalainen P (2004) Regulation of cytoskeletal dynamics by actin-monomer-binding proteins. Trends Cell Biol 14(7):386–394. doi:10.1016/j.tcb.2004.05.002

Penzes P, Cahill ME, Jones KA, VanLeeuwen JE, Woolfrey KM (2011) Dendritic spine pathology in neuropsychiatric disorders. Nat Neurosci 14(3):285–293. doi:10.1038/nn.2741

Peters A, Palay SL, Webster HD (1991) The fine structure of the nervous system. Oxford University Press, New York

Pring M, Weber A, Bubb MR (1992) Profilin-actin complexes directly elongate actin filaments at the barbed end. Biochemistry 31(6):1827–1836

Qualmann B, Koch D, Kessels MM (2011) Let's go bananas: revisiting the endocytic BAR code. EMBO J 30(17):3501–3515. doi:10.1038/emboj.2011.266

Ramakers GJ, Wolfer D, Rosenberger G, Kuchenbecker K, Kreienkamp HJ, Prange-Kiel J, Rune G, Richter K, Langnaese K, Masneuf S, Bosl MR, Fischer KD, Krugers HJ, Lipp HP, van Galen E, Kutsche K (2012) Dysregulation of Rho GTPases in the alphaPix/Arhgef6 mouse model of X-linked intellectual disability is paralleled by impaired structural and synaptic plasticity and cognitive deficits. Hum Mol Genet 21(2):268–286. doi:10.1093/hmg/ddr457

Reid T, Bathoorn A, Ahmadian MR, Collard JG (1999) Identification and characterization of hPEM-2, a guanine nucleotide exchange factor specific for Cdc42. J Biol Chem 274 (47):33587–33593

Riviere JB, van Bon BW, Hoischen A, Kholmanskikh SS, O'Roak BJ, Gilissen C, Gijsen S, Sullivan CT, Christian SL, Abdul-Rahman OA, Atkin JF, Chassaing N, Drouin-Garraud V, Fry AE, Fryns JP, Gripp KW, Kempers M, Kleefstra T, Mancini GM, Nowaczyk MJ, van Ravenswaaij-Arts CM, Roscioli T, Marble M, Rosenfeld JA, Siu VM, de Vries BB, Shendure J, Verloes A, Veltman JA, Brunner HG, Ross ME, Pilz DT, Dobyns WB (2012) De novo mutations in the actin genes ACTB and ACTG1 cause Baraitser-Winter syndrome. Nat Genet 44(4):440–444. doi:10.1038/ng.1091, S441–442

Saarikangas J, Kourdougli N, Senju Y, Chazal G, Segerstråle M, Kuurne J, Minkeviciene R, Mattila PK, Garrett L, Hölter SM, Becker L, Racz I, Hans W, Klopstock T, Wurst W, Zimmer A, Fuchs H, Gailus-Durner V, Hrabě de Angelis M, von Ossowski L, Taira T, Lappalainen P, Rivera C, Hotulainen P (2015) MIM-induced membrane bending promotes dendritic spine initiation. Dev Cell 33:1–16. doi:10.1016/j.devcel.2015.04.014

Sanders MC, Goldstein AL, Wang YL (1992) Thymosin beta 4 (Fx peptide) is a potent regulator of actin polymerization in living cells. Proc Natl Acad Sci U S A 89(10):4678–4682

Schachtele SJ, Losh J, Dailey ME, Green SH (2011) Spine formation and maturation in the developing rat auditory cortex. J Comp Neurol 519(16):3327–3345. doi:10.1002/cne.22728

Schoen CJ, Emery SB, Thorne MC, Ammana HR, Sliwerska E et al (2010) Increased activity of Diaphanous homolog 3 (DIAPH3)/diaphanous causes hearing defects in humans with auditory neuropathy and in Drosophila. Proc Natl Acad Sci U S A 107:13396–13401

Scott EK, Luo L (2001) How do dendrites take their shape? Nat Neurosci 4(4):359–365. doi:10.1038/86006

Sheng M, Hoogenraad CC (2007) The postsynaptic architecture of excitatory synapses: a more quantitative view. Annu Rev Biochem 76:823–847. doi:10.1146/annurev.biochem.76.060805.160029

Shirao T, Gonzalez-Billault C (2013) Actin filaments and microtubules in dendritic spines. J Neurochem 126(2):155–164. doi:10.1111/jnc.12313

Shuib S, McMullan D, Rattenberry E, Barber RM, Rahman F, Zatyka M, Chapman C, Macdonald F, Latif F, Davison V, Maher ER (2009) Microarray based analysis of 3p25-p26 deletions (3p- syndrome). Am J Med Genet A 149A(10):2099–2105. doi:10.1002/ajmg.a.32824

Soares DC, Carlyle BC, Bradshaw NJ, Porteous DJ (2011) DISC1: structure, function, and therapeutic potential for major mental illness. ACS Chem Neurosci 2(11):609–632. doi:10.1021/cn200062k

Soderling SH, Binns KL, Wayman GA, Davee SM, Ong SH, Pawson T, Scott JD (2002) The WRP component of the WAVE-1 complex attenuates Rac-mediated signalling. Nat Cell Biol 4(12):970–975. doi:10.1038/ncb886

Squire L, Berg D, Bloom FE, Lac Sd, Ghosh A, Spitzer NC (2012) Fundamental neuroscience. Elsevier Academic Press, San Diego

St Clair D, Blackwood D, Muir W, Carothers A, Walker M, Spowart G, Gosden C, Evans HJ (1990) Association within a family of a balanced autosomal translocation with major mental illness. Lancet 336(8706):13–16

Stankiewicz T, Linseman D (2014) Rho family GTPases: key players in neuronal development, neuronal survival, and neurodegeneration. Front Cell Neurosci 8:314. doi:10.3389/fncel.2014.00314

Star EN, Kwiatkowski DJ, Murthy VN (2002) Rapid turnover of actin in dendritic spines and its regulation by activity. Nat Neurosci 5(3):239–246. doi:10.1038/nn811

Starr A, Isaacson B, Michalewski HJ, Zeng FG, Kong YY et al (2004) A dominantly inherited progressive deafness affecting distal auditory nerve and hair cells. J Assoc Res Otolaryngol 5:411–426

Steinecke A, Gampe C, Nitzsche F, Bolz J (2014) DISC1 knockdown impairs the tangential migration of cortical interneurons by affecting the actin cytoskeleton. Front Cell Neurosci 8:190. doi:10.3389/fncel.2014.00190

Stiess M, Bradke F (2011) Neuronal polarization: the cytoskeleton leads the way. Dev Neurobiol 71(6):430–444. doi:10.1002/dneu.20849

Suetsugu S, Murayama K, Sakamoto A, Hanawa-Suetsugu K, Seto A, Oikawa T, Mishima C, Shirouzu M, Takenawa T, Yokoyama S (2006) The RAC binding domain/IRSp53-MIM homology domain of IRSp53 induces RAC-dependent membrane deformation. J Biol Chem 281(46):35347–35358. doi:10.1074/jbc.M606814200

Sun HQ, Yamamoto M, Mejillano M, Yin HL (1999) Gelsolin, a multifunctional actin regulatory protein. J Biol Chem 274(47):33179–33182

Tahirovic S, Bradke F (2009) Neuronal polarity. Cold Spring Harb Perspect Biol 1(3):a001644. doi:10.1101/cshperspect.a001644

Takahashi H, Sekino Y, Tanaka S, Mizui T, Kishi S, Shirao T (2003) Drebrin-dependent actin clustering in dendritic filopodia governs synaptic targeting of postsynaptic density-95 and dendritic spine morphogenesis. J Neurosci 23(16):6586–6595

Thomson PA, Malavasi EL, Grunewald E, Soares DC, Borkowska M, Millar JK (2013) DISC1 genetics, biology and psychiatric illness. Front Biol 8(1):1–31. doi:10.1007/s11515-012-1254-7

Threadgill R, Bobb K, Ghosh A (1997) Regulation of dendritic growth and remodeling by Rho, Rac, and Cdc42. Neuron 19(3):625–634

Vindin H, Gunning P (2013) Cytoskeletal tropomyosins: choreographers of actin filament functional diversity. J Muscle Res Cell Motil 34(3–4):261–274. doi:10.1007/s10974-013-9355-8

Vorstman JA, van Daalen E, Jalali GR, Schmidt ER, Pasterkamp RJ, de Jonge M, Hennekam EA, Janson E, Staal WG, van der Zwaag B, Burbach JP, Kahn RS, Emanuel BS, van Engeland H, Ophoff RA (2011) A double hit implicates DIAPH3 as an autism risk gene. Mol Psychiatry 16 (4):442–451. doi:10.1038/mp.2010.26

Waltereit R, Leimer U, von Bohlen Und Halbach O, Panke J, Holter SM, Garrett L, Wittig K, Schneider M, Schmitt C, Calzada-Wack J, Neff F, Becker L, Prehn C, Kutscherjawy S, Endris V, Bacon C, Fuchs H, Gailus-Durner V, Berger S, Schonig K, Adamski J, Klopstock T, Esposito I, Wurst W, de Angelis MH, Rappold G, Wieland T, Bartsch D (2012) Srgap3(-)/(-) mice present a neurodevelopmental disorder with schizophrenia-related intermediate phenotypes. FASEB J 26(11):4418–4428. doi:10.1096/fj.11-202317

Wegner A, Isenberg G (1983) 12-fold difference between the critical monomer concentrations of the two ends of actin filaments in physiological salt conditions. Proc Natl Acad Sci U S A 80 (16):4922–4925

Wegner AM, Nebhan CA, Hu L, Majumdar D, Meier KM, Weaver AM, Webb DJ (2008) N-wasp and the arp2/3 complex are critical regulators of actin in the development of dendritic spines and synapses. J Biol Chem 283(23):15912–15920. doi:10.1074/jbc.M801555200

Willig KI, Barrantes FJ (2014) Recent applications of superresolution microscopy in neurobiology. Curr Opin Chem Biol 20:16–21. doi:10.1016/j.cbpa.2014.03.021

Wilson NK, Lee Y, Long R, Hermetz K, Rudd MK, Miller R, Rapoport JL, Addington AM (2011) A novel microduplication in the neurodevelopmental gene SRGAP3 that segregates with psychotic illness in the family of a COS Proband. Case Rep Genet 2011:585893. doi:10.1155/2011/585893

Wong K, Ren XR, Huang YZ, Xie Y, Liu G, Saito H, Tang H, Wen L, Brady-Kalnay SM, Mei L, Wu JY, Xiong WC, Rao Y (2001) Signal transduction in neuronal migration: roles of GTPase activating proteins and the small GTPase Cdc42 in the Slit-Robo pathway. Cell 107(2): 209–221

Worth DC, Daly CN, Geraldo S, Oozeer F, Gordon-Weeks PR (2013) Drebrin contains a cryptic F-actin-bundling activity regulated by Cdk5 phosphorylation. J Cell Biol 202(5):793–806. doi:10.1083/jcb.201303005

Wu CH, Fallini C, Ticozzi N, Keagle PJ, Sapp PC, Piotrowska K, Lowe P, Koppers M, McKenna-Yasek D, Baron DM, Kost JE, Gonzalez-Perez P, Fox AD, Adams J, Taroni F, Tiloca C, Leclerc AL, Chafe SC, Mangroo D, Moore MJ, Zitzewitz JA, Xu ZS, van den Berg LH, Glass JD, Siciliano G, Cirulli ET, Goldstein DB, Salachas F, Meininger V, Rossoll W, Ratti A, Gellera C, Bosco DA, Bassell GJ, Silani V, Drory VE, Brown RH Jr, Landers JE (2012) Mutations in the profilin 1 gene cause familial amyotrophic lateral sclerosis. Nature 488 (7412):499–503. doi:10.1038/nature11280

Xiang S, Kim EY, Connelly JJ, Nassar N, Kirsch J, Winking J, Schwarz G, Schindelin H (2006) The crystal structure of Cdc42 in complex with collybistin II, a gephyrin-interacting guanine nucleotide exchange factor. J Mol Biol 359(1):35–46. doi:10.1016/j.jmb.2006.03.019

Yamashiro S, Gokhin DS, Kimura S, Nowak RB, Fowler VM (2012) Tropomodulins: pointed-end capping proteins that regulate actin filament architecture in diverse cell types. Cytoskeleton (Hoboken, NJ) 69(6):337–370. doi:10.1002/cm.21031

Yang C, Svitkina T (2011) Filopodia initiation: focus on the Arp2/3 complex and formins. Cell Adh Migr 5(5):402–408. doi:10.4161/cam.5.5.16971

Yu W, Ahmad FJ, Baas PW (1994) Microtubule fragmentation and partitioning in the axon during collateral branch formation. J Neurosci 14(10):5872–5884

Yuste R, Bonhoeffer T (2001) Morphological changes in dendritic spines associated with long-term synaptic plasticity. Annu Rev Neurosci 24:1071–1089. doi:10.1146/annurev.neuro.24.1.1071

Zhao L, Ma QL, Calon F, Harris-White ME, Yang F, Lim GP, Morihara T, Ubeda OJ, Ambegaokar S, Hansen JE, Weisbart RH, Teter B, Frautschy SA, Cole GM (2006) Role of p21-activated kinase pathway defects in the cognitive deficits of Alzheimer disease. Nat Neurosci 9(2):234–242. doi:10.1038/nn1630

Chapter 3
Microtubule Organization and Microtubule-Associated Proteins (MAPs)

Elena Tortosa, Lukas C. Kapitein, and Casper C. Hoogenraad

Abstract Dendrites have a unique microtubule organization. In vertebrates, dendritic microtubules are organized in antiparallel bundles, oriented with their plus ends either pointing away or toward the soma. The mixed microtubule arrays control intracellular trafficking and local signaling pathways, and are essential for dendrite development and function. The organization of microtubule arrays largely depends on the combined function of different microtubule regulatory factors or generally named microtubule-associated proteins (MAPs). Classical MAPs, also called structural MAPs, were identified more than 20 years ago based on their ability to bind to and copurify with microtubules. Most classical MAPs bind along the microtubule lattice and regulate microtubule polymerization, bundling, and stabilization. Recent evidences suggest that classical MAPs also guide motor protein transport, interact with the actin cytoskeleton, and act in various neuronal signaling networks. Here, we give an overview of microtubule organization in dendrites and the role of classical MAPs in dendrite development, dendritic spine formation, and synaptic plasticity.

Keywords Neuron • Dendrite • Cytoskeleton • Microtubule • Microtubule-associated protein • MAP1 • MAP2 • MAP4 • MAP6 • MAP7 • MAP9 • Tau

3.1 Introduction

Microtubules (MTs) are cytoskeletal structures that play essential roles in all eukaryotic cells. MTs are important not only during cell division but also in non-dividing cells, where they are critical structures in numerous cellular processes such as cell motility, migration, differentiation, intracellular transport and organelle positioning. MTs are composed of two proteins, α- and β-tubulin, that form heterodimers and organize themselves in a head-to-tail manner. MTs are dynamic and they can rapidly switch between cycles of growth and shrinkage. This MT

E. Tortosa • L.C. Kapitein • C.C. Hoogenraad (✉)
Cell Biology, Faculty of Science, Utrecht University, Padualaan 8, 3584 CH Utrecht, The Netherlands
e-mail: c.hoogenraad@uu.nl

K. Emoto et al. (eds.), *Dendrites*, DOI 10.1007/978-4-431-56050-0_3

behavior is known as dynamic instability (Desai and Mitchison 1997; Howard and Hyman 2003; Mitchison and Kirschner 1984). Due to the head-to-tail polymerization, MTs are polar structures, designated by "plus" and "minus" ends, each with distinct characteristics. The two MT ends grow and depolymerize at very different rates; the "plus end" is the preferred site for MT assembly and disassembly, while the "minus end" is generally more stable in cells. Although MT minus ends can grow in vitro, minus ends are usually attached and stabilized at the centrosome or MT organizing center (MTOC) in cells (Jiang and Akhmanova 2011). The centrosome is also the primary site for MT nucleation in many cell types. Alternative sites for MT nucleation have been recently described (Bornens 2008). In addition to de novo MT nucleation, new MT ends can also be formed by breakage of preexisting MTs by severing proteins like spastin and katanin (Salinas et al. 2007).

The precise organization of MTs, including their composition, stability, orientation, and spacing, is essential for the correct development, morphology, and function of the neuron. The MT organization depends on several regulatory factors such as various tubulin isoforms, posttranslational modifications, and MT-associated proteins (MAPs). Largely based on their mode of action, the different MAPs can be roughly divided into five groups. The first group contains the MT-based motor proteins that are important for neuronal transport, such as kinesin and dynein motors (Hirokawa et al. 2010; Kardon and Vale 2009; Karki and Holzbaur 1999). The second set consists of regulators of MT dynamics, such as plus-end tracking proteins (+TIPs) and MT depolymerizers (Akhmanova and Steinmetz 2008; Brouhard and Rice 2014; Walczak et al. 2013). The third group contains proteins that modulate MT number, such as regulators of nucleation (Luders and Stearns 2007), enzymes that sever preexisting MTs (Roll-Mecak and McNally 2010; Sharp and Ross 2012), and minus-end targeting proteins (-TIPs), stabilizing minus ends (Akhmanova and Hoogenraad 2015). The fourth set comprises tubulin-modifying enzymes that, through posttranslational modifications, can generate distinct MT subtypes (Hammond et al. 2008; Janke and Bulinski 2011). The fifth group includes cross-linking proteins that align filaments and form MT bundles, such as classical MAPs and some kinesin motors that drive MT sliding (Bratman and Chang 2008; Dehmelt and Halpain 2005; Maccioni and Cambiazo 1995). The actions of many different MAPs together provide the mechanism to spatiotemporal control the architecture of the neuronal MT cytoskeleton during the different steps of development. It is therefore not surprising that, given the importance of these MT regulators, compensation mechanisms exist among different MAPs. This phenomenon is shown in the viability and the absence of strong phenotype in many MAP knockout mice (Table 3.1). However, more severe phenotypes have been observed in double knockouts for MAPs like MAP1B/MAP2 or MAP1B/tau. Double-knockout mice for MAP1B/tau or MAP1B/MAP2 show defects in brain layered structures and fiber track formation (Takei et al. 2000; Teng et al. 2001). Here, we give an overview of the MT organization in neurons and focus on the role of classical MAPs in dendrite development, dendritic spine formation, and synaptic plasticity.

Table 3.1 Animal models – phenotypes associated with classical MAPs

MAP	Knockout animal model phenotypes	References
MAP1A	Perturbation of learning processes Reduced LTP and LTD	Takei et al. (2015)
MAP1B/ MAP5	Body and brain weight loss	Takei et al. (2000), (1997), Edelmann et al. (1996), Gonzalez-Billault et al. (2000, 2005), Meixner et al. (2000), and Benoist et al. (2013)
	Delayed nervous system development	
	Reduced myelination	
	Corpus callosum absent	
	Impaired neuronal migration	
	Altered brain commissures and laminated structures	
	Smaller retina size	
	LTP enhanced and LTD disrupted	
	Motor system abnormalities and lack of exploring activity	
	Reduced motor nerve conduction velocity	
MAP8/ MAP1S	No major defect	Xie et al. (2011)
MAP2	No major defect	Harada et al. (2002), Teng et al. (2001), and Khuchua et al. (2003)
	Decrease in dendritic microtubule (MT) density	
	Disrupted dendritic morphology from CA1 neurons	
	Contextual memory altered	
MAP3/ MAP4	Not reported	
MAP6	Depleted synaptic vesicle pool	Andrieux et al. (2002), Brun et al. (2005), Bouvrais-Veret et al. (2007), (2008), Powell et al. (2007), Fradley et al. (2005), Fournet et al. (2010, (2012b), and Daoust et al. (2014)
	Impaired synaptic plasticity	
	Severe behavioral disorders	
	Abnormalities of glutamatergic, dopaminergic, acetylcholinergic/nicotinic, serotonergic, and noradrenergic neurotransmissions	
	Sensorimotor gating impairment	
	Defects in neuronal transport	
MAP7	Viable. Defects in spermatogenesis	Komada et al. (2000), Sung et al. (2008), Barlan et al. (2013), and Metzger et al. (2012)
	Reduced viability (in *Drosophila*)	
	Myonuclear positioning altered (in *Drosophila*)	
MAP9	Severe developmental defects and embryonic lethality (in zebra fish)	Fontenille et al. (2014)
Tau	Viable	Harada et al. (1994), Dawson et al. (2001), Ikegami et al. (2000), Lei et al. (2012), and Ma et al. (2014)
	Muscle weakness	
	Small caliber axons	
	Behavioral impairments and motor deficits	

3.2 Organization of Axonal and Dendritic Microtubules

The MT cytoskeleton in neurons differs from many other cell types. Even within a single neuron, different compartments have distinct MT properties and organization (Conde and Caceres 2009). For instance, MTs in axon and dendrites differ in their polarity orientations. In mammalian neurons, axons contain MTs with uniform orientation, directed away from the cell body (plus-end out), whereas in dendrites MT orientation is mixed, with about half plus-end out and half minus-end out MTs (Baas et al. 1988; Burton 1988) (Fig. 3.1). In *Drosophila* and *C. elegans* neurons, MTs in axons are arranged with their plus ends distal to the cell body, as in vertebrates. However, in dendrites most MTs are arranged with their minus ends distal to the cell body, although some mixed MTs have also been observed (Stone et al. 2008; Maniar et al. 2011). It is thought that differences in the MT cytoskeleton in axons and dendrites can facilitate polarized cargo trafficking (Kapitein and Hoogenraad 2011; Rolls 2011).

Three different techniques have been used to establish MT orientations in axons and dendrites: the hook-decoration method; fluorescently labeled plus-end tracking proteins, such as GFP-EB3; and second-harmonic generation (SHG) microscopy. In the hook method, exogenous tubulin is added to permeabilized cells and forms curved sheets ("hooks") on existing neuronal MTs. These hooks are visualized in cross section by electron microscopy and provide information about the orientation of axonal and dendritic MTs. This method provides sufficient resolution for counting individual MT orientations. However, the results are usually ambiguous because of the small sample size and the large fraction of non-marked MTs (Baas and Lin 2011). The use of fluorescently labeled plus-end tracking proteins as markers of growing MTs is a relatively easier technique (Stepanova et al. 2003). This method also has limitations because it only detects dynamic MTs and not the stable MT population. Recently, in vivo imaging of growing MTs has confirmed the presence of a mixed MT organization in mature dendrites (Kleele et al. 2014; Yau et al. 2016). SHG microscopy allows label-free imaging without the addition of exogenous probes, but is hard to interpret quantitatively. SHG was used in hippocampal slices to visualize the neuronal MT organization in axon and dendrites (Dombeck et al. 2003; Kwan et al. 2008).

How are the mixed MT arrays in dendrites generated and stabilized? It has been suggested that dendritic MTs are generated by (i) centrosomal nucleation and subsequent release from the centrosome, (ii) breakage of preexisting MTs by katanin or spastin, or (iii) nucleation at noncentrosomal sites, such as cortical γ-tubulin complexes, intracellular membranes, or preexisting MTs (Kuijpers and Hoogenraad 2011). Although the centrosome is the main nucleating point in non-polarized neurons, it loses its function as MTOC during early neuronal development, and only a few MTs emanate from the centrosome in mature neurons (Stiess et al. 2010). Consistently, it has been shown that noncentrosomal MTs are abundantly present in neurons (Yau et al. 2014). Recently, CAMSAP/Patronin/Nezha family proteins have been characterized and found to specifically recognize

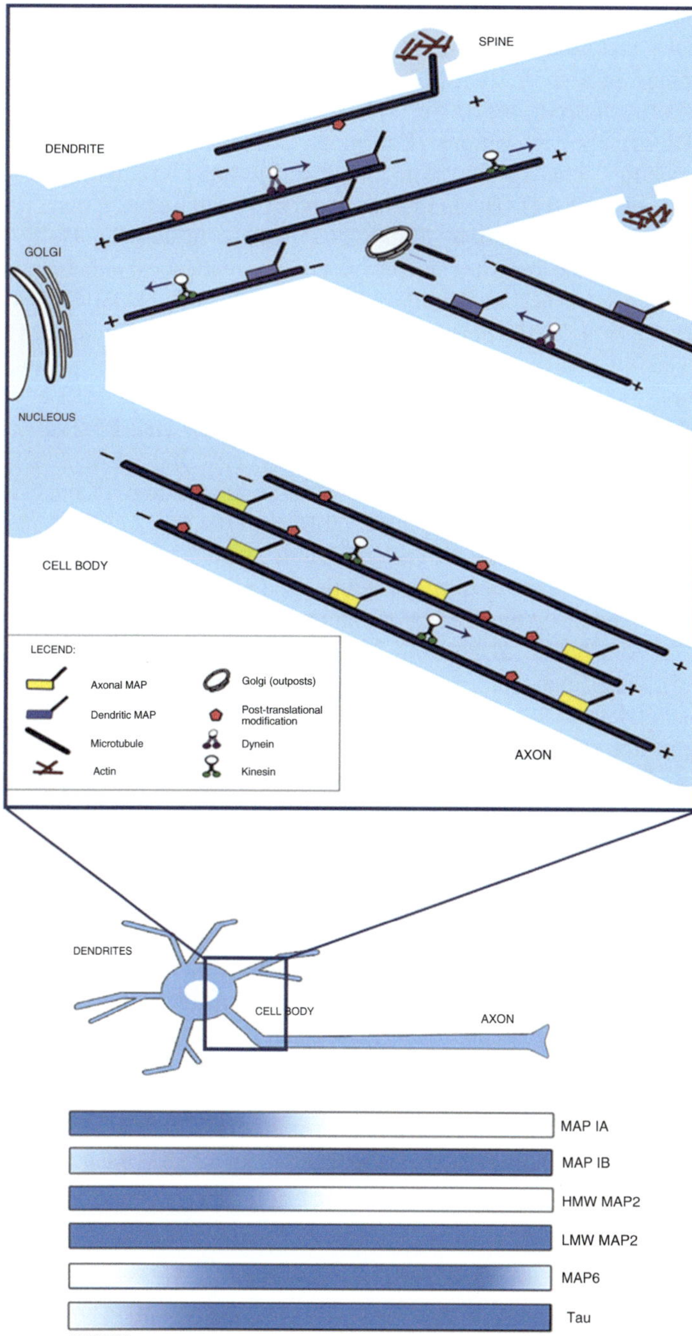

Fig. 3.1 Cytoskeletal organization in neurons. Axons and dendrites present different MT polarity orientations. Axonal MTs are uniformly orientated, whereas MTs in dendrites have mixed orientations. Axonal and dendritic MTs also contain distinct MAPs and differ in their posttranslational modifications. Motor proteins carry out ATP-dependent movements along MTs and either

MT minus ends and stabilize free minus ends against depolymerization (Akhmanova and Hoogenraad 2015). By forming stable CAMSAP stretches at the MT minus ends, these short MT fragments may also serve as "seeds" for new MT regrowth. Indeed, repetitive MT plus-end growth was observed from CAMSAP2 stretches in neurons (Yau et al. 2014). CAMSAP2 is required for neuronal polarity, axon specification, and dendritic branch formation in vitro and in vivo (Yau et al. 2014). The Golgi has also been proposed as a potential site of noncentrosomal MT nucleation (Efimov et al. 2007). In neurons, Golgi outposts have been found in the dendrites and may not only provide local membrane delivery but also act as local MT nucleation points (Horton et al. 2005; Ori-McKenney et al. 2012; Ye et al. 2007). Several other factors are known to regulate MT bundle formation in nonneuronal cells and may play a role in setting up mixed MT organizations in dendrites. For instance, motor proteins, such as MKLP1 (kinesin-6), Eg5 (kinesin-5), and KIF15 (kinesin-12, also called HKLP2), are known to organize antiparallel MT organizations in other systems. Indeed, depletion of each of these factors has been shown to disturb dendrite morphology (Kahn et al. 2015; Lin et al. 2012). In *Drosophila* neurons, EB proteins and kinesin-2 motors are important factors in setting up the uniform minus-end out MT network in dendrites (Mattie et al. 2010). It has been suggested that the minus-end out MT organization in neurons is maintained by steering of polymerizing MTs along the stable MTs by kinesin-2 motors bound to growing MT plus ends (Chen et al. 2014; Doodhi et al. 2014). Other motor proteins, such as kinesin-1, might also be able to cross-link antiparalleled MTs and are critical in forming the characteristic minus-end out MT organization of *C. elegans* dendrites (Yan et al. 2013).

Axonal and dendritic MTs not only differ in their organization but also in their stability. It has been demonstrated that axonal MTs are more resistant to the MT depolymerizing drug nocodazole compared to the dendritic MT population (Baas et al. 1991; Witte et al. 2008). Recently, a new posttranslational modification of tubulin has been identified that directly confers stability to MTs (Song et al. 2013). Biochemical characterization of stable MT fractions demonstrated that polyamination of tubulin is directly involved in stabilizing neuronal MTs. The most commonly studied posttranslational modifications are either acidic (phosphorylation and glutamylation) or charge neutral (acetylation and detyrosination), and they do not directly confer stability to MTs but, rather, accumulate on long-lived MTs (Janke and Kneussel 2010). There is a clear correlation between MT acetylation/detyrosination and stable MTs, and MT tyrosination with dynamic MTs. The different MT posttranslational modifications show a polarized distribution in neurons (Hammond et al. 2010; Kollins et al. 2009; Witte et al. 2008). In growing axons, the ratios of acetylated and

Fig. 3.1 (continued) drive cargo transport toward the *minus end* (dynein) or toward the *plus end* (kinesin). Dynamic MTs can enter dendritic spines and may deliver specific cargos to individual spines. Golgi outposts are present along dendrites and may serve as sites of noncentrosomal MT nucleation. CAMSAP binds to MT *minus ends* and is required for the stabilization of noncentrosomal MTs

detyrosinated tubulin are higher than in the developing neurites, whereas the ratio of glutamylated tubulin does not show differences between these two compartments (Hammond et al. 2010). On the other hand, tyrosinated MTs are more abundant at the tip of the axon and in developing dendrites (Kollins et al. 2009). It has been suggested that the local diversity of posttranslational modifications is due to the specific localization or activity of the modifying enzymes (Hammond et al. 2010). Some posttranslational modifications seem to be essential for proper neuronal development. For example, lack of tyrosinated tubulin in tubulin-tyrosine ligase (TTL) null neurons caused extensive defects in neurite outgrowth and axon development (Erck et al. 2005). The MT posttranslational modifications may also regulate the interaction of other factors with MTs. For example, acetylation of MTs increases the severing activity of katanin (Sudo and Baas 2010). Moreover, posttranslational modifications have been implicated in the regulation of several motor proteins. For instance, kinesin-1 has been reported to prefer stable (acetylated/detyrosinated) MTs (Dunn et al. 2008; Konishi and Setou 2009; Liao and Gundersen 1998; Reed et al. 2006). Additional in vitro studies showed that kinesin-1 motility is increased by polyglutamylation and that detyrosination of α-tubulin promotes kinesin-2 motility (Sirajuddin et al. 2014). It has been hypothesized that posttranslational modifications, together with specific MAP patterns, form a "tubulin code" that can be "read" by factors that interact with MTs, such as motor proteins (Janke and Bulinski 2011; Tischfield and Engle 2010; Verhey and Gaertig 2007). In this way, distinct "tubulin codes" in axons and dendrites may drive polarized cargo sorting of various organelles and proteins into axons and dendrites (Kapitein and Hoogenraad 2011; Rolls 2011).

3.3 Functions of Axon and Dendritic Microtubules

MTs are essential structures in axons and dendrites because they serve as major tracks for long-distance transport and form the basis for stable neuronal morphology. However, MTs are not just passive elements, they also confer plasticity to the neuron and have an active role during different phases of neuronal development. MTs participate in the morphological changes during neuronal migration and differentiation, for instance, by regulating axonal outgrowth, organelle positioning, and dendritic spine dynamics (Hoogenraad and Bradke 2009). Recent studies have found that defects in many MT-related genes lead to a range of nervous system abnormalities and several neurological and neurodegenerative diseases (Gupta et al. 2002; Jaglin and Chelly 2009; Manzini and Walsh 2011; Tischfield et al. 2011). For instance, mutations in genes encoding α- and β-tubulin subunits alter MT dynamics and show defects in axon guidance, neuronal migration, and synaptic connectivity (Jaglin et al. 2009; Keays et al. 2007; Tischfield et al. 2010). In addition, recent studies highlight MTs as a potential target for therapeutic interventions for axon regeneration and neurodegenerative diseases (Baas and Ahmad 2013; Gerdes and Katsanis 2005). Therefore, control of MT organization is of key importance for proper neuronal development and function.

MTs have been extensively studied with respect to axon formation, axon growth, and axon guidance (Hoogenraad and Bradke 2009; Poulain and Sobel 2010). It is known that MTs are also required for dendritic development and dendritic tiling (Grueber and Sagasti 2010; Koleske 2013; Rolls 2011). In many different organisms, proper dendrite morphology depends on MAPs, which regulate MT stabilization, bundling, spacing, and dynamics. MAPs also regulate intracellular transport or link MTs to the actin cytoskeleton, which is another important player in dendrite development (Lansbergen and Akhmanova 2006; Poulain and Sobel 2010; Siegrist and Doe 2007). For instance, classical MAPs like MAP1A and MAP2 are important for dendrite morphogenesis, and upregulation of their expression is correlated with dendrite outgrowth (Harada et al. 2002; Szebenyi et al. 2005; Vaillant et al. 2002). MAP2 knockout mice show altered MT spacing and reduced dendrite arbor size (Harada et al. 2002; Teng et al. 2001). Other MT-related proteins, such as the plus-end binding proteins CLIP-170 and CLASP2, and the minus-end binding protein CAMSAP2, are also involved in dendrite development (Beffert et al. 2012; Swiech et al. 2011; Yau et al. 2014). Moreover, the MT-destabilizing proteins stathmin and SCLIP, or MT-severing proteins, like spastin and katanin p60-like 1, have been reported to regulate dendrite development (Jinushi-Nakao et al. 2007; Lee et al. 2009; Ohkawa et al. 2007; Poulain et al. 2008; Stewart et al. 2012; Ye et al. 2011). In addition, motors that drive dendritic transport are crucial for dendrite morphology (Kapitein et al. 2010; Satoh et al. 2008; Zheng et al. 2008). Therefore, many MT-related factors regulating MT organization, dynamics, and remodeling are critical for proper dendritic development.

In addition to the growth and development of dendrites, MTs are needed for spine morphology and various synaptic processes, including spine formation (Shirao and Gonzalez-Billault 2013), dendritic and synaptic pruning (Kage et al. 2005; Lee et al. 2009; Luo and O'Leary 2005), and synaptic plasticity (Conde and Caceres 2009; Hoogenraad and Bradke 2009). MTs are abundantly present in the dendritic shaft; however, dynamic MTs can also enter actin-rich dendritic spines and regulate synaptic processes (Gu and Zheng 2009; Hu et al. 2008; Jaworski et al. 2009). MT entries in spines may directly regulate spine morphology or provide a way to selectively transport organelles, receptors, and other regulatory factors necessary for synaptic function (Hoogenraad and Bradke 2009). In addition, some classical MAPs like MAP1B and MAP2 have been found in dendritic spines and may regulate their development (Caceres et al. 1983; Collins et al. 2005; Kawakami et al. 2003; Peng et al. 2004; Tortosa et al. 2011). Several other MT regulatory factors such as collapsin response mediator protein (CRMP) members, doublecortin family member DCLK1, and spinophilin/neurabin control dendritic spine maturation (Ryan et al. 2005; Shin et al. 2013; Terry-Lorenzo et al. 2005; Yamashita et al. 2007). Moreover, MAP1B, MAP6, CLASP2, and CRMP family proteins have been implicated to influence synaptic plasticity processes (Andrieux et al. 2002; Beffert et al. 2012; Benoist et al. 2013; Su et al. 2007; Yamashita et al. 2011).

3.4 Classical Maps

Classical MAPs, also called structural MAPs, were isolated more than 20 years ago from mammalian brains by copurification with MTs (Schoenfeld and Obar 1994). A general overview of domain structure of classical MAPs is given in Fig. 3.2. In vitro studies showed that most MAPs bind along the MT lattice and regulate MT polymerization and stabilization. However, the binding characteristics and effect on MTs are different among all MAPs. For instance, MAP2 only needs a single protofilament for MT binding, while other MAPs like MAP4 interact with adjacent protofilaments (Al-Bassam et al. 2002; Kawachi et al. 2003). Some of them, like tau or MAP2, induce MT bundling, whereas others like MAP4 do not have bundling activity (Burgin et al. 1994; Kanai et al. 1989, 1992; Nguyen et al. 1997; Olson et al. 1995). In addition, MAP1B has been shown to control MT dynamic, while MAP4 alters MT surface properties and affects motor protein activity (Bulinski et al. 1997; Samora et al. 2011; Semenova et al. 2014; Tokuraku et al. 2007; Tortosa et al. 2013; Utreras et al. 2008). Interestingly, many of these MAPs not only bind MTs but also interact with actin and participate in numerous signal pathways. Tables 3.2 and 3.3 give an overview of the various phenotypes observed in cultured cells caused by MAP downregulation or overexpression. Given the importance of MAPs for neuronal development and function, it is not surprising that many MAPs are associated with neurological and neurodegenerative diseases. For example, pathological aggregation of tau protein in the human brain leads to neurodegenerative diseases called tauopathies (Zempel and Mandelkow 2014). The best-known tauopathy is Alzheimer's disease, where tau protein is deposited within neurons in the form of neurofibrillary tangles (Avila et al. 2004). Moreover, the deletion of MAP6 in mice leads to severe phenotypes reminiscent of schizophrenia-like symptoms (Andrieux et al. 2002; Fournet et al. 2012b).

3.4.1 MAP1 Family of Microtubule-Associated Proteins

In mammals, MAP1 family proteins include three members: MAP1A, MAP1B, and MAP1S, which are all encoded by different genes. MAP1A and MAP1B are predominantly expressed in neurons and important for the formation and development of axons and dendrites. MAP1S is widespread in murine tissues, and little is known about its function. In *Drosophila*, the single MAP1 homolog is called Futsch, which has been shown to be important in dendritic and axonal development, and regulates synaptic growth. All MAP1 members are multiprotein complexes, formed by a heavy chain and one or two light chains. They are translated as polypeptides and processed by proteolytic cleavage, which leads to the generation of distinct heavy chains (MAP1A-HC, MAP1B-HC, and MAP1S-HC) and light chains (LC2 from MAP1A, LC1 from MAP1B, and MAP1S-LC). Later, HC and LC are assembled together with LC3, which is encoded by a separate gene. Both

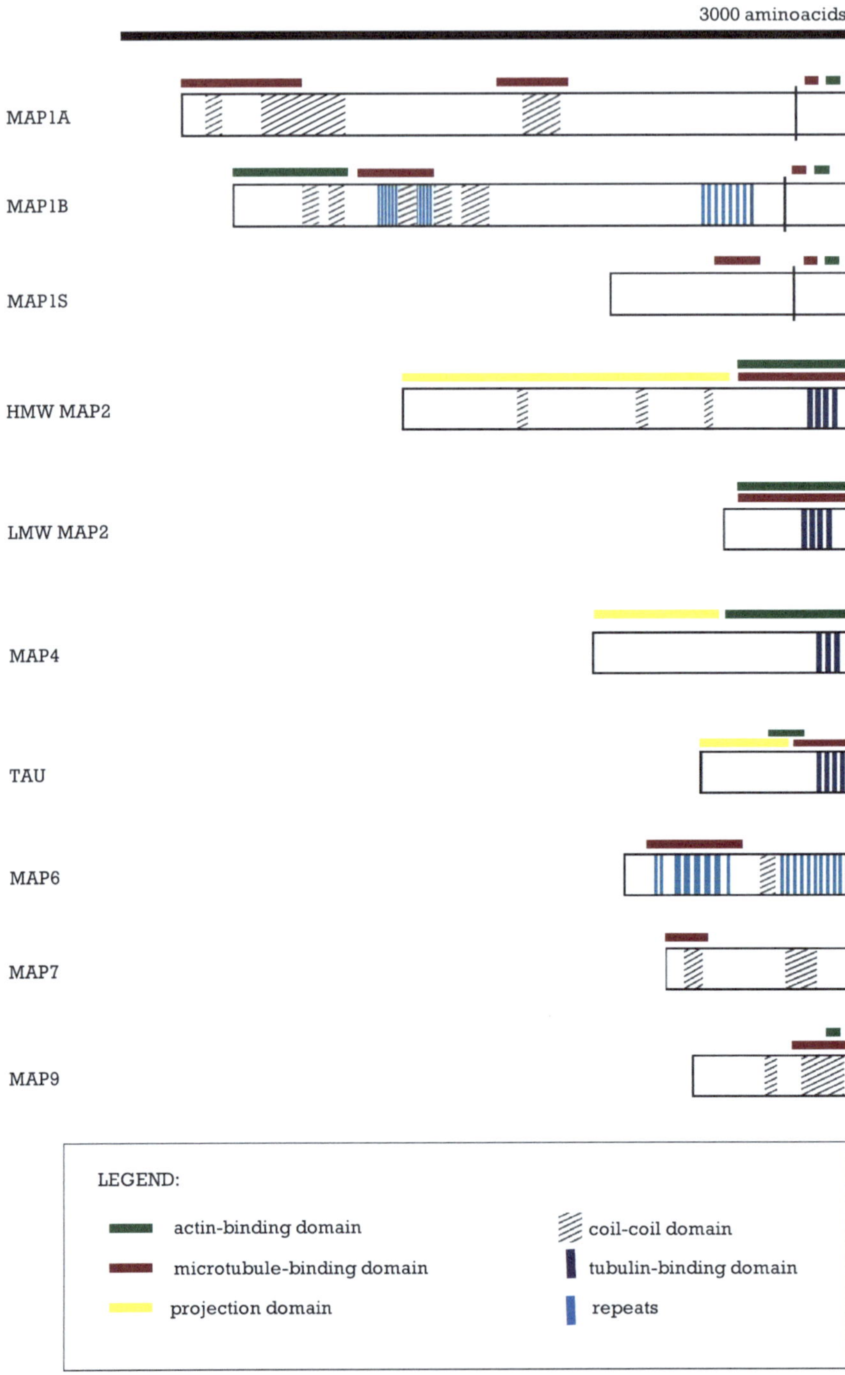

Fig. 3.2 Schematic diagram of classical microtubule-associated proteins (MAPs). The major structural motifs in classical MAPs are illustrated in the diagram. The MAP family proteins contain different MT-binding domains. In MAP2, MAP4, and tau, the MT-binding regions are

HC and LC can bind MTs, filamentous actin, and many other cellular components. Furthermore, their activity is controlled by upstream signaling mechanisms, such as the MAP kinase and glycogen synthase kinase-3 pathways (Halpain and Dehmelt 2006).

MAP1C was originally described as a MAP1 family member with ATPase activity, and further characterization revealed that it represents the microtubule-based motor protein cytoplasmic dynein (Johnson et al. 1984; Paschal et al. 1987; Paschal and Vallee 1987; Vale et al. 1985; Vallee et al. 1988). Dynein is involved in a variety of basic cellular functions, such as the movement of organelles; transport of vesicles, proteins, and mRNA; maintenance of the Golgi apparatus; endosome recycling; cytoskeletal reorientation; and the positioning of the mitotic spindle (McNally 2013; Vallee et al. 2004; Yadav and Linstedt 2011). In neurons, it has a role in neuronal migration, retrograde axonal transport, and polarized trafficking into dendrites (Chevalier-Larsen and Holzbaur 2006; Kapitein and Hoogenraad 2011; Vallee et al. 2009). Mutations in dynein have been directly linked to various neurological and neurodegenerative diseases (Lipka et al. 2013). Here, we will give a general overview of the other three MAP1 family members (MAP1A, MAP1B, and MAP1S) and briefly discuss their role in neuronal development.

3.4.1.1 MAP1 Family Member MAP1A

MAP1A is predominantly expressed in neurons, where it is enriched in dendrites (Schoenfeld et al. 1989) (Fig. 3.1). MAP1A expression increases between 4 and 7 days in hippocampal cultures in vitro and around the second week after birth in the developing mouse brain. During this time, dendrites elongate, branch, and start to make contact with other neurons (Schoenfeld et al. 1989; Szebenyi et al. 2005). MAP1A is a weak MT stabilizer and is important for the functional maintenance and plasticity in mature neurons (Faller and Brown 2009; Takei et al. 2015). MAP1A interacts with actin and postsynaptic components like PSD93/PSD95, and anchors NMDA receptors to the cytoskeleton, supporting their transport along the dendrites (Brenman et al. 1998; Pedrotti et al. 1994b; Reese et al. 2007; Takei et al. 2015). Loss of the MAP1A-PSD95 interaction has been associated with hearing loss, due to defects in synaptic function (Ikeda et al. 2002). MAP1A also

Fig. 3.2 (continued) conserved and well defined (see tubulin-binding domain: pfam00418). Most of the MAPs also contain a coiled-coil region and binding sites for actin. The *light-blue boxes* in the MAP structure indicate repeat domains contributing in some cases to MT binding (amino terminal repeats in MAP1B or repeats in MAP6) or with unknown function in other cases (carboxyl terminal repeats in MAP1B or MAP6). The *yellow lines* indicate the projection domain of some MAPs (MAP2 projection domain: pfam08377). The following rat protein sequences were used for the drawings: MAP1A (NP_112257.1), MAP1B (NP_062090.1), MAP1S (NP_001099540.1), HMWMAP2 (P15146.1), LMWMAP2 (P15146.4), MAP4 (NP_001019449.1), tau (XP_008766496.1), MAP6 (NP_058900.1), MAP7 (NP_001099740.2), and MAP9 (NP_001129188.1)

Table 3.2 Cell culture – alterations caused by classical MAP downregulation

MAP	Inhibition in culture	References
MAP1A	Activity-induced remodeling of the dendritic arbor (dendritic length and branching) blocked	Szebenyi et al. (2005), Takei et al. (2015) and Leenders et al. (2008)
	Dendritic growth inhibited (in mature cultures)	
	Retraction of existing branches	
	Density of active synapses reduced	
	Synaptic surface density of Ca(V)2.2 decreased	
	Enhanced activity-dependent degradation of PSD-93	
	Reduced surface expression and transport of NR2A/2B	
MAP1B/ MAP5	Axon formation delayed	Gonzalez-Billault et al. (2001, 2002b), DiTella et al. (1996), Bouquet et al. (2004), Del Rio et al. (2004), Tortosa et al. (2011, 2013), and Benoist et al. (2013)
	Reduced neurite and axonal length	
	Increased axonal branching	
	Axon guidance altered	
	Formation and maturation of dendritic spines disrupted	
	AMPA receptor-mediated synaptic currents diminished	
	MT dynamics altered	
	Tyrosinated MTs decreased	
MAP8/ MAP1S	Accumulation of dysfunctional mitochondria and autophagosomes (in cardiomyocytes)	Xie et al. (2011)
MAP2	Neurite and axon outgrowth inhibited	Caceres et al. (1992), Gonzalez-Billault et al. (2002a), Sharma et al. (1994), Harada et al. (2002), Dehmelt et al. (2003), and Iriuchijima et al. (2005)
	Reduction in dendritic length	
	cAMP-dependent protein kinase reduced in the dendrites	
	KAP expression reduced	
MAP3/ MAP4	Transport altered (Xenopus melanophores)	Semenova et al. (2014), Samora et al. (2011), and Nguyen et al. (1999)
	Spindle misorientation (epithelia cells)	
	Decreased content of total tubulin (HeLa cells)	
MAP6	Neurite formation impaired (in PC12 cells)	Guillaud et al. (1998), and Andrieux et al. (2002)
	MT loss to cold or nocodazole	
MAP7	Transport altered (in *Drosophila* neurons)	Barlan et al. (2013) and Sung et al. (2008)
	Kinesin-1 recruitment to MTs and motility impaired (ovary extracts and S2 cells, in *Drosophila*)	
MAP9	Severe mitotic defects and cell death (U2OS cells)	Saffin et al. (2005)

(continued)

Table 3.2 (continued)

MAP	Inhibition in culture	References
Tau	Reduced neurite number and length	Liu et al. (1999), Yu et al. (2008), Caceres and Kosik (1990), Caceres et al. (1991), Dawson et al. (2001), and Zempel et al. (2013)
	Decreased axonal elongation and increased axonal branching	
	Neuronal migration inhibited	
	Delayed axonal extension	
	Dendritic length decreased	
	MTs and synapses resistant to Aβ toxicity	

Table 3.3 Cell culture – alterations caused by classical MAP overexpression

MAP	Overexpression in culture	References
MAP1A	MT dynamic altered	Faller and Brown (2009) and Gupta and Yarwood (2005)
	Enhances Rap1 activation by EPAC1	
MAP1B/MAP5	Cell death	Allen et al. (2005), Tortosa et al. (2013), Tymanskyj et al. (2012), and Opal et al. (2003)
	Altered MT dynamics	
	Delayed neuritogenesis (in PC12 cells)	
MAP8/MAP1S	Neurite degeneration and cell death	Ding et al. (2006a)
	MT alterations	
	Axonal transport disruptions	
MAP2	Increased axonal branching	Fukata et al. (2002) and Dehmelt et al. (2003)
	Neurite formation in N2A	
MAP3/MAP4	Transport altered (Ltk cells and Xenopus melanophores)	Semenova et al. (2014) and Bulinski et al. (1997)
MAP6	Reduced dendritic arborization	Schwenk et al. (2014)
	Accelerated retrograde transport	
MAP7	Formation of noncentrosomal MTs (Vero cells)	Masson and Kreis (1993)
MAP9	Aberrant spindles in mitosis (HEK-293 cells)	Saffin et al. (2005)
Tau	Reduced axon elongation	Fukata et al. (2002), Dubey et al. (2008), Chee et al. (2006), Hoover et al. (2010), Stamer et al. (2002), and Zempel et al. (2013)
	Synaptic responses impaired (mutant tau)	
	Dendritic TTLL6 translocation	
	Decreased MT stability	
	Impaired axonal transport	
	Retraction of growing neurites (NB2 cells)	
	Defective synaptic transmission (in *Drosophila* neuromuscular junctions)	

has a presynaptic function. LC2 interacts with the voltage-dependent calcium channels Ca(V)2.2 and mediates its surface localization at presynaptic boutons (Leenders et al. 2008). In addition, MAP1A has been found to localize to clathrin-coated vesicles and to the cargo-adaptor protein AP2 complex (Murakami et al. 2012; Praefcke et al. 2004). Many MAP1A binding partners have been described as signaling factors, suggesting that MAP1A functions as an adaptor to link signaling molecules to MTs. For example, MAP1A can interact with proteins EPAC (exchange protein directly activated by cAMP) (Gupta and Yarwood 2005), the disrupted-in-schizophrenia 1 protein (DISC1) (Morris et al. 2003), the kinase CK1δ (Wolff et al. 2005), BKCa potassium channel (Park et al. 2004), tubby-like protein-1 (Tulp1) (Grossman et al. 2014), the small GTPases RhoB (Lajoie-Mazenc et al. 2008), and a component of the dystrophin-associated protein complex, α1-syntrophin (Fuhrmann-Stroissnigg et al. 2012). Interestingly, disruption of MAP1A could be a very early manifestation of amyloid β-mediated synaptic dysfunction since sublethal doses of soluble Aβ species induce degradation of MAP1A (Clemmensen et al. 2012).

3.4.1.2 MAP1 Family Member MAP1B

In the 1980s, different labs described MAP1B and named the protein MAP1.2, MAP1(x), and MAP5 (Asai et al. 1985; Bloom et al. 1985; Calvert and Anderton 1985; Riederer et al. 1986). MAP1B is strongly expressed in the nervous system during early embryonic development and downregulated during later developmental stages (Diaz-Nido et al. 1990; Garner et al. 1989; Schoenfeld et al. 1989; Tucker et al. 1988b; Tucker and Matus 1988; Viereck et al. 1989). In the mature brain, MAP1B is present in regions with high plasticity, such as the hippocampus (Schoenfeld et al. 1989; Tucker et al. 1989; Viereck et al. 1989). MAP1B is mainly expressed in neurons; however, it is also detected in neuronal progenitors, oligodendrocytes, and astrocytes (Cheng et al. 1999; Fischer and Romano-Clarke 1990; Ulloa et al. 1994a). In developing neurons, MAP1B is present at high levels in growing axons, whereas in mature neurons MAP1B is also present in dendrites and postsynaptic densities (Fig. 3.1) (Black et al. 1994; Collins et al. 2005; Kawakami et al. 2003; Peng et al. 2004; Tortosa et al. 2011). Posttranslational modifications can affect both MAP1B distribution and function. MAP1B can be phosphorylated by many kinases such as casein kinase II (Diaz-Nido et al. 1988; Ulloa et al. 1993), the serine/threonine protein kinase GSK3 (Garcia-Perez et al. 1998; Tymanskyj et al. 2012), the cyclin-dependent kinase 5 together with its regulatory subunit p35 (Kawauchi et al. 2005; Paglini et al. 1998; Pigino et al. 1997), the dual-specificity tyrosine phosphorylation-regulated kinase DYRK1 (Scales et al. 2009), cdc2 (Ulloa et al. 1994b), and members of the mitogen-activated protein kinase family like ERK1/ERK2 (Loeb et al. 1992) and JNK1 (Chang et al. 2003). MAP1B can bind both actin and MTs, and has been suggested to link the two cytoskeletal elements together (Garcia Rocha and Avila 1995; Mansfield et al. 1991; Pedrotti and Islam 1995; Pedrotti et al. 1996). MAP1B, compared with other MAPs, is a weak MT stabilizer (Takemura et al. 1992). Recent studies suggest that MAP1B preferably associates to dynamic, tyrosinated MTs (Tortosa et al. 2013; Tymanskyj et al. 2012; Utreras et al. 2008). MAP1B also interacts with other MT-interacting proteins, including tubulin-tyrosine ligase and EB1/EB3, dynein regulators like LIS1, scaffolding proteins such as dystonin-α2, and motor protein KIF21A (Cheng et al. 2014; Villarroel-Campos and Gonzalez-Billault 2014).

The role of MAP1B in axon formation has been studied for many years (Gordon-Weeks and Fischer 2000). Suppression of MAP1B with antisense oligonucleotides inhibits laminin-enhanced axon growth (DiTella et al. 1996). In addition, hippocampal neurons from MAP1B-deficient mice show a significant delay in axon outgrowth and a decreased axonal elongation (Gonzalez-Billault et al. 2001; Takei et al. 2000). MAP1B is involved in dendritic spine formation and synaptic maturation by regulating the actin cytoskeleton, and has a role in AMPA receptors endocytosis and long-term depression (LTD) in mature neurons (Benoist et al. 2013; Davidkova and Carroll 2007; Lebeau et al. 2011; Tortosa et al. 2011). Moreover, MAP1B can interact with many neurotransmitter receptors such as GABAc receptor, NMDA receptor subunit NR3A, glycine receptor α1 subunit, mGluR receptors, serotonin receptors, and various channels, including the voltage-gated Ca2+ channel Ca(V)2.2 and the sodium channel Nav1.6. It can also interact with receptor regulatory proteins such as glutamate receptor-interacting protein GRIP1 and the AMPA receptors regulating protein stargazin (Villarroel-Campos and Gonzalez-Billault 2014). It is widely believed that MAP1B could act as a scaffold for many of the above mentioned factors by anchoring them to the MT cytoskeleton or controlling their activity and localization. MAP1B has been linked to various neurodegenerative disorders, including Parkinson's disease (Chan et al. 2014; Jensen et al. 2000), Alzheimer's disease (Gevorkian et al. 2008; Good et al. 2004; Hasegawa et al. 1990), giant axonal neuropathy (Allen et al. 2005; Ding et al. 2002), spinocerebellar ataxia type 1 (Opal et al. 2003), fragile X syndrome (Brown et al. 2001; Lu et al. 2004; Zalfa et al. 2003), and the eye-related muscular disorder congenital fibrosis of the extraocular muscles type 1 (CFEOM) (Cheng et al. 2014).

3.4.1.3 MAP1 Family Member MAP1S

MAP1S (also named MAP8, VCY2IP1, and C19ORF5) is the shortest MAP1 protein. MAP1S is expressed in various tissues, including the brain, where it is predominately present in neurons. Compared to other MAPs, this protein has relatively low expression levels (Ding et al. 2006b; Orban-Nemeth et al. 2005). The light chain of MAP1S not only binds and stabilizes MTs but also contains an actin-binding site, suggesting that it may act as a cross-linker between the MT and actin cytoskeletal networks (Ding et al. 2006b; Orban-Nemeth et al. 2005). Recent evidences suggest that MAP1S can also link mitochondria and autophagosomes to MTs (Xie et al. 2011). In the mammalian brain, interaction between NMDA receptor subunit NR3A and MAP1S has been described, and a role in the trafficking and localization of NR3A-containing NMDAR has been proposed (Eriksson et al. 2007). Another neuronal interactor for MAP1S is the WD40 repeat protein nemitin (Wang et al. 2012). No developmental or behavioral defects have been observed in MAP1S-deficient mice (Xie et al. 2011). However, high levels of MAP1S cause excessive MT stabilization, disrupt axonal transport, and lead to neuronal death (Ding et al. 2006a).

3.4.2 *MAP2 Family of Microtubule-Associated Proteins*

MAP2 is the most abundant structural MAP in the brain (Matus 1988; Olmsted 1986; Wiche et al. 1991). MAP2 is mainly expressed in neurons, but it is also detected in nonneuronal cells such as oligodendrocytes and astrocytes (Papasozomenos and Binder 1986; Vouyiouklis and Brophy 1995). All MAP2 isoforms are transcribed from a single gene. MAP2 isoform has been classified into two groups: high molecular weight MAP2 (HMWMAP2), which includes MAP2A and MAP2B, and low molecular weight (LMWMAP2), with MAP2C and MAP2D. MAP2A is mainly present in the adult brain, whereas MAP2B is expressed all along the development of the nervous system (Binder et al. 1984; Burgoyne and Cumming 1984; Nunez 1988). MAP2C is mainly expressed at early developmental stages, but also present in adult brain areas where neuritogenesis occurs like the retina and olfactory bulb (Goedert et al. 1991; Hartel and Matus 1997; Nunez 1988; Przyborski and Cambray-Deakin 1995; Riederer and Matus 1985; Tucker and Matus 1988; Viereck et al. 1989). In contrast, MAP2D is detected in rat brain early after birth and is present at all developmental stages (Doll et al. 1993; Ferhat et al. 1998).

LMWMAP2 is widely distributed in all neuronal compartments (Albala et al. 1995; Meichsner et al. 1993; Tucker et al. 1988a). In contrast, HMWMAP2 is selectively localized in the cell body and dendrites of neurons (Bernhardt and Matus 1982; Caceres et al. 1984, 1986; Chung et al. 1996; Lev and White 1997; Scheetz and Dubin 1994; Shafit-Zagardo and Kalcheva 1998) (Fig. 3.1). MAP2 has also been found in dendritic spines and postsynaptic densities (Caceres et al. 1983; Fifkova and Morales 1992; Hayashi et al. 1996; Langnaese et al. 1996). The specific localization of MAP2 in dendrites is most likely due to a combination of several mechanisms, including dendrite-specific localization of MAP2 mRNA, suppression of MAP2 sorting into axons, and differential protein stability and turnover of MAP2 dendrites (Garner et al. 1988; Hirokawa et al. 1996; Kanai and Hirokawa 1995; Okabe and Hirokawa 1989).

MAP2 forms bundles and determines MT spacing (Avila et al. 1994; Chen et al. 1992; Cunningham et al. 1997; Dhamodharan and Wadsworth 1995; Felgner et al. 1997; Itoh et al. 1997; Kalcheva et al. 1998; Kowalski and Williams 1993; Kurz and Williams 1995; Vandecandelaere et al. 1996; Yamauchi et al. 1993). MAP2 can also bind to actin and to neurofilaments (Bloom and Vallee 1983; Papasozomenos et al. 1985; Pedrotti et al. 1994a; Selden and Pollard 1983). It is highly phosphorylated and this phosphorylation controls the interaction with MTs (Sanchez et al. 2000). MAP2 is thought to be a structural protein important for neurite outgrowth and maintaining the overall MT architecture in dendrites. Moreover, MAP2 is likely to play a role during neuronal plasticity processes (Fanara et al. 2010; Harada et al. 2002; Teng et al. 2001). Interesting in this respect is the observation that chemical LTD in cultured neurons induced the accumulation of EB3 along MAP2-positive MT bundles in the dendritic shaft (Kapitein et al. 2011). In addition, MAP2 is known to effect MT-based transport and has been involved in the association of rough endoplasmic reticulum membranes with MTs (Farah et al. 2005; Heins et al. 1991; Lopez and Sheetz 1993, 1995; Maas et al. 2009; von Massow et al. 1989). MAP2 also participate in the somato-dendritic localization of numerous signaling proteins, such as cAMP-dependent protein kinase (PKA) and its regulatory subunit (Davare et al. 1999; Obar et al. 1989; Rubino et al. 1989; Theurkauf and Vallee 1982), kinase-associated phosphatase (KAP) (Iriuchijima et al. 2005), phosphatase PP2A/Bα (Sontag et al. 2012), calmodulin (Kotani et al. 1985), and tyrosine-protein kinases Src and Fyn (Lim and Halpain 2000; Zamora-Leon et al. 2001). MAP2 can also bind other neuronal proteins, including calcium channels (Davare et al. 1999), MAP2 RNA transacting proteins (MARTA) (Rehbein et al. 2000), CRMP5 (Brot et al. 2010), neural cell adhesion molecule L1 (L1CAM) (Poplawski et al. 2012), and the KIND domain containing RasGEF, very-KIND (Huang et al. 2007). Changes in MAP2 expression levels have been linked to numerous neurological and neurodegenerative diseases such as schizophrenia (Rosoklija et al. 2005) and epilepsy (Jalava et al. 2007; Yan et al. 2012), Alzheimer's disease (Canas et al. 2009; Capetillo-Zarate et al. 2006; Dziewczapolski et al. 2009; Moolman et al. 2004; Takahashi et al. 2013; Wu et al. 2004), spinal cord injury (Abdanipour et al. 2014; Gonzalez et al. 2009), stress (Yan et al. 2010), myotonic dystrophy (Velazquez-Bernardino et al. 2012), and prion diseases (Zhang and Dong 2012).

3.4.3 MAP3/MAP4 Family of Microtubule-Associated Proteins

MAP3 was originally described as a heat-stable protein in a variety of tissues, including the brain, and was later found to be identical to MAP4 (Bulinski and Borisy 1980; Kobayashi et al. 2000; Parysek et al. 1984). MAP4 is encoded by a single gene from which multiple mRNAs are transcribed (Code and Olmsted 1992; West et al. 1991). During development, MAP4 appears transiently in some neurons, while persisting in others, suggesting that MAP4 could be involved in early development. In the adult brain, MAP4 is present in both neurons and glia, but in neurons it is restricted to neurofilament-rich axons (Bernhardt et al. 1985; Huber et al. 1985; Matsushima et al. 2005; Matus et al. 1983; Tokuraku et al. 2010; Voss et al. 1998). Therefore, MAP4 has been considered as a candidate to cross-link MTs and neurofilaments (Huber et al. 1985). In cultures, MAP4 shows patchy staining patterns and localizes at branching points (Tokuraku et al. 2010). In addition, a shorter and neuron-specific isoform has been described, which is restricted to neural ectoderm-derived tissues such as the brain and the adrenal medulla, and its expression is augmented by the addition of nerve growth factor (Matsushima et al. 2005). The short MAP4 isoform still induces MT assembly, but is unable to form bundles (Matsushima et al. 2005). MAP4 promotes tubulin polymerization in vitro and co-localizes with MTs in vivo, binding both glutamylated and tyrosinated MTs (Chapin and Bulinski 1994; Huber et al. 1986; West et al. 1991). The binding of MAP4 to MTs induces conformational changes that promote the overall MT stability (Xiao et al. 2012). MAP4 has also been reported to alter the MT surface properties and affect kinesin-driven movement in vitro (Tokuraku et al. 2007). In nonneuronal cells, overexpression of MAP4 inhibits organelle motility and trafficking, and it inhibits kinesin-driven MT gliding (Bulinski et al. 1997; Tokuraku et al. 2007). In addition, it has been described to interact with p150Glued, part of the dynein-dynactin complex, and to inhibit dynein-mediated MT sliding (Samora et al. 2011). In *Xenopus* melanophores, the binding of XMAP4 to MTs negatively regulates dynein-dependent motility and positively regulates kinesin-2-based cargo movements (Semenova et al. 2014). MAP4 can also regulate the activity of MT-destabilizing factors, such as kinesin-related protein MCAK/XKCM1 and the MAP stathmin (Holmfeldt et al. 2002). Moreover, MAP4 has been suggested to inhibit katanin, by preventing its binding to MTs (McNally et al. 2002). Little is known about the precise role of MAP4 and its interaction partners in neuronal cells. MAP4 has been found upregulated by anti-depressants in rat hippocampus and is linked to chronic stress (Yang et al. 2003). It has also been shown that MAP4 can inhibit muscarinic receptor recovery after agonist exposure in neuroblastoma cells (Cheng et al. 2002). MAP4 is regulated by different kinases, such as protein kinase C (PKC) (Mori et al. 1991), mitogen-activated protein kinase (MAPK) (Hoshi et al. 1992), and protein kinases MARK (Ebneth et al. 1999; Illenberger et al. 1996). Alteration in MAP4 has been linked to cardiac hypertrophy (Cheng et al. 2005; Kumarapeli and Wang 2004), different types of cancer (Bash-Babula et al. 2002; Hait and Yang 2006; Holmfeldt et al. 2003), and Alzheimer's disease (Ray et al. 2008).

3.4.4 MAP6 Family of Microtubule-Associated Proteins

MAP6, also named STOP (stable tubule-only peptides), was discovered as the major factor able to confer cold and nocodazole resistance to neuronal MTs (Andrieux et al. 2002; Bosc et al. 1996; Guillaud et al. 1998). MAP6 is a calmodulin-regulated protein (Job et al. 1981, 1983). Calmodulin-binding sites partially overlap with the MT-binding domain, impairing the binding of MAP6 to MTs. MAP6 also includes four consensus sites for phosphorylation by CaM kinase II that regulates its binding to MTs (Bosc et al. 2003). MAP6 can also bind actin (Baratier et al. 2006). Phosphorylated forms of MAP6 cannot bind MTs and co-localize with actin along the neurites and at branching points. MAP6 is only found in vertebrates and is expressed in several tissues like the brain, heart, muscle, kidney, lung, and testis (Aguezzoul et al. 2003). In the brain, MAP6 is expressed in neurons, astrocytes, and oligodendrocytes (Denarier et al. 1998b; Galiano et al. 2004; Guillaud et al. 1998; Ochoa et al. 2011). Two main splice isoforms are described in neurons; MAP6-E and MAP6-N (Aguezzoul et al. 2003; Denarier et al. 1998a). MAP6-E is the most abundant isoform in embryonic rodent brain and persists in adult brain. MAP6-N appears at birth and its expression is maintained in the adult brain (Bosc et al. 1996; Guillaud et al. 1998).

MAP6 has a strong preference for stable MTs (Bonnet et al. 2002; Slaughter and Black 2003). In neurons from dorsal root ganglia, MAP6 is present in the cell body and throughout the axon, but its expression is reduced in the distal portion of the axon (Fig. 3.1) (Guillaud et al. 1998). MAP6 is also present in dendrites and there are some evidences that localize MAP6 to synapses in mature hippocampal neurons (Andrieux et al. 2002; Baratier et al. 2006; Peng et al. 2004). CaMKII phosphorylation may promote MAP6 translocation from MTs to synaptic compartments where it interacts with actin. This translocation could be important during synaptic plasticity (Baratier et al. 2006). MAP6 is also important for neurite formation and dendritic arborization, where it has been suggested to act as a molecular brake for lysosomal trafficking in dendrites (Guillaud et al. 1998; Schwenk et al. 2014). Initial studies showed that MAP6 knockout mice have reduced number of synaptic vesicles and impaired synaptic plasticity (Andrieux et al. 2002). Later, it was shown that, in addition to changes in glutamatergic synaptic transmission, these mice also present alterations in dopaminergic, acetylcholinergic, nicotinic, serotonergic, and noradrenergic neurotransmission (Bouvrais-Veret et al. 2007, 2008; Brun et al. 2005; Delotterie et al. 2010; Fournet et al. 2010, 2012b; Fradley et al. 2005; Kajitani et al. 2010; Powell et al. 2007). These alterations recapitulate some clinical features observed in schizophrenia disorders (Andrieux et al. 2002; Fournet et al. 2012b). Interestingly, chronic treatments with both typical and atypical antipsychotics improve some defects in MAP6 knockout mice (Andrieux et al. 2002; Begou et al. 2008; Brun et al. 2005; Delotterie et al. 2010; Fradley et al. 2005; Merenlender-Wagner et al. 2010). In addition, treatments with MT-stabilizing compounds like epothilone D and NAP (davunetide) also improved some of the deficits (Andrieux et al. 2006; Fournet et al. 2012a; Merenlender-Wagner et al. 2010). Neuronal transport alterations have also been described in MAP6 knockout neurons, and epothilone D treatments can overcome these defects (Daoust et al. 2014). These results suggest that MT-stabilizing drugs may restore MT stability and subsequent transport functions in neurons.

3.4.5 MAP7 Family of Microtubule-Associated Proteins

MAP7, also called ensconsin and E-MAP-115 (epithelial MAP of 115 kD), was first identified in HeLa cells and purified by its ability to tightly associate with MTs (Bulinski and Bossler 1994; Masson and Kreis 1993). MAP7 is predominantly expressed in epithelial cells where its expression correlates with the presence of stable MTs (Masson and Kreis 1993). Subsequent experiments showed that its association with MTs is very dynamic, and modest expression does not directly affect MT stability (Bulinski et al. 2001; Faire et al. 1999). The MT-binding domain of MAP7 fused to GFP (GFP-EMTB) is used in many labs to label MTs in living cells because it provides a non-perturbing label of the MT network (Bulinski et al. 1999).

MAP7 is widely expressed in the adult mouse; it is predominantly expressed in epithelial cells and in certain neuronal cell types, such as neurons of the trigeminal and dorsal root ganglia (Fabre-Jonca et al. 1998). MAP7 has been suggested to modulate MT functions and control the MT anchoring of other cellular factors (Faire et al. 1999). It has also been demonstrated to be a regulator of kinesin motors. MAP7 is a cofactor of kinesin-1 and required for organelle transport in *Drosophila* neurons (Barlan et al. 2013). MAP7 mutants display defects in motor localization without altering MT cytoskeleton (Sung et al. 2008). In addition, MAP7 has been shown to recruit kinesin-1 to the MT and regulate nuclear positioning, which is essential for skeletal muscle function in myoblasts (Metzger et al. 2012).

3.4.6 MAP9 Family of Microtubule-Associated Proteins

MAP9, also named ASAP (ASter-associated protein), has been characterized as a novel human spindle protein with a role in the correct bipolar spindle assembly and centrosome maintenance (Saffin et al. 2005). MAP9 is phosphorylated and regulated by the mitotic kinases Aurora A and PLK1 (Eot-Houllier et al. 2010; Venoux et al. 2008). It has also been demonstrated that, in response to DNA damage, MAP9 can interact and stabilize p53 (Basbous et al. 2012). In zebra fish, MAP9 is expressed during early embryo development and localizes to the mitotic spindle and centrosomes. MAP9 is expressed in the mammalian nervous system in various brain regions and associates with the mitotic spindle. In addition, both MAP9 knockdown and overexpression produce several developmental defects, leading to early embryonic lethality (Fontenille et al. 2014). However, the precise function of MAP9 remains largely unknown.

3.4.7 Tau/MAPT Family of Microtubule-Associated Proteins

Tau was originally discovered as a MAP that decreases the concentration at which tubulin polymerizes into MTs (Cleveland et al. 1977a, b; Fellous et al. 1977; Weingarten et al. 1975). Tau has many functions; it stabilizes MTs (Bre and Karsenti 1990; Drechsel et al. 1992; Drubin and Kirschner 1986), regulates MT modifications (Perez et al. 2009), alters the mechanical properties of MTs by enhancing their stiffness (Choi et al. 2009; Peck et al. 2011; Samsonov et al. 2004), and functions as a spacer between adjacent MTs (Chen et al. 1992). In neurons, the physiological function of tau is to support assembly and stabilization of axonal MTs to promote neuritogenesis (Caceres and Kosik 1990; Cleveland et al. 1977a; Drubin and Kirschner 1986; Goode et al. 1997). Tau also interacts with the neuronal plasma membrane and anchors enzymes to MTs (Brandt et al. 1995; Lee et al. 1998; Liao et al. 1998; Sontag et al. 1999). Tau can also associate with spectrin (Carlier et al. 1984), actin (Griffith and Pollard 1982), PP1 and PP2A (Liao et al. 1998; Sontag et al. 1999), kinases like CDK5 (Sobue et al. 2000), presenilin 1 (Takashima et al. 1998), α-synuclein (Jensen et al. 1999), phospholipase C (Hwang et al. 1996), the Fyn tyrosine kinase (Klein et al. 2002; Lee et al. 1998), apolipoprotein E (Strittmatter et al. 1994), and calmodulin (Baudier et al. 1987). It can also bind to chaperones such as Hsp70, Hsp90, and Pin-1 (Dou et al. 2003; Lu et al. 1999). A recent study reveals how Hsp90 binds to tau's aggregation-prone MT-binding repeat region (Karagoz et al. 2014). Another possible role for tau is the regulation of kinesin-based transport (Ebneth et al. 1998; Sparacino et al. 2014; Terwel et al. 2002). Tau reduces the attachment of kinesins to MTs and interferes with their transport when is overexpressed, in both in vivo and in vitro experiments (Dixit et al. 2008; Ebneth et al. 1998; Seitz et al. 2002; Stamer et al. 2002; Trinczek et al. 1999; Vershinin et al. 2007).

Tau has several isoforms that are differentially expressed during development. In addition, tau contains many different posttranslational modifications; it can be modified by phosphorylation, glycosylation, ubiquitinylation, glycation, deamidation, oxidation, and truncation (Avila et al. 2004). Tau is mainly present in neurons (Arrasate et al. 1999; Chin and Goldman 1996) and specifically localizes to axons (Binder et al. 1985; Drubin and Kirschner 1986; Kosik and Finch 1987; Mandell and Banker 1996; Migheli et al. 1988), with a strong proximal to distal gradient (Black et al. 1996; Kempf et al. 1996; Mandell and Banker 1996) (Fig. 3.1). It is however unclear how the specific axonal localization of tau is controlled (Scholz and Mandelkow 2014). Under pathological conditions, tau is also present in dendrites, including dendritic spines (Ittner et al. 2010; Kremer et al. 2011). Synaptic activation in cultured cortical neurons and induction of long-term potentiation (LTP) in acute hippocampal slices trigger the translocation of endogenous tau to the postsynaptic compartment. Exposure to amyloid-β oligomers also induces mislocalization of tau to spines (Frandemiche et al. 2014). The synaptic damage induced by amyloid-β oligomers is most likely triggered by the missorting of newly synthesized tau into dendrites. In mouse models for Alzheimer's disease, mutant tau accumulates in dendritic spines, where it suppresses synaptic responses (Hoover et al. 2010). In addition, the expression of mutant tau results in a significant loss of dendritic spines and synapses (Bittner et al. 2010; Mocanu et al. 2008; Rocher et al. 2010). Indeed, in the human brain, tau aggregation also affects dendritic spines (Merino-Serrais et al. 2013). Current models suggest that tau mislocalization recruits the MT polyglutamylation enzyme TTLL6 (tubulin-tyrosine ligase-like-6) into dendrites. Enhanced polyglutamylation of dendritic MTs may subsequently recruit spastin and induces MT breakdown in dendrites (Lacroix et al. 2010; Zempel et al. 2013).

Several tau knockout lines have been generated; all of them are viable and do not display strong phenotype (Dawson et al. 2001; Fujio et al. 2007; Harada et al. 1994; Tucker et al. 2001). However, some muscle weakness and motor and cognitive deficits have been described (Ikegami et al. 2000; Lei et al. 2012; Ma et al. 2014). In addition, age-dependent brain atrophy and neuronal and synapse loss can be found in tau knockout mice (Lei et al. 2012). At cellular level, tau knockout neurons present a decrease in the axon caliber, together with a change in MT stability and organization (Harada et al. 1994). Whereas some models show normal axonal development in culture, a significant delay of the maturation of cultured neurons has been described in others (Dawson et al. 2001; Harada et al. 1994). An increased in MAP1A in tau knockout mice suggests a possible compensation effect in these animal models (Dawson et al. 2001; Fujio et al. 2007; Harada et al. 1994; Ma et al. 2014).

3.5 Discussion and Outlook

For a long time, MTs were considered as static structures distributed along the axon and dendrites. However, many fundamental neurodevelopmental processes such as differentiation, dendritic branching, and synapse formation require a continuous reorganization of the MT cytoskeleton. Moreover, the organization of the MT cytoskeleton largely depends on a tight regulation by many different extracellular and intracellular signaling pathways. Axons and dendrites present different patterns of MT organization that may underlie the different functions of these compartments. The two compartments not only differ in MT orientation but also in stability and posttranslational modifications. These biochemical variations generate distinct MT patterns in neurons and may be directly responsible for sorting cargo transport into axons or dendrites. It will be important to understand how variations in MT patterns can generate MT diversity and drive polarized cargo transport. Future research should help to resolve the basic mechanisms of MT assembly and remodeling. This knowledge will aid us to better understand dendritic development and the alterations that occur in neurological and neurodegenerative diseases.

In this chapter, we have discussed how classical MAPs regulate various MT roles during neuronal development and function. Apart from their traditional role in stabilizing and bundling MTs, MAPs have many additional functions in neurons. Many of the classical MAPs have been described to interact with numerous cellular components and participate in many different signaling pathways. Despite the structural and functional complexity of the MAP family members, the rules governing specific MT organization in axons and dendrites are slowly being revealed. The difficult task ahead will be to sort out how these different MAPs cooperate with each other in time and space to build specific MT arrays. In addition, some very basic questions still remain unknown, such as how do MAPs bind to a subset of dendritic MTs and how MAPs are specifically localized in different compartments. Moreover, what is the spatiotemporal dynamics of MAPs in relationship to MTs dynamics itself and what is the percentage of MTs that is occupied by MAPs. In addition, many different functions of MAPs remain still unexplored. For example, their role in MT nucleation and organization is still largely unknown. In this respect, recent developments in super-resolution imaging are promising techniques to visualize the detailed subcellular localization patterns in developing and mature neurons. The strong implications for MTs and MAPs in many different neurological diseases will stimulate many new research efforts in the near future.

Acknowledgments We thank Phebe Wulf for critically reading the manuscript. This work was supported by the Netherlands Organization for Scientific Research (NOW-ALW-VIDI, LCK; NWO-ALW-VICI, CCH); the Foundation for Fundamental Research on Matter (FOM, LCK, and CCH), which is part of the NWO; the Netherlands Organization for Health Research and Development (ZonMW-TOP, CCH); the European Research Council (ERC) (ERC-starting, LCK; ERC-consolidator, CCH); the Spanish Education Ministry (ET); and an FP7 EU Marie Curie postdoctoral fellowship (ET).

References

Abdanipour A, Schluesener HJ, Tiraihi T, Noori-Zadeh A (2014) Systemic administration of valproic acid stimulates overexpression of microtubule-associated protein 2 in the spinal cord injury model to promote neurite outgrowth. Neurol Res 37(3):223–228. 1743132814Y0000000438

Aguezzoul M, Andrieux A, Denarier E (2003) Overlap of promoter and coding sequences in the mouse STOP gene (Mtap6). Genomics 81:623–627

Akhmanova A, Hoogenraad CC (2015) Microtubule minus-end-targeting proteins. Curr Biol 25: R162–R171

Akhmanova A, Steinmetz MO (2008) Tracking the ends: a dynamic protein network controls the fate of microtubule tips. Nat Rev Mol Cell Biol 9:309–322

Albala JS, Kress Y, Liu WK, Weidenheim K, Yen SH, Shafit-Zagardo B (1995) Human microtubule-associated protein-2c localizes to dendrites and axons in fetal spinal motor neurons. J Neurochem 64:2480–2490

Al-Bassam J, Ozer RS, Safer D, Halpain S, Milligan RA (2002) MAP2 and tau bind longitudinally along the outer ridges of microtubule protofilaments. J Cell Biol 157:1187–1196

Allen E, Ding J, Wang W, Pramanik S, Chou J, Yau V, Yang Y (2005) Gigaxonin-controlled degradation of MAP1B light chain is critical to neuronal survival. Nature 438:224–228

Andrieux A, Salin PA, Vernet M, Kujala P, Baratier J, Gory-Faure S, Bosc C, Pointu H, Proietto D, Schweitzer A et al (2002) The suppression of brain cold-stable microtubules in mice induces synaptic defects associated with neuroleptic-sensitive behavioral disorders. Genes Dev 16:2350–2364

Andrieux A, Salin P, Schweitzer A, Begou M, Pachoud B, Brun P, Gory-Faure S, Kujala P, Suaud-Chagny MF, Hofle G et al (2006) Microtubule stabilizer ameliorates synaptic function and behavior in a mouse model for schizophrenia. Biol Psychiatry 60:1224–1230

Arrasate M, Perez M, Armas-Portela R, Avila J (1999) Polymerization of tau peptides into fibrillar structures. The effect of FTDP-17 mutations. FEBS Lett 446:199–202

Asai DJ, Thompson WC, Wilson L, Dresden CF, Schulman H, Purich DL (1985) Microtubule-associated proteins (MAPs): a monoclonal antibody to MAP 1 decorates microtubules in vitro but stains stress fibers and not microtubules in vivo. Proc Natl Acad Sci U S A 82:1434–1438

Avila J, Dominguez J, Diaz-Nido J (1994) Regulation of microtubule dynamics by microtubule-associated protein expression and phosphorylation during neuronal development. Int J Dev Biol 38:13–25
Avila J, Lucas JJ, Perez M, Hernandez F (2004) Role of tau protein in both physiological and pathological conditions. Physiol Rev 84:361–384
Baas PW, Ahmad FJ (2013) Beyond taxol: microtubule-based treatment of disease and injury of the nervous system. Brain: J Neurol 136:2937–2951
Baas PW, Lin S (2011) Hooks and comets: the story of microtubule polarity orientation in the neuron. Dev Neurobiol 71:403–418
Baas PW, Deitch JS, Black MM, Banker GA (1988) Polarity orientation of microtubules in hippocampal neurons: uniformity in the axon and nonuniformity in the dendrite. Proc Natl Acad Sci U S A 85:8335–8339
Baas PW, Slaughter T, Brown A, Black MM (1991) Microtubule dynamics in axons and dendrites. J Neurosci Res 30:134–153
Baratier J, Peris L, Brocard J, Gory-Faure S, Dufour F, Bosc C, Fourest-Lieuvin A, Blanchoin L, Salin P, Job D et al (2006) Phosphorylation of microtubule-associated protein STOP by calmodulin kinase II. J Biol Chem 281:19561–19569
Barlan K, Lu W, Gelfand VI (2013) The microtubule-binding protein ensconsin is an essential cofactor of kinesin-1. Curr Biol 23:317–322
Basbous J, Knani D, Bonneaud N, Giorgi D, Brondello JM, Rouquier S (2012) Induction of ASAP (MAP9) contributes to p53 stabilization in response to DNA damage. Cell Cycle 11:2380–2390
Bash-Babula J, Toppmeyer D, Labassi M, Reidy J, Orlick M, Senzon R, Alli E, Kearney T, August D, Shih W et al (2002) A Phase I/pilot study of sequential doxorubicin/vinorelbine: effects on p53 and microtubule-associated protein 4. Clin Cancer Res 8:1057–1064
Baudier J, Mochly-Rosen D, Newton A, Lee SH, Koshland DE Jr, Cole RD (1987) Comparison of S100b protein with calmodulin: interactions with melittin and microtubule-associated tau proteins and inhibition of phosphorylation of tau proteins by protein kinase C. Biochemistry 26:2886–2893
Beffert U, Dillon GM, Sullivan JM, Stuart CE, Gilbert JP, Kambouris JA, Ho A (2012) Microtubule plus-end tracking protein CLASP2 regulates neuronal polarity and synaptic function. J Neurosci 32:13906–13916
Begou M, Volle J, Bertrand JB, Brun P, Job D, Schweitzer A, Saoud M, D'Amato T, Andrieux A, Suaud-Chagny MF (2008) The stop null mice model for schizophrenia displays [corrected] cognitive and social deficits partly alleviated by neuroleptics. Neuroscience 157:29–39
Benoist M, Palenzuela R, Rozas C, Rojas P, Tortosa E, Morales B, Gonzalez-Billault C, Avila J, Esteban JA (2013) MAP1B-dependent Rac activation is required for AMPA receptor endocytosis during long-term depression. EMBO J 32:2287–2299
Bernhardt R, Matus A (1982) Initial phase of dendrite growth: evidence for the involvement of high molecular weight microtubule-associated proteins (HMWP) before the appearance of tubulin. J Cell Biol 92:589–593
Bernhardt R, Huber G, Matus A (1985) Differences in the developmental patterns of three microtubule-associated proteins in the rat cerebellum. J Neurosci 5:977–991
Binder LI, Frankfurter A, Kim H, Caceres A, Payne MR, Rebhun LI (1984) Heterogeneity of microtubule-associated protein 2 during rat brain development. Proc Natl Acad Sci U S A 81:5613–5617
Binder LI, Frankfurter A, Rebhun LI (1985) The distribution of tau in the mammalian central nervous system. J Cell Biol 101:1371–1378
Bittner T, Fuhrmann M, Burgold S, Ochs SM, Hoffmann N, Mitteregger G, Kretzschmar H, LaFerla FM, Herms J (2010) Multiple events lead to dendritic spine loss in triple transgenic Alzheimer's disease mice. PLoS One 5:e15477
Black MM, Slaughter T, Fischer I (1994) Microtubule-associated protein 1b (MAP1b) is concentrated in the distal region of growing axons. J Neurosci 14:857–870

Black MM, Slaughter T, Moshiach S, Obrocka M, Fischer I (1996) Tau is enriched on dynamic microtubules in the distal region of growing axons. J Neurosci 16:3601–3619
Bloom GS, Vallee RB (1983) Association of microtubule-associated protein 2 (MAP 2) with microtubules and intermediate filaments in cultured brain cells. J Cell Biol 96:1523–1531
Bloom GS, Luca FC, Vallee RB (1985) Microtubule-associated protein 1B: identification of a major component of the neuronal cytoskeleton. Proc Natl Acad Sci U S A 82:5404–5408
Bonnet C, Denarier E, Bosc C, Lazereg S, Denoulet P, Larcher JC (2002) Interaction of STOP with neuronal tubulin is independent of polyglutamylation. Biochem Biophys Res Commun 297:787–793
Bornens M (2008) Organelle positioning and cell polarity. Nat Rev Mol Cell Biol 9:874–886
Bosc C, Cronk JD, Pirollet F, Watterson DM, Haiech J, Job D, Margolis RL (1996) Cloning, expression, and properties of the microtubule-stabilizing protein STOP. Proc Natl Acad Sci U S A 93:2125–2130
Bosc C, Andrieux A, Job D (2003) STOP proteins. Biochemistry 42:12125–12132
Bouquet C, Soares S, von Boxberg Y, Ravaille-Veron M, Propst F, Nothias F (2004) Microtubule-associated protein 1B controls directionality of growth cone migration and axonal branching in regeneration of adult dorsal root ganglia neurons. J Neurosci 24:7204–7213
Bouvrais-Veret C, Weiss S, Andrieux A, Schweitzer A, McIntosh JM, Job D, Giros B, Martres MP (2007) Sustained increase of alpha7 nicotinic receptors and choline-induced improvement of learning deficit in STOP knock-out mice. Neuropharmacology 52:1691–1700
Bouvrais-Veret C, Weiss S, Hanoun N, Andrieux A, Schweitzer A, Job D, Hamon M, Giros B, Martres MP (2008) Microtubule-associated STOP protein deletion triggers restricted changes in dopaminergic neurotransmission. J Neurochem 104:745–756
Brandt R, Leger J, Lee G (1995) Interaction of tau with the neural plasma membrane mediated by tau's amino-terminal projection domain. J Cell Biol 131:1327–1340
Bratman SV, Chang F (2008) Mechanisms for maintaining microtubule bundles. Trends Cell Biol 18:580–586
Bre MH, Karsenti E (1990) Effects of brain microtubule-associated proteins on microtubule dynamics and the nucleating activity of centrosomes. Cell Motil Cytoskeleton 15:88–98
Brenman JE, Topinka JR, Cooper EC, McGee AW, Rosen J, Milroy T, Ralston HJ, Bredt DS (1998) Localization of postsynaptic density-93 to dendritic microtubules and interaction with microtubule-associated protein 1A. J Neurosci 18:8805–8813
Brot S, Rogemond V, Perrot V, Chounlamountri N, Auger C, Honnorat J, Moradi-Ameli M (2010) CRMP5 interacts with tubulin to inhibit neurite outgrowth, thereby modulating the function of CRMP2. J Neurosci 30:10639–10654
Brouhard GJ, Rice LM (2014) The contribution of alpha/beta-tubulin curvature to microtubule dynamics. J Cell Biol 207:323–334
Brown V, Jin P, Ceman S, Darnell JC, O'Donnell WT, Tenenbaum SA, Jin X, Feng Y, Wilkinson KD, Keene JD et al (2001) Microarray identification of FMRP-associated brain mRNAs and altered mRNA translational profiles in fragile X syndrome. Cell 107:477–487
Brun P, Begou M, Andrieux A, Mouly-Badina L, Clerget M, Schweitzer A, Scarna H, Renaud B, Job D, Suaud-Chagny MF (2005) Dopaminergic transmission in STOP null mice. J Neurochem 94:63–73
Bulinski JC, Borisy GG (1980) Widespread distribution of a 210,000 mol wt microtubule-associated protein in cells and tissues of primates. J Cell Biol 87:802–808
Bulinski JC, Bossler A (1994) Purification and characterization of ensconsin, a novel microtubule stabilizing protein. J Cell Sci 107(Pt 10):2839–2849
Bulinski JC, McGraw TE, Gruber D, Nguyen HL, Sheetz MP (1997) Overexpression of MAP4 inhibits organelle motility and trafficking in vivo. J Cell Sci 110(Pt 24):3055–3064
Bulinski JC, Gruber D, Faire K, Prasad P, Chang W (1999) GFP chimeras of E-MAP-115 (ensconsin) domains mimic behavior of the endogenous protein in vitro and in vivo. Cell Struct Funct 24:313–320

Bulinski JC, Odde DJ, Howell BJ, Salmon TD, Waterman-Storer CM (2001) Rapid dynamics of the microtubule binding of ensconsin in vivo. J Cell Sci 114:3885–3897
Burgin KE, Ludin B, Ferralli J, Matus A (1994) Bundling of microtubules in transfected cells does not involve an autonomous dimerization site on the MAP2 molecule. Mol Biol Cell 5:511–517
Burgoyne RD, Cumming R (1984) Ontogeny of microtubule-associated protein 2 in rat cerebellum: differential expression of the doublet polypeptides. Neuroscience 11:156–167
Burton PR (1988) Dendrites of mitral cell neurons contain microtubules of opposite polarity. Brain Res 473:107–115
Caceres A, Kosik KS (1990) Inhibition of neurite polarity by tau antisense oligonucleotides in primary cerebellar neurons. Nature 343:461–463
Caceres A, Payne MR, Binder LI, Steward O (1983) Immunocytochemical localization of actin and microtubule-associated protein MAP2 in dendritic spines. Proc Natl Acad Sci U S A 80:1738–1742
Caceres A, Banker G, Steward O, Binder L, Payne M (1984) MAP2 is localized to the dendrites of hippocampal neurons which develop in culture. Brain Res 315:314–318
Caceres A, Banker GA, Binder L (1986) Immunocytochemical localization of tubulin and microtubule-associated protein 2 during the development of hippocampal neurons in culture. J Neurosci 6:714–722
Caceres A, Potrebic S, Kosik KS (1991) The effect of tau antisense oligonucleotides on neurite formation of cultured cerebellar macroneurons. J Neurosci 11:1515–1523
Caceres A, Mautino J, Kosik KS (1992) Suppression of MAP2 in cultured cerebellar macroneurons inhibits minor neurite formation. Neuron 9:607–618
Calvert R, Anderton BH (1985) A microtubule-associated protein (MAP1) which is expressed at elevated levels during development of the rat cerebellum. EMBO J 4:1171–1176
Canas PM, Porciuncula LO, Cunha GM, Silva CG, Machado NJ, Oliveira JM, Oliveira CR, Cunha RA (2009) Adenosine A2A receptor blockade prevents synaptotoxicity and memory dysfunction caused by beta-amyloid peptides via p38 mitogen-activated protein kinase pathway. J Neurosci 29:14741–14751
Capetillo-Zarate E, Staufenbiel M, Abramowski D, Haass C, Escher A, Stadelmann C, Yamaguchi H, Wiestler OD, Thal DR (2006) Selective vulnerability of different types of commissural neurons for amyloid beta-protein-induced neurodegeneration in APP23 mice correlates with dendritic tree morphology. Brain: J Neurol 129:2992–3005
Carlier MF, Simon C, Cassoly R, Pradel LA (1984) Interaction between microtubule-associated protein tau and spectrin. Biochimie 66:305–311
Chan SL, Chua LL, Angeles DC, Tan EK (2014) MAP1B rescues LRRK2 mutant-mediated cytotoxicity. Mol Brain 7:29
Chang L, Jones Y, Ellisman MH, Goldstein LS, Karin M (2003) JNK1 is required for maintenance of neuronal microtubules and controls phosphorylation of microtubule-associated proteins. Dev Cell 4:521–533
Chapin SJ, Bulinski JC (1994) Cellular microtubules heterogeneous in their content of microtubule-associated protein 4 (MAP4). Cell Motil Cytoskeleton 27:133–149
Chee F, Mudher A, Newman TA, Cuttle M, Lovestone S, Shepherd D (2006) Overexpression of tau results in defective synaptic transmission in Drosophila neuromuscular junctions. Biochem Soc Trans 34:88–90
Chen J, Kanai Y, Cowan NJ, Hirokawa N (1992) Projection domains of MAP2 and tau determine spacings between microtubules in dendrites and axons. Nature 360:674–677
Chen Y, Rolls MM, Hancock WO (2014) An EB1-kinesin complex is sufficient to steer microtubule growth in vitro. Curr Biol 24:316–321
Cheng A, Krueger BK, Bambrick LL (1999) MAP5 expression in proliferating neuroblasts. Brain Res Dev Brain Res 113:107–113
Cheng G, Iijima Y, Ishibashi Y, Kuppuswamy D, Cooper G (2002) Inhibition of G protein-coupled receptor trafficking in neuroblastoma cells by MAP 4 decoration of microtubules. Am J Physiol Heart Circ Physiol 283:H2379–H2388

Cheng G, Qiao F, Gallien TN, Kuppuswamy D, Cooper G (2005) Inhibition of beta-adrenergic receptor trafficking in adult cardiocytes by MAP4 decoration of microtubules. Am J Physiol Heart Circ Physiol 288:H1193–H1202

Cheng L, Desai J, Miranda CJ, Duncan JS, Qiu W, Nugent AA, Kolpak AL, Wu CC, Drokhlyansky E, Delisle MM et al (2014) Human CFEOM1 mutations attenuate KIF21A autoinhibition and cause oculomotor axon stalling. Neuron 82:334–349

Chevalier-Larsen E, Holzbaur EL (2006) Axonal transport and neurodegenerative disease. Biochim Biophys Acta 1762:1094–1108

Chin SS, Goldman JE (1996) Glial inclusions in CNS degenerative diseases. J Neuropathol Exp Neurol 55:499–508

Choi MC, Raviv U, Miller HP, Gaylord MR, Kiris E, Ventimiglia D, Needleman DJ, Kim MW, Wilson L, Feinstein SC et al (2009) Human microtubule-associated-protein tau regulates the number of protofilaments in microtubules: a synchrotron x-ray scattering study. Biophys J 97:519–527

Chung WJ, Kindler S, Seidenbecher C, Garner CC (1996) MAP2a, an alternatively spliced variant of microtubule-associated protein 2. J Neurochem 66:1273–1281

Clemmensen C, Aznar S, Knudsen GM, Klein AB (2012) The microtubule-associated protein 1A (MAP1A) is an early molecular target of soluble Abeta-peptide. Cell Mol Neurobiol 32:561–566

Cleveland DW, Hwo SY, Kirschner MW (1977a) Physical and chemical properties of purified tau factor and the role of tau in microtubule assembly. J Mol Biol 116:227–247

Cleveland DW, Hwo SY, Kirschner MW (1977b) Purification of tau, a microtubule-associated protein that induces assembly of microtubules from purified tubulin. J Mol Biol 116:207–225

Code RJ, Olmsted JB (1992) Mouse microtubule-associated protein 4 (MAP4) transcript diversity generated by alternative polyadenylation. Gene 122:367–370

Collins MO, Yu L, Coba MP, Husi H, Campuzano I, Blackstock WP, Choudhary JS, Grant SG (2005) Proteomic analysis of in vivo phosphorylated synaptic proteins. J Biol Chem 280:5972–5982

Conde C, Caceres A (2009) Microtubule assembly, organization and dynamics in axons and dendrites. Nat Rev Neurosci 10:319–332

Cunningham CC, Leclerc N, Flanagan LA, Lu M, Janmey PA, Kosik KS (1997) Microtubule-associated protein 2c reorganizes both microtubules and microfilaments into distinct cytological structures in an actin-binding protein-280-deficient melanoma cell line. J Cell Biol 136:845–857

Daoust A, Bohic S, Saoudi Y, Debacker C, Gory-Faure S, Andrieux A, Barbier EL, Deloulme JC (2014) Neuronal transport defects of the MAP6 KO mouse – a model of schizophrenia – and alleviation by Epothilone D treatment, as observed using MEMRI. Neuroimage 96:133–142

Davare MA, Dong F, Rubin CS, Hell JW (1999) The A-kinase anchor protein MAP2B and cAMP-dependent protein kinase are associated with class C L-type calcium channels in neurons. J Biol Chem 274:30280–30287

Davidkova G, Carroll RC (2007) Characterization of the role of microtubule-associated protein 1B in metabotropic glutamate receptor-mediated endocytosis of AMPA receptors in hippocampus. J Neurosci 27:13273–13278

Dawson HN, Ferreira A, Eyster MV, Ghoshal N, Binder LI, Vitek MP (2001) Inhibition of neuronal maturation in primary hippocampal neurons from tau deficient mice. J Cell Sci 114:1179–1187

Dehmelt L, Halpain S (2005) The MAP2/Tau family of microtubule-associated proteins. Genome Biol 6:204

Dehmelt L, Smart FM, Ozer RS, Halpain S (2003) The role of microtubule-associated protein 2c in the reorganization of microtubules and lamellipodia during neurite initiation. J Neurosci 23:9479–9490

Del Rio JA, Gonzalez-Billault C, Urena JM, Jimenez EM, Barallobre MJ, Pascual M, Pujadas L, Simo S, La Torre A, Wandosell F et al (2004) MAP1B is required for netrin 1 signaling in neuronal migration and axonal guidance. Curr Biol 14:840–850
Delotterie D, Ruiz G, Brocard J, Schweitzer A, Roucard C, Roche Y, Suaud-Chagny MF, Bressand K, Andrieux A (2010) Chronic administration of atypical antipsychotics improves behavioral and synaptic defects of STOP null mice. Psychopharmacology (Berl) 208:131–141
Denarier E, Aguezzoul M, Jolly C, Vourc'h C, Roure A, Andrieux A, Bosc C, Job D (1998a) Genomic structure and chromosomal mapping of the mouse STOP gene (Mtap6). Biochem Biophys Res Commun 243:791–796
Denarier E, Fourest-Lieuvin A, Bosc C, Pirollet F, Chapel A, Margolis RL, Job D (1998b) Nonneuronal isoforms of STOP protein are responsible for microtubule cold stability in mammalian fibroblasts. Proc Natl Acad Sci U S A 95:6055–6060
Desai A, Mitchison TJ (1997) Microtubule polymerization dynamics. Annu Rev Cell Dev Biol 13:83–117
Dhamodharan R, Wadsworth P (1995) Modulation of microtubule dynamic instability in vivo by brain microtubule associated proteins. J Cell Sci 108(Pt 4):1679–1689
Diaz-Nido J, Serrano L, Mendez E, Avila J (1988) A casein kinase II-related activity is involved in phosphorylation of microtubule-associated protein MAP-1B during neuroblastoma cell differentiation. J Cell Biol 106:2057–2065
Diaz-Nido J, Serrano L, Hernandez MA, Avila J (1990) Phosphorylation of microtubule proteins in rat brain at different developmental stages: comparison with that found in neuronal cultures. J Neurochem 54:211–222
Ding J, Liu JJ, Kowal AS, Nardine T, Bhattacharya P, Lee A, Yang Y (2002) Microtubule-associated protein 1B: a neuronal binding partner for gigaxonin. J Cell Biol 158:427–433
Ding J, Allen E, Wang W, Valle A, Wu C, Nardine T, Cui B, Yi J, Taylor A, Jeon NL et al (2006a) Gene targeting of GAN in mouse causes a toxic accumulation of microtubule-associated protein 8 and impaired retrograde axonal transport. Hum Mol Genet 15:1451–1463
Ding J, Valle A, Allen E, Wang W, Nardine T, Zhang Y, Peng L, Yang Y (2006b) Microtubule-associated protein 8 contains two microtubule binding sites. Biochem Biophys Res Commun 339:172–179
DiTella MC, Feiguin F, Carri N, Kosik KS, Caceres A (1996) MAP-1B/TAU functional redundancy during laminin-enhanced axonal growth. J Cell Sci 109(Pt 2):467–477
Dixit R, Ross JL, Goldman YE, Holzbaur EL (2008) Differential regulation of dynein and kinesin motor proteins by tau. Science 319:1086–1089
Doll T, Meichsner M, Riederer BM, Honegger P, Matus A (1993) An isoform of microtubule-associated protein 2 (MAP2) containing four repeats of the tubulin-binding motif. J Cell Sci 106(Pt 2):633–639
Dombeck DA, Kasischke KA, Vishwasrao HD, Ingelsson M, Hyman BT, Webb WW (2003) Uniform polarity microtubule assemblies imaged in native brain tissue by second-harmonic generation microscopy. Proc Natl Acad Sci U S A 100:7081–7086
Doodhi H, Katrukha EA, Kapitein LC, Akhmanova A (2014) Mechanical and geometrical constraints control kinesin-based microtubule guidance. Curr Biol 24:322–328
Dou F, Netzer WJ, Tanemura K, Li F, Hartl FU, Takashima A, Gouras GK, Greengard P, Xu H (2003) Chaperones increase association of tau protein with microtubules. Proc Natl Acad Sci U S A 100:721–726
Drechsel DN, Hyman AA, Cobb MH, Kirschner MW (1992) Modulation of the dynamic instability of tubulin assembly by the microtubule-associated protein tau. Mol Biol Cell 3:1141–1154
Drubin DG, Kirschner MW (1986) Tau protein function in living cells. J Cell Biol 103:2739–2746
Dubey M, Chaudhury P, Kabiru H, Shea TB (2008) Tau inhibits anterograde axonal transport and perturbs stability in growing axonal neurites in part by displacing kinesin cargo: neurofilaments attenuate tau-mediated neurite instability. Cell Motil Cytoskeleton 65:89–99

Dunn S, Morrison EE, Liverpool TB, Molina-Paris C, Cross RA, Alonso MC, Peckham M (2008) Differential trafficking of Kif5c on tyrosinated and detyrosinated microtubules in live cells. J Cell Sci 121:1085–1095

Dziewczapolski G, Glogowski CM, Masliah E, Heinemann SF (2009) Deletion of the alpha 7 nicotinic acetylcholine receptor gene improves cognitive deficits and synaptic pathology in a mouse model of Alzheimer's disease. J Neurosci 29:8805–8815

Ebneth A, Godemann R, Stamer K, Illenberger S, Trinczek B, Mandelkow E (1998) Overexpression of tau protein inhibits kinesin-dependent trafficking of vesicles, mitochondria, and endoplasmic reticulum: implications for Alzheimer's disease. J Cell Biol 143:777–794

Ebneth A, Drewes G, Mandelkow EM, Mandelkow E (1999) Phosphorylation of MAP2c and MAP4 by MARK kinases leads to the destabilization of microtubules in cells. Cell Motil Cytoskeleton 44:209–224

Edelmann W, Zervas M, Costello P, Roback L, Fischer I, Hammarback JA, Cowan N, Davies P, Wainer B, Kucherlapati R (1996) Neuronal abnormalities in microtubule-associated protein 1B mutant mice. Proc Natl Acad Sci U S A 93:1270–1275

Efimov A, Kharitonov A, Efimova N, Loncarek J, Miller PM, Andreyeva N, Gleeson P, Galjart N, Maia AR, McLeod IX et al (2007) Asymmetric CLASP-dependent nucleation of noncentrosomal microtubules at the trans-Golgi network. Dev Cell 12:917–930

Eot-Houllier G, Venoux M, Vidal-Eychenie S, Hoang MT, Giorgi D, Rouquier S (2010) Plk1 regulates both ASAP localization and its role in spindle pole integrity. J Biol Chem 285:29556–29568

Erck C, Peris L, Andrieux A, Meissirel C, Gruber AD, Vernet M, Schweitzer A, Saoudi Y, Pointu H, Bosc C et al (2005) A vital role of tubulin-tyrosine-ligase for neuronal organization. Proc Natl Acad Sci U S A 102:7853–7858

Eriksson M, Samuelsson H, Samuelsson EB, Liu L, McKeehan WL, Benedikz E, Sundstrom E (2007) The NMDAR subunit NR3A interacts with microtubule-associated protein 1S in the brain. Biochem Biophys Res Commun 361:127–132

Fabre-Jonca N, Allaman JM, Radlgruber G, Meda P, Kiss JZ, French LE, Masson D (1998) The distribution of murine 115-kDa epithelial microtubule-associated protein (E-MAP-115) during embryogenesis and in adult organs suggests a role in epithelial polarization and differentiation. Differ Res Biol Divers 63:169–180

Faire K, Waterman-Storer CM, Gruber D, Masson D, Salmon ED, Bulinski JC (1999) E-MAP-115 (ensconsin) associates dynamically with microtubules in vivo and is not a physiological modulator of microtubule dynamics. J Cell Sci 112(Pt 23):4243–4255

Faller EM, Brown DL (2009) Modulation of microtubule dynamics by the microtubule-associated protein 1a. J Neurosci Res 87:1080–1089

Fanara P, Husted KH, Selle K, Wong PY, Banerjee J, Brandt R, Hellerstein MK (2010) Changes in microtubule turnover accompany synaptic plasticity and memory formation in response to contextual fear conditioning in mice. Neuroscience 168:167–178

Farah CA, Liazoghli D, Perreault S, Desjardins M, Guimont A, Anton A, Lauzon M, Kreibich G, Paiement J, Leclerc N (2005) Interaction of microtubule-associated protein-2 and p63: a new link between microtubules and rough endoplasmic reticulum membranes in neurons. J Biol Chem 280:9439–9449

Felgner H, Frank R, Biernat J, Mandelkow EM, Mandelkow E, Ludin B, Matus A, Schliwa M (1997) Domains of neuronal microtubule-associated proteins and flexural rigidity of microtubules. J Cell Biol 138:1067–1075

Fellous A, Francon J, Lennon AM, Nunez J (1977) Microtubule assembly in vitro. Purification of assembly-promoting factors. Eur J Biochem/FEBS 78:167–174

Ferhat L, Represa A, Ferhat W, Ben-Ari Y, Khrestchatisky M (1998) MAP2d mRNA is expressed in identified neuronal populations in the developing and adult rat brain and its subcellular distribution differs from that of MAP2b in hippocampal neurones. Eur J Neurosci 10:161–171

Fifkova E, Morales M (1992) Actin matrix of dendritic spines, synaptic plasticity, and long-term potentiation. Int Rev Cytol 139:267–307

Fischer I, Romano-Clarke G (1990) Changes in microtubule-associated protein MAP1B phosphorylation during rat brain development. J Neurochem 55:328–333
Fontenille L, Rouquier S, Lutfalla G, Giorgi D (2014) Microtubule-associated protein 9 (Map9/Asap) is required for the early steps of zebrafish development. Cell Cycle 13:1101–1114
Fournet V, Jany M, Fabre V, Chali F, Orsal D, Schweitzer A, Andrieux A, Messanvi F, Giros B, Hamon M et al (2010) The deletion of the microtubule-associated STOP protein affects the serotonergic mouse brain network. J Neurochem 115:1579–1594
Fournet V, de Lavilleon G, Schweitzer A, Giros B, Andrieux A, Martres MP (2012a) Both chronic treatments by epothilone D and fluoxetine increase the short-term memory and differentially alter the mood status of STOP/MAP6 KO mice. J Neurochem 123:982–996
Fournet V, Schweitzer A, Chevarin C, Deloulme JC, Hamon M, Giros B, Andrieux A, Martres MP (2012b) The deletion of STOP/MAP6 protein in mice triggers highly altered mood and impaired cognitive performances. J Neurochem 121:99–114
Fradley RL, O'Meara GF, Newman RJ, Andrieux A, Job D, Reynolds DS (2005) STOP knockout and NMDA NR1 hypomorphic mice exhibit deficits in sensorimotor gating. Behav Brain Res 163:257–264
Frandemiche ML, De Seranno S, Rush T, Borel E, Elie A, Arnal I, Lante F, Buisson A (2014) Activity-dependent tau protein translocation to excitatory synapse is disrupted by exposure to amyloid-beta oligomers. J Neurosci 34:6084–6097
Fuhrmann-Stroissnigg H, Noiges R, Descovich L, Fischer I, Albrecht DE, Nothias F, Froehner SC, Propst F (2012) The light chains of microtubule-associated proteins MAP1A and MAP1B interact with alpha1-syntrophin in the central and peripheral nervous system. PLoS One 7:e49722
Fujio K, Sato M, Uemura T, Sato T, Sato-Harada R, Harada A (2007) 14-3-3 proteins and protein phosphatases are not reduced in tau-deficient mice. Neuroreport 18:1049–1052
Fukata Y, Itoh TJ, Kimura T, Menager C, Nishimura T, Shiromizu T, Watanabe H, Inagaki N, Iwamatsu A, Hotani H et al (2002) CRMP-2 binds to tubulin heterodimers to promote microtubule assembly. Nat Cell Biol 4:583–591
Galiano MR, Bosc C, Schweitzer A, Andrieux A, Job D, Hallak ME (2004) Astrocytes and oligodendrocytes express different STOP protein isoforms. J Neurosci Res 78:329–337
Garcia Rocha M, Avila J (1995) Characterization of microtubule-associated protein phosphoisoforms present in isolated growth cones. Brain Res Dev Brain Res 89:47–55
Garcia-Perez J, Avila J, Diaz-Nido J (1998) Implication of cyclin-dependent kinases and glycogen synthase kinase 3 in the phosphorylation of microtubule-associated protein 1B in developing neuronal cells. J Neurosci Res 52:445–452
Garner CC, Tucker RP, Matus A (1988) Selective localization of messenger RNA for cytoskeletal protein MAP2 in dendrites. Nature 336:674–677
Garner CC, Matus A, Anderton B, Calvert R (1989) Microtubule-associated proteins MAP5 and MAP1x: closely related components of the neuronal cytoskeleton with different cytoplasmic distributions in the developing brain. Brain Res Mol Brain Res 5:85–92
Gerdes JM, Katsanis N (2005) Small molecule intervention in microtubule-associated human disease. Hum Mol Genet 14(suppl 2):291–300
Gevorkian G, Gonzalez-Noriega A, Acero G, Ordonez J, Michalak C, Munguia ME, Govezensky T, Cribbs DH, Manoutcharian K (2008) Amyloid-beta peptide binds to microtubule-associated protein 1B (MAP1B). Neurochem Int 52:1030–1036
Goedert M, Crowther RA, Garner CC (1991) Molecular characterization of microtubule-associated proteins tau and MAP2. Trends Neurosci 14:193–199
Gonzalez SL, Lopez-Costa JJ, Labombarda F, Gonzalez Deniselle MC, Guennoun R, Schumacher M, De Nicola AF (2009) Progesterone effects on neuronal ultrastructure and expression of microtubule-associated protein 2 (MAP2) in rats with acute spinal cord injury. Cell Mol Neurobiol 29:27–39
Gonzalez-Billault C, Demandt E, Wandosell F, Torres M, Bonaldo P, Stoykova A, Chowdhury K, Gruss P, Avila J, Sanchez MP (2000) Perinatal lethality of microtubule-associated protein 1B-deficient mice expressing alternative isoforms of the protein at low levels. Mol Cell Neurosci 16:408–421

Gonzalez-Billault C, Avila J, Caceres A (2001) Evidence for the role of MAP1B in axon formation. Mol Biol Cell 12:2087–2098

Gonzalez-Billault C, Engelke M, Jimenez-Mateos EM, Wandosell F, Caceres A, Avila J (2002a) Participation of structural microtubule-associated proteins (MAPs) in the development of neuronal polarity. J Neurosci Res 67:713–719

Gonzalez-Billault C, Owen R, Gordon-Weeks PR, Avila J (2002b) Microtubule-associated protein 1B is involved in the initial stages of axonogenesis in peripheral nervous system cultured neurons. Brain Res 943:56–67

Gonzalez-Billault C, Del Rio JA, Urena JM, Jimenez-Mateos EM, Barallobre MJ, Pascual M, Pujadas L, Simo S, Torre AL, Gavin R et al (2005) A role of MAP1B in reelin-dependent neuronal migration. Cereb Cortex 15:1134–1145

Good PF, Alapat D, Hsu A, Chu C, Perl D, Wen X, Burstein DE, Kohtz DS (2004) A role for semaphorin 3A signaling in the degeneration of hippocampal neurons during Alzheimer's disease. J Neurochem 91:716–736

Goode BL, Denis PE, Panda D, Radeke MJ, Miller HP, Wilson L, Feinstein SC (1997) Functional interactions between the proline-rich and repeat regions of tau enhance microtubule binding and assembly. Mol Biol Cell 8:353–365

Gordon-Weeks PR, Fischer I (2000) MAP1B expression and microtubule stability in growing and regenerating axons. Microsc Res Tech 48:63–74

Griffith LM, Pollard TD (1982) The interaction of actin filaments with microtubules and microtubule-associated proteins. J Biol Chem 257:9143–9151

Grossman GH, Beight CD, Ebke LA, Pauer GJ, Hagstrom SA (2014) Interaction of tubby-like protein-1 (Tulp1) and microtubule-associated protein (MAP) 1A and MAP1B in the mouse retina. Adv Exp Med Biol 801:511–518

Grueber WB, Sagasti A (2010) Self-avoidance and tiling: mechanisms of dendrite and axon spacing. Cold Spring Harb Perspect Biol 2:a001750

Gu J, Zheng JQ (2009) Microtubules in dendritic spine development and plasticity. Open Neurosci J 3:128–133

Guillaud L, Bosc C, Fourest-Lieuvin A, Denarier E, Pirollet F, Lafanechere L, Job D (1998) STOP proteins are responsible for the high degree of microtubule stabilization observed in neuronal cells. J Cell Biol 142:167–179

Gupta M, Yarwood SJ (2005) MAP1A light chain 2 interacts with exchange protein activated by cyclic AMP 1 (EPAC1) to enhance Rap1 GTPase activity and cell adhesion. J Biol Chem 280:8109–8116

Gupta A, Tsai LH, Wynshaw-Boris A (2002) Life is a journey: a genetic look at neocortical development. Nat Rev Genet 3:342–355

Hait WN, Yang JM (2006) The individualization of cancer therapy: the unexpected role of p53. Trans Am Clin Climatol Assoc 117:85–101; discussion 101

Halpain S, Dehmelt L (2006) The MAP1 family of microtubule-associated proteins. Genome Biol 7:224

Hammond JW, Cai D, Verhey KJ (2008) Tubulin modifications and their cellular functions. Curr Opin Cell Biol 20:71–76

Hammond JW, Huang CF, Kaech S, Jacobson C, Banker G, Verhey KJ (2010) Posttranslational modifications of tubulin and the polarized transport of kinesin-1 in neurons. Mol Biol Cell 21:572–583

Harada A, Oguchi K, Okabe S, Kuno J, Terada S, Ohshima T, Sato-Yoshitake R, Takei Y, Noda T, Hirokawa N (1994) Altered microtubule organization in small-calibre axons of mice lacking tau protein. Nature 369:488–491

Harada A, Teng J, Takei Y, Oguchi K, Hirokawa N (2002) MAP2 is required for dendrite elongation, PKA anchoring in dendrites, and proper PKA signal transduction. J Cell Biol 158:541–549

Hartel R, Matus A (1997) Cytoskeletal maturation in cultured hippocampal slices. Neuroscience 78:1–5

Hasegawa M, Arai T, Ihara Y (1990) Immunochemical evidence that fragments of phosphorylated MAP5 (MAP1B) are bound to neurofibrillary tangles in Alzheimer's disease. Neuron 4:909–918

Hayashi K, Ishikawa R, Ye LH, He XL, Takata K, Kohama K, Shirao T (1996) Modulatory role of drebrin on the cytoskeleton within dendritic spines in the rat cerebral cortex. J Neurosci 16:7161–7170
Heins S, Song YH, Wille H, Mandelkow E, Mandelkow EM (1991) Effect of MAP2, MAP2c, and tau on kinesin-dependent microtubule motility. J Cell Sci Suppl 14:121–124
Hirokawa N, Funakoshi T, Sato-Harada R, Kanai Y (1996) Selective stabilization of tau in axons and microtubule-associated protein 2C in cell bodies and dendrites contributes to polarized localization of cytoskeletal proteins in mature neurons. J Cell Biol 132:667–679
Hirokawa N, Niwa S, Tanaka Y (2010) Molecular motors in neurons: transport mechanisms and roles in brain function, development, and disease. Neuron 68:610–638
Holmfeldt P, Brattsand G, Gullberg M (2002) MAP4 counteracts microtubule catastrophe promotion but not tubulin-sequestering activity in intact cells. Curr Biol 12:1034–1039
Holmfeldt P, Brattsand G, Gullberg M (2003) Interphase and monoastral-mitotic phenotypes of overexpressed MAP4 are modulated by free tubulin concentrations. J Cell Sci 116:3701–3711
Hoogenraad CC, Bradke F (2009) Control of neuronal polarity and plasticity – a renaissance for microtubules? Trends Cell Biol 19:669–676
Hoover BR, Reed MN, Su J, Penrod RD, Kotilinek LA, Grant MK, Pitstick R, Carlson GA, Lanier LM, Yuan LL et al (2010) Tau mislocalization to dendritic spines mediates synaptic dysfunction independently of neurodegeneration. Neuron 68:1067–1081
Horton AC, Racz B, Monson EE, Lin AL, Weinberg RJ, Ehlers MD (2005) Polarized secretory trafficking directs cargo for asymmetric dendrite growth and morphogenesis. Neuron 48:757–771
Hoshi M, Ohta K, Gotoh Y, Mori A, Murofushi H, Sakai H, Nishida E (1992) Mitogen-activated-protein-kinase-catalyzed phosphorylation of microtubule-associated proteins, microtubule-associated protein 2 and microtubule-associated protein 4, induces an alteration in their function. Eur J Biochem/FEBS 203:43–52
Howard J, Hyman AA (2003) Dynamics and mechanics of the microtubule plus end. Nature 422:753–758
Hu X, Viesselmann C, Nam S, Merriam E, Dent EW (2008) Activity-dependent dynamic microtubule invasion of dendritic spines. J Neurosci 28:13094–13105
Huang J, Furuya A, Furuichi T (2007) Very-KIND, a KIND domain containing RasGEF, controls dendrite growth by linking Ras small GTPases and MAP2. J Cell Biol 179:539–552
Huber G, Alaimo-Beuret D, Matus A (1985) MAP3: characterization of a novel microtubule-associated protein. J Cell Biol 100:496–507
Huber G, Pehling G, Matus A (1986) The novel microtubule-associated protein MAP3 contributes to the in vitro assembly of brain microtubules. J Biol Chem 261:2270–2273
Hwang SC, Jhon DY, Bae YS, Kim JH, Rhee SG (1996) Activation of phospholipase C-gamma by the concerted action of tau proteins and arachidonic acid. J Biol Chem 271:18342–18349
Ikeda A, Zheng QY, Zuberi AR, Johnson KR, Naggert JK, Nishina PM (2002) Microtubule-associated protein 1A is a modifier of tubby hearing (moth1). Nat Genet 30:401–405
Ikegami S, Harada A, Hirokawa N (2000) Muscle weakness, hyperactivity, and impairment in fear conditioning in tau-deficient mice. Neurosci Lett 279:129–132
Illenberger S, Drewes G, Trinczek B, Biernat J, Meyer HE, Olmsted JB, Mandelkow EM, Mandelkow E (1996) Phosphorylation of microtubule-associated proteins MAP2 and MAP4 by the protein kinase p110mark. Phosphorylation sites and regulation of microtubule dynamics. J Biol Chem 271:10834–10843
Iriuchijima N, Sato-Harada R, Takano M, Fujio K, Sato T, Goto F, Harada A (2005) Reduced expression of kinase-associated phosphatase in cortical dendrites of MAP2-deficient mice. Biochem Biophys Res Commun 338:1216–1221
Itoh TJ, Hisanaga S, Hosoi T, Kishimoto T, Hotani H (1997) Phosphorylation states of microtubule-associated protein 2 (MAP2) determine the regulatory role of MAP2 in microtubule dynamics. Biochemistry 36:12574–12582

Ittner LM, Ke YD, Delerue F, Bi M, Gladbach A, van Eersel J, Wolfing H, Chieng BC, Christie MJ, Napier IA et al (2010) Dendritic function of tau mediates amyloid-beta toxicity in Alzheimer's disease mouse models. Cell 142:387–397

Jaglin XH, Chelly J (2009) Tubulin-related cortical dysgeneses: microtubule dysfunction underlying neuronal migration defects. Trends Genet 25:555–566

Jaglin XH, Poirier K, Saillour Y, Buhler E, Tian G, Bahi-Buisson N, Fallet-Bianco C, Phan-Dinh-Tuy F, Kong XP, Bomont P et al (2009) Mutations in the beta-tubulin gene TUBB2B result in asymmetrical polymicrogyria. Nat Genet 41:746–752

Jalava NS, Lopez-Picon FR, Kukko-Lukjanov TK, Holopainen IE (2007) Changes in microtubule-associated protein-2 (MAP2) expression during development and after status epilepticus in the immature rat hippocampus. Int J Dev Neurosci 25:121–131

Janke C, Bulinski JC (2011) Post-translational regulation of the microtubule cytoskeleton: mechanisms and functions. Nat Rev Mol Cell Biol 12:773–786

Janke C, Kneussel M (2010) Tubulin post-translational modifications: encoding functions on the neuronal microtubule cytoskeleton. Trends Neurosci 33:362–372

Jaworski J, Kapitein LC, Gouveia SM, Dortland BR, Wulf PS, Grigoriev I, Camera P, Spangler SA, Di Stefano P, Demmers J et al (2009) Dynamic microtubules regulate dendritic spine morphology and synaptic plasticity. Neuron 61:85–100

Jensen PH, Hager H, Nielsen MS, Hojrup P, Gliemann J, Jakes R (1999) alpha-synuclein binds to Tau and stimulates the protein kinase A-catalyzed tau phosphorylation of serine residues 262 and 356. J Biol Chem 274:25481–25489

Jensen PH, Islam K, Kenney J, Nielsen MS, Power J, Gai WP (2000) Microtubule-associated protein 1B is a component of cortical Lewy bodies and binds alpha-synuclein filaments. J Biol Chem 275:21500–21507

Jiang K, Akhmanova A (2011) Microtubule tip-interacting proteins: a view from both ends. Curr Opin Cell Biol 23:94–101

Jinushi-Nakao S, Arvind R, Amikura R, Kinameri E, Liu AW, Moore AW (2007) Knot/Collier and cut control different aspects of dendrite cytoskeleton and synergize to define final arbor shape. Neuron 56:963–978

Job D, Fischer EH, Margolis RL (1981) Rapid disassembly of cold-stable microtubules by calmodulin. Proc Natl Acad Sci U S A 78:4679–4682

Job D, Rauch CT, Fischer EH, Margolis RL (1983) Regulation of microtubule cold stability by calmodulin-dependent and -independent phosphorylation. Proc Natl Acad Sci U S A 80:3894–3898

Johnson KA, Porter ME, Shimizu T (1984) Mechanism of force production for microtubule-dependent movements. J Cell Biol 99:132s–136s

Kage E, Hayashi Y, Takeuchi H, Hirotsu T, Kunitomo H, Inoue T, Arai H, Iino Y, Kubo T (2005) MBR-1, a novel helix-turn-helix transcription factor, is required for pruning excessive neurites in Caenorhabditis elegans. Curr Biol 15:1554–1559

Kahn OI, Sharma V, Gonzalez-Billault C, Baas PW (2015) Effects of kinesin-5 inhibition on dendritic architecture and microtubule organization. Mol Biol Cell 26:66–77

Kajitani K, Thorne M, Samson M, Robertson GS (2010) Nitric oxide synthase mediates the ability of darbepoetin alpha to improve the cognitive performance of STOP null mice. Neuropsychopharmacology 35:1718–1728

Kalcheva N, Rockwood JM, Kress Y, Steiner A, Shafit-Zagardo B (1998) Molecular and functional characteristics of MAP-2a: ability of MAP-2a versus MAP-2b to induce stable microtubules in COS cells. Cell Motil Cytoskeleton 40:272–285

Kanai Y, Hirokawa N (1995) Sorting mechanisms of tau and MAP2 in neurons: suppressed axonal transit of MAP2 and locally regulated microtubule binding. Neuron 14:421–432

Kanai Y, Takemura R, Oshima T, Mori H, Ihara Y, Yanagisawa M, Masaki T, Hirokawa N (1989) Expression of multiple tau isoforms and microtubule bundle formation in fibroblasts transfected with a single tau cDNA. J Cell Biol 109:1173–1184

Kanai Y, Chen J, Hirokawa N (1992) Microtubule bundling by tau proteins in vivo: analysis of functional domains. EMBO J 11:3953–3961
Kapitein LC, Hoogenraad CC (2011) Which way to go? Cytoskeletal organization and polarized transport in neurons. Mol Cell Neurosci 46:9–20
Kapitein LC, Schlager MA, Kuijpers M, Wulf PS, van Spronsen M, MacKintosh FC, Hoogenraad CC (2010) Mixed microtubules steer dynein-driven cargo transport into dendrites. Curr Biol 20:290–299
Kapitein LC, Yau KW, Gouveia SM, van der Zwan WA, Wulf PS, Keijzer N, Demmers J, Jaworski J, Akhmanova A, Hoogenraad CC (2011) NMDA receptor activation suppresses microtubule growth and spine entry. J Neurosci 31:8194–8209
Karagoz GE, Duarte AM, Akoury E, Ippel H, Biernat J, Moran Luengo T, Radli M, Didenko T, Nordhues BA, Veprintsev DB et al (2014) Hsp90-Tau complex reveals molecular basis for specificity in chaperone action. Cell 156:963–974
Kardon JR, Vale RD (2009) Regulators of the cytoplasmic dynein motor. Nat Rev Mol Cell Biol 10:854–865
Karki S, Holzbaur EL (1999) Cytoplasmic dynein and dynactin in cell division and intracellular transport. Curr Opin Cell Biol 11:45–53
Kawachi A, Ichihara K, Hisanaga S, Iida J, Toyota H, Hotani H, Itoh TJ (2003) Different protofilament-dependence of the microtubule binding between MAP2 and MAP4. Biochem Biophys Res Commun 305:72–78
Kawakami S, Muramoto K, Ichikawa M, Kuroda Y (2003) Localization of microtubule-associated protein (MAP) 1B in the postsynaptic densities of the rat cerebral cortex. Cell Mol Neurobiol 23:887–894
Kawauchi T, Chihama K, Nishimura YV, Nabeshima Y, Hoshino M (2005) MAP1B phosphorylation is differentially regulated by Cdk5/p35, Cdk5/p25, and JNK. Biochem Biophys Res Commun 331:50–55
Keays DA, Tian G, Poirier K, Huang GJ, Siebold C, Cleak J, Oliver PL, Fray M, Harvey RJ, Molnar Z et al (2007) Mutations in alpha-tubulin cause abnormal neuronal migration in mice and lissencephaly in humans. Cell 128:45–57
Kempf M, Clement A, Faissner A, Lee G, Brandt R (1996) Tau binds to the distal axon early in development of polarity in a microtubule- and microfilament-dependent manner. J Neurosci 16:5583–5592
Khuchua Z, Wozniak DF, Bardgett ME, Yue Z, McDonald M, Boero J, Hartman RE, Sims H, Strauss AW (2003) Deletion of the N-terminus of murine map2 by gene targeting disrupts hippocampal ca1 neuron architecture and alters contextual memory. Neuroscience 119:101–111
Kleele T, Marinkovi CP, Williams PR, Stern S, Weigand EE, Engerer P, Naumann R, Hartmann J, Karl RM, Bradke F et al (2014) An assay to image neuronal microtubule dynamics in mice. Nat Commun 5:4827
Klein C, Kramer EM, Cardine AM, Schraven B, Brandt R, Trotter J (2002) Process outgrowth of oligodendrocytes is promoted by interaction of fyn kinase with the cytoskeletal protein tau. J Neurosci 22:698–707
Kobayashi N, Heid HW, Sakai T, Kriz W, Huber G, Mundel P (2000) Molecular characterization reveals identity of microtubule-associated proteins MAP3 and MAP4. Biochem Biophys Res Commun 268:306–309
Koleske AJ (2013) Molecular mechanisms of dendrite stability. Nat Rev Neurosci 14:536–550
Kollins KM, Bell RL, Butts M, Withers GS (2009) Dendrites differ from axons in patterns of microtubule stability and polymerization during development. Neural Dev 4:26
Komada M, McLean DJ, Griswold MD, Russell LD, Soriano P (2000) E-MAP-115, encoding a microtubule-associated protein, is a retinoic acid-inducible gene required for spermatogenesis. Genes Dev 14:1332–1342
Konishi Y, Setou M (2009) Tubulin tyrosination navigates the kinesin-1 motor domain to axons. Nat Neurosci 12:559–567
Kosik KS, Finch EA (1987) MAP2 and tau segregate into dendritic and axonal domains after the elaboration of morphologically distinct neurites: an immunocytochemical study of cultured rat cerebrum. J Neurosci 7:3142–3153

Kotani S, Nishida E, Kumagai H, Sakai H (1985) Calmodulin inhibits interaction of actin with MAP2 and Tau, two major microtubule-associated proteins. J Biol Chem 260:10779–10783
Kowalski RJ, Williams RC Jr (1993) Microtubule-associated protein 2 alters the dynamic properties of microtubule assembly and disassembly. J Biol Chem 268:9847–9855
Kremer A, Maurin H, Demedts D, Devijver H, Borghgraef P, Van Leuven F (2011) Early improved and late defective cognition is reflected by dendritic spines in Tau.P301L mice. J Neurosci 31:18036–18047
Kuijpers M, Hoogenraad CC (2011) Centrosomes, microtubules and neuronal development. Mol Cell Neurosci 48:349–358
Kumarapeli AR, Wang X (2004) Genetic modification of the heart: chaperones and the cytoskeleton. J Mol Cell Cardiol 37:1097–1109
Kurz JC, Williams RC Jr (1995) Microtubule-associated proteins and the flexibility of microtubules. Biochemistry 34:13374–13380
Kwan AC, Dombeck DA, Webb WW (2008) Polarized microtubule arrays in apical dendrites and axons. Proc Natl Acad Sci U S A 105:11370–11375
Lacroix B, van Dijk J, Gold ND, Guizetti J, Aldrian-Herrada G, Rogowski K, Gerlich DW, Janke C (2010) Tubulin polyglutamylation stimulates spastin-mediated microtubule severing. J Cell Biol 189:945–954
Lajoie-Mazenc I, Tovar D, Penary M, Lortal B, Allart S, Favard C, Brihoum M, Pradines A, Favre G (2008) MAP1A light chain-2 interacts with GTP-RhoB to control epidermal growth factor (EGF)-dependent EGF receptor signaling. J Biol Chem 283:4155–4164
Langnaese K, Seidenbecher C, Wex H, Seidel B, Hartung K, Appeltauer U, Garner A, Voss B, Mueller B, Garner CC et al (1996) Protein components of a rat brain synaptic junctional protein preparation. Brain Res Mol Brain Res 42:118–122
Lansbergen G, Akhmanova A (2006) Microtubule plus end: a hub of cellular activities. Traffic 7:499–507
Lebeau G, Miller LC, Tartas M, McAdam R, Laplante I, Badeaux F, DesGroseillers L, Sossin WS, Lacaille JC (2011) Staufen 2 regulates mGluR long-term depression and Map1b mRNA distribution in hippocampal neurons. Learn Mem 18:314–326
Lee G, Newman ST, Gard DL, Band H, Panchamoorthy G (1998) Tau interacts with src-family non-receptor tyrosine kinases. J Cell Sci 111(Pt 21):3167–3177
Lee HH, Jan LY, Jan YN (2009) Drosophila IKK-related kinase Ik2 and Katanin p60-like 1 regulate dendrite pruning of sensory neuron during metamorphosis. Proc Natl Acad Sci U S A 106:6363–6368
Leenders AG, Lin L, Huang LD, Gerwin C, Lu PH, Sheng ZH (2008) The role of MAP1A light chain 2 in synaptic surface retention of Cav2.2 channels in hippocampal neurons. J Neurosci 28:11333–11346
Lei P, Ayton S, Finkelstein DI, Spoerri L, Ciccotosto GD, Wright DK, Wong BX, Adlard PA, Cherny RA, Lam LQ et al (2012) Tau deficiency induces parkinsonism with dementia by impairing APP-mediated iron export. Nat Med 18:291–295
Lev DL, White EL (1997) Organization of pyramidal cell apical dendrites and composition of dendritic clusters in the mouse: emphasis on primary motor cortex. Eur J Neurosci 9:280–290
Liao G, Gundersen GG (1998) Kinesin is a candidate for cross-bridging microtubules and intermediate filaments. Selective binding of kinesin to detyrosinated tubulin and vimentin. J Biol Chem 273:9797–9803
Liao H, Li Y, Brautigan DL, Gundersen GG (1998) Protein phosphatase 1 is targeted to microtubules by the microtubule-associated protein Tau. J Biol Chem 273:21901–21908
Lim RW, Halpain S (2000) Regulated association of microtubule-associated protein 2 (MAP2) with Src and Grb2: evidence for MAP2 as a scaffolding protein. J Biol Chem 275:20578–20587
Lin S, Liu M, Mozgova OI, Yu W, Baas PW (2012) Mitotic motors coregulate microtubule patterns in axons and dendrites. J Neurosci 32:14033–14049

Lipka J, Kuijpers M, Jaworski J, Hoogenraad CC (2013) Mutations in cytoplasmic dynein and its regulators cause malformations of cortical development and neurodegenerative diseases. Biochem Soc Trans 41:1605–1612

Liu CW, Lee G, Jay DG (1999) Tau is required for neurite outgrowth and growth cone motility of chick sensory neurons. Cell Motil Cytoskeleton 43:232–242

Loeb DM, Tsao H, Cobb MH, Greene LA (1992) NGF and other growth factors induce an association between ERK1 and the NGF receptor, gp140prototrk. Neuron 9:1053–1065

Lopez LA, Sheetz MP (1993) Steric inhibition of cytoplasmic dynein and kinesin motility by MAP2. Cell Motil Cytoskeleton 24:1–16

Lopez LA, Sheetz MP (1995) A microtubule-associated protein (MAP2) kinase restores microtubule motility in embryonic brain. J Biol Chem 270:12511–12517

Lu PJ, Wulf G, Zhou XZ, Davies P, Lu KP (1999) The prolyl isomerase Pin1 restores the function of Alzheimer-associated phosphorylated tau protein. Nature 399:784–788

Lu R, Wang H, Liang Z, Ku L, O'Donnell WT, Li W, Warren ST, Feng Y (2004) The fragile X protein controls microtubule-associated protein 1B translation and microtubule stability in brain neuron development. Proc Natl Acad Sci U S A 101:15201–15206

Luders J, Stearns T (2007) Microtubule-organizing centres: a re-evaluation. Nat Rev Mol Cell Biol 8:161–167

Luo L, O'Leary DD (2005) Axon retraction and degeneration in development and disease. Annu Rev Neurosci 28:127–156

Ma QL, Zuo X, Yang F, Ubeda OJ, Gant DJ, Alaverdyan M, Kiosea NC, Nazari S, Chen PP, Nothias F et al (2014) Loss of MAP function leads to hippocampal synapse loss and deficits in the Morris Water Maze with aging. J Neurosci 34:7124–7136

Maas C, Belgardt D, Lee HK, Heisler FF, Lappe-Siefke C, Magiera MM, van Dijk J, Hausrat TJ, Janke C, Kneussel M (2009) Synaptic activation modifies microtubules underlying transport of postsynaptic cargo. Proc Natl Acad Sci U S A 106:8731–8736

Maccioni RB, Cambiazo V (1995) Role of microtubule-associated proteins in the control of microtubule assembly. Physiol Rev 75:835–864

Mandell JW, Banker GA (1996) A spatial gradient of tau protein phosphorylation in nascent axons. J Neurosci 16:5727–5740

Maniar TA, Kaplan M, Wang GJ, Shen K, Wei L, Shaw JE, Koushika SP, Bargmann CI (2011) UNC-33 (CRMP) and ankyrin organize microtubules and localize kinesin to polarize axon-dendrite sorting. Nat Neurosci 15:48–56

Mansfield SG, Diaz-Nido J, Gordon-Weeks PR, Avila J (1991) The distribution and phosphorylation of the microtubule-associated protein MAP 1B in growth cones. J Neurocytol 20:1007–1022

Manzini MC, Walsh CA (2011) What disorders of cortical development tell us about the cortex: one plus one does not always make two. Curr Opin Genet Dev 21:333–339

Masson D, Kreis TE (1993) Identification and molecular characterization of E-MAP-115, a novel microtubule-associated protein predominantly expressed in epithelial cells. J Cell Biol 123:357–371

Matsushima K, Aosaki M, Tokuraku K, Hasan MR, Nakagawa H, Kotani S (2005) Identification of a neural cell specific variant of microtubule-associated protein 4. Cell Struct Funct 29:111–124

Mattie FJ, Stackpole MM, Stone MC, Clippard JR, Rudnick DA, Qiu Y, Tao J, Allender DL, Parmar M, Rolls MM (2010) Directed microtubule growth, +TIPs, and kinesin-2 are required for uniform microtubule polarity in dendrites. Curr Biol 20:2169–2177

Matus A (1988) Microtubule-associated proteins: their potential role in determining neuronal morphology. Annu Rev Neurosci 11:29–44

Matus A, Huber G, Bernhardt R (1983) Neuronal microdifferentiation. Cold Spring Harb Symp Quant Biol 48(Pt 2):775–782

McNally FJ (2013) Mechanisms of spindle positioning. J Cell Biol 200:131–140

McNally KP, Buster D, McNally FJ (2002) Katanin-mediated microtubule severing can be regulated by multiple mechanisms. Cell Motil Cytoskeleton 53:337–349

Meichsner M, Doll T, Reddy D, Weisshaar B, Matus A (1993) The low molecular weight form of microtubule-associated protein 2 is transported into both axons and dendrites. Neuroscience 54:873–880

Meixner A, Haverkamp S, Wassle H, Fuhrer S, Thalhammer J, Kropf N, Bittner RE, Lassmann H, Wiche G, Propst F (2000) MAP1B is required for axon guidance and is involved in the development of the central and peripheral nervous system. J Cell Biol 151:1169–1178

Merenlender-Wagner A, Pikman R, Giladi E, Andrieux A, Gozes I (2010) NAP (davunetide) enhances cognitive behavior in the STOP heterozygous mouse – a microtubule-deficient model of schizophrenia. Peptides 31:1368–1373

Merino-Serrais P, Benavides-Piccione R, Blazquez-Llorca L, Kastanauskaite A, Rabano A, Avila J, DeFelipe J (2013) The influence of phospho-tau on dendritic spines of cortical pyramidal neurons in patients with Alzheimer's disease. Brain: J Neurol 136:1913–1928

Metzger T, Gache V, Xu M, Cadot B, Folker ES, Richardson BE, Gomes ER, Baylies MK (2012) MAP and kinesin-dependent nuclear positioning is required for skeletal muscle function. Nature 484:120–124

Migheli A, Butler M, Brown K, Shelanski ML (1988) Light and electron microscope localization of the microtubule-associated tau protein in rat brain. J Neurosci 8:1846–1851

Mitchison T, Kirschner M (1984) Dynamic instability of microtubule growth. Nature 312:237–242

Mocanu MM, Nissen A, Eckermann K, Khlistunova I, Biernat J, Drexler D, Petrova O, Schonig K, Bujard H, Mandelkow E et al (2008) The potential for beta-structure in the repeat domain of tau protein determines aggregation, synaptic decay, neuronal loss, and coassembly with endogenous Tau in inducible mouse models of tauopathy. J Neurosci 28:737–748

Moolman DL, Vitolo OV, Vonsattel JP, Shelanski ML (2004) Dendrite and dendritic spine alterations in Alzheimer models. J Neurocytol 33:377–387

Mori A, Aizawa H, Saido TC, Kawasaki H, Mizuno K, Murofushi H, Suzuki K, Sakai H (1991) Site-specific phosphorylation by protein kinase C inhibits assembly-promoting activity of microtubule-associated protein 4. Biochemistry 30:9341–9346

Morris JA, Kandpal G, Ma L, Austin CP (2003) DISC1 (Disrupted-In-Schizophrenia 1) is a centrosome-associated protein that interacts with MAP1A, MIPT3, ATF4/5 and NUDEL: regulation and loss of interaction with mutation. Hum Mol Genet 12:1591–1608

Murakami N, Bolton DC, Kida E, Xie W, Hwang YW (2012) Phosphorylation by Dyrk1A of clathrin coated vesicle-associated proteins: identification of the substrate proteins and the effects of phosphorylation. PLoS One 7:e34845

Nguyen HL, Chari S, Gruber D, Lue CM, Chapin SJ, Bulinski JC (1997) Overexpression of full- or partial-length MAP4 stabilizes microtubules and alters cell growth. J Cell Sci 110 (Pt 2):281–294

Nguyen HL, Gruber D, Bulinski JC (1999) Microtubule-associated protein 4 (MAP4) regulates assembly, protomer-polymer partitioning and synthesis of tubulin in cultured cells. J Cell Sci 112(Pt 12):1813–1824

Nunez J (1988) Immature and mature variants of MAP2 and tau proteins and neuronal plasticity. Trends Neurosci 11:477–479

Obar RA, Dingus J, Bayley H, Vallee RB (1989) The RII subunit of cAMP-dependent protein kinase binds to a common amino-terminal domain in microtubule-associated proteins 2A, 2B, and 2C. Neuron 3:639–645

Ochoa CD, Stevens T, Balczon R (2011) Cold exposure reveals two populations of microtubules in pulmonary endothelia. Am J Physiol Lung Cell Mol Physiol 300:L132–L138

Ohkawa N, Fujitani K, Tokunaga E, Furuya S, Inokuchi K (2007) The microtubule destabilizer stathmin mediates the development of dendritic arbors in neuronal cells. J Cell Sci 120:1447–1456

Okabe S, Hirokawa N (1989) Rapid turnover of microtubule-associated protein MAP2 in the axon revealed by microinjection of biotinylated MAP2 into cultured neurons. Proc Natl Acad Sci U S A 86:4127–4131

Olmsted JB (1986) Microtubule-associated proteins. Annu Rev Cell Biol 2:421–457

Olson KR, McIntosh JR, Olmsted JB (1995) Analysis of MAP 4 function in living cells using green fluorescent protein (GFP) chimeras. J Cell Biol 130:639–650
Opal P, Garcia JJ, Propst F, Matilla A, Orr HT, Zoghbi HY (2003) Mapmodulin/leucine-rich acidic nuclear protein binds the light chain of microtubule-associated protein 1B and modulates neuritogenesis. J Biol Chem 278:34691–34699
Orban-Nemeth Z, Simader H, Badurek S, Trancikova A, Propst F (2005) Microtubule-associated protein 1S, a short and ubiquitously expressed member of the microtubule-associated protein 1 family. J Biol Chem 280:2257–2265
Ori-McKenney KM, Jan LY, Jan YN (2012) Golgi outposts shape dendrite morphology by functioning as sites of acentrosomal microtubule nucleation in neurons. Neuron 76:921–930
Paglini G, Pigino G, Kunda P, Morfini G, Maccioni R, Quiroga S, Ferreira A, Caceres A (1998) Evidence for the participation of the neuron-specific CDK5 activator P35 during laminin-enhanced axonal growth. J Neurosci 18:9858–9869
Papasozomenos SC, Binder LI (1986) Microtubule-associated protein 2 (MAP2) is present in astrocytes of the optic nerve but absent from astrocytes of the optic tract. J Neurosci 6:1748–1756
Papasozomenos SC, Binder LI, Bender PK, Payne MR (1985) Microtubule-associated protein 2 within axons of spinal motor neurons: associations with microtubules and neurofilaments in normal and beta, beta'-iminodipropionitrile-treated axons. J Cell Biol 100:74–85
Park SM, Liu G, Kubal A, Fury M, Cao L, Marx SO (2004) Direct interaction between BKCa potassium channel and microtubule-associated protein 1A. FEBS Lett 570:143–148
Parysek LM, Asnes CF, Olmsted JB (1984) MAP 4: occurrence in mouse tissues. J Cell Biol 99:1309–1315
Paschal BM, Vallee RB (1987) Retrograde transport by the microtubule-associated protein MAP 1C. Nature 330:181–183
Paschal BM, Shpetner HS, Vallee RB (1987) MAP 1C is a microtubule-activated ATPase which translocates microtubules in vitro and has dynein-like properties. J Cell Biol 105:1273–1282
Peck A, Sargin ME, LaPointe NE, Rose K, Manjunath BS, Feinstein SC, Wilson L (2011) Tau isoform-specific modulation of kinesin-driven microtubule gliding rates and trajectories as determined with tau-stabilized microtubules. Cytoskeleton (Hoboken) 68:44–55
Pedrotti B, Islam K (1995) Purification of microtubule associated protein MAP1B from bovine brain: MAP1B binds to microtubules but not to microfilaments. Cell Motil Cytoskeleton 30:301–309
Pedrotti B, Colombo R, Islam K (1994a) Interactions of microtubule-associated protein MAP2 with unpolymerized and polymerized tubulin and actin using a 96-well microtiter plate solid-phase immunoassay. Biochemistry 33:8798–8806
Pedrotti B, Colombo R, Islam K (1994b) Microtubule associated protein MAP1A is an actin-binding and crosslinking protein. Cell Motil Cytoskeleton 29:110–116
Pedrotti B, Ulloa L, Avila J, Islam K (1996) Characterization of microtubule-associated protein MAP1B: phosphorylation state, light chains, and binding to microtubules. Biochemistry 35:3016–3023
Peng J, Kim MJ, Cheng D, Duong DM, Gygi SP, Sheng M (2004) Semiquantitative proteomic analysis of rat forebrain postsynaptic density fractions by mass spectrometry. J Biol Chem 279:21003–21011
Perez M, Santa-Maria I, Gomez de Barreda E, Zhu X, Cuadros R, Cabrero JR, Sanchez-Madrid F, Dawson HN, Vitek MP, Perry G et al (2009) Tau – an inhibitor of deacetylase HDAC6 function. J Neurochem 109:1756–1766
Pigino G, Paglini G, Ulloa L, Avila J, Caceres A (1997) Analysis of the expression, distribution and function of cyclin dependent kinase 5 (cdk5) in developing cerebellar macroneurons. J Cell Sci 110(Pt 2):257–270
Poplawski GH, Tranziska AK, Leshchyns'ka I, Meier ID, Streichert T, Sytnyk V, Schachner M (2012) L1CAM increases MAP2 expression via the MAPK pathway to promote neurite outgrowth. Mol Cell Neurosci 50:169–178

Poulain FE, Sobel A (2010) The microtubule network and neuronal morphogenesis: dynamic and coordinated orchestration through multiple players. Mol Cell Neurosci 43:15–32
Poulain FE, Chauvin S, Wehrle R, Desclaux M, Mallet J, Vodjdani G, Dusart I, Sobel A (2008) SCLIP is crucial for the formation and development of the Purkinje cell dendritic arbor. J Neurosci 28:7387–7398
Powell KJ, Hori SE, Leslie R, Andrieux A, Schellinck H, Thorne M, Robertson GS (2007) Cognitive impairments in the STOP null mouse model of schizophrenia. Behav Neurosci 121:826–835
Praefcke GJ, Ford MG, Schmid EM, Olesen LE, Gallop JL, Peak-Chew SY, Vallis Y, Babu MM, Mills IG, McMahon HT (2004) Evolving nature of the AP2 alpha-appendage hub during clathrin-coated vesicle endocytosis. EMBO J 23:4371–4383
Przyborski SA, Cambray-Deakin MA (1995) Developmental regulation of MAP2 variants during neuronal differentiation in vitro. Brain Res Dev Brain Res 89:187–201
Ray M, Ruan J, Zhang W (2008) Variations in the transcriptome of Alzheimer's disease reveal molecular networks involved in cardiovascular diseases. Genome Biol 9:R148
Reed NA, Cai D, Blasius TL, Jih GT, Meyhofer E, Gaertig J, Verhey KJ (2006) Microtubule acetylation promotes kinesin-1 binding and transport. Curr Biol 16:2166–2172
Reese ML, Dakoji S, Bredt DS, Dotsch V (2007) The guanylate kinase domain of the MAGUK PSD-95 binds dynamically to a conserved motif in MAP1a. Nat Struct Mol Biol 14:155–163
Rehbein M, Kindler S, Horke S, Richter D (2000) Two trans-acting rat-brain proteins, MARTA1 and MARTA2, interact specifically with the dendritic targeting element in MAP2 mRNAs. Brain Res Mol Brain Res 79:192–201
Riederer B, Matus A (1985) Differential expression of distinct microtubule-associated proteins during brain development. Proc Natl Acad Sci U S A 82:6006–6009
Riederer B, Cohen R, Matus A (1986) MAP5: a novel brain microtubule-associated protein under strong developmental regulation. J Neurocytol 15:763–775
Rocher AB, Crimins JL, Amatrudo JM, Kinson MS, Todd-Brown MA, Lewis J, Luebke JI (2010) Structural and functional changes in tau mutant mice neurons are not linked to the presence of NFTs. Exp Neurol 223:385–393
Roll-Mecak A, McNally FJ (2010) Microtubule-severing enzymes. Curr Opin Cell Biol 22:96–103
Rolls MM (2011) Neuronal polarity in Drosophila: sorting out axons and dendrites. Dev Neurobiol 71:419–429
Rosoklija G, Keilp JG, Toomayan G, Mancevski B, Haroutunian V, Liu D, Malespina D, Hays AP, Sadiq S, Latov N et al (2005) Altered subicular MAP2 immunoreactivity in schizophrenia. Prilozi 26:13–34
Rubino HM, Dammerman M, Shafit-Zagardo B, Erlichman J (1989) Localization and characterization of the binding site for the regulatory subunit of type II cAMP-dependent protein kinase on MAP2. Neuron 3:631–638
Ryan XP, Alldritt J, Svenningsson P, Allen PB, Wu GY, Nairn AC, Greengard P (2005) The Rho-specific GEF Lfc interacts with neurabin and spinophilin to regulate dendritic spine morphology. Neuron 47:85–100
Saffin JM, Venoux M, Prigent C, Espeut J, Poulat F, Giorgi D, Abrieu A, Rouquier S (2005) ASAP, a human microtubule-associated protein required for bipolar spindle assembly and cytokinesis. Proc Natl Acad Sci U S A 102:11302–11307
Salinas S, Carazo-Salas RE, Proukakis C, Schiavo G, Warner TT (2007) Spastin and microtubules: functions in health and disease. J Neurosci Res 85:2778–2782
Samora CP, Mogessie B, Conway L, Ross JL, Straube A, McAinsh AD (2011) MAP4 and CLASP1 operate as a safety mechanism to maintain a stable spindle position in mitosis. Nat Cell Biol 13:1040–1050
Samsonov A, Yu JZ, Rasenick M, Popov SV (2004) Tau interaction with microtubules in vivo. J Cell Sci 117:6129–6141
Sanchez C, Diaz-Nido J, Avila J (2000) Phosphorylation of microtubule-associated protein 2 (MAP2) and its relevance for the regulation of the neuronal cytoskeleton function. Prog Neurobiol 61:133–168

Satoh D, Sato D, Tsuyama T, Saito M, Ohkura H, Rolls MM, Ishikawa F, Uemura T (2008) Spatial control of branching within dendritic arbors by dynein-dependent transport of Rab5-endosomes. Nat Cell Biol 10:1164–1171

Scales TM, Lin S, Kraus M, Goold RG, Gordon-Weeks PR (2009) Nonprimed and DYRK1A-primed GSK3 beta-phosphorylation sites on MAP1B regulate microtubule dynamics in growing axons. J Cell Sci 122:2424–2435

Scheetz AJ, Dubin MW (1994) Cytochemical polarity in lateral geniculate interneurons. Brain Res 639:181–192

Schoenfeld TA, Obar RA (1994) Diverse distribution and function of fibrous microtubule-associated proteins in the nervous system. Int Rev Cytol 151:67–137

Schoenfeld TA, McKerracher L, Obar R, Vallee RB (1989) MAP 1A and MAP 1B are structurally related microtubule associated proteins with distinct developmental patterns in the CNS. J Neurosci 9:1712–1730

Scholz T, Mandelkow E (2014) Transport and diffusion of Tau protein in neurons. Cell Mol Life Sci 71:3139–3150

Schwenk BM, Lang CM, Hogl S, Tahirovic S, Orozco D, Rentzsch K, Lichtenthaler SF, Hoogenraad CC, Capell A, Haass C et al (2014) The FTLD risk factor TMEM106B and MAP6 control dendritic trafficking of lysosomes. EMBO J 33:450–467

Seitz A, Kojima H, Oiwa K, Mandelkow EM, Song YH, Mandelkow E (2002) Single-molecule investigation of the interference between kinesin, tau and MAP2c. EMBO J 21:4896–4905

Selden SC, Pollard TD (1983) Phosphorylation of microtubule-associated proteins regulates their interaction with actin filaments. J Biol Chem 258:7064–7071

Semenova I, Ikeda K, Resaul K, Kraikivski P, Aguiar M, Gygi S, Zaliapin I, Cowan A, Rodionov V (2014) Regulation of microtubule-based transport by MAP4. Mol Biol Cell 25:3119–3132

Shafit-Zagardo B, Kalcheva N (1998) Making sense of the multiple MAP-2 transcripts and their role in the neuron. Mol Neurobiol 16:149–162

Sharma N, Kress Y, Shafit-Zagardo B (1994) Antisense MAP-2 oligonucleotides induce changes in microtubule assembly and neuritic elongation in pre-existing neurites of rat cortical neurons. Cell Motil Cytoskeleton 27:234–247

Sharp DJ, Ross JL (2012) Microtubule-severing enzymes at the cutting edge. J Cell Sci 125:2561–2569

Shin E, Kashiwagi Y, Kuriu T, Iwasaki H, Tanaka T, Koizumi H, Gleeson JG, Okabe S (2013) Doublecortin-like kinase enhances dendritic remodelling and negatively regulates synapse maturation. Nat Commun 4:1440

Shirao T, Gonzalez-Billault C (2013) Actin filaments and microtubules in dendritic spines. J Neurochem 126:155–164

Siegrist SE, Doe CQ (2007) Microtubule-induced cortical cell polarity. Genes Dev 21:483–496

Sirajuddin M, Rice LM, Vale RD (2014) Regulation of microtubule motors by tubulin isotypes and post-translational modifications. Nat Cell Biol 16:335–344

Slaughter T, Black MM (2003) STOP (stable-tubule-only-polypeptide) is preferentially associated with the stable domain of axonal microtubules. J Neurocytol 32:399–413

Sobue K, Agarwal-Mawal A, Li W, Sun W, Miura Y, Paudel HK (2000) Interaction of neuronal Cdc2-like protein kinase with microtubule-associated protein tau. J Biol Chem 275:16673–16680

Song Y, Kirkpatrick LL, Schilling AB, Helseth DL, Chabot N, Keillor JW, Johnson GV, Brady ST (2013) Transglutaminase and polyamination of tubulin: posttranslational modification for stabilizing axonal microtubules. Neuron 78:109–123

Sontag E, Nunbhakdi-Craig V, Lee G, Brandt R, Kamibayashi C, Kuret J, White CL 3rd, Mumby MC, Bloom GS (1999) Molecular interactions among protein phosphatase 2A, tau, and microtubules. Implications for the regulation of tau phosphorylation and the development of tauopathies. J Biol Chem 274:25490–25498

Sontag JM, Nunbhakdi-Craig V, White CL 3rd, Halpain S, Sontag E (2012) The protein phosphatase PP2A/Balpha binds to the microtubule-associated proteins Tau and MAP2 at a motif also recognized by the kinase Fyn: implications for tauopathies. J Biol Chem 287:14984–14993

Sparacino J, Farias MG, Lamberti PW (2014) Effect of the microtubule-associated protein tau on dynamics of single-headed motor proteins KIF1A. Phys Rev E Stat Nonlin Soft Matter Phys 89:022714

Stamer K, Vogel R, Thies E, Mandelkow E, Mandelkow EM (2002) Tau blocks traffic of organelles, neurofilaments, and APP vesicles in neurons and enhances oxidative stress. J Cell Biol 156:1051–1063

Stepanova T, Slemmer J, Hoogenraad CC, Lansbergen G, Dortland B, De Zeeuw CI, Grosveld F, van Cappellen G, Akhmanova A, Galjart N (2003) Visualization of microtubule growth in cultured neurons via the use of EB3-GFP (end-binding protein 3-green fluorescent protein). J Neurosci 23:2655–2664

Stewart A, Tsubouchi A, Rolls MM, Tracey WD, Sherwood NT (2012) Katanin p60-like1 promotes microtubule growth and terminal dendrite stability in the larval class IV sensory neurons of Drosophila. J Neurosci 32:11631–11642

Stiess M, Maghelli N, Kapitein LC, Gomis-Ruth S, Wilsch-Brauninger M, Hoogenraad CC, Tolic-Norrelykke IM, Bradke F (2010) Axon extension occurs independently of centrosomal microtubule nucleation. Science 327:704–707

Stone MC, Roegiers F, Rolls MM (2008) Microtubules have opposite orientation in axons and dendrites of Drosophila neurons. Mol Biol Cell 19:4122–4129

Strittmatter WJ, Saunders AM, Goedert M, Weisgraber KH, Dong LM, Jakes R, Huang DY, Pericak-Vance M, Schmechel D, Roses AD (1994) Isoform-specific interactions of apolipoprotein E with microtubule-associated protein tau: implications for Alzheimer disease. Proc Natl Acad Sci U S A 91:11183–11186

Su KY, Chien WL, Fu WM, Yu IS, Huang HP, Huang PH, Lin SR, Shih JY, Lin YL, Hsueh YP et al (2007) Mice deficient in collapsin response mediator protein-1 exhibit impaired long-term potentiation and impaired spatial learning and memory. J Neurosci 27:2513–2524

Sudo H, Baas PW (2010) Acetylation of microtubules influences their sensitivity to severing by katanin in neurons and fibroblasts. J Neurosci 30:7215–7226

Sung HH, Telley IA, Papadaki P, Ephrussi A, Surrey T, Rorth P (2008) Drosophila ensconsin promotes productive recruitment of Kinesin-1 to microtubules. Dev Cell 15:866–876

Swiech L, Blazejczyk M, Urbanska M, Pietruszka P, Dortland BR, Malik AR, Wulf PS, Hoogenraad CC, Jaworski J (2011) CLIP-170 and IQGAP1 cooperatively regulate dendrite morphology. J Neurosci 31:4555–4568

Szebenyi G, Bollati F, Bisbal M, Sheridan S, Faas L, Wray R, Haferkamp S, Nguyen S, Caceres A, Brady ST (2005) Activity-driven dendritic remodeling requires microtubule-associated protein 1A. Curr Biol 15:1820–1826

Takahashi RH, Capetillo-Zarate E, Lin MT, Milner TA, Gouras GK (2013) Accumulation of intraneuronal beta-amyloid 42 peptides is associated with early changes in microtubule-associated protein 2 in neurites and synapses. PLoS One 8:e51965

Takashima A, Murayama M, Murayama O, Kohno T, Honda T, Yasutake K, Nihonmatsu N, Mercken M, Yamaguchi H, Sugihara S et al (1998) Presenilin 1 associates with glycogen synthase kinase-3beta and its substrate tau. Proc Natl Acad Sci U S A 95:9637–9641

Takei Y, Kikkawa YS, Atapour N, Hensch TK, Hirokawa N (2015) Defects in synaptic plasticity, reduced NMDA-receptor transport, and instability of postsynaptic density proteins in mice lacking Microtubule-Associated Protein 1A. J Neurosci 35:15539–15554

Takei Y, Kondo S, Harada A, Inomata S, Noda T, Hirokawa N (1997) Delayed development of nervous system in mice homozygous for disrupted microtubule-associated protein 1B (MAP1B) gene. J Cell Biol 137:1615–1626

Takei Y, Teng J, Harada A, Hirokawa N (2000) Defects in axonal elongation and neuronal migration in mice with disrupted tau and map1b genes. J Cell Biol 150:989–1000

Takemura R, Okabe S, Umeyama T, Kanai Y, Cowan NJ, Hirokawa N (1992) Increased microtubule stability and alpha tubulin acetylation in cells transfected with microtubule-associated proteins MAP1B, MAP2 or tau. J Cell Sci 103(Pt 4):953–964

Teng J, Takei Y, Harada A, Nakata T, Chen J, Hirokawa N (2001) Synergistic effects of MAP2 and MAP1B knockout in neuronal migration, dendritic outgrowth, and microtubule organization. J Cell Biol 155:65–76
Terry-Lorenzo RT, Roadcap DW, Otsuka T, Blanpied TA, Zamorano PL, Garner CC, Shenolikar S, Ehlers MD (2005) Neurabin/protein phosphatase-1 complex regulates dendritic spine morphogenesis and maturation. Mol Biol Cell 16:2349–2362
Terwel D, Dewachter I, Van Leuven F (2002) Axonal transport, tau protein, and neurodegeneration in Alzheimer's disease. Neuromolecular Med 2:151–165
Theurkauf WE, Vallee RB (1982) Molecular characterization of the cAMP-dependent protein kinase bound to microtubule-associated protein 2. J Biol Chem 257:3284–3290
Tischfield MA, Engle EC (2010) Distinct alpha- and beta-tubulin isotypes are required for the positioning, differentiation and survival of neurons: new support for the 'multi-tubulin' hypothesis. Biosci Rep 30:319–330
Tischfield MA, Baris HN, Wu C, Rudolph G, Van Maldergem L, He W, Chan WM, Andrews C, Demer JL, Robertson RL et al (2010) Human TUBB3 mutations perturb microtubule dynamics, kinesin interactions, and axon guidance. Cell 140:74–87
Tischfield MA, Cederquist GY, Gupta ML Jr, Engle EC (2011) Phenotypic spectrum of the tubulin-related disorders and functional implications of disease-causing mutations. Curr Opin Genet Dev 21:286–294
Tokuraku K, Noguchi TQ, Nishie M, Matsushima K, Kotani S (2007) An isoform of microtubule-associated protein 4 inhibits kinesin-driven microtubule gliding. J Biochem 141:585–591
Tokuraku K, Okuyama S, Matsushima K, Ikezu T, Kotani S (2010) Distinct neuronal localization of microtubule-associated protein 4 in the mammalian brain. Neurosci Lett 484:143–147
Tortosa E, Montenegro-Venegas C, Benoist M, Hartel S, Gonzalez-Billault C, Esteban JA, Avila J (2011) Microtubule-associated protein 1B (MAP1B) is required for dendritic spine development and synaptic maturation. J Biol Chem 286:40638–40648
Tortosa E, Galjart N, Avila J, Sayas CL (2013) MAP1B regulates microtubule dynamics by sequestering EB1/3 in the cytosol of developing neuronal cells. EMBO J 32:1293–1306
Trinczek B, Ebneth A, Mandelkow EM, Mandelkow E (1999) Tau regulates the attachment/detachment but not the speed of motors in microtubule-dependent transport of single vesicles and organelles. J Cell Sci 112(Pt 14):2355–2367
Tucker RP, Matus AI (1988) Microtubule-associated proteins characteristic of embryonic brain are found in the adult mammalian retina. Dev Biol 130:423–434
Tucker RP, Binder LI, Matus AI (1988a) Differential localization of the high- and low-molecular weight variants of MAP2 in the developing retina. Brain Res 466:313–318
Tucker RP, Binder LI, Matus AI (1988b) Neuronal microtubule-associated proteins in the embryonic avian spinal cord. J Comp Neurol 271:44–55
Tucker RP, Garner CC, Matus A (1989) In situ localization of microtubule-associated protein mRNA in the developing and adult rat brain. Neuron 2:1245–1256
Tucker KL, Meyer M, Barde YA (2001) Neurotrophins are required for nerve growth during development. Nat Neurosci 4:29–37
Tymanskyj SR, Scales TM, Gordon-Weeks PR (2012) MAP1B enhances microtubule assembly rates and axon extension rates in developing neurons. Mol Cell Neurosci 49:110–119
Ulloa L, Diaz-Nido J, Avila J (1993) Depletion of casein kinase II by antisense oligonucleotide prevents neuritogenesis in neuroblastoma cells. EMBO J 12:1633–1640
Ulloa L, Ibarrola N, Avila J, Diez-Guerra FJ (1994a) Microtubule-associated protein 1B (MAP1B) is present in glial cells phosphorylated different than in neurones. Glia 10:266–275
Ulloa L, Montejo de Garcini E, Gomez-Ramos P, Moran MA, Avila J (1994b) Microtubule-associated protein MAP1B showing a fetal phosphorylation pattern is present in sites of neurofibrillary degeneration in brains of Alzheimer's disease patients. Brain Res Mol Brain Res 26:113–122
Utreras E, Jimenez-Mateos EM, Contreras-Vallejos E, Tortosa E, Perez M, Rojas S, Saragoni L, Maccioni RB, Avila J, Gonzalez-Billault C (2008) Microtubule-associated protein 1B interaction with tubulin tyrosine ligase contributes to the control of microtubule tyrosination. Dev Neurosci 30:200–210

Vaillant AR, Zanassi P, Walsh GS, Aumont A, Alonso A, Miller FD (2002) Signaling mechanisms underlying reversible, activity-dependent dendrite formation. Neuron 34:985–998
Vale RD, Schnapp BJ, Reese TS, Sheetz MP (1985) Organelle, bead, and microtubule translocations promoted by soluble factors from the squid giant axon. Cell 40:559–569
Vallee RB, Wall JS, Paschal BM, Shpetner HS (1988) Microtubule-associated protein 1C from brain is a two-headed cytosolic dynein. Nature 332:561–563
Vallee RB, Williams JC, Varma D, Barnhart LE (2004) Dynein: an ancient motor protein involved in multiple modes of transport. J Neurobiol 58:189–200
Vallee RB, Seale GE, Tsai JW (2009) Emerging roles for myosin II and cytoplasmic dynein in migrating neurons and growth cones. Trends Cell Biol 19:347–355
Vandecandelaere A, Pedrotti B, Utton MA, Calvert RA, Bayley PM (1996) Differences in the regulation of microtubule dynamics by microtubule-associated proteins MAP1B and MAP2. Cell Motil Cytoskeleton 35:134–146
Velazquez-Bernardino P, Garcia-Sierra F, Hernandez-Hernandez O, Bermudez de Leon M, Gourdon G, Gomes-Pereira M, Cisneros B (2012) Myotonic dystrophy type 1-associated CTG repeats disturb the expression and subcellular distribution of microtubule-associated proteins MAP1A, MAP2, and MAP6/STOP in PC12 cells. Mol Biol Rep 39:415–424
Venoux M, Basbous J, Berthenet C, Prigent C, Fernandez A, Lamb NJ, Rouquier S (2008) ASAP is a novel substrate of the oncogenic mitotic kinase Aurora-A: phosphorylation on Ser625 is essential to spindle formation and mitosis. Hum Mol Genet 17:215–224
Verhey KJ, Gaertig J (2007) The tubulin code. Cell Cycle 6:2152–2160
Vershinin M, Carter BC, Razafsky DS, King SJ, Gross SP (2007) Multiple-motor based transport and its regulation by Tau. Proc Natl Acad Sci U S A 104:87–92
Viereck C, Tucker RP, Matus A (1989) The adult rat olfactory system expresses microtubule-associated proteins found in the developing brain. J Neurosci 9:3547–3557
Villarroel-Campos D, Gonzalez-Billault C (2014) The MAP1B case: an old MAP that is new again. Dev Neurobiol 74:953–971
von Massow A, Mandelkow EM, Mandelkow E (1989) Interaction between kinesin, microtubules, and microtubule-associated protein 2. Cell Motil Cytoskeleton 14:562–571
Voss AK, Thomas T, Gruss P (1998) Compensation for a gene trap mutation in the murine microtubule-associated protein 4 locus by alternative polyadenylation and alternative splicing. Dev Dyn 212:258–266
Vouyiouklis DA, Brophy PJ (1995) Microtubule-associated proteins in developing oligodendrocytes: transient expression of a MAP2c isoform in oligodendrocyte precursors. J Neurosci Res 42:803–817
Walczak CE, Gayek S, Ohi R (2013) Microtubule-depolymerizing kinesins. Annu Rev Cell Dev Biol 29:417–441
Wang W, Lundin VF, Millan I, Zeng A, Chen X, Yang J, Allen E, Chen N, Bach G, Hsu A et al (2012) Nemitin, a novel Map8/Map1s interacting protein with Wd40 repeats. PLoS One 7: e33094
Weingarten MD, Lockwood AH, Hwo SY, Kirschner MW (1975) A protein factor essential for microtubule assembly. Proc Natl Acad Sci U S A 72:1858–1862
West RR, Tenbarge KM, Olmsted JB (1991) A model for microtubule-associated protein 4 structure. Domains defined by comparisons of human, mouse, and bovine sequences. J Biol Chem 266:21886–21896
Wiche G, Oberkanins C, Himmler A (1991) Molecular structure and function of microtubule-associated proteins. Int Rev Cytol 124:217–273
Witte H, Neukirchen D, Bradke F (2008) Microtubule stabilization specifies initial neuronal polarization. J Cell Biol 180:619–632
Wolff S, Xiao Z, Wittau M, Sussner N, Stoter M, Knippschild U (2005) Interaction of casein kinase 1 delta (CK1 delta) with the light chain LC2 of microtubule associated protein 1A (MAP1A). Biochim Biophys Acta 1745:196–206

Wu CC, Chawla F, Games D, Rydel RE, Freedman S, Schenk D, Young WG, Morrison JH, Bloom FE (2004) Selective vulnerability of dentate granule cells prior to amyloid deposition in PDAPP mice: digital morphometric analyses. Proc Natl Acad Sci U S A 101:7141–7146

Xiao H, Wang H, Zhang X, Tu Z, Bulinski C, Khrapunovich-Baine M, Hogue Angeletti R, Horwitz SB (2012) Structural evidence for cooperative microtubule stabilization by Taxol and the endogenous dynamics regulator MAP4. ACS Chem Biol 7:744–752

Xie R, Nguyen S, McKeehan K, Wang F, McKeehan WL, Liu L (2011) Microtubule-associated protein 1S (MAP1S) bridges autophagic components with microtubules and mitochondria to affect autophagosomal biogenesis and degradation. J Biol Chem 286:10367–10377

Yadav S, Linstedt AD (2011) Golgi positioning. Cold Spring Harbor Perspect Biol 3:a005322

Yamashita N, Morita A, Uchida Y, Nakamura F, Usui H, Ohshima T, Taniguchi M, Honnorat J, Thomasset N, Takei K et al (2007) Regulation of spine development by semaphorin3A through cyclin-dependent kinase 5 phosphorylation of collapsin response mediator protein 1. J Neurosci 27:12546–12554

Yamashita N, Mosinger B, Roy A, Miyazaki M, Ugajin K, Nakamura F, Sasaki Y, Yamaguchi K, Kolattukudy P, Goshima Y (2011) CRMP5 (collapsin response mediator protein 5) regulates dendritic development and synaptic plasticity in the cerebellar Purkinje cells. J Neurosci 31:1773–1779

Yamauchi PS, Flynn GC, Marsh RL, Purich DL (1993) Reduction in microtubule dynamics in vitro by brain microtubule-associated proteins and by a microtubule-associated protein-2 second repeated sequence analogue. J Neurochem 60:817–826

Yan J, Sun XB, Wang HQ, Zhao H, Zhao XY, Xu YX, Guo JC, Zhu CQ (2010) Chronic restraint stress alters the expression and distribution of phosphorylated tau and MAP2 in cortex and hippocampus of rat brain. Brain Res 1347:132–141

Yan XX, Cai Y, Zhang XM, Luo XG, Cai H, Rose GM, Patrylo PR (2012) BACE1 elevation is associated with aberrant limbic axonal sprouting in epileptic CD1 mice. Exp Neurol 235:228–237

Yan J, Chao DL, Toba S, Koyasako K, Yasunaga T, Hirotsune S, Shen K (2013) Kinesin-1 regulates dendrite microtubule polarity in Caenorhabditis elegans. eLife 2:e00133

Yang HJ, Li YF, Zhang HT, Zhang FQ, Zhao N, Gong ZH, Luo ZP (2003) Up-regulation of microtubule-associated protein 4 and drebrin A mRNA expression by antidepressants in rat hippocampus following chronic stress. Neurosci Lett 351:206–208

Yau KW, Schatzle P, Tortosa E, Pages S, Holtmaat A, Kapitein LC, Hoogenraad CC (2016) Dendrites in vitro and in vivo contain microtubules of opposite polarity and axon formation correlates with uniform plus-end-out microtubule orientation. J Neurosci 36:1071–1085

Yau KW, van Beuningen SF, Cunha-Ferreira I, Cloin BM, van Battum EY, Will L, Schatzle P, Tas RP, van Krugten J, Katrukha EA et al (2014) Microtubule minus-end binding protein CAMSAP2 controls axon specification and dendrite development. Neuron 82:1058–1073

Ye B, Zhang Y, Song W, Younger SH, Jan LY, Jan YN (2007) Growing dendrites and axons differ in their reliance on the secretory pathway. Cell 130:717–729

Ye B, Kim JH, Yang L, McLachlan I, Younger S, Jan LY, Jan YN (2011) Differential regulation of dendritic and axonal development by the novel Kruppel-like factor Dar1. J Neurosci 31:3309–3319

Yu W, Qiang L, Solowska JM, Karabay A, Korulu S, Baas PW (2008) The microtubule-severing proteins spastin and katanin participate differently in the formation of axonal branches. Mol Biol Cell 19:1485–1498

Zalfa F, Giorgi M, Primerano B, Moro A, Di Penta A, Reis S, Oostra B, Bagni C (2003) The fragile X syndrome protein FMRP associates with BC1 RNA and regulates the translation of specific mRNAs at synapses. Cell 112:317–327

Zamora-Leon SP, Lee G, Davies P, Shafit-Zagardo B (2001) Binding of Fyn to MAP-2c through an SH3 binding domain. Regulation of the interaction by ERK2. J Biol Chem 276:39950–39958

Zempel H, Mandelkow E (2014) Lost after translation: missorting of Tau protein and consequences for Alzheimer disease. Trends Neurosci 37:721–732

Zempel H, Luedtke J, Kumar Y, Biernat J, Dawson H, Mandelkow E, Mandelkow EM (2013) Amyloid-beta oligomers induce synaptic damage via Tau-dependent microtubule severing by TTLL6 and spastin. EMBO J 32:2920–2937

Zhang J, Dong XP (2012) Dysfunction of microtubule-associated proteins of MAP2/tau family in Prion disease. Prion 6:334–338

Zheng Y, Wildonger J, Ye B, Zhang Y, Kita A, Younger SH, Zimmerman S, Jan LY, Jan YN (2008) Dynein is required for polarized dendritic transport and uniform microtubule orientation in axons. Nat Cell Biol 10:1172–1180

Chapter 4
Membrane Trafficking Mechanisms: Exocytosis and Endocytosis in Dendrites

Bettina Winckler

Abstract The large size of neurons poses great challenges for membrane traffic. Dendrites are often hundreds of micrometers in length, intricately branched, and studded with postsynaptic specializations. How is membrane traffic in dendrites organized to support the large size, the intricate morphology, and the spatial heterogeneity in order to create a functional neuron integrated properly into a circuit? Since dendrite form and function are disrupted when membrane traffic is disturbed, understanding membrane traffic in dendrites is important. In addition, a large number of receptors require accurate trafficking in order to participate in localized neuronal functions. The overall composition of the plasma membrane, which receptor is localized where and at what levels, is also regulated by spatially and temporally precise exo- and endocytosis. This chapter will discuss the basic cell biology of exocytic and endocytic traffic and lay out how the basic mechanisms are the same or are different in dendrites to accommodate the large size and spatial heterogeneity of neurons.

Keywords Endoplasmic reticulum • Golgi • Golgi outpost • Golgi bypass • Unconventional secretion • Endocytosis • Endosome • Neural-enriched proteins • Regulated secretion

4.1 Introduction

4.1.1 Dendrites: Form and Function

Neuronal function is critically and inextricably linked to neuronal form (Lohmann 2013; Polleux and Ghosh 2007; Jan and Jan 2010; Li et al. 2015): axons allow long-distance signaling, whereas dendrites allow integration of many different inputs. The precise function of circuits is thus greatly influenced by the extent and complexity of the dendritic tree. In agreement with this notion, neuronal cell

B. Winckler, Ph.D. (✉)
Department of Neuroscience, University of Virginia, 409 Lane Rd. Extd., Charlottesville, VA MR4-6116, USA
e-mail: bwinckler@virginia.edu

K. Emoto et al. (eds.), *Dendrites*, DOI 10.1007/978-4-431-56050-0_4

types participating in different circuits elaborate highly distinct dendritic arbors: striking examples are Purkinje neurons in the cerebellum with their characteristic near-planar and highly branched dendritic arbors which any neuroscientist recognizes by sight. Not surprisingly, many diseases of the nervous system, both neurodevelopmental and neurodegenerative, manifest with alterations in dendrite complexity (Jan and Jan 2010; Lohmann 2013; Ehlers 2013). It is thus important for advancing our understanding of neuronal function in health and disease to uncover the mechanisms that regulate the elaboration of dendrites. Among the key mechanisms that enable dendrite growth and dendrite function are the cytoskeleton (see Chaps. 2 and 3) and membrane trafficking (see also Chap. 4). Membrane transport in dendrites critically underlies the elaboration of dendrites during development. New membrane is added via exocytosis for growth of dendrites and branching, whereas membrane is removed via endocytosis for pruning of dendrites. In addition, important signaling events take place in endosomes (Cosker and Segal 2014; Irannejad et al. 2013; Irannejad and Zastrow 2014; Murphy et al. 2009). Endosomal trafficking and endosomal signaling of a multitude of membrane receptors thus also contribute to dendrite development (Yap and Winckler 2012). Later on, maintenance of dendrites requires regulated turnover of membrane protein and continued insertion of newly synthesized receptors (see Chap. 15). In addition, membrane trafficking of neurotransmitter receptors at synapses is essential for synaptic plasticity (Ehlers 2013; Greger and Esteban 2007). This chapter will not primarily focus on the complex regulation of membrane trafficking at synapses, which is covered elsewhere in this book, but will focus mostly on the overall organization of membrane trafficking in dendrites generally.

4.1.2 Membrane Traffic in Dendrites: Challenges of Large Size and Spatial Heterogeneity

During development, many newborn neurons migrate to their final sites and then elaborate axonal and dendritic processes (Flynn et al. 2013; Ma and Gibson 2013; Polleux and Ghosh 2007). Initial processes are often extended during the migratory phase, but extensive morphological changes take place post-migration. Many migratory neurons have a bipolar shape with an extended leading process and a trailing process. In the cortex, the trailing process of a radially migrating cell usually becomes the axon, whereas the leading process becomes a dendrite, often the major apical dendrite (Noctor et al. 2013). Depending on the neuronal cell type, additional primary dendrites extend from the soma or from the proximal portion of the axon. Primary dendrites frequently grow very long (hundreds of micrometers) and develop secondary and tertiary branches that cover a large area. Growth and branching of dendrites require cytoskeleton and membrane traffic to support the enormous size of the dendritic arbor (Lewis et al. 2013). Just as axons, growing dendrites are tipped by growth cones, actin-rich motile structures that are responsible for exploring the extracellular environment and enabling growth. Microtubules are similarly essential for dendrite growth (Franker and Hoogenraad 2013). In mature dendrites, microtubules serve as tracks for vesicular traffic, which carry cargos long distances to supply new material to distant sites of this enormous cell. The size of the dendritic tree thus poses unique challenges for cytoskeleton and membrane traffic. Cytoskeletal regulation of dendrite growth is discussed in Chaps. 2 and 3.

Membrane trafficking in dendrites plays crucial roles in multiple ways (Bonifacino 2014; Ehlers 2013; Winckler and Yap 2011). Initially, adding more membrane via exocytosis is needed to expand the plasma membrane to support dendrite growth and branching. Subsequently to dendritic growth, many branches and dendrites are extensively remodeled later on and even completely pruned (Schuldiner and Yaron 2015). Pruning requires the removal of membrane via endocytosis (Zhang et al. 2014). One role of membrane traffic in dendrites is thus to add dendritic membrane mass during initial growth and to remove it during the subsequent pruning. Secondly, membrane traffic in dendrites regulates the exact membrane composition of the dendritic plasma membrane (Ehlers 2013): what receptors are where on the surface, at what quantities, and for how long? Many receptors are locally exocytosed in precise locations and removed by endocytosis in a temporally tightly regulated fashion. Since many receptors elicit intracellular signaling cascades and since signal strength is correlated to receptor levels on the surface, regulating membrane composition is crucial to neuronal function, both during development and later at mature synapses (Ehlers 2013; Yap and Winckler 2012). Dendrites are expansive and show great spatial heterogeneity: membrane composition at proximal and distal portion of dendrites differs, membrane composition at synaptic vs. non-synaptic regions differs, membrane composition at inhibitory vs. excitatory synaptic sites differs, membrane composition at specialized dendritic protrusions (i.e., spines) differs, and so forth. Regulating membrane traffic in dendrites is thus truly a formidable challenge.

4.1.3 Ubiquitous and Neuron-Specific Regulators of Membrane Traffic

Membrane traffic has been – and still is – widely studied in many cell types. Most of the fundamental insights and identification of large families of traffic regulators stem from work in yeast and in non-neuronal mammalian cell lines, such as HeLa cells, COS cells, or CHO cells. Neurons express most if not all of the same families of proteins, but some proteins (or specific splice variants) are highly expressed only in neurons with low or undetectable levels in non-neuronal cells. This raises the question as to whether neurons modify canonical compartments with neuronal-specific/neuronal-enriched regulators or whether they create unique compartments specialized to carry out neuronal-specific functions, such as trafficking neurotransmitter receptors in and out of dendritic spines. These are not mutually exclusive possibilities, and all of them likely contribute to regulating membrane traffic in the large and spatially heterogeneous neuron. For example, important roles have been identified for ubiquitously expressed regulators in neurons. These proteins appear to play comparable roles in neurons to the functions that have been ascribed to these same proteins in non-neuronal cells (e.g., Bonifacino 2014; Brown et al. 2007; Kawauchi et al. 2010; Park 2004; Ye et al. 2007). Surprisingly, some ubiquitous regulators play modified, cell-type-specific roles in neurons which are different from their roles described in non-neuronal cells (e.g., Yap et al. 2010). Lastly, more and more examples are appearing in the literature for roles of neuronal-specific/neuronal-enriched trafficking proteins in neuronal function (e.g., Ha et al. 2008; Hoogenraad et al. 2010; Lasiecka et al. 2014; Muthusamy et al. 2014; Myers et al. 2012; Newell-Litwa et al. 2010; Steiner et al. 2002). This is just the tip of the iceberg, and many fundamental questions remain unanswered in neurons about how ubiquitous regulators of membrane traffic in conjunction with neuronal-specific/neuronal-enriched proteins generate and maintain dendritic morphology and support dendritic functions.

4.1.4 Intrinsic and Extrinsic Regulation of Dendrite Morphogenesis

When cortical or hippocampal neurons are dissociated and plated singly, they execute an intrinsic program of cellular differentiation that includes the specification of a single axon and multiple dendrites (Dotti et al. 1988; Higgins et al. 1997; Polleux and Ghosh 2007). Dendrite number and some aspects of growth, branching, and synapse formation can also be recapitulated in culture (e.g., McAllister 2007; Williams et al. 2011), but the full extent of dendritic morphogenesis requires the complete context of extracellular and cellular information (McAllister 2007; Puram and Bonni 2013). It should be kept in mind, though, that cultured neurons are also exposed to a multitude of secreted and contact-mediated signals, so that many of the processes studied in cultured neurons are also mediated by extrinsic factors (e.g., Baptista et al. 1994). In fact, cell density in culture affects differentiation of neurons (e.g., Boiko et al. 2007; Dunn 1971), and the presence of astrocytes in the culture provides crucial support and modulates the characteristics of neurons grown in a dish (Christopherson et al. 2005). Extrinsic regulation thus is essential to proper dendritic growth and for pruning.

Since one of the characteristics of neurons is the ability to signal electrically, the question of if and how activity might influence dendritic development has intrigued many researchers for quite some time. There is ample evidence that activity is required for dendrite sculpting (Cline 2001; McAllister 2007). In addition, extracellular ligands and signaling molecules, such as BDNF, reelin, and Semaphorin3A, have been shown to affect dendrite growth or orientation (Kwon et al. 2011; Matsuki et al. 2010; Polleux et al. 2000; Sotelo et al. 2014). Interestingly, activity might in fact operate via the stimulated secretion of proteins (such as NT3 or BDNF) that might then affect dendrite development (Joo et al. 2014; Park and Poo 2013). It is noteworthy that often distinct aspects of dendrite morphogenesis are separable, such that an extracellular cue specifically affects dendrite orientation, number of primary dendrites, extent and pattern of branching, extent and pattern of pruning, dendrite diameter, etc., but does not affect all these facets of dendrite morphology to the same extent. Ultimately, the cascade downstream of extracellular signals and activity must impinge on both the cytoskeleton and membrane trafficking to affect changes in dendrites. It is this link between receptors in the plasma membrane and specific cytoskeletal and trafficking regulation that is still largely unknown.

4.2 The Biosynthetic Membrane Compartments in Dendrites: Endoplasmic Reticulum, Golgi, and Trans-Golgi Network

4.2.1 The Biosynthetic Pathways in Non-neuronal Cells

Dendrites are highly compartmentalized and spatially heterogeneous in terms of their membrane composition and their functional properties (Ehlers 2013). How is membrane traffic orchestrated in a spatially defined manner to set up this exquisite compartmentalization? In order to start asking this question, it is important to take a look at what is known about biosynthetic membrane traffic in non-neuronal cells. Then we will ask how dendrites make use of ubiquitous as well as cell-type-specific machinery and compartments in order to establish and maintain the spatial heterogeneity of membrane composition (Sect. 4.2.2).

The secretory pathway was first described by George Palade who ultimately won the Nobel Prize for his work (Palade 1975). The basic organization of the secretory pathway is thus well established: secreted and transmembrane proteins are inserted into the endoplasmic reticulum (ER) where initial glycosylation takes place, then transported to the Golgi complex where extensive modification of glycosylation takes place as secretory cargos sequentially transit the Golgi stacks. The trans-Golgi network (TGN) then sorts and packages secretory cargos into carriers with different final destinations, such as the plasma membrane or endosomes.

4.2.1.1 The Endoplasmic Reticulum

The endoplasmic reticulum (ER) is an extensive membrane network that extends throughout the cell and is continuous with the nuclear membrane (Shibata et al. 2010; Westrate et al. 2014). It consists of peripheral membrane sheets as well as networks of interconnected tubules. Markers found on the ER are often enriched on either sheets or tubular ER. For instance, calreticulin is found more prominently on sheets. ER tubules can be formed by the action of microtubule motors, but tubular ER can ultimately be maintained without microtubules. Sheets and tubules are dynamically interrelated, and distinct membrane proteins are responsible for formation of either sheets or tubules. For instance, the reticulon and the DP1/Reep family of proteins shape ER tubules, whereas CLIMP63 overexpression increases sheet formation. Fission and fusion events might additionally contribute to sculpting the ER. The ER can also form branched tubules, a morphology requiring the dynamin-like membrane protein atlastin (Hu et al. 2009; Shibata et al. 2010).

It has been known for a long time that some ER is smooth, whereas other regions appear rough by electron microscopy. The rough appearance is due to ribosomes attached to the ER. These ER-attached ribosomes are in the process of co-translationally inserting secretory cargos into the ER lumen. An initial set of mannose-rich oligosaccharides is added in the ER to asparagine residues of many secretory cargos in a single reaction (reviewed in Bieberich 2014; Shrimal et al. 2014). This ER-resident glycosylation can be removed experimentally by treating samples with endoglycosidase H (EndoH). EndoH sensitivity is thus a hallmark that a protein has been modified in the ER, but has not yet transited through the Golgi (Rothman and Fine 1980). The ER also plays important roles in quality control of correctly folded proteins and is thus an important organelle for proteostasis and cellular stress responses (reviewed in Benyair et al. 2014; Koenig and Ploegh 2014). Smooth ER does not have translating ribosomes attached and plays instead important roles in calcium homeostasis and lipid metabolism (reviewed in Bieberich 2014; Redondo and Rosado 2015).

4.2.1.2 The ER-Golgi Transitional Compartments

After passing quality control in the ER, secretory cargos leave the ER (for recent reviews see Budnik and Stephens 2009; Lorente-Rodriguez and Barlowe 2011). The transport of secretory cargos from the ER to the Golgi is a complex process involving transitional compartments. It is now clear that most secretory cargos leave the ER from specialized ER exit sites (ERES) which appear as fairly stable punctate structure and that cargo exit does not occur all along the rough ER surface. ERES secrete cargo by fission of vesicles coated with COPII coats (Jensen and Schekman 2011). COPII coat assembly requires the activation of the small GTPase sar1 which is recruited to the ER via the ER-resident membrane protein sec12 which serves as a GTP exchange factor (GEF) for sar1, thus activating it on the ER surface. Activated sar1 then initiates the assembly of a complete COPII coat. Expression of a dominant-negative sar1 protein inhibits COPII formation and is used experimentally to block ER exit. The COPII-coated vesicles then fuse heterotypically with a preexisting Golgi-near compartment called the ERGIC (ER-Golgi intermediate compartment) or might also fuse homotypically with each other to create new ERGICs. The ERGIC can generate COPI-coated vesicles for retrieval of ER-resident proteins back to the ER. Forward transport of cargo from the ERGIC to the Golgi is still not well understood, but might include homotypic fusion and/or maturation of ERGIC to a tubular cis-Golgi network (CGN) or heterotypic fusion into the preexisting cis-Golgi stack. Tubular carriers can also sometimes be observed to emerge from the ER/ERGIC. The trafficking between post-ER and pre-Golgi compartments is very dynamic, and the exact relationships remain to be further established.

4.2.1.3 The Golgi Apparatus

After leaving the ER, secretory cargos traverse the Golgi apparatus where they are modified (for recent review see Martínez-Menárguez 2013). By electron microscopy, the Golgi apparatus in mammalian cells is comprised of three to six membrane stacks surrounded by tubules and vesicles. The stacks are frequently connected at the rims into continuous ribbons. Cargo enters the Golgi from the cis side and is modified as it moves through the stacks. It takes about 20 min for a typical cargo to traverse the Golgi stacks. The Golgi has well-established roles in protein transport and glycosylation. A set of modifying enzymes resident in different Golgi stacks modifies the mannose-rich oligosaccharide chains, which remain EndoH sensitive until α-mannosidase II removes two mannose residues. Additional roles for the Golgi besides transport and glycosylation are becoming apparent as well. These include nucleation of Golgi-associated microtubules as well as calcium homeostasis. The Golgi stacks are surrounded by a matrix made up of golgins and GRASP proteins. These components associate dynamically with the Golgi and can be recruited from the cytosol. The Golgi matrix plays multiple roles, including stacking of membranes, tethering of vesicles (Wong and Munro 2014), and nucleation of non-centrosomal microtubules (Rios 2014). Both the cis and the trans side of the Golgi can nucleate non-centrosomal microtubules via different matrix proteins. Golgi-associated microtubules might be involved in maintaining the stack and ensuring directionality of cargo transit. In non-polarized cells, the Golgi is positioned near the centrosome and requires dynein for its perinuclear localization. The perinuclear localization of the Golgi depends on centrosomal microtubules, and Golgi microtubules are not involved. In other cell types, the Golgi can be dispersed or exist in multiple mini-stacks. This will be discussed for neuronal cells below.

4.2.1.4 The Trans-Golgi Network (TGN)

The trans side of the Golgi is surrounded by an extensive network of membrane tubules, the trans-Golgi network. The TGN is a distinct organelle from the trans-Golgi because the Golgi and the TGN behave differently after treatment with the drug Brefeldin A (BFA) which inhibits the activation of the small GTPase Arf1 at the Golgi: Golgi membranes are absorbed retrogradely into the ER, effectively blocking secretion from the Golgi (Lippincott-Schwartz et al. 1989). TGN membranes, in contrast, fuse with the recycling endosomal system in the presence of BFA and lead to disruption of transport to lysosomes (Lippincott-Schwartz et al. 1991). The TGN has been shown to be an important sorting station for secretory cargo, likely budding out multiple carriers with distinct cargo composition and target routes. There is extensive transport from the TGN not only to the plasma membrane but also to the endolysosomal system. In fact, some plasma membrane proteins travel to the cell surface via endosomes (Ang et al. 2004; Fölsch et al. 2009; Rodriguez-Boulan and Müsch 2005). In polarized cells, multiple distinct carriers are formed at the TGN that transport different sets of cargos to distinct plasma membrane domains (Park et al. 2011).

4.2.1.5 Unconventional Secretion

In addition to the conventional secretory pathway described above, a subset of proteins is secreted via unconventional pathways (reviewed in Grieve and Rabouille 2011; Rabouille et al. 2013). These pathways are frequently associated with stress and immune surveillance (for review see Rabouille et al. 2013), but we know still relatively little about these pathways, and many more cargos using unconventional secretion will likely be identified in the future. For membrane proteins, there are at least two pathways that use unconventional secretion. One involves a COPII-independent pathway of ER exit. Cargos using this pathway can be identified because they are insensitive to expression of dominant-negative sar1. The second pathway is a Golgi-bypass pathway in which secretory cargos leave the ER in COPII vesicles but are transported to the cell surface without transiting the Golgi. Golgi-bypass cargos can be identified because their transport is BFA insensitive and the proteins remain EndoH sensitive, i.e., their glycosylation patterns have not been modified by α-mannosidase II. In many cases, it is not 100 % of the protein that becomes BFA insensitive, but there are both BFA-insensitive and BFA-sensitive populations. An example of a cargo that exits the ER via a non-COPII pathway is the potassium channel Kv4, which requires KChIP binding for transport. Examples of cargos that have been shown to bypass the Golgi are connexin26 and connexin30, pannexins, CD45, and cystic fibrosis-associated chloride channel CFTR. In the case of CFTR, it can under certain conditions and in a cell-type-specific manner exit the ER in a COPII-independent manner and traffic independently of the Golgi SNARE syntaxin5, but there is still some modification of glycans to achieve EndoH resistance. The exact trafficking route for CFTR is not known, but the SNARE syntaxin13 has been implicated (reviewed in Prydz et al. 2012). Syntaxin13 is associated with the TGN and endosomes. Golgi-bypass mechanisms in neurons are not well studied, but a complex of F3/contactin and Caspr/paranodin arrives at the plasma membrane with EndoH-sensitive glycans (Bonnon et al. 2003).

4.2.2 The Secretory Pathways in Dendrites

Neurons contain all of the compartments of the secretory pathway described in non-neuronal cells (Cooney et al. 2002; Krijnse-Locker et al. 1995; Spacek and Harris 1997), but significant adaptations are in place to accommodate the large size and spatial heterogeneity of axons and dendrites. Axonal compartments will not be discussed in this chapter. Ribosomes, rough ER, and Golgi have all been described as prominent in the neuronal soma, but can also be clearly detected in dendrites (Krijnse-Locker et al. 1995). How is the ER organized in soma and dendrites and how does it relate to the Golgi?

4.2.2.1 Dendritic ER

ER is found extensively in dendrites and extends to the distal portions of them (Cooney et al. 2002; Krijnse-Locker et al. 1995; Spacek and Harris 1997). Soma and proximal dendrites contain rough ER with many ribosomes attached, whereas more smooth ER and fewer attached ribosomes are present on ER in distal dendrites (Krijnse-Locker et al. 1995; Valenzuela et al. 2011). There is also some smooth ER present in dendritic spines, which can release calcium after stimulation of mGluRs (Holbro et al. 2009; Ng and Toresson 2011). A fine structure reconstruction of dendritic ER was undertaken at the electron microscopy level (Cui-Wang et al. 2012). An anostomizing network of ER tubules is found throughout the dendrites with increased complexity at dendritic branchpoints. At branchpoints, cisternal ER (rather than tubular ER) is also apparent, with an increase in attached ribosomes. Similarly, more complex ER is found near spines. The presence of rough ER in distal dendrites and at branchpoints raises several questions. Is mRNA localized to dendrites translated on dendritic ER and not in the soma? Where are the Golgi compartments localized through which distally translated membrane cargo would traffic on its way to the plasma membrane? Secondly, can secretory cargos inserted into the ER in the soma redistribute laterally to enter distal dendrites? Is this regulated? Where are the ER exit sites, the ERGIC, and the Golgi?

Several labs have looked at the motility in the ER using labeled ER-resident proteins or proteins which were mutated to be retained in the ER. Many of these proteins diffuse rapidly and freely throughout the dendritic ER, including ER-retained AMPA receptor subunits GluA1 and GluA2 (Cui-Wang et al. 2012), GABA receptor subunit GABAB1R (Valenzuela et al. 2014) (which remains in the ER and diffuses widely in the absence of co-expressed GABAB2R), as well as ER luminal proteins (Bannai 2004). Strikingly, diffusion throughout the dendritic ER is not uniformly fast everywhere. Valenzuela et al. described two populations of GABAB1R with different speeds (Valenzuela et al. 2014). Cui-Wang et al. observed that the diffusion was slowed at branchpoints and near spines, regions where there was increased ER complexity by 3D serial EM. Even more important, the diffusion rates were regulated downstream of mGluR signaling which caused slowing by a regulatory mechanism involving PKC and the ER-microtubule-binding protein CLIMP63 (Cui-Wang et al. 2012). The ER might thus provide a non-vesicular transport system, continuous through wide region of the soma and dendrites (Valenzuela et al. 2011). Could ER-based diffusion contribute to distributing proteins into dendrites? Again, this begs the question of where the post-ER compartments are localized to support a potential local secretory pathway near branchpoints or synapses. Bannai used fluorescent proteins fused to the KDEL ER-retrieval sequence (Bannai 2004). The KDEL sequence is found on ER-resident proteins and allows retrograde retrieval of escaped ER proteins back to the ER via the KDEL receptor. Live imaging of GFP- or RFP-KDEL shows motile carriers in dendrites. These carriers contain several other ER markers and are capable of calcium release. Bannai (2004) propose that these motile carriers correspond to a specialized vesicular ER subcompartment. Since post-ER compartment markers (such as ERGIC53) were not co-stained, the exact relationship of the motile ER vesicular carriers to the ERGIC or to retrograde carriers remains to be clarified. In any case, it is clear that there is extensive trafficking of ER-/post-ER-associated membranes throughout the dendritic tree, and secretory cargos can enter distal dendrites either by lateral diffusion within the continuous reticular ER or in vesicular ER subcompartments/post-ER carriers. The distribution of secretory cargos in the distal ER cannot be only explained by passive diffusion since interference with active motor-based transport affects this distribution (Valenzuela et al. 2014). To what degree motor dependence is due to a dependence of distal ER tubule extension on motors or of motor-driven vesicular ER carriers remains to be established more firmly.

4.2.2.2 Transitional ER-Golgi Compartments in Dendrites

Where is secretory cargo exiting the ER in neurons? Several labs have visualized ER exit using a number of different approaches. Aridor and co-workers visualized directly where COPII coats assembled in dendrites by using a digitonin-permeabilization assay to selectively permeabilize the plasma membrane but not the ER (Aridor 2004). Using this assay in DIV21 (days in vitro) cultured neurons, they observed extensive formation of ERES (i.e., sites of COPII assembly) all over the soma and dendritic tree, including distal portions of dendrites. Newly synthesized NMDA receptors were seen to exit the ER in distal dendrites as well as proximally. mGluR stimulation increased NMDA receptor exit at ERES all over the dendritic tree. Since mRNA for NMDA receptors is known to be localized to dendrites and its translation stimulated by activity, these distal ERES could be crucially involved in local insertion near active synapses.

Ehlers and colleagues (Hanus et al. 2014; Horton and Ehlers 2003) directly visualized cargo exiting the ER and its subsequent post-ER trafficking using a temperature-sensitive mutant of the viral glycoprotein VSV-G, called VSV-G-ts045. At the restrictive temperature (39.5 °C), VSV-G-ts045 is misfolded and does not pass ER quality control for exit. At 32 °C, VSV-G-ts045 rapidly folds, trimerizes, and exits the ER toward the Golgi. VSV-G-ts045-GFP can be accumulated in the dendritic ER at the restrictive temperature and then released as a synchronous wave upon temperature shift to the permissive temperature (Presley et al. 1997). After temperature shifting, VSV-G-ts045-GFP exits the ER and accumulates in stationary puncta all over the dendritic tree and the soma in a matter of minutes. Sec24-YFP co-localizes with the emerging VSV-G-ts-045-GFP puncta, indicating that they correspond to ERES (Horton and Ehlers 2003). The stationary puncta also greatly co-localize with ERGIC markers (Hanus et al. 2014) and are often clustered at branchpoints where more complex ER was observed by EM (Cui-Wang et al. 2012). Motile carriers are observed to emerge after 15 min at the permissive temperature from the dendritic ERES/ERGIC and to traffic bidirectionally in dendrites (Horton and Ehlers 2003). A longer residence time of post-ER cargos at dendritic branchpoints was also observed by the Kennedy lab (Chen et al. 2013) using a novel way to induce ER exit by making fusion proteins containing light-sensitive oligomerization modules. ER exit can thus be locally induced by illumination. This technique is uniquely suited to ask questions about spatial regulation of secretion and will likely provide important insights in the future.

4.2.2.3 Golgi Compartments in Dendrites

As discussed in the previous section, innovative new techniques show that secretory cargos are found far distally in dendrites while still in the ER and exit the ER all throughout the dendritic tree. Branchpoints in dendrites play a special role in that the ER is more complex there and post-ER cargos reside in ERGIC-positive compartments at branchpoints for many minutes. Where then is the Golgi apparatus to which the post-ER secretory cargo will have to transit on its way to the plasma membrane? Staining of primary neurons against Golgi-resident enzymes reveals a prominent somatic Golgi apparatus perinuclearly with multiple stacks reminiscent of Golgi in non-neuronal cell types (Horton and Ehlers 2004; Krijnse-Locker et al. 1995). In addition, primary neurons with a major dendrite (probably corresponding to the apical dendrite in vivo) show Golgi membranes extending tens of micrometers into the proximal portion of the longest dendrite (Horton et al. 2005; Matsuki et al. 2010). These Golgi extensions still appear continuous with the somatic Golgi ribbons. Work from the 1990s indicated that glycosylation can take place in experimentally severed dendrites that lack somatic ER and Golgi suggesting the existence of a satellite secretory system spatially separated from the somatic compartments (Torre and Steward 1996). In addition, markers for Golgi compartments could be found in dendrites and even spines by immunofluorescence and electron microscopy (Pierce et al. 2000, 2001; Torre and Steward 1996). Unlike the extended Golgi ribbons in the proximal major dendrite, these "Golgi outposts" appear discontinuous from the somatic Golgi and can be found up to 100 μm in distal regions of the major dendrites in cultured neurons (Horton et al. 2005; Horton and Ehlers 2003). Both the Golgi polarized into the proximal portion of the major dendrites and the discontinuous Golgi outposts can be observed in vivo in the cortex by immunostaining and electron microscopy (Horton et al. 2005). Using VSV-G-ts045-GFP as an experimental cargo, Ehlers group demonstrated that these distal "Golgi outposts" in fact receive biosynthetic cargo subsequent to their ER exit (Horton and Ehlers 2003). Dual live imaging of VSV-G-ts045-GFP together with the Golgi-specific enzyme GalTase-CFP shows that motile carriers positive for VSV-G-ts045-GFP undergo either long-range retrograde transport to enter the soma or undergo presumptive fusion with distally localized Golgi outposts after temperature shift to the permissive temperature. BDNF-GFP similarly was transported to both somatic Golgi and Golgi outposts after ER exit (Horton and Ehlers 2003). Couve and colleagues showed that GABAB1 receptor is also secreted from somatic and dendritic Golgi compartments since blocking Golgi exit by expression of constitutive active Arf1 leads to accumulation of GABAB1 receptors in both somatic Golgi and Golgi outposts (Valenzuela et al. 2014). Interestingly, Golgi outposts were subsequently also identified in *Drosophila* neurons, and their localization affected dendrite branching (Ye et al. 2007). Similar correlations of Golgi outposts and increased dendrite length and branching were also noted in mammalian neurons (Horton et al. 2005). Interestingly, the position and extent of the extended Golgi are regulated downstream of Reelin signaling and might contribute to polarization of neurons by directing Golgi-derived traffic preferentially into the large apical dendrite (Matsuki et al. 2010). Most recently, Caceres and colleagues discovered how Golgi outposts are generated in major dendrites (Quassollo et al. 2015). They asked if Golgi outposts were locally generated from distally localized ERGICs or if they were generated from the somatic Golgi. Live imaging revealed that long tubules are generated from the somatic Golgi that enter the major dendrite, undergo fission, and then condense into Golgi outposts. These newly deployed Golgi outposts contain markers for all Golgi stacks as well as for the TGN. Interestingly, a RhoA cascade was shown to be upstream of regulating the deployment of Golgi-derived tubules into major dendrites. It will be interesting to determine if RhoA might be involved in the Reelin-dependent reorganization of the Golgi. In addition, other signaling pathways that converge onto RhoA might also regulate the extent of Golgi outpost formation into major dendrites.

4.2.2.4 Bypassing the Somatic Golgi: Secretion from Golgi Outposts

Are there cargos that traffic majorly or even exclusively via Golgi outposts? Jeyifous et al. made the surprising finding that AMPA receptors (GluA1) and NMDA receptors (NR1 and NR2) behave very differently after Golgi block using constitutively active Arf1, 20 °C temperature block, or Brefeldin A: GFP-GluA1 accumulates in the somatic Golgi, whereas NR1-GFP does not accumulate in the somatic Golgi but instead accumulates in dendritic Golgi outposts (Jeyifous et al. 2009). NR1-GFP was found to be transported into dendrites in motile carriers that also contained DsRed-ER and might thus correspond to a distinct ER subcompartment. The NR-binding protein SAP-97 and its binding partner CASK are also present in the motile subcompartment together with the microtubule motor KIF17. Interestingly, downregulation of SAP-97 or CASK leads to reduced trafficking of NR1 via Golgi outposts, increased trafficking via somatic Golgi, and reduced synaptic NR1. NR1 is thus preferentially included in an ER subcompartment by interaction with CASK and SAP-97 for delivery to dendritic Golgi outposts for more efficient delivery to synapses. Some of the mechanisms underlying the differential trafficking of GluA1 and NR1 have been discovered. SAP-97 can take on a compact or an extended configuration, which have differential affinities for AMPA receptors and NMDA receptors: SAP-97 in the extended configuration binds preferentially to NMDA receptors. Interestingly, CASK binding to SAP-97 stabilizes the extended configuration, thus favoring SAP-97 binding to NMDA receptors (Lin et al. 2013). Since CASK is important for binding to KIF17, cargos capable of associating with KIF17 via CASK might be preferentially included into the motile ER subcompartment that sends them to dendrites for secretion (Guillaud et al. 2008). Kir2 channels also bind to CASK and SAP-97 and might thus also be subject to bypassing the somatic Golgi (Leonoudakis et al. 2004). Clearly, the biosynthetic routes of glutamate receptors are diverse and regulated in complex ways which likely include somatic Golgi bypass for a specific subset (see Horak et al. 2014 for review).

What is the evidence that Golgi outposts secrete cargos for local exocytosis? The development of pH-sensitive GFP variants (such as superecliptic pHluorin SEP) enabled the visualization of SEP-tagged receptors arriving at the plasma membrane. The SEP tag becomes brighter as it is exposed to the neutral pH of the extracellular medium and is increasingly quenched in acidified compartments. This allowed visualization of exocytic events in dendrites, at/near synapses, and at spines (Kennedy and Ehlers 2011). Secretion from dendrites has been shown by using VAMP2-SEP (Cui-Wang et al. 2012). Since secretory compartments are mildly acidified, VAMP2-SEP becomes brighter upon exocytosis and exposure of the SEP tag on the cell surface. VAMP2-SEP can be seen to appear preferentially on the cell surface near branchpoints, suggesting the possibility that it is secreted from Golgi outposts (Cui-Wang et al. 2012). Alternatively, post-Golgi vesicles generated in the somatic Golgi might also have a preference for exocytosis at branchpoints. GABAB1 receptors appear with identical kinetics on the somatic and distal dendritic cell surface, consistent with secretion from Golgi outposts to the dendrite surface (Valenzuela et al. 2014). In order to definitively demonstrate that the local secretion in dendrites was via Golgi outposts rather than by long-range transport of post-Golgi carriers from the somatic Golgi, a local labeling approach will be needed. The light-dependent secretory block developed by Kennedy and co-workers will be powerful in combination with monitoring surface appearance. So far, Chen and Kennedy showed that VSV-G-Eos can be locally photoconverted to red and released from dendritic ER with a UV pulse and is then seen to accumulate preferentially at branchpoints (Chen et al. 2013). These secretory hotspots are presumed to correspond to Golgi outposts.

Work by Bonifacino and colleagues showed that several dendritic cargos, including NMDA receptors, become mislocalized to axons and dendrites when the function of the clathrin adaptor AP-1 is impacted by expression of a mutant μ1 chain (Farías et al. 2012). AP-1 adaptors have been shown to operate at the level of the TGN and endosomes in non-neuronal cells (Bonifacino 2014; Fölsch et al. 2009). This might also be their site of action in neurons. The sensitivity of NMDA receptor surface distribution to disturbance of AP-1 function suggests that sorting at the TGN is required for its trafficking. If AP-1 is acting at Golgi outposts, it has not been shown and awaits more work.

4.2.2.5 Golgi Bypass in Neurons

Current evidence strongly favors a view that many secretory cargos can enter dendrites by diffusing in the ER and exit the ER near branchpoints where ERGICs is concentrated. Surprisingly, distal ER exit is observed in all dendrites, even those that do not contain Golgi outposts. Golgi outposts are not found in all neurons and usually only in the major dendrite in those neurons that do have Golgi outposts (Horton and Ehlers 2003). In cultured hippocampal neurons, 30 % of neurons do not contain Golgi outposts, and they tend to correspond to neurons, which do not have a single major dendrite and are GABAergic. Even in the 70 % of neurons that do contain Golgi outposts, it is usually only one of the dendrites, which has Golgi outposts. Other dendrites do not have Golgi outposts, but still have distal ER exit sites secreting cargo. Further experiments using VSV-G-ts045 live imaging show that VSV-G exits the ER and then rapidly accumulates atERGIC53-positive branchpoints for second to minutes (Hanus et al. 2014). Motile carriers can be observed to emerge from ERGIC-positive compartments and travel long distance before apparently fusing with other ERGIC-positive compartments. The motility of post-ER carriers in dendrites is regulated by activity, such that increased activity immobilizes post-ER carriers and leads to faster appearance of VSV-G on the surface. Suppressing activity on the other hand increases long-range transport. Given the relative scarcity of Golgi outposts, the best hypothesis is that post-ER carriers are immobilized at branchpoints in a post-ER/pre-Golgi ERGIC53-positive secretory compartment which is secretion competent without transport to somatic Golgi (Ehlers 2013; Hanus et al. 2014). Golgi bypass might thus be used by synaptic receptors to change synaptic composition in an activity-dependent manner. This is an exciting possibility that needs to be further explored.

4.3 The Dendritic Endosomal Compartments

4.3.1 The Endosomal Pathways in Non-neuronal Cells

4.3.1.1 The Centrality of Endosomes to Cell Function

Endocytosis and endosomal trafficking is well studied in non-neuronal cell types (Burd and Cullen 2014; Doherty and Mcmahon 2009; Huotari and Helenius 2011; Naslavsky and Caplan 2011; Rink et al. 2005; Traub and Bonifacino 2013; Zoncu et al. 2009) Understanding the organization and regulation of endosomal pathways remains a topic of high interest because endocytosis is a powerful and rapid mechanism to regulate receptor levels and distribution (Yap and Winckler 2012). It is therefore relevant to any process that is dependent on surface receptors. In addition, many ligand-receptor systems signal not only at the plasma membrane but also from endosomes after they have undergone endocytosis (Bokel and Brand 2014; Cosker and Segal 2014; Murphy et al. 2009; Sann et al. 2009; Scott et al. 2014; Yogev et al. 2008). For instance, Sandra Schmidt and co-workers showed several decades ago that inhibition of endocytosis of the EGF receptor prolonged the downstream signaling of several pathways (Vieira et al. 1996). This was the expected result since endocytosis of the EGF receptor was thought to downregulate signaling. Inhibition of endocytosis was thus expected to increase signal duration. Surprisingly though, several downstream signaling cascades were not initiated in the absence of EGF receptor endocytosis. This surprising result suggested that distinct signaling cascades are initiated at the plasma membrane and at endosomes downstream of ligand-receptor binding. Other examples of receptors signaling from endosomes are becoming apparent. Neurotrophin signaling is probably the best-known example in neuronal cells (Cosker and Segal 2014; Harrington and Ginty 2013; Marlin and Li 2015). Neurotrophins bind their cognate receptor (such as TrkA for NGF) at the axon terminal leading to endocytosis of the activated ligand-receptor complex and subsequent transport of the endosome to the cell body for delivery of long-range survival signals to the nucleus. The endosomes traveling retrogradely with activated Trk receptors are commonly referred to as "signaling endosomes." In neurons, endosomes generally are less well understood than in fibroblasts, but are a topic of keen interest (for reviews see Lasiecka and Winckler 2011; Yap and Winckler 2012). Many processes in neurons are regulated by endocytosis and endosomal trafficking, including neuronal migration (Kawauchi 2012), axon and dendrite outgrowth (Lasiecka et al. 2014; Schuldiner and Yaron 2015; Zhang et al. 2014), and synaptic function (Ehlers 2013). Endocytosis and endosomes are thus at the crossroads of neuronal function.

4.3.1.2 The Canonical Endosomal Compartments

Endosomes come in several "flavors" which are commonly divided into early endosomes, recycling endosomes, late endosomes, and lysosomes. Most endocytosed cargos coalesce into early endosomes marked by Rab5 and EEA1 (early endosome antigen 1). Rab5 is a crucial player in early endosome function, and downregulation of Rab5 leads to loss of EEA1-positive early endosomes (Zeigerer et al. 2012; Wandinger-Ness and Zerial 2014). Because of this common entry point into the endosomal system, the early endosome is sometimes referred to as the "common endosome" (Jovic et al. 2010). From the early endosome, cargos can take different routes, i.e., they are sorted away from one another into distinct domains and distinct carriers with distinct destinations. The early endosome is therefore often also referred to as the "sorting endosome." There are three fundamental sorting directions out of the early endosome: recycling back to the plasma membrane, retrograde transport back to the TGN, or degradation. Cargos that recycle back to the plasma membrane can do so directly from the early endosome using Rab4-positive domains. Recycling cargos can also traffic via the recycling endosome, marked by Rab11. Recycling via recycling endosomes is slower, and cargos can reside in recycling endosomes for 20 min or more. In fibroblasts, early endosomes emerge in the cell periphery and grow in size by homotypic fusion events (Perini et al. 2014). With time, there is centripetal movement of early endosomes to more perinuclear locations. Recycling endosomes are mostly clustered perinuclearly in fibroblasts and are found in close proximity to the TGN (Thompson et al. 2007). Recycling endosomes are more tubular compared to the more spherical early endosomes. In fact, early endosomes generate tubular carriers that carry specific cargos to the recycling endosome or back to the TGN. As early endosomes spew out tubular carriers (see D'Souza et al. 2014; Yudowski et al. 2009 for striking live imaging), some cargos start to bud into the lumen of the endosome generating intraluminal vesicles (ILVs) (Scott et al. 2014). A host of regulatory protein complexes is responsible for sorting cargos into diverse tubular carriers or into ILVs. This process is not fully understood and much is still being learned. Generation and regulation of ILV formation are downstream of another Rab cascade, namely, Rab7 (Huotari and Helenius 2011). As more ILVs bud into the endosome lumen, the endosome takes on the appearance of a multivesicular body (MVB), which can be easily distinguished by electron microscopy from early endosomes (few ILVs and tubules emerging from the periphery) and recycling endosomes (tubules) (Klumperman and Raposo 2014). MVBs are viewed by some as an intermediate carrier to the late endosome (Falguières et al. 2009) and by others as a precursor to the late endosome itself. There are still sorting events taking place in this arm of the endosomal pathway, and late endosomes are dynamic. Some cargos can still be retrieved from late endosomes and returned to the TGN for another round of trafficking. The transition from early to late endosomes is accompanied by progressive acidification as well as a switch from Rab5 to Rab7 and concomitant switch to Rab7 effectors (Poteryaev et al. 2010; Rink et al. 2005). Late endosomes can ultimately fuse with lysosomes, which have the most acidic pH of any intracellular membrane compartment and are filled with electron-dense material as seen by electron microscopy (Klumperman and Raposo 2014). Lysosomes contain many hydrolases that degrade incoming cargos. The endosomal system was initially viewed exclusively as a set of stable compartments that exchange content via vesicular and tubular carriers. It is now clear that in addition to this system of carriers budding and fusing from preexisting compartments, the compartments themselves undergo progressive maturation and convert from an earlier to a later type of endosome (Rink et al. 2005). These conversions are accompanied and likely triggered by changes in Rabs and Rab effectors and in internal pH. The molecular details of endosomal maturation are still being elucidated. Because there are gradual as well as abrupt changes in molecular composition, some workers refer to additional subtypes of endosomes, such as "early late" endosomes and "late late" endosomes. This nomenclature is helpful in order to keep in mind the maturational nature of endosomes with many intermediate compartments on the way. It is nevertheless helpful to divide endosomes into Rab5-positive early endosomes vs. Rab11-positive recycling endosomes vs. Rab7-positive late endosomes, keeping in mind that dually positive compartments exist in the transitions.

4.3.2 *The Endosomal Pathway in Neurons*

4.3.2.1 Spatial Heterogeneity in Endosomal Populations

Neurons contain all of the canonical endosomal compartments described in non-neuronal cells, but the size and spatial and functional heterogeneity in dendrites and axons necessitate additional points of diversification and regulation. Surprisingly, some of the canonical regulators of endosomes identified in non-neuronal cells are localized in a highly regionalized manner. A striking example of this is EEA1, one of the most widely used markers for early endosomes in non-neuronal cells: EEA1 is only detectable in the soma and dendrites, but absent from axons (Wilson et al. 2000). This is surprising given the crucial roles attributed to EEA1 in tethering and fusion of early endosomes in fibroblasts (Christoforidis et al. 1999; Lawe et al. 2000). Since there is copious amount of endocytosis in axon terminals, it is certainly not the case that there are no early endosomes in axons. There is, for instance, plenty of Rab5 localized to axons (Wilson et al. 2000), but these axonal Rab5-positive early endosomes lack EEA1. Are there alternative tethering/fusion factors for axonal early endosomes, or do they not fuse as extensively as in the somatodendritic domain, thus not requiring EEA1 for their maturation? This is not known, and it is therefore necessary to investigate the dynamics and regulation of neuronal endosomes keeping spatial parameters in mind. Given that synapses are unique features on neurons, they represent a particularly salient neuronal-specific sublocale with active endocytosis and exocytosis taking place. Endosomes are found widely distributed in dendrites out to near the distal tips. EM reconstruction and live imaging have revealed that endosomes are present in the dendritic shaft, at the spine neck, and even in the spine head itself (Cooney et al. 2002). The spine-associated endosomes share markers (such as Rab11 or the endosomal SNARE syntaxin13) with canonical endosomes (Park 2004; Park et al. 2006). One intriguing open question thus is whether spine-near endosomes have special roles and might be regulated differently, especially by activity. Given the extent of the literature on endosomal contributions to synapse function, I refer the reader to excellent chapters in this volume covering trafficking at synapses.

4.3.2.2 Dynamics of Dendritic Endosomes

In fibroblasts, early endosomes are found throughout the cytoplasm including near the cell periphery where they are initially generated after endocytosis. Recycling endosomes, on the other hand, tend to form a cluster of tubules in the center of the cell near the centrosome (Thompson et al. 2007). Similarly, late endosomes and lysosomes are more concentrated in the center of the cell near the centrosome. In fact, microtubule motors are required to maintain this perinuclear distribution. In particular, dynein motor activity is responsible for the perinuclear localization of endosomes (Jordens et al. 2001; Neefjes and van der Kant 2014). Where are early, recycling, late endosome, and lysosomes localized in dendrites? In neurons, in contrast to fibroblasts and polarized epithelial cells (MDCK cells), all populations of endosomes are found both in the soma and dispersed far into the dendrites (Thompson et al. 2007). Similarly to fibroblasts, neurons have dispersed early endosomes (EEA1-positive) throughout the length of the dendrite (Choy et al. 2014; Cooney et al. 2002; Lasiecka et al. 2010). In fact, EEA1-positive endosomes are largely stationary in dendrites, and little retrograde flux toward the soma is evident in a 10-min period (Lasiecka et al. 2014). In fibroblasts, EEA1-positive endosomes move centripetally and mature as they start to accumulate near the nucleus. In dendrites, EEA1-positive endosomes mature and lose EEA1 while they are stationary and dispersed (Lasiecka et al. 2010, 2014). In agreement with this observation, LAMP1-positive late endosomes/lysosomes can be found throughout the dendrite, but they are often more clustered in the soma (Schwenk et al. 2014). Therefore, both early and late endosomes and even lysosomes are dispersed throughout dendrites and likely carry out local functions at these more distal locations. In fact, dysregulation of bidirectional trafficking of lysosomes in dendrites leads to defects in dendrite branching (Schwenk et al. 2014).

In addition, live imaging has revealed dynamic populations of endosomes that travel dynamically in both directions. For instance, Rab11-positive recycling endosomes and EHD1-positive endosomes undergo long-range and persistent movements along dendrites (Lasiecka et al. 2010). These movements are bidirectional with no particular bias toward anterograde or retrograde movements within dendrites. Many of the Rab11-positive endosomes move at very high speeds (>1 μm/s) and need high-resolution imaging to track fully. In our own imaging of DIV8 rat hippocampal neurons, for example, we have observed that the overwhelming majority of Rab11 endosomes are motile in any given two-minute window with only a small stationary pool (Lasiecka et al. 2010). This might change later in development as synapses and spines become prominent, and regional specializations develop along the dendritic shaft (Jullie et al. 2014; Lazo et al. 2013).

Live imaging of recycling endosomes in dendrites shows that at least some cargos (the transferrin receptor (TfR), the G protein-coupled receptor, β2-adrenergic receptor (B2AR), GluA2) undergo little long-range trafficking and appear to rapidly recycle locally (Choy et al. 2014; Jullie et al. 2014). At least two endocytosed cargos, on the other hand, additionally travel bidirectionally and long range in endosomes: transferrin and endocytosed NgCAM/L1 cell adhesion molecule (Lasiecka et al. 2010; Yap et al. 2008). These two cargos travel at different speeds with different efficiencies and in distinct endosomal populations (Yap et al. 2008). Where they are transported to, and where they might exocytose, is an unresolved question. Even locally recycling cargos do not completely share endosomal carriers: TfR and B2AR only co-localize in ~65 % of endosomes in dendrites (Jullie et al. 2014), suggesting that B2AR endosomes might be specialized. In striking contrast to TfR and NgAM, endocytosed BACE travels only retrogradely in dendritic Rab11 endosomes (Buggia-Prévot et al. 2013), and this motility is dependent on EHD1 function. Exactly what EHD1 is doing to promote retrograde motility of BACE-containing endosomes is not known. Clearly, we do not yet have a firm understanding of the heterogeneity of recycling compartments and to what degree cargos regulate their transport and recycling.

4.3.2.3 Directionality of Dendritic Vesicle Transport

The unexpected observation of unidirectional motility of endocytosed BACE in dendrites raises intriguing questions about which motors are responsible for motility of endosomes. A recent paper by Gary Banker and colleagues adapted an inducible split motor construct approach to investigate which motors are in principle capable of moving endosomes in fibroblasts. They found that KIF13A and KIF13B preferentially bind to early endosomes, whereas KIF1A and KIF1Bβ preferentially move late endosomes (Bentley et al. 2015). This assay can be used to screen through known kinesins and ask which ones move what kind of intracellular organelle, a question of great interest for neurons as well. The inducible motor assay has been used powerfully before to study roles of dyneins and kinesins in neurons and other cell types (Kapitein et al. 2010; van Spronsen et al. 2013; Yau et al. 2014) and to study myosin function in restricting access of dendritic proteins to the axon (Al-Bassam et al. 2012; Lewis et al. 2011). Most recently, Kapitein and Hoogenraad further optimized this system to allow dimerization upon illumination (van Bergeijk et al. 2015). This will for the first time allow investigators to ask not only temporal but also spatial questions about motor function in cellular behaviors.

In fact, how endosome motility is regulated is still a wide open question. It is still more difficult to explain regulation of directional motility in dendrites compared to axons, since microtubules in axons are of overwhelmingly uniform polarity (plus ends toward the distal end), so dynein motors are responsible for retrograde transport, and kinesin motors are responsible for anterograde movements. Dendrites, in contrast, contain microtubules of mixed polarity, as first shown by Peter Baas using a labor-intensive microtubule hook assay that requires electron microscopy (Baas et al. 1989). Mixed microtubule polarity is a confounding feature when it comes to regulating directional microtubule-based transport and raises many questions about regulating directional vesicle transport in dendrites. Dynein is in fact able to transport cargos to dendrites and might additionally remove aberrantly transported dendritic cargos out of the axon (Kapitein et al. 2010). New fluorescence-based techniques have made it easier to determine the orientation of microtubules in cells. These assays are based on expressing fluorescent versions of proteins that preferentially associate with the growing plus end of microtubules ("+ tip proteins"), such as EB1 or EB3. Live imaging of EB1-GFP reveals "comets" of GFP fluorescence in the movies that indicate the direction of plus ends (Baas and Lin 2011). Using this assay, it was found that in *Drosophila* dendrites microtubules are overwhelmingly of the "minus end"-out orientation (Mattie et al. 2010). Further analysis revealed that it was long dendrites (>40 μm) that showed overwhelmingly retrograde EB1 comets, diagnostic of minus end-out microtubule orientation (Ori-McKenney et al. 2012). In contrast, medium-sized dendrites had mixed microtubule polarity, and short terminal branches even showed mostly anterograde EB1 comets. How these distinctly oriented microtubule populations are generated and regulated is an important and fascinating question (Mattie et al. 2010; Ori-McKenney et al. 2012; Zhou et al. 2014). More detailed discussions on what is known about the microtubule motors powering transport in dendrites are presented elsewhere in this volume.

4.3.2.4 Exocytosis from Endosomes/Recycling in Dendrites

Exocytosis in dendrites was shown in the 1990s, some of it from TGN-derived vesicles as well as from endosomes. In addition, it was shown that calcium increase promotes these exocytic events (Huang and Neher 1996; Maletic-Savatic and Malinow 1998). As imaging techniques became more advanced, direct visualization of fusion events became feasible. In particular, total internal reflection (TIRF) microscopy allowed monitoring of membrane-near events (within 100 nm) and detection of fluorescent proteins as they approached the plasma membrane in vesicles and then fused. The development of pH-sensitive GFP variants (such as SEP) enabled the visualization of SEP-tagged receptors arriving at the plasma membrane. Exocytosis in dendrites was shown using SEP-tagged TfR which largely resides in endosomes. These exocytic events therefore likely correspond to endosomal exocytosis (Park 2004; Park et al. 2006). Rab11 and Rab11 effectors (such as Rab11-FIP2) were implicated in exocytosis in spines using SEP-TfR as a readout (Wang et al. 2008), arguing that exocytosis was occurring from a recycling endosome compartment. Subsequently, synaptic cargos such as GluA2 were visualized as SEP-fusion proteins and shown to undergo exocytosis in dendrites, including spines (Kennedy et al. 2010). In contrast, exocytosis in spines was rarely, if ever, observed by Jullie et al. (2014). More work is needed to understand under what conditions spine exocytosis takes place and how it is regulated. Recycling endosomes are clearly very important for spine formation and/or maintenance since interfering with recycling endosomes leads to disappearance of spines (Hoogenraad and Van Der Sluijs 2010; Park 2004).

Looking at β2-adrenergic receptor (B2AR), von Zastrow and colleagues identified two distinct modes of dendritic exocytosis: a rapid "burst" mode and a slow "persistent" mode (Yudowski et al. 2006). Using SEP-B2AR and TIRF microscopy, Yudowski found that exocytic events of SEP-B2AR were often characterized by a transient puff of increased SEP fluorescence followed by rapid diffusional dispersion of the fluorescence on the plasma membrane. This exocytic mode is consistent with full fusion of the vesicle with the plasma membrane and diffusion of the receptor after the fusion event. The second type of event was characterized by a sudden increase in SEP fluorescence, which persisted brightly in one spot for many seconds. How should these persistent events be interpreted? The persistent events might represent cases where the receptors remained clustered after the fusion and did not diffuse away. Alternatively, the fusion pore might not have expanded fully to allow diffusion of the receptor but remained open to allow fluorescence of theSEP tag. This potential mode of fusion was termed "kiss and wait." Two papers in 2014 shed further light on these persistent events and demonstrated that the persistent events corresponded to "kiss-and-run" fusions from endosomes (Jullie et al. 2014; Roman-Vendrell et al. 2014). "Kiss-and-run" fusion events are characterized by rapid opening of the fusion pore and partial discharge of cargo onto the plasma membrane, followed by closing of the fusion pore without full fusion. The reason that the SEP fluorescence appears persistent in these events is that acidification takes place at variable rates and can take >30 s before SEP fluorescence is quenched. This was shown elegantly by Jullie et al. using rapid switching of external pH buffers. This allowed them to estimate that the fusion pore closed on average after 2.6 s (Jullie et al. 2014). The "kiss-and-run" mode of endosomal exocytosis was found to be particularly prevalent in neurons (as opposed to astrocytes and fibroblasts) and to become more frequent as neurons matured in culture (Jullie et al. 2014; Roman-Vendrell et al. 2014). Furthermore, "kiss-and-run" endosomes can undergo multiple rounds of partial fusion and resealing (Jullie et al. 2014; Roman-Vendrell et al. 2014). Interestingly, an endosome that underwent a round of kiss-and-run exocytosis might undergo a burst full fusion event in the next round (Jullie et al. 2014). These observations argue that a particular endosome is not restricted to one mode of exocytosis, but mode can be switched dynamically, presumably by the recruitment of the appropriate machinery. This dual mode of exocytosis might allow tight temporal and spatial control of recycling particular receptors to the plasma membrane and is clearly distinct from full fusion where the entire content of the endosome is externalized. The observation that fluorescence from SEP can persist in slowly acidifying endosomes raises a cautionary flag as to the interpretation of SEP fluorescence (Jullie et al. 2014): it is not necessarily on the plasma membrane, but could be in a non-acidified endosome after endocytosis.

What is known about the molecular machinery for dendritic endosomal exocytosis? Both calcium ionophores and addition of KCl increase exocytosis, in particular the "kiss-and-run" mode (Roman-Vendrell et al. 2014), suggesting specific regulation. Interestingly, the myosin motor MyoVb mediates movements of recycling endosomes into spines downstream of Rab11 effectors in a calcium-regulated manner (Wang et al. 2008), and the SNARE syntaxin4 was implicated in SEP-GluA2 exocytosis in spines (Kennedy et al. 2010). Syntaxin3 has also been implicated in glutamate receptor exocytosis at postsynaptic sites (Jurado et al. 2013). Are Rab11-positive endosomes the major source of endosomal exocytosis in dendrites? Rab11 has clearly been implicated in endosomal exocytosis (Jullie et al. 2014; Wang et al. 2008). Other endosomal Rabs, such as Rab4, on the other hand show little association with exocytosing endosomes (Jullie et al. 2014). Rab11 was also found on exocytosing endosomes and interestingly disappeared rapidly after a burst exocytic event. Expression of a dominant-negative Rab11 did not affect "kiss-and-run" exocytosis, but did decrease burst exocytosis, implicating Rab11 in burst events (Jullie et al. 2014). The machinery mediating burst vs. "kiss-and-run" exocytosis might thus differ and awaits more studies.

Other endosomal carriers might also undergo exocytosis in dendrites. In particular, retromer-containing endosomes were shown to exocytose SEP-B2AR in dendrites (Choy et al. 2014). Retromer is a complex of proteins which was initially implicated in retrograde transport from endosomes to the TGN (Burd and Cullen 2014). It is now well established that retromer subunits can also mediate retrieval into recycling pathways away from degradation at the stage of the early/late transitional endosome. B2AR can bind the retromer-associated sorting nexin SNX27 to undergo recycling. Choy et al. show that the retromer subunit VPS35 is maintained on a subdomain of the early (EEA1-positive) endosome in neurons (Choy et al. 2014), similar to observations by others (Bhalla et al. 2012). The VPS35-containing secretory endosomes were abundant in dendrites (found every 2 μm), and ~60 % of SEP-B2AR exocytic events were associated with retromer-containing endosomes (Choy et al. 2014). In addition, these fusion events were frequently partial, i.e., might be of the "kiss-and-run" variety. Interestingly, dendritic endosomal exocytosis occurred frequently from elongated tubules rather than spherical carriers (Choy et al. 2014; Jullie et al. 2014). To what extent the tubules generated on early endosomes in a VPS35-dependent manner are the same or distinct from Rab11 tubules is not entirely clear. It is of note that the majority of TfR was in the same endosomes as B2AR with only 35 % of TfR being on its own. Again, the diversity and heterogeneity of endosomal carriers are still not fully worked out, and it is an open question how the cargo orchestrates the formation and subsequent fate of its carrier (Burd and Cullen 2014; D'Souza et al. 2014; Puthenveedu et al. 2010; Yudowski et al. 2009).

4.3.2.5 Neuronal-Enriched Endosomal Proteins: Two Examples

Observations such as the one discussed above, namely, that neurons specifically use a dual mode of endosomal exocytosis in dendrites, raise the question of how such neuronal-specific regulation could be achieved. Neuronally enriched endosomal proteins might in fact provide the cellular toolbox for implementing neuronal adaptations to trafficking. Even though many trafficking proteins are highly expressed and even enriched in the brain, their precise functions are still being uncovered. Two examples of such neuronal-enriched endosomal proteins are GRASP-1 (GRIP-associated protein-1) and NEEP21 (neuronal-enriched endosomal protein 21kd). Knockdown of either one impairs LTP (Alberi et al. 2005; Hoogenraad et al. 2010), but in somewhat distinct ways. What is known about GRASP-1 and NEEP21?

GRASP-1 is a cytosolic PDZ domain-containing protein that binds Rab4-GTP as well as the endosomal SNARE syntaxin13 (stx13) (Hoogenraad et al. 2010; Hoogenraad and Van Der Sluijs 2010). GRASP-1 is enriched on somatodendritic endosomes that are positive for Rab4, but not Rab5, or Rab7. It is also co-localized with Rab11. Live imaging and immuno-EM show that GRASP-1 is present on tubules associated with early endosomes (Rab4-positive), moves on the tip of tubular Rab4-positive carriers, and associates with large Rab11-positive endosomes, but not with tubular Rab11 compartments. Overexpression and knock-down of GRASP-1 show that GRASP-1 regulates the transition from Rab4- to Rab11-positive recycling endosomes and might in fact be associated with carriers that are generated at the early endosome and fuse with the recycling endosome in a stx13-dependent manner. Interestingly, downregulation of GRASP-1 leads to fewer TfR-containing endosomes in spines, reduction in spine density, and impaired recycling of AMPA receptors. Recycling into spines and maintenance of spines might thus require neuron-specific machinery, such as GRASP-1.

Interestingly, stx13 also binds another neuron-enriched endosomal protein, namely, NEEP21. NEEP21's function is even less understood than GRASP-1's (Muthusamy et al. 2014). Similarly to GRASP-1, NEEP21 is largely restricted to somatodendritic endosomes and mostly excluded from the axon. NEEP21 is a small transmembrane protein that regulates the recycling of several neuronal cargos, including GluA2 (Steiner et al. 2005), and neurotensin receptor (Debaigt et al. 2004). In addition, it binds to βAPP and regulates its proteolytic processing: downregulation of NEEP21 leads to increased amyloidogenic processing of βAPP (Norstrom et al. 2010). NEEP21 also plays a surprising role in trafficking of the cell adhesion molecule L1 to the axon. This role of NEEP21 is carried out at somatodendritic endosomes (Yap et al. 2008). Somatodendritic endosomes can sort cargos toward axons and not just toward local somatodendritic domains (Ascaño et al. 2009; Eva et al. 2010; Wisco et al. 2003). This pathway is called transcytosis. Transit of L1 through NEEP21 somatodendritic endosomes is required for L1-mediated axon growth in primary neurons (Lasiecka et al. 2014). NEEP21 serves to restrict easy access of endocytosed cargos to LAMP1-positive late endosomes/lysosomes, since knockdown of NEEP21 leads to increased trafficking of GluA2 (Steiner et al. 2005), L1 (Yap et al. 2008), and neurotensin receptors (Debaigt et al. 2004) to lysosomes. NEEP21-containing endosomes share markers with early endosomes (such as EEA1) (Hoogenraad et al. 2010; Lasiecka et al. 2014; Steiner et al. 2002), but NEEP21 is also found in post-EEA1 endosomes after EEA1 dissociation (Lasiecka et al. 2014). Since GRASP-1 does not co-localize with NEEP21 (Hoogenraad et al. 2010), these two regulators act in different compartments/subdomains to affect AMPA receptor recycling and to keep cargos out of lysosomes. Regulating recycling vs. degradative pathways can thus likely take place both along the EEA1/Rab5/Rab7 axis and along the Rab4/Rab11 axis of the endosomal system.

4.4 Conclusion

Neurons have an extensive dendritic arbor that has many spatially distinct domains, including postsynaptic densities. Special adaptations to trafficking are in use in dendrites to ensure that there is rapid endo- and exocytosis even in distal regions of the dendrite that is far removed from somatic compartments. In particular, local trafficking pathways are employed to regulate membrane composition both by exocytosis and by endocytosis. In addition, long-range trafficking needs to be coordinated to supply the appropriate membrane proteins throughout the dendrite and to communicate with the soma. Much of this trafficking is regulated, by growth factors (such as neurotrophins), by developmental cues (such as reelin), and by activity (downstream of calcium and metabotropic glutamate receptor signaling). Neurons are using conserved machinery to carry out trafficking, but often the prevalence and regulation of this conserved machinery are adapted to the spatial challenges of neurons. For instance, unconventional secretion can be observed in all cell types, but might be particularly prevalent for synapse-related trafficking at distal site in dendrites. Similarly, endosomal secretion in neurons uses "kiss-and-run" exocytosis at a much greater rate than in other cell types. New imaging advances have opened new frontiers, and the field is poised for exciting new discoveries. In addition, rapid progress in trafficking research in non-neuronal cell types continues to uncover new mechanisms that might also have been adapted in neuronal trafficking. These are exciting times for neuronal trafficking.

References

Al-Bassam S, Xu M, Wandless TJ, Arnold DB (2012) Differential trafficking of transport vesicles contributes to the localization of dendritic proteins. Cell Rep 2:89–100. doi:10.1016/j.celrep.2012.05.018

Alberi S, Boda B, Steiner P, Nikonenko I, Hirling H, Muller D (2005) The endosomal protein NEEP21 regulates AMPA receptor-mediated synaptic transmission and plasticity in the hippocampus. Mol Cell Neurosci 29:313–319. doi:10.1016/j.mcn.2005.03.011

Ang AL, Taguchi T, Francis S, Fölsch H, Murrells LJ, Pypaert M, Warren G, Mellman I (2004) Recycling endosomes can serve as intermediates during transport from the Golgi to the plasma membrane of MDCK cells. J Cell Biol 167:531–543. doi:10.1083/jcb.200408165

Aridor M (2004) Endoplasmic reticulum export site formation and function in dendrites. J Neurosci 24:3770–3776. doi:10.1523/JNEUROSCI.4775-03.2004

Ascaño M, Richmond A, Borden P, Kuruvilla R (2009) Axonal targeting of Trk receptors via transcytosis regulates sensitivity to neurotrophin responses. J Neurosci 29:11674–11685. doi:10.1523/JNEUROSCI.1542-09.2009

Baas PW, Lin S (2011) Hooks and comets: the story of microtubule polarity orientation in the neuron. Dev Neurobiol 71:403–418. doi:10.1002/dneu.20818

Baas PW, Black MM, Banker GA (1989) Changes in microtubule polarity orientation during the development of hippocampal neurons in culture. J Cell Biol 109:3085–3094

Bannai H (2004) Kinesin dependent, rapid, bi-directional transport of ER sub-compartment in dendrites of hippocampal neurons. J Cell Sci 117:163–175. doi:10.1242/jcs.00854

Baptista CA, Hatten ME, Blazeski R, Mason CA (1994) Cell-cell interactions influence survival and differentiation of purified Purkinje cells in vitro. Neuron 12:243–260

Bentley M, Decker H, Luisi J, Banker G (2015) A novel assay reveals preferential binding between Rabs, kinesins, and specific endosomal subpopulations. J Cell Biol 208:273–281. doi:10.1021/pr200179r

Benyair R, Ogen-Shtern N, Lederkremer GZ (2014) ARTICLE IN PRESS. Semin Cell Dev Biol:1–11. doi:10.1016/j.semcdb.2014.11.006

Bhalla A, Vetanovetz CP, Morel E, Chamoun Z, Di Paolo G, Small SA (2012) Neurobiology of disease. Neurobiol Dis 47:126–134. doi:10.1016/j.nbd.2012.03.030

Bieberich E (2014) Synthesis, processing, and function of N-glycans in N-glycoproteins. In: Advances in neurobiology. Springer, New York, pp 47–70

Boiko T, Vakulenko M, Ewers H, Yap CC, Norden C, Winckler B (2007) Ankyrin-dependent and -independent mechanisms orchestrate axonal compartmentalization of L1 family members neurofascin and L1/neuron-glia cell adhesion molecule. J Neurosci 27:590–603. doi:10.1523/JNEUROSCI.4302-06.2007

Bokel C, Brand M (2014) Endocytosis and signaling during development. Cold Spring Harb Perspect Biol 6:a017020–a017020. doi:10.1101/cshperspect.a017020

Bonifacino JS (2014) Adaptor proteins involved in polarized sorting. J Cell Biol 204:7–17. doi:10.1074/jbc.274.9.5385

Bonnon C, Goutebroze L, Denisenko-Nehrbass N, Girault J-A, Faivre-Sarrailh C (2003) The paranodal complex of F3/contactin and caspr/paranodin traffics to the cell surface via a Non-conventional pathway. J Biol Chem 278:48339–48347. doi:10.1074/jbc.M309120200

Brown TC, Correia SS, Petrok CN, Esteban JA (2007) Functional compartmentalization of endosomal trafficking for the synaptic delivery of AMPA receptors during long-term potentiation. J Neurosci 27:13311–13315. doi:10.1523/JNEUROSCI.4258-07.2007

Budnik A, Stephens DJ (2009) ER exit sites – localization and control of COPII vesicle formation. FEBS Lett 583:3796–3803. doi:10.1016/j.febslet.2009.10.038

Buggia-Prévot V, Fernandez CG, Udayar V, Vetrivel KS, Elie A, Roseman J, Sasse VA, Lefkow M, Meckler X, Bhattacharyya S, George M, Kar S, Bindokas VP, Parent AT, Rajendran L, Band H, Vassar R, Thinakaran G (2013) A function for EHD family proteins in unidirectional retrograde dendritic transport of BACE1 and Alzheimer's disease Ab production. Cell Rep 5:1552–1563. doi:10.1016/j.celrep.2013.12.006

Burd C, Cullen PJ (2014) Retromer: a master conductor of endosome sorting. Cold Spring Harb Perspect Biol 6:a016774–a016774. doi:10.1101/cshperspect.a016774

Chen D, Gibson ES, Kennedy MJ (2013) A light-triggered protein secretion system. J Cell Biol 201:631–640. doi:10.1038/nmeth.2021

Choy RW-Y, Park M, Temkin P, Herring BE, Marley A, Nicoll RA, Zastrow MV (2014) Retromer mediates a discrete route of local membrane delivery to dendrites. Neuron 82:55–62. doi:10.1016/j.neuron.2014.02.018

Christoforidis S, McBride HM, Burgoyne RD, Zerial M (1999) The Rab5 effector EEA1 is a core component of endosome docking. Nature 397:621–625. doi:10.1038/17618

Christopherson KS, Ullian EM, Stokes CCA, Mullowney CE, Hell JW, Agah A, Lawler J, Mosher DF, Bornstein P, Barres BA (2005) Thrombospondins are astrocyte-secreted proteins that promote CNS synaptogenesis. Cell 120:421–433. doi:10.1016/j.cell.2004.12.020

Cline HT (2001) Dendritic arbor development and synaptogenesis. Curr Opin Neurobiol 11:118–126

Cooney JR, Hurlburt JL, Selig DK, Harris KM, Fiala JC (2002) Endosomal compartments serve multiple hippocampal dendritic spines from a widespread rather than a local store of recycling membrane. J Neurosci 22:2215–2224

Cosker KE, Segal RA (2014) Neuronal signaling through endocytosis. Cold Spring Harb Perspect Biol 6:a020669–a020669. doi:10.1101/cshperspect.a020669

Cui-Wang T, Hanus C, Cui T, Helton T, Bourne J, Watson D, Harris KM, Ehlers MD (2012) Local zones of endoplasmic reticulum complexity confine cargo in neuronal dendrites. Cell 148:309–321. doi:10.1016/j.cell.2011.11.056

Debaigt C, Hirling H, Steiner P, Vincent JP, Mazella J (2004) Crucial role of neuron-enriched endosomal protein of 21 kDa in sorting between degradation and recycling of internalized G-protein-coupled receptors. J Biol Chem 279:35687–35691. doi:10.1074/jbc.M402751200

Doherty GJ, Mcmahon HT (2009) Mechanisms of endocytosis. Annu Rev Biochem 78:857–902. doi:10.1146/annurev.biochem.78.081307.110540

Dotti CG, Sullivan CA, Banker GA (1988) The establishment of polarity by hippocampal neurons in culture. J Neurosci 8:1454–1468

Dunn GA (1971) Mutual contact inhibition of extension of chick sensory nerve fibres in vitro. J Comp Neurol 143:491–507. doi:10.1002/cne.901430406

Ehlers MD (2013) Dendritic trafficking for neuronal growth and plasticity. Biochem Soc Trans 41:1365–1382. doi:10.1042/BST20130081

Eva R, Dassie E, Caswell PT, Dick G, Ffrench-Constant C, Norman JC, Fawcett JW (2010) Rab11 and its effector Rab coupling protein contribute to the trafficking of 1 integrins during axon growth in adult dorsal root ganglion neurons and PC12 cells. J Neurosci 30:11654–11669. doi:10.1523/JNEUROSCI.2425-10.2010

Falguières T, Luyet P-P, Gruenberg J (2009) Molecular assemblies and membrane domains in multivesicular endosome dynamics. Exp Cell Res 315:1567–1573. doi:10.1016/j.yexcr.2008.12.006

Farías GG, Cuitino L, Guo X, Ren X, Jarnik M, Mattera R, Bonifacino JS (2012) Signal-mediated, AP-1/clathrin-dependent sorting of transmembrane receptors to the somatodendritic domain of hippocampal neurons. Neuron 75:810–823. doi:10.1016/j.neuron.2012.07.007

Flynn KC, Stiess M, Bradke F (2013) Role of the cytoskeleton and membrane trafficking in axon–dendrite morphogenesis. Cell Migr Form Neuro Connect:19–50. doi:10.1016/B978-0-12-397266-8.00036-3

Fölsch H, Mattila PE, Weisz OA (2009) Taking the scenic route: biosynthetic traffic to the plasma membrane in polarized epithelial cells. Traffic 10:972–981. doi:10.1111/j.1600-0854.2009.00927.x

Franker MAM, Hoogenraad CC (2013) Microtubule-based transport – basic mechanisms, traffic rules and role in neurological pathogenesis. J Cell Sci 126:2319–2329. doi:10.1242/jcs.115030

Greger IH, Esteban JA (2007) AMPA receptor biogenesis and trafficking. Curr Opin Neurobiol 17:289–297. doi:10.1016/j.conb.2007.04.007

Grieve AG, Rabouille C (2011) Golgi bypass: skirting around the heart of classical secretion. Cold Spring Harb Perspect Biol 3:a005298–a005298. doi:10.1101/cshperspect.a005298

Guillaud L, Wong R, Hirokawa N (2008) Disruption of KIF17–Mint1 interaction by CaMKII-dependent phosphorylation: a molecular model of kinesin–cargo release. Nat Cell Biol 10:19–29. doi:10.1038/ncb1665

Ha J, Lo KW-H, Myers KR, Carr TM, Humsi MK, Rasoul BA, Segal RA, Pfister KK (2008) A neuron-specific cytoplasmic dynein isoform preferentially transports TrkB signaling endosomes. J Cell Biol 181:1027–1039. doi:10.1083/jcb.200803150

Hanus C, Kochen L, Tom Dieck S, Racine V, Sibarita J-B, Schuman EM, Ehlers MD (2014) Synaptic control of secretory trafficking in dendrites. Cell Rep 7:1771–1778. doi:10.1016/j.celrep.2014.05.028

Harrington AW, Ginty DD (2013) Long-distance retrograde neurotrophic factor signalling in neurons. 1–11. doi:10.1038/nrn3253

Higgins D, Burack M, Lein P, Banker G (1997) Mechanisms of neuronal polarity. Curr Opin Neurobiol 7:599–604

Holbro N, Grunditz A, Oertner TG (2009) Differential distribution of endoplasmic reticulum controls metabotropic signaling and plasticity at hippocampal synapses. Proc Natl Acad Sci 106:15055–15060. doi:10.1073/pnas.0905110106

Hoogenraad CC, Van Der Sluijs P (2010) GRASP-1 regulates endocytic receptor recycling and synaptic plasticity. Commun Integr Biol 3:433–435. doi:10.4161/cib.3.5.12209

Hoogenraad CC, Popa I, Futai K, Sanchez-Martinez E, Wulf PS, Van Vlijmen T, Dortland BR, Oorschot V, Govers R, Monti M, Heck AJR, Sheng M, Klumperman J, Rehmann H, Jaarsma D, Kapitein LC, Van Der Sluijs P (2010) Neuron specific Rab4 effector GRASP-1 coordinates membrane specialization and maturation of recycling endosomes. PLoS Biol 8:e1000283. doi:10.1371/journal.pbio.1000283.t001

Horak M, Petralia RS, Kaniakova M, Sans N (2014) ER to synapse trafficking of NMDA receptors. Front Cell Neurosci 8:394. doi:10.3389/fncel.2014.00394

Horton AC, Ehlers MD (2003) Dual modes of endoplasmic reticulum-to-Golgi transport in dendrites revealed by live-cell imaging. J Neurosci 23:6188–6199

Horton AC, Ehlers MD (2004) Secretory trafficking in neuronal dendrites. Nat Cell Biol 6:585–591. doi:10.1038/ncb0704-585

Horton AC, Rácz B, Monson EE, Lin AL, Weinberg RJ, Ehlers MD (2005) Polarized secretory trafficking directs cargo for asymmetric dendrite growth and morphogenesis. Neuron 48:757–771. doi:10.1016/j.neuron.2005.11.005

Hu J, Shibata Y, Zhu P-P, Voss C, Rismanchi N, Prinz WA, Rapoport TA, Blackstone C (2009) A class of dynamin-like GTPases involved in the generation of the tubular ER network. Cell 138:549–561. doi:10.1016/j.cell.2009.05.025

Huang LY, Neher E (1996) Ca(2+)-dependent exocytosis in the somata of dorsal root ganglion neurons. Neuron 17:135–145

Huotari J, Helenius A (2011) Endosome maturation. EMBO J 30:3481–3500. doi:10.1038/emboj.2011.286

Irannejad R, Zastrow MV (2014) ScienceDirectGPCR signaling along the endocytic pathway. Curr Opin Cell Biol 27:109–116. doi:10.1016/j.ceb.2013.10.003

Irannejad R, Tomshine JC, Tomshine JR, Chevalier M, Mahoney JP, Steyaert J, Rasmussen SGF, Sunahara RK, El-Samad H, Huang B, Zastrow MV (2013) Conformational biosensors reveal GPCR signalling from endosomes. Nature 495:534–538. doi:10.1038/nature12000

Jan Y-N, Jan LY (2010) Branching out: mechanisms of dendritic arborization. Nat Rev Neurosci 11:316–328. doi:10.1038/nrn2836

Jensen D, Schekman R (2011) COPII-mediated vesicle formation at a glance. J Cell Sci 124:1–4. doi:10.1242/jcs.069773

Jeyifous O, Waites CL, Specht CG, Fujisawa S, Schubert M, Lin EI, Marshall J, Aoki C, de Silva T, Montgomery JM, Garner CC, Green WN (2009) SAP97 and CASK mediate sorting of NMDA receptors through a previously unknown secretory pathway. Nat Neurosci 12:1011–1019. doi:10.1038/nn.2362

Joo W, Hippenmeyer S, Luo L (2014) Dendrite morphogenesis depends on relative levels of NT-3/TrkC signaling. Science 346:626–629. doi:10.1126/science.1258996

Jordens I, Fernandez-Borja M, Marsman M, Dusseljee S, Janssen L, Calafat J, Janssen H, Wubbolts R, Neefjes J (2001) The Rab7 effector protein RILP controls lysosomal transport by inducing the recruitment of dynein-dynactin motors. Curr Biol 11:1680–1685

Jovic M, Sharma M, Rahajeng J, Caplan S (2010) The early endosome: a busy sorting station for proteins at the crossroads. Histol Histopathol 25:99–112

Jullie D, Choquet D, Perrais D (2014) Recycling endosomes undergo rapid closure of a fusion pore on exocytosis in neuronal dendrites. J Neurosci 34:11106–11118. doi:10.1523/JNEUROSCI.0799-14.2014

Jurado S, Goswami D, Zhang Y, Molina AJM, Südhof TC, Malenka RC (2013) LTP requires a unique postsynaptic SNARE fusion machinery. Neuron 77:542–558. doi:10.1016/j.neuron.2012.11.029

Kapitein LC, Schlager MA, Kuijpers M, Wulf PS, van Spronsen M, MacKintosh FC, Hoogenraad CC (2010) Mixed microtubules steer dynein-driven cargo transport into dendrites. Curr Biol 20:290–299. doi:10.1016/j.cub.2009.12.052

Kawauchi T (2012) Cell adhesion and its endocytic regulation in cell migration during neural development and cancer metastasis. Int J Mol Sci 13:4564–4590. doi:10.3390/ijms13044564

Kawauchi T, Sekine K, Shikanai M, Chihama K, Tomita K, Kubo K-I, Nakajima K, Nabeshima Y-I, Hoshino M (2010) Rab GTPases-dependent endocytic pathways regulate neuronal migration and maturation through N-cadherin trafficking. Neuron 67:588–602. doi:10.1016/j.neuron.2010.07.007

Kennedy MJ, Ehlers MD (2011) Mechanisms and function of dendritic exocytosis. Neuron 69:856–875. doi:10.1016/j.neuron.2011.02.032

Kennedy MJ, Davison IG, Robinson CG, Ehlers MD (2010) Syntaxin-4 defines a domain for activity-dependent exocytosis in dendritic spines. Cell 141:524–535. doi:10.1016/j.cell.2010.02.042

Klumperman J, Raposo G (2014) The complex ultrastructure of the endolysosomal system. Cold Spring Harb Perspect Biol. doi:10.1101/cshperspect.a016857

Koenig P-A, Ploegh HL (2014) Protein quality control in the endoplasmic reticulum. F1000Prime Rep. doi:10.12703/P6-49

Krijnse-Locker J, Parton RG, Fuller SD, Griffiths G, Dotti CG (1995) The organization of the endoplasmic reticulum and the intermediate compartment in cultured rat hippocampal neurons. Mol Biol Cell 6:1315–1332

Kwon M, Fernandez JR, Zegarek GF, Lo SB, Firestein BL (2011) BDNF-promoted increases in proximal dendrites occur via CREB-dependent transcriptional regulation of cypin. J Neurosci 31:9735–9745. doi:10.1523/JNEUROSCI.6785-10.2011

Lasiecka ZM, Winckler B (2011) Mechanisms of polarized membrane trafficking in neurons – focusing in on endosomes. Mol Cell Neurosci 48:278–287. doi:10.1016/j.mcn.2011.06.013

Lasiecka ZM, Yap CC, Caplan S, Winckler B (2010) Neuronal early endosomes require EHD1 for L1/NgCAM trafficking. J Neurosci 30:16485–16497. doi:10.1523/JNEUROSCI.3127-10.2010

Lasiecka ZM, Yap CC, Katz J, Winckler B (2014) Maturational conversion of dendritic early endosomes and their roles in L1-mediated axon growth. J Neurosci 34:14633–14643. doi:10.1523/JNEUROSCI.1837-14.2014

Lawe DC, Patki V, Heller-Harrison R, Lambright D, Corvera S (2000) The FYVE domain of early endosome antigen 1 is required for both phosphatidylinositol 3-phosphate and Rab5 binding. Critical role of this dual interaction for endosomal localization. J Biol Chem 275:3699–3705

Lazo OM, Gonzalez A, Ascaño M, Kuruvilla R, Couve A, Bronfman FC (2013) BDNF regulates Rab11-mediated recycling endosome dynamics to induce dendritic branching. J Neurosci 33:6112–6122. doi:10.1523/JNEUROSCI.4630-12.2013

Leonoudakis D, Conti LR, Radeke CM, McGuire LMM, Vandenberg CA (2004) A multiprotein trafficking complex composed of SAP97, CASK, veli, and Mint1 is associated with inward

rectifier Kir2 potassium channels. J Biol Chem 279:19051–19063. doi:10.1074/jbc.M400284200
Lewis TL, Mao T, Arnold DB (2011) A role for myosin VI in the localization of axonal proteins. PLoS Biol 9:e1001021. doi:10.1371/journal.pbio.1001021
Lewis TL, Courchet J, Polleux F (2013) Cell biology in neuroscience: cellular and molecular mechanisms underlying axon formation, growth, and branching. J Cell Biol 202:837–848. doi:10.1101/cshperspect.a009886
Li L, Gervasi N, Girault J-A (2015) Dendritic geometry shapes neuronal cAMP signalling to the nucleus. Nat Commun 6:6319. doi:10.1038/ncomms7319
Lin EI, Jeyifous O, Green WN (2013) CASK regulates SAP97 conformation and its interactions with AMPA and NMDA receptors. J Neurosci 33:12067–12076. doi:10.1523/JNEUROSCI.0816-13.2013
Lippincott-Schwartz J, Yuan LC, Bonifacino JS, Klausner RD (1989) Rapid redistribution of Golgi proteins into the ER in cells treated with brefeldin A: evidence for membrane cycling from Golgi to ER. Cell 56:801–813
Lippincott-Schwartz J, Yuan L, Tipper C, Amherdt M, Orci L, Klausner RD (1991) Brefeldin A's effects on endosomes, lysosomes, and the TGN suggest a general mechanism for regulating organelle structure and membrane traffic. Cell 67:601–616
Lohmann C (2013) Dendritic development. Cell Migr Form Neur Connect:213–227. doi:10.1016/B978-0-12-397266-8.00007-7
Lorente-Rodriguez A, Barlowe C (2011) Entry and exit mechanisms at the cis-face of the Golgi complex. Cold Spring Harb Perspect Biol 3:a005207–a005207. doi:10.1101/cshperspect.a005207
Ma L, Gibson DA (2013) Axon growth and branching. Cell Migr Form Neur Connect:51–68. doi:10.1016/B978-0-12-397266-8.00056-9
Maletic-Savatic M, Malinow R (1998) Calcium-evoked dendritic exocytosis in cultured hippocampal neurons. Part I: trans-Golgi network-derived organelles undergo regulated exocytosis. J Neurosci 18:6803–6813
Marlin MC, Li G (2015) Biogenesis and function of the NGF/TrkA signaling endosome. Int Rev Cell Mol Biol 314:239–257. doi:10.1016/bs.ircmb.2014.10.002
Martínez-Menárguez JA (2013) Intra-Golgi transport: roles for vesicles, tubules, and cisternae. ISRN Cell Biol 2013:1–15. doi:10.1016/j.cell.2008.04.044
Matsuki T, Matthews RT, Cooper JA, van der Brug MP, Cookson MR, Hardy JA, Olson EC, Howell BW (2010) Reelin and Stk25 have opposing roles in neuronal polarization and dendritic Golgi deployment. Cell 143:826–836. doi:10.1016/j.cell.2010.10.029
Mattie FJ, Stackpole MM, Stone MC, Clippard JR, Rudnick DA, Qiu Y, Tao J, Allender DL, Parmar M, Rolls MM (2010) Directed microtubule growth, +TIPs, and kinesin-2 are required for uniform microtubule polarity in dendrites. Curr Biol 20:2169–2177. doi:10.1016/j.cub.2010.11.050
McAllister AK (2007) Dynamic aspects of CNS synapse formation. Annu Rev Neurosci 30:425–450. doi:10.1146/annurev.neuro.29.051605.112830
Murphy JE, Padilla BE, Hasdemir B, Cottrell GS, Bunnett NW (2009) Endosomes: a legitimate platform for the signaling train. Proc Natl Acad Sci 106:17615–17622. doi:10.1073/pnas.0906541106
Muthusamy N, Chen Y-J, Yin D-M, Mei L, Bergson C (2014) Complementary roles of the neuron-enriched endosomal proteins NEEP21 and calcyon in neuronal vesicle trafficking. J Neurochem 132:20–31. doi:10.1111/jnc.12989
Myers KR, Wang G, Sheng Y, Conger KK, Casanova JE, Zhu JJ (2012) Arf6-GEF BRAG1 regulates JNK-mediated synaptic removal of GluA1-containing AMPA receptors: a new mechanism for nonsyndromic X-linked mental disorder. J Neurosci 32:11716–11726. doi:10.1523/JNEUROSCI.1942-12.2012
Naslavsky N, Caplan S (2011) EHD proteins: key conductors of endocytic transport. Trends Cell Biol 21:122–131. doi:10.1016/j.tcb.2010.10.003

Neefjes J, van der Kant R (2014) Stuck in traffic: an emerging theme in diseases of the nervous system. Trends Neurosci 37:66–76. doi:10.1016/j.tins.2013.11.006

Newell-Litwa K, Chintala S, Jenkins S, Pare J-F, Mcgaha L, Smith Y, Faundez V (2010) Hermansky-pudlak protein complexes, AP-3 and BLOC-1, differentially regulate presynaptic composition in the striatum and hippocampus. J Neurosci 30:820–831. doi:10.1523/JNEUROSCI.3400-09.2010

Ng AN, Toresson H (2011) Endoplasmic reticulum dynamics in hippocampal dendritic spines induced by agonists of type I metabotropic glutamate but not by muscarinic acetylcholine receptors. Synapse 65:351–355. doi:10.1002/syn.20887

Noctor SC, Cunningham CL, Kriegstein AR (2013) Radial migration in the developing cerebral cortex. Cell Migr Form Neuro Connect:299–316. doi:10.1016/B978-0-12-397266-8.00024-7

Norstrom EM, Zhang C, Tanzi R, Sisodia SS (2010) Identification of NEEP21 as a -amyloid precursor protein-interacting protein in vivo that modulates amyloidogenic processing in vitro. J Neurosci 30:15677–15685. doi:10.1523/JNEUROSCI.4464-10.2010

Ori-McKenney KM, Jan LY, Jan Y-N (2012) Golgi outposts shape dendrite morphology by functioning as sites of acentrosomal microtubule nucleation in neurons. Neuron 76:921–930. doi:10.1016/j.neuron.2012.10.008

Palade G (1975) Intracellular aspects of the process of protein synthesis. Science 189:347–358

Park M (2004) Recycling endosomes supply AMPA receptors for LTP. Science 305:1972–1975. doi:10.1126/science.1102026

Park H, Poo M-M (2013) Neurotrophin regulation of neural circuit development and function. 1–17. doi:10.1038/nrn3379

Park M, Salgado JM, Ostroff L, Helton TD, Robinson CG, Harris KM, Ehlers MD (2006) Plasticity-induced growth of dendritic spines by exocytic trafficking from recycling endosomes. Neuron 52:817–830. doi:10.1016/j.neuron.2006.09.040

Park JJ, Gondré-Lewis MC, Eiden LE, Loh YP (2011) A distinct trans-Golgi network subcompartment for sorting of synaptic and granule proteins in neurons and neuroendocrine cells. J Cell Sci 124:735–744. doi:10.1242/jcs.076372

Perini ED, Schaefer R, Stöter M, Kalaidzidis Y, Zerial M (2014) Mammalian CORVET is required for fusion and conversion of distinct early endosome sub-populations. Traffic. doi:10.1111/tra.12232

Pierce JP, van Leyen K, McCarthy JB (2000) Translocation machinery for synthesis of integral membrane and secretory proteins in dendritic spines. Nat Neurosci 3:311–313. doi:10.1038/73868

Pierce JP, Mayer T, McCarthy JB (2001) Evidence for a satellite secretory pathway in neuronal dendritic spines. Curr Biol 11:351–355

Polleux F, Ghosh A (2007) Dendrites, 10th edn. pp 97–115. doi:10.1093/acprof:oso/9780198566564.001.0001

Polleux F, Morrow T, Ghosh A (2000) Semaphorin 3A is a chemoattractant for cortical apical dendrites. Nature 404:567–573. doi:10.1038/35007001

Poteryaev D, Datta S, Ackema K, Zerial M, Spang A (2010) Identification of the switch in early-to-late endosome transition. Cell 141:497–508. doi:10.1016/j.cell.2010.03.011

Presley JF, Cole NB, Schroer TA, Hirschberg K, Zaal KJ, Lippincott-Schwartz J (1997) ER-to-Golgi transport visualized in living cells. Nature 389:81–85. doi:10.1038/38001

Prydz K, Tveit H, Vedeler A, Saraste J (2012) Arrivals and departures at the plasma membrane: direct and indirect transport routes. Cell Tissue Res 352:5–20. doi:10.1007/s00441-012-1409-5

Puram SV, Bonni A (2013) Cell-intrinsic drivers of dendrite morphogenesis. Development 140:4657–4671. doi:10.1242/dev.087676

Puthenveedu MA, Lauffer B, Temkin P, Vistein R, Carlton P, Thorn K, Taunton J, Weiner OD, Parton RG, Zastrow MV (2010) Sequence-dependent sorting of recycling proteins by actin-stabilized endosomal microdomains. Cell 143:761–773. doi:10.1016/j.cell.2010.10.003

Quassollo G, Wojnacki J, Salas DA, Gastaldi L, Marzolo M-P, Conde C, Bisbal M, Couve A, Caceres A (2015) A RhoA signaling pathway regulates dendritic Golgi outpost formation. Curr Biol 25:1–12. doi:10.1016/j.cub.2015.01.075

Rabouille C, Malhotra V, Nickel W (2013) Diversity in unconventional protein secretion. J Cell Sci 125:5251–5255. doi:10.1242/jcs.103630

Redondo PC, Rosado JA (2015) Store-operated calcium entry: unveiling the calcium handling signalplex. Int Rev Cell Mol Biol 316:183–226. doi:10.1016/bs.ircmb.2015.01.007

Rink J, Ghigo E, Kalaidzidis Y, Zerial M (2005) Rab conversion as a mechanism of progression from early to late endosomes. Cell 122:735–749. doi:10.1016/j.cell.2005.06.043

Rios RM (2014) The centrosome-Golgi apparatus nexus. Philos Trans R Soc B Biol Sci 369:20130462–20130462. doi:10.1038/ncb1777

Rodriguez-Boulan E, Müsch A (2005) Protein sorting in the Golgi complex: shifting paradigms. Biochim Biophys Acta (BBA) Mol Cell Res 1744:455–464. doi:10.1016/j.bbamcr.2005.04.007

Roman-Vendrell C, Chevalier M, Acevedo-Canabal AM, Delgado-Peraza F, Flores-Otero J, Yudowski GA (2014) Imaging of kiss-and-run exocytosis of surface receptors in neuronal cultures. Front Cell Neurosci 8:363. doi:10.3389/fncel.2014.00363

Rothman JE, Fine RE (1980) Coated vesicles transport newly synthesized membrane glycoproteins from endoplasmic reticulum to plasma membrane in two successive stages. Proc Natl Acad Sci U S A 77:780–784

Sann S, Wang Z, Brown H, Jin Y (2009) Roles of endosomal trafficking in neurite outgrowth and guidance. Trends Cell Biol 19:317–324. doi:10.1016/j.tcb.2009.05.001

Schuldiner O, Yaron A (2015) Mechanisms of developmental neurite pruning. CMLS Cell Mol Life Sci 72:101–119. doi:10.1007/s00018-014-1729-6

Schwenk BM, Lang CM, Hogl S, Tahirovic S, Orozco D, Rentzsch K, Lichtenthaler SF, Hoogenraad CC, Capell A, Haass C, Edbauer D (2014) The FTLD risk factor TMEM106B and MAP6 control dendritic trafficking of lysosomes. EMBO J 33:450–467. doi:10.1002/embj.201385857

Scott CC, Vacca F, Gruenberg J (2014) Seminars in cell & developmental biology. Semin Cell Dev Biol 31:2–10. doi:10.1016/j.semcdb.2014.03.034

Shibata Y, Shemesh T, Prinz WA, Palazzo AF, Kozlov MM, Rapoport TA (2010) Mechanisms determining the morphology of the peripheral ER. Cell 143:774–788. doi:10.1016/j.cell.2010.11.007

Shrimal S, Cherepanova NA, Gilmore R (2014) ARTICLE IN PRESS. Sem Cell Dev Biol:1–8. doi:10.1016/j.semcdb.2014.11.005

Sotelo P, Farfán P, Benitez ML, Bu G, Marzolo M-P (2014) Sorting nexin 17 regulates ApoER2 recycling and reelin signaling. PLoS ONE 9:e93672. doi:10.1371/journal.pone.0093672.s006

D'Souza RS, Semus R, Billings EA, Meyer CB, Conger K, Casanova JE (2014) Rab4 Orchestrates a small GTPase cascade for recruitment of adaptor proteins to early endosomes. Curr Biol:1–12. doi:10.1016/j.cub.2014.04.003

Spacek J, Harris KM (1997) Three-dimensional organization of smooth endoplasmic reticulum in hippocampal CA1 dendrites and dendritic spines of the immature and mature rat. J Neurosci 17:190–203

Steiner P, Sarria J-CF, Glauser L, Magnin S, Catsicas S, Hirling H (2002) Modulation of receptor cycling by neuron-enriched endosomal protein of 21 kD. J Cell Biol 157:1197–1209. doi:10.1083/jcb.200202022

Steiner P, Alberi S, Kulangara K, Yersin A, Sarria J-CF, Regulier E, Kasas S, Dietler G, Muller D, Catsicas S, Hirling H (2005) Interactions between NEEP21, GRIP1 and GluR2 regulate sorting and recycling of the glutamate receptor subunit GluR2. EMBO J 24:2873–2884. doi:10.1038/sj.emboj.7600755

Thompson A, Nessler R, Wisco D, Anderson E, Winckler B, Sheff D (2007) Recycling endosomes of polarized epithelial cells actively sort apical and basolateral cargos into separate subdomains. Mol Biol Cell 18:2687–2697. doi:10.1091/mbc.E05-09-0873

Torre ER, Steward O (1996) Protein synthesis within dendrites: glycosylation of newly synthesized proteins in dendrites of hippocampal neurons in culture. J Neurosci 16:5967–5978
Traub LM, Bonifacino JS (2013) Cargo recognition in clathrin-mediated endocytosis. Cold Spring Harb Perspect Biol 5:a016790–a016790. doi:10.1101/cshperspect.a016790
Valenzuela JI, Jaureguiberry-Bravo M, Couve A (2011) Molecular and cellular neuroscience. Mol Cell Neurosci 48:269–277. doi:10.1016/j.mcn.2011.07.001
Valenzuela JI, Jaureguiberry-Bravo M, Salas DA, Ramirez OA, Cornejo VH, Lu HE, Blanpied TA, Couve A (2014) Transport along the dendritic endoplasmic reticulum mediates the trafficking of GABAB receptors. J Cell Sci 127:3382–3395. doi:10.1242/jcs.151092
van Bergeijk P, Adrian M, Hoogenraad CC, Kapitein LC (2015) Optogenetic control of organelle transport and positioning. Nature 518:111–114. doi:10.1038/nature14128
van Spronsen M, Mikhaylova M, Lipka J, Schlager MA, van den Heuvel DJ, Kuijpers M, Wulf PS, Keijzer N, Demmers J, Kapitein LC, Jaarsma D, Gerritsen HC, Akhmanova A, Hoogenraad CC (2013) TRAK/Milton motor-adaptor proteins steer mitochondrial trafficking to axons and dendrites. Neuron 77:485–502. doi:10.1016/j.neuron.2012.11.027
Vieira AV, Lamaze C, Schmid SL (1996) Control of EGF receptor signaling by clathrin-mediated endocytosis. Science 274:2086–2089
Wandinger-Ness A, Zerial M (2014) Rab proteins and the compartmentalization of the endosomal system. Cold Spring Harb Perspect Biol 6:a022616–a022616. doi:10.1101/cshperspect.a022616
Wang Z, Edwards JG, Riley N, Jr DWP, Karcher R, Li X-D, Davison IG, Ikebe M, Mercer JA, Kauer JA, Ehlers MD (2008) Myosin Vb mobilizes recycling endosomes and AMPA receptors for postsynaptic plasticity. Cell:1–14. doi:10.1016/j.cell.2008.09.057
Westrate LM, Lee JE, Prinz WA, Voeltz GK (2014) Form follows function: the importance of endoplasmic reticulum shape. Annu Rev Biochem 84:150112144356008. doi:10.1146/annurev-biochem-072711-163501
Williams ME, Wilke SA, Daggett A, Davis E, Otto S, Ravi D, Ripley B, Bushong EA, Ellisman MH, Klein G, Ghosh A (2011) Cadherin-9 regulates synapse-specific differentiation in the developing hippocampus. Neuron 71:640–655. doi:10.1016/j.neuron.2011.06.019
Wilson JM, de Hoop M, Zorzi N, Toh BH, Dotti CG, Parton RG (2000) EEA1, a tethering protein of the early sorting endosome, shows a polarized distribution in hippocampal neurons, epithelial cells, and fibroblasts. Mol Biol Cell 11:2657–2671
Winckler B, Yap CC (2011) Endocytosis and endosomes at the crossroads of regulating trafficking of axon outgrowth-modifying receptors. Traffic 12:1099–1108. doi:10.1111/j.1600-0854.2011.01213.x
Wisco D, Anderson ED, Chang MC, Norden C, Boiko T, Fölsch H, Winckler B (2003) Uncovering multiple axonal targeting pathways in hippocampal neurons. J Cell Biol 162:1317–1328. doi:10.1083/jcb.200307069
Wong M, Munro S (2014) The specificity of vesicle traffic to the Golgi is encoded in the golgin coiled-coil proteins. Science 346:1256898–1256898. doi:10.1126/science.1256898
Yap CC, Winckler B (2012) Harnessing the power of the endosome to regulate neural development. Neuron 74:440–451. doi:10.1016/j.neuron.2012.04.015
Yap CC, Wisco D, Kujala P, Lasiecka ZM, Cannon JT, Chang MC, Hirling H, Klumperman J, Winckler B (2008) The somatodendritic endosomal regulator NEEP21 facilitates axonal targeting of L1/NgCAM. J Cell Biol 180:827–842. doi:10.1083/jcb.200707143
Yap CC, Lasiecka ZM, Caplan S, Winckler B (2010) Alterations of EHD1/EHD4 protein levels interfere with L1/NgCAM endocytosis in neurons and disrupt axonal targeting. J Neurosci 30:6646–6657. doi:10.1523/JNEUROSCI.5428-09.2010
Yau KW, van Beuningen SFB, Cunha-Ferreira I, Cloin BMC, van Battum EY, Will L, Schätzle P, Tas RP, van Krugten J, Katrukha EA, Jiang K, Wulf PS, Mikhaylova M, Harterink M, Pasterkamp RJ, Akhmanova A, Kapitein LC, Hoogenraad CC (2014) Microtubule minus-end binding protein CAMSAP2 controls axon specification and dendrite development. Neuron 82:1058–1073. doi:10.1016/j.neuron.2014.04.019

Ye B, Zhang Y, Song W, Younger SH, Jan LY, Jan Y-N (2007) Growing dendrites and axons differ in their reliance on the secretory pathway. Cell 130:717–729. doi:10.1016/j.cell.2007.06.032

Yogev S, Schejter ED, Shilo B-Z (2008) Drosophila EGFR signalling is modulated by differential compartmentalization of Rhomboid intramembrane proteases. EMBO J 27:1219–1230. doi:10.1038/emboj.2008.58

Yudowski GA, Puthenveedu MA, Zastrow MV (2006) Distinct modes of regulated receptor insertion to the somatodendritic plasma membrane. Nat Neurosci 9:622–627. doi:10.1038/nn1679

Yudowski GA, Puthenveedu MA, Henry AG, Zastrow MV (2009) Cargo-mediated regulation of a rapid Rab4-dependent recycling pathway. Mol Biol Cell 20:2774–2784. doi:10.1091/mbc.E08-08-0892

Zeigerer A, Gilleron J, Bogorad RL, Marsico G, Nonaka H, Seifert S, Epstein-Barash H, Kuchimanchi S, Peng CG, Ruda VM, Del Conte-Zerial P, Hengstler JG, Kalaidzidis Y, Koteliansky V, Zerial M (2012) Rab5 is necessary for the biogenesis of the endolysosomal system in vivo. Nature 485:465–470. doi:10.1038/nature11133

Zhang H, Wang Y, Wong JJL, Lim K-L, Liou Y-C, Wang H, Yu F (2014) Endocytic pathways downregulate the L1-type cell adhesion molecule neuroglian to promote dendrite pruning in drosophila. Dev Cell 30:463–478. doi:10.1016/j.devcel.2014.06.014

Zhou W, Chang J, Wang X, Savelieff MG, Zhao Y, Ke S, Ye B (2014) GM130 is required for compartmental organization of dendritic Golgi outposts. Curr Biol:1–7. doi:10.1016/j.cub.2014.04.008

Zoncu R, Perera RM, Balkin DM, Pirruccello M, Toomre D, Camilli PD (2009) A phosphoinositide switch controls the maturation and signaling properties of APPL endosomes. Cell 136:1110–1121. doi:10.1016/j.cell.2009.01.032

Chapter 5
Activity Dependent Protein Transport from the Synapse to the Nucleus

Sujoy Bera, Gonca Bayraktar, Katarzyna M. Grochowska, Michelle Melgarejo da Rosa, and Michael R. Kreutz

Abstract It is widely accepted that signaling between synapses and neuronal nuclei regulates activity-dependent gene transcription that plays a crucial role in synaptic plasticity. However, despite many years of research, it is still essentially unclear how signals from distal synapses are transduced to the nucleus. Several studies in the past decade have proposed mechanisms of activity-dependent transport of synaptic proteins to the nucleus. In this chapter we will discuss these mechanisms and how this transport might couple in particular NMDAR activation to specific aspects of gene transcription.

Keywords Synapto-nuclear protein messengers (SNPs) • Synaptic and extra-synaptic NMDARs • Synaptic plasticity • Nuclear localization signal (NLS) • Active retrograde transport • Importin nuclear import machinery • CREB

5.1 Introduction

The N-methyl-D-aspartate receptor (NMDAR) plays a central role in synaptic plasticity (Malenka and Nicoll 1993; Hunt and Castillo 2012). Signals generated from this receptor are transduced to neuronal nuclei and regulate gene transcription that is required for long-lasting changes in synapto-dendritic inputs (West et al. 2002; Deisseroth et al. 2003; Greer and Greenberg 2008; Flavell and Greenberg 2008; Cohen and Greenberg 2008; Alberni 2009). Along these lines the late phase of long-term potentiation (LTP) at hippocampal CA1 synapses, a synaptic adaptation that is believed to be part of the molecular basis of learning and memory consolidation in rodents, requires NMDAR-dependent gene expression (Nguyen et al. 1994; Vickers et al. 2005). Although several studies have demonstrated the importance of NMDARs in these processes, the underlying mechanism of NMDAR signaling to the nucleus is largely unknown. The complex

S. Bera • G. Bayraktar • K.M. Grochowska • M.M. da Rosa • M.R. Kreutz (✉)
RG Neuroplasticity, Leibniz Institute for Neurobiology, Brenneckestr 6, 39118 Magdeburg, Germany
e-mail: kreutz@lin-magdeburg.de

K. Emoto et al. (eds.), *Dendrites*, DOI 10.1007/978-4-431-56050-0_5

cytoarchitecture of pyramidal neurons poses a unique challenge for cellular signaling mechanisms that couple synaptic activity in distant dendrites to activity-dependent gene transcription in the nucleus. Synapses can be several hundred micrometers away from the nucleus, which imposes a number of problems with respect to long-distance communication. The classical view invokes that intracellular Ca^{2+} signals that are generated upon activation of NMDARs are instrumental for communication between synapses and neuronal nuclei. These signals are believed to propagate to the nucleus as either regenerative Ca^{2+} waves or back-propagating dendritic action potentials (Jaffe and Brown 1994; Berridge 1998; Kapur et al. 2001; Berridge et al. 2003; Adams and Dudek 2005; Watanabe et al. 2006). The question arises how these two mechanisms alone can generate specific genomic responses with respect to the diversity of synaptic signals that can be transduced to the nucleus. In fact, several studies in recent years have gathered evidence for an alternative mechanism where proteins that are located at synapses translocate to the nucleus upon enhanced neuronal activity (i.e., Abi-1, AIDA-1D, CREB2, CRTC1, Jacob, LAPSER1, NFκB, and the Wnt receptors Frizzled and Ryk/Derailed) (Jordan and Kreutz 2009; Budnik and Salinas 2011; Kaushik et al. 2014). This in turn regulates specific aspects of gene expression programs, which are in control of a variety of processes including cell survival, cell death, development, and plasticity (Jordan and Kreutz 2009; Ch'ng and Martin 2011; Fainzilber et al. 2011). Although this concept is appealing, it still lacks ultimate proof. Protein transport from the synapse to the nucleus is challenged by several space and time constraints. The number of molecules that can be released from individual synapses is probably low, the distance between synapse and nucleus is large, and synaptic signals have a short lifetime. Taken together this might result in a rapid decay of signals if it is based on the transport of molecules between both compartments. In this chapter we will summarize the current knowledge about the underlying transport mechanisms and put forward the hypothesis that the NMDAR complex is a rich source of protein messengers and that different NMDAR signals might induce the nuclear translocation of different proteins. It should, however, also be mentioned that a number of transmembrane proteins undergo proteolytic processing, and as a result their intracellular domains (ICDs) can migrate to the nucleus to regulate transcription. Indeed, ICDs from Notch, APP, CD44, and ErbB4 have been detected in the cell nucleus where they mediate key cellular responses to extracellular stimuli (Jordan and Kreutz 2009). This alternative mode of synapse-to-nucleus signaling has been covered by previous reviews (Jordan and Kreutz 2009; Ch'ng and Martin 2011).

5.2 Protein Transport from the Synapse to the Nucleus

The current evidence for the existence of protein transport from the synapse to the nucleus is largely indirect. Synapto-nuclear protein messengers (SNPs) exhibit a dual localization at synapses and in neuronal nuclei. In addition an activity-dependent dissociation of these proteins from synapses with a subsequent accumulation in the nucleus that cannot be blocked by protein synthesis inhibitors has been shown (Jordan and Kreutz 2009). However, direct evidence for long-distance transport of such proteins is still lacking. Furthermore, the mechanisms of transport have not been addressed yet in any detail. Proteomic studies have revealed that many protein components of the post-synaptic density (PSD) contain a nuclear localization signal (NLS), which points toward the possibility that they can serve as nuclear cargos for the importin nuclear import machinery (Jordan et al. 2004). Interestingly components of this machinery like importin-α and importin-β are located at synapses and translocate to the nucleus in an NMDAR-dependent manner (Thompson et al. 2004; Dieterich et al. 2008; Brill et al. 2009). Several synapto-nuclear protein messengers contain an NLS, and evidence was provided that they can bind to importin-α (Jordan and Kreutz 2009; Kaushik et al. 2014). Of note importin-α can also directly bind an NLS in the intracellular domain of GluN1 from where it then dissociates in an activity-dependent manner (Jeffrey et al. 2009). In previous work, it was shown that neuronal importins also directly associate with a dynein motor for retrograde transport to the nucleus (Hanz et al. 2003). Interestingly, the NMDAR complex, which contains probably more than 70 different proteins, is also a rich source of different protein messengers. Many of these proteins translocate to the nucleus in a strictly NMDAR-dependent manner (Fig. 5.1).

5.3 NMDAR-Dependent Protein Transport from the Synapse to the Nucleus

NMDARs in the forebrain are tri- or diheterotetrameric ligand and voltage-dependent sodium and calcium channels that consist of two GluN1 and one or two GluN2A and GluN2B subunits (Paoletti et al. 2013). The C-terminal intracellular domains of GluN2A and GluN2B are several hundred amino acids long and provide the binding interface for a multitude of proteins. All receptor subunits harbor carboxy-terminal PDZ ligands for the interaction with PDZ domains of MAGUKs like PSD95 or SAP102, and NMDARs are tethered to the PSD via binding to these scaffolding proteins. The association with these scaffolding proteins can conceivably further enrich the repertoire of SNPs that can be released from synaptic sites for nuclear trafficking. A larger signaling complex is assembled that will arguably differ between different receptor subtypes, and it might contain different SNPs. NMDARs are present at both synaptic and extra-synaptic sites, and the subcellular localization of each receptor profoundly and differentially affects the nuclear response to its activation. Several studies suggest that in particular diheteromeric GluN2B containing NMDARs are abundant at both sites (Paoletti et al. 2013). Other reports indicate that triheteromeric GluN1/GluN2A/GluN2B NMDARs are the most abundant NMDARs at synaptic sites, followed by

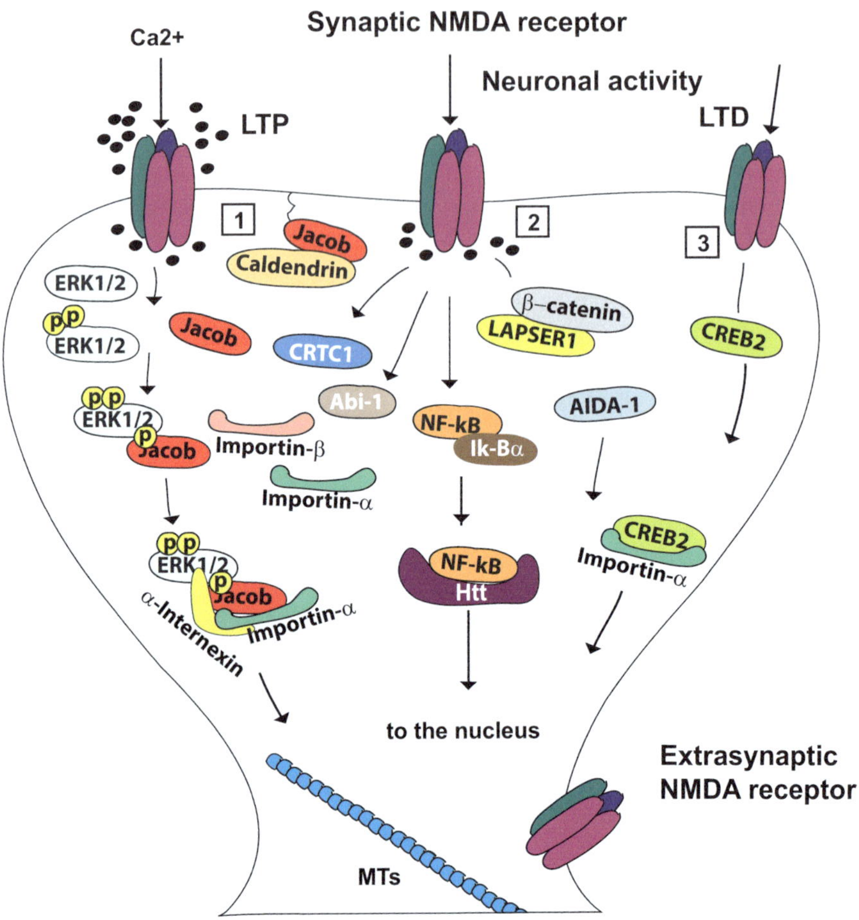

Fig. 5.1 Assembly of synapto-nuclear protein messenger at the synapse. (*1*) Following activation of synaptic NMDARs, Jacob is phosphorylated at a serine at position 180 by active ERK1/2. Phospho-Jacob subsequently associates with a soluble fragment of the neurofilament protein α-internexin and forms a stable trimeric complex. In this complex, α-internexin hinders the dephosphorylation of Jacob during long-distance trafficking to the nucleus. (*2*) Different protein messengers (CRTC1, Abi-1, NF-κB, LAPSER1) downstream of NMDAR activation translocate to the nucleus in an activity-dependent manner. (*3*) Induction of LTD but not LTP leads to translocation of CREB2 and vice versa of Jacob from the synapse to the nucleus

diheteromeric GluN2A containing receptors, whereas diheteromeric GluN2B NMDARs are the least abundant (reviewed in Paoletti et al. (2013)). Thus, the diversity of NMDARs and cross talk with other synaptic receptor pathways can generate a diversity of signals that might be transduced to the nucleus. Although it is still essentially unclear how NMDAR signals induce dissociation of synapto-nuclear protein messengers from the synapse, one can speculate that different signals will induce the nuclear import of different proteins.

Several lines of evidence suggest that this might be indeed the case. For instance, the transcriptional corepressor cAMP response element-binding protein-2 (CREB-2) also known as ATF-4 only transits to the nucleus after the induction of NMDAR-dependent long-term depression (LTD) but not long-term potentiation (LTP) (Lai et al. 2008). The opposite was found for Jacob, a protein that might operate as a mobile hub, which docks an NMDAR-derived signalosome to nuclear target sites like CREB (Behnisch et al. 2011; Karpova et al. 2013). Abelson interacting protein-1 (Abi-1) is another protein that can relay NMDAR signals to the nucleus. At the synapse it directly binds to the post-synaptic scaffolding protein ProSAP2/Shank3 (Proepper et al. 2007). Nuclear trafficking requires the convergence of both NMDAR activation and abelson kinase activity. Recently it was reported that ProSAP2/Shank3 itself might accumulate in the nucleus following NMDAR activation (Grabrucker et al. 2014). Since in contrast to ProSAP2/Shank3, Abi-1 lacks a nuclear localization signal (NLS), it is possible that a macromolecular complex transits to the nucleus. Another ProSAP2/Shank3 binding partner is LAPSER1 (Schmeisser et al. 2009), which is in one complex with the members of the canonical Wnt pathway including beta-catenin at post-synaptic sites. Following NMDAR stimulation, LAPSER1 and beta-catenin traffic to the nucleus suggesting a cross talk between NMDAR signaling and the Wnt pathway (Schmeisser et al. 2009).

Another example for the encoding of signals at the site of origin and decoding in the nucleus is the protein Jacob. Activation of synaptic NMDARs induces the expression of cell survival and plasticity-related genes, while their extra-synaptic counterparts primarily drive the expression of cell death genes, linking the pathway to disease (Hardingham and Bading 2010). An unresolved issue is how can the distant nucleus discriminate between synaptic and extra-synaptic NMDAR-induced signals? Previous work showed that extra-synaptic NMDAR activation induces nuclear import of Jacob, which results in sustained dephosphorylation and transcriptional inactivation of the transcription factor CREB, a loss of synaptic contacts, a retraction of dendrites, and eventually cell death (Dieterich et al. 2008). However, Jacob also transits to the nucleus of CA1 hippocampal neurons following induction of Schaffer collateral-dependent LTP but not LTD (Behnisch et al. 2011). Hence it acts as a messenger for both synaptic and extra-synaptic NMDAR pathways (Behnisch et al. 2011). In more recent work, it was found that Jacob following its nuclear import can encode the synaptic and extra-synaptic origin of NMDAR signals via differential ERK1/2-kinase binding and ERK-dependent phosphorylation of serine 180 which encodes synaptic but not extra-synaptic NMDAR activation (Karpova et al. 2013) (Fig. 5.1). A stable trimeric complex with proteolytically cleaved fragments of the neurofilament α-internexin is subsequently formed, and α-internexin binding protects Jacob and active ERK against phosphatase activity during retrograde transport to the nucleus. In the nucleus, this signalosome-like complex enhances plasticity-related and CREB-dependent gene expression as well as synaptic strength. Following extra-synaptic NMDAR activation, ERK is not activated and remains outside of the nucleus, while non-phosphorylated Jacob will dock yet unknown protein components to CREB that eventually induce CREB shutoff and cell death. Therefore, it appears that Jacob operates as a mobile hub that docks NMDAR-derived signalosomes that might arguably differ between synaptic and extra-synaptic NMDARs to nuclear target sites (Karpova et al. 2013). Phosphorylation-dependent transport and nucleocytoplasmic shuttling upon activation of synaptic NMDARs have also been shown for CREB-regulated transcription coactivator 1 (CRTC1) (Ch'ng et al. 2012). Increased NMDAR activity leads to calcineurin-dependent dephosphorylation of CRTC1, which then promotes its dissociation from the synaptic anchor 14-3-3ε and enables nuclear import (Ch'ng et al. 2012). Following nuclear import of CRTC1, the nuclear residing time can be tightly regulated by subsequent neuronal activity (Ch'ng et al. 2012). Thus, as a protein transiting to the nucleus, CRTC1 is another example for the activation of different synaptic kinase and phosphatase pathways that are encoded locally by phosphorylation of messenger molecules (Ch'ng et al. 2012) (Fig. 5.1).

5.4 Retrograde Transport of Protein Messenger to the Nucleus

One of the main issues for a better understanding of synapse-to-nucleus communication is the limited knowledge about the mechanisms underlying this particular type of long-distance protein transport. Fundamental mechanisms of protein redistribution following enhanced synaptic activity are passive and facilitated diffusion as opposed to active transport. Passive diffusion is random, does not require ATP, and is only efficient in case of small molecules and short distances (less than 200 nm) (Howe 2005). It is therefore unlikely that such mechanisms are sufficient to effectively transport larger protein complexes over long distances. A more plausible scenario is the two-dimensional diffusion along tracks, like the endoplasmic reticulum (ER), which is continuous with the outer nuclear membrane and extends throughout dendrites into dendritic spines (Jordan and Kreutz 2009).

In contrast to diffusion, active transport requires ATP to generate force by molecular motor proteins and it can provide fast, directional, and diffusion-independent targeting of synaptic proteins to the distant nucleus (Fig. 5.2). Active retrograde transport might also explain the incredibly efficient spatiotemporal integration of synaptic signals in the nucleus demonstrated in a recent paper by Zhai et al. (2013). After LTP induction in only 3–7 spines located in different, distal (more than 200 μm away from the nucleus) dendritic branches, the authors observed nuclear ERK1/2 activation with a delay of 30 min. This late nuclear ERK1/2 activation was independent of voltage-dependent calcium channel (VDCC) activation and membrane depolarization. Zhai et al. suggest that ERK1/2 phosphorylation can spread by PKC-dependent positive feedback loops (Zhai et al. 2013). On the other hand, directional, active transport could be an alternative mechanism

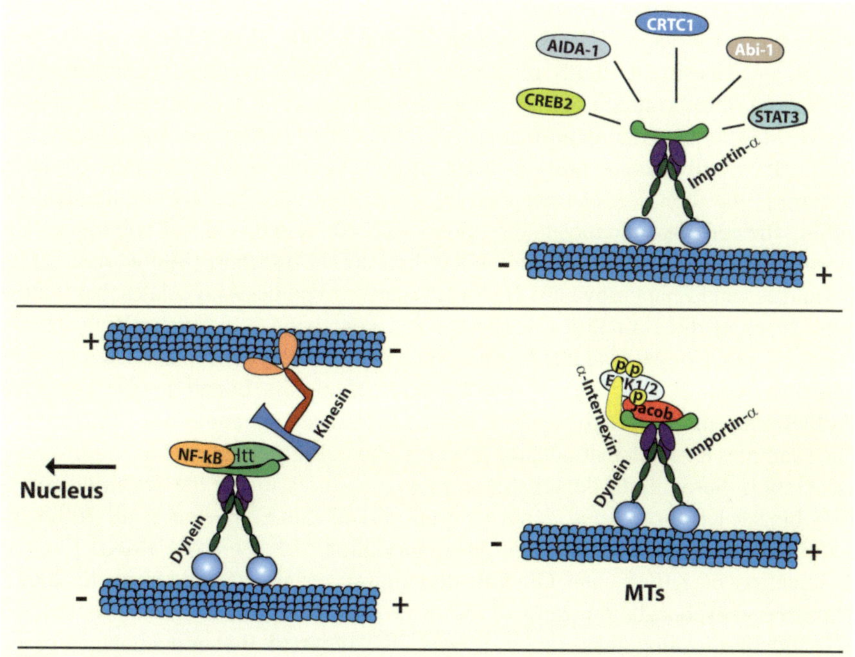

Fig. 5.2 Active retrograde transport along microtubule to the nucleus. Due to the mixed polarity of MTs, both dynein and kinesin can mediate transport to the nucleus. It has been shown that NFκB exits the synapse for nuclear import in a complex with htt. In addition, it was shown that the phosphorylation of huntingtin at S421 could switch transport from a dynein to a kinesin motor. Jacob is transported to the nucleus in a complex with active ERK1/2 and α-internexin, and nuclear trafficking is dynein dependent. AIDA-1D, CREB2, and STAT3 were shown to associate with importin-α; thus, it is possible that the translocation to the nucleus is also dynein dependent

explaining the fast accumulation of ERK1/2 in the nucleus. Mechanisms for signal retention of phosphorylation events in spine synapses were already demonstrated in several studies. Namely, binding of proteins like 14-3-3 (Dent et al. 1995; Wang et al. 2004), vimentin (Perlson et al. 2005), or α-internexin (Karpova et al. 2013) masks phosphorylated residues, thus hindering dephosphorylation that can occur in the phosphatase-rich neuronal environment. For example, active ERK1/2 translocates to the nucleus in a dynein-dependent manner in association with an intermediate filament vimentin, which is structurally related to α-internexin. Binding to vimentin links pERK1/2 to importin-β and preserves the phosphorylation signal (Perlson et al. 2005).

Active retrograde transport via molecular motors can provide high-speed and directional translocation of cargo. The motor proteins dynein, kinesins, and myosins require ATP to translocate along microtubule or actin tracts (Liu et al. 2005; Perlson et al. 2005; Marcora and Kennedy 2010; Ben-Yaakov et al. 2012; Dent and Baas 2014; Karpova et al. 2013). Whereas myosins mediate actin-based, short-distance transport, dyneins and kinesins are involved in long-distance, microtubule-based trafficking (Hirokawa and Takemura 2004; Roberts et al. 2013). Both dynein and kinesin are highly processive motors. While dyneins translocate to the minus end of microtubules (MTs), kinesins translocate to the plus end. Unlike in axons, MTs in proximal dendrites align bidirectionally (Kapitein and Hoogenraad 2011). It is still unclear how different motors bound to one transport complex determine the direction of transport. It is plausible that they act simultaneously, pulling the complex in opposite directions. According to this tug-of-war model, the most numerous motor will determine the direction of transport (Müller et al. 2010; Bryantseva and Zhapparova 2012). An alternative hypothesis postulates that motors with different directionalities cooperate and coactivate each other (Welte 2010; Harterink and Hoogenraad 2013). This could provide a highly sophisticated level of transport regulation depending on the type of motors bound to the complex. Furthermore, motor-based transport can be modulated by numerous mechanisms. For example, it seems that adaptor proteins play an important role in motor-cargo association. Setou et al. demonstrated that overexpression of the scaffold protein JIP3 blocks KIF5-mediated dendritic transport of GluA2 (Setou et al. 2002). In addition, it was demonstrated that phosphorylation of KIF17 by CaMKII leads to dissociation of KIF17- and GluN2B-containing vesicles (Guillaud et al. 2008). Thus, the phosphorylation status of motor proteins can provide an additional level of regulation. Along these lines, it was also reported that mitochondria-based transport is regulated by Ca^{2+} (Glater et al. 2006; Stowers et al. 2002; Guo et al. 2005) and small G proteins were implicated in regulation of transport of vesicles and organelles. Although described mechanisms were not demonstrated for non-vesicular retrograde transport, it is plausible that they play an important regulatory role together with posttranslational modifications of MTs (Echard et al. 1998; Nielsen et al. 1999). Molecular motors might also be involved in the translocation of proteins from the synapse to the dendrite. Actin-based myosins are involved in bidirectional transport to synaptic membranes (Kneussel and Wagner 2013). They can thereby provide a link between microfilaments- and MT-based transports. Alternatively, assembled transport complexes could leave the synapse upon NMDAR activation and calcium-dependent entry of dynamic MTs (Jaworski et al. 2009; Merriam et al. 2011, 2013).

Numerous proteins transported to the nucleus are reported to associate with at least one member of the importin family of proteins (e.g., Jacob, CREB2, AIDA-1, STAT3) (Karpova et al. 2013; Kindler et al. 2009; Dieterich et al. 2008; Lai et al. 2008; Jordan et al. 2007; Liu et al. 2005) (Fig. 5.2). The helix-loop-helix armadillo repeat of importin-α binds to NLS-containing proteins, which bind to importin-β forming a trimeric complex. Although importin-β does not have to be necessarily part of this complex, it can increase the affinity of importin-α to the NLS-containing molecule (Thompson et al. 2004; Otis et al. 2006).

A molecule that has a predicted tertiary structure like importin-β is huntingtin (htt) (Andrade and Bork 1995). Htt was reported to be involved in both vesicular and non-vesicular trafficking in axons and dendrites (Caviston and Holzbaur 2006).

In dendrites of striatal medium spiny neurons htt is essential for dynein-dependent retrograde transport of TrkB-signaling endosomes (Liot et al. 2013). Htt also forms a transport complex with importin-α and Htt-associated protein1 (HAP-1) in which NFκB is transported to the nucleus (Takano and Gusella 2002; Fagerlund et al. 2005; Marcora and Kennedy 2010). HAP1 interacts directly with p150Glued of dynactin, dynein intermediate chain, and kinesin, which could serve as a general docking platform (Colin et al. 2008; Marcora et al. 2003). In this regard Colin et al. (2008) demonstrated an interesting regulatory mechanism of directionality of vesicle transport. Phosphorylation of htt on serine 421 promotes interaction with kinesin-1 and anterograde transport, whereas non-phosphorylated htt translocates in a dynein-dependent manner to the nucleus (Colin et al. 2008). Such a mechanism could enable more efficient cargo trafficking in dendrites where MTs have mixed polarity.

It is important to note that there are at least 6 importin-α isoforms (Hosokawa et al. 2008), which together with importin-β and htt may form numerous transport complexes. Furthermore, several proteins were reported to contain an atypical NLS (Christophe et al. 2000), where nuclear transport is mediated by importin-β-like molecules (Flores and Seger 2013). For example, importin-7 mediates the translocation of active ERK1/2, MEK, or Smad3 (Chuderland et al. 2008). It is therefore likely that different import pathways coexist in neurons providing numerous ways of regulation of transport directionality and efficiency (Fig. 5.2).

5.5 Regulation of Gene Transcription by Protein Messenger

Activity-dependent remodeling of dendrites and synapses requires gene transcription, which occurs within a few minutes following stimulation of NMDARs (Matamales 2012; Deisseroth et al. 2003; Saha and Dudek 2013). Several different transcription factors (TFs) are regulated by NMDAR signaling, and the published studies show that different synapto-nuclear protein messengers might indeed transfer information about specific synaptic signals to distinct nuclear target sites (Fig. 5.3). Abi-1, for instance, is associated with Myc/Max transcription factor complex where it mediates E-box-regulated gene transcription (Proepper et al. 2007). LAPSER1 is a messenger on the Wnt pathway (Schmeisser et al. 2009), and AIDA-1D might regulate RNA splicing (Jordan et al. 2007). A key TF for NMDAR-induced gene expression is probably CREB. Gene expression by CREB is required in a variety of cellular processes that include cell survival, development, synaptic plasticity, learning and memory, cell death, and neurodegenaration (Silva et al. 1998; Hardingham et al. 2001, 2002; Deisseroth et al. 2006; Benito and Barco 2010; Hardingham and Bading 2010; Bading 2013). Several proteins that translocate from the synapse to the nucleus either directly or

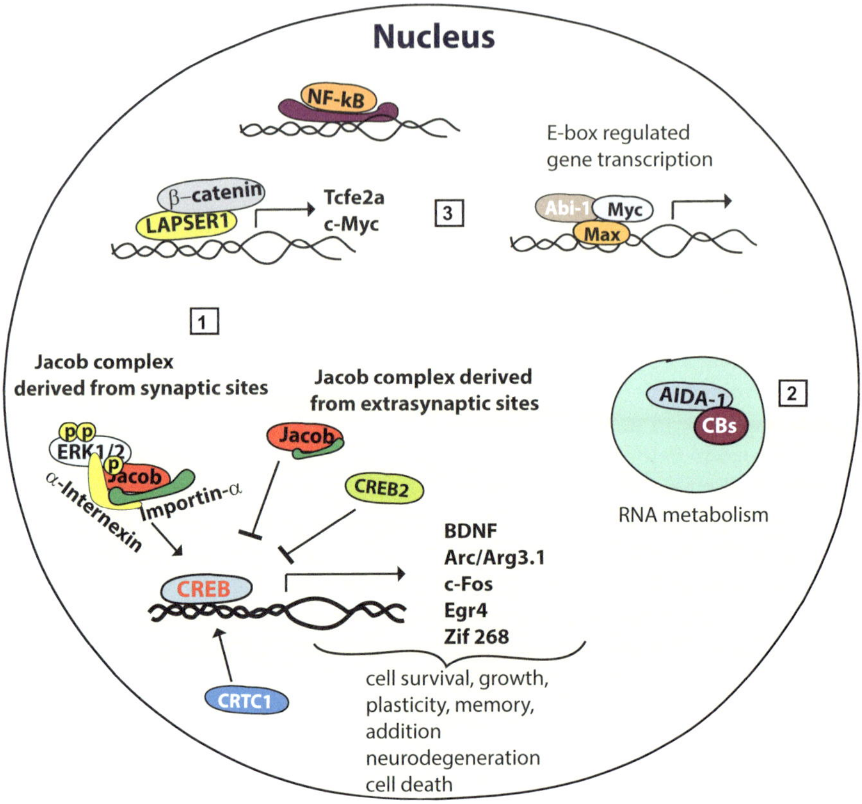

Fig. 5.3 Regulation of gene transcription by protein messenger. (*1*) Translocation of phospho-Jacob from synapses to the nucleus leads to CREB-dependent gene transcription required for cell survival and plasticity. Following stimulation of extra-synaptic NMDARs the non-phosphorylated form of Jacob triggers CREB shutoff pathway leading to stripping of synaptic contacts and simplification of dendritic architecture. Nuclear CRTC1 regulates CREB-dependent gene transcription and plays an important role in synaptic plasticity while the TF CREB2 represses CREB-dependent gene transcription. (*2*) The synapto-nuclear messenger AIDA-1 is associated with Cajal bodies and regulates global protein synthesis. (*3*) Protein messengers transported from synapses to the nucleus that regulate specific aspects of gene expression required for dendritic development, synaptogenesis, and synaptic maturation

indirectly associate with CREB or the CREB complex and regulate the transcriptional activity of CREB (Kaushik et al. 2014). Interestingly, however, binding of these proteins can have very different consequences resulting in transcriptional inactivation of CREB (Dieterich et al. 2008), differential phosphorylation of CREB (Ch'ng et al. 2012; Karpova et al. 2013), and regulation of a different set of genes (Ch'ng et al. 2012; Karpova et al. 2013). This further suggests that protein messengers cannot only carry an input specific signal from distal synapses but can also provide high specificity in terms of transcriptional regulation (Fig. 5.3).

References

Adams JP, Dudek SM (2005) Late-phase long-term potentiation: getting to the nucleus. Nat Rev Neurosci 6:737–743

Alberni CM (2009) Transcription factors in long-term memory and synaptic plasticity. Physiol Rev 89:1221–145

Andrade MA, Bork P (1995) HEAT repeats in the Huntington's disease protein. Nat Genet 11 (2):115–116

Bading H (2013) Nuclear calcium signaling in the regulation of brain function. Nat Rev Neurosci 14:593–608

Behnisch T, Yuanxiang P et al (2011) Nuclear translocation of jacob in hippocampal neurons after stimuli inducing long-term potentiation but not long- term depression. PLoS One 6(2):e17276

Benito E, Barco A (2010) CREB's control of intrinsic and synaptic plasticity: implications of CREB dependent memory models. Trends Neurosci 33(5):230–240

Ben-Yaakov K, Dagan SY et al (2012) Axonal transcription factors signal retrogradely in lesioned peripheral nerve. EMBO J 31(6):1350–1363

Berridge MJ (1998) Neuronal calcium signaling. Neuron 21:13–26

Berridge MJ, Bootman MD, Roderick HL (2003) Calcium signaling: dynamics, homeostasis and remodelling. Nat Rev Mol Cell Biol 4:517–529

Brill MS, Ninkovic J, Winpenny E et al (2009) Adult generation of glutamatergic olfactory bulb interneurons. Nat Neurosci 12:1524–1533

Bryantseva SA, Zhapparova ON (2012) Bidirectional transport of organelles: unity and struggle of opposing motors. Cell Biol Invest 36(1):1–6

Budnik V, Salinas PC (2011) Wnt signaling during synaptic development and plasticity. Curr Opin Neurobiol 3(1):151–159

Caviston JP, Holzbaur EL (2006) Microtubule motors at the intersection of trafficking and transport. Trends Cell Biol 16(10):530–537

Ch'ng TH, Martin KC (2011) Synapse-to-nucleus signaling. Curr Opin Neurobiol 21(2):345–352

Ch'ng TH, Uzgil B, Lin P et al (2012) Activity-dependent transport of transcriptional coactivator CRTC1 from synapse to nucleus. Cell 150(1):207–221

Christophe D, Christophe-Hobertus C, Pichon B (2000) Nuclear targeting of proteins: how many different signals? Cell Signal 12(5):337–341

Chuderland D, Konson A, Seger R (2008) Identification and characterization of a general nuclear translocation signal in signaling proteins. Mol Cell 31(6):850–861

Cohen S, Greenberg ME (2008) Communication between the synapse and the nucleus in neuronal development, plasticity, and disease. Annu Rev Cell Dev Biol 24:183–209

Colin E, Zala D et al (2008) Huntingtin phosphorylation acts as a molecular switch for anterograde/retrograde transport in neurons. EMBO J 27(15):2124–2134

Deisseroth K, Feng G, Majewska AK et al (2006) Next-generation optical technologies for illuminating genetically targeted brain circuits. J Neurosci 26(41):10380–10386

Deisseroth K, Mermelstein PG et al (2003) Signaling from synapse to nucleus: the logic behind the mechanisms. Curr Opin Neurobiol 13:354–365

Dent EW, Bass PW (2014) Microtubules in neurons as information carriers. J Neurochem 192(2):235–239

Dent P, Jelinek T, Morrison DK et al (1995) Reversal of Raf-1 activation by purified and membrane-associated protein phosphatases. Science 268(5219):1902–1906

Dieterich DC, Karpova A, Mikhaylova M et al (2008) Caldendrin-Jacob: a protein liaison that couples NMDA receptor signaling to the nucleus. PLoS Biol 6:e34

Echard A, Jollivet F, Martinez O, Lacapere J, Rousselet A, Janoueix-Lerosey I, Goud B (1998) Interaction of a Golgi-associated kinesin-like protein with Rab6. Science 279:580–585

Fagerlund R, Kinnunen L, Köhler M et al (2005) NF-{kappa}B is transported into the nucleus by importin {alpha}3 and importin {alpha}4. J Biol Chem 280(16):15942–15951

Fainzilber M, Budnik V, Segal RA et al (2011) From synapse to nucleus and back again-communication over distance within neurons. J Neurosci 31(45):16045–16048
Flavell SW, Greenberg ME (2008) Signaling mechanisms linking neuronal activity to gene expression and plasticity of the nervous system. Annu Rev Neurosci 31:563–590
Flores K, Seger R (2013) Stimulated nuclear import by β-like importins. F1000Prime Rep 5:41
Glater EE, Megeath LJ, Stowers RS, Schwarz TL (2006) Axonal transport of mitochondria requires milton to recruit kinesin heavy chain and is light chain independent. J Cell Biol 173:545–557
Grabrucker S, Proepper C, Mangus K, Eckert M, Chhabra R, Schmeisser MJ, Boeckers TM, Grabrucker AM (2014) The PSD protein ProSAP2/Shank3 displays synapto-nuclear shuttling which is deregulated in a schizophrenia-associated mutation. Exp Neurol 253:126–137
Greer PL, Greenberg ME (2008) From synapse to nucleus: calcium dependent gene transcription in control of synapse development and function. Neuron 59:846–860
Guillaud L, Wong R, Hirokawa N (2008) Disruption of KIF17–Mint1 interaction by CaMKII-dependent phosphorylation: a molecular model of kinesin-cargo release. Nat Cell Biol 10:19–29
Guo X, Macleod GT, Wellington A, Hu F, Panchumarthi S, Schoenfield M, Martin L, Charlton MP, Atwood HL, Zinsmaier KE (2005) The GTPase dMiro is required for axonal transport of mitochondria to Drosophila synapses. Neuron 47:379–393
Hanz S, Perlson E, Willis D et al (2003) Axoplasmic importins enable retrograde injury signaling in lesioned nerve. Neuron 40(6):1095–1104
Hardingham GE, Bading H (2010) Synaptic versus extrasynaptic NMDARs signalling: implications for neurodegenerative disorders. Nat Rev Neurosci 11:682–696
Hardingham GE, Arnold FJ, Bading H (2001) Nuclear calcium signalling controls CREB-mediated gene expression triggered by synaptic activity. Nat Neurosci 4:261–267
Hardingham GE, Fukunaga Y, Bading H (2002) Extrasynaptic NMDARs oppose synaptic NMDARS by triggering CREB shut-off and cell death pathways. Nat Neurosci 5:405–414
Harterink M, Hoogenraad CC (2013) Slide to the left and slide to the right: motor coordination in neurons. Dev Cell 26(4):326–328
Hirokawa N, Takemura R (2004) Kinesin superfamily proteins and their various functions and dynamics. Exp Cell Res 301(1):50–59
Hosokawa K, Nishi M, Sakamoto H et al (2008) Regional distribution of importin subtype mRNA expression in the nervous system: study of early postnatal and adult mouse. Neuroscience 157:864–877
Howe CL (2005) Modeling the signaling endosome hypothesis: why a drive to the nucleus is better than a (random) walk. Theor Biol Med Model 2:43
Hunt DL, Castillo PE (2012) Synaptic plasticity of NMDARs: mechanisms and functional implications. Curr Opin Neurobiol 22(3):496–508
Jaffe DB, Brown TH (1994) Metabotropic glutamate receptor activation induces calcium waves within hippocampal dendrites. J Neurophysiol 72:471–474
Jaworski J, Kapitein LC, Gouveia SM et al (2009) Dynamic microtubules regulate dendritic spine morphology and synaptic plasticity. Neuron 61(1):85–100
Jeffrey RA, Ch'ng TH, O'Dell TJ et al (2009) Activity-dependent anchoring of importin alpha at the synapse involves regulated binding to the cytoplasmic tail of the NR1-1a subunit of the NMDA receptor. J Neurosci 29(50):15613–15620
Jordan BA, Fernholz BD, Boussac M et al (2004) Identification and verification of novel rodent postsynaptic density proteins. Mol Cell Proteomics 3:857–871
Jordan BA, Fernholz BD, Khatri L et al (2007) Activity-dependent AIDA-1 nuclear signaling regulates nucleolar numbers and protein synthesis in neurons. Nat Neurosci 10:427–435
Jordan BA, Kreutz MR (2009) Nucleocytoplasmic protein shuttling: the direct route in synapse-to-nucleus signaling. Trends Neurosci 32(7):392–401
Kapitein LC, Hoogenraad CC (2011) Which way to go? Cytoskeletal organization and polarized transport in neurons. Mol Cell Neurosci 46(1):9–20

Kapur A, Yeckel M, Johnston D (2001) Hippocampal mossy fiber activity evokes Ca2+ release in CA3 pyramidal neurons via a metabotropic glutamate receptor pathway. Neuroscience 107:59–69

Karpova A, Mikhaylova M, Bera S et al (2013) Encoding and transducing the synaptic or extrasynaptic origin of NMDA receptor signals to the nucleus. Cell 152(5):1119–1133

Kaushik R, Grochowska KM, Butnaru I et al (2014) Protein trafficking from synapse to nucleus in control of activity-dependent gene expression. Neuroscience 280C:340–350

Kindler S, Dieterich DC, Schütt J et al (2009) Dendritic mRNA targeting of Jacob and N-methyl-D-aspartate-induced nuclear translocation after calpain-mediated proteolysis. J Biol Chem 284:25431–25440

Kneussel M, Wagner W (2013) Myosin motors at neuronal synapses: drivers of membrane transport and actin dynamics. Nat Rev Neurosci 14(4):233–247

Lai KO, Zhao Y, Ch'ng TH et al (2008) Importin-mediated retrograde transport of CREB2 from distal processes to the nucleus in neurons. Proc Natl Acad Sci U S A 105:17175–17180

Liot G, Zala D, Pla P et al (2013) Mutant Huntingtin alters retrograde transport of TrkB receptors in striatal dendrites. J Neurosci 33(15):6298–6309

Liu L, McBride KM, Reich NC (2005) STAT3 nuclear import is independent of tyrosine phosphorylation and mediated by importin-alpha3. Proc Natl Acad Sci U S A 102 (23):8150–8155

Malenka RC, Nicoll RA (1993) NMDAR receptors dependent synaptic plasticity: multiple forms and mechanisms. Trends Neurosci 16:521–527

Marcora E, Kennedy MB (2010) The Huntington's disease mutation impairs Huntingtin's role in the transport of NF-κB from the synapse to the nucleus. Hum Mol Genet 19(22):4373–4384

Marcora E, Gowan K, Lee JE (2003) Stimulation of NeuroD activity by huntingtin and huntingtin-associated proteins HAP1 and MLK2. Proc Natl Acad Sci U S A 100(16):9578–9583

Matamales M (2012) Neuronal activity-regulated gene transcription: how are distant synaptic signals conveyed to the nucleus? F1000Res 1:69

Merriam EB, Lumbard DC, Viesselmann C et al (2011) Dynamic microtubules promote synaptic NMDA receptor-dependent spine enlargement. PLoS One 6(11):e27688

Merriam EB, Millette M, Lumbard DC et al (2013) Synaptic regulation of microtubule dynamics in dendritic spines by calcium, F-actin, and drebrin. J Neurosci 33(42):16471–16482

Müller MJ, Klumpp S, Lipowsky R (2010) Bidirectional transport by molecular motors: enhanced processivity and response to external forces. Biophys J 98(11):2610–2618

Nguyen PV, Abel T, Kandel ER (1994) Requirement of critical period of transcription for induction of late phase LTP. Science 265:1104–1107

Nielsen E, Severin F, Backer JM, Hyman AA, Zerial M (1999) Rab5 regulates motility of early endosomes on microtubules. Nat Cell Biol 1:376–382

Otis KO, Thompson KR, Martin KC (2006) Importin-mediated nuclear transport in neurons. Curr Opin Neurobiol 16(3):329–335

Paoletti P, Bellone C, Zhou Q (2013) NMDA receptor subunit diversity: impact on receptor properties, synaptic plasticity and disease. Nat Rev Neurosci 14:383–400

Perlson E, Hanz S, Ben-Yaakov K et al (2005) Vimentin-dependent spatial translocation of an activated MAP kinase in injured nerve. Neuron 45(5):715–726

Proepper C, Johannsen S, Liebau S et al (2007) Abelson interacting protein 1 (Abi-1) is essential for dendrite morphogenesis and synapse formation. EMBO J 26:7188–7196

Roberts AJ, Kon T, Knight PJ et al (2013) Functions and mechanisms of dynein motor proteins. Nat Cell Mol Bio 14:713–726

Saha RN, Dudek SM (2013) Splitting hares and tortoises: a classification of neuronal immediate early gene transcription based on poised RNA polymerase II. Neuroscience 247:175–181

Schmeisser MJ, Grabrucker AM, Bockmann J et al (2009) Synaptic cross-talk between N-methyl-D-aspartate receptors and LAPSER1-beta-catenin at excitatory synapses. J Biol Chem 284:29146–29157

Setou M, Seog D-H, Tanaka Y et al (2002) Glutamate-receptor-interacting protein GRIP1 directly steers kinesin to dendrites. Nature 417:83–87
Silva AJ, Kogan JH, Frankland PW et al (1998) CREB and memory. Annu Rev Neurosci 21:127–148
Stowers RS, Megeath LJ, Górska-Andrzejak J, Meinertzhagen IA, Schwarz TL (2002) Axonal transport of mitochondria to synapses depends on milton, a novel Drosophila protein. Neuron 36:1063–1077, 2002
Takano H, Gusella JF (2002) The predominantly HEAT-like motif structure of huntingtin and its association and coincident nuclear entry with dorsal, an NF-kB/Rel/dorsal family transcription factor. BMC Neurosci 3:15
Thompson KR, Otis KO, Chen DY et al (2004) Synapse to nucleus signalling during long-term synaptic plasticity: a role for classical active nuclear import pathway. Neuron 44:997–1009
Vickers CA, Dickson KS, Wyllie DA (2005) Induction and maintenance of late-phase long-term potentiation in isolated dendrites of rat hippocampal CA1 pyramidal neurones. J Physiol 568:803–813
Wang X, Grammatikakis N, Siganou A et al (2004) Interactions between extracellular signal-regulated protein kinase 1, 14-3-3epsilon, and heat shock factor 1 during stress. J Biol Chem 279:49460–49469
Watanabe S, Hong M, Lasser-Ross N et al (2006) Modulation of calcium wave propagation in the dendrites and to the soma of rat hippocampal pyramidal neurons. J Physiol 575:455–468
Welte MA (2010) Bidirectional transport: matchmaking for motors. Curr Biol 20(9):R410–R413
West AE, Griffith EC, Greenberg ME (2002) Regulation of transcription factors by neuronal activity. Nat Rev Neurosci 3:921–931
Zhai S, Ark ED, Parra-Bueno P et al (2013) Long-distance integration of nuclear ERK signaling triggered by activation of a few dendritic spines. Science 342(6162):1107–1111

Chapter 6
Ultrastructure Methods

Nuno Maçarico da Costa

Abstract Information is delivered to neurons through thousands of synapses, most of them located on the branches of treelike structures we call dendrites. These dendrites are quite remarkable since on them lies the responsibility of collecting most of the inputs to a neuron and channeling this information to the soma and axon hillock where a decision is made to fire or not an action potential, to participate or not in the active circuit.

Dendrites are therefore the place where structure interacts with information in a neuron. Since dendrites are not sphere-like structures, which would render the neurons iso-potential units, but look like a tree, their morphology and the distribution of synapses on their branches reflects the nature of this interaction and therefore are some of the most crucial components defining the computations performed within each neuron through development to adulthood. This makes dendritic structure and connectivity two of the most important defining features of cell types.

Electron microscopy (EM) is perfectly poised to investigate connectivity as well as the fine morphology of the dendrites since it is widely recognized as the gold standard to identify synapses and can image at the resolution to precisely quantify their morphology and if required their intracellular organelles and components.

Keywords Dendrites • Synapses • Electron microscope • Connectomics • Code • Cortex

6.1 Dendrites as Seen by the Electron Microscope

Dendrites take up to 35 % of cortical neuropil (49 % if one includes spines) (Braitenberg and Schüz 1991). The extensive cataloguing of these different morphological components of the dendrites and their appendages have been masterly done by Peters and Palay (Peters et al. 1991) and extensively reviewed elsewhere (e.g., Fiala and Harris 1999) and therefore will not be pursued in detail here.

N.M. da Costa (✉)
The Allen Institute for Brain Sciences, Seattle, WA, USA
e-mail: nunod@alleninstitute.org

K. Emoto et al. (eds.), *Dendrites*, DOI 10.1007/978-4-431-56050-0_6

Nevertheless it is worthwhile to start by mentioning some of the defining morphological features of cortical dendrites and synapses since such local morphological characteristics are crucial to map dendrites and their connections.

At the EM level, the presence of microtubules and mitochondria in dendrites' cytoplasm is very noticeable, and they also carry neurofilaments, agranular endoplasmic reticulum, Golgi apparatus, and granular endoplasmic reticulum in the proximal regions. In electron micrographs, the dendrites are usually recognized as the processes that receive synapses, and these synapses can be formed either directly on their shaft (Fig. 6.1b, c) or on protrusions named dendritic spines (Fig. 6.1a). Though somata and initial segments of axon also receive synapses, their shape and/or cytoplasmic contend is different. It is however worth noting that there is no hard border between dendrite and soma. The initial segment of the axon is different from dendrites and is usually recognized at the EM level by a low amount of granular endoplasmic reticulum, fasciculated microtubules, and an electron dense layer under the plasma membrane (Peters et al. 1991). The synapses in the initial portion of the axon are usually symmetric (putative GABAergic), while the ones on dendrites are asymmetric (Fig. 6.1a, b) and symmetric (Fig. 6.1c).

Dendrites of excitatory and inhibitory neurons also have morphological differences that allow for their classification on structural grounds, especially if serial sections are available. Inhibitory neurons usually have smooth dendrites that lack spines, though there are numerous exceptions to this rule (e.g., Kubota et al. 2011). The smooth dendrites also receive multiple synapses on the shaft (Fig. 6.1b), mostly excitatory, and are usually more densely packed with mitochondria (of large cross section) then the dendrites of pyramidal neurons (Somogyi et al. 1983; Peters and Saint Marie 1984; Kisvárday et al. 1986; Beaulieu 1993). If a long enough series of sections is obtained, the varicosities present along the dendrites can also be identified (Ahmed et al. 1997). Dendrites of excitatory cells, besides being spiny, have a lighter cytoplasm and most asymmetric synapses occur in spines (Fig. 6.1a); symmetric synapses are mostly present on the shaft (and soma) but occasionally also appear on spines (Saint Marie and Peters 1985; Kisvárday et al. 1986).

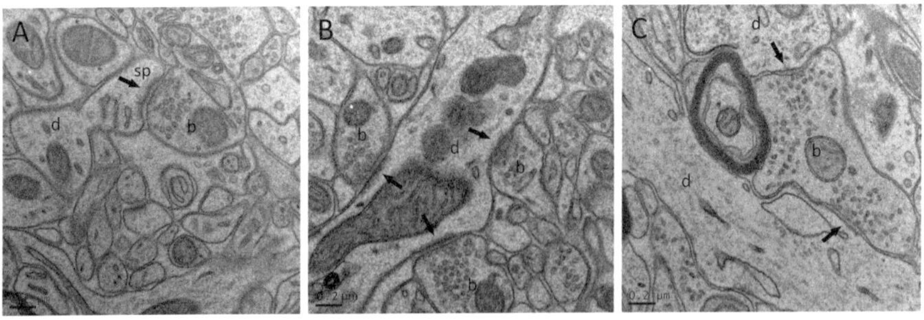

Fig. 6.1 Electron micrographs of dendrites (dt) forming asymmetric (**a**, **b**) or symmetric (**c**) synapses (*arrows*) with boutons (*b*). The dendrite in (**a**) is likely excitatory since the synapse is formed with a spine. The dendrite in (**b**) is likely inhibitory since it receives many asymmetric synapses in the shaft

Dendrites from spiny stellate neurons in the cat (Fig. 6.2) are one exception to this rule, with many of its asymmetric synapses formed with the dendritic shaft (Ahmed et al. 1994; Anderson et al. 1994).

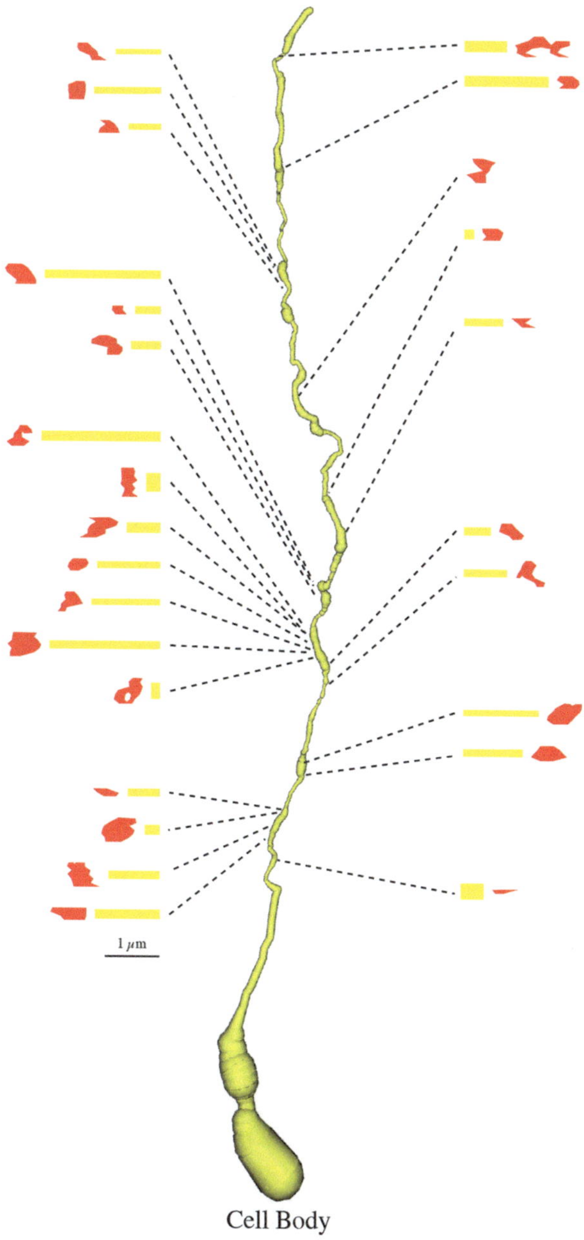

Fig. 6.2 3D reconstruction of a dendritic branch from spiny stellate neurons from the visual cortex of the cat (Adapted from Da Costa 2013). Spines located on the dendrite are shown on the right and left size with spine necks shown by *yellow lines* and reconstructions en face of the postsynaptic density shown in *red*. The closest spine to the soma was located at 34 μm distance and the furthest away at 124 μm. Scale bar relates to the thickness and length of the spine necks

Spines are one of the most recognizable features of excitatory dendrites and are the site of most synapses in the neocortex. Though they surely serve to connect (Peters and Kaiserman-Abramof 1970), there were other proposed roles that range from biochemical and electrical compartmentalization to plasticity (Koch and Zador 1993; Yuste and Denk 1995; Yuste 2011; Chen and al 2012). Even though not all spines have a thin neck (see review of spine types by Fiala and Harris (1999)), many of these roles depend on the existence and morphology of such a neck that connects the head (and its synapse) to the dendritic shaft. Precise ultrastructure measurements have helped constrain the possible role of these short and thin cables (Harris and Stevens 1988; Harris and Stevens 1989; Segev et al. 1995a; Da Costa 2013).

6.2 Dendritic Cable and Topology

The shape of dendrites imposes that neurons are not iso-potential units and therefore the treelike shape of a dendrite will affect dendritic integration, spatiotemporal summation of the synaptic input that is distributed along its branches, and the firing properties of neurons. The role of these structural, more commonly known as passive, properties of dendrites was pioneered by Rall (1959; 1964; see also Segev et al. (1995b) for a compendium of selected works of Rall, and Segev and London (2000) and London and Häusser (2005) for reviews).

Most of the theory and modeling done on dendritic computation has been successful without the need for complete or partial EM reconstructions. Instead they take advantage of the topology and estimations of diameter obtained from light microscopy reconstructions to model the passive dendritic properties (e.g., Mainen and Sejnowski 1996; Jaffe and Carnevale 1999; Cuntz et al. 2007; Banitt et al. 2007; Wang et al. 2010). While this provides less accurate measurements of membrane surface area, they are nevertheless sufficient. A drawback is that one can't relate the synaptic map of individual neurons to the topology of their individual dendritic trees, but since those synaptic maps are presently not available, one has to use average distributions of inputs anyway (Banitt et al. 2007). These average distributions come in part from sparse sampling of connections (e.g., Da Costa and Martin 2011), but it also originates from reconstructions of long stretches of single dendrites and even trees (e.g., Davis and Sterling 1979; Ahmed et al. 1994; 1997; Anderson et al. 1994; Da Costa 2013).

Dendrites represent approximately 10 % of all the cable in the neocortex (Braitenberg and Schüz 1991). While their shorter and thicker trees make them much easier to reconstruct than axons, it is still a phenomenal effort to collect thousands of ultrathin serial sections to reconstruct the several millimeters that constitute a complete dendritic tree and image it at nanometer resolution.

The first extensive reconstruction of a dendritic tree of a cortical neuron from electron micrographs come in the form of a wooden sculpture instead of a digital 3D rendition (White and Rock 1979; White and Rock 1980). This impressive reconstruction is still one of the largest ever performed and includes several dendrites of a spiny stellate from mouse somatosensory cortex. Given the lack of 3D rendering in 1979, the authors built a wooden reconstruction from which they measured with a ruler the dimensions of the dendrites as well as the neck lengths and head diameters of spines. These were then used in biophysical detailed models by Segev and colleagues, investigating how the fine morphology of the spines and dendrites affected the electrical impact of synapses, both locally on the dendrite and at the soma (Segev et al. 1995a).

Though the task of reconstructing a complete dendritic tree might be easier in the present day, it is still a formidable task. Depending on the cell type and animal, the volume required to enclose a complete cortical dendritic tree can be as "small" as 0.05 mm^3 (approximately 9000 serial ultrathin sections at 40 nm and 8×10^{13} voxels at 4 nm X,Y resolution) and as large as 0.5 mm^3. With the advent of fast automated sectioning techniques as the ATUM (Hayworth et al. 2014) and fast EM imaging systems as the TEMCA (transmission electron microscope with camera arrays Bock et al. 2011) or the newly developed multi-beam SEM, these numbers appear less scary, however the imaging will still take a considerable amount of time and it has yet not been achieved in any published work on neocortex. The value of these reconstructions will be on the combination of the detail metrics of the cable, which alone would just serve to calibrate the light microscopy reconstructions, with the detailed knowledge of the connectivity.

6.3 The Morphology and Syntax of Dendrites

Dendrites collect almost all of the synapses received by a neuron, which can vary from a few thousands like in the monkey primate visual cortex (Cragg 1967; O'Kusky and Colonnier 1982; Beaulieu et al. 1992) to more than ten thousand as in some of the estimations for the rodent visual cortex (Cragg 1967; Warren and Bedi 1982; Turner and Greenough 1985; Schüz and Palm 1989; Miki et al. 1997). One way to look at the distribution of these synapses and their weights along the dendrites of neurons is that they form a structural code that through nature and nurture is written into the dendrites and that provides the set of instructions to make that neuron fire in a purposive manner.

Unfortunately we don't know the message written in a complete dendritic tree or even a single dendrite, but the general consensus is that ultrastructure methods will play a crucial role in the understanding of its grammar since the fine structural detail provided by the electron microscope (EM) allows the identification of such synapses on morphological grounds (Gray 1959a, b; Colonnier 1968).

Two strategies have been applied for understanding this language. The first one looks into the average distribution of connections between different cell types, their dendritic organization, their weights, etc. (e.g., Binzegger et al. 2004; Lefort et al. 2009; Da Costa and Martin 2011; Schoonover et al. 2014). This resembles the study of syntax in a language, since it investigates the general principles and rules of the sentence structure. The second one looks at neurons as individuals and investigates the precise connections between each one of those individuals (e.g., Bock et al. 2011; Briggman et al. 2011; Helmstaedter et al. 2013; Takemura et al. 2013). Such studies provide connectivity graphs, and the raw data is the distribution of connections (though incomplete) along the dendritic trees. It is tempting to see these individual projections as the fundamental grammatical unit in this code resembling what in linguistics is known as a morpheme. The identification, analysis, and description of the structure of a given language's morphemes is called morphology (Wikipedia).

Most of the research in connectivity has focused on describing the syntax of this code. The work of Gray showed that one could identify synapses in ultrathin sections. However, to determine the pre- and postsynaptic neurons that come together to form that synapse, one would have to follow both that axon and dendrite back to their respective somata, which might be hundreds of microns (in the case of dendrites) to millimeters (in the case of axons) away and therefore require thousands to tens of thousands of serial ultrathin sections.

Since tracing the cable of axons and dendrites with light microscopy is comparably a much simpler process, the possibility of correlated light and electron microscopy emerged as a much more inviting technique. The gross morphology of labeled neurons can be described at light level, and then regions of interest are blocked out of the tissue to be measured and investigated at synapse detail with the EM. Stell (1965) pioneered this strategy using the Golgi technique to label neurons and visualize them at both light and electron microscopy levels. While this allowed for the identification of the dendrite, it lets the identity of the presynaptic partners undisclosed, a problem that was solved by Blackstad who combined the Golgi technique with identification of afferent degenerating axons in the rat hippocampus (Blackstad 1965, 1975). While the Golgi method has been mostly substituted by other forms of labeling and degeneration is not used anymore as a method to mark afferent pathways, the strategy pioneered by Stell and Blackstad has been a very successful strategy to map the statistics of connectivity in cortical circuits from in vitro to in vivo preparations and in a wide range of animal models. In parallel with the light and EM correlation, there was also an effort to reconstruct neurons that have not been labeled. This provided the possibility of a more unbiased sample of neurons and dendrites (Vaughan and Peters 1973; Peters et al. 1976; Peters and Feldman 1977; Davis and Sterling 1979). The problem of unlabelled tissue was that the reconstructions either started in the soma, allowing for cell-type identification but missed distal dendrites or started at distal dendrites but missed the soma and other dendrites, and therefore it was very difficult to identify the parent cell type.

One way to avoid the problem of orphan processes (axons or dendrites) would be to develop a procedure to identify different cell types on short series of sections. One approach is an EM brainbow in which an EM genetic tag can be targeted to different compartments in different cell types, allowing for their labeling in the same way that in light microscopy one can use different colors to identify different cells in the same volume of tissue. Such EM tags (Fig. 6.3d–f) are now available (2011; Shu et al. 2011; Martell et al. 2012; Atasoy et al. 2014; Lam et al. 2015), and

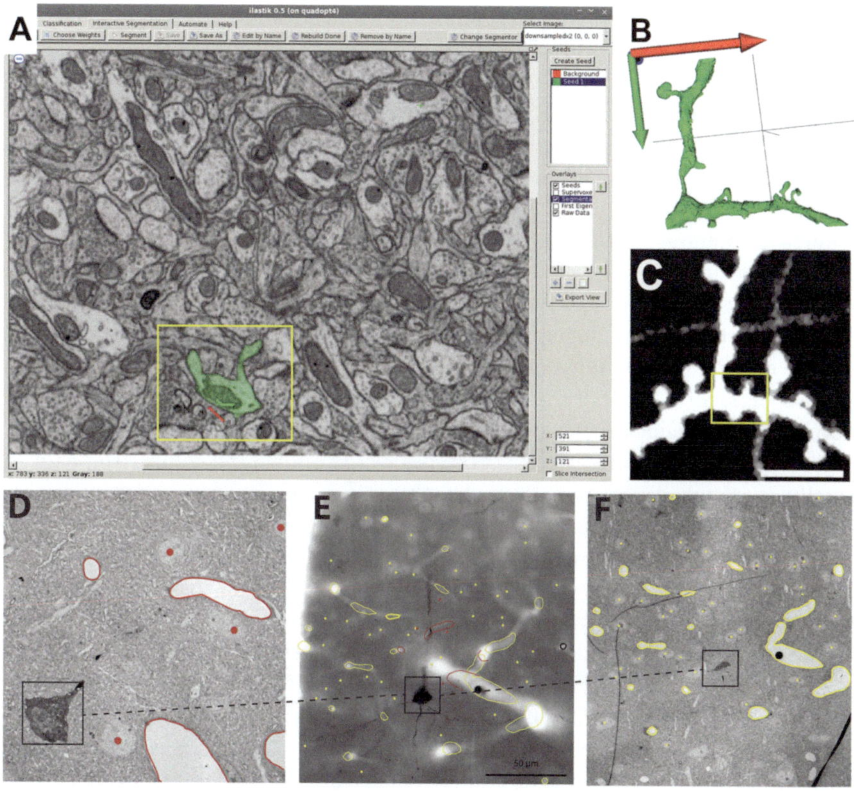

Fig. 6.3 Examples of correlated light and microscopy (CLEM). (**a**–**c**) (Images reproduced from Maco et al. 2013, under the Creative Commons Attribution License, http://creativecommons.org/licenses/by/2.5/) CLEM between in vivo fluorescent microscopy and EM. (**a**) EM visualization with ilastik. (**b**) 3D EM reconstruction. (**c**) Fluorescent micrograph from the same section of a layer 5 dendrite shown in (**b**). (**d**–**f**) CLEM of a neuron expressing APEX visualized using a Nickel-DAB subtract. The electron micrographs (**d**, **f**) are correlated with the light micrographs (**e**) using blood vessels (contours) and cell bodies (*dots*). Apical dendrites and myelinated fibers can also be used as structural markers for co-registration

there is intense work on discovering cell type-specific promoters that would allow their differential expression (Gong et al. 2007; Taniguchi et al. 2011; Harris et al. 2014). A second approach uses correlated light and electron microscopy (Fig. 6.3a–c) either with a modern version of Stell and Blackstad strategy (Knott et al. 2009; Da Costa and Martin 2011; Bleckert et al. 2013; Maco et al. 2013; Schoonover et al. 2014) or using array tomography (Micheva and Smith 2007; Micheva et al. 2010) or using a modern view onto the old strategy of photoactivation (Paez-Segala et al. 2015). A third strategy requires no labels and is based on the identification of synapses from different cell types on morphological grounds. The classical example of this is separation of synapses in Gray type I (or asymmetric synapse) as putative excitatory synapses and Gray type II (or symmetric synapse) as putative inhibitory synapses (Gray 1959a, b; Colonnier 1968). Another example is the classification of presynaptic terminal in the thalamus as large terminals with round vesicles (RL), small terminals with round vesicles (RS), presynaptic dendrites (PSD), and terminals with flattened or pleomorphic vesicles forming symmetric synapses (F) (reviewed by Jones 2007). Similar work has also been performed in the cortex (Saint Marie and Peters 1985; Peters and Harriman 1990, 1992; Peters et al. 1990; Ahmed et al. 1994), investigating mostly features of boutons that concerned the bouton size and location. These features were then used to map the input to dendrites of excitatory and inhibitory neurons. This is a strategy that deserves further investigation for two reasons. Firstly, because morphological fingerprinting of bouton would allow for decoding the general rules of connectivity of the cortex using nature's own brainbow without the use of labels. Secondly, having a thorough quantitative description of the morphological characteristics of synapses (e.g., synapse size, density of vesicles, target type, etc.) between different cell types would allow for a comparison with an equally thorough description from the physiology of the same connections. If correlations are found, then this would be an invaluable tool to provide estimations of physiology from ultrastructure reconstructions and help interpret the EM work in the context of synaptic physiology. An example of dendrites in layer 4 receiving synapses from different synaptic terminals is shown in Fig. 6.4.

The above-described investigations into the dendrites of neurons will go as far as to describe this syntax or average connectivity. But since neurons respond to stimulus in specific ways, it is expected that their input is also specific, or at least it is arranged along its dendrites in a specific way. In the same way that different morphemes form different words and different words form different sentences all following the same syntax rules, so will specific sets of inputs, or different spatial arrangement of inputs, form different instructions on the dendrites of neurons.

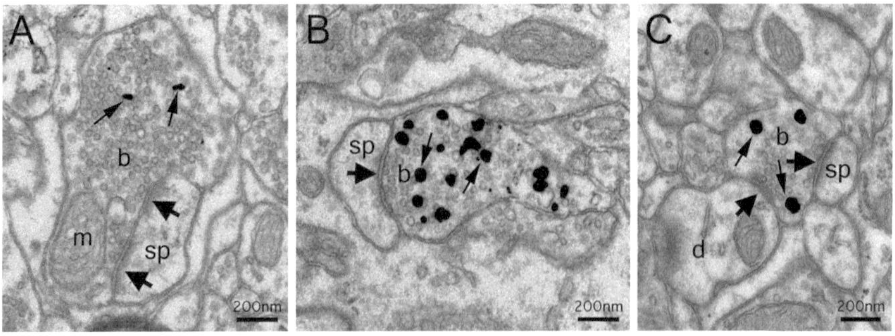

Fig. 6.4 Electron micrographs of dendrites in layer 4 of mouse V1 receiving asymmetric synapses from three different presynaptic partners. Presynaptic partners labeled with silver-enhanced gold particles (*small arrows*), synapses indicated by large arrows. (**a**) VGluT2 terminal from a putative thalamic neuron forming a perforated synapse with a spine (sp.). (**b**) Nr5a terminal from a putative layer 4 neuron forming a synapse a spine (sp.). (**c**) Ntsr1 terminal from a putative layer 6 corticothalamic neuron forming small synapses with both a dendritic shaft (**d**) and a spine (sp.)

However, to know the source of a specific input to a dendrite, one has to trace back the axon that forms that specific synapse. For that to happen, one needs to image at synapse resolution a volume that is sufficiently large to include the presynaptic neuron(s) we are interested in. To know the source of all the inputs to a single cortical cell might require imaging large volumes of the brain covering several areas on both hemispheres, but for the local circuit input, the volume might be close to a cubic millimeter or millimeters depending on the animal model.

The electron microscope surely has the resolution to investigate these problems, but while unbeatable in resolution, the Achilles heel of ultrastructure methods has always been its lack of capacity to scale. This was reflected not just at the level of sectioning and imaging but also at the level of data storage and ultimately reconstruction, with consequently reduction in the number of samples and size of the dendritic reconstructions.

In recent years, not only the classic studies of the worm connectomes (Albertson and Thomson 1976; White et al. 1976, 1986) underwent a renaissance (e.g., Jarrell et al. 2012; Bumbarger et al. 2013), but large-scale EM projects have been pursued in fly, retina, and cortex (Bock et al. 2011; Briggman et al. 2011; Helmstaedter et al. 2013; Takemura et al. 2013). Such studies have demonstrated that scaling up is possible, though the size and speed of the scaling is still a matter of discussion; EM imaging is evolving to make possible data acquisition of large volumes of tissue at considerable voxel speeds (MHz to GHz) (see review by Briggman and Bock 2012). Automated serial sectioning, either in the form of serial block face (Denk and Horstmann 2004) or tape collection (Hayworth et al. 2014), also offers the possibility of scaling serial sectioning to volumes and at speeds that are not possible manually and can reach more than thousand sections per day. Reconstruction is likely the biggest challenge on this process of scaling the ultrastructure techniques (Da Costa and Martin 2013; Plaza et al. 2014; Lichtman et al. 2014). While humans are more accurate, they are too slow and a mouse brain will take 4.5 million person years using current strategies according to Plaza et al. (2014), and the automated methods are still far from reaching human performance.

However, we should not forget that in the journey to large-scale reconstructions, one could choose to reconstruct first the processes that will have the most impact in our understanding of general principles. And in this regard, knowing the complete map of the local circuit synapses to the complete dendritic tree of a single neuron might be highly insightful, especially if combined with some knowledge of the function of its presynaptic partner (see Bock et al. 2011). Because, at the very least, it will give us for the first time a view of the complete writings of that circuit into that dendrite, and therefore a better chance to decode its syntax and morphology.

Acknowledgments I would like to thank Agnes Bodor, JoAnn Buchanan, and Marc Takeno for their help preparing the figures and to Agnes Bodor, Hannah Krakauer, Jill Scott, and Elizabeth Whitney for their comments on the manuscript. I also would like to thank Ali Cetin and Tanya Daigle-Ting for the generation and optimization of Apex expressing recombinant adeno-associated viral vectors for in vivo applications. I am grateful to the Allen Institute founders, P.G. Allen and J. Allen, for their vision, encouragement, and support.

References

Ahmed B, Anderson J, Douglas R et al (1994) Polyneuronal innervation of spiny stellate neurons in cat visual cortex. J Comp Neurol 341:39–49

Ahmed B, Anderson J, Martin K, Nelson J (1997) Map of the synapses onto layer 4 basket cells of the primary visual cortex of the cat. J Comp Neurol 380:230–242

Albertson DG, Thomson JN (1976) The pharynx of Caenorhabditis elegans. Philos Trans R Soc Lond B Biol Sci 275:299–325

Anderson J, Douglas R, Martin K, Nelson J (1994) Map of the synapses formed with the dendrites of spiny stellate neurons of cat visual cortex. J Comp Neurol 341:25–38

Atasoy D, Betley JN, Li W-P et al (2014) A genetically specified connectomics approach applied to long-range feeding regulatory circuits. Nat Neurosci 17:1830–1839. doi:10.1038/nn.3854

Banitt Y, Martin K, Segev I (2007) A biologically realistic model of contrast invariant orientation tuning by thalamocortical synaptic depression. J Neurosci 27:10230–10239. doi:10.1523/JNEUROSCI.1640-07.2007

Beaulieu C (1993) Numerical data on neocortical neurons in adult rat, with special reference to the GABA population. Brain Res 609:284–292

Beaulieu C, Kisvarday Z, Somogyi P et al (1992) Quantitative distribution of GABA-immunopositive and-immunonegative neurons and synapses in the monkey striate cortex (area 17). Cereb Cortex 2:295–309

Binzegger T, Douglas R, Martin K (2004) A quantitative map of the circuit of cat primary visual cortex. J Neurosci 24:8441–8453

Blackstad TW (1965) Mapping of experimental axon degeneration by electron microscopy of Golgi preparations. Z Zellforsch Mikrosk Anat 67:819–834

Blackstad TW (1975) Electron microscopy of experimental axonal degeneration in photochemically modified Golgi preparations: a procedure for precise mapping of nervous connections. Brain Res 95:191–210

Bleckert A, Parker ED, Kang Y et al (2013) Spatial relationships between GABAergic and glutamatergic synapses on the dendrites of distinct types of mouse retinal ganglion cells across development. PLoS One 8:e69612. doi:10.1371/journal.pone.0069612

Bock DD, Lee W-CA, Kerlin AM et al (2011) Network anatomy and in vivo physiology of visual cortical neurons. Nature 471:177–182. doi:10.1038/nature09802

Braitenberg V, Schüz A (1991) Anatomy of the cortex, 2nd edn. Springer, Berlin

Briggman KL, Bock DD (2012) Volume electron microscopy for neuronal circuit reconstruction. Curr Opin Neurobiol 22:154–161. doi:10.1016/j.conb.2011.10.022

Briggman KL, Helmstaedter M, Denk W (2011) Wiring specificity in the direction-selectivity circuit of the retina. Nature 471:183–188. doi:10.1038/nature09818

Bumbarger DJ, Riebesell M, Rödelsperger C, Sommer RJ (2013) System-wide rewiring underlies behavioral differences in predatory and bacterial-feeding nematodes. Cell 152:109–119. doi:10.1016/j.cell.2012.12.013

Chen Y et al (2012) Signaling in dendritic spines and spine microdomains. Curr Opin Neurobiol 22:389

Colonnier M (1968) Synaptic patterns on different cell types in the different laminae of the cat visual cortex. An electron microscope study. Brain Res 9:268–287

Cragg B (1967) The density of synapses and neurones in the motor and visual areas of the cerebral cortex. J Anat 101:639–654

Cuntz H, Borst A, Segev I (2007) Optimization principles of dendritic structure. Theor Biol Med Model 4:21. doi:10.1186/1742-4682-4-21

Da Costa NM (2013) Diversity of thalamorecipient spine morphology in cat visual cortex and its implication for synaptic plasticity. J Comp Neurol 521:2058–2066. doi:10.1002/cne.23272

Da Costa NM, Martin KAC (2011) How thalamus connects to spiny stellate cells in the cat's visual cortex. J Neurosci 31:2925–2937. doi:10.1523/JNEUROSCI.5961-10.2011

Da Costa NM, Martin KAC (2013) Sparse reconstruction of brain circuits: or, how to survive without a microscopic connectome. NeuroImage:1–10. doi:10.1016/j.neuroimage.2013.04.054
Davis T, Sterling P (1979) Microcircuitry of cat visual cortex: classification of neurons in layer IV of area 17, and identification of the patterns of lateral geniculate input. J Comp Neurol 188:599–627
Denk W, Horstmann H (2004) Serial block-face scanning electron microscopy to reconstruct three-dimensional tissue nanostructure. PLoS Biol 2:e329
Fiala JC, Harris KM (1999) Dendrite structure. In: Stuart G, Spruston N, Häusser M (eds) Dendrites. Oxford University Press, Oxford
Gong S, Doughty M, Harbaugh CR et al (2007) Targeting Cre recombinase to specific neuron populations with bacterial artificial chromosome constructs. J Neurosci 27:9817–9823. doi:10.1523/JNEUROSCI.2707-07.2007
Gray EG (1959a) Axo-somatic and axo-dendritic synapses of the cerebral cortex: an electron microscope study. J Anat 93:420–433
Gray EG (1959b) Electron microscopy of synaptic contacts on dendrite spines of the cerebral cortex. Nature 183:1592–1593
Harris KM, Stevens JK (1988) Dendritic spines of rat cerebellar Purkinje cells: serial electron microscopy with reference to their biophysical characteristics. J Neurosci 8:4455–4469
Harris KM, Stevens JK (1989) Dendritic spines of CA 1 pyramidal cells in the rat hippocampus: serial electron microscopy with reference to their biophysical characteristics. J Neurosci 9:2982–2997
Harris JA, Hirokawa KE, Sorensen SA et al (2014) Anatomical characterization of Cre driver mice for neural circuit mapping and manipulation. Front Neural Circ 8:76. doi:10.3389/fncir.2014.00076
Hayworth KJ, Morgan JL, Schalek R et al (2014) Imaging ATUM ultrathin section libraries with WaferMapper: a multi-scale approach to EM reconstruction of neural circuits. Front Neural Circ 8:68. doi:10.3389/fncir.2014.00068
Helmstaedter M, Briggman KL, Turaga SC et al (2013) Connectomic reconstruction of the inner plexiform layer in the mouse retina. Nature 500:168–174. doi:10.1038/nature12346
Jaffe DB, Carnevale NT (1999) Passive normalization of synaptic integration influenced by dendritic architecture. J Neurophysiol 82:3268–3285
Jarrell TA, Wang Y, Bloniarz AE et al (2012) The connectome of a decision-making neural network. Science 337:437–444. doi:10.1126/science.1221762
Jones EG (2007) Thalamic neurons, synaptic organization, and functional properties. In: The Thalamus, 2nd edn. Cambridge University Press, Cambridge, pp 192–204
Kisvárday ZF, Martin KA, Freund TF et al (1986) Synaptic targets of HRP-filled layer III pyramidal cells in the cat striate cortex. Exp Brain Res 64:541–552
Knott GW, Holtmaat A, Trachtenberg JT et al (2009) A protocol for preparing GFP-labeled neurons previously imaged in vivo and in slice preparations for light and electron microscopic analysis. Nat Protoc 4:1145–1156. doi:10.1038/nprot.2009.114
Koch C, Zador A (1993) The function of dendritic spines: devices subserving biochemical rather than electrical compartmentalization. J Neurosci 13:413–422
Kubota Y, Karube F, Nomura M et al (2011) Conserved properties of dendritic trees in four cortical interneuron subtypes. Sci Rep 1:89. doi:10.1038/srep00089
Lam SS, Martell JD, Kamer KJ et al (2015) Directed evolution of APEX2 for electron microscopy and proximity labeling. Nat Methods 12:51–54. doi:10.1038/nmeth.3179
Lefort S, Tomm C, Sarria JF, Petersen C (2009) The excitatory neuronal network of the C2 barrel column in mouse primary somatosensory cortex. Neuron 61:301
Lichtman JW, Pfister H, Shavit N (2014) The big data challenges of connectomics. Nat Neurosci 17:1448–1454. doi:10.1038/nn.3837
London M, Häusser M (2005) Dendritic computation. Annu Rev Neurosci 28:503–532. doi:10.1146/annurev.neuro.28.061604.135703

Maco B, Holtmaat A, Cantoni M et al (2013) Correlative in vivo 2 photon and focused ion beam scanning electron microscopy of cortical neurons. PLoS One 8:e57405. doi:10.1371/journal.pone.0057405
Mainen ZF, Sejnowski TJ (1996) Influence of dendritic structure on firing pattern in model neocortical neurons. Nature 382:363–366. doi:10.1038/382363a0
Martell JD, Deerinck TJ, Sancak Y et al (2012) Engineered ascorbate peroxidase as a genetically encoded reporter for electron microscopy. Nat Biotechnol 30:1143–1148. doi:10.1038/nbt.2375
Micheva KD, Smith SJ (2007) Array tomography: a new tool for imaging the molecular architecture and ultrastructure of neural circuits. Neuron 55:25–36. doi:10.1016/j.neuron.2007.06.014
Micheva KD, Busse B, Weiler NC et al (2010) Single-synapse analysis of a diverse synapse population: proteomic imaging methods and markers. Neuron 68:639–653. doi:10.1016/j.neuron.2010.09.024
Miki T, Fukui Y, Itoh M et al (1997) Estimation of the numerical densities of neurons and synapses in cerebral cortex. Brain Res Brain Res Protoc 2:9–16
O'Kusky J, Colonnier M (1982) A laminar analysis of the number of neurons, glia, and synapses in the adult cortex (area 17) of adult macaque monkeys. J Comp Neurol 210:278–290. doi:10.1002/cne.902100307
Paez-Segala MG, Sun MG, Shtengel G et al (2015) Fixation-resistant photoactivatable fluorescent proteins for CLEM. Nat Methods. doi:10.1038/nmeth.3225
Peters A, Feldman M (1977) The projection of the lateral geniculate nucleus to area 17 of the rat cerebral cortex. IV. Terminations upon spiny dendrites. J Neurocytol 6:669–689
Peters A, Harriman KM (1990) Different kinds of axon terminals forming symmetric synapses with the cell bodies and initial axon segments of layer II/III pyramidal cells. I. Morphometric analysis. J Neurocytol 19:154–174
Peters A, Harriman KM (1992) Different kinds of axon terminals forming symmetric synapses with the cell bodies and initial axon segments of layer II/III pyramidal cells. III. Origins and frequency of occurrence of the terminals. J Neurocytol 21:679–692
Peters A, Kaiserman-Abramof I (1970) The small pyramidal neuron of the rat cerebral cortex. The perikaryon, dendrites and spines. Am J Anat 127:321–355
Peters A, Saint Marie RL (1984) Smooth and sparsely spinous nonpyramidal cells forming local axonal plexuses. In: Peters A, Jones EG (eds) Cerebral cortex, vol 1. Cellular components of the cerebral cortex. Plenum, New York, pp 419
Peters A, Feldman M, Saldanha J (1976) The projection of the lateral geniculate nucleus to area 17 of the rat cerebral cortex. II. Terminations upon neuronal perikarya and dendritic shafts. J Neurocytol 5:85–107
Peters A, Sethares C, Harriman KM (1990) Different kinds of axon terminals forming symmetric synapses with the cell bodies and initial axon segments of layer II/III pyramidal cells. II. Synaptic junctions. J Neurocytol 19:584–600
Peters A, Palay S, Webster F (1991) The fine structure of the nervous system: neurons and their supporting cells, 3rd edn. Oxford University Press, New York
Plaza SM, Scheffer LK, Chklovskii DB (2014) Toward large-scale connectome reconstructions. Curr Opin Neurobiol 25:201–210. doi:10.1016/j.conb.2014.01.019
Rall W (1959) Branching dendritic trees and motoneuron membrane resistivity. Exp Neurol 1:491–527. doi:10.1016/0014-4886(59)90046-9
Rall W (1964) Theoretical significance of dendritic trees for neuronal input-output relations. In Reiss RF (ed) Neural theory and modeling. Stanford University Press, Stanford, pp 73–97
Saint Marie R, Peters A (1985) The morphology and synaptic connections of spiny stellate neurons in monkey visual cortex (area 17): a Golgi-electron microscopic study. J Comp Neurol 233:213–235
Schoonover CE, Tapia J-C, Schilling VC et al (2014) Comparative strength and dendritic organization of thalamocortical and corticocortical synapses onto excitatory layer 4 neurons. J Neurosci 34:6746–6758. doi:10.1523/JNEUROSCI.0305-14.2014

Schüz A, Palm G (1989) Density of neurons and synapses in the cerebral cortex of the mouse. J Comp Neurol 286:442–455. doi:10.1002/cne.902860404
Segev I, London M (2000) Untangling dendrites with quantitative models. Science 290:744–750
Segev I, Friedman A, White E, Gutnick M (1995a) Electrical consequences of spine dimensions in a model of a cortical spiny stellate cell completely reconstructed from serial thin sections. J Comput Neurosci 2:117–130
Segev I, Shepherd G, Rinzel J (1995b) The theoretical foundation of dendritic functions. Bradford Book, The MIT Press, Cambridge, Mass
Shu X, Lev-Ram V, Deerinck TJ et al (2011) A genetically encoded tag for correlated light and electron microscopy of intact cells, tissues, and organisms. PLoS Biol 9:e1001041. doi:10.1371/journal.pbio.1001041.g005
Somogyi P, Kisvárday ZF, Martin KA, Whitteridge D (1983) Synaptic connections of morphologically identified and physiologically characterized large basket cells in the striate cortex of cat. Neuroscience 10:261–294
Stell WK (1965) Correlation of retinal cytoarchitecture and ultrastructure in Golgi preparations. Anat Rec 153:389–397
Takemura S-Y, Bharioke A, Lu Z et al (2013) A visual motion detection circuit suggested by Drosophila connectomics. Nature 500:175–181. doi:10.1038/nature12450
Taniguchi H, He M, Wu P et al (2011) A resource of Cre driver lines for genetic targeting of GABAergic neurons in cerebral cortex. Neuron 71:995–1013. doi:10.1016/j.neuron.2011.07.026
Turner AM, Greenough WT (1985) Differential rearing effects on rat visual cortex synapses. I. Synaptic and neuronal density and synapses per neuron. Brain Res 329:195–203
Vaughan DW, Peters A (1973) A three dimensional study of layer I of the rat parietal cortex. J Comp Neurol 149:355–370. doi:10.1002/cne.901490305
Wang H-P, Spencer D, Fellous J-M, Sejnowski TJ (2010) Synchrony of thalamocortical inputs maximizes cortical reliability. Science 328:106–109. doi:10.1126/science.1183108
Warren MA, Bedi KS (1982) Synapse-to-neuron ratios in the visual cortex of adult rats undernourished from about birth until 100 days of age. J Comp Neurol 210:59–64. doi:10.1002/cne.902100107
White E, Rock M (1979) Distribution of thalamic input to different dendrites of a spiny stellate cell in mouse sensorimotor cortex. Neurosci Lett 15:115–119
White E, Rock M (1980) Three-dimensional aspects and synaptic relationships of a Golgi-impregnated spiny stellate cell reconstructed from serial thin sections. J Neurocytol 9:615–636
White JG, Southgate E, Thomson JN, Brenner S (1976) The structure of the ventral nerve cord of Caenorhabditis elegans. Philos Trans R Soc Lond B Biol Sci 275:327–348
White J, Southgate E, Thomson J, Brenner S (1986) The structure of the nervous system of the nematode Caenorhabditis elegans. Philos Trans R Soc Lond B Biol Sci 314:1–340
Wikipedia Morphology (linguistics). In: httpen.wikipedia.orgwikiMorphologylinguistics. Accessed 20 Jan 2015
Yuste R (2011) Dendritic spines and distributed circuits. Neuron 71:772–781. doi:10.1016/j.neuron.2011.07.024
Yuste R, Denk W (1995) Dendritic spines as basic functional units of neuronal integration. Nature 375:682–684. doi:10.1038/375682a0

Chapter 7
Imaging Signal Transduction in Dendrites Using Genetically Encoded Biosensors

Hideji Murakoshi and Ryohei Yasuda

Abstract Activity-dependent synaptic plasticity is the basis of circuit plasticity and, ultimately, behavioral plasticity, learning, and memory. Intracellular signal transduction, mediated by hundreds of species of signaling proteins in dendrites, plays an important role in linking postsynaptic neuronal activity with synaptic plasticity. Recent advances in the development of fluorescent protein-based biosensors and optical techniques have enabled us to image signal transduction with the resolution of single synapses, revealing the mechanism by which spatiotemporal dynamics of biochemical signaling are regulated in dendrites and synapses. In this chapter, we will review genetically encoded fluorescent protein-based sensors, Förster resonance energy transfer (FRET) imaging, and recent discoveries based on these techniques.

Keywords 2pFLIM • FRET • GFP • Small GTPase • Synapse

7.1 Introduction

Thousands of intracellular signaling proteins, including ~500 kinases (Manning et al. 2002), ~150 small GTPase proteins (Wennerberg et al. 2005), and ~800 G-protein-coupled receptors (Bjarnadottir et al. 2006), are known to be essential components of biochemical reactions in a cell. These molecules form a complicated signaling network through their extensive interactions. In dendrites, biochemical reactions are spatiotemporally regulated by complicated subcellular structures such as dendritic spines, small postsynaptic protrusion emanating from dendritic surface (Augustine et al. 2003; Murakoshi and Yasuda 2012). A dendritic spine consists of

---

H. Murakoshi
Supportive Center for Brain Research, National Institute for Physiological Sciences, Okazaki, Aichi 444-8585, Japan

Japan Science and Technology Agency (JST), Kawaguchi, Saitama 332-0012, Japan

R. Yasuda (✉)
Max Planck Florida Institute for Neuroscience, Jupiter, FL 33458, USA
e-mail: ryohei.yasuda@mpfi.org

K. Emoto et al. (eds.), *Dendrites*, DOI 10.1007/978-4-431-56050-0_7

a spherical "head," typically the size of ~0.1 fl, and a narrow (diameter ~100 nm) "neck" that connects the head with the dendritic surface. The neck acts as diffusion barrier and compartmentalizes biochemical reaction in the spine (Alvarez and Sabatini 2007). Each dendritic spine has sub-spine structures such as the postsynaptic density (PSD) and endosomes (Ehlers 2013; Colgan and Yasuda 2014). These nanostructures provide additional layers of signal compartmentalization. While biochemical reactions in dendritic spine are required for input specificity, signaling initiated in a single spine can spread beyond the spine and affect surrounding dendrites and spines over 5–50 μm or more (Harvey et al. 2008b; Govindarajan et al. 2011). Thus, biochemical signaling in dendrites appears to be regulated in several levels of spatial scale. In order to understand such biochemical computation that occurs in dendrites, it is necessary to measure the spatiotemporal dynamics of a large set of signaling molecules. For a long time, intracellular signaling has been studied using biochemical methods such as Western blotting to detect phosphorylation, activation, molecular interaction, and protein modifications. However, because these methods require cell homogenate, they lose the subcellular spatial information. For the complete understanding of the intracellular signaling system, spatial and temporal information in living cells should be acquired in addition to the identity of the active proteins.

Since the discovery of green fluorescent protein and its variants (Tsien 1998), many researchers have developed biosensors based on fluorescent proteins for monitoring various cellular events such as molecular translocation, mobility (Lippincott-Schwartz et al. 2001), and second messenger concentration changes (Whitaker 2010; Oldach and Zhang 2014). The advantages of fluorescent protein-based sensors include (1) compatibility with living cells by ectopic DNA expression, (2) cell-type-specific expression by utilization of specific promoters, and (3) relative ease of engineering because only gene engineering is required. These advantages have led researchers to develop various types of genetically encoded fluorescent sensors. In particular, variants of GFPs with circular permuted configurations are useful in converting signaling events into fluorescence intensity changes (Baird et al. 1999). In addition, by taking advantage of the distance dependency of FRET at the nanometer length scale, a wide variety of FRET biosensors for detecting protein-protein interactions and conformation changes of signaling proteins have been developed (Zhang et al. 2002; Miyawaki 2003; Newman et al. 2011; Ueda et al. 2013).

These sensors have greatly improved our understanding of biochemical signaling in dendrites and synapses. For example, imaging activation of signaling molecules using 2-photon fluorescence lifetime imaging microscopy (2pFLIM) revealed that while activated, Ca^{2+}-/CaM-dependent kinase II (CaMKII), Cdc42, and cofilin are restricted to the stimulated spine (Lee et al. 2009; Murakoshi et al. 2011; Bosch et al. 2014), activated Ras and RhoA spread from the stimulated spine into its parent dendritic shaft (Harvey et al. 2008b; Murakoshi et al. 2011), and subsequent ERK phosphorylation occurs in the nucleus (Zhai et al. 2013). These studies demonstrate that the intracellular dendritic signaling events do not occur homogeneously but rather in a spatiotemporally regulated manner. In this review, we introduce various fluorescent protein-based signaling sensors, the related fluorescence techniques, and new insights into molecular mechanisms underlying synaptic plasticity provided by these sensors.

7.2 Sensing Cellular States with Single-Fluorescent Protein-Based Sensors

To develop signaling sensors based on fluorescent proteins, it is necessary to convert signaling events into a fluorescence signature. A breakthrough came from the development of circularly permuted GFP (cpGFP) (Baird et al. 1999), which is made by connecting both ends of GFP to create new terminals. Using cpGFP, calcium-sensitive fluorescent protein G-CaMP was developed by fusing calmodulin (CaM) and M13 peptide derived from myosin light-chain kinase to C- and N-terminus of cpGFP, respectively (Nakai et al. 2001) (Fig. 7.1). This sensor exhibits little fluorescence in the absence of Ca^{2+}, but the binding of calmodulin and M13 in response to a Ca^{2+} increase leads to the closure of the two ends of cpGFP, increasing its fluorescence (Fig. 7.1). This probe and its variants have been used widely to monitor calcium concentration changes in various types of living cells. The signal-to-noise ratio of optimized G-CaMP is extremely high, allowing researchers to image single action potentials in individual neurons and synaptic inputs to single dendritic spines in vivo (Ohkura et al. 2005; Chen et al. 2013). In addition, color variants of cpGFP-based calcium sensors including blue and red versions have been developed, facilitating multicolor imaging (Akerboom et al. 2013; Inoue et al. 2014). CpGFP has also been utilized to develop other sensors including glutamate sensors (Marvin et al. 2013) and ATP: ADP ratio sensors (Tantama et al. 2013).

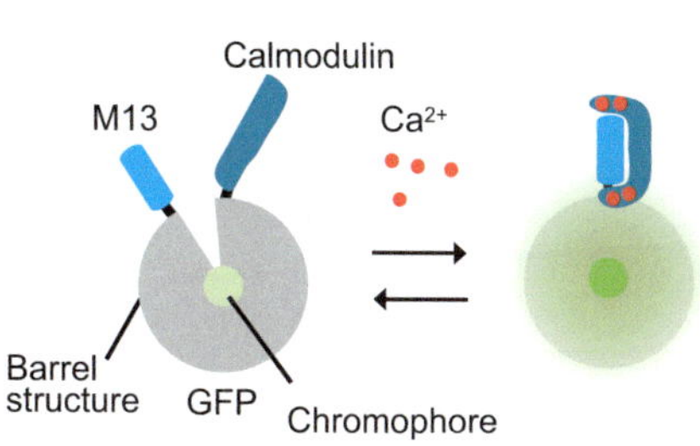

Fig. 7.1 A Ca^{2+}-sensitive green fluorescent protein (G-CaMP). Binding of Ca^{2+} to calmodulin causes binding between calmodulin and M13 peptide, leading to a conformation change of GFP. This process increases fluorescence from GFP

7.3 Sensing Molecular Interaction and Activation by FRET

Another way to measure signaling events is to use FRET (Lakowicz 2006). FRET is a phenomenon in which the energy of the excited donor is transferred to the acceptor molecule in close proximity (within ~10 nm), leading to an increased fluorescence emission from the acceptor molecule (Fig. 7.2a) (Miyawaki 2003; Lakowicz 2006). For example, when a GFP molecule is excited, it emits green fluorescence. However, if a red fluorescent protein (RFP) is located close to the GFP, the energy of the excited GFP is transferred to the RFP, and the RFP emits red fluorescence (Fig. 7.2a). The efficiency of the energy transfer is a steep function of the distance between the donor and the acceptor (Fig. 7.2b). The Förster distance is defined as the distance at which 50 % of the energy is transferred. For commonly used FRET pairs, the Förster distance is approximately 5 nm. FRET occurrence

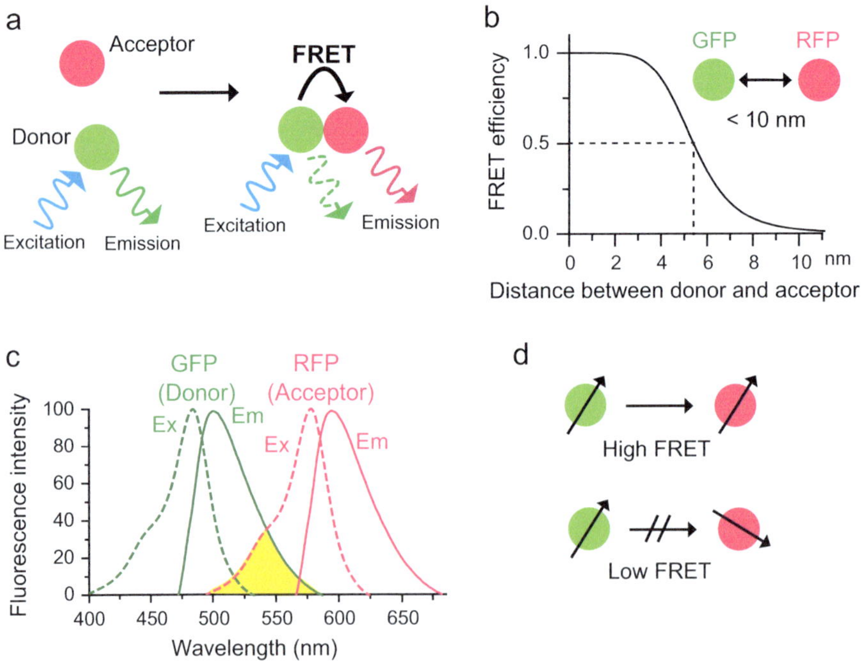

Fig. 7.2 Criteria for FRET occurrence. (**a**) A donor fluorescent molecule absorbs excitation light and emits fluorescence in the absence of FRET. When an acceptor is close to the donor, the excitation energy of the donor is transferred to the acceptor due to FRET, causing increased emission from the donor and decreased emission from the acceptor. (**b**) A typical distance dependency of FRET efficiency. FRET occurs when a donor and an acceptor are located within ~10 nm. The energy transfer efficiency is inversely proportional to the sixth power of the distance between donor and acceptor molecules. The distance that causes 50 % of FRET is called the Förster distance (*dotted lines*). (**c**) For FRET occurrence, spectral overlap between donor emission (Em) and acceptor excitation (Ex) spectra is required. (**d**) FRET also depends on the angle between the emission dipole moment of the donor (*arrow*) and absorption dipole moment (*arrow*). The efficiency is higher when they are parallel and lower when they are vertical to each other

requires the following conditions: (1) The emission spectrum of the donor overlaps with the excitation spectrum of the acceptor (larger overlap increases Förster distance) (Fig. 7.2c), (2) the donor and the acceptor are located within ~twofold Förster distance (typically 10 nm) (Fig. 7.2b), and (3) the relative orientation of the donor emission dipole moment and the acceptor absorption dipole moment must not be vertical (FRET efficiency is the maximum when those are in parallel) (Fig. 7.2d).

7.4 Biosensors Optimized for Ratiometric FRET Imaging

The most common imaging method for FRET is to measure the ratio between the acceptor fluorescence and the donor fluorescence (ratiometric imaging). Since FRET results in an increase in acceptor fluorescence and a decrease in donor fluorescence, the ratio of the acceptor and donor fluorescence provides a measure of the relative strength of FRET. Many signaling sensors have been developed based on this mechanism (Newman et al. 2011; Ueda et al. 2013).

A typical ratiometric FRET sensor consists of a pair of interacting molecules attached to a donor and an acceptor. For example, Cameleon, a FRET-based calcium sensor, is a fusion protein that consists of enhanced cyan fluorescent protein (ECFP), calmodulin, M13, and enhanced yellow fluorescent protein (EYFP) in that order (Fig. 7.3a) (Miyawaki et al. 1997). Ca^{2+} elevation causes binding between calmodulin and M13, bringing ECFP and EYFP close to each other and increasing FRET from ECFP to EYFP. In another example, the Ras activity sensor Raichu is made of Ras and Ras-binding domain (RBD) from Raf-1 (Fig. 7.3b) (Mochizuki et al. 2001). When Ras is activated, it binds to RBD, resulting in a change in the distance between ECFP and Venus (Fig. 7.3b). This design has been further generalized to many other small GTPase proteins such as Cdc42, Rac1, Rab5, and RhoA (Itoh et al. 2002; Yoshizaki et al. 2003; Pertz et al. 2006; Kitano et al. 2008). Several kinase sensors, including AKAR, a protein kinase A sensor (Zhang et al. 2001), CKAR, a protein kinase C sensor (Violin et al. 2003), EKAR, an extracellular signal-regulated kinase sensor (Harvey et al. 2008a), and a focal adhesion kinase sensor (Seong et al. 2011), use a similar design made of a kinase substrate peptides and a phospho-binding domain such as forkhead-associated domain (FHA) instead of pairs of M13-CaM or Ras-RBD.

In addition to protein-protein interactions, a conformation change of a protein associated with its activation has been measured using ratiometric imaging. For example, Camuiα, a FRET-based CaMKII sensor, has ECFP as a donor and Venus as an acceptor on its N- and C-termini of CaMKIIα subunit (Fig. 7.3c) (Takao et al. 2005). When the CaMKII is activated and changes its conformation from the closed to open form, FRET decreases. This design has been used to make sensors for other molecules, including PKC, ERK, WASP, and Raf-1 (Ward et al. 2004; Braun et al. 2005; Terai and Matsuda 2005; Fujioka et al. 2006).

Although ratiometric imaging provides a simple and easy way to measure FRET, it is prone to artifacts due to local concentration changes of the donor and acceptor

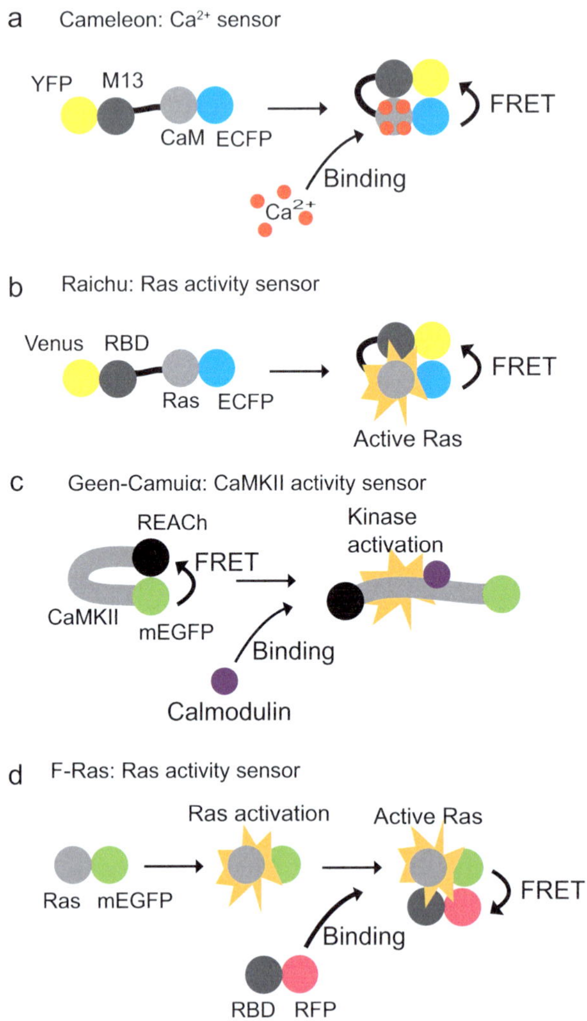

Fig. 7.3 Schematic drawing of FRET sensors. (**a**) Cameleon, an intermolecular FRET sensor for Ca^{2+} concentration (Miyawaki et al. 1997). Binding of Ca^{2+} to calmodulin (CaM) causes interaction between CaM and M13 peptide derived from myosin light-chain kinase, increasing FRET. (**b**) Raichu, an intermolecular FRET sensor for Ras activity (Mochizuki et al. 2001). Ras activation leads to binding of Ras and Ras-binding domain (RBD) of Raf, increasing FRET. (**c**) Green-Camuiα, a FRET sensor for CaMKII activity (Takao et al. 2005; Lee et al. 2009). The sensor measures changes in the conformation of CaMKII upon its activation. (**d**) F-Ras, an intermolecular FRET sensor for Ras activity (Yasuda et al. 2006). The mechanism is similar to Raichu, but the sensor uses intermolecular interaction

(Piston and Kremers 2007). In particular, when the donor and acceptor are in different compartments, it is difficult to isolate the changes in the ratio due to FRET from that due to differential translocation of the proteins. Therefore, biosensors for ratiometric imaging are usually designed so that the donor and acceptor are on the same polypeptide. One disadvantage in the development of an intramolecular FRET sensor is that it is difficult to gain sufficient dynamic range because the FRET efficiency at the unbound state is often high (Fig. 7.3). Thus, it requires an extensive optimization of the linkers between two fluorescent proteins and signaling proteins to maximize the sensitivity (Komatsu et al. 2011). Furthermore, photobleaching and wavelength-dependent light scattering can cause artifacts, since they change the ratio of the donor and acceptor.

7.5 Biosensors Optimized for FLIM-FRET Imaging

In addition to the ratiometric method, FRET can also be detected by measuring fluorescence lifetime, the duration of the excited state (i.e., the time elapsed between fluorophore excitation and fluorescence photon emission) (Fig. 7.4). Fluorescent molecules have their own unique fluorescence lifetime, typically at the timescale of nanoseconds (Lakowicz 2006). FRET shortens the fluorescence lifetime of the donor in proportion to the FRET efficiency, due to competing processes between the photon emission of the donor and FRET (Fig. 7.4a). For fluorescence lifetime imaging of neurons in brain slices, the time-correlated single-photon counting (TCSPC) method in combination with 2-photon laser scanning microscopy (2pLSM) has been commonly used (Yasuda et al. 2006). TCSPC measures the time elapsed from an exciting laser pulse to the fluorescence photon detection (Fig. 7.4b). By repeating this process, a histogram of the fluorescence decay curve at each pixel can be generated (Fig. 7.4c). It is relatively straightforward to combine TCSPC with 2pLSM, since 2pLSM uses a pulsed laser for the excitation source. TCSPC combined with 2pLSM, i.e., 2pFLIM, is used for imaging signaling activity in dendritic spines in brain slices (Becker 2005; Yasuda et al. 2006). Unlike ratiometric FRET imaging, fluorescence lifetime measurement typically uses only

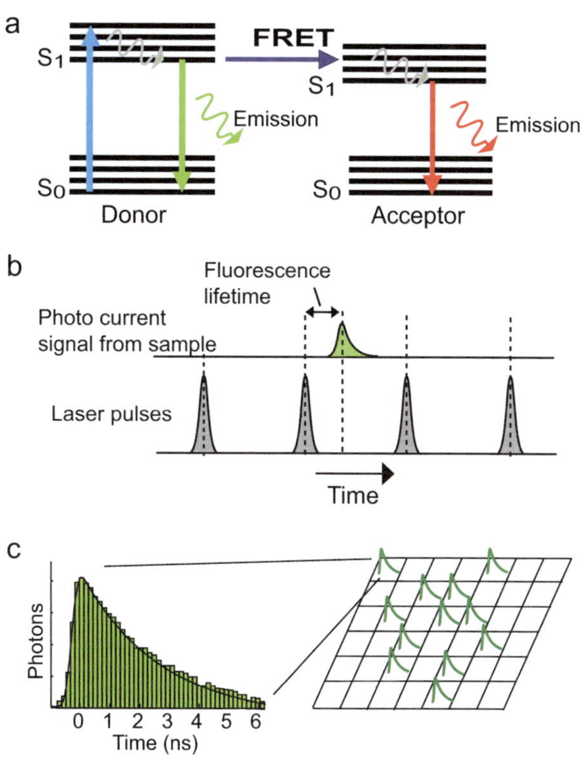

Fig. 7.4 Fluorescence lifetime measurement. (**a**) Jablonski diagram. When a fluorescent molecule is excited by light, the energy state of the molecule changes to the excited state (S1). Following thermal relaxation (*gray*), the state spontaneously comes back to the ground state (S0) with emission of a fluorescence photon. The time spent in the excited state is called fluorescence lifetime. (**b**) The time lags between the arrival time of a fluorescence photon and the excitation laser pulses are repeatedly measured to create a histogram of fluorescence lifetime. (**c**) Repeating fluorescence lifetime measurements in all pixels can generate an image (**c**)

donor fluorescence. Thus, a non- or dim-fluorescent acceptor with a high extinction coefficient and a low quantum yield can be used (Ganesan et al. 2006; Murakoshi et al. 2008). For example, Green-Camuiα, a Camuiα variant specifically designed for FLIM-FRET, uses a dim EYFP variant called resonance energy transfer acceptor chromophore (REACh) as an acceptor and monomeric EGFP (mEGFP) as a donor (Fig. 7.3c) (Kwok et al. 2008; Lee et al. 2009). The fluorescence decay is fast in the closed conformation because of high FRET and becomes slower when the protein takes the open conformation upon protein activation (Fig. 7.5b). Because

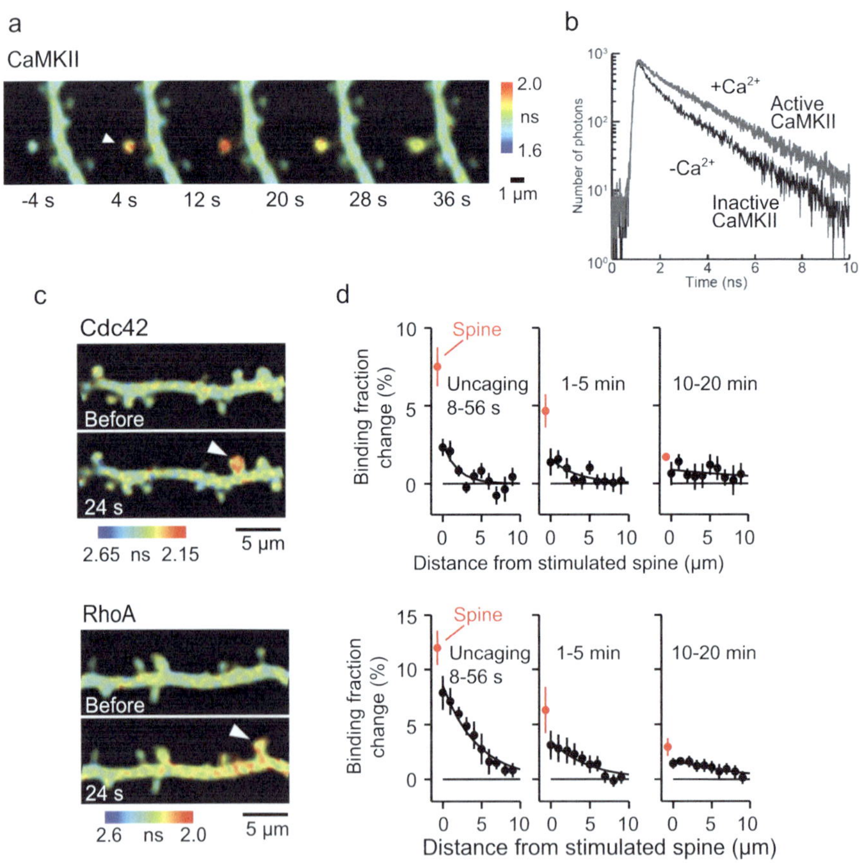

Fig. 7.5 Imaging activity of signaling proteins. (**a**) Fluorescence lifetime images of CaMKII activity in a secondary dendrite of hippocampal CA1 neuron in response to stimulation of a single spine (*white arrowhead*) with 2-photon glutamate uncaging. The activity was measured with Green-Camuiα, a FRET-based CaMKII sensor (see Fig. 7.3c) (From (Lee et al. 2009) with permission). (**b**) Fluorescence decay curves of Green-Camuiα before and after activation measured in cuvette. (**c**) Fluorescence lifetime images of Cdc42 and RhoA after stimulation of a single spine with 2-photon glutamate uncaging (*white arrowhead*). The activity was measured with intermolecular sensors (see Fig. 7.3d) (Murakoshi et al. 2011 with permission). (**d**) Spatial profiles of Cdc42 and RhoA activations after stimulation (Murakoshi et al. 2011 with permission)

the measurement only requires the donor fluorescence, three major advantages arise from the use of FLIM-FRET: (1) the measurement is independent of the local concentration ratio of the donor and acceptor, (2) the measurement is independent of wavelength-dependent light scattering, and (3) the measurement is relatively insensitive to photobleaching. The independence of FLIM on local concentration changes enables the measurement of protein-protein interactions using intermolecular FRET sensors. For example, a Ras sensor based on FLIM-FRET is made of two modules: mEGFP-Ras and RBD fused with two mRFPs at each end of RBD (mRFP-RBD-mRFP) (Fig. 7.3d). By measuring the fluorescence lifetime of mEGFP, one can measure the fraction of mEGFP-Ras that binds to mRFP-RBD-mRFP (Yasuda et al. 2006). FRET sensors with this kind of bimolecular (intermolecular) designs tend to have higher signal compared to unimolecular (intramolecular) FRET sensors because the FRET efficiency is essentially zero when the molecules are separated (Yasuda et al. 2006) (Fig. 7.2b). One limitation of FLIM is that the sensitivity of FLIM is generally lower than ratiometric imaging for the same biosensor, because only donor fluorescence is used in FLIM, while the ratiometric method uses both donor and acceptor fluorescence (Harvey et al. 2008a).

7.6 Bioluminescence-Based Sensors

Bioluminescence has been used mostly for nonimaging signal detection such as gene transcription reporter assays (Fan and Wood 2007). However, the recent development of bright luminophores opened new methods to image biochemical signaling using luminescence (Saito et al. 2012). For example, protein-protein interactions can be detected by the ratiometric imaging of bioluminescence resonance energy transfer (BRET) between a luminophore and a fluorophore as a donor and an acceptor, respectively (Boute et al. 2002). Since acceptor fluorescence depends only on BRET, ratiometric-BRET signal is not affected by the concentration ratio between the donor and the acceptor, making binding analyses simple. BRET also has been used to create color variants of luminophores and BRET-based sensors for intracellular signaling (De et al. 2013). A variant of *Renilla* luciferase optimized for BRET with EYFP showed further improvement in the brightness (Saito et al. 2012). In addition, it has been demonstrated that the fusion of circularly permuted luminophore and chemical sensing domain, the design similar to G-CaMP, can be used to sense Ca^{2+} and other messengers in cells using luminescence (Saito et al. 2012). These improvements have made the temporal resolution of luminescence imaging comparable with fluorescent-based imaging. Bioluminescence imaging has many advantages over fluorescence-based assays. The technique has essentially no background and no phototoxicity. However, there are the following two drawbacks: (1) the continuous addition of expensive cofactors is required and (2) the high-resolution imaging in the tissue is difficult because of light scattering.

7.7 Imaging Signal Transduction in Single Dendritic Spines During Synaptic Plasticity

Signal transduction that induces long-term potentiation (LTP) and long-term depression (LTD) is spatiotemporally regulated in a complicated manner in dendrites, influencing the adjacent dendritic spines as well as the stimulated spines. The recent development of 2-photon glutamate uncaging technique has greatly improved our understanding of the spatiotemporal control of electrophysiological and structural plasticity of dendritic spines (Matsuzaki et al. 2004). For example, it has been demonstrated that LTP is associated with spine enlargements (Matsuzaki et al. 2004), whereas LTD is associated with spine shrinkage (Zhou et al. 2004; Hayama et al. 2013; Oh et al. 2013). Interestingly, LTD appears to be either spine specific or nonspecific depending on induction protocols. When LTD is induced with a simple low-frequency glutamate uncaging, LTD is spine specific (Oh et al. 2013). However, when LTD is induced by pairing back-propagating action potentials with subsequent 2-photon glutamate uncaging and GABA uncaging, the depression spreads over ~5–10 μm (Hayama et al. 2013).

LTP induction in a single spine causes the activation of diverse signaling cascades. The reaction starts from glutamate binding to NMDA receptors, which triggers a Ca^{2+} influx through the receptors (Lisman et al. 2012). This leads to the formation of Ca^{2+}/CaM complexes in the spine. Subsequently, the Ca^{2+}/CaM complexes bind to CaMKII induces the activation of CaMKII (Lisman et al. 2012). CaMKII is one of the most abundant signaling proteins in neurons and consists of 12 subunits (Lisman et al. 2012). Each subunit is activated by the binding of Ca^{2+}/CaM. When active calmodulin molecules bind to CaMKII subunits located next to each other in the complex, those CaMKII molecules undergo cooperative trans-autophosphorylation to phosphorylate the adjacent subunit at threonine-286 and lead to the autonomous and Ca^{2+}-independent activation of CaMKII. Because of this property, it has been thought that CaMKII activity in spines lasts for a long time to maintain LTP. However, the duration of CaMKII in spines had not been measured because of the lack of a method to measure the temporal pattern of CaMKII activity in dendritic spines. Recently, CaMKII activation during LTP and associated spine structural long-term potentiation (sLTP) induced by 2-photon glutamate uncaging was monitored using Green-Camuiα in combination with 2pFLIM-FRET (Fig. 7.5a, b) (Lee et al. 2009). This revealed that the activation of CaMKII after single-spine glutamate stimulation is spine specific and short-lived (~1 min).

Since LTP and sLTP last more than ~1 h, short-lived CaMKII activation should be relayed to long-lasting signals. Rho GTPase proteins, particularly RhoA, Rac1, and Cdc42, are good candidates for such signaling because these proteins are known to regulate actin cytoskeleton and spine morphology (Luo 2000; Tashiro and Yuste 2004; Saneyoshi et al. 2008). RhoA and Cdc42 FRET sensors have been developed and their activities were imaged with 2pFLIM-FRET (Fig. 7.5c). This study demonstrated that the activation of Cdc42 and RhoA is CaMKII dependent and long lasting (Murakoshi et al. 2011). However, their spatial profiles are different: while Cdc42 activity is localized in the stimulated spine, RhoA spreads into the dendritic shaft and adjacent spines (Fig. 7.5d). Thus, Cdc42 appears to play an important role in relaying short Ca^{2+}-CaMKII signaling into spine-specific and long-lasting signaling (Murakoshi and Yasuda 2012). The spreading of Rho may have specific roles in the surrounding area, such as heterosynaptic metaplasticity as discussed in the following section. As a downstream signaling of Rho GTPase proteins, LIMK-1 and cofilin have been thought to be major regulators of actin during LTP and LTD (Meng et al. 2002; Rust et al. 2010). Recently, FLIM-FRET sensors for inter-cofilin and cofilin-actin interactions demonstrated that cofilin is recruited into the stimulated spine and forms tight complex with actin filaments, likely playing a role in the consolidation of structural plasticity (Bosch et al. 2014).

7.8 Signal Spreading

While the input-specific biochemical reactions are required for synapse-specific plasticity, there is evidence suggesting that signals initiated in a single spine can diffuse out of the stimulated spine and invade adjacent spines, activating the downstream molecules. For example, LTP induction in a single spine facilitates LTP in surrounding spines located within 5 μm for ~5 min (Harvey et al. 2008b). In addition, similar stimulation can facilitate spinogenesis in adjacent dendritic shaft (Kwon and Sabatini 2011). In contrast, when LTP is induced in several adjacent spines, spine shrinkages are induced in non-stimulated spines located within 3 μm (heterosynaptic LTD) (Oh et al. 2015).

Imaging protein activity with FLIM-FRET has revealed that there are signals that spread from the stimulated spines. For example, small GTPase proteins RhoA and Ras are activated in a stimulated spine and then spread into adjacent spines over 5–10 μm (Harvey et al. 2008b; Murakoshi et al. 2011). Activated Ras spreading lowers the threshold of LTP induction in neighboring spines (Harvey et al. 2008b) and increases the probability of spinogenesis (Kwon and Sabatini 2011). LTP-inducing stimuli can also activate calcineurin (CaN), a Ca^{2+}-/CaM-dependent phosphatase important for LTD, in the stimulated spines and adjacent dendrites (Fujii et al. 2013). The spreading of CaN could be important for heterosynaptic LTD (Oh et al. 2015).

Since many subcellular compartments required for synaptic plasticity exist outside of the spine, the spreading of biochemical signaling may be required for communication to the extra-spine compartments such as recycling endosomes containing AMPARs (Park et al. 2006) and protein synthesis machinery (Steward and Levy 1982; Ostroff et al. 2002; Buxbaum et al. 2014). In addition, various forms of heterosynaptic plasticity are likely to be induced by signal spreading as discussed above for cross talk and heterosynaptic LTD (see above). Finally, the spread of signaling may also be important for clustered plasticity, which creates local accumulation of synaptic inputs that in turn results in functional compartmentalization of dendritic segments (Govindarajan et al. 2006; Larkum and Nevian 2008; Branco et al. 2010). The clustered plasticity has been found in several paradigms in vitro and in vivo. For example, calcium imaging in combination with electrophysiological recording suggests that neighboring synapses are likely to be coactivated, and such clustering formation depends on neuronal activity and NMDA receptor activation (Kleindienst et al. 2011; Takahashi et al. 2012). Furthermore, imaging of the GluA1 subunit of AMPA receptors suggests that synaptic plasticity occurs in a clustered manner on dendrites (Makino and Malinow 2011). In addition, clustered spinogenesis in the motor cortex is found to be correlated with motor learning, further supporting its physiological significance (Fu et al. 2012).

Some intracellular signals are not restricted to short stretch of dendritic segments and spread over much longer distance. FLIM-FRET imaging of ERK activity revealed that the spreading of Ras-ERK signaling from stimulated spine reaches into the nucleus (Zhai et al. 2013). Interestingly, ERK activation in the nucleus occurs when only a few (3–7) spines are stimulated (Harvey et al. 2008a; Zhai et al. 2013). Furthermore, ERK in the nucleus was more efficiently activated when multiple dendrites were stimulated. Thus, neuronal signaling appears to integrate multiple dendritic branches to activate gene transcription. Overall, these studies provided many insights into biochemical computations in spines, dendritic branches, and the nucleus.

7.9 Conclusion

Genetically encoded biosensors based on fluorescent and luminescent proteins allow researchers to probe intracellular biochemical reactions in live cells. Many of them show high sensitivity that is sufficient for imaging signals in neuronal subcompartments with high spatiotemporal resolution. Diverse signal sensing probes including circular permuted, luminophore-/fluorophore-based, and BRET-/FRET-based probes have been developed. It has also been demonstrated that FRET sensor in combination with 2pFLIM-FRET enables the monitoring of protein activity at the single synapse level in brain slices. By combining optical manipulation techniques such as 2-photon glutamate uncaging, the spatiotemporal dynamics of signaling proteins in dendrites and spines in response to local activation of signaling can be studied.

While the imaging and optical manipulation techniques have been proven to be useful for the study of intracellular signaling, the activity of only several species of proteins among thousands of signaling proteins has been monitored. The rate-limiting steps for sensor creation appear to be the optimization and validation of each sensor. Establishing automated, high-throughput screening system will be important for developing many sensors (Chen et al. 2013; Slattery and Hahn 2014). Although it will take a long time to make sensitive biosensors to cover a majority of signaling sensors, the efforts will provide many insights into the principles of biochemical computation in neurons.

Acknowledgments We thank the members of Murakoshi's lab and Yasuda's lab for the discussion and T. Yasuda for the discussion and critical reading.

References

Akerboom J, Carreras Calderon N, Tian L et al (2013) Genetically encoded calcium indicators for multi-color neural activity imaging and combination with optogenetics. Front Mol Neurosci 6:2

Alvarez VA, Sabatini BL (2007) Anatomical and physiological plasticity of dendritic spines. Annu Rev Neurosci 30:79–97

Augustine GJ, Santamaria F, Tanaka K (2003) Local calcium signaling in neurons. Neuron 40 (2):331–346

Baird GS, Zacharias DA, Tsien RY (1999) Circular permutation and receptor insertion within green fluorescent proteins. Proc Natl Acad Sci U S A 96(20):11241–11246

Becker W (2005) In: Castleman AW, Toennies JP, Zinth W (eds) In advanced time-correlated single photon counting techniques, vol 81. Springer, New York

Bjarnadottir TK, Gloriam DE, Hellstrand SH et al (2006) Comprehensive repertoire and phylogenetic analysis of the G protein-coupled receptors in human and mouse. Genomics 88 (3):263–273

Bosch M, Castro J, Saneyoshi T et al (2014) Structural and molecular remodeling of dendritic spine substructures during long-term potentiation. Neuron 82(2):444–459

Boute N, Jockers R, Issad T (2002) The use of resonance energy transfer in high-throughput screening: BRET versus FRET. Trends Pharmacol Sci 23(8):351–354

Branco T, Clark BA, Hausser M (2010) Dendritic discrimination of temporal input sequences in cortical neurons. Science 329(5999):1671–1675

Braun DC, Garfield SH, Blumberg PM (2005) Analysis by fluorescence resonance energy transfer of the interaction between ligands and protein kinase Cdelta in the intact cell. J Biol Chem 280 (9):8164–8171

Buxbaum AR, Wu B, Singer RH (2014) Single beta-actin mRNA detection in neurons reveals a mechanism for regulating its translatability. Science 343(6169):419–422

Chen TW, Wardill TJ, Sun Y et al (2013) Ultrasensitive fluorescent proteins for imaging neuronal activity. Nature 499(7458):295–300

Colgan LA, Yasuda R (2014) Plasticity of dendritic spines: subcompartmentalization of signaling. Annu Rev Physiol 76:365–385

De A, Jasani A, Arora R et al (2013) Evolution of BRET biosensors from live cell to tissue-scale in vivo imaging. Front Endocrinol (Lausanne) 4:131

Ehlers MD (2013) Dendritic trafficking for neuronal growth and plasticity. Biochem Soc Trans 41 (6):1365–1382

Fan F, Wood KV (2007) Bioluminescent assays for high-throughput screening. Assay Drug Dev Technol 5(1):127–136

Fu M, Yu X, Lu J et al (2012) Repetitive motor learning induces coordinated formation of clustered dendritic spines in vivo. Nature 483(7387):92–95

Fujii H, Inoue M, Okuno H et al (2013) Nonlinear decoding and asymmetric representation of neuronal input information by CaMKIIalpha and calcineurin. Cell Rep 3(4):978–987

Fujioka A, Terai K, Itoh RE et al (2006) Dynamics of the Ras/ERK MAPK cascade as monitored by fluorescent probes. J Biol Chem 281(13):8917–8926

Ganesan S, Ameer-Beg SM, Ng TT et al (2006) A dark yellow fluorescent protein (YFP)-based Resonance Energy-Accepting Chromoprotein (REACh) for Forster resonance energy transfer with GFP. Proc Natl Acad Sci U S A 103(11):4089–4094

Govindarajan A, Israely I, Huang SY et al (2011) The dendritic branch is the preferred integrative unit for protein synthesis-dependent LTP. Neuron 69(1):132–146
Govindarajan A, Kelleher RJ, Tonegawa S (2006) A clustered plasticity model of long-term memory engrams. Nat Rev Neurosci 7(7):575–583
Harvey CD, Ehrhardt AG, Cellurale C et al (2008a) A genetically encoded fluorescent sensor of ERK activity. Proc Natl Acad Sci U S A 105(49):19264–19269
Harvey CD, Yasuda R, Zhong H et al (2008b) The spread of Ras activity triggered by activation of a single dendritic spine. Science 321(5885):136–140
Hayama T, Noguchi J, Watanabe S et al (2013) GABA promotes the competitive selection of dendritic spines by controlling local Ca^{2+} signaling. Nat Neurosci 16(10):1409–1416
Inoue M, Takeuchi A, Horigane SI et al (2014) Rational design of a high-affinity, fast, red calcium indicator R-CaMP2. Nat Methods 12:64–70
Itoh RE, Kurokawa K, Ohba Y et al (2002) Activation of rac and cdc42 video imaged by fluorescent resonance energy transfer-based single-molecule probes in the membrane of living cells. Mol Cell Biol 22(18):6582–6591
Kitano M, Nakaya M, Nakamura T et al (2008) Imaging of Rab5 activity identifies essential regulators for phagosome maturation. Nature 453(7192):241–245
Kleindienst T, Winnubst J, Roth-Alpermann C et al (2011) Activity-dependent clustering of functional synaptic inputs on developing hippocampal dendrites. Neuron 72(6):1012–1024
Komatsu N, Aoki K, Yamada M et al (2011) Development of an optimized backbone of FRET biosensors for kinases and GTPases. Mol Biol Cell 22(23):4647–4656
Kwok S, Lee C, Sanchez SA et al (2008) Genetically encoded probe for fluorescence lifetime imaging of CaMKII activity. Biochem Biophys Res Commun 369(2):519–525
Kwon HB, Sabatini BL (2011) Glutamate induces de novo growth of functional spines in developing cortex. Nature 474(7349):100–104
Lakowicz JR (2006) Principles of fluorescence spectroscopy. Springer, New York
Larkum ME, Nevian T (2008) Synaptic clustering by dendritic signalling mechanisms. Curr Opin Neurobiol 18(3):321–331
Lee SJ, Escobedo-Lozoya Y, Szatmari EM et al (2009) Activation of CaMKII in single dendritic spines during long-term potentiation. Nature 458(7236):299–304
Lippincott-Schwartz J, Snapp E, Kenworthy A (2001) Studying protein dynamics in living cells. Nat Rev Mol Cell Biol 2(6):444–456
Lisman J, Yasuda R, Raghavachari S (2012) Mechanisms of CaMKII action in long-term potentiation. Nat Rev Neurosci 13(3):169–182
Luo L (2000) Rho GTPases in neuronal morphogenesis. Nat Rev Neurosci 1(3):173–180
Makino H, Malinow R (2011) Compartmentalized versus global synaptic plasticity on dendrites controlled by experience. Neuron 72(6):1001–1011
Manning G, Whyte DB, Martinez R et al (2002) The protein kinase complement of the human genome. Science 298(5600):1912–1934
Marvin JS, Borghuis BG, Tian L et al (2013) An optimized fluorescent probe for visualizing glutamate neurotransmission. Nat Methods 10(2):162–170
Matsuzaki M, Honkura N, Ellis-Davies GC et al (2004) Structural basis of long-term potentiation in single dendritic spines. Nature 429(6993):761–766
Meng Y, Zhang Y, Tregoubov V et al (2002) Abnormal spine morphology and enhanced LTP in LIMK-1 knockout mice. Neuron 35(1):121–133
Miyawaki A (2003) Visualization of the spatial and temporal dynamics of intracellular signaling. Dev Cell 4(3):295–305
Miyawaki A, Llopis J, Heim R et al (1997) Fluorescent indicators for Ca^{2+} based on green fluorescent proteins and calmodulin. Nature 388(6645):882–887
Mochizuki N, Yamashita S, Kurokawa K et al (2001) Spatio-temporal images of growth-factor-induced activation of Ras and Rap1. Nature 411(6841):1065–1068
Murakoshi H, Lee SJ, Yasuda R (2008) Highly sensitive and quantitative FRET-FLIM imaging in single dendritic spines using improved non-radiative YFP. Brain Cell Biol 36(1-4):31–42

Murakoshi H, Wang H, Yasuda R (2011) Local, persistent activation of Rho GTPases during plasticity of single dendritic spines. Nature 472(7341):100–104
Murakoshi H, Yasuda R (2012) Postsynaptic signaling during plasticity of dendritic spines. Trends Neurosci 35(2):135–143
Nakai J, Ohkura M, Imoto K (2001) A high signal-to-noise Ca^{2+} probe composed of a single green fluorescent protein. Nat Biotechnol 19(2):137–141
Newman RH, Fosbrink MD, Zhang J (2011) Genetically encodable fluorescent biosensors for tracking signaling dynamics in living cells. Chem Rev 111(5):3614–3666
Oh WC, Hill TC, Zito K (2013) Synapse-specific and size-dependent mechanisms of spine structural plasticity accompanying synaptic weakening. Proc Natl Acad Sci U S A 110(4): E305–E312
Oh WC, Parajuli LK, Zito K (2015) Heterosynaptic structural plasticity on local dendritic segments of hippocampal CA1 neurons. Cell Rep 10(2):162–169
Ohkura M, Matsuzaki M, Kasai H et al (2005) Genetically encoded bright Ca^{2+} probe applicable for dynamic Ca^{2+} imaging of dendritic spines. Anal Chem 77(18):5861–5869
Oldach L, Zhang J (2014) Genetically encoded fluorescent biosensors for live-cell visualization of protein phosphorylation. Chem Biol 21(2):186–197
Ostroff LE, Fiala JC, Allwardt B et al (2002) Polyribosomes redistribute from dendritic shafts into spines with enlarged synapses during LTP in developing rat hippocampal slices. Neuron 35 (3):535–545
Park M, Salgado JM, Ostroff L et al (2006) Plasticity-induced growth of dendritic spines by exocytic trafficking from recycling endosomes. Neuron 52(5):817–830
Pertz O, Hodgson L, Klemke RL et al (2006) Spatiotemporal dynamics of RhoA activity in migrating cells. Nature 440(7087):1069–1072
Piston DW, Kremers GJ (2007) Fluorescent protein FRET: the good, the bad and the ugly. Trends Biochem Sci 32(9):407–414
Rust MB, Gurniak CB, Renner M et al (2010) Learning, AMPA receptor mobility and synaptic plasticity depend on n-cofilin-mediated actin dynamics. EMBO J 29(11):1889–1902
Saito K, Chang YF, Horikawa K et al (2012) Luminescent proteins for high-speed single-cell and whole-body imaging. Nat Commun 3:1262
Saneyoshi T, Wayman G, Fortin D et al (2008) Activity-dependent synaptogenesis: regulation by a CaM-kinase kinase/CaM-kinase I/betaPIX signaling complex. Neuron 57(1):94–107
Seong J, Ouyang M, Kim T et al (2011) Detection of focal adhesion kinase activation at membrane microdomains by fluorescence resonance energy transfer. Nat Commun 2:406
Slattery SD, Hahn KM (2014) A high-content assay for biosensor validation and for examining stimuli that affect biosensor activity. Curr Protoc Cell Biol 65:14.15.11–14.15.31
Steward O, Levy WB (1982) Preferential localization of polyribosomes under the base of dendritic spines in granule cells of the dentate gyrus. J Neurosci 2(3):284–291
Takahashi N, Kitamura K, Matsuo N et al (2012) Locally synchronized synaptic inputs. Science 335(6066):353–356
Takao K, Okamoto K, Nakagawa T et al (2005) Visualization of synaptic Ca^{2+}/calmodulin-dependent protein kinase II activity in living neurons. J Neurosci 25(12):3107–3112
Tantama M, Martinez-Francois JR, Mongeon R et al (2013) Imaging energy status in live cells with a fluorescent biosensor of the intracellular ATP-to-ADP ratio. Nat Commun 4:2550
Tashiro A, Yuste R (2004) Regulation of dendritic spine motility and stability by Rac1 and Rho kinase: evidence for two forms of spine motility. Mol Cell Neurosci 26(3):429–440
Terai K, Matsuda M (2005) Ras binding opens c-Raf to expose the docking site for mitogen-activated protein kinase kinase. EMBO Rep 6(3):251–255
Tsien RY (1998) The green fluorescent protein. Annu Rev Biochem 67:509–544
Ueda Y, Kwok S, Hayashi Y (2013) Application of FRET probes in the analysis of neuronal plasticity. Front Neural Circ 7:163
Violin JD, Zhang J, Tsien RY et al (2003) A genetically encoded fluorescent reporter reveals oscillatory phosphorylation by protein kinase C. J Cell Biol 161(5):899–909

Ward ME, Wu JY, Rao Y (2004) Visualization of spatially and temporally regulated N-WASP activity during cytoskeletal reorganization in living cells. Proc Natl Acad Sci U S A 101 (4):970–974

Wennerberg K, Rossman KL, Der CJ (2005) The Ras superfamily at a glance. J Cell Sci 118 (Pt 5):843–846

Whitaker M (2010) Genetically encoded probes for measurement of intracellular calcium. Methods Cell Biol 99:153–182

Yasuda R, Harvey CD, Zhong H et al (2006) Supersensitive Ras activation in dendrites and spines revealed by two-photon fluorescence lifetime imaging. Nat Neurosci 9(2):283–291

Yoshizaki H, Ohba Y, Kurokawa K et al (2003) Activity of Rho-family GTPases during cell division as visualized with FRET-based probes. J Cell Biol 162(2):223–232

Zhai S, Ark ED, Parra-Bueno P et al (2013) Long-distance integration of nuclear ERK signaling triggered by activation of a few dendritic spines. Science 342(6162):1107–1111

Zhang J, Campbell RE, Ting AY et al (2002) Creating new fluorescent probes for cell biology. Nat Rev Mol Cell Biol 3(12):906–918

Zhang J, Ma Y, Taylor SS et al (2001) Genetically encoded reporters of protein kinase A activity reveal impact of substrate tethering. Proc Natl Acad Sci U S A 98(26):14997–15002

Zhou Q, Homma KJ, Poo MM (2004) Shrinkage of dendritic spines associated with long-term depression of hippocampal synapses. Neuron 44(5):749–757

Part III
Patterning Dendritic Architecture of Neurons and Their Populations

Chapter 8
Dendritic Self-Avoidance

Peter Soba

Abstract Self-avoidance allows a neuron's dendrites to recognize and repel each other. It constitutes one of the basic mechanisms facilitating even patterning of the receptive field. Self-avoidance enables neurons to distinguish self- from nonself-dendrites in complex systems so their receptive fields can coexist in overlapping territories. The phenomenon was first described in leech and later in *Drosophila* and *C. elegans*, with recent studies in vertebrate systems revealing self-avoidance as a universally conserved mechanism required to pattern dendrites of many if not all neurons. Molecularly, members of the immunoglobulin superfamily (IgSF), in particular Down's syndrome cell adhesion molecule (Dscam), and clustered protocadherins (Pcdhs) play a prominent role in mediating self-avoidance. Here, I will summarize the findings and advances in our understanding of the conceptual and molecular nature of dendrite self-avoidance from the cellular to organismal level and discuss its contribution to dendritic patterning.

Keywords Dendritic self-avoidance • Down's syndrome cell adhesion molecule (Dscam) • Clustered protocadherins (Pcdhs) • Dendrite repulsion • Homotypic interaction

8.1 Introduction

Neurons have to grow and pattern their processes in appropriate ways that allow them to connect and build functional circuits. Notably, most types of excitatory neurons develop very stereotyped dendritic fields that make them instantly recognizable as a neuron of a specific class, recognized already by Cajal as a hallmark for a given neuronal subtype (e.g., Purkinje cells, pyramidal neurons, or retinal ganglion cells). Dendritic fields in vivo feature stereotypy in spite of their extraordinary complexity. A key characteristic of most dendritic fields is the remarkable non-overlap of isoneuronal dendrites, which is particularly evident in neurons

P. Soba (✉)
Center for Molecular Neurobiology Hamburg (ZMNH), University Medical Campus Hamburg-Eppendorf (UKE), University of Hamburg, Hamburg, Germany
e-mail: peter.soba@zmnh.uni-hamburg.de

K. Emoto et al. (eds.), *Dendrites*, DOI 10.1007/978-4-431-56050-0_8

that grow in an almost two-dimensional shape, e.g., ganglion cells in the mammalian retina or a subset of *Drosophila* sensory neurons, so-called dendritic arborization (da) neurons (Fig. 8.1a, b). How growing neurites of a given neuron achieve such stereotyped and non-overlapping patterning remained enigmatic for a long time. The phenomenon however was already described in 1968 by Nicholls and Baylor who studied the development of sensory neurons in medicinal leech. They observed that receptive fields of peripheral sensory touch receptors displayed highly stereotyped and non-overlapping organization and suggested that "the presence of innervation may in some way prevent additional nerve fibers from invading a territory and forming additional endings in it" (Nicholls and Baylor 1968). These studies were later extended by observations from Kramer, Kuwada, and Stent using the same system (Fig. 8.1c), showing that growing branches seemed to actively avoid each other (Kramer and Kuwada 1983; Kramer and Stent 1985). Their work coined the term self-avoidance, and they speculated that "the functional role for a 'self-avoidance' phenomenon could be, for instance, to ensure an optimal coverage of the available target territory" (Kramer and Kuwada 1983).

More than two decades past before the conceptual revival of self-avoidance which started with the elucidation of its molecular basis. By now we have a partial understanding of the molecular mechanisms required for dendrite self-avoidance in several model systems converging on the use of cellular recognition, repulsion, and self- vs. nonself-discrimination. Dendritic patterning through self-avoidance seems to be functionally conserved throughout the animal kingdom and plays a pivotal role in generating evenly spaced dendritic fields.

8.2 Molecular Basis of Dendrite Self-Avoidance

8.2.1 Dendrite Self-Avoidance at the Single-Cell Level

The first evidence that self-avoidance is a cell-autonomous molecular mechanism emanated again from leech studies. Sensory branches that were severed from the neuron by laser surgery quickly lost their repulsive nature and failed to be recognized by their former sister neurites (Wang and Macagno 1997). Later studies over the last decade in different model organisms have implicated several cell surface-related signals in self-avoidance. Most prominently, Down's syndrome cell adhesion molecule (Dscam) in *Drosophila* and clustered protocadherins (Pcdhs) in mammals will be discussed in detail as they are the most universal self-avoidance molecules known to date. A number of reviews cover additional aspects of self-avoidance in neuronal patterning (Cameron and Rao 2010; Grueber and Sagasti 2010; Lawrence Zipursky and Grueber 2013; Zipursky and Sanes 2010) and functions of Dscam (Fuerst and Burgess 2009; Garrett et al. 2012a) and Pcdhs (Chen and Maniatis 2013; Keeler et al. 2015; Morishita and Yagi 2007; Yagi 2012).

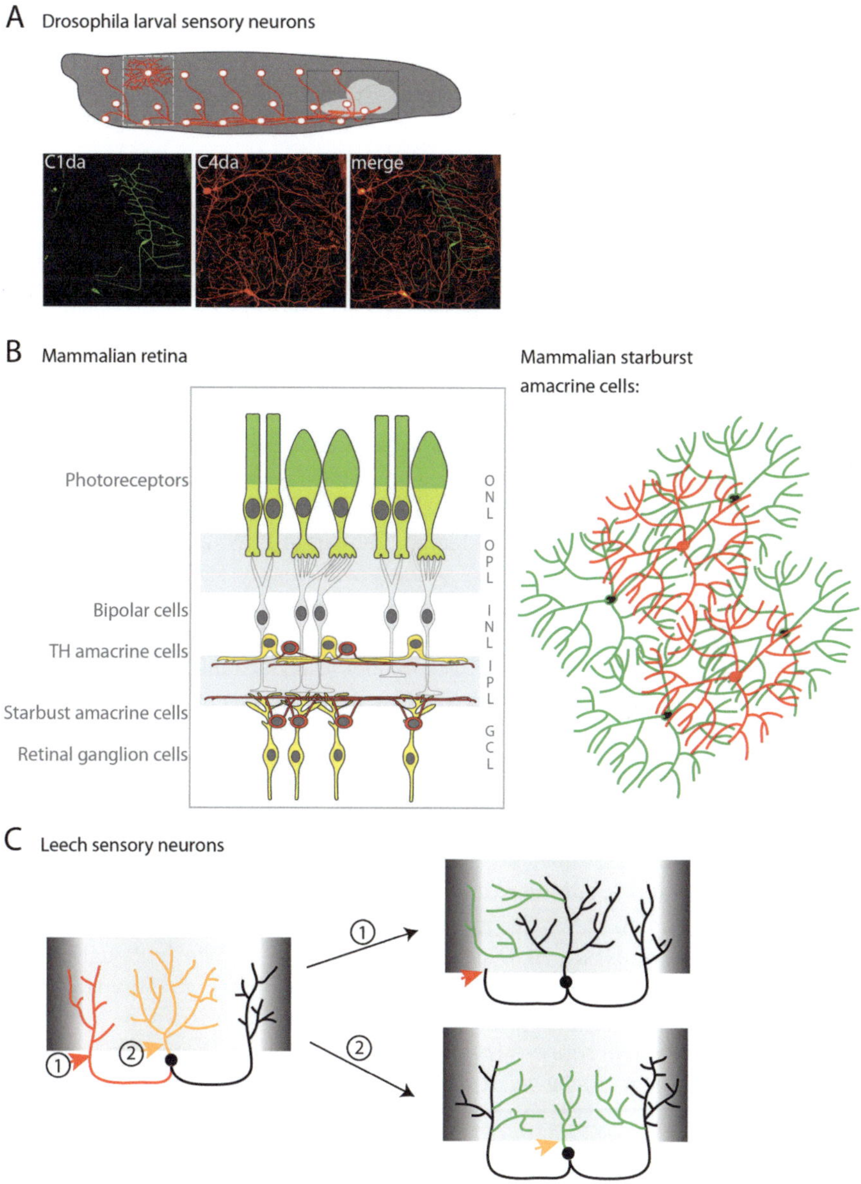

Fig. 8.1 Dendrite self-avoidance in different cell types across species. (**a**) *Drosophila* larval sensory da neurons innervate the body wall in a segmentally repeated pattern. All types of da neurons (shown are exemplary class I (C1) and class IV (C4) da neurons) exhibit dendrite self-avoidance and evenly spaced non-overlapping innervation of their dendritic fields. C1da and C4da neurons occupy overlapping dendritic territories by being able to distinguish self-avoidance of isoneuronal dendrites and coexistence of heteroneuronal dendrites. (**b**) Schematic organization of the mammalian retina with the different cell types and layers (*ONL*/*INL* outer/inner nuclear layer, *OPL*/*IPL* outer/inner plexiform layer, *GCL* ganglion cell layer). The IPL consists of stratified dendrites of different subtypes of amacrine and retinal ganglion cells. Starburst amacrine cell dendritic fields typically display complete yet nonredundant coverage within individual subtypes (tiling) and little or no overlap of isoneuronal dendrites (self-avoidance). Different subtypes have

Here I will focus on the role of self-avoidance in establishing ordered receptive fields and its molecular and mechanistic basis.

8.2.1.1 *Drosophila* Down's Syndrome Cell Adhesion Molecule (Dscam)

The first identified mediator of dendrite self-avoidance was *Drosophila Down's syndrome cell adhesion molecule* (*Dscam*), which was originally found as an axon guidance molecule in larval photoreceptor neurons (Schmucker et al. 2000) and shown to mediate axon branch segregation (Wang et al. 2002a). Dscam belongs to the IgSF of surface receptors and is able to interact in a homotypic manner via its extracellular domain (Wojtowicz et al. 2004, 2007). *Dscam* was subsequently shown to play a role in dendrite development of *Drosophila* projection neurons in the antennal lobe, which display condensed and underdeveloped dendritic fields without Dscam (Zhu et al. 2006). Definitive proof that Dscam is a major mediator of dendrite self-avoidance came from several subsequent studies on the peripheral nervous system (PNS) of *Drosophila* larvae. The larval PNS features four distinct classes of dendritic arborization (da) neurons with highly stereotyped dendritic fields (class I–IV, Grueber et al. 2002). The key advantage of this system is the ability to visualize and genetically manipulate individual neurons in the living animal which made it an extensively used model to understand dendrite development (Fig. 8.1a and Jan and Jan 2010). The absence of *Dscam* in da neurons results in crossing and fasciculation of isoneuronal dendrites as they fail to recognize and repel each other (Fig. 8.3a, Soba et al. 2007; Hughes et al. 2007; Matthews et al. 2007). Live imaging of da neuron dendrites showed that Dscam likely mediates repulsive signaling between isoneuronal branches in a contact-dependent manner (Matthews et al. 2007; Soba et al. 2007). Removal of the Dscam intracellular domain results in adhesion of dendrites instead of repulsion suggesting that downstream signaling pathways trigger the retraction and reorientation of growing branches. Thus the presence of Dscam on the dendrite surface is thought to be sufficient for recognition and repulsion of growing branches (Fig. 8.3b). However, the signaling pathways regulating Dscam's repulsive action remain to be identified. Although p21-activated serine-threonine kinase (Pak) has been implicated in Dscam-mediated axon guidance, it is not involved in dendrite self-avoidance (Hughes et al. 2007).

Fig. 8.1 (continued) overlapping dendritic territories similarly as *Drosophila* sensory da neurons (coexistence). Adapted from Zipursky and Grueber (2013). (**c**) Findings by Kramer and Stent described self-avoidance in mechanosensory neurons in medicinal leech. Sensory processes innervate the leech skin and display dynamic growth during development with self-avoidance of isoneuronal branches. Surgical removal of individual branches (indicated by numbered *arrows*) results in compensatory growth by neighboring branches (shown in *green*) to close the gap left behind by the removed branch (shown in *red*, Adapted from Grueber and Sagasti 2010)

8.2.1.2 Dscam in Mammals

Dscam is structurally well conserved in mammals, and subsequent studies showed that Dscam and its paralog DscamL1 have functions in the mouse retina apparently analogous to its role in *Drosophila* (Fuerst et al. 2008, 2009). Many retinal subtypes of ganglion and amacrine cells develop and maintain regular mosaic spacing, and their radial dendritic fields have minimal overlap with their neighboring cells (see tiling chapter). In addition, their stratified dendritic fields show uniform mosaic patterning. Dscam and DscamL1 are expressed in non-overlapping subsets of retinal ganglion cells (RGCs) and starburst amacrine cells (SACs), and loss of function analyses showed that the dendrites of the affected cell types were strongly fasciculated. The aggregation and crossing of isoneuronal dendrites without Dscam or DscamL1 are highly similar to the previous observations in *Drosophila* sensory neurons lacking Dscam. Later studies confirmed that mouse Dscam indeed seems to be required for self-avoidance of retinal cells in a cell-type autonomous manner, which was shown by conditional deletion of Dscam in specific retinal cell subsets (Fuerst et al. 2012). Interestingly, analyses of mosaic animals with a small number of retinal cells being mutant for Dscam showed that self-avoidance defects of these cells also affected the development of wild-type cells of the same subtype. Moreover, mouse Dscam is also required for cell number regulation, regular mosaic spacing, lamination, and synaptogenesis of some retinal cell types suggesting it has multiple developmental functions that are temporally regulated.

Nonetheless, these studies provided the first in vivo evidence that the function of *Dscam* in self-avoidance is likely conserved in mammals. It also suggested that mammals use a similar self-avoidance mechanism to assemble their nervous system. Due to the restricted expression of Dscam in mammals and its limited capacity to mediate self-avoidance in an entire nervous system, it is however likely that additional self-avoidance molecules exist.

8.2.1.3 Clustered Protocadherins (Pcdhs)

More recent studies implicated members of the cadherin family of adhesion proteins, in particular clustered protocadherins (Pcdhs), in self-avoidance in mammals. These phylogenetically related Pcdhs are organized in three gene clusters (14 α-, 22 β-, and 22 γ-Pcdhs in mouse) which are highly conserved in vertebrates (Morishita and Yagi 2007; Wu and Maniatis 1999). Clustered Pcdhs play an important role in neuronal development and survival (rewiewed in Hayashi et al. 2015; Keeler et al. 2015). While β-Pcdhs have not been studied functionally so far, loss of α-Pcdh function leads to decreased dendritic arborization and spine loss in hippocampal neurons (Suo et al. 2012). Additionally, α-Pcdh knockout mice diplay aggregated retinogeniculate terminals and reduced visual acuity (Meguro et al. 2015). γ-Pcdhs have been shown to play important roles in neuronal development including spinal cord neuronal survival and synaptogenesis (Lefebvre et al. 2008; Wang et al. 2002b; Weiner et al. 2005). However, mice in which γ-Pcdhs were knocked out constitutively displayed widespread neuronal apoptosis and lethality which precluded earlier analyses of their function in self-avoidance. Consequently, Lefebvre and colleagues then used a conditional knockout approach to study γ-Pcdhs in retinal SACs and in Purkinje neurons, both of which normally exhibit dendrite self-avoidance (Lefebvre et al. 2012). Specific removal of all 22 γ-Pcdhs in these cell types resulted in striking self-avoidance defects, which are characterized by extensive isoneuronal dendrite crossing and fasciculation. Unlike removal of mammalian Dscam, cell-type-specific knockout of γ-Pcdhs did not affect overall organization of the retina but only self-avoidance of their dendrites. The described phenotypes strongly resemble the findings for *Drosophila Dscam* and indicate that γ-Pcdhs are major self-avoidance molecules in the mammalian system.

Although several downstream components have been identified for clustered Pcdhs, the signal transduction cascade required for dendrite self-avoidance is currently unknown. Proline-rich tyrosine kinase 2 (Pyk2) and focal adhesion kinase (FAK) have been shown to interact with the intracellular domains of both α- and γ-Pcdhs (Chen et al. 2009), and their functional interaction has been linked to dendritic patterning defects and spine morphogenesis in cortical neurons (Garrett et al. 2012b; Suo et al. 2012).It is however unclear if these pathways contribute to Pcdh-mediated dendrite self-avoidance.

8.2.1.4 Additional Candidate Molecules Regulating Self-Avoidance

Interestingly, dendrite self-avoidance is also found in *C. elegans* where Dscam or Pcdhs are not conserved. The PVD sensory neuron, which features an elaborate dendritic arbor, was shown to utilize UNC-6/netrin and its receptors UNC-40/DCC (deleted in colorectal cancer) and UNC-5 to repel contacting dendrites and prevent dendrite crossing (Smith et al. 2012). Similarly, cell-type-specific self-avoidance molecules have been identified in various model organisms. The IgCAM Turtle (Long et al. 2009) and the atypical protocadherin Flamingo (Matsubara et al. 2011) might play a specific role in dendrite self-avoidance in *Drosophila* class IV da neurons, similarly to Dscam/DscamL1 being required in retinal subtypes in mouse. A recent study also implicated the axon guidance receptor Robo2 and its ligand Slit2 in dendrite self-avoidance (Gibson et al. 2014). Here, loss of either the Robo2 receptor or its ligand Slit2 resulted in isoneuronal crossing of developing Purkinje neuron dendrites. Thus it seems that several molecular signals can act in the same neuronal subtypes to mediate self-avoidance, like γ-Pcdhs and Robo2/Silt2 in Purkinje neurons. This seeming redundancy of signals might be required to increase the robustness of self-avoidance or reflect molecular cross talk between different cell surface receptors. To what extent each of these signals is used in other neurons remains to be investigated.

In summary, several different cell surface molecules have been shown to be mediators of self-avoidance in different species. Several additional yet unidentified molecules might be used in specific subtypes of neurons to create non-overlapping isoneuronal dendritic fields. Independent of the molecules used, dendrites of all neuronal populations examined so far exhibit self-avoidance behavior suggesting it to be a major nervous system patterning mechanism utilized across the animal kingdom. The most compelling evidence for a universal function of Dscam in *Drosophila* and γ-Pcdhs in mammals in dendrite self-avoidance stems from their unique ability to produce a large number of unique isoforms, which will be discussed below.

8.2.2 Self- vs. Nonself-Recognition: How to Avoid Yourself But Not Others

In many systems including leech, dendritic fields of distinct neuronal subtypes typically cover overlapping territories. As one conclusion of their studies, Kramer and Kuwada already stated that "self-avoidance would require that processes projected by a neuron be able to distinguish between self and non-self" (Kramer and Kuwada 1983). A single self-avoidance molecule would thus not be able to pattern dendrites of all neurons in an organism, due to the resulting extensive cross recognition between different types of developing neurons which share overlapping dendritic territories. In simple systems individual molecules might be sufficient self-avoidance signals as shown for unc-6/Netrin in *C. elegans* PVD neurons (Smith et al. 2012). However, it is hard to envision how a single molecule can mediate self-avoidance considering the complexity of the nervous system, particularly in higher organisms. Likely every neuron has to ensure self-avoidance of its dendrites while encountering numerous other growing dendrites from their neighbors. In addition, for self-avoidance to work with high fidelity, dendrites have to use a common growth substrate allowing them to encounter and repel each other to achieve an evenly spaced nonredundant receptive field.

This raises a major problem: how can a neuron's dendrites develop independently from their neighbors using the same signal and the same growth substrate? As a consequence at least three requirements have to be met for dendrite self-avoidance to work in a complex developing nervous system:

1. Isoneuronal recognition and repulsion of "self"-branches
2. Coexistence of heteroneuronal dendrites by distinguishing "self" from "nonself"
3. A common growth substrate which ensures "self""-encounters of growing dendrites to enable self-avoidance

8.2.2.1 Creating Unique Neuronal Identity Allows Coexistence of Dendritic Fields

Drosophila Dscam Alternative Splicing and Isoform-Specific Interaction

Drosophila Dscam as the first identified prototypic self-avoidance receptor was also the first to be recognized to have the properties needed to ensure self-avoidance and to distinguish self- from nonself-dendrites. Dscam is expressed in virtually all *Drosophila* neurons and as a single molecule would not be able to mediate the distinction between self- and nonself-dendrites. However, it was noticed early on that its complex genomic locus contains numerous alternatively spliced exons that can give rise to more than 38,000 isoforms (Schmucker et al. 2000). The *Drosophila Dscam* gene encodes a single-pass transmembrane protein with conserved immunoglobulin and fibronectin domains. Three of Dscam's exons (exons 4, 6, 9) contain numerous alternatively spliced variants (12, 48, and 33 variants, respectively). In turn, each of these exons encodes a critical portion of a corresponding Ig domain (Ig2, Ig3, and Ig7), which gives rise to 19,008 ($12 \times 48 \times 33$) potential Dscam ectodomain variants (Fig. 8.2a). In addition, two alternative exons encoding the Dscam transmembrane domain (exon 17) determine the preferential targeting of Dscam to dendrites (17.1) or axons (17.2) (Wang et al. 2004), with additional alternative splicing of the intracellular domain raising the number of potential isoforms even more (Yu et al. 2009).

Remarkably, the variable exons encoding the three Ig domain segments are also critically involved in Dscam homotypic interaction (Wojtowicz et al. 2004, 2007). Identity of each of the alternatively spliced Ig domains is generally required for strong molecular interaction between two Dscam molecules (Fig. 8.2b). This was shown in elegant biochemical studies testing the interaction of Dscam ectodomain variants in vitro, which revealed striking isoform-specific interaction: each Dscam isoform tested so far exhibits almost exclusive homophilic interaction between identical but not divergent isoforms (Fig. 8.2c, Wojtowicz et al. 2004, 2007). In this manner, the 19,008 ectodomain isoforms encoded by the Dscam gene give rise to 19,008 homotypic interaction pairs.

Subsequent structural analysis of Dscam explained these stringent binding properties that lead to extremely high specificity (Meijers et al. 2007; Sawaya et al. 2008). Two interacting Dscam ectodomains form a double-S-like structure in which the alternatively spliced regions within Ig2, Ig3, and Ig7 are directly involved in the interaction (Sawaya et al. 2008). Each of these Ig domains binds to its cognate counterpart of the opposing Dscam molecule and thereby forms the contact surface critical for Dscam homodimerization. Dscam homophilic interaction relies on the electrostatic and sterical properties of Ig2/3/7 isoforms, as only identical isoforms offer the proper conformation and charge to promote strong interaction (Fig. 8.2b). This complex and sterically constrained structural interaction offers an explanation for why even the substitution of a single Ig variant generally leads to the disruption of stable binding.

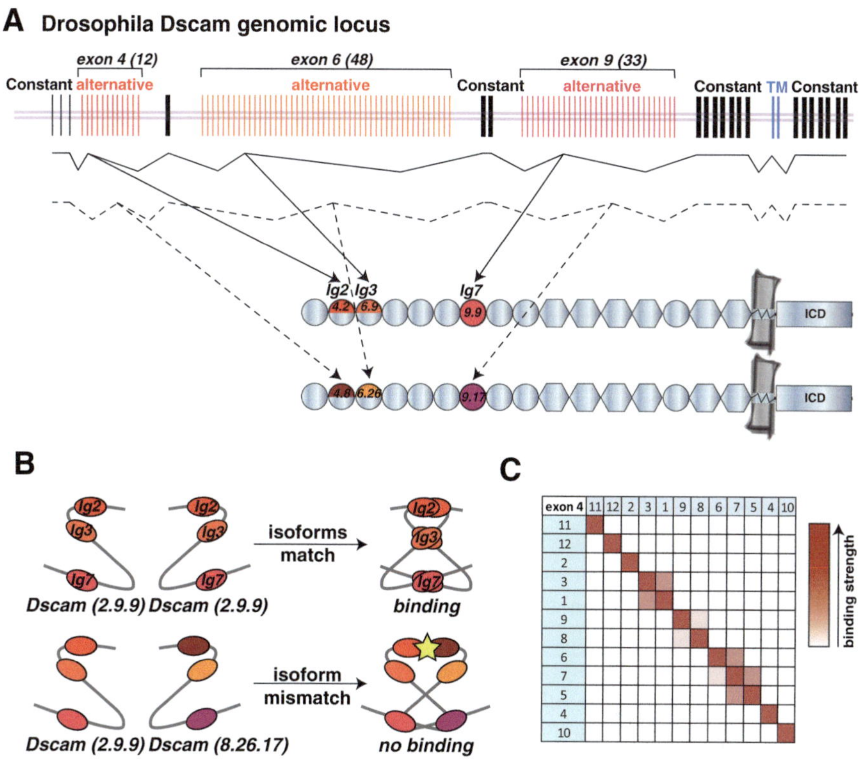

Fig. 8.2 *Drosophila* Dscam genomic organization and isoform-specific interaction. (**a**) Dscam genomic locus with constant and alternatively spliced exons. Exons 4, 6, and 9 contain 12, 48, and 33 alternative exons, respectively, which can give rise to 19,008 different isoforms by mutually exclusive splicing. Additionally, the transmembrane (TM) domain is encoded by two alternative exons determining preferential axonal or dendritic localization of Dscam. Exons 4, 6, and 9 encode portions of Dscam Ig2, Ig3, and Ig7 domains, which are critical for homotypic interaction. (**b**) Dscam isoforms display virtually exclusive homotypic interaction. Identical isoforms form a molecular complex by stable interaction between the Ig2, Ig3, and Ig7 domains. Structural analysis revealed that the Dscam ectodomain adopts a double-S-shaped conformation to accommodate specific Ig domain interactions. Nonidentical isoforms do not form complementary surfaces within the alternatively spliced Ig domains and thus generally do not show strong binding affinities. (**c**) Exemplary interaction matrix of Dscam exon 4 splice variants based on Wojtowicz et al. (2007). Binding properties of Dscam ectodomains containing the 12 different exon 4 variants were tested in vitro. In general, only identical isoforms showed high binding affinities, with some lower-affinity binding between more closely related isoforms. Adapted from Zipursky and Sanes (2010).

The structural knowledge of the Dscam ectodomain subsequently allowed the modeling and generation of chimeric Dscam molecules which exhibit heterophilic instead of homophilic binding properties (Wu et al. 2012). Interestingly, these chimeric molecules did not support self-avoidance in vivo but were able to mediate recognition and repulsion of heterotypic neurons expressing complementary pairs of chimeric Dscam. Overall, the structural and functional analysis showed that isoform-specific homotypic interaction of Dscam is strictly required to support self-avoidance.

Dscam Isoform Diversity Is Required in *Drosophila* In Vivo

How much does *Dscam* function actually rely on its diversity? The requirement of Dscam alternative splicing has been tested extensively in *Drosophila* in vivo.

In the larval PNS, expression of single Dscam isoforms is able to rescue dendrite morphology of da neurons at the single-cell level (Hughes et al. 2007; Matthews et al. 2007; Soba et al. 2007). All isoforms tested so far support dendrite self-avoidance at the single-cell level suggesting there is no intrinsic difference in the activity of Dscam ectodomain splice variants. However, as soon as the same isoform is expressed in two neighboring neurons which normally have overlapping dendritic territories, it results in recognition and repulsion of heteroneuronal dendrites (Fig. 8.3c). Thus, neurons normally not recognizing each other partition their receptive fields in a tiled manner simply by expressing the same Dscam isoform.

This raises the question of how many Dscam isoforms are actually needed in vivo. Animals expressing only one Dscam isoform under a *Dscam* promoter showed strong recognition and repulsion of heteroneuronal dendrites (Soba et al. 2007). This finding illustrates that Dscam diversity is required for the discrimination between self- and nonself-dendrites to allow coexistence of heteroneuronal dendritic fields. In subsequent studies this idea was explored in more detail. Reducing Dscam ectodomain diversity in flies from 19,008 to less than 5000 or several hundreds revealed that most neuronal populations examined require at least several thousand isoforms to prevent developmental defects due to inappropriate recognition of neighboring neurons (Hattori et al. 2009). Even in a relatively simple system like the *Drosophila* PNS, where less than ten neurons share overlapping dendritic territories, significant Dscam isoform diversity is needed: reducing diversity to about 1100 possible isoforms already resulted in strong coexistence defects of heteroneuronal dendritic fields (Fig. 8.3c).

Taken together, Dscam isoform diversity is critical for *Drosophila* neurons to discriminate self- from nonself-dendrites during development.

Drosophila Dscam Is Probabilistically Spliced and Creates a Stochastic Neuronal Identity Code

Drosophila Dscam is expressed throughout the nervous system, yet its remarkable isoform diversity seems to be sufficient to ensure dendrite self-avoidance in each individual neuron. This also implies that most if not all possible splice variants need to be utilized and each neuron should only express a subset of isoforms to give it a unique identity based on its Dscam profile.

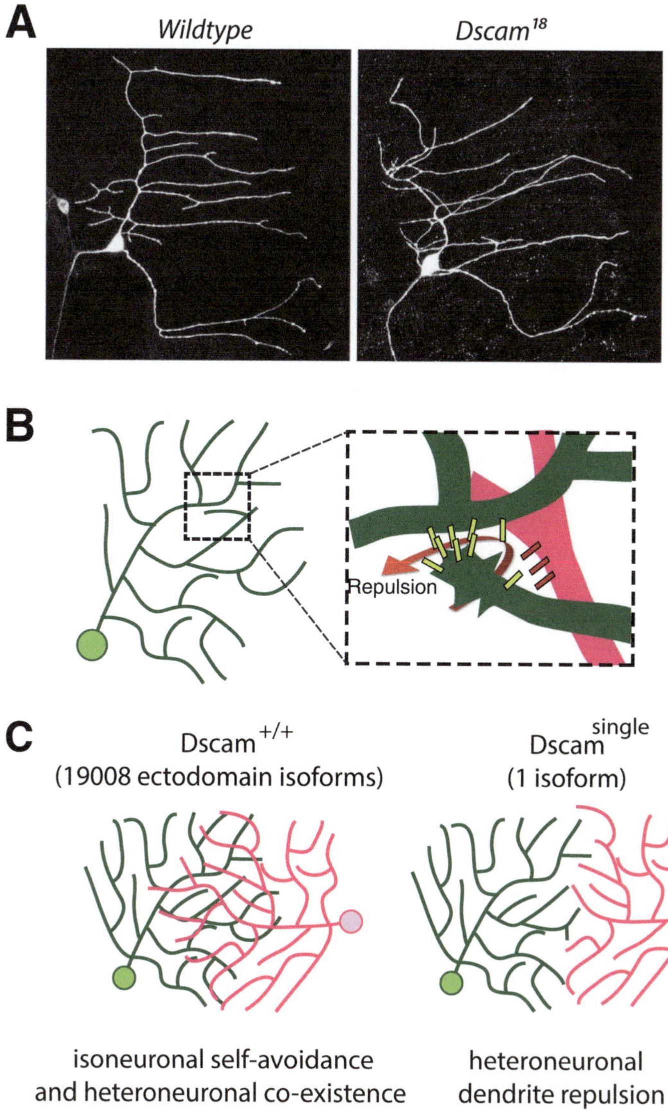

Fig. 8.3 *Drosophila* Dscam-mediated dendrite self-avoidance in vivo. (**a**) Exemplary images of wild-type and Dscam mutant class I (C1) da neurons show dendrite self-avoidance defects when Dscam function is lost, evident by extensive isoneuronal dendrite crossing. (**b**) Model for Dscam-mediated dendrite self-avoidance. Isoneuronal dendrites contain an identical Dscam isoform repertoire, which results in recognition and repulsion of dendrites growing close enough. Heteroneuronal dendrites (shown in *magenta*) express distinct Dscam isoforms and are therefore not repelled. (**c**) Wild-type *Drosophila* neurons probabilistically express a small subset of the 19,008 possible Dscam ectodomain variants thus creating a unique Dscam-based identity. This diversity is sufficient to distinguish self- from nonself-dendrites, resulting in proper self-avoidance and coexistence. Reducing the number of possible isoforms results in cross recognition and repulsion of heterotypic dendrites, which would preclude the coexistence of dendritic fields. Adapted from Grueber and Sagasti (2010).

Indeed, microarray analyses revealed that most Dscam isoforms are actually produced in vivo, both in neurons and in immune cells (Neves et al. 2004; Watson et al. 2005; Zhan et al. 2004). Interestingly, analysis of individual neuronal populations or even single cells revealed that Dscam alternative exons are largely stochastically spliced. Typically, each neuron expresses a random subset of 10–50 isoforms, although a cell-type-specific bias for certain exons could be observed (Neves et al. 2004; Zhan et al. 2004).

Later studies in *Drosophila* da neurons exposed the temporal and cellular specificity of Dscam expression. Da neurons offer the unique opportunity to follow and compare individual cells and cell types in the same animal throughout larval development. In a well-designed study, it was shown that individual sensory da neurons express Dscam probabilistically and even neurons of the same subtype in the same animal show highly divergent exon usage (Miura et al. 2013). Strikingly, exon utilization changed during the course of development in individual neurons suggesting random and temporally not restricted splicing of Dscam isoforms. As a particular neuron expresses only a small subset of Dscam isoforms at any given time, probabilistic splicing of Dscam ensures a unique neuronal identity during the temporal window of dendrite development.

In summary, *Drosophila Dscam* proved to be the prototypic molecule enabling dendrite self-avoidance and self-/nonself-discrimination by probabilistic splicing, stochastic isoform expression, and isoform-specific recognition and repulsion of dendrites.

Protocadherins Create Neuronal Uniqueness in Mammals

Unlike *Drosophila Dscam*, its mammalian counterpart lacks isoform diversity. Despite the important role of mouse Dscam in retinal cell lamination and self-avoidance, it likely cannot support self- vs. nonself-discrimination of dendrites. Recent studies point to clustered Pcdhs as functional analogs of *Drosophila* Dscam in mammals (Lefebvre et al. 2012; Thu et al. 2014). They do not share any structural similarity with *Drosophila Dscam*, yet they seem to have evolved in parallel as functional counterparts.

One of the three prerequisites for dendrite self-avoidance is to create neuronal uniqueness in order to avoid cross recognition between neighboring cells. Clustered Pcdhs bring molecular diversity and stochastic expression to the plate. It was first shown for α-Pcdhs, originally identified as cadherin-related neuronal receptors (CNRs), that they are differentially expressed in individual neurons (Kohmura et al. 1998). Subsequently, analysis of the genomic organization of the Pcdh cluster led to the identification of all α-, β-, and γ-Pcdhs in human and mouse (Wu and Maniatis 1999; Wu et al. 2001). The three clustered Pcdh subtypes are phylogenetically highly related and share a similar genomic organization: for α- and γ-Pcdhs, variable exons encode the entire extracellular and transmembrane domain, which are spliced with constant exons encoding a common portion of the intracellular domain (Fig. 8.4a). β-Pcdhs in turn consist of 22 individual exons encoding the

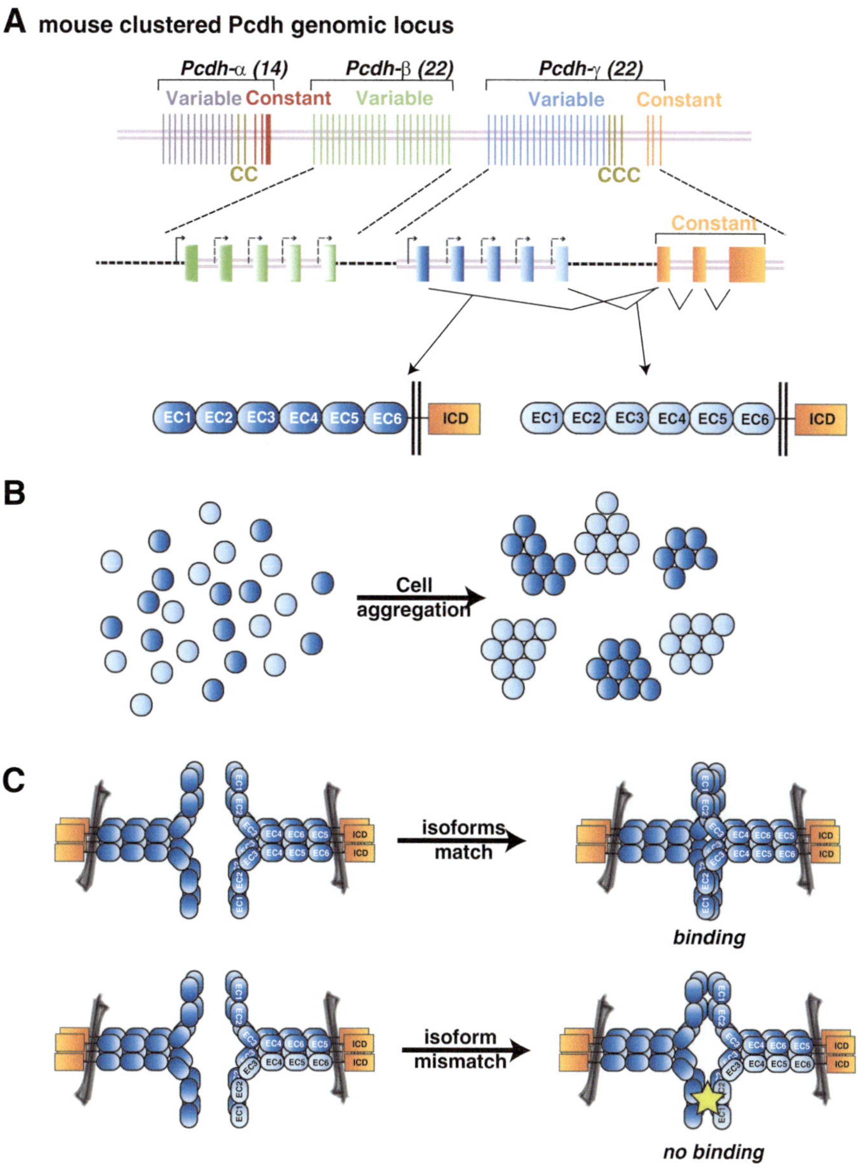

Fig. 8.4 Clustered Pcdh genomic organization and isoform-specific interaction. (**a**) Clustered Pcdhs reside in the same genomic locus encoding 14 α-, 22 β-, and 22 γ-Pcdhs in mouse. α- and γ-Pcdhs contain variable exons encoding the entire ectodomain and constitutively spliced exons encoding the intracellular domains. Individual β-Pcdh exons encode the entire protein. Each variable exon of Pcdhs is preceded by its own promoter. Monoallelic and stochastic expression of only small subset of Pcdhs gives rise to neurons with a unique Pcdh-based identity (with the exception of C-type Pcdhs, which are constitutively expressed (indicated by "C")). Adapted from Zipursky and Grueber (2013). (**b**) Isoform-specific interaction properties of Pcdhs were shown using cellular aggregation assays (based on Schreiner and Weiner 2010; Thu et al. 2014). Only cells expressing the same γ-Pcdhs were able to interact and form clusters showing that Pcdhs interact in an isoform-specific manner. (**c**) Model of Pcdh-based interaction (based on Schreiner and Weiner 2010; Thu et al. 2014).

entire transmembrane proteins with unique intracellular domains. In addition, α- and γ-Pcdh genomic clusters contain divergent isoforms (2 and 3, respectively) termed C-type, which are more closely related to each other than to Pcdhs of their respective cluster.

Strikingly, each individual variable exon of clustered Pcdhs is preceded by its own promoter, which was shown for α- and γ-Pcdh to result in promoter choice-based expression rather than *trans*-splicing (Ribich et al. 2006; Tasic et al. 2002; Wang et al. 2002c). In this way clustered Pcdhs in mouse can give rise to 58 related proteins with distinct extracellular domains (Fig. 8.4a).

Pcdh promoter choice in vivo seems to be highly stochastic. Strong evidence was provided by elegant experiments using in situ hybridization of mouse brain sections and profiling of single isolated neurons by RT-PCR (Esumi et al. 2005; Hirano et al. 2012; Kaneko et al. 2006; Noguchi et al. 2009). Taking advantage of mouse strain polymorphisms, Esumi and colleagues used crossing and F1 analysis of two distinct inbred mouse strains to reveal monoallelic and combinatorial expression of all α-Pcdhs in individual Purkinje cells. Later studies revealed similar stochastic expression properties for β- (Hirano et al. 2012) and γ-Pcdhs (Kaneko et al. 2006). The exception are C-type Pcdhs, which are constitutively and biallelically expressed and seem to have a very general role in neuronal survival rather than self-avoidance, at least in the case of C-type γ-Pcdhs (Chen et al. 2012).

How expression of individual Pcdh isoforms is regulated and restricted is not precisely known. However, the CTCF/cohesin complex was proposed to play a crucial role in facilitating expression of α-Pcdhs by DNA looping and interaction with several individual promoters (Guo et al. 2012; Monahan et al. 2012). In addition, *de novo* methylation of Pcdh promoters by Dnmt3b seems to be an important regulatory mechanism during early development to restrict the number of expressed Pcdh isoforms in individual neurons (Toyoda et al. 2014). Although the precise number of expressed Pcdh isoforms per neuron is unknown, it was estimated that individual cells express 10 random Pcdh isoforms, plus the 5 C-type isoforms which are constitutively expressed in all neurons (Yagi 2012).

Protocadherin-Based Interaction and Neuronal Identity Code

How can a set of 58 Pcdh molecules in mouse create molecular diversity comparable to *Drosophila* Dscam with its 19,008 ectodomain isoforms, which would presumably be required to create neuronal uniqueness during mammalian brain development? Several recent studies provide evidence that Pcdhs form multimeric complexes that result in unique interaction properties.

Fig. 8.4 (continued) Pcdhs form cis-tetramers and can interact with an identical complex via its EC1–3 domains. Substituting one isoform is sufficient to prevent interaction with an isoform-mismatched complex showing strong preference for homotypic interaction of isoform-matched complexes

All clustered Pcdhs contain six extracellular cadherin domains (EC1–6) that can engage in cis- and trans-interactions. Schreiner and Weiner were the first to provide evidence that g-Pcdh isoforms can form cis-tetramers, which then promote homophilic interaction with identical tetramers on opposing cells (Fig. 8.4b, c (Schreiner and Weiner 2010). The EC2-EC3 domain interface was determined to mediate homophilic binding specificity distinguishing γ-Pcdhs from classical cadherins, where the EC1 domain determines binding specificity. Interestingly, the EC2 and EC3 domains also show the highest sequence diversity within the clustered Pcdh isoforms suggesting they are critically involved in isoform-specific binding. Schreiner and Weiner proposed that this could lead to 234,256 distinct adhesive interfaces assembled by g-Pcdh cis-tetramers with homophilic trans-interaction properties.

Extending these results, a comprehensive study by Thu and colleagues analyzed the interaction properties of all clustered Pcdhs in a cell culture model (Thu et al. 2014). With one exception, all β- and γ-Pcdhs exhibit highly specific homophilic trans-interaction as shown by promoting aggregation of cells expressing the same isoform. Even highly similar isoforms displaying up to 97 % sequence identity within the critical EC2-EC3 domains did not show binding. Thus β- and γ-Pcdhs show similarly strict isoform-specific binding properties as *Drosophila* Dscam (Fig. 8.4b, c).

In contrast, α-Pcdhs are not able to mediate cellular interactions by themselves, as they are not transported to the cell surface without the help of γ-Pcdhs (Murata et al. 2004). However, coexpression of β- or γ-Pcdhs is sufficient to transport α-Pcdhs to the cell surface, a process that requires Pcdh interaction via the EC6 domain (Thu et al. 2014). In a recent study, which solved the structure of the EC1-EC3 region of cPcdhs, it was shown that intercellular *trans*-interaction between Pcdhs relies on antiparallel interaction between their EC1-EC4 domains (Rubinstein et al. 2015).

Strikingly, α-/β-/γ-Pcdhs can also form hetero-multimeric cell surface complexes which are able to mediate cell recognition. When coexpressed, the formed α-/β-/γ-Pcdh complex allows recognition of cells containing the same set of isoforms only (Thu et al. 2014). Even the substitution of a single isoform resulted in loss of cross recognition between those cells suggesting that a tetrameric complex of different Pcdh isoforms creates a unique interaction interface able to recognize only itself. These findings might explain how a set of only about 50 molecules can generate a combinatorial diversity code that extends in the range of billions across all clustered Pcdhs assuming they form functional hetero-tetramers (Yagi 2012).

Thus the combinatorial and stochastic expression of α-, β-, and γ-Pcdhs by probabilistic promoter choice and resulting recognition complex formation might be able to create a unique cell surface code in mammalian neurons. To date, clustered Pcdhs are the best candidates as general mediators of dendrite self-avoidance in vertebrates.

Protocadherins Promote Dendrite Self-Avoidance and Self-/Nonself-Discrimination In Vivo

As detailed above, clustered Pcdhs are likely able to provide a unique cell surface code of receptors which combine the properties of stochastic expression and isoform-specific cell surface recognition required of true self-avoidance receptors. Although the generality of their function in dendrite self-avoidance remains to be shown, exemplary key findings in SACs and Purkinje neurons suggest so. To fulfill the requirements, Pcdhs should not only mediate recognition and repulsion of isoneuronal dendrites but also allow discrimination of self- from nonself-dendrites. Lefebvre and colleagues showed that the cell-type-specific loss of all 22 γ-Pcdhs results in dendrite self-avoidance defects in these neuronal populations suggesting that the lack of recognition and repulsion of self-dendrites is due to the absence of γ-Pcdhs (Lefebvre et al. 2012). These defects could be fully rescued when a single γ-Pcdh isoform was expressed in SACs suggesting that γ-Pcdhs mediate self-avoidance cell autonomously. Importantly, however, this study also showed that expression of the same γ-Pcdh isoform in neighboring SACs that usually share overlapping dendritic territories results in decreased interaction of heteroneuronal dendrites (Fig. 8.5). These findings imply that SAC dendrites expressing the same γ-Pcdh isoforms are able to recognize and repel each other. As γ-Pcdhs normally display probabilistic expression of individual isoforms in SACs, it supports the requirement of Pcdh isoform diversity that allows dendrites of interacting neighboring SACs to distinguish self from nonself.

Taken together, the results of this study point to a pivotal role of Pcdhs in dendrite self-avoidance in mammals analogously to that of Dscam in *Drosophila*.

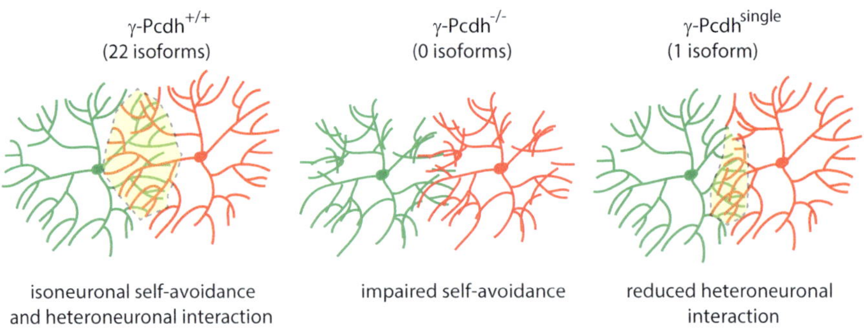

Fig. 8.5 γ-Pcdhs and their diversity are required for dendrite self-avoidance and heteroneuronal interaction (based on Lefebvre et al. 2012). SACs in the mammalian retina display clear dendrite self-avoidance but also specific heteroneuronal dendrite interaction. Removal of all γ-Pcdhs from SACs results in dendrite self-avoidance defects. The expression of single variants of γ-Pcdhs rescues dendrite self-avoidance but results in reduced heteroneuronal dendrite interaction between neighboring SACs

8.2.2.2 Ensuring Self-Encounters of Growing Dendrites by Substrate Interaction

Self-avoidance relies on transmembrane proteins like Dscam or Pcdhs encountering each other on the surface of two incoming "self"-dendrites and inducing recognition and repulsion. How are robust encounters of dendrites ensured considering their growth in a complex three-dimensional tissue environment?

Some evidence was again provided by studies on *Drosophila* sensory da neurons. Da neuron dendrites are confined to a two-dimensional growth surface defined by the extracellular matrix (ECM) and an underlying epithelium (Han et al. 2012; Kim et al. 2012). In this system, neuronal integrins are required to maintain dendrite-ECM interaction and thus ensure the common growth substrate for all da neuron dendrites. Integrins are receptors for laminins, which are the ECM components required for dendrite adhesion in this system. Loss of the α-integrin *mew* (*multiple edematous wing*) or β-integrin *mys* (*myospheroid*) results in dendrite detachment and growth within the epithelial cell layer, which encloses growing dendrites. Thus the use of different growth substrates (ECM and epithelial cells) precludes self-avoidance since isoneuronal dendrites do not encounter each other anymore. Integrins function independently of *Dscam* in this system yet they illustrate the requirement of creating a common growth substrate for dendritic encounters to happen. Additional signals and cell surface receptors like Ret (Soba et al. 2015) and Raw (Lee et al. 2014) seem to regulate integrin-based dendrite-ECM adhesion in da neurons suggesting that numerous molecular players are in place that ensure structured growth on the same substrate, thus facilitating self-avoidance of dendrites (Fig. 8.6).

How a uniform growth substrate for dendrites is created in other systems is so far not entirely clear. Many types of neurons including retinal ganglion cells form a

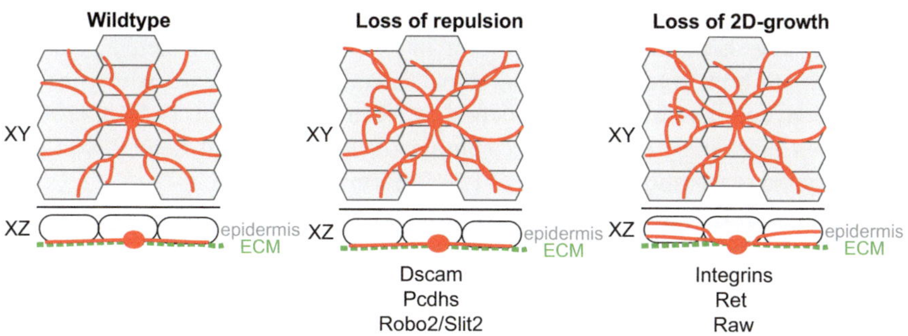

Fig. 8.6 Coherent substrate growth contributes to robust dendrite self-avoidance. Dendrite self-avoidance requires robust encounters of growing branches. All the neuronal subtypes studied so far (e.g., *Drosophila* da neurons, SACs, Purkinje neurons) display highly stratified dendritic fields. Dendrites of *Drosophila* da neurons in particular are confined to a 2D growth space between the ECM and an epithelial cell layer. Loss of isoneuronal dendrite repulsion (e.g., Dscam, Pcdhs) results in self-avoidance defects; however, loss of substrate adhesion (e.g., integrins, Ret) results in similar disorganization of dendritic fields due to the lack of dendritic encounters

single-cell layer and exhibit laminated dendritic fields that are restricted to a specific layer. Interestingly, laminins play an important role in ganglion cell positioning during development (Edwards et al. 2010; Pinzón-Duarte et al. 2010). Thus integrins are likely candidates mediating uniform growth substrate usage, as they are universally conserved cell-matrix adhesion proteins and laminin receptors. A recent study indeed showed that β1-integrin is developmentally required for the formation of the single ganglion cell layer through p130Cas signaling (Riccomagno et al. 2014). Semaphorin-plexin signaling might serve a related purpose and has been shown to be required for the separation and stratification of ON and OFF SAC dendrites (Sun et al. 2013). Interestingly, ON SAC dendrites show reduced fields and an increased number of self-crossing suggesting that similarly to *Drosophila* da neurons, failure to properly use a common growth substrate impairs self-avoidance. Thus it is likely that the ECM and specific molecular interactions ensure robust self-encounters of growing dendrites by creating a 2D growth environment (Fig. 8.6). This is likely also true for highly complex dendritic arbors like those of Purkinje neurons which also exhibit self-avoidance (Gibson et al. 2014; Lefebvre et al. 2012). How a common growth substrate for robust dendrite-dendritic encounters is ensured in these and other neuronal populations remains to be investigated.

8.3 Conclusion and Outlook

The phenomenon of dendrite self-avoidance seems to be generally conserved in all species analyzed to date, from *C. elegans*, medicinal leech, and *Drosophila to* mouse. However, only organisms with more complex nervous systems including *Drosophila* and mammals encode proteins that can create unique cellular identities through isoform diversity. Both *Drosophila* and mouse use a combinatorial code of self-avoidance molecules which are stochastically or probabilistically expressed with only few isoforms per cell. Dscam in *Drosophila* and clustered Pcdhs in mammals are both able to form homotypic interaction complexes which lead to recognition and repulsion of self-dendrites. Due to the strict isoform-specific interaction properties, unwanted recognition of heteroneuronal dendrites is precluded by the stochastic expression of few isoforms per neuron.

It seems that the strategy of creating neuronal uniqueness by stochastic expression of cell surface recognition receptors is conserved, even if the molecules utilized differ in insects and vertebrates. Nonetheless, mechanistically they serve the same purpose allowing the repulsion of "self"-dendrites in a strictly cell-autonomous manner. Creating a unique neuronal identity seems to be a key feature for self-avoidance and dendritic patterning in general and should provide the basis for future studies on how self-avoidance of dendrites in conjunction with other signals sculpts dendritic fields and circuits.

Although the recent progress detailed above has firmly established self-avoidance as a universal mechanism in nervous system development, numerous open questions remain.

First, several potential self-avoidance molecules have been identified in the same systems, e.g., Robo2/Slit2 and γ-protocadherins in Purkinje neurons. Both are required for dendrite self-avoidance and have been shown to act independently (Gibson et al. 2014). Similarly, multiple self-avoidance molecules besides Dscam have been proposed for *Drosophila* sensory dendritic patterning, including the IgSF Turtle (Long et al. 2009) and protocadherin Flamingo (Matsubara et al. 2011). It is possible that several avoidance machineries are required to shape dendritic fields with high fidelity. Alternatively, additional mechanisms like substrate interaction and nonautonomous protein functions contribute to dendrite self-avoidance in individual neuronal populations, similarly to directed guidance of dendrite growth by interaction with pre-patterned ligand complexes in the surrounding tissue as shown in *C. elegans* sensory neurons (Dong et al. 2013; Salzberg et al. 2013).

Second, very little is known about the downstream signaling required for self-avoidance. Dscam can signal via the Pak; however, this seems to be restricted to Dscam's role in axon guidance but not self-avoidance (Hughes et al. 2007). Similarly, several downstream components including Pyk2 and FAK have been identified for clustered Pcdhs (Chen et al. 2009; Garrett et al. 2012b; Suo et al. 2012), yet the signal transduction cascade required for dendrite self-avoidance is currently unknown. It remains to be addressed which downstream signals convey the repulsive signals required for self-avoidance.

Third, how is contact-dependent homophilic binding translated into repulsion of isoneuronal dendrites? It seems counterintuitive that proteins like Dscam and protocadherins, which exhibit adhesive properties in vitro, actually mediate dendrite repulsion in vivo. So far it is unclear how an initially attractive interaction of transmembrane proteins is translated into dendrite repulsion. The current data indicates that mammalian Dscam (Schramm et al. 2012) and also Pcdhs (Hambsch et al. 2005) can be cleaved extracellularly by proteases, which could be a possible mechanism to release interaction and initiate cytoskeletal rearrangement required for dendrite repulsion.

Fourth, robust encounters of isoneuronal dendrites require a uniform substrate that ensures self-encounters of growing dendrites. Our understanding of the interaction of dendrites with its in vivo growth substrate is just at the beginning. However, it is likely a major contributor to self-avoidance and dendritic patterning in general and might involve additional adhesion and signaling mechanism besides integrins and semaphorin/plexin.

Taken together, layering of several mechanisms and intricate cellular regulation of specific homotypic interaction complexes contributes to dendrite self-avoidance in all systems investigated so far. Future studies will likely uncover additional requirements and cell-type-specific differences, which should reveal the full impact of dendrite self-avoidance on brain development.

Acknowledgment I would like to thank Dörte Clausen for the artwork and J. Parrish for the critical reading of the manuscript.

References

Cameron S, Rao Y (2010) Molecular mechanisms of tiling and self-avoidance in neural development. Mol Brain 3:28

Chen WV, Maniatis T (2013) Clustered protocadherins. Development 140:3297–3302

Chen J, Lu Y, Meng S, Han M-H, Lin C, Wang X (2009) Alpha- and gamma-protocadherins negatively regulate PYK2. J Biol Chem 284:2880–2890

Chen WV, Alvarez FJ, Lefebvre JL, Friedman B, Nwakeze C, Geiman E, Smith C, Thu CA, Tapia JC, Tasic B et al (2012) Functional significance of isoform diversification in the protocadherin gamma gene cluster. Neuron 75:402–409

Dong X, Liu OW, Howell AS, Shen K (2013) An extracellular adhesion molecule complex patterns dendritic branching and morphogenesis. Cell 155:296–307

Edwards MM, Mammadova-Bach E, Alpy F, Klein A, Hicks WL, Roux M, Simon-Assmann P, Smith RS, Orend G, Wu J et al (2010) Mutations in Lama1 disrupt retinal vascular development and inner limiting membrane formation. J Biol Chem 285:7697–7711

Esumi S, Kakazu N, Taguchi Y, Hirayama T, Sasaki A, Hirabayashi T, Koide T, Kitsukawa T, Hamada S, Yagi T (2005) Monoallelic yet combinatorial expression of variable exons of the protocadherin-alpha gene cluster in single neurons. Nat Genet 37:171–176

Fuerst PG, Burgess RW (2009) Adhesion molecules in establishing retinal circuitry. Curr Opin Neurobiol 19:389–394

Fuerst PG, Koizumi A, Masland RH, Burgess RW (2008) Neurite arborization and mosaic spacing in the mouse retina require DSCAM. Nature 451:470–474

Fuerst PG, Bruce F, Tian M, Wei W, Elstrott J, Feller MB, Erskine L, Singer JH, Burgess RW (2009) DSCAM and DSCAML1 function in self-avoidance in multiple cell types in the developing mouse retina. Neuron 64:484–497

Fuerst PG, Bruce F, Rounds RP, Erskine L, Burgess RW (2012) Cell autonomy of DSCAM function in retinal development. Dev Biol 361:326–337

Garrett AM, Tadenev ALD, Burgess RW (2012a) DSCAMs: restoring balance to developmental forces. Front Mol Neurosci 5:86

Garrett AM, Schreiner D, Lobas MA, Weiner JA (2012b) γ-Protocadherins control cortical dendrite arborization by regulating the activity of a FAK/PKC/MARCKS signaling pathway. Neuron 74:269–276

Gibson DA, Tymanskyj S, Yuan RC, Leung HC, Lefebvre JL, Sanes JR, Chédotal A, Ma L (2014) Dendrite self-avoidance requires cell-autonomous slit/robo signaling in cerebellar purkinje cells. Neuron 81:1040–1056

Grueber WB, Sagasti A (2010) Self-avoidance and tiling: mechanisms of dendrite and axon spacing. Cold Spring Harb Perspect Biol 2:a001750

Grueber WB, Jan LY, Jan YN (2002) Tiling of the drosophila epidermis by multidendritic sensory neurons. Development 129:2867–2878

Guo Y, Monahan K, Wu H, Gertz J, Varley KE, Li W, Myers RM, Maniatis T, Wu Q (2012) CTCF/cohesin-mediated DNA looping is required for protocadherin α promoter choice. Proc Natl Acad Sci U S A 109:21081–21086

Hambsch B, Grinevich V, Seeburg PH, Schwarz MK (2005) {gamma}-Protocadherins, presenilin-mediated release of C-terminal fragment promotes locus expression. J Biol Chem 280: 15888–15897

Han C, Wang D, Soba P, Zhu S, Lin X, Jan LY, Jan Y-N (2012) Integrins regulate repulsion-mediated dendritic patterning of drosophila sensory neurons by restricting dendrites in a 2D space. Neuron 73:64–78

Hattori D, Chen Y, Matthews BJ, Salwinski L, Sabatti C, Grueber WB, Zipursky SL (2009) Robust discrimination between self and non-self neurites requires thousands of Dscam1 isoforms. Nature 461:644–648

Hayashi S, Takeichi M (2015) Emerging roles of protocadherins: from self-avoidance to enhancement of motility. J Cell Sci 128:1–10

Hirano K, Kaneko R, Izawa T, Kawaguchi M, Kitsukawa T, Yagi T (2012) Single-neuron diversity generated by Protocadherin-β cluster in mouse central and peripheral nervous systems. Front Mol Neurosci 5:90

Hughes ME, Bortnick R, Tsubouchi A, Bäumer P, Kondo M, Uemura T, Schmucker D (2007) Homophilic Dscam interactions control complex dendrite morphogenesis. Neuron 54:417–427

Jan YN, Jan LY (2010) Branching out: mechanisms of dendritic arborization. Nat Rev Neurosci 11:316–328

Kaneko R, Kato H, Kawamura Y, Esumi S, Hirayama T, Hirabayashi T, Yagi T (2006) Allelic gene regulation of Pcdh-α and Pcdh-γ clusters involving both monoallelic and biallelic expression in single Purkinje cells. J Biol Chem 281:30551–30560

Keeler AB, Molumby MJ, Weiner JA (2015) Protocadherins branch out: multiple roles in dendrite development. Cell Adh Migr 9:214–226

Kim ME, Shrestha BR, Blazeski R, Mason CA, Grueber WB (2012) Integrins establish dendrite-substrate relationships that promote dendritic self-avoidance and patterning in drosophila sensory neurons. Neuron 73:79–91

Kohmura N, Senzaki K, Hamada S, Kai N, Yasuda R, Watanabe M, Ishii H, Yasuda M, Mishina M, Yagi T (1998) Diversity revealed by a novel family of cadherins expressed in neurons at a synaptic complex. Neuron 20:1137–1151

Kramer AP, Kuwada JY (1983) Formation of the receptive fields of leech mechanosensory neurons during embryonic development. J Neurosci 3:2474–2486

Kramer AP, Stent GS (1985) Developmental arborization of sensory neurons in the leech Haementeria ghilianii. II. Experimentally induced variations in the branching pattern. J Neurosci 5:768–775

Lawrence Zipursky S, Grueber WB (2013) The molecular basis of self-avoidance. Annu Rev Neurosci 36:547–568

Lee J, Peng Y, Lin W-Y, Parrish JZ (2014) Coordinate control of terminal dendrite patterning and dynamics by the membrane protein Raw. Development 142:1–12

Lefebvre JL, Zhang Y, Meister M, Wang X, Sanes JR (2008) Gamma-Protocadherins regulate neuronal survival but are dispensable for circuit formation in retina. Development 135: 4141–4151

Lefebvre JL, Kostadinov D, Chen WV, Maniatis T, Sanes JR (2012) Protocadherins mediate dendritic self-avoidance in the mammalian nervous system. Nature 488:517–521

Long H, Ou Y, Rao Y, van Meyel DJ (2009) Dendrite branching and self-avoidance are controlled by turtle, a conserved IgSF protein in drosophila. Development 136:3475–3484

Matsubara D, Horiuchi S-Y, Shimono K, Usui T, Uemura T (2011) The seven-pass transmembrane cadherin Flamingo controls dendritic self-avoidance via its binding to a LIM domain protein, Espinas, in Drosophila sensory neurons. Genes Dev 25:1982–1996

Matthews BJ, Kim ME, Flanagan JJ, Hattori D, Clemens JC, Zipursky SL, Grueber WB (2007) Dendrite self-avoidance is controlled by Dscam. Cell 129:593–604

Meguro R, Hishida R, Tsukano H, Yoshitake K, Imamura R, Tohmi M, Kitsukawa T, Hirabayashi T, Yagi T, Takebayashi H, Shibuki K (2015) Impaired clustered protocadherin-α leads to aggregated retinogeniculate terminals and impaired visual acuity in mice. J Neurochem 133:66–72

Meijers R, Puettmann-Holgado R, Skiniotis G, Liu J, Walz T, Wang J, Schmucker D (2007) Structural basis of Dscam isoform specificity. Nature 449:487–491

Miura SK, Martins A, Zhang KX, Graveley BR, Zipursky SL (2013) Probabilistic splicing of Dscam1 establishes identity at the level of single neurons. Cell 155:1166–1177

Monahan K, Rudnick ND, Kehayova PD, Pauli F, Newberry KM, Myers RM, Maniatis T (2012) Role of CCCTC binding factor (CTCF) and cohesin in the generation of single-cell diversity of Protocadherin- gene expression. Proc Natl Acad Sci 109:9125–9130

Morishita H, Yagi T (2007) Protocadherin family: diversity, structure, and function. Curr Opin Cell Biol 19:584–592

Murata Y, Hamada S, Morishita H, Mutoh T, Yagi T (2004) Interaction with protocadherin-gamma regulates the cell surface expression of protocadherin-alpha. J Biol Chem 279: 49508–49516

Neves G, Zucker J, Daly M, Chess A (2004) Stochastic yet biased expression of multiple Dscam splice variants by individual cells. Nat Genet 36:240–246

Nicholls JG, Baylor DA (1968) Specific modalities and receptive fields of sensory neurons in CNS of the leech. J Neurophysiol 31:740–756

Noguchi Y, Hirabayashi T, Katori S, Kawamura Y, Sanbo M, Hirabayashi M, Kiyonari H, Uchimura A, Yagi T (2009) Total expression and dual gene-regulatory mechanisms maintained in deletions and duplications of the Pcdha cluster. J Biol Chem 284:32002–32014

Pinzón-Duarte G, Daly G, Li YN, Koch M, Brunken WJ (2010) Defective formation of the inner limiting membrane in laminin beta2- and gamma3-null mice produces retinal dysplasia. Invest Ophthalmol Vis Sci 51:1773–1782

Ribich S, Tasic B, Maniatis T (2006) Identification of long-range regulatory elements in the protocadherin-alpha gene cluster. Proc Natl Acad Sci U S A 103:19719–19724

Riccomagno MM, Sun LO, Brady CM, Alexandropoulos K, Seo S, Kurokawa M, Kolodkin AL (2014) Cas adaptor proteins organize the retinal ganglion cell layer downstream of integrin signaling. Neuron 81:779–786

Rubinstein R, Thu CA, Goodman KM, Wolcott HN, Bahna F, Mannepalli S, Ahlsen G, Chevee M, Halim A, Clausen H, Maniatis T, Shapiro L, Honig B (2015) Molecular logic of neuronal self-recognition through protocadherin domain interactions. Cell 163:629–642

Salzberg Y, Díaz-Balzac CA, Ramirez-Suarez NJ, Attreed M, Tecle E, Desbois M, Kaprielian Z, Bülow HE (2013) Skin-derived cues control arborization of sensory dendrites in Caenorhabditis elegans. Cell 155:308–320

Sawaya MR, Wojtowicz WM, Andre I, Qian B, Wu W, Baker D, Eisenberg D, Zipursky SL (2008) A double S shape provides the structural basis for the extraordinary binding specificity of Dscam isoforms. Cell 134:1007–1018

Schmucker D, Clemens JC, Shu H, Worby CA, Xiao J, Muda M, Dixon JE, Zipursky SL (2000) Drosophila Dscam is an axon guidance receptor exhibiting extraordinary molecular diversity. Cell 101:671–684

Schramm RD, Li S, Harris BS, Rounds RP, Burgess RW, Ytreberg FM, Fuerst PG (2012) A novel mouse Dscam mutation inhibits localization and shedding of DSCAM. PLoS One 7:e52652

Schreiner D, Weiner JA (2010) Combinatorial homophilic interaction between gamma-protocadherin multimers greatly expands the molecular diversity of cell adhesion. Proc Natl Acad Sci U S A 107:14893–14898

Smith CJ, Watson JD, VanHoven MK, Colón-Ramos DA, Miller DM (2012) Netrin (UNC-6) mediates dendritic self-avoidance. Nat Neurosci 15:731–737

Soba P, Zhu S, Emoto K, Younger S, Yang S-JJ, Yu H-HH, Lee T, Jan LY, Jan Y-NN (2007) Drosophila sensory neurons require Dscam for dendritic self-avoidance and proper dendritic field organization. Neuron 54:403–416

Soba P, Han C, Zheng Y, Perea D, Miguel-Aliaga I, Jan LY, Jan YN (2015) The ret receptor regulates sensory neuron dendrite growth and integrin mediated adhesion. Elife 4:e05491

Sun LO, Jiang Z, Rivlin-Etzion M, Hand R, Brady CM, Matsuoka RL, Yau K-W, Feller MB, Kolodkin AL (2013) On and off retinal circuit assembly by divergent molecular mechanisms. Science 342:1241974

Suo L, Lu H, Ying G, Capecchi MR, Wu Q (2012) Protocadherin clusters and cell adhesion kinase regulate dendrite complexity through Rho GTPase. J Mol Cell Biol 4:362–376

Tasic B, Nabholz CE, Baldwin KK, Kim Y, Rueckert EH, Ribich SA, Cramer P, Wu Q, Axel R, Maniatis T (2002) Promoter choice determines splice site selection in protocadherin alpha and gamma pre-mRNA splicing. Mol Cell 10:21–33

Thu CA, Chen WV, Rubinstein R, Chevee M, Wolcott HN, Felsovalyi KO, Tapia JC, Shapiro L, Honig B, Maniatis T (2014) Single-cell identity generated by combinatorial homophilic interactions between α, β, and γ Protocadherins. Cell 158:1045–1059

Toyoda S, Kawaguchi M, Kobayashi T, Tarusawa E, Toyama T, Okano M, Oda M, Nakauchi H, Yoshimura Y, Sanbo M et al (2014) Developmental epigenetic modification regulates stochastic expression of clustered protocadherin genes, generating single neuron diversity. Neuron 82:94–108

Wang H, Macagno ER (1997) A detached branch stops being recognized as self by other branches of a neuron. J Neurobiol 35:53–64

Wang J, Zugates CT, Liang IH, Lee CH, Lee T (2002a) Drosophila Dscam is required for divergent segregation of sister branches and suppresses ectopic bifurcation of axons. Neuron 33:559–571

Wang X, Weiner JA, Levi S, Craig AM, Bradley A, Sanes JR (2002b) Gamma protocadherins are required for survival of spinal interneurons. Neuron 36:843–854

Wang X, Su H, Bradley A (2002c) Molecular mechanisms governing Pcdh-γ gene expression: evidence for a multiple promoter and cis-alternative splicing model. Genes Dev 16:1890–1905

Wang J, Ma X, Yang JS, Zheng X, Zugates CT, Lee CH, Lee T (2004) Transmembrane/juxtamembrane domain-dependent Dscam distribution and function during mushroom body neuronal morphogenesis. Neuron 43:663–672

Watson FL, Püttmann-Holgado R, Thomas F, Lamar DL, Hughes M, Kondo M, Rebel VI, Schmucker D (2005) Extensive diversity of Ig-superfamily proteins in the immune system of insects. Science 309:1874–1878

Weiner JA, Wang X, Tapia JC, Sanes JR (2005) Gamma protocadherins are required for synaptic development in the spinal cord. Proc Natl Acad Sci U S A 102:8–14

Wojtowicz WM, Flanagan JJ, Millard SS, Zipursky SL, Clemens JC (2004) Alternative splicing of Drosophila Dscam generates axon guidance receptors that exhibit isoform-specific homophilic binding. Cell 118:619–633

Wojtowicz WM, Wu W, Andre I, Qian B, Baker D, Zipursky SL (2007) A vast repertoire of Dscam binding specificities arises from modular interactions of variable Ig domains. Cell 130:1134–1145

Wu Q, Maniatis T (1999) A striking organization of a large family of human neural cadherin-like cell adhesion genes. Cell 97:779–790

Wu Q, Zhang T, Cheng JF, Kim Y, Grimwood J, Schmutz J, Dickson M, Noonan JP, Zhang MQ, Myers RM et al (2001) Comparative DNA sequence analysis of mouse and human protocadherin gene clusters. Genome Res 11:389–404

Wu W, Ahlsen G, Baker D, Shapiro L, Zipursky SL (2012) Complementary chimeric isoforms reveal Dscam1 binding specificity in vivo. Neuron 74:261–268

Yagi T (2012) Molecular codes for neuronal individuality and cell assembly in the brain. Front Mol Neurosci 5:45

Yu HH, Yang JS, Wang J, Huang Y, Lee T (2009) Endodomain diversity in the Drosophila Dscam and its roles in neuronal morphogenesis. J Neurosci 29:1904–1914

Zhan X-LL, Clemens JC, Neves G, Hattori D, Flanagan JJ, Hummel T, Vasconcelos ML, Chess A, Zipursky SL (2004) Analysis of Dscam diversity in regulating axon guidance in Drosophila mushroom bodies. Neuron 43:673–686

Zhu H, Hummel T, Clemens JC, Berdnik D, Zipursky SL, Luo L (2006) Dendritic patterning by Dscam and synaptic partner matching in the Drosophila antennal lobe. Nat Neurosci 9:349–355

Zipursky SL, Sanes JR (2010) Chemoaffinity revisited: dscams, protocadherins, and neural circuit assembly. Cell 143:343–353

Chapter 9
Tiling and Mosaic Spacing of Dendrites

Jay Z. Parrish

Abstract Dendrites exhibit a diverse array of complex but often stereotyped morphologies that influence the inputs a neuron can receive and how these inputs are processed. While dendrite morphology is a hallmark of neuronal type, with neurons of a given type elaborating type-specific dendrite arbors, certain organizing principles govern patterning and spacing of dendrite and axon arbors. In many cases, particularly in sensory systems, dendrites of same-type neurons are arranged in ordered mosaics. In some of these mosaics, neurites of a single functional class of neurons are arranged to allow complete and nonredundant representation of territory – a phenomenon referred to as tiling. Tiling dendrites (and axons) minimally overlap with dendrites of the same functional class, but freely overlap with dendrites of different neurons, therefore tiling dendrites must be able to recognize and avoid "self"-dendrites while coexisting with "nonself"-dendrites. Tiled mosaics have been observed in non-neuronal cells as well, suggesting that tiling is a general mechanism for cell spacing. Here I will provide an overview of the cellular mechanism of tiling in neurons, with a focus on studies in model systems that have provided information about signaling systems underlying the establishment and maintenance of tiled mosaics.

Keywords Tiling • Dendrite mosaics • Somatosensory neuron • Ganglion cell • Homotypic repulsion

9.1 Introduction

Dendrites exhibit a diverse array of complex but often stereotyped morphologies that influence the inputs a neuron can receive and how these inputs are processed. While dendrite morphology is a hallmark of neuronal type, with neurons of a given type elaborating type-specific dendrite arbors, certain organizing principles govern patterning and spacing of dendrite and axon arbors. In many cases, particularly in sensory systems, dendrites of same-type neurons are arranged in ordered mosaics.

J.Z. Parrish (✉)
Department of Biology, University of Washington, Seattle, WA 98195, USA
e-mail: jzp2@uw.edu

K. Emoto et al. (eds.), *Dendrites*, DOI 10.1007/978-4-431-56050-0_9

In some of these mosaics, neurites of a single functional class of neurons are arranged to allow complete and nonredundant representation of territory – a phenomenon referred to as tiling.

The term tiling is intentionally evocative of floor tiles, and the metaphor illustrates several of the principle features of tiling. Individual components (floor tiles, representing arbors of individual neurons) are recognizably of the same type, have uniformity in size and shape, exhibit minimal overlap, and cover a surface with little empty space in a mostly two-dimensional fashion. Tiling is particularly prevalent in sensory systems, where complete and nonredundant coverage of receptive fields ensures efficient sampling of sensory inputs while additionally allowing for spatial acuity. One of the most elaborate manifestations of tiling can be found in the vertebrate retina, where axons and dendrites of several different types of neurons tile the retina (Amthor and Oyster 1995). Different classes of sensory neurons likewise tile the skin in invertebrates and vertebrates (Gallegos and Bargmann 2004; Grueber et al. 2001, 2002; Sagasti et al. 2005; Smith et al. 2010), with insect multidendritic sensory neurons and more recently *C. elegans* mechanosensory neurons figuring prominently in genetic analysis of tiling.

Tiling has been additionally used to describe the even spacing of distinct arbors within individual neurons (Vonhoff and Duch 2010), mosaics of non-neuronal cells such as astrocytes (Bushong et al. 2002; Ogata and Kosaka 2002), and leech comb cells (Gershon et al. 1998), as well as the evenly spaced arrangement of synapses (Mizumoto and Shen 2013). While these variations exhibit shared features of tiling – spacing and, in some cases, recognition – and likely rely on some of the same machinery, this review will focus on tiling as it relates to the nonoverlapping arrangement of arbors of same-type neurons innervating a target. Recent reviews have provided comprehensive overviews of the phenomenology of tiling and self-avoidance (Grueber and Sagasti 2010) as well as cell adhesion molecules that facilitate the self-recognition and avoidance in tiling (Cameron and Rao 2010). Here I will provide an overview of the cellular mechanism of tiling in neurons, with a focus on studies in model systems that have provided information about signaling systems underlying the establishment and maintenance of tiled mosaics.

Studies of tiled neuronal mosaics over the last several decades have revealed several key features of tiling. First, tiling is particularly prevalent in sensory systems, suggesting that it may be generally used to provide complete, nonredundant sampling of sensory fields and hence allow for a high degree of spatial acuity. Second, the tiled orientation is suggestive of several forms of cellular recognition. Tiling dendrites minimally overlap with dendrites of the same functional class, but freely overlap with dendrites of different neurons. Thus, tiling dendrites must be able to recognize and avoid "self"-dendrites from the same neuron (isoneuronal tiling) or neighboring neurons of the same type (heteroneuronal tiling) while coexisting with "nonself"-dendrites. Third, tiling can occur transiently during development of particular types of neurons, suggesting that tiling signals or receptors for the signals are developmentally regulated in these neurons. Fourth, different types of neurons utilize different cellular and molecular mechanisms to tile territory, but many tiling neurites are confined to two-dimensional territories to facilitate homotypic interactions. Finally, after establishing tiled mosaics, neurons often maintain them for long periods of time, doing so in some cases during periods of animal growth. These neurons must therefore integrate extrinsic signals to facilitate maintenance of receptive field coverage.

9.2 Phenomenology of Tiling

9.2.1 Tiling in Somatosensory Neurons

One of the most conspicuous examples of "perfect" tiling, complete coverage with no overlap, is manifest in somatosensory neurons of invertebrates and vertebrates (Fig. 9.1). Early observations that leech somatosensory neurons of a given modality (T, touch; P, pressure; N, nociceptive) completely and nonredundantly covered the skin in a class-specific manner (Blackshaw 1981; Nicholls and Baylor 1968) illustrate some of the first principles of tiling. Similarly, dendrites of different functional classes of multidendritic dendrite arborization (da) neurons tile the body wall in *Manduca* and *Drosophila* (Grueber et al. 2001, 2002), dendrites of mechanosensory neurons tile the body wall in *C. elegans* (Gallegos and Bargmann 2004; Smith et al. 2010), and peripheral axonal arbors of somatosensory neurons tile the body wall in *Xenopus* (Hayes and Roberts 1983), salamanders (Kitson and Roberts 1983), and fish (Sagasti et al. 2005). Similar rules appear to govern tiling in each of these systems: multiple distinct classes of neurons tile the body wall, with dendrites (or axons) of different classes of neurons overlapping extensively, whereas dendrites of cells of the same class exhibit minimal overlap (Fig. 9.2).

9.2.2 Tiling and Cellular Mosaics in the Retina

Some of the earliest examples of tiling were identified in the retina, which is a highly organized ensemble of cellular mosaics. The mammalian retina is populated by ~60 different types of cells (Masland 2012; Rockhill et al. 2002), each of which is nonrandomly distributed over the retina, with cell bodies of same-type neurons more evenly spaced than would be expected by chance to form regular mosaics (Cook 1996; Reese and Keeley 2014; Wässle and Riemann 1978). These cellular mosaics provide uniformity in coverage of the visual field by a given type of neuron, but mosaics of different subtypes are independent of each other; thus, cells of different types of neurons are more or less randomly distributed with respect to each other (Rockhill et al. 2000).

Dendrites of a given retinal type cover the spaces between the regularly spaced cell bodies and, in many cases, are arranged to minimize overlap (Devries and

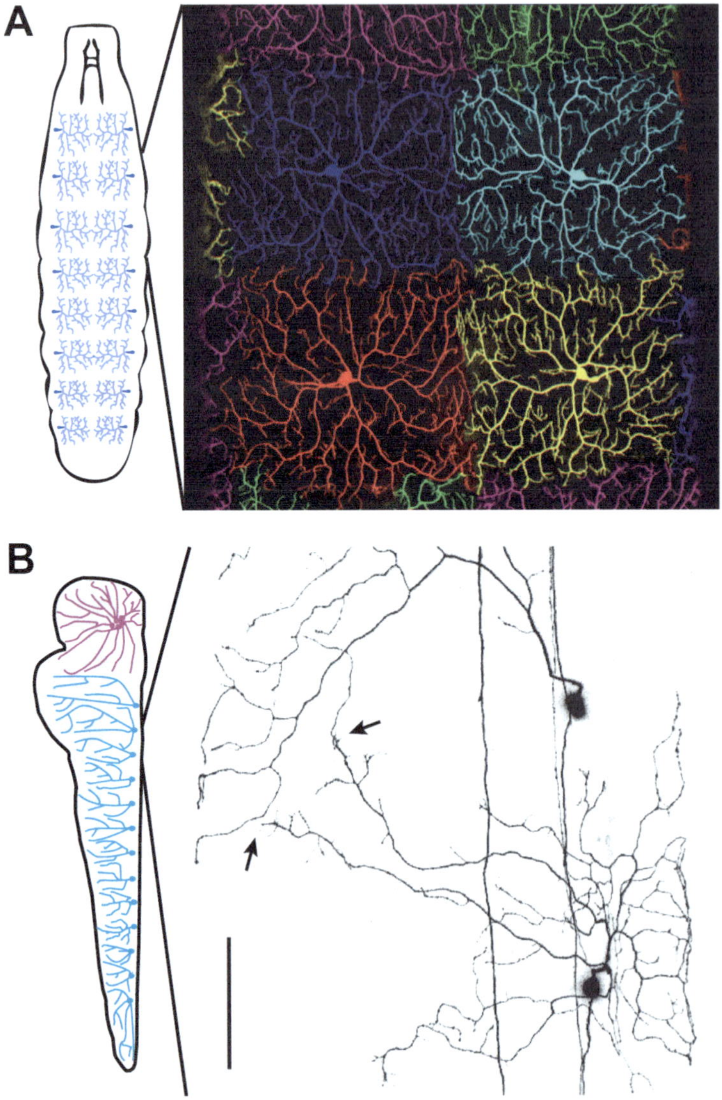

Fig. 9.1 Tiling in somatosensory neurons. (**a**) *Left*, schematic representation of *Drosophila* larval class IV da neurons, whose dendrites completely and nonredundantly cover the body wall. *Drosophila* larvae have six class IV da neurons per segment, two of which are positioned dorsally and depicted in the schematic. *Right*, confocal image of dendritic arbors of class IV neurons labeled with a membrane-targeted fluorescent protein (CD4-tdTomato). Arbors of different neurons are pseudo-colored to facilitate tracking of receptive field boundaries. Note the regularity of size and position of dendritic fields and the minimal interdigitation of dendrites where arbors of different neurons meet (Photo credit: Wen-Yang Lin). (**b**) *Left*, schematic representation of tiling by peripheral axons of trigeminal neurons (*magenta*) and Rohon-Beard neurons (*blue*) in zebrafish larvae. *Right*, confocal image of two GFP-expressing Rohon-Beard neurons at ~48 h postfertilization (Photo credit: Donald Julien)

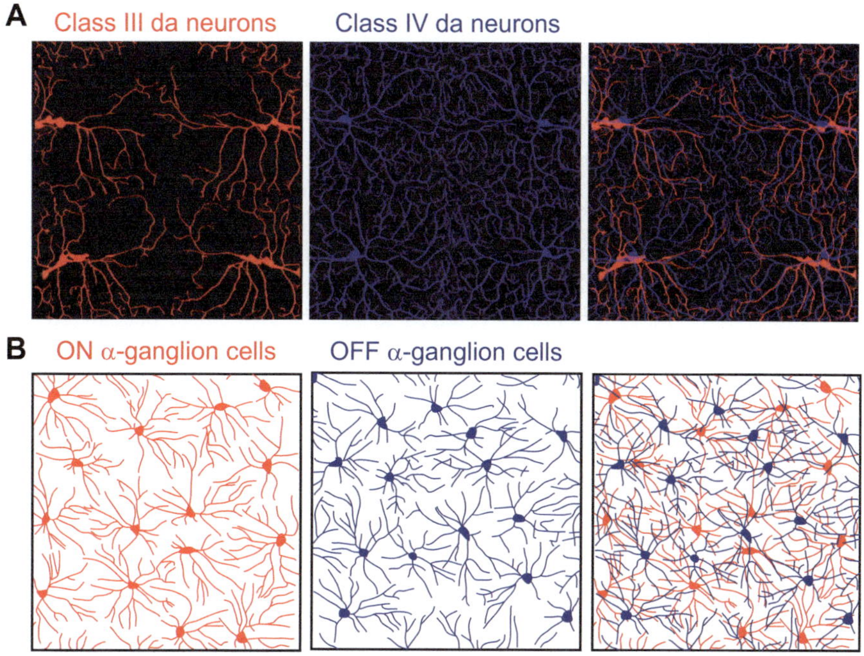

Fig. 9.2 Coexistence of different tiled mosaics. (**a**) Confocal image depicting dendritic tiling by two distinct functional classes of sensory neurons in *Drosophila*. Class III (*red*) and class IV (*blue*) dendrites tile the larval body wall, with dendrites of each class exhibiting minimal overlap, whereas dendrites of class III and class IV neurons readily overlap with one another (Photo credit: Jay Parrish). (**b**) Schematic representation of dendritic mosaics in the cat retina. Dendrites of ON ganglion cells (*red*) or OFF ganglion cells (*blue*) show minimal overlap with homotypic dendrites but extensive overlap with heterotypic dendrites, suggesting that dendritic territories of these neurons are shaped by interactions with neighboring ganglion cells of the same type (Adapted from (Wässle et al. 1981))

Baylor 1997; Wässle et al. 1981). Dendrites of some types of RGCs display almost perfect tiling, with no empty spaces in the mosaic and no redundant coverage (Amthor and Oyster 1995; Dacey 1993). As with the spacing of the cell bodies, dendrites of different types of RGCs show no ordered arrangement relative to one another (Fig. 9.2b). Tiling is used to partition retinal axon territories as well, with neighboring same-type axons filling available space evenly without overlap in some bipolar cell types (Wässle et al. 2009). However, it is important to note that perfect tiling is not a general feature of all retinal cell types. In fact, most retinal neurons exhibit overlap in their arbors. For example, cholinergic amacrine cells have greater than 25-fold redundancy in their coverage (Keeley et al. 2007). While a relatively small number of retinal cell types exhibit perfect tiling, some retinal cell types that exhibit dendritic overlap likely use related mechanisms to organize their receptive fields.

9.3 Cellular Mechanisms of Tiling

9.3.1 Homotypic Recognition and Avoidance Behavior

In many tiled mosaics, experimental evidence demonstrates that cell-cell interactions, especially homotypic repulsion, likely contribute to establishment of neurite spacing. Repulsive interactions that contribute to tiling take two basic forms: isoneuronal interactions (sister branches of the same neuron) and heteroneuronal interactions (branches from different neurons of the same type). Isoneuronal repulsive interactions, more commonly referred to as self-avoidance, depend on the ability of an individual neuron to recognize “self” from “nonself” and function to ensure uniformity of coverage within the arbor. The phenomenon of self-avoidance was first described in studies of isoneural homotypic interactions in leech mechanosensory neurons (Kramer and Kuwada 1983). Each leech mechanosensory neuron innervates a particular territory in the skin, with the receptive field divided into subfields, contiguous but nonoverlapping territories innervated by different peripheral axons. Notably, these subfields overlap extensively with the receptive fields of neighboring same-type mechanosensory neurons (Nicholls and Baylor 1968), suggesting that homotypic interactions between axons of the same neuron but not between axons of homologous neurons influence arborization patterns in these neurons. Indeed, surgical manipulations that eliminated subfields led to expansion of the territories occupied by spared axons in the same neuron without affecting arborization of homologous neurons (Kramer and Stent 1985), demonstrating that homotypic repulsive interactions limited growth of peripheral axons in the same neuron. Although self-recognition is an important component of tiling, tiling neurons must additionally avoid a subset of heteroneuronal contacts, namely, same-type neurons. Here, we will focus on heteroneuronal contacts; a detailed discussion of self-avoidance can be found in an accompanying chapter.

Some of the first indications of heteroneuronal repulsive signaling came from observations of irregularities in arbor boundaries between neighboring same-type neurons, suggesting that local interactions shaped arbor boundaries. For example, in ON- and OFF-transient ganglion cells, dendritic trees are irregularly shaped to maximize coverage of the retina while producing minimal overlap with other like dendrites (Fig. 9.2) (Wässle et al. 1981). Similarly, *Xenopus* neurons from the right and left trigeminal ganglia normally terminate at the dorsal midline where they meet other same-type neurons. However, arbors frequently make small incursions across the midline, but when they do so, contralateral arbors tend to stop short of the midline, minimizing overlap between same-type arbors (Kitson and Roberts 1983). Thus, interactions between same-type axons appear to determine growth behavior of arbors at the midline. In the growing retina, orientation of dendrite outgrowth in the peripheral zone changes during development (Hitchcock and Easter 1986), suggesting that as the central retina becomes populated with dendrites, dendrite growth into the central retina from the periphery becomes impeded by dendrites present in the central retina.

These early studies were suggestive of recognition and repulsion between same-type dendrites, and live-imaging studies provided further support for homotypic repulsion in defining boundaries of tiled arbors. In the *Drosophila* peripheral nervous system, nociceptive class IV da neurons establish complete but nonredundant dendrite coverage of the body wall early in larval development (Grueber et al. 2002; Parrish et al. 2009). In vivo time-lapse imaging of dendrite growth in developing class IV da neurons showed that when growing dendrites encounter dendrite branches of the same neuron or a neighboring class IV neuron, the branches either turn to avoid contact or retract following contact (Emoto et al. 2004; Grueber et al. 2003; Sugimura et al. 2003). Similarly, time-lapse imaging of sensory axon growth in zebrafish demonstrated that when a growth cone encountered a same-type neuron, it collapsed or changed direction (Fig. 9.1) (Sagasti et al. 2005). In some cases, tiling dendrites freely cross one another at certain stages during development, suggesting that homotypic avoidance signaling is developmentally regulated or dispensable for tiling in these neurons (Gallegos and Bargmann 2004; Lee et al. 2011). Finally, it is important to note that tiling can, in principle, be achieved in the absence of homotypic avoidance behavior, for example, by properly spacing cells that elaborate arbors of a fixed size (Lin et al. 2004; Montague and Friedlander 1991) or by relying on environmental signals to define boundaries between tiling neurons (discussed below).

9.3.2 Homotypic Avoidance Patterns Tiled Arbors

Cell ablation and duplication studies have provided compelling evidence for the roles that homotypic avoidance interactions play in establishing tiled arrangements of neurites. Embryonic ablation of *Drosophila* class IV da neurons prior to establishment of tiling causes invasion of the vacated territories by neighboring class IV dendrites (Fig. 9.3a) (Gao et al. 2000; Grueber et al. 2003; Sugimura et al. 2003). Conversely, duplication of class IV neurons results in a partitioning of the receptive field (Grueber et al. 2003). In *C. elegans*, dendrites of PVD and FLP nociceptive neurons tile the body wall (Albeg et al. 2011; Smith et al. 2010), and as with *Drosophila* da neurons, dendrites of supernumerary nociceptive neurons (generated via mutations that alter mechanosensory neuron cell fates) partition the body wall (Fig. 9.3b) (Smith et al. 2013). Finally, transplantation studies in zebrafish demonstrate that homotypic repulsive interactions define territories and constrain growth of peripheral somatosensory arbors (Sagasti et al. 2005). Transplants of wild-type cells into *neuregenin-1* mutant fish, which are depleted of Rohon-Beard cells, yielded donor-derived Rohon-Beard cells with expanded peripheral arbors, likely because contact inhibition that normally curtails growth of these arbors was absent. Indeed, removal of contralateral neurons similarly led to expanded peripheral arbors in Rohon-Beard and trigeminal sensory neurons. Although the components of the signaling system responsible for homotypic repulsion have not been fully identified in any of these systems (some molecules involved in this signaling are

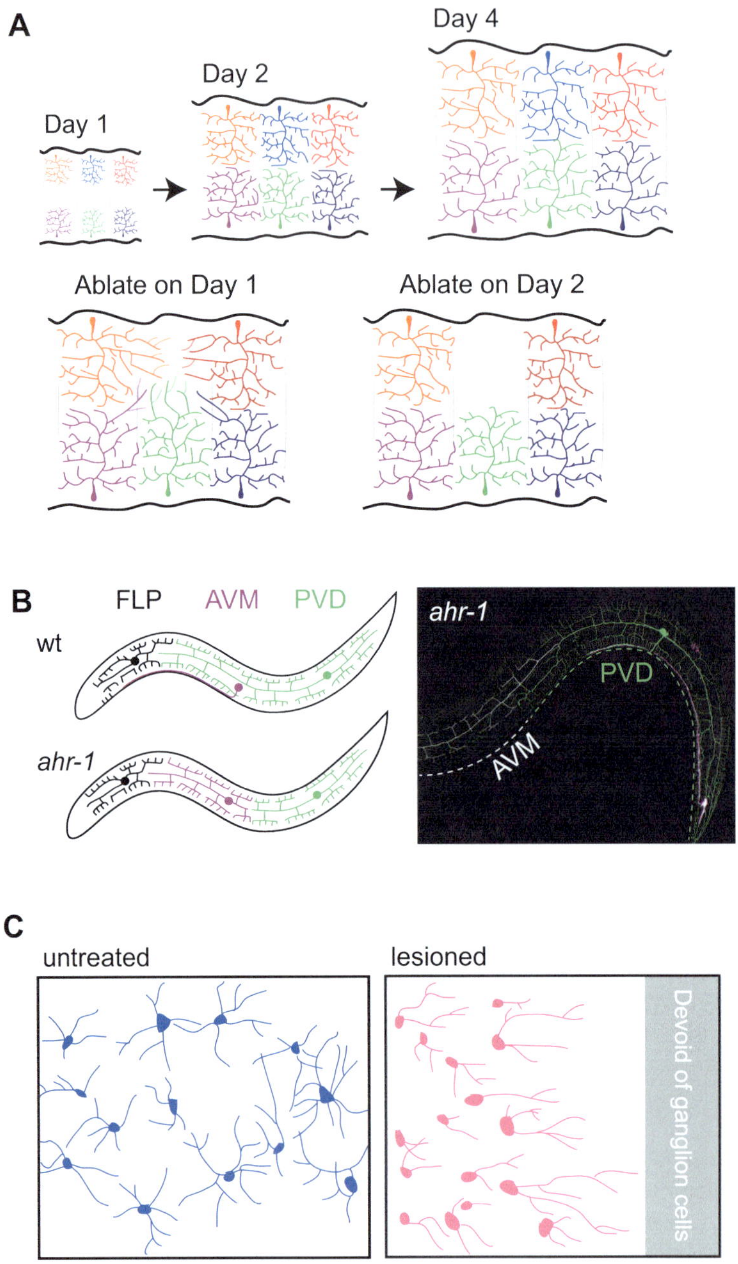

Fig. 9.3 Competitive interactions between homotypic dendrites. (**a**) Like-repels like homotypic repulsion defines dendritic territories of *Drosophila* class IV da neurons. Dendrites of these neurons grow until they encounter same-type dendrites ("space filling") and subsequently maintain their tiled orientation by growing in precise synchrony with the animal as a whole. Ablation of a neuron early in development (Day 1) leads to exuberant dendrite growth in spared neighboring neurons. After dendritic tiling has been established (Day 2), dendrites lose their ability to grow into vacated territory, suggesting that signals other than homotypic repulsion constrain dendritic

discussed below), features of the signaling can be extrapolated from these studies. First, tiling neurites of same-type neighboring somatosensory neurons can "sense" one another. Second, receptive field boundaries of these neurons appear to be established by homotypic repulsion. Whether homotypic repulsion is continuously required to maintain tiling is currently unknown. Third, exclusionary interactions curtail the growth potential of tiling neurites in many contexts.

Exclusionary neighbor interactions likely contribute to tiling in invertebrate and vertebrate visual systems as well. The *Drosophila* medulla is a laminated structure that is organized into columns containing axons from photoreceptor neurons (R7–R8), lamina neurons (L1–L5), and several other cell types. Cells in columns make synaptic connections with target cells within the same column, but not with target cells in neighboring columns, creating a tiled array of axon terminals across the medulla (Millard and Zipursky 2008; Nern et al. 2015). In genetic mosaics in which ommatidia containing supernumerary R7 cells are surrounded by ommatidia lacking all R7 cells, R7 axons invade neighboring columns (Ashley and Katz 1994). By contrast, supernumerary R7 cells surrounded by ommatidia containing R7 cells have axons that are confined to the appropriate column. Therefore, repulsive interactions underlie tiling of R7 axons and, as discussed below, tiling of other cells in the *Drosophila* visual system.

Lesion studies in several vertebrate systems including rats (Perry and Linden 1982), cats (Eysel et al. 1985; Leventhal et al. 1988), and goldfish (Hitchcock 1989; Negishi et al. 1982) demonstrate that when a patch of ganglion cells is destroyed, RGCs near the wound to reorient their arbors toward the lesion (Fig. 9.3c). It therefore appears that these neurons are predisposed to fill in the space lacking innervation. While these results could indicate that the ablated neurons normally provide a repulsive cue, it is also conceivable that the denervated region provides an attractive cue to the spared neuron. In either case, these results indicate that neighbor interactions influence dendritic tiling in retinal neurons, as in other systems. However, the plasticity of retinal neurons varies substantially depending on cell type, inputs, and position in the retina, suggesting that a complex interplay of intrinsic and extrinsic factors shapes dendritic territories of retinal neurons (Leventhal et al. 1989; Reese et al. 2011).

Fig. 9.3 (continued) growth after tiling is established. (**b**) In wild-type (wt) *C. elegans* larvae, dendrites of FLP and PVD nociceptive mechanosensory neurons tile, whereas AVM neurons extend an unbranched dendrite that has no obvious influence on FLP or AVM dendrite arborization. AHR-1, the *C. elegans* aryl hydrocarbon receptor transcription factor, prevents AVM from adopting a PVD-like cell fate; in *ahr-1* mutants, AVM adopts a PVD-like cell fate and tiles with FLP and PVD, limiting the anterior elaboration of PVD dendrites. Schematics depict results presented in (Smith et al. 2013), and the confocal image depicts the tiling. Confocal image showing tiling between AVM dendrites (*white*) and PVD dendrites (*green*) in an *ahr-1* mutant L3 larva (Photo credit: David Miller and Cody Smith). (**c**) Drawing of cell bodies and dendrites of ganglion cells in the absence or presence of optic nerve lesion. Whereas ganglion cells are radially oriented in the untreated retina, dendrites orient toward the region devoid of ganglion cells in the lesioned retina (Adapted from Perry and Linden (1982))

9.3.3 Subclasses Tile Independent of One Another

As noted above, different types of neurons often innervate the same territory and tile independent of one another (Fig. 9.2). Pioneering studies in leech and insects demonstrated that functionally and morphologically distinct classes of sensory neurons tile independent of one another. Ablation of leech mechanosensory or nociceptive neurons that innervate the epidermis caused other neurons of the same subclass to expand territories, whereas other subclasses of neurons showed no effect (Blackshaw et al. 1982). Likewise, two different morphological classes of da neurons tile the larval body wall in *Manduca* and *Drosophila* (Fig. 9.2a) (Grueber et al. 2001, 2002). Whereas same-type dendrites repel one another and exhibit minimal overlap, cell ablation and cell duplication experiments revealed no evidence for instructive or exclusionary interactions between dendrites of different classes (Grueber et al. 2003; Sugimura et al. 2003). Similarly, different classes of retinal neurons exhibit highly ordered arrangements relative to neurons of the same class but not neurons of different functional classes.

In situations where multiple classes of neurons tile the same territory, each type of neuron must employ a unique form of recognition. In some cases, for example, in the *Drosophila* visual system, the different tiled mosaics are largely dependent on homotypic repulsion mediated by distinct cell surface molecules (Cameron and Rao 2010). In other situations, different cell types utilize distinct cellular mechanisms to tile the same territory. For example, two morphologically and functionally distinct classes of sensory neuron dendrites tile the body wall in *Manduca* and *Drosophila* (Grueber et al. 2001, 2002). In contrast to class IV da neurons, ablation of class III da neurons has no effect on dendrites of spared neurons, and duplication of class III da neurons leads to extensive overlap between same-type neurons (Grueber et al. 2003; Sugimura et al. 2003), suggesting that tiling in class III neurons does not rely on homotypic repulsion. These observations illustrate two important points about tiling: neurons that tile the same surface can utilize distinct signaling to establish tiling, and mechanisms other than homotypic repulsion contribute to tiling.

9.3.4 Temporally Regulated Tiling

Although dendrites of many types of retinal neurons exhibit extensive overlap, tiling interactions could contribute to formation of retinal mosaics even in neurons with extensive dendrite overlap. For example, transient interactions between dendrites early in development could govern cell spacing, after which repulsive interactions could be down regulated, allowing dendrites to grow into neighboring territories. Indeed, studies of horizontal cell development suggest that transient tiling may contribute to the formation of horizontal cell mosaics (Fig. 9.4). In

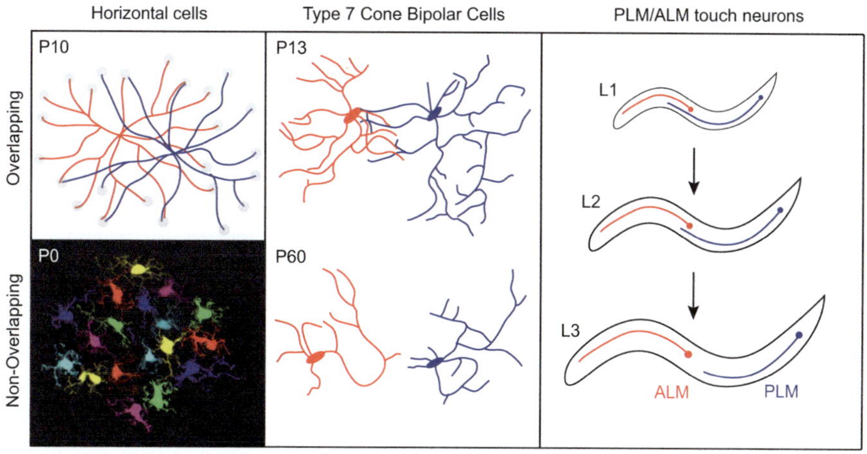

Fig. 9.4 Transient tiling interactions. Schematic representations of three different manifestations of transient tiling are shown. For each cell type, dendrite arbors of extensively overlap with same-type neurons during a portion of development and exhibit transient tiling interactions during another portion of development. Whereas dendrites of horizontal cells overlap extensively (an *en face* view of horizontal cell dendrites is depicted according to (Reese et al. 2005)), horizontal cells exhibit columnar tiling via homotypic interactions early in development (Photo credit: Rachel Huckfeldt). By contrast, dendrites of type 7 cone bipolar cells intermingle with same-type neighbors early in development but exhibit tiling after a period of dendritic refinement (Schematics adapted from Lee et al. (2011)). Similarly, dendrites of PLM and ALM touch neurons initially cross, but growth of these dendrites is slowed during larval development such that larval growth outpaces dendrite growth, and overlap between these dendrites is eliminated (Gallegos and Bargmann 2004)

mature retinas, dendrites of horizontal cells overlap extensively, with any point in space covered by dendrites of six or more horizontal cells (Raven et al. 2005). Nevertheless, horizontal cell spacing and dendrite field size correlate with cell density, suggesting that homotypic interactions may influence horizontal cell arrangement in the retina (Reese et al. 2005). Horizontal cells become evenly spaced prior to appearance of laminar dendrites, suggesting that dendrites of these neurons do not influence cell spacing. Instead, horizontal cells elaborate transient, vertically oriented neurites that tile, forming minimally overlapping columnar arbors (Huckfeldt et al. 2009). These tiling neurites exhibit homotypic exclusionary interactions, as ablation of embryonic horizontal cells results in arbor expansion in the spared neighboring neurons. Although the functional significance of these interactions is still not known, one plausible hypothesis is that repulsive interactions between these transient arbors could facilitate spacing.

Developmentally regulated homotypic interactions between neurites contribute to retinal tiling in other cells as well. In many instances of tiling discussed thus far, for example, in somatosensory neurons, space-filling neurites grow until they encounter same-type neurites that constrain arbor growth and define arbor boundaries. By contrast, dendrites of some neurons initially overgrow and subsequently tile (Fig. 9.4). For example, dendrites of type 7 cone bipolar cells initially overlap extensively but subsequently tile (Lee et al. 2011; Wässle et al. 2009). Homotypic interactions likely contribute to tiling in these neurons as altering cell number leads to alterations in dendritic field size that preserve tiling (Lee et al. 2011). However, these homotypic interactions must be developmentally regulated, as dendrites of juvenile type 7 bipolar cells show no signs of repulsive interactions. Similarly, dendrites of *C. elegans* mechanosensory neurons PLM and ALM tile after initially overlapping (Gallegos and Bargmann 2004), but tiling appears to depend on a different cellular mechanism in the two cases. Type 7 cone bipolar cells extend exuberant juvenile arbors whose size is reduced in maturity; whether this is achieved by dendrite retraction, degeneration, or pruning remains to be seen. By contrast, tiling of PLM and ALM depends on tuning of growth rates: the neurons initially grow to overlap, but dendrite growth slows, while elongation of the body wall continues, eventually eliminating the overlap between PLM and ALM dendrites.

9.4 Molecular Mechanisms of Tiling

9.4.1 Tiling by Adhesion Molecules

Cell ablation and duplication experiments demonstrated that in many tiled mosaics, exclusionary interactions between homotypic neurites play key roles in establishing tiled mosaics. How are these interactions mediated? While there appear to be a variety of molecules involved, one commonly used mechanism involves interactions between adhesion molecules that mediate avoidance signaling. The best-characterized family of cell surface molecules with roles in neurite self-recognition is the Dscam (Down syndrome cell adhesion molecules) family of immunoglobulin-domain containing cell adhesion molecules (IgCAMs) (Hattori et al. 2008). The founding member of the Dscam family in *Drosophila*, Dscam1, exhibits extreme isoform diversity as a result of alternative splicing, and this diversity forms the basis of self-recognition and self-avoidance in the *Drosophila* nervous system, with each neuron expressing a unique repertoire of Dscam1 isoforms which interact homotypically to mediate repulsion (Hattori et al. 2008; Schmucker et al. 2000). Though vertebrate Dscams lack the extreme diversity of Dscam1, an analogous system appears to mediate self-recognition in vertebrates. In mice, the clustered protocadherins (Pcdh) are expressed stochastically and combinatorially and interact homophilically to mediate self-avoidance (Lefebvre et al. 2012). The Pcdh proteins are encoded by 58 genes, many of which are highly alternatively spliced, yielding extreme diversity. As with Dscam, no single *pcdh* isoform is necessary for self-avoidance, but a single isoform is sufficient for self-avoidance. However, when all starburst amacrine cells express the same isoform, heteroneuronal dendrites avoid one another, demonstrating that isoform diversity allows different neurons to distinguish isoneuronal from heteroneuronal dendrites. Thus, similar mechanisms form the basis of self-/nonself-discrimination in vertebrates and invertebrates.

In *Drosophila*, Dscam1 does not play a major role in heteroneuronal contacts (Hughes et al. 2007; Matthews et al. 2007; Soba et al. 2007). Instead, different cell surface molecules mediate reciprocal repulsion in different cell types, and the best-characterized paradigms are in the fly visual system (Fig. 9.5a, b). For example, Dscam2 mediates homotypic repulsion of L1 axons to achieve columnar tiling in the medulla (Millard et al. 2007). In L1 neurons lacking functional *Dscam2*, axons invade neighboring columns (Fig. 9.5c). Similarly, wild-type L1 neurons adjacent to *Dscam2* mutant neurons exhibit tiling defects, invading the column containing the *Dscam2*-deficient L1 neuron. Dscam2 is therefore required autonomously and nonautonomously for tiling of lamina neurons. Dscam2 is additionally expressed in L2 neurons, where it regulates synaptic recognition, but L1 and L2 neurons do not repel one another. Although *Dscam2* does not exhibit extreme diversity, Dscam2 has two splice isoforms that exhibit homophilic binding in an isoform-specific manner, and these isoforms are differentially expressed in L1 and L2. When neighboring L1 cells express different Dscam2 isoforms, they fail to recognize and repel each other (Lah et al. 2014).

As in L1 neurons, adhesive interactions regulate tiling of R7 and R8 axons. In R7 photoreceptor axons, the IgCAM Turtle (Tutl) functions as a tiling receptor (Ferguson et al. 2009), but instead of directly mediating repulsion, Tutl appears to modulate other adhesive interactions. In *tutl* mutants, R7 axon terminals invade neighbor columns (Fig. 9.5), and as with Dscam2 in L1 neurons, Tutl functions both autonomously and nonautonomously in R7 cells to constrain R7 axons to the appropriate column. However, ablating wild-type R7 cells neighboring R7 cells lacking *tutl* ameliorates tiling defects of *tutl* mutant neurons (Fig. 9.5d), and R7 axon terminals from neighboring cartridges appear to fuse in homozygous *Tutl* mutant flies, suggesting that Tutl may modulate the activity of other adhesion complexes. Indeed, Tutl genetically and physically interacts with another IgCAM, Borderless (Bdl) to regulate R7 tiling (Cameron et al. 2013). Loss of *bdl* suppresses the axon tiling phenotype of *tutl* mutants, whereas *bdl* overexpression causes tiling defects similar to *tutl* mutants. Tutl and Bdl can physically interact, suggesting Tutl may directly modulate the adhesive activity of Bdl to allow for even spacing of R7 axons. Thus, heterophilic interactions between cell surface molecules likely contribute to establishment of tiling R7 interactions. In addition to its function in R7 axon tiling, Tutl appears to contribute to self-avoidance of *Drosophila* class IV da neuron dendrites as *tutl* mutant class IV dendrites exhibit isoneuronal crossing defects (Sulkowski et al. 2011). Intriguingly, Tutl is expressed in all da neurons but only required for self-avoidance in class IV neurons, suggesting that its role in self-avoidance may depend on heterophilic binding partners present only in a subset of da neurons.

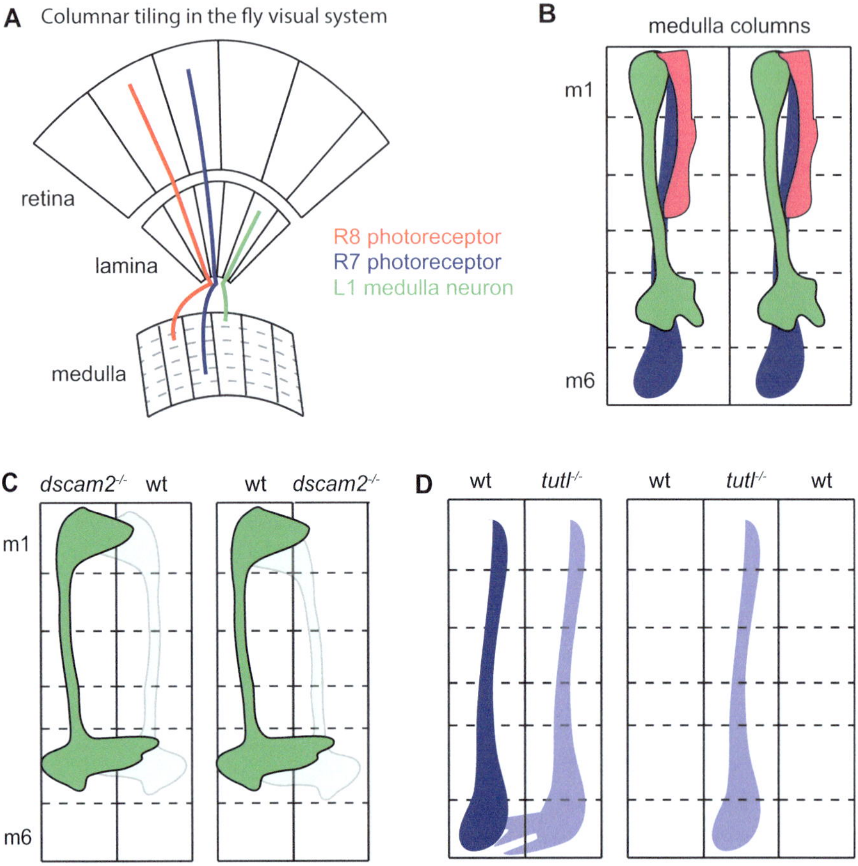

Fig. 9.5 Tiling by homophilic adhesive interactions in the fly visual system. (**a**) Columnar arrangement of the *Drosophila* visual system. The medulla is a laminated structure that is organized into columns containing axons from photoreceptor neurons (R7–R8), lamina neurons (L1), and several other cell types. Cells in columns make connections with target cells in the same column, but not with cells in neighboring columns, creating a tiled array of axon terminals across the medulla. (**b**) Schematic representation of axon terminals of R7, R8, and L1 neurons in two neighboring columns in the medulla. (**c**) Dscam2 regulates L1 axon tiling. Axons of L1 neurons lacking *dscam2* function (*dscam*$^{-/-}$) or wild-type L1 neurons neighboring *dscam*$^{-/-}$ mutant L1 neurons invade neighboring columns, demonstrating that Dscam2 is required for mutual repulsion in these neurons. The lightly shaded neurons were present but not visualized (Millard et al. 2007). (**d**) Tutl regulates R7 axon tiling. R7 neurons lacking *tutl* function (*tutl*$^{-/-}$) or wild-type R7 neurons neighboring columns with mutant R7 neurons (not shown) invade neighboring columns. Genetic ablation of R7 cells in columns neighboring *tutl* mutant R7 neurons suppresses axon tiling defects, suggesting that Tutl functions by antagonizing R7–R7 adhesion (Illustrations depict results from Ferguson et al. (2009))

9.4.2 *Homotypic Interactions Regulate Mosaic Spacing in the Retina*

As described above, although dendrites of many retinal cell types provide complete coverage of the retina, few of these cells exhibit perfect tiling. For example, dendrites of starburst amacrine cells (SACs) readily overlap with their neighbors (Tauchi and Masland 1984; Vaney 1984), but terminal dendrites of neighboring SACs are highly ordered with respect to one another while being nonrandomly distributed over the retina (Tauchi and Masland 1985). How are these mosaics established? One plausible mechanism would be to have transient tiling interactions facilitate cell spacing early in development; however, SAC mosaics are established early in development and persist even as cells are continuously added (Galli-Resta et al. 1997). Further, the mosaics are relatively insensitive to changes in cell number, and patterning of SAC dendrites do not appear to be influenced by neighbor interactions (Keeley and Reese 2010). While these observations distinguish the mosaic spacing in SACs (and many other retinal neurons) from the tiled orientation seen in some ganglion cells and elsewhere, one recent report suggests that mosaic spacing of SACs is regulated by transmembrane proteins that mediate homotypic interactions, analogous to the molecular design of tiling receptors (Kay et al. 2012). In SACs, the single-pass transmembrane protein MEGF10 (multiple epidermal growth factor-like domains) is required for mosaic spacing, as loss of *Megf10* reduces the size of the exclusion zone (the region around a SAC in which other SACs are rarely found), and ectopic expression of Megf10 in patches of cells led to exclusion of SACs from the edge of the patch, whereas SAC spacing was normal at the center of the patch. Similarly, clusters of MEGF10-expressing cells form sharp borders in cell culture. These results are consistent with MEGF10 functioning as both receptor and ligand for repellant signaling, positioning the soma so as to equalize repulsive signaling on all sides. However, unlike the exclusionary interactions that mediate tiling, MEGF10 repellant signaling affects cell spacing of SACs and horizontal cells without affecting dendrite spacing of these cell types; therefore, it will be intriguing to see how these different types of repellant signals are transduced intracellularly.

9.4.3 *Tiling in Two Dimensions*

Tiling neurites that have been studied to date, including retinal neurons and somatosensory neurons, primarily occupy a two-dimensional plane. Early anatomical studies of somatosensory neurons noted that growing neurites cross over regions of skin that are already innervated and only insert into the basal lamina when they reach areas of skin that have not yet been innervated (Hayes and Roberts 1983; Kitson and Roberts 1983; Scott et al. 1981), suggesting that repulsive interactions between neurites required interactions with the extracellular matrix

(ECM), confinement to a two-dimensional plane, or both. Indeed, recent studies in invertebrates and vertebrates confirm that direct and indirect contacts between sensory neurons and ECM are required to position neurites for avoidance signaling. For example, dendrites of *Drosophila* class IV dendrite arborization neurons are normally confined to a mostly two-dimensional space on the basal surface of epidermal cells, positioned by direct interactions between dendritic integrins and the collagen-rich ECM (Han et al. 2012a; Kim et al. 2012) that is likewise tethered to the epidermis by epidermal integrins (Jiang et al. 2014) (Fig. 9.6a, b). Compromising dendrite-ECM attachment by reducing neuronal expression of integrins causes dendrites to reorient in 3D space and become embedded inside epithelial cells (Han et al. 2012a; Kim et al. 2012). Dendrite adhesion in these neurons is additionally regulated by the receptor tyrosine kinase Ret, which

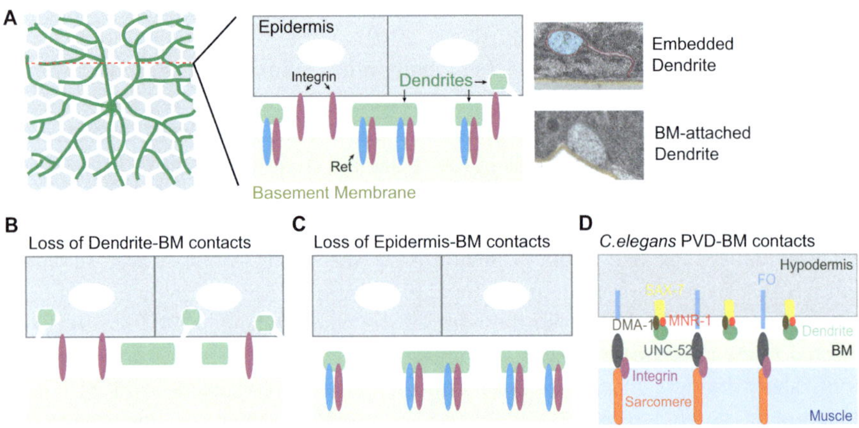

Fig. 9.6 Dendrite-ECM interactions potentiate homotypic interactions. (**a**) *Drosophila* larval sensory dendrites (*green*) grow over a monolayer of epithelial cells (*gray*) in a mostly two-dimensional plane, attached to a collagen-rich basement membrane (BM, tan) along the basal surface of epithelial cells by virtue of neuronal integrin and Ret (Han et al. 2012b; Kim et al. 2012; Soba et al. 2015). The basement membrane is likewise tethered to the epidermis by epithelial integrins (Jiang et al. 2014). The majority of dendrites are positioned at the basal plasma membrane of epithelial cells (BM-attached dendrites, shown at high magnification in a TEM micrograph at right), but a small portion of dendrites become embedded in the epidermis. Image credit: Nan Jiang. Somatosensory dendrites are similarly embedded in zebrafish (O'Brien et al. 2012). (**b**) Compromising dendrite-ECM interaction by blocking integrin or Ret function in neurons leads to an increase in epidermal embedding of dendrites, suggesting that dendrite-BM contacts counteract attractive cues from the epidermis. (**c**) Compromising integrin function in the epidermis blocks epidermal embedding of dendrites and increases the spacing between epidermis and the BM. (**d**) Interactions with the BM position dendrites of the *C. elegans* mechanosensory neuron PVD. A tripartite complex between PVD-expressed DMA-1 (a leucine-rich repeat transmembrane protein), hypodermal SAX-7 (L1CAM), and MNR-1 (a novel membrane protein) spatially patterns PVD dendrite growth (Dong et al. 2013; Salzberg et al. 2013). Muscle integrins interact with the BM protein UNC-52/perlecan and influence positioning of hemidesmosomes (the fibrous organelle, FO) in epidermal cells, indirectly regulating dendrite positioning by locally excluding SAX-7 (Liang et al. 2015)

interacts with integrins to promote dendrite-ECM attachment (Soba et al. 2015). Similarly, targeting of zebrafish somatosensory neuron peripheral axons to the skin depends on interactions between a neuronal receptor (LAR) and Heparin-sulfate proteoglycans that are enriched in the epidermal ECM (Wang et al. 2012).

In addition to these direct interactions between dendrites and ECM, indirect interactions with the ECM contribute to positioning of sensory dendrites in both *Drosophila* and *C. elegans* (Jiang et al. 2014; Liang et al. 2015). In *Drosophila*, dendrite positioning of class IV da neurons is influenced by epidermis-ECM adhesion mediated by epidermal integrins, in addition to dendrite-ECM interactions (Fig. 9.6c). In *C. elegans*, dendrites of mechanosensory neurons tile the body wall, and dendrites of these neurons are spatially patterned by direct interactions with the hypodermis and by hypodermis-muscle interactions (Fig. 9.6d). First, neuronal DMA-1 (a leucine-rich repeat transmembrane protein) interacts with hypodermal SAX-7 (L1CAM) and MNR-1 (a novel membrane protein) to spatially pattern primary, secondary, and tertiary PVD dendrites (Dong et al. 2013; Salzberg et al. 2013). Second, terminal dendrites are patterned by exclusionary interactions between SAX-7 and UNC-52 (heparan sulfate proteoglycan) (Liang et al. 2015). UNC-52 is a major component of the ECM that covers muscle, with UNC-52 tethered to muscle in a striped pattern by virtue of integrin-based contacts. UNC-52 regulates the position of hypodermal hemidesmosomes, which connect hypodermal cells to the ECM, and SAX-7 interdigitates between UNC-52 stripes, which in turn directs growth of PVD 4° branches by interaction with DMA-1 and MNR-1. Altogether, these studies suggest that for a large number of tiling neurons, neurites are confined to a two-dimensional plane to facilitate homotypic avoidance.

9.4.4 Tiling by Environmental Cues

Receptive field boundaries of many invertebrate somatosensory neurons correspond to the position of non-neuronal structures, suggesting that nonautonomous positional cues likely assist in defining boundaries of dendritic territories (Fig. 9.7). For example, segmental boundaries demarcate anterior and posterior boundaries of receptive fields in *Drosophila* class III and IV da neurons (Grueber et al. 2002), suggesting that positional cues exist at segment boundaries to demarcate the borders of these dendritic fields. During metamorphosis, dendrites of *Drosophila* class IV neurons are removed by pruning, and adult-specific processes establish tiling of the abdomen (Shimono et al. 2009). Shortly after eclosion, dendrite arbors of these neurons are remodeled and subsequently reorient to interdigitate lateral tergosternal muscles (Yasunaga et al. 2010). The remodeling of these dendrites depends on the activity of matrix metalloproteinase 2, which degrades the ECM and allows the dendrites to reorient along muscle fibers, but the nature of instructive cues that orient the dendrites has not yet been identified. Similarly, as described above, dendrites of *C. elegans* PVD mechanosensory neurons interdigitate muscle

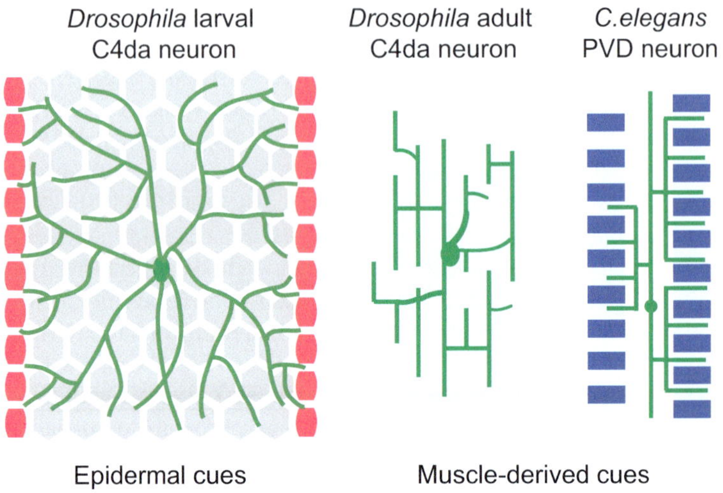

Fig. 9.7 Tiling by environmental cues. (**a**) Dendrites of dorsal *Drosophila* larval class IV da (C4da) neurons (*green*) completely cover the dorsal larval hemisegment, terminating at segment boundaries, where apodemes (*magenta*) form myotendinous junctions. Dendrites occasionally pass between apodemes but not over them. Dendrites are additionally asymmetrically oriented over body wall epithelial cells (*gray*), preferentially growing along the periphery of epithelial cells (Parrish et al. 2009). (**b**) Dendrites of the *Drosophila* adult mechanosensory neuron v'ada (*green*) interdigitate along the lateral tergosternal muscle fibers (*blue*) shortly after eclosion. Signals that position the dendrites are unknown; however, ECM remodeling by matrix metalloproteinases is necessary for remodeling of the dendrites from a radial array to this parallel alignment (Yasunaga et al. 2010). (**c**) Terminal dendrite (4°) branches (*green*) of the *C. elegans* mechanosensory neuron PVD interdigitate with the dense body and M-line (*red*) of muscle sarcomeres (Liang et al. 2015)

(Liang et al. 2015), and while positional information in muscle shapes muscle-hypodermis interactions that influence dendrite positioning, it remains to be determined whether direct interactions with muscle are required for proper dendrite spacing.

9.4.5 Tiling by Secreted Signals

Although there is growing evidence that adhesive mechanisms regulate tiling in many neurons, secreted signals regulate tiling as well. In a genetic screen for photoreceptor axon-targeting defects in *Drosophila*, mutations in TGF-β signaling were found to disrupt R7 axonal tiling. Specifically, mutations in the genes encoding the TGF-β ligand activin, its receptor baboon (Babo), or the downstream effector dSmad2 caused R7 axons to invade neighboring territories (Ting et al. 2007). Each of these genes is required cell-autonomously for R7 tiling, indicating that autocrine signaling from photoreceptor neurons is required for R7 tiling. Interestingly, this signaling appears to be distinct from the homotypic repulsion mediated by Tutl/Btl as axons of *babo* mutant R7 neurons are still responsive to repulsion from neighboring wild-type neurons, and mutations in *dSmad2* and *tutl* have additive effects on tiling (Ferguson et al. 2009). Instead of directly mediating repulsion signals, activin signaling appears to regulate axon growth such that R7 growth cones are positioned in the appropriate layer within the medulla, where repulsive interactions regulate axon spacing. Consistent with this notion, Smad2 and importin-α3, which are required for nuclear import of Smad2 and activin signaling in R7 cells, localize to axon terminals, and blocking retrograde transport causes tiling defects similar to mutations in the activin pathway (Ting et al. 2007).

Although the effectors of activin signaling in this context are not known, one recent study suggests that activin signaling may be temporally regulated in R7 cells (Kniss et al. 2013). The transcription factor Ttk69 is expressed in R7 cells concomitant with axon arrival at the medulla, and Ttk69 appears to function upstream of activin to regulate axonal tiling. Ttk69 overexpression can suppress the tiling defects of R7 cells defective in activin signaling, suggesting that Ttk69 may regulate additional factors involved in tiling control. Temporally regulated expression programs likely regulate transient tiling in other contexts, although no such programs have been identified to date.

In addition to this autocrine signaling in R7 tiling control, photoreceptor-derived activin functions as a paracrine signal to restrict dendritic fields of medulla neurons (Ting et al. 2014). Medulla projection neurons (called transmedulla [Tm] neurons) confine their dendrite arbors to their cognate columns, but in *babo* mutants, dendrites of Tm20 neurons invade neighboring columns, where they form aberrant synapses. Wild-type Tm20 neurons neighboring *babo* mutant Tm20 neurons have dendrites confined to the appropriate column, demonstrating that TGF-β signaling is required autonomously in Tm20 neurons for tiling. In this context, activin ligand is supplied at least partially from photoreceptors, as knocking down activin expression in photoreceptors and preventing R8 cells from targeting the appropriate layer in the medulla lead to Tm20 dendrite expansion into neighboring columns. Thus, activin is required to constrain dendrite expansion, providing an example of paracrine signaling that can limit the extent of arbor expansion.

Aside from activin, secreted growth inhibitory or repulsive cues are notably absent from the list of tiling regulators identified thus far, although modeling studies suggest that such factors could account for dendritic tiling (Shimono et al. 2014). Reminiscent of the activin autocrine signaling, the secreted molecule Slit2 and its receptor Robo2 are required in Purkinje cells for dendrite self-avoidance (Gibson et al. 2014). In dissociated cultures, Purkinje neurons adopt a tiled orientation with dendrites of Purkinje neurons avoiding overlap with neighboring Purkinje neurons but overlapping frequently with granule cells (Fujishima et al. 2012). It has been reported that Purkinje neurons may exhibit some form of tiling in vivo, but it remains to be seen whether Slit/Robo or other signaling cues regulate this tiling.

9.4.6 Intracellular Signaling

While several cell surface molecules have been identified that transduce avoidance signaling, the intracellular signaling cascades responsible for shaping dendrite growth in response to these signals are less well understood. Given that avoidance behavior involves turning, retraction, or degeneration of neurites, the output of this signaling must converge on the cytoskeleton. Indeed, one of the first signaling molecules implicated in tiling control, the NDR (nuclear dbf2 related) kinase tricornered (Trc), plays evolutionarily conserved role in branching morphogenesis that involves control of the actin cytoskeleton (Emoto et al. 2004; Hergovich et al. 2006). In *Drosophila*, loss of Trc or its activator, the adaptor protein Furry, leads to increased isoneuronal and heteroneuronal crossings in class IV neurons (Emoto et al. 2004). The ability of dendrites to turn and avoid homotypic dendrites depends on Trc kinase activity, suggesting that Trc kinase signaling regulates the ability of neurons to either receive repulsive signals or to respond to the signal. Arguing in favor of the latter scenario, Trc regulates actin-based processes in neurons and non-neuronal cells and additionally regulates dendritic branching by negatively regulating Rac1 signaling, though this latter activity is separable from its function in dendrite repulsion.

Although the targets of Trc in regulating dendrite avoidance remain unknown, several factors have been identified that associate with Trc and regulate its activity in neurons. First, the Ste20 kinase hippo (Hpo) activates Trc via phosphorylation of Trc threonine 449 (Emoto et al. 2006), which is associated with maximal kinase activation (Hergovich et al. 2005), and *hpo* mutants have dendritic tiling defects similar to *trc* mutants. Second, the target of rapamycin complex 2 (TORC2) promotes Trc activation, both by phosphorylation of Trc and by promoting membrane association of Trc (Koike-Kumagai et al. 2009). Interestingly, TORC2 and Trc have synergistic effects on actin assembly in cultured cells, suggesting that TORC2 may similarly function to affect actin assembly in neurons. Finally, a *Drosophila* membrane protein, Raw, was recently described that interacts with Trc and is required for Trc phosphorylation in dendrites (Lee et al. 2015). Raw promotes cell adhesion to collagen, and neurons lacking raw have defects in axial positioning of terminal dendrites that lead to out-of-plane dendrite-dendrite crossings. Notably, the dendrite-dendrite crossing defects of *raw* mutant neurons can be mitigated by expression of a membrane-targeted form of Trc or overexpression of integrins. Likewise, many of the dendrite-dendrite crossing defects in *trc* mutant neurons are out-of-plane crossings, and these crossing defects can be suppressed by overexpression of integrins (Han et al. 2012a). Taken together with a recent report that Ndr2 regulates integrin trafficking in hippocampal neurons (Rehberg et al. 2014), these findings suggest that Trc may function locally at sites of contact to modulate dendrite adhesion to the ECM and cytoskeletal remodeling.

Although the mechanism by which Trc regulates the cytoskeleton in response to avoidance signals is not yet known, cell surface proteins that regulate dendrite-ECM attachment in *Drosophila* class IV da neurons do so in part by signaling through Rac-like proteins. The receptor tyrosine kinase Ret interacts with Rac and regulates F-actin distribution in dendrites (Soba et al. 2015), and similar to *Ret* mutants, class IV da neurons with compromised Rac function are no longer confined to a two-dimensional plane and hence exhibit out-of-plane dendrite crossing over. Ret functions together with integrin to regulate dendrite-ECM attachment; hence, it appears that ECM adhesion by Ret/integrin regulates the actin cytoskeleton via Rac. Given that *Drosophila* has three Rac-like proteins that appear to have an additive effect on dendrite adhesion (Soba et al. 2015), it seems plausible that Trc may also signal through one of the Rac-like proteins other than Rac1.

Trc appears to play an evolutionarily conserved role in tiling, as dendritic tiling of *C. elegans* mechanosensory neurons PLM and ALM require the function of Trc (Sax-1) and Fry (Sax-2) homologues (Gallegos and Bargmann 2004). However, the cellular mechanisms of tiling differ in these neurons and class IV da neurons. PLM and ALM tile the long axis of the *C. elegans* body wall with nonoverlapping dendritic fields, but the tiled arrangement of PLM and ALM involves an early period of overgrowth, leading to PLM/ALM dendrite overlap, and a subsequent period of refinement in which dendrite outgrowth is modulated to eliminate overlap (Fig. 9.5). Sax-1 and Sax-2 are required for this latter refinement step, as dendrite overlap persists in neurons lacking the activity of either *sax-1* or *sax-2*. Though tiling in this developmental context does not appear to involve repulsion and the cellular output is distinct (rate of neurite extension instead of neurite turning), it remains to be determined whether Trc and Sax-1 are nevertheless regulating tiling through a common set of effectors.

9.5 Maintenance of Tiling

9.5.1 Scalar Expansion of Tiling Arbors

Tiling is established early in development in many systems, and the tiled arrangement is often maintained for prolonged periods of time. This presents tiling neurons with two distinct problems. First, the structural integrity of dendrite arbors must be maintained for long periods of time. Long-lived neurons throughout the nervous system face the same problem, and the solutions are likely to operate independent of neurite spacing (Koleske 2013). Second, in many cases, arbor expansion must be coordinated with animal growth to maintain receptive field coverage. This scalar expansion of dendrite arbors during animal growth has been most extensively documented in the retina. For example, small ganglion cells in crucian carp maintain constant coverage of the growing retina, precisely counterbalancing the decrease in cell density that occurs during growth by increasing dendrite length (Kock 1982). Similarly, certain classes of ganglion cells in goldfish (type 1.2 ganglion cells) (Bloomfield and Hitchcock 1991), cats (alpha- and beta-type ganglion cells) (Dann et al. 1988; Ramoa et al. 1987, 1988), and mice (alpha cells) (Ren et al. 2010) establish dendritic coverage of the retina early in development, after which they expand their dendrites synchronously with the several-fold expansion of the retina that occurs postnatally (Fig. 9.8a). While this scalar expansion of dendrite arbors is not a general feature of ganglion cells, as some classes of ganglion cells exhibit little growth after birth, exuberant growth followed by retraction, or other modes of growth during retinal expansion (Ault and Leventhal 1994; Ren et al. 2010), for at least some classes of ganglion cells, extrinsic signals likely coordinate expansion of these dendrite arbors with overall expansion of the eye. Likewise, scalar expansion of dendrite (and axonal) arbors occurs to sustain proper connectivity and maintain arbor geometry in non-tiling neurons (Bentley and Toroian-Raymond 1981; Bucher and Pflüger 2000; Truman and Reiss 1988).

Some of the first clues into the nature of signals involved in coordinating dendrite arbor expansion with animal growth came from studies of dendrite growth control in *Drosophila* sensory neurons. Dendrites of class III and class IV da neurons establish tiling of the body wall early in larval development and maintain this tiling for the remainder of larval development (Parrish et al. 2009). Whereas dendrites of these neurons expand faster than their substrate to establish tiling, dendrite arbors expand synchronously with body wall epithelial cells to maintain tiling during larval growth, doing so while expanding more than tenfold in linear dimensions (Fig. 9.8b). Prior to establishment of tiling, ablating C4da neurons leads to exuberant dendrite growth into unoccupied territory by spared neurons (Grueber et al. 2003; Parrish et al. 2009; Sugimura et al. 2003). This invasive growth potential is lost during the maintenance phase, and although dendrite arbors continue to grow, growth occurs only to maintain proportional receptive field coverage (Fig. 9.3) (Parrish et al. 2009), showing that signals constrain late-stage growth of these dendrites. Notably, this signaling does not involve the homotypic repulsion required to establish tiling as mutants with defects in homotypic avoidance exhibit normal scalar expansion of their dendrites. Similarly, growth of vertebrate RGCs and somatosensory neurons into vacated territory is progressively limited during development (Eysel et al. 1985; Frank and Westerfield 1982; O'Brien et al. 2009), suggesting that extrinsic signals constrain late-stage growth of these dendrites as well.

What signals are responsible for synchronous expansion of dendrite and substrate? In *Drosophila*, the microRNA bantam *(ban)* regulates epithelia-dendrite signaling that coordinates dendrite and substrate growth to maintain tiling (Parrish et al. 2009). Dendrites of *ban* mutants tile the body wall properly, but dendrite growth subsequently outpaces substrate growth, causing dendrites to cross normal receptive field boundaries and occupy larger proportions of the body wall. *ban* functions in epithelial cells, where it regulates developmental coupling of dendrites and epithelial cells by controlling epithelial endoreplication (Jiang et al. 2014). Endoreplication is a modified cell cycle that entails genome amplification without cell division to facilitate cell growth and, in certain cases, differentiation (Orr-Weaver 2015). In the *Drosophila* epidermis, endoreplication begins during

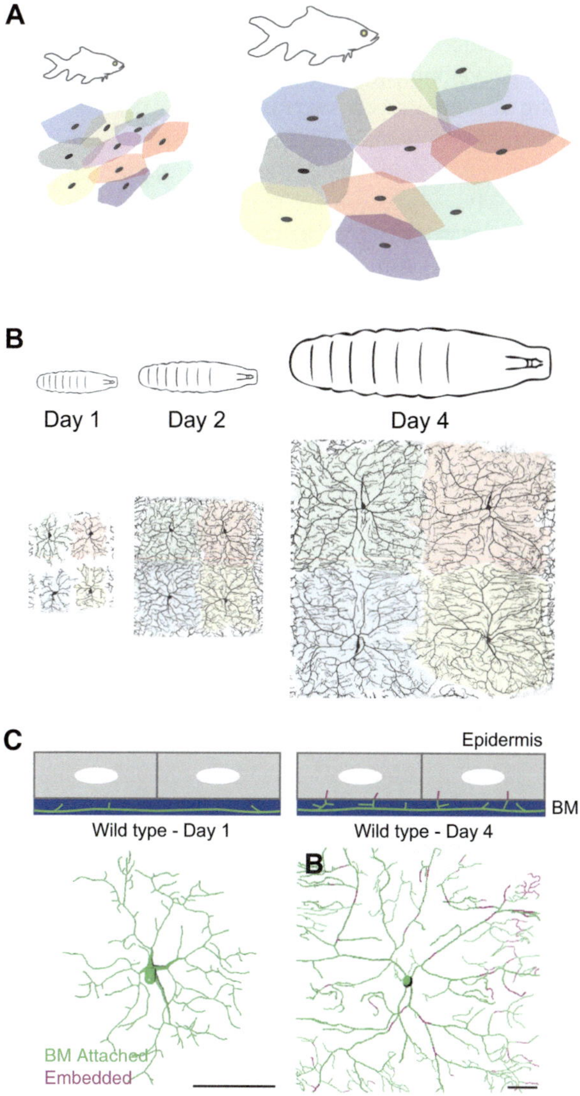

Fig. 9.8 Scalar expansion of dendrite arbors. During animal growth, dendrite arbor expansion must be coordinated with substrate expansion to maintain receptive field coverage. (**a**) Goldfish ganglion cells maintain constant coverage of the growing retina. Dendritic territories (*shaded polygons*) and cell body position (*black oval*) are depicted for type 1.2 ganglion cells from two fish, one 3.9 cm in length, the other 14 cm in length (Adapted from Hitchcock (1987)). (**b**) *Drosophila* class IV da neurons establish dendritic coverage of the body wall by Day 2 of development and maintain this coverage even as larvae grow more than ten-fold in linear dimensions. Confocal images show dendrite arbors of class IV neurons on Days 1, 2, and 4 of development; dendritic territories of individual neurons are shaded (Image credit: Jay Parrish). (**c**) As development progresses, dendrites of *Drosophila* class IV da neurons become more closely associated with epithelial cells, and some dendrites become embedded inside epithelial cells (*magenta*). The

larval development, when class IV dendrites have already established tiling of the body wall, and influences the adhesive properties of the epidermis (Jiang et al. 2014). Early in development, dendrites are rarely embedded in the epidermis, but this embedding is progressively increased during development as a result of epithelial β-integrin (myospheroid, Mys) expression, which is upregulated by endoreplication (Fig. 9.8c). Additionally, endoreplication and the accompanying changes in epithelial Mys expression are required to constrain late-stage dendrite growth and structural plasticity. Hence, modulating epithelial-ECM attachment likely influences substrate permissivity for dendrite growth and contributes to dendrite-substrate coupling that ensures proportional expansion of the two cell types.

Although endoreplication has not been widely studied in nervous system growth, these studies highlight endoreplication as a solution for cell/tissue growth in circumstances where cell division might disrupt patterning or connectivity. In a related line of work, it was shown that glia which wrap the *Drosophila* ventral nerve cord likewise rely on endoreplicate to maintain integrity of the blood-brain barrier during larval growth (Unhavaithaya and Orr-Weaver 2012). The epidermis in *C. elegans*, mouse, and humans endoreplicates (Gandarillas and Freije 2014; Hedgecock and White 1985; Zanet et al. 2010), and in *C. elegans*, this endoreplication drives body wall expansion (Lozano et al. 2006), so it will be intriguing to see whether endoreplication similarly contributes to coordination of somatosensory dendrite-substrate expansion in those systems.

In addition to these local signals that coordinate dendrite and substrate expansion, systemic growth cues likely contribute to scalar growth of dendrite arbors. Starvation during larval development affects *Drosophila* adult body size, and dendrite arbors of sensory neurons are scaled appropriately to tile the abdomen of the miniaturized flies. These neurons must integrate nutritional cues to tune their growth, and a recent study has identified some of the molecular machinery involved in this form of growth control. In flies bearing mutations in the *CHORD/morgana* gene, which encodes an evolutionarily conserved cysteine- and histidine-rich protein that functions as an HSP90 co-chaperone, sensory dendrites are miniaturized and cannot respond to changes in body size (Shimono et al. 2014). Though the mechanism of CHORD action in dendrite growth control is unknown, *CHORD* genetically interacts with components of target of rapamycin complex 2 (TORC2), and CHORD regulates Rho kinase in other contexts, so one model is that CHORD functions together with TORC2 to regulate dendrite outgrowth via modulation of Rho activity. These results suggest that these neurons, by default, develop small dendrite arbors whose growth is potentiated by nutritional cues.

Fig. 9.8 (continued) increased association is driven by a developmental transition in epithelial cells (endoreplication) that alters the adhesive properties of the epidermis (Jiang et al. 2014). Traces depict the distribution of BM-attached and embedded dendrites (Image credit: Nan Jiang)

9.5.2 *Coordinate Control of Establishment and Maintenance of Dendritic Tiling*

In addition to signaling that coordinates dendrite and substrate growth, an independent signaling pathway has been identified that cell-autonomously regulates maintenance of dendrite coverage in class IV da neurons (Emoto et al. 2006; Parrish et al. 2007). This genetic pathway consists of the NDR kinase warts, polycomb group (PcG) transcriptional repressors, and Bithorax complex (BX-C) homeobox transcription factors that cell-autonomously regulate maintenance of tiling. Mutants lacking the function of any of each of these factors initially tile the body wall but then fail to maintain this tiling, with terminal dendrite retraction leading to large gaps in receptive field coverage. BX-C transcription factors are expressed in all larval da neurons but are downregulated specifically in class IV neurons after dendritic tiling is established. PcG proteins and warts can physically associate, and these genes are required for proper regulation of BX-C expression in class IV neurons and in other developmental contexts. These findings provided some of the first evidence for cell-autonomous, neuron-type-specific, and dendrite-specific mechanisms for maintenance of tiling, but the mechanism by which this pathway regulates maintenance is still not known. Other classes of sensory neurons similarly establish coverage of the receptive field early in development and maintain the coverage through larval stages, including class III da neurons which tile the body wall, but the warts-polycomb pathway is dispensable for maintenance of dendrite coverage in these neurons, suggesting that other cell-type-specific pathways may control dendrite maintenance in these neurons.

9.6 Summary and Future Prospects

While the phenomenon of tiling is widespread in sensory systems, presumably to ensure efficient sampling of sensory inputs while additionally allowing for spatial acuity, it remains to be seen how widespread tiling is in other types of neurons. The observation that some types of neurons transiently tile suggests that tiling may be even more widespread than previously appreciated. Several major questions remain to be answered about tiling, in particular regarding the molecular mechanisms of tiling. Homotypic repulsion appears to be a major mechanism for establishing tiling, but how these signals are transduced intracellularly and whether homotypic signaling is constantly required to maintain tiled mosaics remain to be determined. Likewise, molecular mechanisms for transient tiling and tiling by environmental cues have not been described, though it would seem that the former likely involves developmental control of tiling signals or receptors for the signals. Finally, it remains to be seen whether tiling in neuronal and non-neuronal cells involves similar cellular and molecular mechanisms.

Acknowledgments I would like to thank Akira Chiba, Nan Jiang, Wen-Yang Lin, David Miller, Alvaro Sagasti, and Rachel Wong for contributing unpublished images. Thanks also to Peter Soba and Michael Kim for helpful discussions. This work was supported by National Institutes of Health grant NINDS R01-NS076614.

References

Albeg A, Smith CJ, Chatzigeorgiou M, Feitelson DG, Hall DH, Schafer WR, Miller DM, Treinin M (2011) C. elegans multi-dendritic sensory neurons: morphology and function. Mol Cell Neurosci 46:308–317. doi:10.1016/j.mcn.2010.10.001

Amthor FR, Oyster CW (1995) Spatial organization of retinal information about the direction of image motion. Proc Natl Acad Sci U S A 92:4002–4005

Ashley JA, Katz FN (1994) Competition and position-dependent targeting in the development of the Drosophila R7 visual projections. Dev Camb Engl 120:1537–1547

Ault SJ, Leventhal AG (1994) Postnatal development of different classes of cat retinal ganglion cells. J Comp Neurol 339:106–116. doi:10.1002/cne.903390110

Bentley D, Toroian-Raymond A (1981) Embryonic and postembryonic morphogenesis of a grasshopper interneuron. J Comp Neurol 201:507–518. doi:10.1002/cne.902010404

Blackshaw SE (1981) Morphology and distribution of touch cell terminals in the skin of the leech. J Physiol 320:219–228

Blackshaw SE, Nicholls JG, Parnas I (1982) Expanded receptive fields of cutaneous mechanoreceptor cells after single neurone deletion in leech central nervous system. J Physiol 326:261–268

Bloomfield SA, Hitchcock PF (1991) Dendritic arbors of large-field ganglion cells show scaled growth during expansion of the goldfish retina: a study of morphometric and electrotonic properties. J Neurosci Off J Soc Neurosci 11:910–917

Bucher, Pflüger (2000) Directional sensitivity of an identified wind-sensitive interneuron during the postembryonic development of the locust. J Insect Physiol 46:1545–1556

Bushong EA, Martone ME, Jones YZ, Ellisman MH (2002) Protoplasmic astrocytes in CA1 stratum radiatum occupy separate anatomical domains. J Neurosci Off J Soc Neurosci 22:183–192

Cameron S, Rao Y (2010) Molecular mechanisms of tiling and self-avoidance in neural development. Mol Brain 3:28. doi:10.1186/1756-6606-3-28

Cameron S, Chang W-T, Chen Y, Zhou Y, Taran S, Rao Y (2013) Visual circuit assembly requires fine tuning of the novel Ig transmembrane protein borderless. J Neurosci Off J Soc Neurosci 33:17413–17421. doi:10.1523/JNEUROSCI.1878-13.2013

Cook JE (1996) Spatial properties of retinal mosaics: an empirical evaluation of some existing measures. Vis Neurosci 13:15–30

Dacey DM (1993) The mosaic of midget ganglion cells in the human retina. J Neurosci Off J Soc Neurosci 13:5334–5355

Dann JF, Buhl EH, Peichl L (1988) Postnatal dendritic maturation of alpha and beta ganglion cells in cat retina. J Neurosci Off J Soc Neurosci 8:1485–1499

Devries SH, Baylor DA (1997) Mosaic arrangement of ganglion cell receptive fields in rabbit retina. J Neurophysiol 78:2048–2060

Dong X, Liu OW, Howell AS, Shen K (2013) An extracellular adhesion molecule complex patterns dendritic branching and morphogenesis. Cell 155:296–307. doi:10.1016/j.cell.2013.08.059

Emoto K, He Y, Ye B, Grueber WB, Adler PN, Jan LY, Jan Y-N (2004) Control of dendritic branching and tiling by the Tricornered-kinase/Furry signaling pathway in Drosophila sensory neurons. Cell 119:245–256. doi:10.1016/j.cell.2004.09.036

Emoto K, Parrish JZ, Jan LY, Jan Y-N (2006) The tumour suppressor Hippo acts with the NDR kinases in dendritic tiling and maintenance. Nature 443:210–213. doi:10.1038/nature05090
Eysel UT, Peichl L, Wässle H (1985) Dendritic plasticity in the early postnatal feline retina: quantitative characteristics and sensitive period. J Comp Neurol 242:134–145. doi:10.1002/cne.902420109
Ferguson K, Long H, Cameron S, Chang W-T, Rao Y (2009) The conserved Ig superfamily member turtle mediates axonal tiling in Drosophila. J Neurosci Off J Soc Neurosci 29:14151–14159. doi:10.1523/JNEUROSCI.2497-09.2009
Frank E, Westerfield M (1982) The formation of appropriate central and peripheral connexions by foreign sensory neurones of the bullfrog. J Physiol 324:495–505
Fujishima K, Horie R, Mochizuki A, Kengaku M (2012) Principles of branch dynamics governing shape characteristics of cerebellar Purkinje cell dendrites. Dev Camb Engl 139:3442–3455. doi:10.1242/dev.081315
Gallegos ME, Bargmann CI (2004) Mechanosensory neurite termination and tiling depend on SAX-2 and the SAX-1 kinase. Neuron 44:239–249. doi:10.1016/j.neuron.2004.09.021
Galli-Resta L, Resta G, Tan SS, Reese BE (1997) Mosaics of islet-1-expressing amacrine cells assembled by short-range cellular interactions. J Neurosci Off J Soc Neurosci 17:7831–7838
Gandarillas A, Freije A (2014) Cycling up the epidermis: reconciling 100 years of debate. Exp Dermatol 23:87–91. doi:10.1111/exd.12287
Gao FB, Kohwi M, Brenman JE, Jan LY, Jan YN (2000) Control of dendritic field formation in Drosophila: the roles of flamingo and competition between homologous neurons. Neuron 28:91–101
Gershon TR, Baker MW, Nitabach M, Macagno ER (1998) The leech receptor protein tyrosine phosphatase HmLAR2 is concentrated in growth cones and is involved in process outgrowth. Dev Camb Engl 125:1183–1190
Gibson DA, Tymanskyj S, Yuan RC, Leung HC, Lefebvre JL, Sanes JR, Chédotal A, Ma L (2014) Dendrite self-avoidance requires cell-autonomous slit/robo signaling in cerebellar purkinje cells. Neuron 81:1040–1056. doi:10.1016/j.neuron.2014.01.009
Grueber WB, Sagasti A (2010) Self-avoidance and tiling: mechanisms of dendrite and axon spacing. Cold Spring Harb Perspect Biol. doi:10.1101/cshperspect.a001750
Grueber WB, Graubard K, Truman JW (2001) Tiling of the body wall by multidendritic sensory neurons in Manduca sexta. J Comp Neurol 440:271–283. doi:10.1002/cne.1385
Grueber WB, Jan LY, Jan YN (2002) Tiling of the Drosophila epidermis by multidendritic sensory neurons. Dev Camb Engl 129:2867–2878
Grueber WB, Ye B, Moore AW, Jan LY, Jan YN (2003) Dendrites of distinct classes of Drosophila sensory neurons show different capacities for homotypic repulsion. Curr Biol CB 13:618–626
Han C, Wang D, Soba P, Zhu S, Lin X, Jan LY, Jan Y-N (2012a) Integrins regulate repulsion-mediated dendritic patterning of drosophila sensory neurons by restricting dendrites in a 2D space. Neuron 73:64–78. doi:10.1016/j.neuron.2011.10.036
Han C, Wang D, Soba P, Zhu S, Lin X, Jan LY, Jan Y-N (2012b) Integrins regulate repulsion-mediated dendritic patterning of drosophila sensory neurons by restricting dendrites in a 2D space. Neuron 73:64–78. doi:10.1016/j.neuron.2011.10.036
Hattori D, Millard SS, Wojtowicz WM, Zipursky SL (2008) Dscam-mediated cell recognition regulates neural circuit formation. Annu Rev Cell Dev Biol 24:597–620. doi:10.1146/annurev.cellbio.24.110707.175250
Hayes BP, Roberts A (1983) The anatomy of two functional types of mechanoreceptive "free" nerve-ending in the head skin of Xenopus embryos. Proc R Soc Lond B Biol Sci 218:61–76
Hedgecock EM, White JG (1985) Polyploid tissues in the nematode Caenorhabditis elegans. Dev Biol 107:128–133
Hergovich A, Bichsel SJ, Hemmings BA (2005) Human NDR kinases are rapidly activated by MOB proteins through recruitment to the plasma membrane and phosphorylation. Mol Cell Biol 25:8259–8272. doi:10.1128/MCB.25.18.8259-8272.2005
Hergovich A, Stegert MR, Schmitz D, Hemmings BA (2006) NDR kinases regulate essential cell processes from yeast to humans. Nat Rev Mol Cell Biol 7:253–264. doi:10.1038/nrm1891

Hitchcock PF (1987) Constant dendritic coverage by ganglion cells with growth of the goldfish's retina. Vision Res 27:17–22
Hitchcock PF (1989) Exclusionary dendritic interactions in the retina of the goldfish. Dev Camb Engl 106:589–598
Hitchcock PF, Easter SS (1986) Retinal ganglion cells in goldfish: a qualitative classification into four morphological types, and a quantitative study of the development of one of them. J Neurosci Off J Soc Neurosci 6:1037–1050
Huckfeldt RM, Schubert T, Morgan JL, Godinho L, Di Cristo G, Huang ZJ, Wong ROL (2009) Transient neurites of retinal horizontal cells exhibit columnar tiling via homotypic interactions. Nat Neurosci 12:35–43. doi:10.1038/nn.2236
Hughes ME, Bortnick R, Tsubouchi A, Bäumer P, Kondo M, Uemura T, Schmucker D (2007) Homophilic Dscam interactions control complex dendrite morphogenesis. Neuron 54:417–427. doi:10.1016/j.neuron.2007.04.013
Jiang N, Soba P, Parker E, Kim CC, Parrish JZ (2014) The microRNA bantam regulates a developmental transition in epithelial cells that restricts sensory dendrite growth. Dev Camb Engl 141:2657–2668. doi:10.1242/dev.107573
Kay JN, Chu MW, Sanes JR (2012) MEGF10 and MEGF11 mediate homotypic interactions required for mosaic spacing of retinal neurons. Nature 483:465–469. doi:10.1038/nature10877
Keeley PW, Reese BE (2010) Morphology of dopaminergic amacrine cells in the mouse retina: independence from homotypic interactions. J Comp Neurol 518:1220–1231. doi:10.1002/cne.22270
Keeley PW, Whitney IE, Raven MA, Reese BE (2007) Dendritic spread and functional coverage of starburst amacrine cells. J Comp Neurol 505:539–546. doi:10.1002/cne.21518
Kim ME, Shrestha BR, Blazeski R, Mason CA, Grueber WB (2012) Integrins establish dendrite-substrate relationships that promote dendritic self-avoidance and patterning in drosophila sensory neurons. Neuron 73:79–91. doi:10.1016/j.neuron.2011.10.033
Kitson DL, Roberts A (1983) Competition during innervation of embryonic amphibian head skin. Proc R Soc Lond B Biol Sci 218:49–59
Kniss JS, Holbrook S, Herman TG (2013) R7 photoreceptor axon growth is temporally controlled by the transcription factor Ttk69, which inhibits growth in part by promoting transforming growth factor-β/activin signaling. J Neurosci Off J Soc Neurosci 33:1509–1520. doi:10.1523/JNEUROSCI.2023-12.2013
Kock JH (1982) Dendritic tree structure and dendritic hypertrophy during growth of the crucian carp eye. J Comp Neurol 209:275–286. doi:10.1002/cne.902090306
Koike-Kumagai M, Yasunaga K, Morikawa R, Kanamori T, Emoto K (2009) The target of rapamycin complex 2 controls dendritic tiling of Drosophila sensory neurons through the Tricornered kinase signalling pathway. EMBO J 28:3879–3892. doi:10.1038/emboj.2009.312
Koleske AJ (2013) Molecular mechanisms of dendrite stability. Nat Rev Neurosci 14:536–550. doi:10.1038/nrn3486
Kramer AP, Kuwada JY (1983) Formation of the receptive fields of leech mechanosensory neurons during embryonic development. J Neurosci Off J Soc Neurosci 3:2474–2486
Kramer AP, Stent GS (1985) Developmental arborization of sensory neurons in the leech Haementeria ghilianii. II. Experimentally induced variations in the branching pattern. J Neurosci Off J Soc Neurosci 5:768–775
Lah GJ-E, Li JSS, Millard SS (2014) Cell-specific alternative splicing of Drosophila Dscam2 is crucial for proper neuronal wiring. Neuron 83:1376–1388. doi:10.1016/j.neuron.2014.08.002
Lee SCS, Cowgill EJ, Al-Nabulsi A, Quinn EJ, Evans SM, Reese BE (2011) Homotypic regulation of neuronal morphology and connectivity in the mouse retina. J Neurosci Off J Soc Neurosci 31:14126–14133. doi:10.1523/JNEUROSCI.2844-11.2011
Lee J, Peng Y, Lin W-Y, Parrish JZ (2015) Coordinate control of terminal dendrite patterning and dynamics by the membrane protein Raw. Dev Camb Engl 142:162–173. doi:10.1242/dev.113423

Lefebvre JL, Kostadinov D, Chen WV, Maniatis T, Sanes JR (2012) Protocadherins mediate dendritic self-avoidance in the mammalian nervous system. Nature 488:517–521. doi:10.1038/nature11305

Leventhal AG, Schall JD, Ault SJ (1988) Extrinsic determinants of retinal ganglion cell structure in the cat. J Neurosci Off J Soc Neurosci 8:2028–2038

Leventhal AG, Ault SJ, Vitek DJ, Shou T (1989) Extrinsic determinants of retinal ganglion cell development in primates. J Comp Neurol 286:170–189. doi:10.1002/cne.902860204

Liang X, Dong X, Moerman DG, Shen K, Wang X (2015) Sarcomeres pattern proprioceptive sensory dendritic endings through UNC-52/Perlecan in C. elegans. Dev Cell 33:388–400. doi:10.1016/j.devcel.2015.03.010

Lin B, Wang SW, Masland RH (2004) Retinal ganglion cell type, size, and spacing can be specified independent of homotypic dendritic contacts. Neuron 43:475–485. doi:10.1016/j.neuron.2004.08.002

Lozano E, Sáez AG, Flemming AJ, Cunha A, Leroi AM (2006) Regulation of growth by ploidy in Caenorhabditis elegans. Curr Biol CB 16:493–498. doi:10.1016/j.cub.2006.01.048

Masland RH (2012) The neuronal organization of the retina. Neuron 76:266–280. doi:10.1016/j.neuron.2012.10.002

Matthews BJ, Kim ME, Flanagan JJ, Hattori D, Clemens JC, Zipursky SL, Grueber WB (2007) Dendrite self-avoidance is controlled by Dscam. Cell 129:593–604. doi:10.1016/j.cell.2007.04.013

Millard SS, Zipursky SL (2008) Dscam-mediated repulsion controls tiling and self-avoidance. Curr Opin Neurobiol 18:84–89. doi:10.1016/j.conb.2008.05.005

Millard SS, Flanagan JJ, Pappu KS, Wu W, Zipursky SL (2007) Dscam2 mediates axonal tiling in the Drosophila visual system. Nature 447:720–724. doi:10.1038/nature05855

Mizumoto K, Shen K (2013) Interaxonal interaction defines tiled presynaptic innervation in C. elegans. Neuron 77:655–666. doi:10.1016/j.neuron.2012.12.031

Montague PR, Friedlander MJ (1991) Morphogenesis and territorial coverage by isolated mammalian retinal ganglion cells. J Neurosci Off J Soc Neurosci 11:1440–1457

Negishi K, Teranishi T, Kato S (1982) New dopaminergic and indoleamine-accumulating cells in the growth zone of goldfish retinas after neurotoxic destruction. Science 216:747–749

Nern A, Pfeiffer BD, Rubin GM (2015) Optimized tools for multicolor stochastic labeling reveal diverse stereotyped cell arrangements in the fly visual system. Proc Natl Acad Sci U S A. doi:10.1073/pnas.1506763112

Nicholls JG, Baylor DA (1968) Specific modalities and receptive fields of sensory neurons in CNS of the leech. J Neurophysiol 31:740–756

O'Brien GS, Martin SM, Söllner C, Wright GJ, Becker CG, Portera-Cailliau C, Sagasti A (2009) Developmentally regulated impediments to skin reinnervation by injured peripheral sensory axon terminals. Curr Biol CB 19:2086–2090. doi:10.1016/j.cub.2009.10.051

O'Brien GS, Rieger S, Wang F, Smolen GA, Gonzalez RE, Buchanan J, Sagasti A (2012) Coordinate development of skin cells and cutaneous sensory axons in zebrafish. J Comp Neurol 520:816–831. doi:10.1002/cne.22791

Ogata K, Kosaka T (2002) Structural and quantitative analysis of astrocytes in the mouse hippocampus. Neuroscience 113:221–233

Orr-Weaver TL (2015) When bigger is better: the role of polyploidy in organogenesis. Trends Genet TIG. doi:10.1016/j.tig.2015.03.011

Parrish JZ, Emoto K, Jan LY, Jan YN (2007) Polycomb genes interact with the tumor suppressor genes hippo and warts in the maintenance of Drosophila sensory neuron dendrites. Genes Dev 21:956–972. doi:10.1101/gad.1514507

Parrish JZ, Xu P, Kim CC, Jan LY, Jan YN (2009) The microRNA bantam functions in epithelial cells to regulate scaling growth of dendrite arbors in drosophila sensory neurons. Neuron 63:788–802. doi:10.1016/j.neuron.2009.08.006

Perry VH, Linden R (1982) Evidence for dendritic competition in the developing retina. Nature 297:683–685

Ramoa AS, Campbell G, Shatz CJ (1987) Transient morphological features of identified ganglion cells in living fetal and neonatal retina. Science 237:522–525
Ramoa AS, Campbell G, Shatz CJ (1988) Dendritic growth and remodeling of cat retinal ganglion cells during fetal and postnatal development. J Neurosci Off J Soc Neurosci 8:4239–4261
Raven MA, Stagg SB, Nassar H, Reese BE (2005) Developmental improvement in the regularity and packing of mouse horizontal cells: implications for mechanisms underlying mosaic pattern formation. Vis Neurosci 22:569–573. doi:10.1017/S095252380522504X
Reese BE, Keeley PW (2014) Design principles and developmental mechanisms underlying retinal mosaics. Biol Rev Camb Philos Soc. doi:10.1111/brv.12139
Reese BE, Raven MA, Stagg SB (2005) Afferents and homotypic neighbors regulate horizontal cell morphology, connectivity, and retinal coverage. J Neurosci Off J Soc Neurosci 25:2167–2175. doi:10.1523/JNEUROSCI.4876-04.2005
Reese BE, Keeley PW, Lee SCS, Whitney IE (2011) Developmental plasticity of dendritic morphology and the establishment of coverage and connectivity in the outer retina. Dev Neurobiol 71:477–490. doi:10.1002/dneu.20903
Rehberg K, Kliche S, Madencioglu DA, Thiere M, Müller B, Meineke BM, Freund C, Budinger E, Stork O (2014) The serine/threonine kinase Ndr2 controls integrin trafficking and integrin-dependent neurite growth. J Neurosci Off J Soc Neurosci 34:5342–5354. doi:10.1523/JNEUROSCI.2728-13.2014
Ren L, Liang H, Diao L, He S (2010) Changing dendritic field size of mouse retinal ganglion cells in early postnatal development. Dev Neurobiol 70:397–407. doi:10.1002/dneu.20777
Rockhill RL, Euler T, Masland RH (2000) Spatial order within but not between types of retinal neurons. Proc Natl Acad Sci U S A 97:2303–2307. doi:10.1073/pnas.030413497
Rockhill RL, Daly FJ, MacNeil MA, Brown SP, Masland RH, (2002) The diversity of ganglion cells in a mammalian retina. J Neurosci Off J Soc Neurosci 22:3831–3843
Sagasti A, Guido MR, Raible DW, Schier AF (2005) Repulsive interactions shape the morphologies and functional arrangement of zebrafish peripheral sensory arbors. Curr Biol CB 15:804–814. doi:10.1016/j.cub.2005.03.048
Salzberg Y, Díaz-Balzac CA, Ramirez-Suarez NJ, Attreed M, Tecle E, Desbois M, Kaprielian Z, Bülow HE (2013) Skin-derived cues control arborization of sensory dendrites in Caenorhabditis elegans. Cell 155:308–320. doi:10.1016/j.cell.2013.08.058
Schmucker D, Clemens JC, Shu H, Worby CA, Xiao J, Muda M, Dixon JE, Zipursky SL (2000) Drosophila Dscam is an axon guidance receptor exhibiting extraordinary molecular diversity. Cell 101:671–684
Scott SA, Cooper E, Diamond J (1981) Merkel cells as targets of the mechanosensory nerves in salamander skin. Proc R Soc Lond B Biol Sci 211:455–470
Shimono K, Fujimoto A, Tsuyama T, Yamamoto-Kochi M, Sato M, Hattori Y, Sugimura K, Usui T, Kimura K, Uemura T (2009) Multidendritic sensory neurons in the adult Drosophila abdomen: origins, dendritic morphology, and segment- and age-dependent programmed cell death. Neural Dev 4:37. doi:10.1186/1749-8104-4-37
Shimono K, Fujishima K, Nomura T, Ohashi M, Usui T, Kengaku M, Toyoda A, Uemura T (2014) An evolutionarily conserved protein CHORD regulates scaling of dendritic arbors with body size. Sci Rep 4:4415. doi:10.1038/srep04415
Smith CJ, Watson JD, Spencer WC, O'Brien T, Cha B, Albeg A, Treinin M, Miller DM (2010) Time-lapse imaging and cell-specific expression profiling reveal dynamic branching and molecular determinants of a multi-dendritic nociceptor in C. elegans. Dev Biol 345:18–33. doi:10.1016/j.ydbio.2010.05.502
Smith CJ, O'Brien T, Chatzigeorgiou M, Spencer WC, Feingold-Link E, Husson SJ, Hori S, Mitani S, Gottschalk A, Schafer WR, Miller DM (2013) Sensory neuron fates are distinguished by a transcriptional switch that regulates dendrite branch stabilization. Neuron 79:266–280. doi:10.1016/j.neuron.2013.05.009
Soba P, Zhu S, Emoto K, Younger S, Yang S-J, Yu H-H, Lee T, Jan LY, Jan Y-N (2007) Drosophila sensory neurons require Dscam for dendritic self-avoidance and proper dendritic field organization. Neuron 54:403–416. doi:10.1016/j.neuron.2007.03.029

Soba P, Han C, Zheng Y, Perea D, Miguel-Aliaga I, Jan LY, Jan YN (2015) The ret receptor regulates sensory neuron dendrite growth and integrin mediated adhesion. eLife 4:e05491. doi:10.7554/eLife.05491

Sugimura K, Yamamoto M, Niwa R, Satoh D, Goto S, Taniguchi M, Hayashi S, Uemura T (2003) Distinct developmental modes and lesion-induced reactions of dendrites of two classes of Drosophila sensory neurons. J Neurosci Off J Soc Neurosci 23:3752–3760

Sulkowski MJ, Iyer SC, Kurosawa MS, Iyer EPR, Cox DN (2011) Turtle functions downstream of cut in differentially regulating class specific dendrite morphogenesis in Drosophila. PLoS One 6:e22611. doi:10.1371/journal.pone.0022611

Tauchi M, Masland RH (1984) The shape and arrangement of the cholinergic neurons in the rabbit retina. Proc R Soc Lond B Biol Sci 223:101–119

Tauchi M, Masland RH (1985) Local order among the dendrites of an amacrine cell population. J Neurosci Off J Soc Neurosci 5:2494–2501

Ting C-Y, Herman T, Yonekura S, Gao S, Wang J, Serpe M, O'Connor MB, Zipursky SL, Lee C-H (2007) Tiling of r7 axons in the Drosophila visual system is mediated both by transduction of an activin signal to the nucleus and by mutual repulsion. Neuron 56:793–806. doi:10.1016/j.neuron.2007.09.033

Ting C-Y, McQueen PG, Pandya N, Lin T-Y, Yang M, Reddy OV, O'Connor MB, McAuliffe M, Lee C-H (2014) Photoreceptor-derived activin promotes dendritic termination and restricts the receptive fields of first-order interneurons in Drosophila. Neuron 81:830–846. doi:10.1016/j.neuron.2013.12.012

Truman JW, Reiss SE (1988) Hormonal regulation of the shape of identified motoneurons in the moth Manduca sexta. J Neurosci Off J Soc Neurosci 8:765–775

Unhavaithaya Y, Orr-Weaver TL (2012) Polyploidization of glia in neural development links tissue growth to blood-brain barrier integrity. Genes Dev 26:31–36. doi:10.1101/gad.177436.111

Vaney DI (1984) "Coronate" amacrine cells in the rabbit retina have the "starburst" dendritic morphology. Proc R Soc Lond B Biol Sci 220:501–508

Vonhoff F, Duch C (2010) Tiling among stereotyped dendritic branches in an identified Drosophila motoneuron. J Comp Neurol 518:2169–2185. doi:10.1002/cne.22380

Wang F, Wolfson SN, Gharib A, Sagasti A (2012) LAR receptor tyrosine phosphatases and HSPGs guide peripheral sensory axons to the skin. Curr Biol CB 22:373–382. doi:10.1016/j.cub.2012.01.040

Wässle H, Riemann HJ (1978) The mosaic of nerve cells in the mammalian retina. Proc R Soc Lond B Contain Pap Biol Charact R Soc G B 200:441–461

Wässle H, Peichl L, Boycott BB (1981) Dendritic territories of cat retinal ganglion cells. Nature 292:344–345

Wässle H, Puller C, Müller F, Haverkamp S (2009) Cone contacts, mosaics, and territories of bipolar cells in the mouse retina. J Neurosci Off J Soc Neurosci 29:106–117. doi:10.1523/JNEUROSCI.4442-08.2009

Yasunaga K, Kanamori T, Morikawa R, Suzuki E, Emoto K (2010) Dendrite reshaping of adult Drosophila sensory neurons requires matrix metalloproteinase-mediated modification of the basement membranes. Dev Cell 18:621–632. doi:10.1016/j.devcel.2010.02.010

Zanet J, Freije A, Ruiz M, Coulon V, Sanz JR, Chiesa J, Gandarillas A (2010) A mitosis block links active cell cycle with human epidermal differentiation and results in endoreplication. PLoS One 5:e15701. doi:10.1371/journal.pone.0015701

Chapter 10
Mosaics and Lamination in the Retina

Peter G. Fuerst

Abstract The highly ordered architecture of the neural retina makes it an excellent system to study neuron spacing and dendrite organization. Adult retinal neurites and cell bodies are organized in a highly stereotyped manner in formation of the radial circuitry of the retina, from photoreceptors to retinal ganglion cells. The circuitry formed by these cells is functionally separated into ON and OFF pathways in the retinal outer plexiform layer. These pathways are then spatially divided within the retinal inner plexiform layer, in which coding for additional aspects of vision, such as movement and edge detection, are added. Many cell types and neural circuits in the retina are also spaced across the horizontal plane of the retina so as to sample these and other features of vision evenly across the visual field. The mechanisms and molecules underpinning organization of the retina have been the subjects of intensive study. These studies have found that retinal circuitry is largely hardwired by transmembrane receptors, many of them cell adhesion molecules, and their ligands, with activity helping to further refine development. In this chapter, we focus on developmental mechanisms, such as programmed cell death, migration, and neurite refinement that increase organization of retinal cell bodies and neurites, as well as the molecules by which cells recognize and respond to appropriate contacts and domains.

Keywords Mosaic • Lamination • Refinement • Neurite • Dendrite • Apoptosis • Retina

Terms

Agene	Vertebrate DNA or RNA sequence
APROTEIN	Vertebrate protein
CAD	Cadherin

P.G. Fuerst (✉)
Department of Biological Sciences, University of Idaho, Moscow, ID 83844, USA

WWAMI Medical Education Program, University of Washington School of Medicine, Moscow, ID 83844, USA
e-mail: fuerst@uidaho.edu

K. Emoto et al. (eds.), *Dendrites*, DOI 10.1007/978-4-431-56050-0_10

CAR	Cone arrestin
CBC	Cone bipolar cell
ChAT	Choline acetyltransferase
CNTN	Contactin
DAC	Dopaminergic amacrine cell
DCD	Developmental cell death
DSCAM	Down syndrome cell adhesion molecule
ECM	Extracellular matrix
HC	Horizontal cell
INL	Inner nuclear layer
IPL	Inner plexiform layer
ipRGC	Intrinsically photoresponsive retinal ganglion cell
MEGF	Multiple EGF-like domain protein
NGL2	Netrin-G ligand 2/leucine-rich repeat-containing protein 4
ONL	Outer nuclear layer
OPL	Outer plexiform layer
PCDHG	γ-Protocadehrin
PLXN	Plexin
PR	Photoreceptor
PTTG	Pituitary tumor-transforming gene
RBC	Rod bipolar cell
RGC	Retinal ganglion cell
RGL	Retinal ganglion layer
SAC	Starburst amacrine cell
SDK	Sidekick
SEMA	Semaphorin

10.1 Introduction: Mosaic and Neurite Lamination in the Retina

The neural retina offers an excellent system to consider the structure and function of dendrites. The vertebrate retina is clearly divided into three cellular layers and two synaptic layers, composed of a limited number of cell types that make highly stereotyped connections (Masland 2012). The organized structure of the retina and its clear physiological output has made this organ an excellent model to study principles by which dendrites are organized. In this chapter, the mechanisms and molecular underpinnings of retinal organization will be discussed, including (1) recognition of self, like, and non-like cells, (2) mosaic spacing of cells, and (3) laminar targeting. By necessity, we were unable to include much equally relevant and important research, especially with respect to non-vertebrate systems.

10.2 Anatomy of the Retina

Retinal neurons are organized along multiple axes, principally the radial axis (in this chapter, this is considered from the sclera to vitreous) from photoreceptor to retinal ganglion cell (RGC), and the horizontal axis, running perpendicular to the radial axis. The mature retina is organized in three cellular layers, the outer nuclear layer (ONL), which contains the soma of rod and cone photoreceptors; the inner nuclear layer (INL), which contains the soma of horizontal cells (HCs), Müller glia, bipolar cells, and amacrine cells; and the retinal ganglion cell layer (RGL), which contains the soma of RGCs and amacrine cells (Fig. 10.1a, b). In addition to these cell populations, which are all produced by retinal progenitor cells, the retina contains populations of astrocytes, microglia, and cells of the retinal vasculature. Retinal neurons form synapses within two layers of the retina, the outer plexiform layer (OPL), which primarily contains the neurites and synapses between rods or cones and HCs and bipolar cells, and the inner plexiform layer (IPL) which contains the neurites and synapses between bipolar, amacrine, and RGCs. The IPL is conventionally divided into five sublamina, labeled S1–S5, in the direction from the outer retina (S1) to the inner retina (S5). The IPL can also be divided functionally between circuits that encode synapses that receive input from bipolar cells that depolarize in the presence of light and are therefore responsive to light (ON) and circuits that receive input primarily from bipolar cells that depolarize in the absence of light-driven stimulus and are therefore responsive to darkness (OFF) (Fig. 10.1c) (Nelson et al. 1978). The organization of retinal neurons into circuits across the radial axis of the retina is essential for the functional physiology of the retina, but the individual circuits are also spaced across the horizontal axis of the retina to maximize sampling and output of different aspects of vision, for example, color or motion detection.

Organization of the retina across the horizontal plane of the retina involves large-scale changes in cellular composition as well as shorter-range organization of cells and circuits. Dorsal-ventral changes in photoreceptor distribution have been well documented in mice, with the latter transitioning from primarily green cones dorsally to blue/green cones ventrally (Szel et al. 1992). A dramatic change in cell composition has also been well characterized in the primate retina, in which the cone dominant fovea is centered in the cone-rich macula, surrounded by the rod-dominant peripheral retina. On a more local scale, many cell types in the retina are spaced with respect to cells of the same subtype in patterns called mosaics, in which cells avoid areas close to cells of the same subtype (Fig. 10.1d) (Morris 1970; Wassle and Riemann 1978). Organization of retinal neurons into mosaics is common enough in the retina that mosaic organization is considered evidence that these cells constitute a distinct subtype of cell. Even those cells that do not have soma organized into mosaics, such as bipolar cells, have tiled dendritic and axonal fields, so as to sample and represent a given receptive field evenly across the retina.

Distribution of retinal axons and dendrites, like the distribution of neural soma, is highly stereotyped and can be described by a few simple principles, although

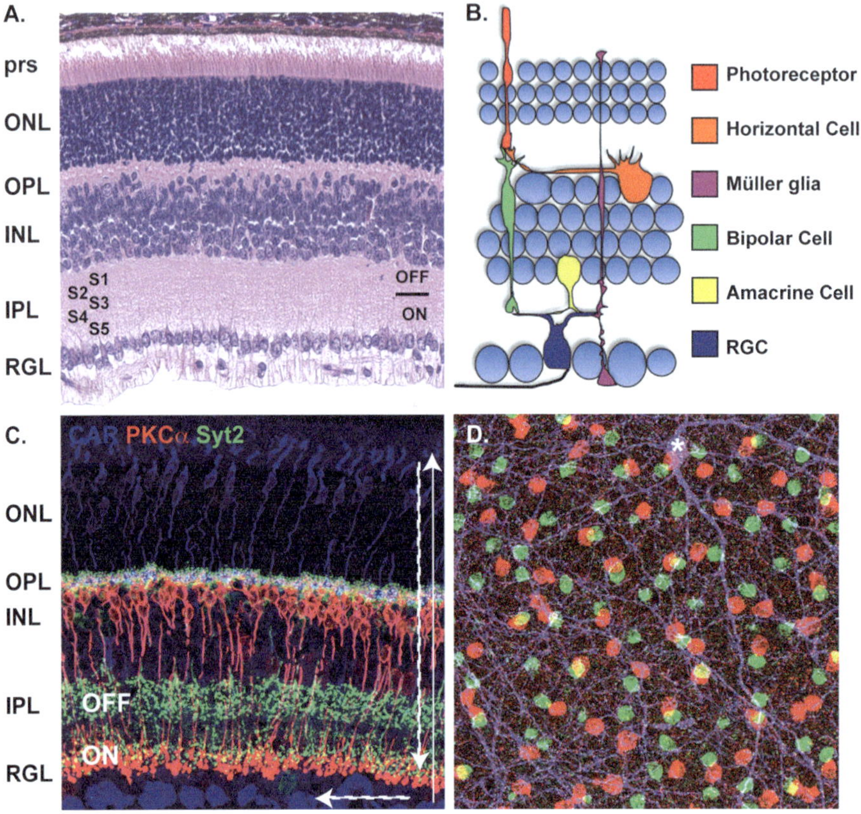

Fig. 10.1 Organization of the neural retina. (**a**) Hematoxylin- and eosin-stained section of mouse retina. The retina is organized in the radial plane from the outer most retina, which contains the photoreceptor inner and outer segments (prs) to the retinal ganglion cell layer (RGL), located in the inner retina. The cellular layers include the outer nuclear layer (ONL), inner nuclear layer (INL), and retinal ganglion cell layer (RGL), while the synaptic outer plexiform layer (OPL) and inner plexiform layer (IPL) contain the synapses connecting the cells of the neural retina. The IPL is functionally divided into an ON and OFF half and conventionally divided into five strata, S1-S5. (**b**) Cartoon of the retina depicting the location of the five principal types of retinal neuron and Müller glia. The ONL contains the soma of rods and cones. The INL contains the soma of horizontal cells (HCs), amacrine cells, bipolar cells, and Müller glia. The RGL contains the soma of retinal ganglion cells (RGCs) and amacrine cells. (**c**) Section of mouse retina stained with antibodies to cone arrestin (CAR), labeling cones; PKCα, labeling (ON) rod bipolar cells; and Syt2, which labels type 2 (OFF) and type 6 (ON) cone bipolar cells. Light passes through the inner retina (*arrow*) to the photoreceptors. Information is sent and processed from photoreceptors (*vertical dashed line*) through the retina to the retinal ganglion cells (RGCs), where visual information is transmitted to the rest of the brain. (**d**) Confocal image of the mouse retina stained with antibodies to ChAT, which labels OFF and ON starburst amacrine cells (SACs), and tyrosine hydroxylase, which labels a single dopaminergic amacrine cell in the field (DAC, *asterisk*). OFF SACs (labeled in *green/light gray*), located in the INL, are organized such that they are spaced away from like cells. ON SACs (labeled in *red/dark gray*), located in the RGL, are also spaced away from other SACs, but this is disrupted somewhat by growth of the retinal vasculature

defining what process is an axon versus a dendrite is not as straightforward in the retina as in other regions of the brain. Many neurons in the retina, especially amacrine cells, do not have what might classically be considered axons and dendrites, and these processes, which receive and transmit synaptic information, are often referred to as neurites. Unlike the use of this term in reference to immature neural outgrowths, the neurites of the adult retina are mature processes, and the ambiguity in their description reflects their dual function. Other cell types, such as bipolar cells, have what could be recognized as dendrites, projecting into the OPL, and axons, projecting into the IPL, while RGCs clearly have receptive dendrites and axons that project out of the retina and into the rest of the brain.

10.3 Self, Like, and Non-like Recognition in Retinal Organization

Organization of retinal neurons and neurites across the radial and horizontal planes of the retina occurs within cell types and appears to be insensitive to even widespread ablation of a given cell type's synaptic partners. Studies in which the various classes of cells (RGCs, amacrine cells, bipolar cells, Müller glia) were depleted or ablated found that mosaic spacing and formation of the IPL could occur in the absence of many given cell types and even in the absence of multiple cell types (Rossi et al. 2003; Lin et al. 2004; Randlett et al. 2013; Kay et al. 2004; Williams et al. 2001). Regulation of cell number and dendrite complexity appears to be regulated within cell types and independent of afferent input or target number (Keeley et al. 2014a, b; Lee et al. 2011; Raven and Reese 2003). It therefore appears that retinal neurons can be organized with a limited number of cues coming from heterotypic cells. In this chapter, we will first consider requirements that individual cells need in order to adopt their morphology, then the requirements that a cell type would have in order to adopt its organization exclusively within that cell type, and finally cues that help to organize the organ as a whole.

Development of the retina involves multiple levels of organization, from the cell, to the cell type, and finally to integrating cell types into functional circuitry. First, consider organization of a cell's neurites with respect to self: do the cell's neurites avoid crossing or bundling with its own neurites, are a cell's neurites indifferent to crossing with its own neurites, or are the cell's neurites attracted to other neurites projecting from the same cell (Fig. 10.2a)? This describes the cell autonomous organization of the cell. Next, consider the interactions between the neurites of a cell and the neurites of other cells of the same subtype: do the cell's neurites avoid the neurites of like cells, are they indifferent to the neurites of like cells, or are they attracted to the neurites of like cells? This describes the cell-type level of organization within the retina. Answering these two questions results in nine different possible outcomes, before even considering interactions with heterotypic cells, and these can be used to describe dendrite organization in considering

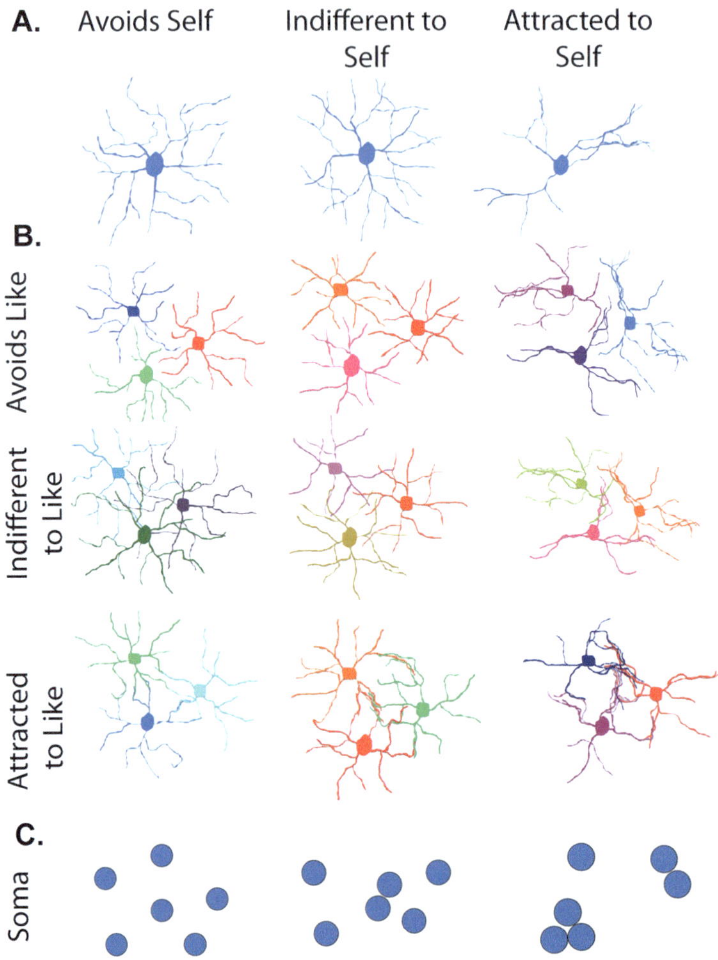

Fig. 10.2 Organization of retinal neurites and cells across the horizontal plane of the retina. Cartoon depictions of neurite and soma organization in the retina. The degree to which neurites avoid self, are indifferent to self or are attracted to self can be assayed by monitoring the number of times neurites cross like neurites, which is minimized in the case of avoidance. (**a**) A cartoon depiction of the neurite organization patterns of single cells. The neurites of a single cell can either avoid other neurites extending from the same cell (avoids self), be indifferent to neurites extending from the same cell (indifferent to self), or adhere to neurites extending from the same cell (attracted to self). (**b**) A cartoon depicting the organization of neurites within a single cell type. Based on the organization in (**a**), and how the neurites of a cell interact with neurites extending from other cells of the same subtype, nine possible organizational strategies observed during development, in the adult and in disease. (**c**) A cartoon depicting different spacing strategies observed with respect to single populations of retinal neurons. The cell bodies of neurons of the same type can organize to avoid cells of the same type (avoids self); they can be indifferent to the position of cells of the same type (indifferent to self), or they can adhere to cells of the same subtype (attracted to self)

both the development and disorders of retinal neurites (Fig. 10.2b). Several terms are widely used to describe several of these outcomes. Tiling describes a situation in which a cell's neurites avoid crossing the neurites of cells of the same subtype, as is the case for the dendrites and axons of bipolar cells (Wassle et al. 2009). Tiling is also used to describe situations where cells such as RGCs have spaced receptive fields, in which case there is often overlap at the edges of a cell's receptive fields. In contrast to tiling, many other cells project neurites that overlap extensively with the neurites of other cells of the same type (Reese and Keeley 2014). This is described with the term coverage, and reflects the number of cells that sample a given area within the retina, which can be as high as 30 for some cell types, such as starburst amacrine cells (SACs). Other cells, especially after degeneration and remodeling, or as the result of a genetic deficiency, have clumped neurites (Marc et al. 2007, 2008).

A similar framework is appropriate for considering spacing of the soma of retinal neurons, which may exhibit self-avoidance, self-indifference, or self-attraction (Fig. 10.2c). The spacing of cell soma in mosaics is often referred to as tiling, although in the mouse retina, this can introduce confusion because the tiled soma of a cell type may project neurites that overlap extensively and therefore do not tile.

10.4 Mechanisms of Organization

Given the organized structure of the retina, and the functional consequences of disrupting this organization, identifying the mechanisms by which the retina develops has been an active area of research. Two general mechanisms by which organization of retinal cell bodies and neurites are refined are developmental cell death (DCD) and cell and neurite motility.

Development of the retina and brain in general involves the overproduction of neurons that project neurites through a broader area of the plexiform layers than what is observed in the adult (White et al. 1998; Kuida et al. 1996). About half of these cells are eventually eliminated, and this is thought to reflect the elimination of cells that are misplaced or do not make appropriate synaptic contacts (Fig. 10.3a, b). Studies using mice deficient in cell death pathway proteins such as BAX have demonstrated that the additional cells rescued from undergoing cell death include those that are misplaced, therefore providing evidence that DCD can be triggered to remove cells that do not integrate into normal retinal circuits (Chen et al. 2013; Keeley et al. 2012; Wong and Hughes 1987).

Other mechanisms that promote organization of retinal soma include tangential migration of cell bodies across the horizontal plane of the retina and the refinement of neurites (Fig. 10.3c) (Reese et al. 1999; Reese and Galli-Resta 2002). Evidence for horizontal migration includes the varying tendency of cells to migrate out of the clonal columns in which they were born. By tracking cells during migration,

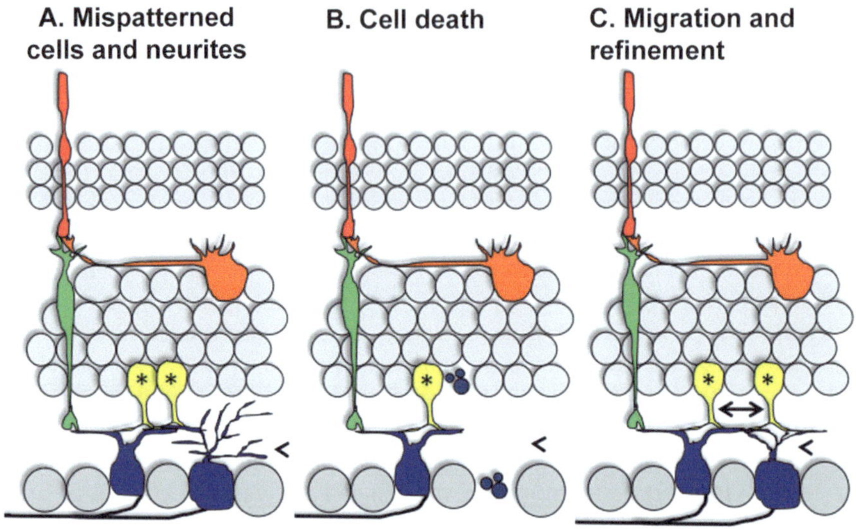

Fig. 10.3 Cell death, cell movement, and neurite refinement as retinal organizers. (**a**) A cartoon depiction of a retina with mispatterned cell bodies (*asterisks*) and unrefined neurites (*arrowhead*). The mispatterning of cell somata, such that they are spaced close to cells of the same subtype or in inappropriate lamina, and the misplacement of neurites are observed to a greater degree in the developing retina than in the mature retina. (**b**) A cartoon of the retina depicting apoptosis as a mechanism to resolve disorganization shown in (**a**). Apoptotic death of misplaced cells has been demonstrated to increase spacing within cell types and reduce instances of misplaced synapses. Apoptotic cells are depicted as condensed nuclear bodies. (**c**) A cartoon depicting soma migration (*arrows*) and neurite refinement as a solution to disorganization depicted in (**a**). Tangential migration of soma across the horizontal plane of the retina has been demonstrated to increase the distance between cells of the same type, and molecules that act in recognition and migration have been identified that promote this process

tangential dispersal has been demonstrated to guide the spacing of SACs as they migrate into their respective mosaics (Galli-Resta et al. 1997).

Movement not only increases the regularity of soma spacing but also helps to restrict and refine the organization of retinal neurites. RGCs, and other retinal neurons, initially project neurites in a broader pattern than is observed in the adult (Sernagor et al. 2001; Wong 1990). The initial bushy distribution of neurites likely involves two distinct requirements: that cells find the correct synaptic partners and that cells space themselves properly. This can involve multiple stages of neurite outgrowth and retraction and radically change the organization of a given cell type, for example, HC neurites tile during early development of the mouse retina, which is thought to account for the highly spaced organization of these cells' soma (Huckfeldt et al. 2009). Later in development, however, these HC neurites retract, and the new dendrites and axons that HCs develop overlap extensively, although there is evidence that they regulate arbor size in response to HC density (Poche et al. 2008).

10.5 Molecular Mechanisms of Retinal Neurite and Soma Organization

Much current work in developmental biology of the retina is focused on identifying the molecules that promote the processes listed above. The widespread adoption of transgenic approaches, for visualization, targeted cell depletion and mutagenesis, has rapidly expanded our understanding of the molecules that drive these processes and are discussed below. Unless otherwise noted, this research was performed in mice.

10.6 Genetics and Organization of the Starburst Amacrine Cell

The cell type for which organization is best understood on the molecular level is currently the SAC (Famiglietti 1983). Early expression of *islet1* and choline acetyltransferase (*ChAT*) in these cells, their critical role in direction selectivity and retinal waves, the ease at which they can be selectively ablated, their highly ordered spatial organization, and their abundance in the retina have all contributed to the scientific community's considerable understanding of this cell type's development and function. The next section will outline what is known about the molecular regulation of this cell type in development to illuminate progress that has been made, as well as to identify current knowledge gaps with respect to this cell type and organization of retinal neurites and soma in general.

SACs are located in separate mosaics in both the INL and RGL (Famiglietti 1983). SACs migrating into their respective layers are not initially organized into mosaics but undergo tangential migration as they enter the mosaic to maximize spacing (Galli-Resta et al. 1997). Unlike many retinal neurons, this spacing does not require normal developmental cell death, as elimination of the cell death protein BAX does not reduce the regularity of their spacing, as is observed for several other types of retinal neurons (Keeley et al. 2012). After entering their respective mosaics, SACs project neurites, which are visible in two distinct bands as early as postnatal day 0 (P0). SACs play a central role in establishing several aspects of retinal organization and function. The early developmental time at which SACs laminate is thought to set guideposts by which other cells organize their neurites, although these may be redundant with other cues, given that SACs can be ablated without disrupted lamination of other cell types (Farajian et al. 2004; Reese et al. 2001), and the angle at which starburst amacrine cell neurites intersect with direction-selective retinal ganglion cells results in the asymmetric delivery of inhibition, which is critical for their direction selectivity (Stacy and Wong 2003; Grzywacz et al. 1998; Wei et al. 2011; Briggman et al. 2011; Yoshida et al. 2001). Multiple genes, each discussed and defined in their respective sections, have been identified that are required for organization of SACs with respect to self neurites

(*Pcdhg*, *Sema6a*, and *Plxna2*) (Lefebvre et al. 2012; Sun et al. 2013), within respective mosaics (*Megf10*, *Pttg1*, and *Barhl2*) (Kay et al. 2012; Ding et al. 2009; Keeley et al. 2014c), and with respect to the two individual SAC mosaics (*Sox2*, *Sema6a*, and *Plxna2*) (Sun et al. 2013; Whitney et al. 2014).

The first question we will consider in how SACs adopt their morphology and spacing is: how do they adopt the characteristic starburst morphology of their neurites, in which proximal neurites avoid self-crossing and distal neurites fasciculate with other SAC neurites? In addressing this question, it is important to consider that a mechanism could be entirely intrinsic, that is, the SAC could project neurites, which grow a determined amount and then branch. In fact, the SAC requires molecules that permit it to recognize and avoid like neurites while still exhibiting the high coverage that these cells have at any given location. Genetic studies have found that the self-avoidance observed in SAC neurites is mediated by differential expression of γ-protocadherins. The γ-protocadherin (*Pcdhg*) complex is a genetic locus that encodes 22 different proteins by alternative inclusion of different variable exons (Tasic et al. 2002). Genetic deletion of the γ-protocadherin complex was initially thought to result primarily in an increase in developmental cell death (Lefebvre et al. 2008), but a more thorough examination found that the γ-protocadherin complex is sufficient to prevent self-avoidance, resulting in clumping of SAC neurites (Fig. 10.4a, b) (Lefebvre et al. 2012). The number of γ-protocadherin isoforms specified by the cluster, 22, roughly matches the coverage of SACs, estimated between 17 and 39 (Farajian et al. 2004; Chun et al. 1996), suggesting that the γ-protodcadherins are a mechanism by which SACs can avoid self but still overlap extensively with other SACs. This hypothesis was tested by driving a single isoform of γ-protocadherin in all SACs, and indeed a reduction in interactions between the neurites of adjacent SACs was observed. The

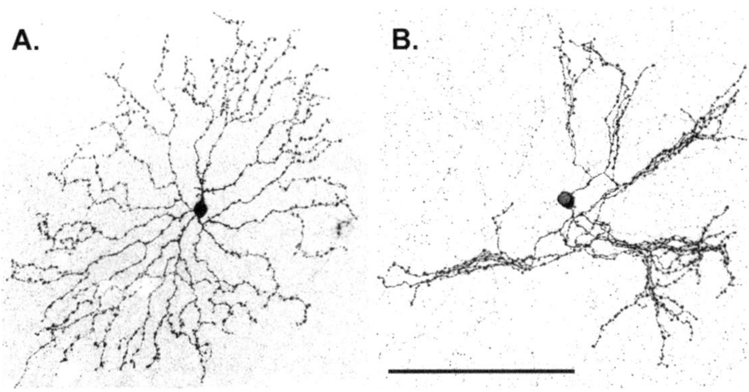

Fig. 10.4 Self-avoidance mediated by γ-protocadherin in SACs. (**a**) A single wild-type SAC. The neurites of this cell project out from the soma with minimal overlap. (**b**) A single γ-protocadherin (*pcdhg*) null SAC. Extensive adhesion between neurites is observed. The scale bar is equivalent to 125 μm (Reprinted by permission from Macmillan Publishers Ltd: Nature. Protocadherins mediate dendritic self-avoidance in the mammalian nervous system. Lefebvre et al. (2012), copyright 2012)

γ-protocadherins are widely expressed in the retina, including in Müller glia, and a major unanswered question is: how widely is this mechanism utilized in other cell types? Outside of the retina, the requirement of γ-protocadherin for self-avoidance has been clearly demonstrated in Purkinje cells, and they are also required for synapse formation in the spinal cord (Garrett and Weiner 2009; Lefebvre et al. 2012) and dendrite arborization in the cortex (Garrett et al. 2012), suggesting that the γ-protocadherin roles in dendrite organization may be widespread in the brain at large, but to date, the primary phenotype demonstrated in other cell types within the retina is to inhibit developmental cell death.

In addition to the γ-protocadherins, the transmembrane protein semaphorin 6a (SEMA6A) and its receptor plexin A2 (PLXNA2) are both required for self-avoidance at the distal tips of SAC neurites in vivo (Sun et al. 2013). Interestingly, SACs in these mutant backgrounds also failed to adopt a symmetrical distribution of neurites. These proteins also play an important role in establishing the ON and OFF symmetry of the IPL, which is discussed in more detail in the next section of this chapter. If γ-protocadherins, semaphorins, and plexins contribute to self-avoidance and yet permit extensive overlap, what are mechanisms that mediate spacing of SACs, given this overlap? Given that biased production of SACs is not sufficient to space these cells (Galli-Resta et al. 1997), efforts to undercover molecules that mediate this process are ongoing.

In order to space the soma across the horizontal plane of the retina, SACs require a system to recognize other SACs and move in response to these. The clearest part of this puzzle involves recognition mediated by the multiple EGF-like domain 10 (MEGF10) protein, which exhibits homotypic binding and is expressed by SACs during the embryonic times during which SACs are still migrating into position (Kay et al. 2012). Deletion of *Megf10* results in randomization of SAC spacing, and its ectopic expression can induce SAC avoidance in vitro and in vivo (Fig. 10.5a, b). A model wherein SAC somata migrate to equalize contact with proximally localized MEGF10 protein would provide a mechanism for SACs to sense density and reposition themselves to maximize spacing. Expression of *Megf10* is limited to SACs and HCs, all of which stratify in spatially distinct domains of the retina, therefore allowing a single gene to mediate this function in multiple cell populations. The temporal dynamics of SAC mosaic spacing suggests that MEGF10-mediated signaling occurs during early development of SACs: the ON SAC mosaic is initially spaced with a similar regularity as the OFF mosaic, but blood vessel growth across the RGL pushes cells out of their mosaic and degrades this spacing (Whitney et al. 2008). That MEGF10 does not correct this spacing therefore suggests a temporal limit on its function in this capacity. Therefore, MEGF10 appears to be a mechanism by which SACs can detect local concentrations of like cells and reposition themselves. It is currently unknown how SACs physically reposition themselves, but recent work identifying the pituitary tumor-transforming gene (*Pttg1*) as a player in this process is helping to fill this gap. The overexpression of *Pttg1* is associated with multiple types of tumor, and like *Megf10*, *Pttg1* is required to promote spacing of SACs (Fig. 10.5c, d) (Keeley et al. 2014c). While the mechanism by which *Pttg1* maintains SAC spacing remains

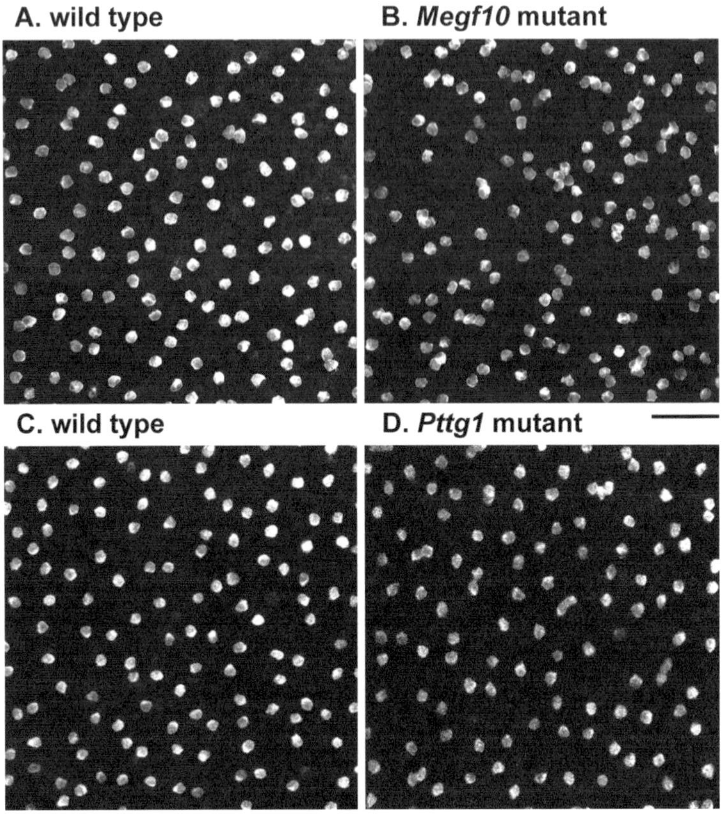

Fig. 10.5 *Megf10* and *Pttg1* are required for SAC spacing. (**a**) SACs in the INL of the mouse retina. In the wild-type retina, SACs space themselves across the horizontal plane of the retina. (**b**) SACs in the INL of the multiple EGF-like domains protein 10 (*Megf10*) mutant retina. SACs exhibit an avoidance defect in the *Megf10* mutant retina such that they are more often located directly next to other SACs. (**c**) SACs in the INL of the mouse retina. In the wild-type retina, SACs space themselves across the horizontal plane of the retina. (**d**) SACs in the INL of the pituitary tumor-transforming gene (*Pttg1*) mutant retina. SACs exhibit an avoidance defect in the *Pttg1* mutant retina. The scale bar is equivalent to 50 μm (Reprinted by permission from Macmillan Publishers Ltd: Nature. MEGF10 and MEGF11 mediate homotypic interactions required for mosaic spacing of retinal neurons. Kay et al. (2012). We thank Drs. Ben Reese and Patrick Keeley for providing images of the *Pttg1* mutant and control retinas)

to be determined, one possibility is that it influences migration by modulating signaling through the AP-1 complex. Interestingly, *Pttg1* is also involved in the spacing of HCs, which like SAC utilize MEGF proteins for recognition and spacing (Keeley et al. 2014c; Kay et al. 2012).

Another promising clue to the spacing of SACs involves the transcription factor BARHL2. Deletion of *Barhl2* results in a marked increase in SAC number and an apparent defect in the spacing of SACs (Ding et al. 2009). This suggests that genes regulated by BARHL2 are required to prevent adhesion that occurs in the absence of BARHL2-mediated regulation. Identifying genes that are not expressed or are misregulated in the *Barhl2* mutant retina offers a promising avenue to the identification of these.

How well the mechanisms by which SACs and HCs organize themselves across the horizontal plane of the retina will hold with respect to other neuron populations is not clear. SACs and HCs are unusual populations of retinal neurons in that they do not require developmental cell death in order to space themselves. This is evidenced by studies of the *Bax* null retina, in which cell death is greatly reduced. The spacing organization of many types of retinal neurons is degraded in the *Bax* null retina, but SAC and HC spacing remains intact. Some of the proteins that facilitate spacing of SACs and HCs, such as MEGF10 and MEGF11, are also unique to these neuron populations, although they are later expressed by Müller glia. *Pcdhg*, *Pttg1*, and *Barhl2* are more broadly expressed, but their deletion does not result in soma spacing defects, apart from SACs and HCs in the *Pttg1* mutant retina. SACs and HCs are also unusual in that they do not express any of the homotypic cell adhesion molecule encoding sidekick (*Sdk*) or Down syndrome cell adhesion molecule (*Dscam*) genes, which are expressed in different combinations by most, and possible all, other populations of retinal neurons (de Andrade et al. 2014; Yamagata and Sanes 2008; Fuerst et al. 2008, 2009). *Dscam* genes are required for normal developmental cell death in the mouse retina, and in their absence, excessive ectopic adhesion is observed between cells and neurites of the same subtype, including bipolar, amacrine, and ganglion cells, greatly disrupting the normal mosaic organization of these cells (Fig. 10.6). This suggests that these proteins could act to help eliminate homotypic cells that are too closely spaced and reduce adhesion between remaining cells, although given the broad expression of both genes in spatially overlapping cell types in the mouse retina, other proteins must act as cell-type identifiers. The function of SDK proteins in the mouse may be more analogous to the role of both SDK and DSCAM proteins in the chick retina, where they guide lamination, the topic of the next section of this chapter.

10.7 Lamination in the Retina: The Outer Plexiform Layer

The two synaptic layers of the retina, the OPL and IPL, contain the electrical and chemical synapses that connect and process visual input. In one sense, the OPL is simpler than the IPL, in that it contains only two types of large multi-synaptic signaling hubs, those at the rod spherule and those at the cone pedicle (Fig. 10.7) (Hopkins and Boycott 1992; Boycott and Hopkins 1991; Wassle et al. 1978). The OPL synapses themselves, while limited in number of type, are highly complex and are considered by some to be the most complex in the brain, involving the assembly of numerous cells of different types at each signaling unit. In the mouse retina, the synaptic terminal of rods contains an invagination in which two HC axon tips enfold one to three rod bipolar cell (RBC) dendrite tips, converging on a rod synaptic ribbon (Carter-Dawson and LaVail 1979). Rod spherules are also

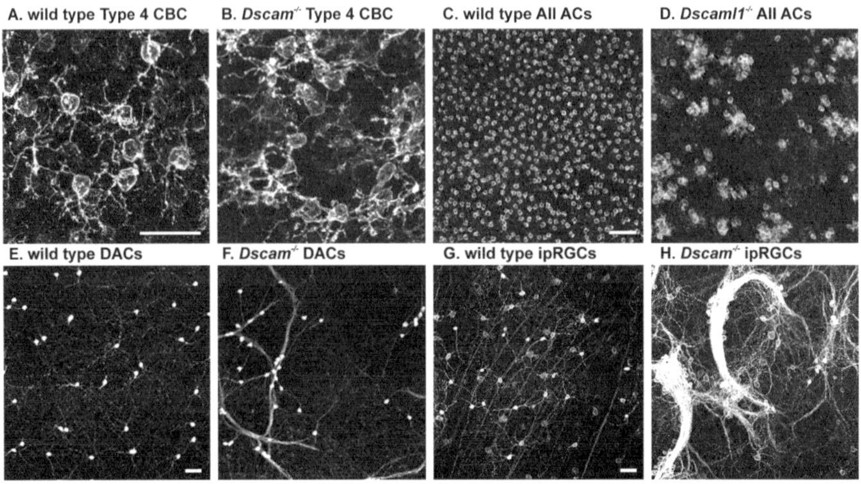

Fig. 10.6 Clumping in the *Dscam* or *Dscaml1* mutant retinas. (**a**, **b**) Type 4 cone bipolar cells (CBCs) in the wild-type and Down syndrome cell adhesion molecule (*Dscam*) mutant retina. Clumping of type 4 CBC dendrites was observed in the *Dscam* mutant retina compared to wild-type dendrites. (**c**, **d**) AII amacrine cells in the wild-type and *Dscaml1* mutant retina. Clumping of AII amacrine is observed in the *Dscaml1* mutant retina compared to wild-type cells. (**e**, **f**), DACs in the wild-type and *Dscam* mutant retina. Clumping of DACs and their neurites is observed in the *Dscam* mutant retina compared to wild type. (**g**, **h**) Melanopsin-positive RGCs in the wild-type and *Dscam* mutant retina. Clumping of melanopsin-positive RGCs and their dendrites is observed in the *Dscam* mutant retina compared to wild type (The scale bars in each set of images are equivalent to 25 μm)

contacted by the telodendria of cone terminals, which can also project to other cone terminals (Normann et al. 1984). In addition to the conventional rod spherule, a retracted morphology, in which the synapse directly abuts the soma, is observed in the wild-type retina in rods on the inner surface of the ONL and in disease, in which the synapse is located amidst the ONL. The cone synapse is organized in a similar fashion to the rod synapse but contains multiple ribbons. Like the rod synapse, the cone synapse is built around ribbons that release glutamate onto invaginating bipolar cell dendrite tips and, in this case, invaginating HC dendrite tips. Unlike the rod, each cone is sampled by multiple invaginating ON types of bipolar cells, in addition to OFF bipolar cells, which make flat contacts at the base of each invagination. Cone terminals in the primate retina have been reported to contain up to 42 ribbons and corresponding invaginations and are thought to transmit signal to each distinct type of bipolar cell (Boycott and Hopkins 1991; Chun et al. 1996). The large size of the OPL synapses and their easily identified structures have made them an attractive target for studies into the mechanisms and molecules driving retinal lamination, detailed below.

Development of the rod synapse requires pairing of rod terminals with HC axon tips and the dendrites of RBCs. Cell ablation, genetic targeting of physiology, and studies of the aging retina suggest that horizontal and bipolar cells are guided to

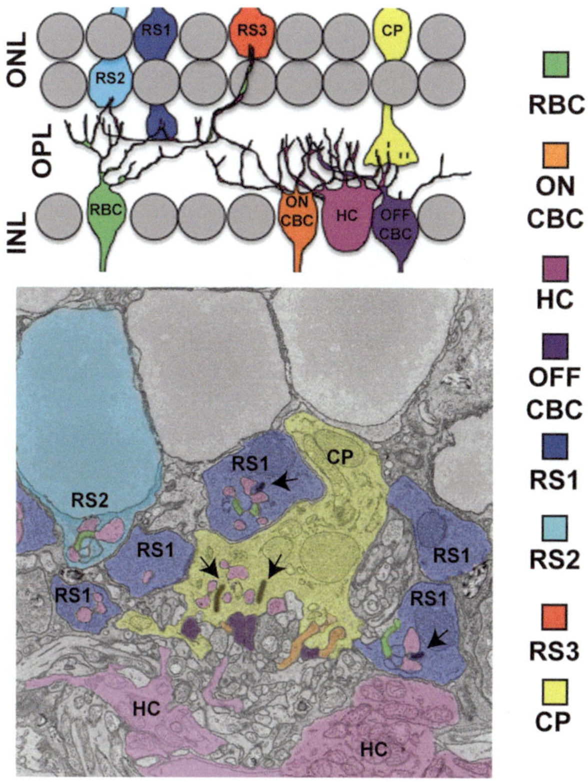

Fig. 10.7 Organization of the retinal OPL synapses. (**a**) A cartoon depiction of the outer plexiform layer, and cells that project neurites into it. (**b**) Electron micrograph image with cell types and synaptic structures labeled. (**a** and **b**) Rod and cone photoreceptors located in the ONL make synapses with bipolar and HCs, located in the INL. Rods generally project a spherule containing their synapses into the OPL (RS1), but some rods on the surface of the ONL have synapses directly abutting the nucleus (RS2). These superficially resemble the retracted synapses observed in disease models (RS3). Cones also project an axon-like terminal into the OPL, termed the cone pedicle (CP). Rod bipolar cells (RBC) and HCs project dendrites and axons tips, respectively, to rods and can be seen invaginating into the rod spherule. ON CBCs and horizontal cells (HC) project invaginating dendrites into the cone pedicle (CP). OFF CPCs make flat contacts at the base of the cone pedicle. Ribbon structures (*arrows*) terminate at the invaginating processes of HCs and ON bipolar cells

rods by a combination of intrinsic and extrinsic factors. The outcome of the modulations listed above is a mixture of rod spherule retraction into the ONL, HC sprouting and or bipolar cell sprouting into the ONL, and ectopic synaptogenesis in the ONL, and principles of OPL organization can be gleaned by comparing the similarities and differences between these.

In cases where synaptic activity is disrupted by mutation of cyclic nucleotide-gated channels (Claes et al. 2004), or in aging, or the genetic equivalent induced by knockout of the serine/threonine kinase Lkb1 (Samuel et al. 2011, 2014; Liets et al. 2006), both HCs and RBCs project neurites into the ONL and form synapses adjacent to rod soma. In mice in which activity is eliminated, HCs project into the ONL, followed by bipolar cells. In this case, it is not clear if rod spherules are retracting into the ONL, and then subsequently attracting the neurites of other cells, or if these cells' neurites explore the ONL and induce formation of ectopic synapses (Michalakis et al. 2012). In the aging retina, a similar phenomenon is observed, but here it is clear that the rods initially make synapses in the OPL, and then retract their spherules, which are targeted by HCs and rod bipolar cells (Samuel et al. 2011). This suggests that rods are able to secrete factors that attract the neurites of HCs and rod bipolar cells, each of which is making numerous other synaptic connections that remain in place. This has been further demonstrated in retinas that lack horizontal cells, in which case rod spherules retract and then, much later, bipolar cells sprout and make ectopic synaptic connections with the retracted spherules (Keeley et al. 2013). This study also indicates that the delayed projection of bipolar cell dendrites into the ONL, observed in other studies, is not dependent on HCs. Taken together, these studies indicate that rods and their synaptic partners produce molecules that attract and promote formation of synapses, and several promising candidates have been identified that may mediate this process.

Several adhesion and scaffolding molecules have been identified that promote formation of the rod synapse, although no single identified factor is absolutely required for connectivity, arguing for the presence of some redundancy in the system by which this synapse is programmed. Studies in mouse and chick have localized DSCAML1 and SDK2 (Sidekick-2), respectively, at the photoreceptor synapse. In the *Dscaml1* null mouse retina, a reduction in rod ribbon synapses has been reported (Fuerst et al. 2009). SDK2 and the MAGI PDZ protein that is tethered to are localized to this synapse in chick and can promote synaptic vesicle clustering in an in vitro photoreceptor model (Yamagata and Sanes 2010). *Plexa4* and *Sema6a* are both expressed by HCs, and in either mutant background, HCs project neurites into the ONL (Matsuoka et al. 2012). These HC neurites do not form ectopic synapses with rods or induce sprouting of bipolar cell neurites but rather fail to form the normally paired set of HC axon tips surrounding the ribbon, suggesting that these proteins may help to stabilize HC neurites within the rod synapse. This also supports the hypothesis that retracted rods secrete factors that attract HC and RBC neurites, as opposed to HC or RBC neurites inducing presynaptic specialization within the ONL. Studies of molecules that target HCs to the rod spherule have also been productive. HCs express the protein netrin-G ligand protein 2 (NGL2), which is localized to their axon tips (Soto et al. 2013). In retinas lacking NGL2, the axons of HC cells project into the ONL, but do not form synapses in this layer. Rods express the NGL2 ligand netrin-G2 and these proteins can facilitate formation of transynaptic adhesions, but this has not yet been demonstrated in the retina.

Considerably, less is known about organization of the cone synapses, but expression of similar molecules that organize the rod circuit and contribute to IPL organization have been reported, suggesting that similar mechanisms may be at play. An added layer of complexity when considering the cone pedicle is the large number of different bipolar cells that sample each pedicle, often in a nonoverlapping fashion. Therefore, a mechanism by which each invaginating bipolar cell could recognize and avoid like neurites is necessary, and the refinement of this synapse suggests that its organization may involve competition between neurites for targets that is not observed in other retinal circuits (Okawa et al. 2014).

10.8 Lamination in the Retina: The Inner Plexiform Layer

Unlike the OPL, where the individual synapses are complex but stereotyped and easy to identify, the IPL contains numerous synapses organized in diverse assemblages throughout this layer (Anderson et al. 2011). During development of the IPL, neurites initially occupy a broad territory, which is then refined. The identification of multiple cell adhesion and transmembrane signaling molecules that contribute to avoidance and attraction in the IPL, and the absence of phenotypes in the IPL of mice and fish lacking synaptic transmission or reared in the dark, suggests that differential adhesion plays a leading role in this process, with the role of activity, an important player in refining this initial organization (Nevin et al. 2008; Okawa et al. 2014).

IPL neurites express a complex code of adhesion and transmembrane signaling molecules, and the next section of this chapter will focus on a number of those that have been functionally demonstrated to influence lamination. Some of these adhesion molecules mediate homotypic recognition, followed by either adhesion or avoidance, while others, such as the semaphorins and their plexin receptors, act as heterotypic ligands and receptors. In fact, given that cell adhesion molecules can act through repulsion (*Drosophila* Dscam1 being an excellent example), it might be best to consider even homotypic adhesion molecules in the context of ligands and receptors. The localization of a large number of these molecules in the IPL suggests that the adhesive environment within the IPL is highly complex, even when considering a small number of proteins, several of which are described in more detail below.

In the chick retina, DSCAM and SDK proteins are concentrated in distinct IPL sublamina (Yamagata and Sanes 2008) and represent an elegant solution to the problem of laminar specificity (Fig. 10.8a). The pattern of DSCAMs in the less complex mouse retina is much more broad, with the DSCAM protein, for example, localized throughout the mouse IPL (Fig. 10.8a). While these proteins spatially overlap in the mouse retina, their expression appears to be cell-type exclusive, like the expression of *Dscam* and *Sdk* genes in the chick. Adding to this level of complexity, the contactins (CNTNs), a homologous family of homophilic Ig-superfamily proteins, are also expressed in overlapping domains of the IPL (Fig. 10.8b) (Yamagata and Sanes 2012). Non-Ig-superfamily proteins also have complex localization patterns and are essential players in organization of the IPL. Transmembrane SEMA6A is concentrated in the ON half of the IPL, with its PLEXA4 receptor concentrated in the OFF sublamina and PLEXA2 concentrated in the neurites of SACs (Fig. 10.8c) (Matsuoka et al. 2011b). Another set of PLXN

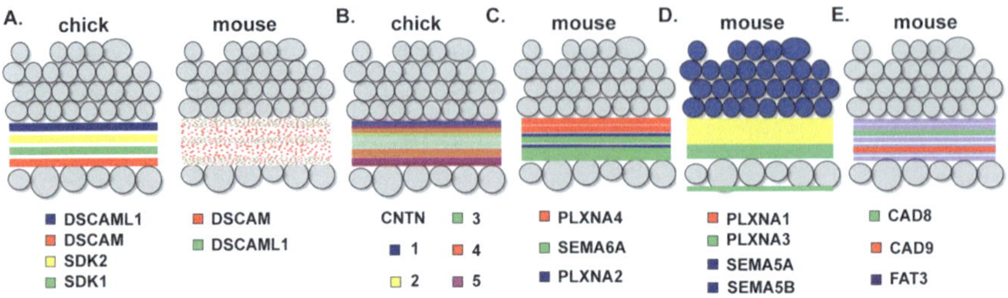

Fig. 10.8 Distribution of adhesion and signaling molecules in the chick or mouse IPL. The distribution of transmembrane signaling molecules and their ligands in the IPL is depicted. (**a**) A cartoon depicting localization of DSCAM and Sidekick (SDK) proteins in the chick IPL. DSCAM and SDK proteins are concentrated in distinct nonoverlapping bands within the chick IPL, and the genes are expressed in a nonoverlapping fashion. DSCAM proteins are more broadly distributed in the mouse retina and spatially overlap but like in chick are expressed in nonoverlapping cell types. (**b**) A cartoon of the chick IPL depicting contactin (CNTN) protein localization. CNTNs are localized in a complex, partially overlapping fashion in the chick IPL. (**c**) A cartoon of the mouse IPL showing distribution of plexin (PLXN) and semaphorin (SEMA) proteins. PLXNA4 protein is localized roughly in the OFF half of the mouse IPL, while SEMA6A protein is localized to the ON half of the IPL. PLXNA2 is concentrated in the ON and OFF SAC neurites. (**d**) A cartoon depicting localization of PLXNA1, PLXNA3, SEMA5A, and SEMA5B in the mouse INL and IPL. SEMA5A and SEMA5B are localized to the developing neuroblast layer. PLXNA1 and PLXNA3 are localized in a largely overlapping fashion in the IPL, although PLXNA3 projects more closely to the RGL and in the RGC axons compared to PLXNA1. (**e**) A cartoon of the mouse IPL showing localization of several cadherin proteins. FAT3 protein is localized throughout the IPL. Cadherin-8 (*Cad8*) is expressed in type 2 CBCs, which laminate in S1 and S2 of the IPL. Cadherin-9 (*Cad9*) is expressed in type 5 CBCs, which laminate in S4 of the IPL

and SEMA proteins that help to establish the IPL are PLXNA1/3, localized in the IPL, and SEMA5A/B localized in the developing inner neuroblast layer (Fig. 10.8d). Two types of cadherins have also been shown to direct lamination in the IPL, the atypical cadherin FAT3 (Deans et al. 2011) and the type 2 cadherins 8 and 9 (CAD8 and CAD9) (Fig. 10.9e) (Duan et al. 2014). In addition to the overlap of these molecules' distribution in the adult retina, some of their expression patterns are dynamic during development, adding yet another level of complexity to consider (Sun et al. 2013). The molecules listed above, which will be discussed below, represent only a fraction of adhesion molecules localized in the IPL, and other adhesion molecules, including multiple cadherins, plexins, semaphorins, and other Ig-superfamily CAMs, are localized in complex patterns in the IPL, although their function is still currently being ascertained.

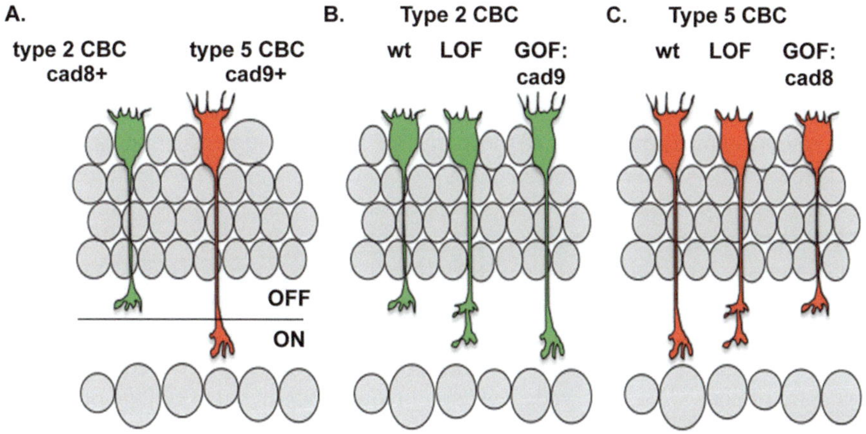

Fig. 10.9 Type 2 cadherins stabilize CBCs in specific lamina of the IPL. (**a**) A cartoon of type 2 and 5 CBC axon terminals in the mouse IPL. Type 2 CBCs maintain axons in S1 and S2 of the OFF sublamina and express *Cad8*. Type 5 CBCs maintain axon terminals in S4 of the ON sublamina and express *Cad9*. (**b**) Cartoon of *Cad8* loss-of-function and *Cad9* gain-of-function study results for type 2 CBCs. Loss of *Cad8* results in type 2 CBCs adopting a bistratified morphology. Expressing *Cad9* in these cells retargets their neurites into domains normally occupied by type 5 CBCs. (**c**) Cartoon of *Cad9* loss-of-function and *Cad8* gain-of-function study results in type 5 CBCs. Loss of *Cad9* results in type 5 CBCs adopting a bistratified morphology. Expressing *Cad8* in these cells retargets their neurites into domains normally occupied by type 2 CBCs

10.9 Inner Plexiform Laminar Targeting Through Adhesion

Much of development, including major events like gastrulation, is mediated by differential adhesion provided by the expression of cell adhesion molecules. Several studies have demonstrated that differential adhesion facilitates organization of the neural retina. According to this model, the differential adhesion hypothesis, neurites are targeted or retained in locations in which they encounter the greatest degree of adhesion. Several large gene families provide likely candidates to test this hypothesis, including members of the immunoglobulin (Ig) superfamily of cell adhesion molecules and cadherins. Expression and gain- and loss-of-function studies have demonstrated that adhesion provided by these molecules is sufficient in many cases to target neurites to specific lamina of the retina.

The first proteins for which this was demonstrated are the SDK proteins, members of the Ig superfamily. The *Sdk1* and *Sdk2* genes were shown to be expressed by cells on either side of the IPL and in distinct, nonoverlapping cell populations (Yamagata et al. 2002). The authors of this work expanded their analysis to include the DSCAM proteins (DSCAM and DSCAML1), which are homologous to the SDK proteins. They found that the four genes were expressed in nonoverlapping cell types and that they were concentrated in distinct lamina and further found that each protein engaged in homotypic binding (Yamagata and Sanes 2008). Both gain- and loss-of-function studies in the chick retina support a model in which these proteins promote neurite lamination through adhesion; when protein was depleted, cells that laminate in a given region have a less refined organization, and ectopic expression of a given molecule was observed to retarget the neurites of cells toward the lamina in which a given protein was concentrated. Identification of specific cell types and prospective synaptic partners is more challenging in chick than in mouse, and additional work in this model has permitted the mechanism of these genes to be further explored. *Dscaml1* in the mouse is expressed in many cells of the rod circuit, including rods, rod bipolar cells, and AII amacrine cells, consistent with a model wherein these genes organize cells into distinct circuits (Fuerst et al. 2009). Physiology within this circuit was intact, but a reduced number of synaptic ribbons were observed, suggesting redundancy in the specification of this circuit in mouse. *Dscam* is very broadly expressed in cells contributing to the mouse. The broad expression pattern of the *Dscam* genes in the mouse retina suggests that they will overlap with *Sdk* gene expression in the mouse. An interesting possibility is that the loss of two circuits during the evolution of the mammalian retina, compared to the chick, allowed the genes to specialize in function, with the DSCAMs promoting refinement and cell death, while the SDKs function in mammals in a fashion more similar to that of their chick orthologs. Together, these studies in mouse and chick support a model wherein homotypic binding of DSCAMs or SDKs can promote targeting and refinement of neurites in the IPL. Future studies of SDK function in the mouse retina will be key in confirming these roles. Like the DSCAM and SDK proteins in chick, the homotypic CNTN proteins function in guiding neurites to their appropriate lamina. Some CNTN proteins were found to laminate in a specific fashion, while others are localized to multiple IPL bands, suggesting that understanding their interactions will be more complicated than interactions between DSCAM and SDK proteins in this model.

The first gene identified as required for targeting of retinal neurites is a member of the cadherin family of cell adhesion molecule. Cadherins are classically thought of as homotypic adhesion molecules (adherence in the presence of Ca2+), although this may need to be revised in light of recent experiments, outlined below (Takeichi et al. 1986). In a classic experiment, cadherin-2 was shown to promote laminar targeting of chick RGC axons to the tectum in an ex vivo assay system (Yamagata and Sanes 1995; Yamagata et al. 1995). More recent work by the same group has demonstrated that the type 2 cadherins 8 and 9 are required for organization of neurites in the IPL and for the generation of functional circuitry (Duan et al. 2014). *Cad8* and *Cad9* are expressed by type 2 and type 5 cone bipolar cells (CBCs), respectively. Type 2 CBCs project within a band of the OFF half of the IPL, while type 5 CBCs project to a band within the ON half of the IPL. In the absence of each respective factor (*Cad8* in type 2 CBCs and *Cad9* in type 5 CBCs), the axon terminals of either cell type are rerouted to occupy space in bands within both the ON and OFF half of the IPL (Fig. 10.9). Interestingly, ectopic expression of *Cad9* in type 2 CBCs or *Cad8* in type 5 CBCs retargeted the axon of these cells to either the

ON (type 2) or OFF (type 5) sublamina of the IPL. Further experiments in which either cadherin was introduced in mutant backgrounds indicated, surprisingly, that the cadherins appear to be responding to heterotypic binding with as yet unknown partners. Identification of these ligands will be an exciting follow-up to this elegant study. The bipolar cells that cadherins promote the organization of are a developmentally interesting class of retinal neurons in the mouse in that they do not target axons or dendrites into the lamina in which they ultimately reside. Rather, the postmitotic cell retains apical and basal connections to the basement membrane in both the inner and outer retina (Morgan et al. 2006). The apical end differentiates into an axon and laminates after sampling all possible strata in the IPL, while the basal end differentiates into the cell's dendrites. Given this developmental organization, it will be interesting to see if the cadherins are uniquely utilized by this class of neuron or if they also promote lamination of other cell types as well.

10.10 Inner Plexiform Layer Laminar Targeting Through Avoidance

While the requirement for adhesion and recognition between interacting cells is required to specify cell circuits, repulsion and avoidance has emerged as the major factor establishing large-scale organization of the retina, such as plexiform layer formation. Targeting of neurites in the IPL involves projecting neurites into the IPL, localizing in approximately the correct area, pairing with the appropriate synaptic partners, and refining inappropriate contacts. The first two of these, identifying the location of the IPL and targeting to the correct lamina, are mediated at least in part by avoidance induced by binding of transmembrane SEMA ligands to their PLXN receptors. PLXN/SEMA signaling has emerged as a major factor in neurodevelopment. In the retina, both transmembrane semaphorins and diffusible semaphorins promote organization of neurites. Transmembrane SEMA proteins signal through PLXN receptors, while diffusible SEMA proteins require both PLXN receptors and neuropilin coreceptors (Jongbloets and Pasterkamp 2014).

Signaling between transmembrane SEMA proteins and PLXNs is required to target neurites to the IPL, as opposed to the nuclear layers or OPL (Matsuoka et al. 2011a). PLXNA1 and PLXNA3 are localized in a largely overlapping fashion in the IPL, while both SEMA5A and SEMA5B are localized to the developing neuroblast layer. In mutant backgrounds lacking *Sema5a* and *Sema5b* or *Plxna1* and *Plxna3*, three remarkable phenotypes were observed: first, neurites of some cell types retargeted within the IPL; second, neurites projected through the INL, into the OPL, and in some cases into the ONL; and third, projection of neurites into the INL was so extensive that a second IPL was established (Fig. 10.10a, b). These phenotypes were observed in many different amacrine cell types, and retargeting was also observed in bipolar cells and RGCs, which may also avoid SEMA5A/B protein in the INL or may follow patterning of amacrine cells. In vitro experiments confirmed

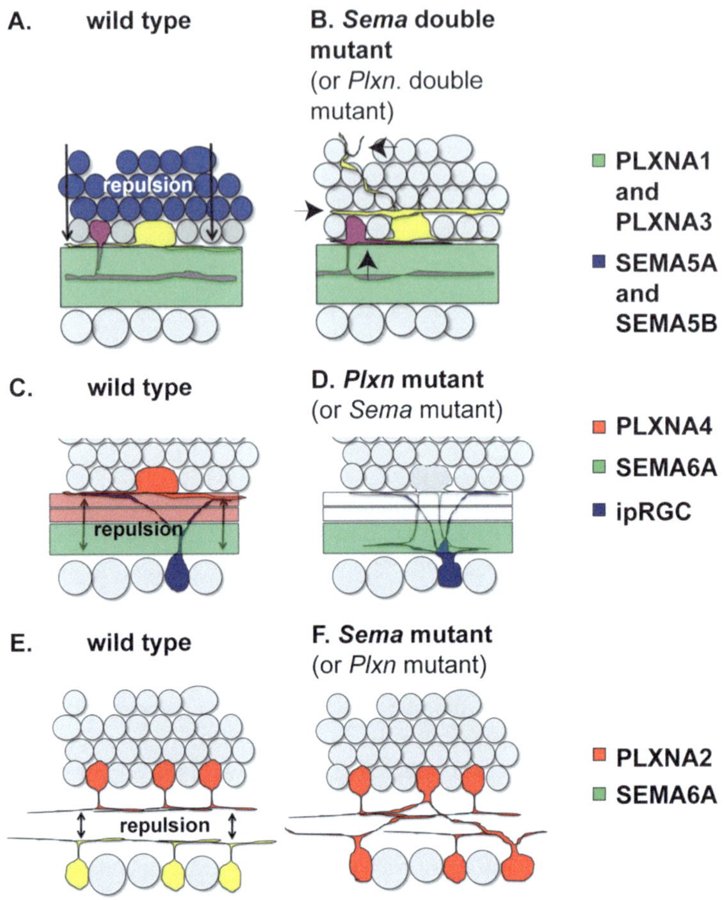

Fig. 10.10 Plexin-semaphorin signaling establishes ON/OFF symmetry in the IPL. (**a**) A cartoon of PLXNA1/3 and SEMA5A/B protein distribution in the mouse retina. PLXNA1/3 are localized within the IPL, while SEMA5A/B are localized to the developing cells of the neuroblast layer. (**b**) In mutant retinas in which both *Plxna1* and *Plxna3* or *Sema5a* and *Sema5b* have been knocked out, three phenotypes are observed. First, cells, such as vglut3-positive amacrine cells, project neurites to novel lamina within the IPL (*lower arrow*). Second, neurites of cells that normally target to the IPL project neurites in the expected lamina of the IPL and also project neurites into the INL, OPL, and occasionally into the ONL (*top arrow*). Finally, a second IPL is established within the INL (*middle arrow*). (**c**) A cartoon depicting SEMA6A and PLXNA4 protein localization in the mouse IPL. SEMA6A is concentrated in the ON half of the retina, while PLXNA4 is concentrated in the OFF half of the IPL, including in DACs, which express the *Plxna4* gene. (**d**) In retinas in which either factor are mutant, DACs project neurites into S5 of the IPL, which are then targeted by the dendrites of type I ipRGCs. (**e**) A cartoon of the mouse IPL depicting ON and OFF SACs. *Plxna2* is expressed by ON and OFF SACs, and the PLXNA2 protein is concentrated in their neurites. *Sema6a* is expressed by ON SACs only, and the SEMA6A protein is localized to their neurites. (**f**) In retina mutant for either *Sema6a* or *Plxna2*, SAC neurites originating in either the ON or OFF sublamina project across the center of the IPL and adhere with each other

that signaling between these molecules was mediating neurite avoidance. Therefore, these results indicate that the developing neuroblast layer (which resides in the same location as the ONL, OPL, and outer INL do in the adult retina) produces factors that repel the neurites of amacrine cells and suggests that these cells project neurites in multiple directions. While a much more limited phenotype was observed in the *Sema5b* mutant retina, this study is also remarkable because it demonstrated that redundancy in specifying developmental patterns is an actual phenomenon and demonstrated that this problem is not insurmountable. In addition to these proteins, multiple other transmembrane and secreted semaphorins and their plexin receptors are localized in the IPL, and it will be interesting to see if they also act in a redundant fashion (Matsuoka et al. 2013).

Other factors that help to specify targeting and formation of the IPL include the atypical cadherin FAT3. *Fat3* is required for delimiting the IPL and in its absence, a second IPL forms within the INL (Deans et al. 2011). *Fat3* is expressed while amacrine cells are migrating and prevents these cells from projecting neurites in between the soma of the INL, resulting in a mutant phenotype very similar to the development of an ectopic plexiform layer in the *Sema5a/b* or *Plxna1/3* double mutants described above.

While it is unclear if the next example involves avoidance, an example of a gene, in this case a transcription factor, which is required for targeting neurites to the OPL by developing HCs, has been identified in *Lim1*. This was demonstrated by deletion of *Lim1*, which results in HCs targeting neurites to the IPL, instead of to the OPL (Poche et al. 2007). Transcriptional profiling of HCs in the wild-type and *Lim1* null background could help to identify factors that establish the OPL, possibly additional semaphorin and plexin proteins required to prevent HC neurites from targeting the IPL, in a similar fashion to those that inhibit projection into the ONL.

Once targeting to the IPL is established, additional transmembrane SEMA protein ligands and PLXN receptors establish the ON/OFF symmetry of the IPL. These molecules are essential for establishing the ON and OFF boundaries of the IPL. This has been demonstrated by loss-of-function studies of *Sema6a* and either *Plxna4* or *Plxna2* (Sun et al. 2013; Matsuoka et al. 2011b). The localization of these two proteins is essential for their function; PlXNA4 protein is roughly localized to the OFF half of the IPL, while SEMA6A is localized to the ON half of the IPL. Repulsive interactions between PLXNA4 and SEMA6A are required for normal lamination of type I dopaminergic amacrine cells (DACs). DACs, which express *Plxna4*, normally laminate in S1 of the retina. In retinas deficient for *Plxna4* or the gene encoding its ligand *Sema6a*, DACs extend neurites into S5 (Matsuoka et al. 2011b) (Fig. 10.10c, d). In response, the intrinsically photoresponsive RGCs (ipRGCs) that make synaptic connections with DACs (Zhang et al. 2008) route into both S1 and S5. In addition to identifying cues that guide neurites to specific lamina, this elegant study also illustrates two other important features of retinal organization. First, DACs rerouted many neurites to S5 of the IPL but avoided S2-S4. This result suggests that the ON and OFF sublamina of the IPL are organized in a manner akin to mirror reflections of each other. Second, the retargeting of type I ipRGC dendrites to S5 suggests that DACs provide cues that are targeted by RGCs, akin to

SACs serving as a scaffold for direction-selective RGCs, thus providing a second example of this phenomena and suggesting that it may be a general manner in which amacrine and ganglion cells interact, surprising in light of RGC earlier birthdate.

The second example of plexins and their semaphorin ligands establishing the ON/OFF boundaries of the IPL involves interactions between PLXNA2 and SEMA6A in the two populations of SACs (Sun et al. 2013). ON SACs express both *Plxna2* and *Sema6a*, while OFF SACs express *Plxna2* only. In the absence of either of these factors, merging and misprojection of SAC neurites were observed, and in vitro experiments demonstrated that cells expressing *Plxna2* avoid exogenous SEMA6A protein. In both of these cases, the data strongly support a model wherein SEMA6A serves as a ligand that mediates repulsion of *Plxna4*-expressing cells (Fig. 10.10e, f).

Semaphorins and plexins are therefore responsible for the major subdivisions of the retinal IPL, first by demarcating the IPL and second by establishing the ON/OFF symmetry of the IPL.

The recent advances outlined in this section demonstrate how rapidly understanding of neurite organization within the retina has progressed. These results show that avoidance mediated by semaphorins and plexins serve to subdivide the retina, first by boxing neurites into the IPL and then by restricting neurites into either the ON or OFF sublamina of the IPL. The reprojection of neurites in the *Sema*, *Plxn*, and *Cad* mutant retinas, for example, in which DAC neurites project to both S1 and S5, but not S2–S4, predicts that additional molecules will further subdivide the ON and OFF lamina. A major unresolved question is the mechanism by which cells refine their projections within the IPL. Developing neurons initially project a bushy arbor that is presumably limited by PLXN/SEMA restriction, but how these neurites are further refined into the very sharp banding observed in the adult retina is currently unknown. Another largely unanswered question is how cells choose the correct synaptic partners, given the dense and overlapping projections of neurites within the IPL.

10.11 Other Mechanisms in Neurite and Soma Organization

Several mechanisms that may contribute to this process, apart from transmembrane signaling molecules, are briefly outlined in the next section of this chapter, and despite their brief discussion here, they are each deserving of further treatment that was not possible here.

10.12 Programmed Dendrite Organization

Apart from adhesion and repulsion triggered by interactions at the surface of cells and their neurites, complex intrinsic programming of dendrite organization is sufficient in many cases to produce a dendrite arbor, in isolation, that approximates what is observed in vivo (Jan and Jan 2010). Presumably, this is due to the regulatory elements expressed by a given cell type that result in specified branching patterns, and while assaying, cell adhesion molecules have been a helpful exercise in determining how the retinal IPL is organized; other clues to the organization of the IPL can be gleaned from studies of transcription factors, such as the POU4Fs (Badea et al. 2009) and SOX2 (Whitney et al. 2014). In addition to specifying cell type, transcription factors can also specify aspects of the cells that express them, for example, dendrite morphology. The POU4F (BRN3A, B, and C) transcription factors are combinatorially expressed by RGCs and required for their survival (Badea et al. 2009). Studies of *Pou4f1* (*Brn3a*) have demonstrated a conversion of RGCs that would normally express this factor from a monostratified morphology to a bistratified morphology. Strain-specific polymorphisms in *Sox2* or deletion of *Sox2* results in a decrease in the number of SACs that successfully migrate into the RGL, compared to the number in the INL. Interestingly, in an analogous fashion to *Pou4f1*-negative RGCs, the SACs that do successfully migrate into the RGL in the *Sox2* null retina now have some affinity for the OFF SAC neurite band and projected a subset of their neurites to both layers. While these factors are considered under intrinsic factors that regulate dendrite morphology, it would also not be surprising if they regulated expression of cell surface molecules that control neurite morphogenesis. Identification of differentially regulated genes in *Sox2* null SACs, comparison of ON versus OFF SACs, or changes in gene expression in *Pou4f* null RGCs compared to wild type offers a straightforward pathway to better understand the genes and processes driving neurite organization.

10.13 Glia and Retinal Neurite Organization

The retina contains three populations of glia: Müller glia, astrocytes, and microglia. An emerging body of literature has identified expression of what are classically considered immune system proteins at the synapse and has identified a role for these proteins in synapse elimination and dendrite remodeling. A landmark study in this field was the identification of the complement protein C1q at the synapse of RGCs, both in the IPL and in the tectum. C1q was found to act as a signal by which synapses were engulfed and eliminated and this process was found to be increased in glaucoma (Stevens et al. 2007). Another immune protein involved in RGC dendrite remodeling is CD3-zeta; in the absence of which, RGC dendrite density is increased, consistent with this protein functions in refinement of RGC dendrites (Xu et al. 2010). Interestingly, the MEGF10 protein has recently been implicated in refinement of RGC axons at their target, indicating that proteins active in neuron-neuron interactions can also facilitate glia-mediated interactions (Chung et al. 2013). These studies indicate that consideration of how the retina's resident glia contribute to refinement and organization of neurites will be an important avenue of future research.

10.14 Activity in Retinal Neurite and Soma Organization

While much of retinal circuitry appears to be hardwired by receptors and their ligands, synaptic activity also plays an important role in organization and refinement of neurites within the retina (Bleckert and Wong 2011). The role of synaptic activity and retinal waves has been documented to drive refinement within the central targets of retinal ganglion cell axons, at the rod synapse (Dunn et al. 2013), and new studies suggest that the activity could also promote refinement within the retina in a similar manner, but in the reverse direction. Inactivation of the synaptic activity at the bipolar to ganglion cell synapse was recently shown to influence not the neurite arbors in the IPL but rather the bipolar cell dendrites, in a feed-backward manner (Johnson and Kerschensteiner 2014). These results suggest that bipolar cell dendrite refinement is influenced by activity and highlights the need for studies that incorporate concurrent analysis of the inner and outer retina.

10.15 The Extracellular Matrix in Retinal Neurite and Soma Organization

Another important area that deserves additional study in identifying drivers of soma spacing and neurite lamination within the retina is the interactions between neurons and components of the extracellular matrix (ECM). ECM components play a central role in targeting and lamination of RGC targets in the tectum, but less is known about their functions within the retina proper. Classic experiments have demonstrated that ECM components promote RGC dendrite outgrowth, and their inclusion in in vitro culture media is often required to support survival and neurite outgrowth (Meyer-Franke et al. 1995). Because of their widespread expression and necessity in development, studying the role of the ECM has been challenging, but initial results support the role of ECM components in neurite organization. RGC axons project toward the inner retina in response to laminin-α1, and dendrites may polarize opposite of this (Randlett et al. 2011), although recent work in *Xenopus* suggests that RGCs polarize dendrites in response to gradients of diffusible SEMA3A and SEMA3F, which signal through neuropilin1 and PLXNA1 (Kita et al. 2013). Organization of the RGL is dependent on interactions between laminin and integrin and downstream signaling, identifying another area where ECM components are essential (Riccomagno et al. 2014). Within the IPL, laminins have been shown to help regulate outgrowth of DAC neurites and are required for normal banding of ON SACs, a phenotype similar to the initial phenotype reported for these cells in the γ-protocadherin null retina (Denes et al. 2007). In the OPL, dystroglycan and pikachurin have been shown to promote development of the rod synapse and are required for normal visual function (Han and Townes-Anderson 2012; Omori et al. 2012; Sato et al. 2008; Satz et al. 2009). An important additional emerging role of the ECM in lamination has been its ability to capture and present diffusible signaling molecules (Xiao et al. 2011). Further identification of ECM components that are differentially localized, followed by attempts to target these components, may be an avenue to further address the role of ECM in organization of retinal neurites and soma.

10.16 Summary

Our understanding of the organization of soma and neurites in the retina has undergone a transformation from essentially an unknown process to one in which basic rules and molecules have been identified. Much work remains in order to understand the interactions between various factors, to uncover the roles of activity in retinal organization and in translating these findings to the rest of the brain.

References

Anderson JR, Jones BW, Watt CB, Shaw MV, Yang JH, Demill D, Lauritzen JS, Lin Y, Rapp KD, Mastronarde D, Koshevoy P, Grimm B, Tasdizen T, Whitaker R, Marc RE (2011) Exploring the retinal connectome. Mol Vis 17:355–379

Badea TC, Cahill H, Ecker J, Hattar S, Nathans J (2009) Distinct roles of transcription factors brn3a and brn3b in controlling the development, morphology, and function of retinal ganglion cells. Neuron 61:852–864

Bleckert A, Wong RO (2011) Identifying roles for neurotransmission in circuit assembly: insights gained from multiple model systems and experimental approaches. Bioessays 33:61–72

Boycott BB, Hopkins JM (1991) Cone bipolar cells and cone synapses in the primate retina. Vis Neurosci 7:49–60

Briggman KL, Helmstaedter M, Denk W (2011) Wiring specificity in the direction-selectivity circuit of the retina. Nature 471:183–188

Carter-Dawson LD, Lavail MM (1979) Rods and cones in the mouse retina. I. Structural analysis using light and electron microscopy. J Comp Neurol 188:245–262

Chen SK, Chew KS, Mcneill DS, Keeley PW, Ecker JL, Mao BQ, Pahlberg J, Kim B, Lee SC, Fox MA, Guido W, Wong KY, Sampath AP, Reese BE, Kuruvilla R, Hattar S (2013) Apoptosis regulates ipRGC spacing necessary for rods and cones to drive circadian photoentrainment. Neuron 77:503–515

Chun MH, Grunert U, Martin PR, Wassle H (1996) The synaptic complex of cones in the fovea and in the periphery of the macaque monkey retina. Vision Res 36:3383–3395

Chung WS, Clarke LE, Wang GX, Stafford BK, Sher A, Chakraborty C, Joung J, Foo LC, Thompson A, Chen C, Smith SJ, Barres BA (2013) Astrocytes mediate synapse elimination through MEGF10 and MERTK pathways. Nature 504:394–400

Claes E, Seeliger M, Michalakis S, Biel M, Humphries P, Haverkamp S (2004) Morphological characterization of the retina of the CNGA3(-/-)Rho(-/-) mutant mouse lacking functional cones and rods. Invest Ophthalmol Vis Sci 45:2039–2048

de Andrade GB, Long SS, Fleming H, Li W, Fuerst PG (2014) DSCAM localization and function at the mouse cone synapse. J Comp Neurol 522:2609–2633

Deans MR, Krol A, Abraira VE, Copley CO, Tucker AF, Goodrich LV (2011) Control of neuronal morphology by the atypical cadherin Fat3. Neuron 71:820–832

Denes V, Witkovsky P, Koch M, Hunter DD, Pinzon-Duarte G, Brunken WJ (2007) Laminin deficits induce alterations in the development of dopaminergic neurons in the mouse retina. Vis Neurosci 24:549–562

Ding Q, Chen H, Xie X, Libby RT, Tian N, Gan L (2009) BARHL2 differentially regulates the development of retinal amacrine and ganglion neurons. J Neurosci 29:3992–4003

Duan X, Krishnaswamy A, de la Huerta I, Sanes JR (2014) Type II cadherins guide assembly of a direction-selective retinal circuit. Cell 158:793–807

Dunn FA, Della Santina L, Parker ED, Wong RO (2013) Sensory experience shapes the development of the visual system's first synapse. Neuron 80:1159–1166

Famiglietti EV Jr (1983) On and off pathways through amacrine cells in mammalian retina: the synaptic connections of "starburst" amacrine cells. Vision Res 23:1265–1279

Farajian R, Raven MA, Cusato K, Reese BE (2004) Cellular positioning and dendritic field size of cholinergic amacrine cells are impervious to early ablation of neighboring cells in the mouse retina. Vis Neurosci 21:13–22

Fuerst PG, Koizumi A, Masland RH, Burgess RW (2008) Neurite arborization and mosaic spacing in the mouse retina require DSCAM. Nature 451:470–474

Fuerst PG, Bruce F, Tian M, Wei W, Elstrott J, Feller MB, Erskine L, Singer JH, Burgess RW (2009) DSCAM and DSCAML1 function in self-avoidance in multiple cell types in the developing mouse retina. Neuron 64:484–497

Galli-Resta L, Resta G, Tan SS, Reese BE (1997) Mosaics of islet-1-expressing amacrine cells assembled by short-range cellular interactions. J Neurosci 17:7831–7838

Garrett AM, Weiner JA (2009) Control of CNS synapse development by {gamma}-protocadherin-mediated astrocyte-neuron contact. J Neurosci 29:11723–11731

Garrett AM, Schreiner D, Lobas MA, Weiner JA (2012) Gamma-protocadherins control cortical dendrite arborization by regulating the activity of a FAK/PKC/MARCKS signaling pathway. Neuron 74:269–276

Grzywacz NM, Amthor FR, Merwine DK (1998) Necessity of acetylcholine for retinal directionally selective responses to drifting gratings in rabbit. J Physiol 512(Pt 2):575–581

Han J, Townes-Anderson E (2012) Cell specific post-translational processing of pikachurin, a protein involved in retinal synaptogenesis. PLoS ONE 7:e50552

Hopkins JM, Boycott BB (1992) Synaptic contacts of a two-cone flat bipolar cell in a primate retina. Vis Neurosci 8:379–384

Huckfeldt RM, Schubert T, Morgan JL, Godinho L, Di Cristo G, Huang ZJ, Wong RO (2009) Transient neurites of retinal horizontal cells exhibit columnar tiling via homotypic interactions. Nat Neurosci 12:35–43

Jan YN, Jan LY (2010) Branching out: mechanisms of dendritic arborization. Nat Rev Neurosci 11:316–328

Johnson RE, Kerschensteiner D (2014) Retrograde plasticity and differential competition of bipolar cell dendrites and axons in the developing retina. Curr Biol 24:2301–2306

Jongbloets BC, Pasterkamp RJ (2014) Semaphorin signalling during development. Development 141:3292–3297

Kay JN, Roeser T, Mumm JS, Godinho L, Mrejeru A, Wong RO, Baier H (2004) Transient requirement for ganglion cells during assembly of retinal synaptic layers. Development 131:1331–1342

Kay JN, Chu MW, Sanes JR (2012) MEGF10 and MEGF11 mediate homotypic interactions required for mosaic spacing of retinal neurons. Nature 483:465–469

Keeley PW, Sliff BJ, Lee SC, Fuerst PG, Burgess RW, Eglen SJ, Reese BE (2012) Neuronal clustering and fasciculation phenotype in Dscam- and Bax-deficient mouse retinas. J Comp Neurol 520:1349–1364

Keeley PW, Luna G, Fariss RN, Skyles KA, Madsen NR, Raven MA, Poche RA, Swindell EC, Jamrich M, Oh EC, Swaroop A, Fisher SK, Reese BE (2013) Development and plasticity of outer retinal circuitry following genetic removal of horizontal cells. J Neurosci 33:17847–17862

Keeley PW, Madsen NR, St John AJ, Reese BE (2014a) Programmed cell death of retinal cone bipolar cells is independent of afferent or target control. Dev Biol 394:191–196

Keeley PW, Whitney IE, Madsen NR, St John AJ, Borhanian S, Leong SA, Williams RW, Reese BE (2014b) Independent genomic control of neuronal number across retinal cell types. Dev Cell 30:103–109

Keeley PW, Zhou C, Lu L, Williams RW, Melmed S, Reese BE (2014c) Pituitary tumor-transforming gene 1 regulates the patterning of retinal mosaics. Proc Natl Acad Sci U S A 111:9295–9300

Kita EM, Bertolesi GE, Hehr CL, Johnston J, McFarlane S (2013) Neuropilin-1 biases dendrite polarization in the retina. Development 140:2933–2941

Kuida K, Zheng TS, Na S, Kuan C, Yang D, Karasuyama H, Rakic P, Flavell RA (1996) Decreased apoptosis in the brain and premature lethality in CPP32-deficient mice. Nature 384:368–372

Lee SC, Cowgill EJ, Al-Nabulsi A, Quinn EJ, Evans SM, Reese BE (2011) Homotypic regulation of neuronal morphology and connectivity in the mouse retina. J Neurosci 31:14126–14133

Lefebvre JL, Zhang Y, Meister M, Wang X, Sanes JR (2008) Gamma-protocadherins regulate neuronal survival but are dispensable for circuit formation in retina. Development 135:4141–4151

Lefebvre JL, Kostadinov D, Chen WV, Maniatis T, Sanes JR (2012) Protocadherins mediate dendritic self-avoidance in the mammalian nervous system. Nature 488:517–521

Liets LC, Eliasieh K, van der List DA, Chalupa LM (2006) Dendrites of rod bipolar cells sprout in normal aging retina. Proc Natl Acad Sci U S A 103:12156–12160

Lin B, Wang SW, Masland RH (2004) Retinal ganglion cell type, size, and spacing can be specified independent of homotypic dendritic contacts. Neuron 43:475–485

Marc RE, Jones BW, Anderson JR, Kinard K, Marshak DW, Wilson JH, Wensel T, Lucas RJ (2007) Neural reprogramming in retinal degeneration. Invest Ophthalmol Vis Sci 48:3364–3371

Marc RE, Jones BW, Watt CB, Vazquez-Chona F, Vaughan DK, Organisciak DT (2008) Extreme retinal remodeling triggered by light damage: implications for age related macular degeneration. Mol Vis 14:782–806

Masland RH (2012) The neuronal organization of the retina. Neuron 76:266–280

Matsuoka RL, Chivatakarn O, Badea TC, Samuels IS, Cahill H, Katayama K, Kumar SR, Suto F, Chedotal A, Peachey NS, Nathans J, Yoshida Y, Giger RJ, Kolodkin AL (2011a) Class 5 transmembrane semaphorins control selective mammalian retinal lamination and function. Neuron 71:460–473

Matsuoka RL, Nguyen-Ba-Charvet KT, Parray A, Badea TC, Chedotal A, Kolodkin AL (2011b) Transmembrane semaphorin signalling controls laminar stratification in the mammalian retina. Nature 470:259–263

Matsuoka RL, Jiang Z, Samuels IS, Nguyen-Ba-Charvet KT, Sun LO, Peachey NS, Chedotal A, Yau KW, Kolodkin AL (2012) Guidance-cue control of horizontal cell morphology, lamination, and synapse formation in the mammalian outer retina. J Neurosci 32:6859–6868

Matsuoka RL, Sun LO, Katayama K, Yoshida Y, Kolodkin AL (2013) Sema6B, Sema6C, and Sema6D expression and function during mammalian retinal development. PLoS ONE 8: e63207
Meyer-Franke A, Kaplan MR, Pfrieger FW, Barres BA (1995) Characterization of the signaling interactions that promote the survival and growth of developing retinal ganglion cells in culture. Neuron 15:805–819
Michalakis S, Schaferhoff K, Spiwoks-Becker I, Zabouri N, Koch S, Koch F, Bonin M, Biel M, Haverkamp S (2012) Characterization of neurite outgrowth and ectopic synaptogenesis in response to photoreceptor dysfunction. Cell Mol Life Sci 70(10):1831–1847
Morgan JL, Dhingra A, Vardi N, Wong RO (2006) Axons and dendrites originate from neuroepithelial-like processes of retinal bipolar cells. Nat Neurosci 9:85–92
Morris VB (1970) Symmetry in a receptor mosaic demonstrated in the chick from the frequencies, spacing and arrangement of the types of retinal receptor. J Comp Neurol 140:359–398
Nelson R, Famiglietti EV Jr, Kolb H (1978) Intracellular staining reveals different levels of stratification for on- and off-center ganglion cells in cat retina. J Neurophysiol 41:472–483
Nevin LM, Taylor MR, Baier H (2008) Hardwiring of fine synaptic layers in the zebrafish visual pathway. Neural Dev 3:36
Normann RA, Perlman I, Kolb H, Jones J, Daly SJ (1984) Direct excitatory interactions between cones of different spectral types in the turtle retina. Science 224:625–627
Okawa H, Hoon M, Yoshimatsu T, Della Santina L, Wong RO (2014) Illuminating the multifaceted roles of neurotransmission in shaping neuronal circuitry. Neuron 83:1303–1318
Omori Y, Araki F, Chaya T, Kajimura N, Irie S, Terada K, Muranishi Y, Tsujii T, Ueno S, Koyasu T, Tamaki Y, Kondo M, Amano S, Furukawa T (2012) Presynaptic dystroglycan-pikachurin complex regulates the proper synaptic connection between retinal photoreceptor and bipolar cells. J Neurosci 32:6126–6137
Poche RA, Kwan KM, Raven MA, Furuta Y, Reese BE, Behringer RR (2007) Lim1 is essential for the correct laminar positioning of retinal horizontal cells. J Neurosci 27:14099–14107
Poche RA, Raven MA, Kwan KM, Furuta Y, Behringer RR, Reese BE (2008) Somal positioning and dendritic growth of horizontal cells are regulated by interactions with homotypic neighbors. Eur J Neurosci 27:1607–1614
Randlett O, Poggi L, Zolessi FR, Harris WA (2011) The oriented emergence of axons from retinal ganglion cells is directed by laminin contact in vivo. Neuron 70:266–280
Randlett O, MacDonald RB, Yoshimatsu T, Almeida AD, Suzuki SC, Wong RO, Harris WA (2013) Cellular requirements for building a retinal neuropil. Cell Rep 3:282–290
Raven MA, Reese BE (2003) Mosaic regularity of horizontal cells in the mouse retina is independent of cone photoreceptor innervation. Invest Ophthalmol Vis Sci 44:965–973
Reese BE, Galli-Resta L (2002) The role of tangential dispersion in retinal mosaic formation. Prog Retin Eye Res 21:153–168
Reese BE, Keeley PW (2014) Design principles and developmental mechanisms underlying retinal mosaics. Biol Rev Camb Philos Soc 90(3):854–876
Reese BE, Necessary BD, Tam PP, Faulkner-Jones B, Tan SS (1999) Clonal expansion and cell dispersion in the developing mouse retina. Eur J Neurosci 11:2965–2978
Reese BE, Raven MA, Giannotti KA, Johnson PT (2001) Development of cholinergic amacrine cell stratification in the ferret retina and the effects of early excitotoxic ablation. Vis Neurosci 18:559–570
Riccomagno MM, Sun LO, Brady CM, Alexandropoulos K, Seo S, Kurokawa M, Kolodkin AL (2014) Cas adaptor proteins organize the retinal ganglion cell layer downstream of integrin signaling. Neuron 81:779–786
Rossi C, Strettoi E, Galli-Resta L (2003) The spatial order of horizontal cells is not affected by massive alterations in the organization of other retinal cells. J Neurosci 23:9924–9928
Samuel MA, Zhang Y, Meister M, Sanes JR (2011) Age-related alterations in neurons of the mouse retina. J Neurosci 31:16033–16044

Samuel MA, Voinescu PE, Lilley BN, de Cabo R, Foretz M, Viollet B, Pawlyk B, Sandberg MA, Vavvas DG, Sanes JR (2014) LKB1 and AMPK regulate synaptic remodeling in old age. Nat Neurosci 17:1190–1197

Sato S, Omori Y, Katoh K, Kondo M, Kanagawa M, Miyata K, Funabiki K, Koyasu T, Kajimura N, Miyoshi T, Sawai H, Kobayashi K, Tani A, Toda T, Usukura J, Tano Y, Fujikado T, Furukawa T (2008) Pikachurin, a dystroglycan ligand, is essential for photoreceptor ribbon synapse formation. Nat Neurosci 11:923–931

Satz JS, Philp AR, Nguyen H, Kusano H, Lee J, Turk R, Riker MJ, Hernandez J, Weiss RM, Anderson MG, Mullins RF, Moore SA, Stone EM, Campbell KP (2009) Visual impairment in the absence of dystroglycan. J Neurosci 29:13136–13146

Sernagor E, Eglen SJ, Wong RO (2001) Development of retinal ganglion cell structure and function. Prog Retin Eye Res 20:139–174

Soto F, Watkins KL, Johnson RE, Schottler F, Kerschensteiner D (2013) NGL-2 regulates pathway-specific neurite growth and lamination, synapse formation, and signal transmission in the retina. J Neurosci 33:11949–11959

Stacy RC, Wong RO (2003) Developmental relationship between cholinergic amacrine cell processes and ganglion cell dendrites of the mouse retina. J Comp Neurol 456:154–166

Stevens B, Allen NJ, Vazquez LE, Howell GR, Christopherson KS, Nouri N, Micheva KD, Mehalow AK, Huberman AD, Stafford B, Sher A, Litke AM, Lambris JD, Smith SJ, John SW, Barres BA (2007) The classical complement cascade mediates CNS synapse elimination. Cell 131:1164–1178

Sun LO, Jiang Z, Rivlin-Etzion M, Hand R, Brady CM, Matsuoka RL, Yau KW, Feller MB, Kolodkin AL (2013) On and off retinal circuit assembly by divergent molecular mechanisms. Science 342:1241974

Szel A, Rohlich P, Caffe AR, Juliusson B, Aguirre G, van Veen T (1992) Unique topographic separation of two spectral classes of cones in the mouse retina. J Comp Neurol 325:327–342

Takeichi M, Shirayoshi Y, Hatta K, Nose A (1986) Cadherins: their morphogenetic role in animal development. Prog Clin Biol Res 217B:17–27

Tasic B, Nabholz CE, Baldwin KK, Kim Y, Rueckert EH, Ribich SA, Cramer P, Wu Q, Axel R, Maniatis T (2002) Promoter choice determines splice site selection in protocadherin alpha and gamma pre-mRNA splicing. Mol Cell 10:21–33

Wassle H, Riemann HJ (1978) The mosaic of nerve cells in the mammalian retina. Proc R Soc Lond B Biol Sci 200:441–461

Wassle H, Boycott BB, Peichl L (1978) Receptor contacts of horizontal cells in the retina of the domestic cat. Proc R Soc Lond B Biol Sci 203:247–267

Wassle H, Puller C, Muller F, Haverkamp S (2009) Cone contacts, mosaics, and territories of bipolar cells in the mouse retina. J Neurosci 29:106–117

Wei W, Hamby AM, Zhou K, Feller MB (2011) Development of asymmetric inhibition underlying direction selectivity in the retina. Nature 469:402–406

White FA, Keller-Peck CR, Knudson CM, Korsmeyer SJ, Snider WD (1998) Widespread elimination of naturally occurring neuronal death in Bax-deficient mice. J Neurosci 18:1428–1439

Whitney IE, Keeley PW, Raven MA, Reese BE (2008) Spatial patterning of cholinergic amacrine cells in the mouse retina. J Comp Neurol 508:1–12

Whitney IE, Keeley PW, St John AJ, Kautzman AG, Kay JN, Reese BE (2014) Sox2 regulates cholinergic amacrine cell positioning and dendritic stratification in the retina. J Neurosci 34:10109–10121

Williams RR, Cusato K, Raven MA, Reese BE (2001) Organization of the inner retina following early elimination of the retinal ganglion cell population: effects on cell numbers and stratification patterns. Vis Neurosci 18:233–244

Wong RO (1990) Differential growth and remodelling of ganglion cell dendrites in the postnatal rabbit retina. J Comp Neurol 294:109–132

Wong RO, Hughes A (1987) Role of cell death in the topogenesis of neuronal distributions in the developing cat retinal ganglion cell layer. J Comp Neurol 262:496–511

Xiao T, Staub W, Robles E, Gosse NJ, Cole GJ, Baier H (2011) Assembly of lamina-specific neuronal connections by slit bound to type IV collagen. Cell 146:164–176

Xu HP, Chen H, Ding Q, Xie ZH, Chen L, Diao L, Wang P, Gan L, Crair MC, Tian N (2010) The immune protein CD3zeta is required for normal development of neural circuits in the retina. Neuron 65:503–515

Yamagata M, Sanes JR (1995) Lamina-specific cues guide outgrowth and arborization of retinal axons in the optic tectum. Development 121:189–200

Yamagata M, Sanes JR (2008) Dscam and sidekick proteins direct lamina-specific synaptic connections in vertebrate retina. Nature 451:465–469

Yamagata M, Sanes JR (2010) Synaptic localization and function of sidekick recognition molecules require MAGI scaffolding proteins. J Neurosci 30:3579–3588

Yamagata M, Sanes JR (2012) Expanding the Ig superfamily code for laminar specificity in retina: expression and role of contactins. J Neurosci 32:14402–14414

Yamagata M, Herman JP, Sanes JR (1995) Lamina-specific expression of adhesion molecules in developing chick optic tectum. J Neurosci 15:4556–4571

Yamagata M, Weiner JA, Sanes JR (2002) Sidekicks: synaptic adhesion molecules that promote lamina-specific connectivity in the retina. Cell 110:649–660

Yoshida K, Watanabe D, Ishikane H, Tachibana M, Pastan I, Nakanishi S (2001) A key role of starburst amacrine cells in originating retinal directional selectivity and optokinetic eye movement. Neuron 30:771–780

Zhang DQ, Wong KY, Sollars PJ, Berson DM, Pickard GE, Mcmahon DG (2008) Intraretinal signaling by ganglion cell photoreceptors to dopaminergic amacrine neurons. Proc Natl Acad Sci U S A 105:14181–14186

Chapter 11
Modifying Dendritic Structure After Function

Yuan Wang and Edwin W. Rubel

Abstract Neurons develop highly specialized dendritic architecture for certain operations of information processing in the brain. In this chapter, we review the development and regulation of characterized patterns of dendritic morphology and arrangement of bipolar neurons in the auditory brainstem, an excellent example of highly specialized dendritic architecture for their function in temporal coding and coincidence detection. We describe dramatic dynamics of the dendrites in both developing and mature systems and discuss the role of afferent synaptic input in influencing both the size of dendritic tree and the pattern of dendritic arborizations. The unique dendritic structure of these neurons provides an advantageous model for further understanding of the specific roles of neurotransmission, calcium signaling, protein synthesis, and cytoskeletal regulation in this important form of brain dynamics. Importantly, these neurons are highly conserved structurally and functionally across vertebrates including humans, emphasizing stereotyped dendritic regulation that is evolutionarily conserved for fundamental information processing operations in the vertebrate brain.

Keywords Neuronal morphology • Dendritic regulation • Afferent influence • Auditory brainstem • Sound localization • Bipolar neurons • Development • Nucleus laminaris • Medial superior olive

Functional neural circuits require that neurons develop their dendritic architecture to maximize information processing specific to that circuit. For a distinct neuronal cell type, dendritic architecture is determined not only by dendritic morphology of individual neurons but also by the way in which dendrites of all neurons within a neural circuit are arranged as a population. The former underlies how information is

---

Y. Wang
Department of Biomedical Sciences, Florida State University College of Medicine, Tallahassee, FL 32306, USA

E.W. Rubel (✉)
Virginia Merrill Bloedel Hearing Research Center, Department of Otolaryngology-Head and Neck Surgery, University of Washington School of Medicine, Box 357923, Seattle, WA 98195, USA
e-mail: rubel@uw.edu

K. Emoto et al. (eds.), *Dendrites*, DOI 10.1007/978-4-431-56050-0_11

received and integrated within individual neurons, while the latter enables effective communication between neurons of the same or different types within a local network. Specialized dendritic architectures usually form gradually during development and may remain flexible after they mature. Such structural dynamics is presumed to be important for changes in perception, cognition, learning, and memory, but may also result in neurological dysfunction and degeneration at any stage of the life cycle.

An excellent example of highly specialized dendritic architecture with dramatic dynamics in both developing and mature systems is observed in bipolar neurons of the auditory brainstem. These neurons are located in the nucleus laminaris (NL) of reptiles and birds and in the medial superior olive (MSO) of mammals (Fig. 11.1). Across vertebrate species, these bipolar neurons are highly comparable in their dendritic morphology and arrangement, afferent and efferent connectivity, biochemistry particularly specialized levels of expression of a number of functionally specialized proteins, as well as physiology. They act as coincidence detectors for processing temporal properties of the incoming signals from the two ears and are exquisitely sensitive to interaural time differences (ITDs), the time disparities in the arrival of low-frequency information between the two ears. This ITD sensitivity is believed to be essential for normal sound localization in low-frequency hearing species including reptiles, birds, some rodents, primates, and humans (reviewed in Grothe 2000; Grothe and Pecka 2014). In addition, bipolar neurons in NL and MSO respond in a comparable manner to chronic changes in their environment, particularly to altered afferent input through manipulations of their excitatory input from the ears. These similarities emphasize stereotyped dendritic regulation that is evolutionarily conserved for fundamental information processing operations in the vertebrate brain.

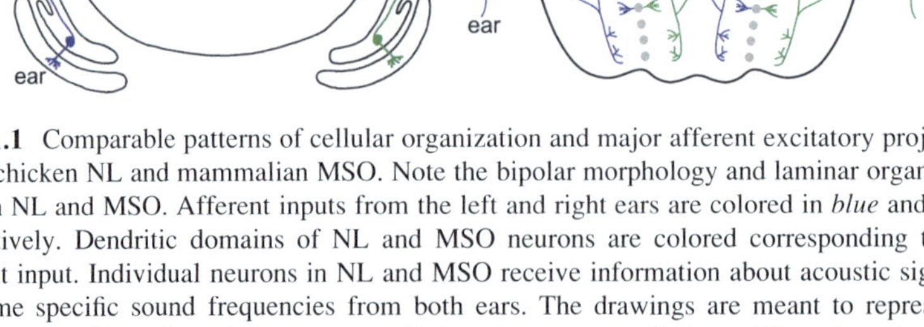

Fig. 11.1 Comparable patterns of cellular organization and major afferent excitatory projections in the chicken NL and mammalian MSO. Note the bipolar morphology and laminar organization of both NL and MSO. Afferent inputs from the left and right ears are colored in *blue* and *green*, respectively. Dendritic domains of NL and MSO neurons are colored corresponding to their afferent input. Individual neurons in NL and MSO receive information about acoustic signals at the same specific sound frequencies from both ears. The drawings are meant to represent an iso-frequency dimension. Abbreviations: *NM* nucleus magnocellularis, *NL* nucleus laminaris, *AVCN* anterior ventral cochlear nucleus, *MSO* medial superior olive

This chapter focuses on the development and regulation of dendritic architecture in the chicken NL, which has been most extensively studied owing to its simple anatomy and well-defined organization. We first describe a number of structural and organizational specializations of NL dendrites that are exquisitely tailored for its function as coincidence detectors and precise temporal processing. This is followed by a discussion of how these specializations are established during development and how they are regulated after onset of function, with a focus on the role of excitatory afferent input in these structural dynamics.

11.1 Structural and Organizational Specializations of NL Dendrites

11.1.1 Bipolar Configuration and Differential Innervation

The following description is of the chicken NL, and it should be recognized that variations of this basic plan are seen in other avian species (Carr and Code 2000). The somata of NL neurons are aligned into a single sheet, resulting in separate dorsal and ventral dendritic neuropil layers (Figs. 11.2a and 11.3a, b; Rubel et al. 1976; Seidl et al. 2010). These dendritic layers are largely free of glial cells but surrounded by a high density of glial cells just beyond the border of the dendritic arbors (Fig. 11.2c). At the single-cell level, the most characteristic structural property of the chicken NL neurons is their bipolar morphology with two segregated dendritic domains (Fig. 11.2b; Ramón y Cajal 1909; Smith and Rubel 1979; Wang and Rubel 2012). The two domains are largely comparable in their size and shape, as well as in their fine structure such as synapse distribution and organelle organization (Deitch and Rubel 1989a, b). Dendritic arbors are generally spineless but sometimes roughly surfaced. Distal dendritic endings are often characterized with an enlarged bulge with or without narrow filopodial-like extensions (Fig. 11.2d).

This bipolar configuration and the single-cell body layer provide the architecture for the organization of afferent inputs, which is critical for coincidence detection. NL receives glutamatergic excitatory inputs solely from the nucleus magnocellularis (NM) on both sides of the brain (Fig. 11.1). NM is a primary target of the cochlear ganglion cells and homologous to the mammalian anterior ventral cochlear nucleus (AVCN). This projection is arranged in a strict topographic manner resulting in a precise map of sound frequencies (tonotopic) in NL (Rubel and Parks 1975). NL neurons responsible for encoding high-frequency information are located in the rostromedial pole, and neurons that are optimally activated by lower-frequency sound are positioned in progressively caudolateral regions of the nucleus. Ipsilateral and contralateral terminals from the same parent neuron in NM target identical tonotopic positions in NL on the two sides of the brain, but segregate onto dorsal and ventral dendritic domains, respectively (Parks and

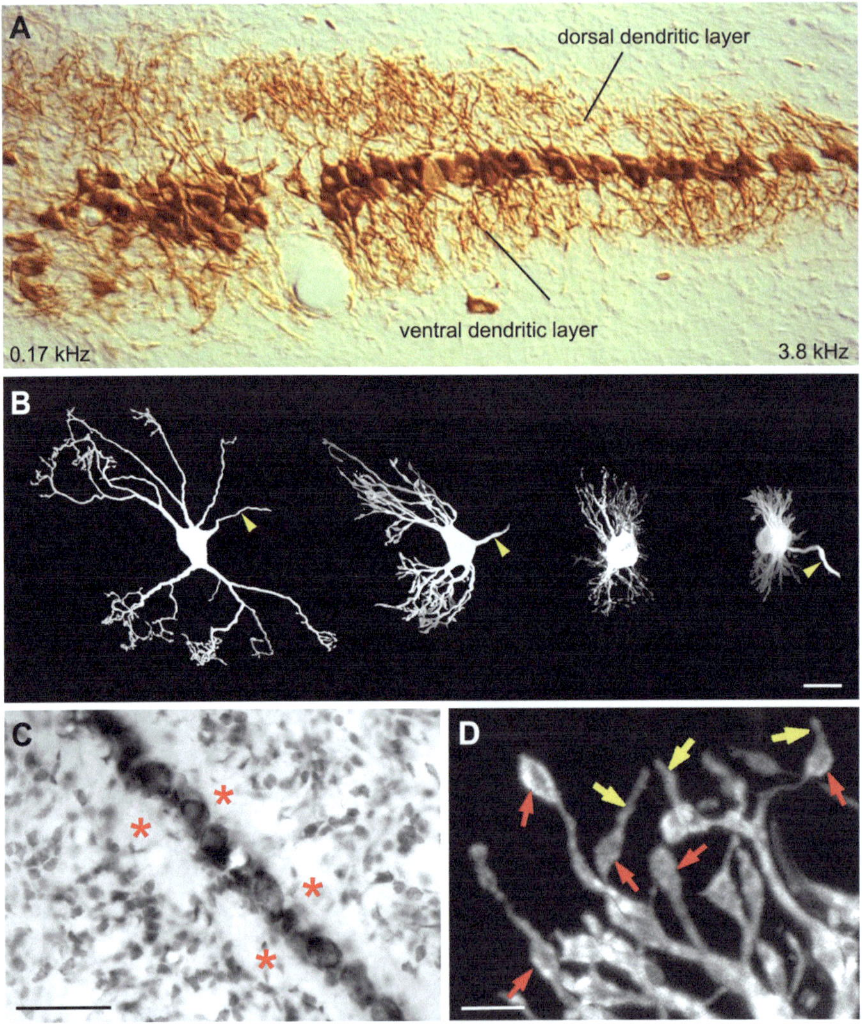

Fig. 11.2 Dendritic morphology and arrangement in the chicken NL. (**a**) MAP2 immunoreactivity (a somatodendritic marker) in NL. The plane of the section is approximately parallel to the caudolateral-rostromedial axis of the nucleus. Thus, this photo covers a large, although not the entire, range of the tonotopic axis. Note the spatial gradient of dendritic extension along this axis, as well as the three-layer configuration of the nucleus. The lowest and highest characterized frequencies recorded from the chicken NL are indicated (Rubel and Park 1975). (**b**) Examples of individual dye-filled NL neurons. These bipolar neurons vary their dendritic length and arborization pattern as a function of the location. A neuron responding optimally to low-frequency sounds is shown on the left, with neurons responding to progressively higher frequencies to the right. Axons are indicated by *yellow arrowheads*. (**c**) Nissl staining demonstrating that NL dendritic neuropil regions are largely glia-free (*red stars*). (**d**) Morphology of distal dendritic endings of NL neurons, characterized by an enlarged bulge (*red arrows*) with or without narrow extensions (*yellow arrows*). Scale bar: 10 μm in (**b**) (applies to **a** and **b**), 50 μm in (**c**), and 5 μm in (**d**)

Rubel 1975; Young and Rubel 1983). With this arrangement, individual NL neurons receive information about acoustic signals at the same specific sound frequencies from both ears.

To perform coincidence detection of action potential input from the two ears, ipsilateral and contralateral terminal arbors of each NM neuron form radically different and highly stereotyped morphologies. The ipsilateral axon bifurcates to provide equivalent axon length to dorsal dendrites of NL neurons along an iso-frequency dimension (Young and Rubel 1983). The contralateral axon, however, extends across the midline and bifurcates several times to create an orderly, serial set of axonal branches to ventral dendrites along a matching iso-frequency lamina of NL on the opposite side of the brain. This arrangement of innervation, predominately along the medial to lateral dimension, results in medial NL neurons receiving contralateral inputs from the shortest input axons and lateral NL neurons receiving from the longest axons. This systematic increase in axonal length across the ventral dendritic field effectively establishes a series of delay lines. The combination of these delay lines formed by the serial branching and differential conduction velocities of the ipsilateral and contralateral circuit compensates for time delays between the two ears as sound emanates from different positions in azimuth (Jeffress 1948; Young and Rubel 1983, 1986; Overholt et al. 1992; Seidl et al. 2010, 2014).

11.1.2 Spatial Gradient

In addition to their bipolar morphology, another dendritic specialization of NL neurons is its highly stereotyped dendritic gradient (Fig. 11.2a, b; Smith and Rubel 1979; Smith 1981; Deitch and Rubel 1984). This gradient is evident not only in the total dendritic branch length of individual neurons but also in the distance from the most distal dendritic branches to the soma, i.e., the "width" or dorsal-ventral extent of the dendritic field. The dendritic gradient conforms precisely to the tonotopic axis with an 11-fold increase in total dendritic branch length and approximately fivefold increase in the width of the dendritic band from high- to low-frequency neurons. Along with these parameters, the number of primary dendrites decreases about 13-fold from high-frequency to low-frequency coding regions of NL. While there is no evidence of spatial gradient in dendritic size and shape in the mammalian MSO, both dendritic extension and arbor density exhibit a dramatic gradient in the alligator NL (Wang et al. 2014).

This dendritic length gradient in NL appears to be an adaptation for ITD processing for particular sound frequencies. Low- and middle-frequency neurons, which have the longer dendrites, may exhibit dendritic filtering resulting from their large surface area (Kuba et al. 2005). A possible advantage of this filtering property is that it may enhance the electrical isolation of dorsal and ventral dendrites and, thus, the inputs from each ear. Indeed, basic biophysical modeling of NL neurons demonstrates that NL dendrites act as current sinks, improving coincidence detection by allowing a nonlinear summation between the segregated inputs from the ipsilateral and contralateral NM (Agmon-Snir et al. 1998).

11.1.3 High Dendritic Activity

NL dendrites are highly active. In quite conditions, average multiunit spike rate is usually over 1000 per second in NM and NL (Born et al. 1991), while single units in NM spike around 100 times per second (Fukui et al. 2010). With acoustic stimulation, the maximal instantaneous spike rate of NM and NL neurons reaches up to several folds of their spontaneous activity. As required by this high spiking activity,

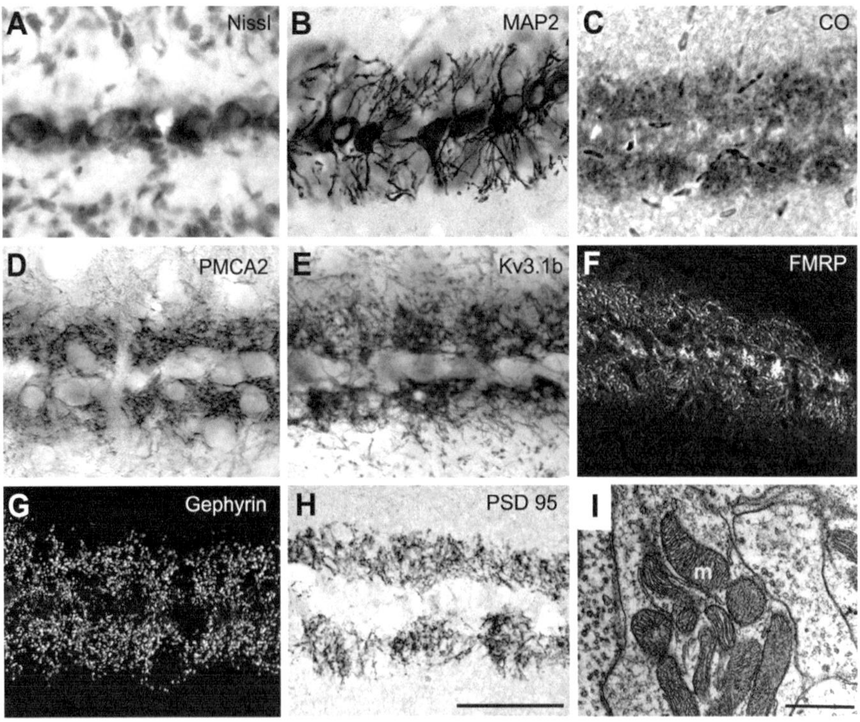

Fig. 11.3 Specialized proteins and cellular organelles in neuronal dendrites of the chicken NL. Images are taken from sections stained for Nissl (**a**), immunoreactivity for microtubule-associated protein 2 (MAP2, **b**), cytochrome oxidase (CO, **c**), and immunoreactivities for plasma membrane calcium ATPase type 2 (PMCA2, **d**), high voltage-gated potassium channel 3.1b (Kv3.1b, **e**), fragile X mental retardation protein (FMRP, **f**), gephyrin (**g**), and PSD 95 (**h**). NL dendrites contain high levels of CO, PMCA2, Kv3.1b, and FMRP and both excitatory and inhibitory synapses, indicated by PSD 95 and gephyrin staining. In **b**–**h** note that all of these highly expressed dendritic proteins are symmetrically distributed in the two dendritic domains. **i** An ultrastructural image illustrating a high density of mitochondria (m) in NL dendrites. Scale bar: 50 μm in (**h**) (applies to **a**–**h**) and 1 μm in (**i**)

NL neurons contain a high density of mitochondria throughout their dendrites (Fig. 11.3i; Deitch and Rubel 1989a). In addition, as in other central auditory neurons, NL is characterized by a high level of cytochrome oxidase (CO, Fig. 11.3c), a mitochondrial enzyme supplying ATP to the cell and thus an indicator for the amount of energy consumption (Dezsö et al. 1993). CO activity in NL is most intense in dendritic layers and adjacent glial regions containing NM axons. Similarly, NL exhibits strong glucose uptake in these regions as measured by 2-deoxyglucose method (Lippe et al. 1980; Heil and Scheich 1986). Studies with cochlea destruction reveal that this high metabolic level of NL is partially due to neuronal spiking which completely depends on cochlea integrity (Born et al. 1991) and is partially due to intrinsic cellular events as indicated by still higher CO level in NL (and other auditory nuclei) than most other regions in the brain even after bilateral cochlea removal (Dezsö et al. 1993). In addition, the brainstem region containing NM and NL is rich in blood vessels, and the amount of blood flow to this region is reduced following cochlea removal (Richardson and Durham 1990).

Consistently, these highly active NL dendrites exhibit strong localization of ion channels including high voltage-gated potassium channel 3.1b (Kv3.1b) (Fig. 11.3e), local protein translation regulators such as fragile X mental retardation protein (FMRP, Fig. 11.3f; see more details in 3.1—Molecular signals and pathways), as well as both excitatory and inhibitory synapses, indicated by PSD 95 and gephyrin staining (Fig. 11.3g, h). Importantly, NL dendrites adopt a number of cellular mechanisms to handle the challenges associated with this high activity. One such challenge is extensive calcium influx during action potentials (spikes), which could lead to calcium toxicity if overaccumulated. As one important mechanism for maintaining basal calcium homeostasis, NL dendrites are heavily loaded with the plasma membrane calcium ATPase type 2 (PMCA2, Fig. 11.3d), the fastest calcium efflux pump that expels calcium out of the cell and sets the resting calcium concentration (Wang et al. 2009).

11.2 Formation and Regulation of NL Dendrites During Development

11.2.1 Normal Development

Chickens can hear well before they hatch (about 21 days of incubation). Similar to other auditory neurons, the major developmental events in NL occur *in ovo* (Fig. 11.4). Detailed reviews of these developmental events are provided elsewhere (Rubel and Parks 1988; Wang et al. 2010). Below we highlight the main time course of dendritic and axonal development within NL.

NL neurons are formed at embryonic days 3–4 (E3–4) from the rhombomeres 5–6 of the neural tube (Rubel et al. 1976; Cramer et al. 2000). Following migration from the ventricular surface into the caudal brainstem, NL and NM are first

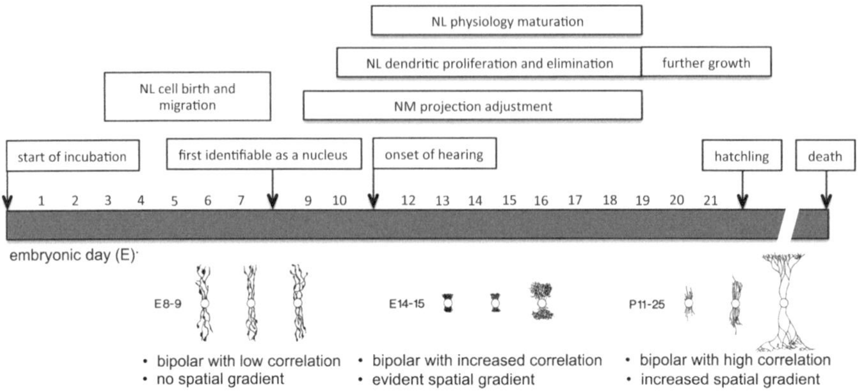

Fig. 11.4 Developmental time course of the chicken NL dendrites. Schematic drawings of dendritic morphology at three time points are illustrated below. For each time point, the three cells from left to right are examples from the caudal, middle, and rostral NL (These drawings are modified from Smith (1981))

identifiable as a nucleus at E8–9. At this time, the morphology and organization of NL and its input from NM have exhibited a number of principles of their mature architecture. Although the somata of NL neurons have not yet aligned into a single layer at this time, the glia-free margin surrounding it has already became apparent (Rubel et al. 1976). Individual NL neurons are recognizably bipolar, often with dorsally and ventrally oriented dendrites, characterized by dense branching of thin processes and wispy filopodia (Smith 1981). The connection between NM and NL is formed and organized topographically, although NM terminals appear immature in morphology and size of terminal field (Young and Rubel 1986). One important aspect of immaturity, though, is the apparent absence of spatial gradient of dendritic length and extension across the nucleus (Smith 1981).

Subsequently, NL neurons undergo dramatic changes in dendritic morphology (Smith 1981). Starting at E10, there is a spatiotemporal gradient of proliferation of numerous fine dendritic processes from rostromedial to caudolateral. This is followed by a spatiotemporal gradient of the elimination of dendritic processes along the same spatial direction. Such changes along with growth of dendritic trees as a function of the location result in the emergence of the distinct spatial gradient of remaining dendrites. This gradient is evident at E15 and becomes prominent by E19. During the same period of dendritic proliferation and elimination, the incoming NM axons gradually adjust their dorsal and ventral terminal fields to achieve the tonotopic specificity (Young and Rubel 1986). In addition, the maturation of intrinsic and synaptic properties of NL neurons also takes place primarily between E9–17 through extensive recomposition and relocalization of synaptic receptors and ion channels in dendrites (Kuba et al. 2002; Gao and Lu 2008; Sanchez et al. 2010, 2012). By E19, NL dendrites and their NM inputs are mature like structurally and physiologically, although the dendritic gradient continues to develop after hatchling, as does the correlation of dendritic length between the two domains of individual neurons (Smith 1981).

11.2.2 Role of Afferent Input

In the chicken, onset of hearing and synaptogenesis in NL take place at E11–14 (Saunders, et al. 1973; Jackson et al. 1982; Rebillard and Rubel 1981). The development of these highly specialized structural features of NL dendrites appears temporally correlated with the establishment of synaptically driven neuronal activity and may be influenced by the level and pattern of this activity. Interestingly, spontaneous activity in the embryonic auditory brainstem emerges before the onset of sound-evoked activity (Lippe 1994, 1995). Possible roles of afferent activity, synaptically driven and/or spontaneous, in dendritic development of NL neurons are supported by significant effects on NL dendrites of otocyst (embryonic precursor of the inner ear) removal at E3. This manipulation prevents the formation of the inner ear and thus completely removes excitatory afferent input to the ipsilateral NM and then to the dorsal NL dendrites ipsilaterally and ventral NL dendrites contralaterally. As compared to normal development, unilateral otocyst removal produces 44 % reduction in total dendritic length of NL neurons at E17–18 (Parks 1981). Unexpectedly, bilateral otocyst removal produces a smaller (14 %) reduction (Parks et al. 1987). On the other hand, the spatial gradient of NL dendrites is unaffected following either unilateral or bilateral manipulation, indicating molecular factors inherent to the nucleus.

The more dramatic effect of unilateral otocyst removal on NL dendrites than that induced by the bilateral manipulation may suggest that symmetrical afferent input is more important to the regulation of NL dendritic length than the absolute level of this input during development. However, the underlying mechanism might be complicated if one considers potential involvement of newly formed neuronal connection induced by the manipulation. Early removal of one otocyst induces formation of a functional aberrant axonal projection to the ipsilateral NM from the contralateral NM (Jackson and Parks 1988). This projection is excitatory and maintains the tonotopic map in NM and NL on both sides of the brain (Lippe et al. 1992; Zhou and Parks 1993; Parks 1997) and thus may compensate, at least partially, the lost afferent input to the initially deprived NL dendrites. The formation of this aberrant projection is not unique to birds; a similar projection from the contralateral AVCN to the deprived ipsilateral AVCN is formed following early ablation of one cochlea in gerbils (Kitzes et al. 1995). Another new source of excitatory afferent input to the deprived NL dendrites may arise directly from axons of the contralateral NM, which are normally restricted to the opposite domain of NL dendrites (Rubel et al. 1981).

In support of the importance of afferent input to NL development, acoustic enrichment by overstimulation of species-specific and music sound during the prenatal period increases neuronal and glial cell numbers in the chicken NM and

NL, as well as the volume of NL neuropil regions (Wadhwa et al. 1999), although exact effect on NL dendritic architecture is not studied. In summary, it is clear that molecular markers, whose expression is genetically controlled, determine the principles of NL dendritic architecture. It is also evident that excitatory afferent input affects dendritic growth and/or maintenance during development.

11.3 Afferent Influence of NL Dendrites Following Maturation

After NL neurons become structurally and functionally mature, the dendritic arborizations maintain their ability to rapidly and dramatically change their architecture in response to changes in their environment. One such change is altered excitatory afferent input, evident by dramatic effects of afferent deprivation on the behavior, structure, and biochemistry of mature NL dendrites in young chickens.

11.3.1 Effects of Afferent Deprivation on Dendritic Structure

Direct Deafferentation The most direct way of depriving a postsynaptic target of afferent physiological input is to physically disconnect incoming axons to their postsynaptic targets. Of course deafferentation initiates a series of events beyond the elimination of activation of postsynaptic receptors. Nevertheless, it is a direct, repeatable, and productive manipulation and often reveals important information on the roles of a specific presynaptic input on postsynaptic cellular and subcellular elements. The location and organization of NL neurons and incoming NM axons allow complete elimination of excitatory to ventral NL dendrites in vivo by surgical transection of the crossed dorsal cochlear tract (XDCT) at the midline, which contains contralateral NM axons (Fig. 11.5a). Since dorsal and ventral NL dendrites receive excitatory input almost exclusively from ipsilateral or contralateral NM axons, respectively, XDCT transection does not directly alter the excitatory input to dorsal dendrites of the same NL neurons. In addition, there is no evidence of retargeting of damaged contralateral NM axons or sprouting of fibers from the ipsilateral NM to the ventral NL dendrites within 3 months (Rubel et al. 1981). Thus, this differential manipulation allows examination of the subcellular spatial resolution of the response to deafferentation.

XDCT transection produces rapid and progressive retraction of deafferented ventral dendrites in NL (Benes et al. 1977; Deitch and Rubel 1984; Wang and Rubel 2012). Within just 2 hours (h), total branch length of the ventral dendrites is 20 % shorter than those on the dorsal side of the same neurons or the ventral dendrites in control animals. Loss of ventral dendrites continues over the time, such that by 16 days

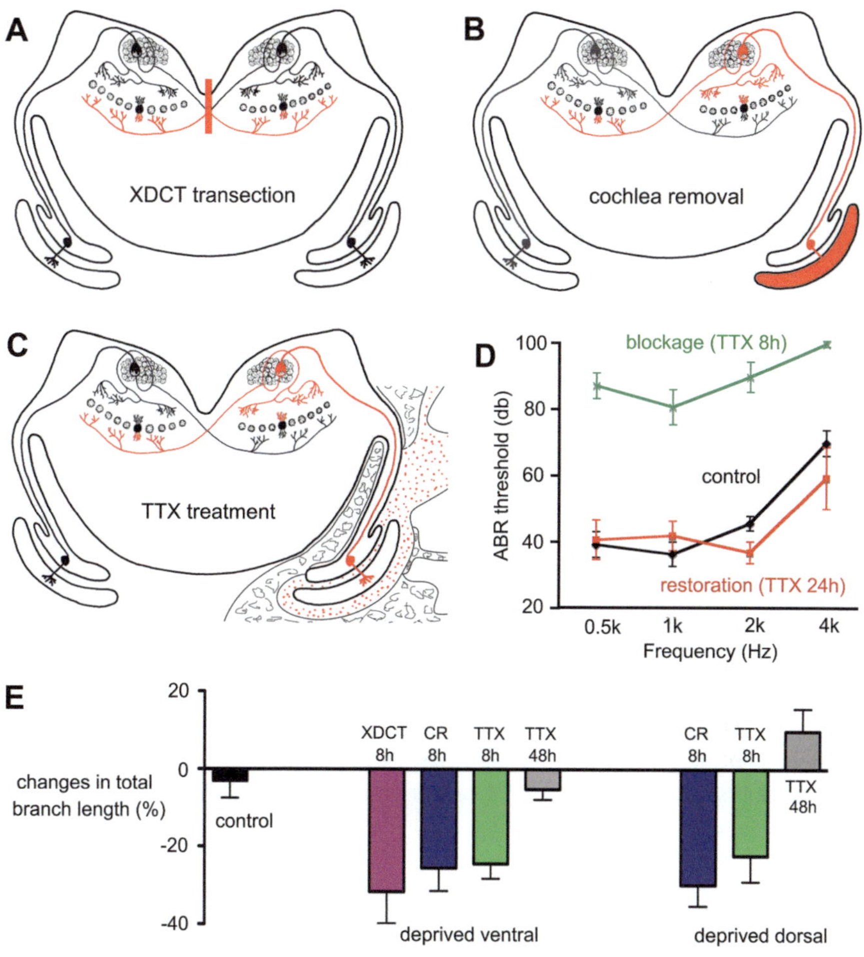

Fig. 11.5 Effects of afferent manipulations on dendritic length of NL neurons. (**a–c**) Schematic drawings of the chick brainstem illustrate the affected dendritic fields following XDCT transection (**a**), unilateral cochlea removal (**b**), and unilateral TTX treatment (**c**). *Red* color indicates the surgical site or drug application location for each manipulation, as well as deafferented axons and dendrites influenced by each manipulation. For TTX treatment, TTX is injected into the perilymph of the vestibular system to avoid disrupting the integrity of the middle ear or cochlear fluids (Born and Rubel 1988; Canady and Rubel 1992). (**d**) Effects and time course of TTX treatment on auditory brainstem response (ABR). ABR threshold is plotted as a function of the frequency of sound stimulation. ABR thresholds in response to sounds presented to the injected ear increase dramatically across chicken's entire hearing frequency range immediately after the TTX injection and remain so at 8 hours (h) (TTX 8 h, *green*) and then return to the control levels (*black*) by 24 h (TTX 24 h, *red*). (**e**) Change in total branch length of the deprived dendrites of individual NL neurons is plotted as a function of the manipulation. These changes are calculated as the differences of the deprived dendritic domain as compared to the intact domain of the same neurons. Error bar is standard deviation. At 8 h following XDCT transection (XDCT 8 h), unilateral cochlea removal (CR 8 h), or local TTX treatment (TTX 8 h), deprived dendrites exhibit 20–30 % reduction in total branch length (pink, blue, and green bars). Neurons at 48 h following TTX treatment (TTX 48 h) have more than 24 h with restored afferent activity, and the dorsal and ventral domains are no longer significantly different (*gray bars*), similar to control neurons (*black bar*)

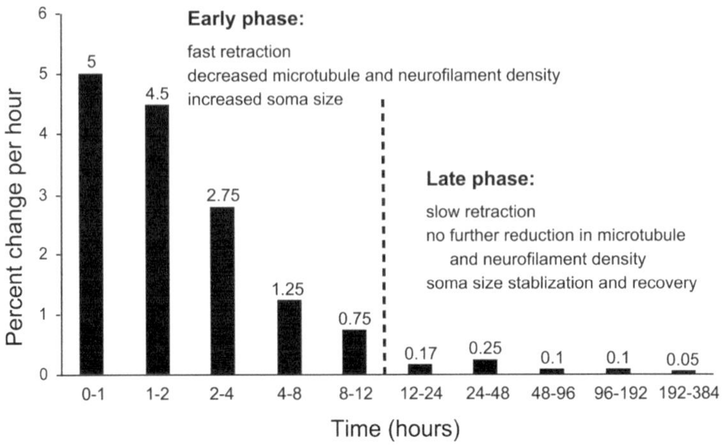

Fig. 11.6 Time course of dendritic retraction following XDCT transection in the chicken NL. The y-axis is the percent change per hour of total dendritic branch length of the deprived ventral dendrites of individual NL neurons as compared to the intact dorsal dendrites of the same neurons. The time course of dendritic change appears to have two phases, which differ in the retraction rate of dendritic branches, in cytoskeletal reorganization and in somatic alternations. The retraction rates are recalculated from Deitch and Rubel (1984). Changes in soma and cytoskeleton are described by Deitch and Rubel (1989a, b)

following XDCT transection, 60 % of the ventral dendrites have disappeared. Quantitative analyses reveal that the changes in dendritic branches are very fast during the first 12 h (on average 2 % reduction per hour) and slow down later (less than 0.25 % reduction per hour, Fig. 11.6). The short-term changes appear characterized by a rapid decrease in the density of dendritic cytoskeleton (microtubule and neurofilament), while the late phase exhibits no further changes in cytoskeletal density but shows a reduced density of cellular organelles (Deitch and Rubel 1989a, b). Importantly, at any given time point within 16 days following the manipulation, the percentage, not the absolute amount, of dendritic lost is the same across the tonotopic axis of the nucleus. Hence, the tonotopic gradient of ventral dendrites remains, but this gradient is "flattened" because the large (low-frequency coding) neurons loose more total dendrite than the smaller (high-frequency coding) dendrites. This observation is consistent with the notion that both the formation and maintenance of dendritic gradient in NL are largely independent of excitatory afferent input, but the overall size of any specific dendritic arbor is dramatically regulated by presynaptic activity.

Careful ultrastructural examination of synapses and axonal structure was unable to reveal degeneration during the early stage of ventral dendritic reorganization following XDCT transection (Deitch and Rubel 1989a). This result is consistent with the interpretation that the early changes in dendritic structure are due to the loss of excitatory synaptic input onto the ventral dendrite and not the result of degeneration of the afferent elements, per se. Surprisingly, no evidence of degeneration of dendritic plasma membrane is evident until at least 2 days following XDCT transection. Examination of the soma, however, reveals an increase in volume within hours, which is maintained up to 8 days following the lesion (Deitch and Rubel 1989b). This result suggests that resorption of dendritic plasma membrane is apart of the process of dendritic remodeling, at least the early decrease in ventral dendritic length.

The most important observation from these experiments is that the changes in dendrites are confined to the ventral (deafferented) domain (Benes et al. 1977; Deitch and Rubel 1984, 1989a). There is no evidence of altered morphology or structure in the dorsal dendrites of the same NL neurons, indicating that the cellular response is confined to the dendritic surface which is directly deafferented. This specificity provides a clear example of the fact that afferent excitatory inputs can independently regulate different parts of the same neuron, an observation that has now been confirmed in a variety of other neuronal systems. Detailed quantitative studies in NL have demonstrated that structural difference between the two domains of the same neuron reliably predicts the actual change in the deprived domain following differential afferent deprivation. Thus NL provides a convenient model to study local dynamics of dendrites with the benefit of a matched control for structural properties within the same neuron.

Transneuronal Afferent Deprivation The rapidity and domain specificity of dendritic retraction following XDCT transection raise the possibility that these postsynaptic changes, at least the short-term changes, are exclusively or primarily due to the cessation of presynaptic action potentials. This possibility is confirmed by comparable changes in NL dendrites following transneuronal deprivation of afferent activity. As the cochlea provides the only excitatory input to the ipsilateral NM, unilateral cochlea removal eliminates presynaptic action potentials to the ipsilateral NM and then to the dorsal dendrites of the ipsilateral NL and ventral dendrites of the contralateral NL, without directly damaging NM axons (Fig. 11.5b; Born et al. 1991). By 8 h following unilateral cochlea removal, the total dendritic length of deprived NL dendrites decreases by ~30 % as compared to the intact dendritic domain of the same neurons, comparable to that following XDCT transection (Wang and Rubel 2012). The similar retraction rate following the two manipulations strongly suggests that presynaptic activity, instead of lesion-related events, is responsible for maintaining dendritic arbors at the early stage.

The short-term modifications of deprived NL dendrites are fully reversible, as evidenced by complete recovery of normal dendritic length and dorsal-ventral symmetry following restoration of normal afferent activity (Wang and Rubel 2012). Local exposure of one ear with tetrodotoxin (TTX) completely blocks action potentials in the auditory nerve, totally silences action potential generation of NM neurons, and is completely reversible (Fig. 11.5c; Born and Rubel 1988; Canady and Rubel 1992). TTX treatment with carefully adjusted amount of the drug allows complete blockage of activity for at least 8 h, followed by rapid recovery; the level of activity and hearing thresholds return to control levels by 24 h and are maintained thereafter (Fig. 11.5d; Wang and Rubel 2012). NL dendrites deprived of excitatory input for 8 h by this manipulation show comparable retraction to that seen following cochlea removal or XDCT transection (Fig. 11.5e), further confirming presynaptic action potentials as a key afferent component in regulating postsynaptic dendrites. These dendrites then regrow to their normal size by 48 h following the onset of TTX treatment, by which time NM neurons have had at least 24 h with restored normal activity. This rapid recovery of dendritic structure demonstrates that retracting NL dendrites, following afferent deprivation, maintain a high degree of plasticity and actively respond to changes in afferent input in either direction.

In the paragraphs above, we only consider the early phase of dendritic changes following afferent deprivation and argue that these changes are mostly or entirely due to the deprivation of action potentials, presumably glutamate release and activation of postsynaptic receptors on the local postsynaptic elements (dendrites). On the other hand, integrity of presynaptic axons, in addition to their activity, may contribute to dendritic regulation during the later phases of dendritic remodeling following afferent axon elimination. To assess this hypothesis, we further studied the effects of unilateral cochlea removal. It is well known that unilateral cochlea removal in young chickens leads to extensive, but not complete, neuronal death in the ipsilateral NM (Born and Rubel 1985). Therefore, this manipulation produces partial deafferentation of specific dendritic domains in both ipsilateral and contralateral NL. The variation between animals in the amount of neuronal loss allowed us to compare the amount of dendritic retraction in the affected dendritic domains of NL neurons with the amount of NM cell loss in each animal. Across animals, the degree of dendritic retraction in NL exhibits a highly significant correlation ($r = 0.76$, $p < 0.01$) with the percentage of cell death in NM at 4 days following unilateral cochlea removal (Fig. 11.7). One interesting possibility is that the maintenance of remaining dendrites is affected by the integrity of NM neurons that continue to innervate these dendrites. Alternatively, cell survival in NM and long-term dendritic maintenance in NL may be influenced by similar mechanisms that depend on action potential-mediated signals.

While the MSO of low-frequency hearing mammals presents many of the same advantages for dendritic analyses as NL, detailed structural and temporal analyses are not yet available. Nonetheless, similar dendritic changes have been observed in MSO of young rats (Feng and Rogowski 1980) and adult gerbils (Russell and Moore 1999). Unilateral cochlea ablation also produces domain-specific changes in dendritic structure days and months later; whether these relatively long-term changes correlate with neuronal survival in AVCN and whether there are earlier changes in MSO dendrites have not been examined.

Dendritic Properties Regulated by Afferent Input To understand functional significance or consequence of afferent-regulated dendritic dynamics in NL, we need to identify the major dendritic properties that are both sensitive to changes in afferent input and essential to the function of the neuron. Two such properties have been identified, the size of dendritic surface which helps define the number of synapses

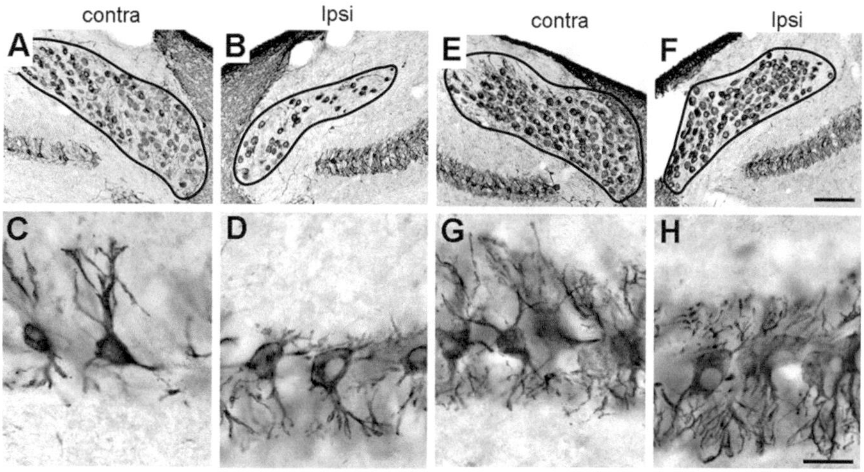

Fig. 11.7 Correlation of cell survival in NM with dendritic maintenance in NL. All images were taken at 4 days following unilateral cochlea removal. (**a**–**h**) Photomicrographs of coronal sections labeled with MAP2 immunoreactivity through NM (**a**, **b**, **e**, **f**) and NL (**c**, **d**, **g**, **h**). (**a**–**d**) Were taken from one animal, and **e**–**h** from the second animal. Left (**a**, **c**, **e**, **g**) and right (**b**, **d**, **f**, **h**) columns are from the contralateral and ipsilateral side, respectively. In both cases, the ipsilateral NM has a notable reduction in cell number as compared to the contralateral NM of the same case. Within NL, the ventral dendrites of the contralateral side and the dorsal dendrites of the ipsilateral side show clear retraction as compared to the other domain of the same side. The difference between the two cases is that the first case (**a**–**d**) exhibits a much larger degree of cell death in the ipsilateral NM as compared to the second case (**e**–**h**). Correlated to this difference in transneuronal NM cell death, dendritic retractions in NL are more dramatic in the first case than in the second case. *Solid lines* in **a**, **b** and **e**, **f** outline the borders of NM (Images are modified from Wang and Rubel (2012)). Scale bar: 100 μm in (**f**) (applies to **a**, **b** and **e**, **f**) and 20 μm in (**h**) (applies to **c**, **d** and **g**, **h**)

that can be formed and the dendritic arborization pattern which contributes to the way synapses are distributed and how inputs are integrated.

Quantitative analyses at the single-cell level following XDCT transection reveal no changes in the average diameter of dendritic branches (Deitch and Rubel 1984; Wang and Rubel 2012). Consistently, the overall size of dendritic surface correlates with total dendritic length in both control and deprived dendrites (Smith and Rubel 1979; Deitch and Rubel 1984). As a result, afferent activity influences dendritic surface area by dynamically regulating dendritic length.

Afferent activity also influences the arborization pattern of NL dendrites by reorganizing the dendritic tree in a branch-specific manner. First, the number of primary dendrites is unaffected by afferent deprivation (Deitch and Rubel 1984; Wang and Rubel 2012), suggesting that retraction mostly takes place in higher-order dendritic arbors. This suggestion is also supported by in vitro time-lapse imaging studies showing that deafferentation-induced dendritic retraction is observed primarily at the distal portion of the NL dendritic arborizations (Sorensen and Rubel 2006, 2011). Second, the number of dendritic endings or segments is regulated by afferent activity in such a way that it reliably predicts changes in total branch length of both retracing and regrowing dendrites (Wang and Rubel 2012). This finding indicates that selective elimination of a number of distal arbors and addition of new branches contribute substantially to the overall changes in total branch length. In vitro time-lapse imaging further demonstrates that individual dendritic arbors of a particular neuron are always dynamic (Sorensen and Rubel 2006). At any time over any 30-min period, about 50 % of the dendritic branches remain stable (less than 10 % change) in length. This statement holds true under both normal (balanced) and imbalanced conditions of afferent activity. The other 50 % of the dendritic branches either grow or retract by over 10 %. Over a 7-h imaging period, most or all terminal branches could be found growing or retracting. Changes in total branch length result from changes in the *balance* of arbors that are growing, retracting, or remaining stable at any time, not a sudden surge of growth or retraction. Thus, afferent-regulated dendritic reorganization does not take place uniformly along all affected dendritic branches and may occur in a branch-specific manner by which the fate (growing, retracting, or remaining stable) of individual branches is influenced by a combination of afferent inputs and intrinsic properties of each branch. This suggestion is consistent with the compartmentalization of electrical, calcium, and other biochemical signaling with single branch specificity in neuronal dendrites (Rinzel and Rall 1974; Rinzel 1975; Sjöström et al. 2008; Branco and Häusser 2010; Siegel and Lohmann 2013). Similar regulatory patterns have been reported in visual neurons for dendritic growth induced by sensory stimulation and for dendritic retraction following neurotransmission inactivation (Rajan and Cline 1998; Rajan et al. 1999; Sin et al. 2002).

Responsible Components of Afferent Activity Both XDCT transection and unilateral transneuronal manipulations deprive presynaptic activity to one, but not both, dendritic domain of NL neurons. Thus, changes in either the balance or the overall level of activity, or both, could be responsible for induced changes in postsynaptic NL dendrites.

In vitro studies using acute slice preparations of the chicken NL support the importance of the balance of afferent activity to the two dendritic domains of the same neurons (Sorensen and Rubel 2006, 2011). Slices of the brainstem were maintained in artificial CSF for several hours with bipolar stimulating electrodes on both auditory nerves, while individual fluorescently labeled NL neurons were images with a multiphoton microscope (Fig. 11.8). Two control conditions were included: (1) both auditory nerves were stimulated at a rate of 10 Hz for 7 h and at a level that reliably evoked action potentials in NL neurons or (2) neither auditory nerve was stimulated for the same period of time. In both of these conditions, NL neurons exhibited very little changes in the total dendritic length over the 7 h, although the arbors were continuously growing and retracting. This observation indicates that NL neurons maintain their dendrites in the absence of presynaptic action potentials to both dendritic domains *in vitro*. In contrast, unilateral electrical stimulation, activating NM axons to only one set of NL dendrites (either dorsal or ventral), results in the unstimulated dendrites losing total dendritic branch length

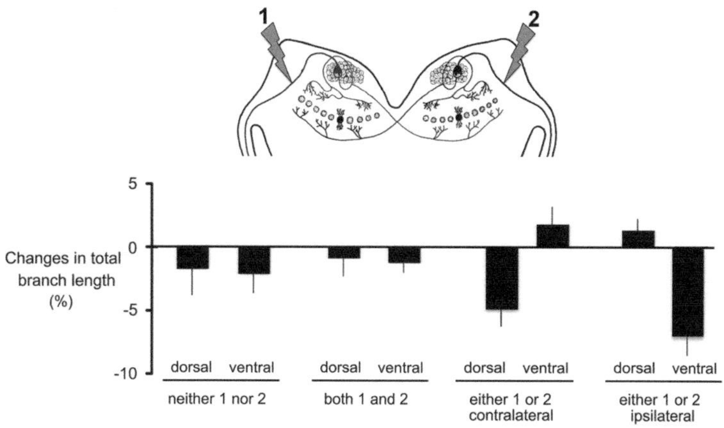

Fig. 11.8 Differential effects of balanced and imbalanced afferent inputs on dendritic length of the chicken NL. The y-axis is the changes in total branch length of NL dendrites. Control conditions on the left show that there is no significant change in the dorsal or ventral dendrites in response to no stimulation (neither 1 nor 2) or bilateral stimulation (both 1 and 2) at 10 Hz for 7 h. In contrast the data on the right show that unilateral (either 1 or 2) stimulation using the same parameters of afferent input from NM causes dramatic and opposite changes in branch lengths of the dorsal and ventral dendrites (Data are derived from Sorensen and Rubel (2006, 2011))

and the stimulated dendrites of the same NL neurons exhibiting a net growth. This result suggests that imbalanced input results in differential changes in the two sets of dendrites in a way that dendritic real estate can rapidly be shifted from inactive inputs to active inputs.

On one hand, these studies are consistent with the importance of the balance of afferent activity. On the other hand, dendritic growth or retraction during changes in afferent activity may depend on differences in the exact level and pattern of presynaptic activity elicited by electrodes as compared to that derived from the intact ears *in vivo*. That is, it is possible that both the balance between the two afferent inputs and the specific properties (level and/or pattern) of each input influence the maintenance of NL dendrites.

Molecular Signals and Pathways Examining effects of afferent alternations on dendritic biochemistry helps identify intracellular signals and pathways involved in afferent-regulated dendritic reorganization, especially if such effects show similar domain specificity as exhibited by changes in structure following differential afferent manipulations.

One such pathway is calcium-mediated microtubule regulation (Fig. 11.9). Following XDCT transection, one of the first changes in the fine structure of the deafferented ventral dendrites is dramatic reduction in the density of microtubules (Deitch and Rubel 1989b), a major cytoskeletal component of neuronal dendrites. This reduction may be associated with rapid decrease in the microtubule-associated

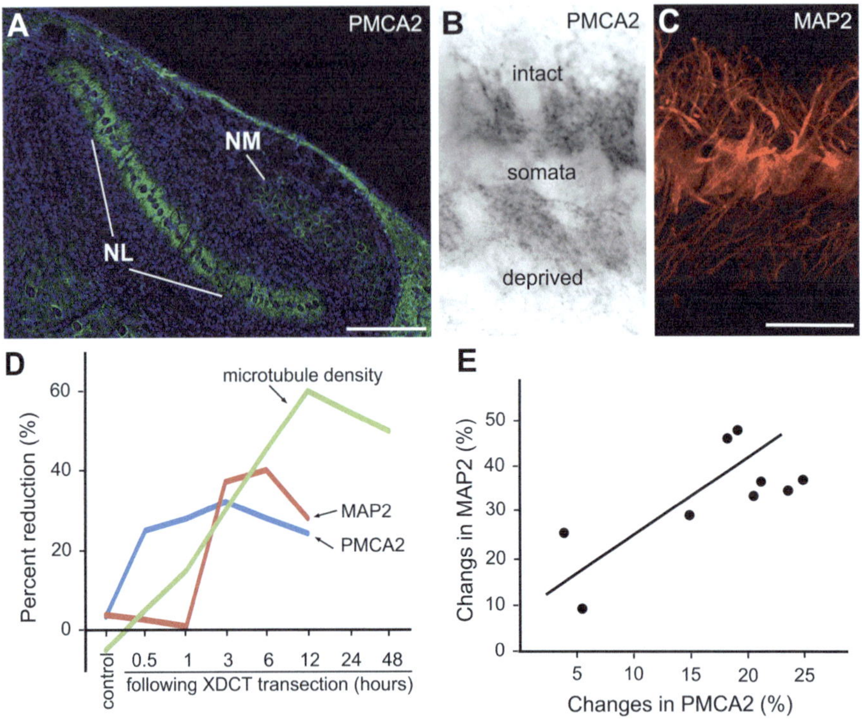

Fig. 11.9 Effects of afferent deprivation on dendritic levels of PMCA2 and MAP2 in the chicken NL. (**a**) A low-magnification photomicrograph showing the overall pattern of PMCA2 immunoreactivity in NM and NL. Note high PMCA2 level in NL dendritic layers. Also shown in Fig. 11.3d. (**b**, **c**) Immunoreactivities for PMCA2 (**b**) and MAP2 (**c**) in NL neurons at 6 h following XDCT transection. The levels of PMCA2 and MAP2 are reduced dramatically in deprived ventral dendrites as compared to the intact dorsal dendrites of the same NL neurons. (**d**) Time courses of percent reductions in dendritic levels of PMCA2 (*blue line*), MAP2 (*red line*), and microtubule density (*green line*) following XDCT transection (Recalculated from Wang and Rubel (2008), Wang et al. (2009), and Deitch and Rubel (1989a)). (**e**) Scatter plot showing relationship between changes in PMCA2 and MAP2 in deprived NL dendrites following unilateral cochlea removal. Each data point represents an individual case. Animals with a larger change in PMCA2 immunoreactivity tend to have a larger change in MAP2 immunoreactivity. Abbreviations: *PMCA2* plasma membrane calcium ATPase type 2, *MAP2* microtubule-associated protein 2. Scale bar: 200 μm in (**a**) and 25 μm in (**c**) (applies to **b** and **c**)

protein 2 (MAP2) that is confined to afferent-deprived NL dendrites (Wang and Rubel 2008). MAP2 is essential for microtubule stabilization and is believed to play a fundamental role in regulating dendritic morphology in developing and adult brains (reviewed in Dehmelt and Halpain 2005). Importantly, changes in MAP2 are correlated with changes in the plasma membrane calcium ATPase type 2 (PMCA2), a high-affinity calcium efflux protein, in deprived NL dendrites (Wang et al. 2009). Neurons with a larger change in PMCA2 immunoreactivity tend to have a larger change in MAP2 immunoreactivity. Changes in PMCA2 start as early as 30 min following afferent deprivation, prior to detectable changes in MAP2 and the structure of these dendrites. This temporal property provides a potential mechanism by which deprivation can change calcium transport that, in turn, may be important for rapid, domain-specific dendritic remodeling. This possibility is strongly supported by the well-known regulation of MAP2 and microtubule by calcium signaling (Vaillant et al. 2002), as well as with the pivotal role of intradendritic calcium signals in activity-dependent dendritic regulation (Lohmann and Wong 2005; Redmond and Ghosh 2005). Further support for an important role of this pathway is provided by the observations that afferent or sensory deprivation leads to changes in the expression of a number of calcium-binding proteins (parvalbumin, calretinin, and calbindin) in the mammalian MSO in adults (Caicedo et al. 1997; Alvarado et al. 2004). These data naturally lead to the speculation that deprivation of afferent excitatory activity leads to chronic changes in dendritic Ca^{++} regulation in the affected dendrite, a hypothesis supported by the observation that $[Ca^{++}]_i$ appears to be independently regulated in the dorsal and ventral dendritic domains (Blackmer et al. 2009). This is an important area for further investigation using modern imaging methods.

Another intracellular signaling pathway that may be involved in afferent regulation of dendritic structure is fragile X mental retardation protein (FMRP, Fig. 11.10). FMRP is an mRNA-binding protein, associated with a highly selected group of mRNAs (Darnell et al. 2011). A major function of FMRP is thought to be regulating local protein synthesis of its mRNA targets, particularly in neuronal dendrites (reviewed in Santoro et al. 2012). Loss of FMRP protein leads to abnormal dendritic structure and reduced dendritic dynamics in response to neuronal activity (Galvez et al. 2003, 2005; Restivo et al. 2005; Zarnescu et al. 2005; Thomas et al. 2008; Santoro et al. 2012). We have recently shown that *dendritic* FMRP is expressed at unusually high levels in NL and MSO across a wide range of low-frequency hearing vertebrates (alligator, chicken, gerbil, and human; Wang et al. 2014). Remarkably, dendritic FMRP in NL and MSO neurons accumulates at branch points and enlarged distal tips, loci known to be critical for branch-specific dendritic arbor dynamics. This localization pattern raises the possibility that FMRP may play an important role in determining the fate of individual branches either branching out, retracting, or staying unchanged and/or provides quick supplies of required proteins for such cellular events on an individual branch basis. In support of this possibility, in neurotrophin-stimulated neurites of PC12 cells, FMRP granules are predominantly localized to swellings along developing neurite, often locations where new branches are generated (De Diego Otero et al. 2002). Finally, it is interesting to note that FMRP mechanisms may be linked with calcium and microtubule regulation in NL and MSO dendrites as both mRNAs encoding PMCA2 and MAP2 show a very high binding affinity with FMRP (Darnell et al. 2011). In addition, Kv3.1b, a functionally specialized protein in NL dendrites (Fig. 11.3e), is also an identified target of FMRP (Darnell et al. 2011).

Given the activity dependency of afferent-regulated dendritic structure, it is reasonable to argue for a critical role of neurotransmission in this regulation.

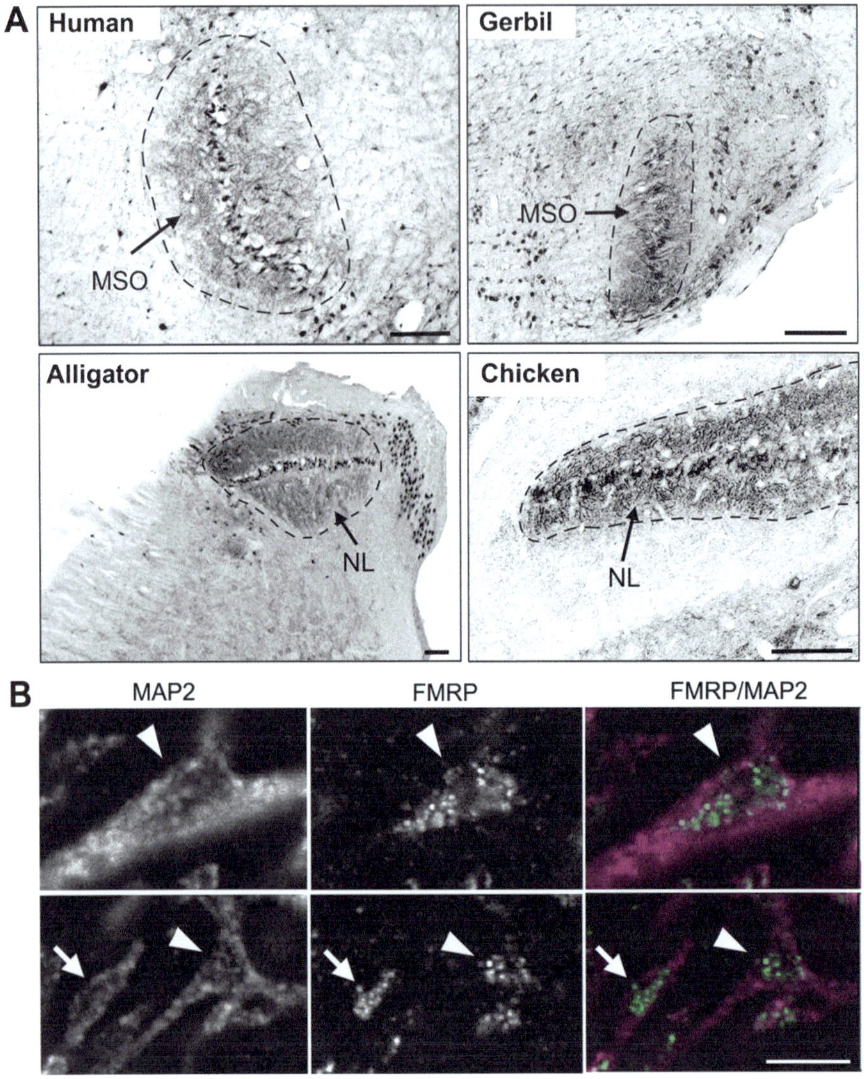

Fig. 11.10 Intense and specialized dendritic localization of FMRP in NL and MSO neurons. (**a**) High expression levels of dendritic FMRP immunoreactivity in human and gerbil MSO, as well as in alligator and chicken NL, as compared to the surrounding regions. (**b**) Complimentary pattern of MAP2 and FMRP in NL dendrites of the chicken. Note the strong accumulation of FMRP immunoreactivity in branch points (*arrowheads*) and terminal endings (*arrows*) and the relatively low level of MAP2 in these locations. Abbreviations: *FMRP* fragile X mental retardation protein. Scale bar: 100 μm in (**a**) (all panels) and 5 μm in (**b**)

Indeed, in the chicken NL, blocking excitatory neurotransmission using glutamate receptor inhibitors produces dendritic retraction (Sorensen and Rubel 2006), and unilateral cochlea removal also affects GABAa receptor expression (Code and

Churchill 1991). In the mammalian MSO, unilateral cochlea removal leads to significant changes in GABA and glycine release and uptake, as well as in the regulatory signaling of neurotransmitter release (Suneja et al. 1998a, b, 2000; Mo et al. 2006; Yan et al. 2007).

11.3.2 Other Regulatory Factors

NL and MSO neurons are among neuronal cell types with a high sensitivity to cues in the environment, including but not limited to changes in afferent input from the ear. For example, alcohol exposure affects dendritic length and structure in the chicken NL (Pettigrew and Hutchinson 1984) and neuronal number of the monkey MSO (Mooney and Miller 2001). Interestingly, this change in neuronal number may be exclusive to MSO, as compared to relatively smaller changes in other auditory regions in the brainstem. As another example, both the number and extension of MSO dendrites are reduced in underfed rats as compared to rats normally nourished (Salas et al. 1994). Similarly, aging-associated neurochemical changes are observed in the MSO but not other auditory groups in the monkey brainstem (Gray et al. 2014), and in the brainstem of humans with autism spectrum disorder (ASD), neuronal loss and disorganization appear most dramatic in MSO as compared to other auditory neurons (Kulesza 2007). This high vulnerability of MSO neuronal morphology and neuronal survival is consistent with reduced sound localization ability as a hallmark of age-related hearing deficits and as a common phenotype in a number of neurological disorders.

11.4 Conclusion

Afferent synaptic input is one important factor regulating neuronal structure including dendritic architecture in developing and mature brains. Bipolar neurons in NL and MSO present an example of dendritic architecture highly specialized for their function in temporal coding and coincidence detection. These neurons are also highly sensitive to cues in their environment, particularly variations in afferent input during both development and in the mature brain. Relatively brief changes in presynaptic glutamate-driven synaptic activity dramatically influences both the size of dendritic surface and the pattern of dendritic arborizations, and these changes appear largely localized to the precise regions of the altered input, providing unparalleled access to the membrane compartments undergoing plasticity with appropriate control compartments in the same neurons. The unique dendritic structure of NL and MSO neurons provides an advantageous model for further understanding of the specific roles of neurotransmission, calcium signal, protein synthesis, and cytoskeletal regulation in this important form of brain dynamics.

Acknowledgments Sponsored by National Institute on Deafness and Other Communication Disorders grants DC-013074, DC-03829, DC-02739, DC-04661, and DC-00018.

References

Agmon-Snir H, Carr CE, Rinzel J (1998) The role of dendrites in auditory coincidence detection. Nature 393:268–272

Alvarado JC, Fuentes-Santamaria V, Henkel CK, Brunso-Bechtold JK (2004) Alterations in calretinin immunostaining in the ferret superior olivary complex after cochlear ablation. J Comp Neurol 470:63–79

Benes FM, Parks TN, Rubel EW (1977) Rapid dendritic atrophy following deafferentation: an EM morphometric analysis. Brain Res 122:1–13

Blackmer T, Kuo SP, Bender KJ, Apostolides PF, Trussell LO (2009) Dendritic calcium channels and their activation by synaptic signals in auditory coincidence detector neurons. J Neurophysiol 102:1218–1226

Born DE, Rubel EW (1985) Afferent influences on brain stem auditory nuclei of the chicken: neuron number and size following cochlea removal. J Comp Neurol 231:435–445

Born DE, Rubel EW (1988) Afferent influences on brain stem auditory nuclei of the chicken: presynaptic action potentials regulate protein synthesis in nucleus magnocellularis neurons. J Neurosci 8:901–919

Born DE, Durham D, Rubel EW (1991) Afferent influences on brainstem auditory nuclei of the chick: nucleus magnocellularis neuronal activity following cochlea removal. Brain Res 557:37–47

Branco T, Häusser M (2010) The single dendritic branch as a fundamental functional unit in the nervous system. Curr Opin Neurobiol 20:494–502

Caicedo A, d'Aldin C, Eybalin M, Puel JL (1997) Temporary sensory deprivation changes calcium-binding proteins levels in the auditory brainstem. J Comp Neurol 378:1–15

Canady KS, Rubel EW (1992) Rapid and reversible astrocytic reaction to afferent activity blockade in chick cochlear nucleus. J Neurosci 12:1001–1009

Carr CE, Code RA (2000) The central auditory systems of reptiles and birds. In: Dooling RJ, Fay RR, Popper AN (eds) Comparative hearing: birds and reptiles. Springer, New York

Code RA, Churchill L (1991) GABAA receptors in auditory brainstem nuclei of the chick during development and after cochlea removal. Hear Res 54:281–295

Cramer KS, Rosenberger MH, Frost DM, Cochran SL, Pasquale EB, Rubel EW (2000) Developmental regulation of EphA4 expression in the chick auditory brainstem. J Comp Neurol 426:270–278

Darnell JC, Van Driesche SJ, Zhang C, Hung KY, Mele A, Fraser CE, Stone EF, Chen C, Fak JJ, Chi SW, Licatalosi DD, Richter JD, Darnell RB (2011) FMRP stalls ribosomal translocation on mRNAs linked to synaptic function and autism. Cell 146:247–261

De Diego Otero Y, Severijnen LA, van Cappellen G, Schrier M, Oostra B, Willemsen R (2002) Transport of fragile X mental retardation protein via granules in neurites of PC12 cells. Mol Cell Biol 22:8332–8341

Dehmelt L, Halpain S (2005) The MAP2/Tau family of microtubule-associated proteins. Genome Biol 6:204

Deitch JS, Rubel EW (1984) Afferent influences on brain stem auditory nuclei of the chicken: time course and specificity of dendritic atrophy following deafferentation. J Comp Neurol 229:66–79

Deitch JS, Rubel EW (1989a) Rapid changes in ultrastructure during deafferentation-induced dendritic atrophy. J Comp Neurol 281:234–258

Deitch JS, Rubel EW (1989b) Changes in neuronal cell bodies in N. laminaris during deafferentation-induced dendritic atrophy. J Comp Neurol 281:259–268
Dezsö A, Schwarz DW, Schwarz IE (1993) A survey of auditory brainstem nuclei in the chicken (Gallus domesticus) with cytochrome oxidase histochemistry. J Otolaryngol 22:385–390
Feng AS, Rogowski BA (1980) Effects of monaural and binaural occlusion on the morphology of neurons in the medial superior olivary nucleus of the rat. Brain Res 189:530–534
Fukui I, Burger RM, Ohmori H, Rubel EW (2010) GABAergic inhibition sharpens the frequency tuning and enhances phase locking in chicken nucleus magnocellularis neurons. J Neurosci 30:12075–12083
Galvez R, Gopal AR, Greenough WT (2003) Somatosensory cortical barrel dendritic abnormalities in a mouse model of the fragile X mental retardation syndrome. Brain Res 971:83–89
Galvez R, Smith RL, Greenough WT (2005) Olfactory bulb mitral cell dendritic pruning abnormalities in a mouse model of the Fragile-X mental retardation syndrome: further support for FMRP's involvement in dendritic development. Brain Res Dev Brain Res 157:214–216
Gao H, Lu Y (2008) Early development of intrinsic and synaptic properties of chicken nucleus laminaris neurons. Neuroscience 153:131–143
Gray DT, Engle JR, Recanzone GH (2014) Age-related neurochemical changes in the rhesus macaque superior olivary complex. J Comp Neurol 522:573–591
Grothe B (2000) The evolution of temporal processing in the medial superior olive, an auditory brainstem structure. Prog Neurobiol 61:581–610
Grothe B, Pecka M (2014) The natural history of sound localization in mammals – a story of neuronal inhibition. Front Neural Circ 8:116
Heil P, Scheich H (1986) Effects of unilateral and bilateral cochlea removal on 2-deoxyglucose patterns in the chick auditory system. J Comp Neurol 252:279–301
Jackson H, Parks TN (1988) Induction of aberrant functional afferents to the chick cochlear nucleus. J Comp Neurol 271:106–114
Jackson H, Hackett JT, Rubel EW (1982) Organization and development of brain stem auditory nuclei in the chick: ontogeny of postsynaptic responses. J Comp Neurol 210:80–86
Jeffress LA (1948) A place theory of sound localization. J Comp Physiol Psychol 41:35–39
Kitzes LM, Kageyama GH, Semple MN, Kil J (1995) Development of ectopic projections from the ventral cochlear nucleus to the superior olivary complex induced by neonatal ablation of the contralateral cochlea. J Comp Neurol 353:341–363
Kuba H, Koyano K, Ohmori H (2002) Synaptic depression improves coincidence detection in the nucleus laminaris in brainstem slices of the chick embryo. Eur J Neurosci 15:984–990
Kuba H, Yamada R, Fukui I, Ohmori H (2005) Tonotopic specialization of auditory coincidence detection in nucleus laminaris of the chick. J Neurosci 25:1924–1934
Kulesza RJ (2007) Cytoarchitecture of the human superior olivary complex: medial and lateral superior olive. Hear Res 225:80–90
Lippe WR (1994) Rhythmic spontaneous activity in the developing avian auditory system. J Neurosci 14:1486–1495
Lippe WR (1995) Relationship between frequency of spontaneous bursting and tonotopic position in the developing avian auditory system. Brain Res 703:205–213
Lippe WR, Steward O, Rubel EW (1980) The effect of unilateral basilar papilla removal upon nuclei laminaris and magnocellularis of the chick examined with [3H]2-deoxy-D-glucose autoradiography. Brain Res 196:43–58
Lippe WR, Fuhrmann DS, Yang W, Rubel EW (1992) Aberrant projection induced by otocyst removal maintains normal tonotopic organization in the chick cochlear nucleus. J Neurosci 12:962–969
Lohmann C, Wong RO (2005) Regulation of dendritic growth and plasticity by local and global calcium dynamics. Cell Calcium 37:403–409
Mo Z, Suneja SK, Potashner SJ (2006) Phosphorylated cAMP response element-binding protein levels in guinea pig brainstem auditory nuclei after unilateral cochlear ablation. J Neurosci Res 83:1323–1330

Mooney SM, Miller MW (2001) Episodic exposure to ethanol during development differentially affects brainstem nuclei in the macaque. J Neurocytol 30:973–982
Overholt EM, Rubel EW, Hyson RL (1992) A circuit for coding interaural time differences in the chick brainstem. J Neurosci 12:1698–1708
Parks TN (1981) Changes in the length and organization of nucleus laminaris dendrites after unilateral otocyst ablation in chick embryos. J Comp Neurol 202:47–57
Parks TN (1997) Effects of early deafness on development of brain stem auditory neurons. Ann Otol Rhinol Laryngol Suppl 168:37–43
Parks TN, Rubel EW (1975) Organization and development of brain stem auditory nuclei of the chicken: organization of projections from n. magnocellularis to n. laminaris. J Comp Neurol 164:435–448
Parks TN, Gill SS, Jackson H (1987) Experience-independent development of dendritic organization in the avian nucleus laminaris. J Comp Neurol 260:312–319
Pettigrew AG, Hutchinson I (1984) Effects of alcohol on functional development of the auditory pathway in the brainstem of infants and chick embryos. Ciba Found Symp 105:26–46
Rajan I, Cline HT (1998) Glutamate receptor activity is required for normal development of tectal cell dendrites in vivo. J Neurosci 18:7836–7846
Rajan I, Witte S, Cline HT (1999) NMDA receptor activity stabilizes presynaptic retinotectal axons and postsynaptic optic tectal cell dendrites in vivo. J Neurobiol 38:357–368
Ramón y Cajal S (1909) Histologie du Systeme Nerveux, vol I (1952 reprint). Instituto Ramon y Cajal, Madrid
Rebillard G, Rubel EW (1981) Electrophysiological study of the maturation of auditory responses from the inner ear of the chick. Brain Res 229:15–23
Redmond L, Ghosh A (2005) Regulation of dendritic development by calcium signaling. Cell Calcium 37:411–416
Restivo L, Ferrari F, Passino E, Sgobio C, Bock J, Oostra BA, Bagni C, Ammassari-Teule M (2005) Enriched environment promotes behavioral and morphological recovery in a mouse model for the fragile X syndrome. Proc Natl Acad Sci U S A 102:11557–11562
Richardson BE, Durham D (1990) Blood flow changes in chicken brain stem auditory nuclei following cochlea removal. Hear Res 46:53–61
Rinzel J (1975) Voltage transients in neuronal dendritic trees. Fed Proc 34:1350–1356
Rinzel J, Rall W (1974) Transient response in a dendritic neuron model for current injected at one branch. Biophys J 14:759–790
Rubel EW, Parks TN (1975) Organization and development of brain stem auditory nuclei of the chicken: tonotopic organization of n. magnocellularis and n. laminaris. J Comp Neurol 164:411–433
Rubel EW, Parks TN (1988) Organization and development of the avian brain-stem auditory system. In: Edelman GM, Gall WE, Cowan WM (eds) Auditory function: neurobiological bases of hearing. Wiley, New York, pp 3–92
Rubel EW, Smith DJ, Miller LC (1976) Organization and development of brain stem auditory nuclei of the chicken: ontogeny of n. magnocellularis and n. laminaris. J Comp Neurol 166:469–489
Rubel EW, Smith ZD, Steward O (1981) Sprouting in the avian brainstem auditory pathway: dependence on dendritic integrity. J Comp Neurol 202:397–414
Russell FA, Moore DR (1999) Effects of unilateral cochlear removal on dendrites in the gerbil medial superior olivary nucleus. Eur J Neurosci 11:1379–1390
Salas M, Torrero C, Regalado M, Martínez-Gómez M, Pacheco P (1994) Dendritic arbor alterations in the medial superior olivary neurons of neonatally underfed rats. Acta Anat (Basel) 151:180–187
Sanchez JT, Wang Y, Rubel EW, Barria A (2010) Development of glutamatergic synaptic transmission in binaural auditory neurons. J Neurophysiol 104:1774–1789

Sanchez JT, Seidl AH, Rubel EW, Barria A (2012) Control of neuronal excitability by NMDA-type glutamate receptors in early developing binaural auditory neurons. J Physiol 590:4801–4818

Santoro MR, Bray SM, Warren ST (2012) Molecular mechanisms of fragile X syndrome: a twenty-year perspective. Annu Rev Pathol 7:219–245

Saunders JC, Coles RB, Gates GR (1973) The development of auditory evoked responses in the cochlea and cochlear nuclei of the chick. Brain Res 63:59–74

Seidl AH, Rubel EW, Harris DM (2010) Mechanisms for adjusting interaural time differences to achieve binaural coincidence detection. J Neurosci 30:70–80

Seidl AH, Rubel EW, Barría A (2014) Differential conduction velocity regulation in ipsilateral and contralateral collaterals innervating brainstem coincidence detector neurons. J Neurosci 34:4914–4919

Siegel F, Lohmann C (2013) Probing synaptic function in dendrites with calcium imaging. Exp Neurol 242:27–32

Sin WC, Haas K, Ruthazer ES, Cline HT (2002) Dendrite growth increased by visual activity requires NMDA receptor and Rho GTPases. Nature 419:475–480

Sjöström PJ, Rancz EA, Roth A, Häusser M (2008) Dendritic excitability and synaptic plasticity. Physiol Rev 88:769–840

Smith ZD (1981) Organization and development of brain stem auditory nuclei of the chicken: dendritic development in N. laminaris. J Comp Neurol 203:309–333

Smith DJ, Rubel EW (1979) Organization and development of brain stem auditory nuclei of the chicken: dendritic gradients in nucleus laminaris. J Comp Neurol 186:213–239

Sorensen SA, Rubel EW (2006) The level and integrity of synaptic input regulates dendrite structure. J Neurosci 26:1539–1550

Sorensen SA, Rubel EW (2011) Relative input strength rapidly regulates dendritic structure of chick auditory brainstem neurons. J Comp Neurol 519:2838–2851

Suneja SK, Benson CG, Potashner SJ (1998a) Glycine receptors in adult guinea pig brain stem auditory nuclei: regulation after unilateral cochlear ablation. Exp Neurol 154:473–488

Suneja SK, Potashner SJ, Benson CG (1998b) Plastic changes in glycine and GABA release and uptake in adult brain stem auditory nuclei after unilateral middle ear ossicle removal and cochlear ablation. Exp Neurol 151:273–288

Suneja SK, Potashner SJ, Benson CG (2000) AMPA receptor binding in adult guinea pig brain stem auditory nuclei after unilateral cochlear ablation. Exp Neurol 165:355–369

Thomas CC, Combe CL, Dyar KA, Inglis FM (2008) Modest alterations in patterns of motor neuron dendrite morphology in the Fmr1 knockout mouse model for fragile X. Int J Dev Neurosci 26:805–811

Vaillant AR, Zanassi P, Walsh GS, Aumont A, Alonso A, Miller FD (2002) Signaling mechanisms underlying reversible, activity-dependent dendrite formation. Neuron 34:985–998

Wadhwa S, Anand P, Bhowmick D (1999) Quantitative study of plasticity in the auditory nuclei of chick under conditions of prenatal sound attenuation and overstimulation with species specific and music sound stimuli. Int J Dev Neurosci 17:239–253

Wang Y, Rubel EW (2008) Rapid regulation of microtubule-associated protein 2 in dendrites of nucleus laminaris of the chick following deprivation of afferent activity. Neuroscience 154:381–389

Wang Y, Rubel EW (2012) In vivo reversible regulation of dendritic patterning by afferent input in bipolar auditory neurons. J Neurosci 32:11495–11504

Wang Y, Cunningham DE, Tempel BL, Rubel EW (2009) Compartment-specific regulation of plasma membrane calcium ATPase type 2 in the chick auditory brainstem. J Comp Neurol 514:624–640

Wang Y, Sanchez J, Rubel EW (2010) Nucleus laminaris. In: Shepherd G, Addona D (eds) Handbook of brain microcircuits. Oxford University Press, New York, pp 224–233

Wang Y, Sakano H, Beebe K, Brown MR, de Laat R, Bothwell M, Kulesza RJ, Rubel EW (2014) Intense and specialized dendritic localization of the fragile X mental retardation protein in binaural brainstem neurons: a comparative study in the alligator, chicken, gerbil, and human. J Comp Neurol 522:2107–2128

Yan L, Suneja SK, Potashner SJ (2007) Protein kinases regulate glycine receptor binding in brain stem auditory nuclei after unilateral cochlear ablation. Brain Res 1135:102–106

Young SR, Rubel EW (1983) Frequency-specific projections of individual neurons in chick brainstem auditory nuclei. J Neurosci 3:1373–1378

Young SR, Rubel EW (1986) Embryogenesis of arborization pattern and topography of individual axons in N. laminaris of the chicken brain stem. J Comp Neurol 254:425–459

Zarnescu DC, Shan G, Warren ST, Jin P (2005) Come FLY with us: toward understanding fragile X syndrome. Genes Brain Behav 4:385–392

Zhou N, Parks TN (1993) Maintenance of pharmacologically-immature glutamate receptors by aberrant synapses in the chick cochlear nucleus. Brain Res 628:149–156

Part IV
Cellular and Molecular Control of Dendrite Development and Maintenance

Chapter 12
Molecular Control of Dendritic Remodeling

Kazuya Togashi, Hiroyuki Koizumi, Takahiro Kanamori, and Kazuo Emoto

Abstract Dendrites are the entry site of neural signals into neurons. Once formed, dendrites do not maintain their structure, rather are dynamically remodeled in vivo: some dendrites are pruned away, while others lengthen and branch out. Dendritic remodeling occurs not only during neural development but also in mature dendrites under both physiological and pathological conditions, suggesting its contribution to neural plasticity. The underlying cellular and molecular mechanisms remain poorly understood; however, they are just beginning to be elucidated from studies in vertebrate and invertebrate model systems. Here, we review recent advances in our understanding of dendritic remodeling by focusing particularly on insights obtained from layer IV spiny stellate neurons in the rodent somatosensory cortex and *Drosophila* sensory neurons.

Keywords Pruning • Drosophila sensory neurons • Barrel cortex

12.1 Introduction

Dendrites are the entry site of neural signals. In most types of neurons, neural signals, such as external stimuli for sensory neurons and synaptic inputs for postsynaptic neurons, are processed and converted to electrical signals in dendritic branches. Early anatomical studies pioneered by Ramón y Cajal revealed that dendritic structures could vary considerably between neurons (Cajal 1911). This led researchers to believe that dendritic divergence might explain, at least in part, why neurons show distinct functional properties in the brain. The fact that functional properties of neurons can often change in a plastic manner in vivo raises an intriguing question: do neurons remodel their dendrites structurally?

It has become increasingly clear that dendrites are remodeled in developmental, physiological, and pathological contexts. During neural development, neurons selectively eliminate exuberant dendritic branches to refine neural circuits. For

K. Togashi • H. Koizumi • T. Kanamori • K. Emoto, Ph.D. (✉)
Department of Biological Sciences, Graduate School of Science, The University of Tokyo,
7-3-1 Hongo, Bunkyo-ku, Tokyo 113-0033, Japan
e-mail: emoto@bs.s.u-tokyo.ac.jp

K. Emoto et al. (eds.), *Dendrites*, DOI 10.1007/978-4-431-56050-0_12

example, in the developing olfactory system of rats, mitral cells initially form multiple primary dendrites that contact adjacent glomeruli. However, they eventually prune all but one dendritic branch that remains in contact with a single glomerulus (Malun and Brunjes 1996). Developmental refinement of dendritic branches is also observed in other types of neurons, including retinal ganglion cells (Wang et al. 2001), spiny stellate cells in layer IV of the primary somatosensory cortex (Mizuno et al. 2014), and the cerebellar Purkinje cells (Kaneko et al. 2011). In a physiological condition, the superficial layer two-thirds interneurons in cerebral cortex dynamically add and eliminate dendritic branches, and the fraction of dynamic dendrites increases by three-fold after sensory deprivation, suggesting that neural activity influences the dynamics of dendritic remodeling (Chen et al. 2011; Lee et al. 2006). Pathologically, chronic stress and drug exposure can induce dendritic remodeling in a wide range of cerebral neurons in rodents (Ehlinger et al. 2012; Gourley et al. 2013; Li et al. 2012; Liston et al. 2006). Thus, increasing evidence indicates that both developing and mature neurons have the ability to remodel their dendritic branches. However, due to the lack of an appropriate model, the cellular and molecular mechanisms of dendritic remodeling have remained poorly understood.

Structural remodeling of dendrites reportedly occurs also in invertebrate nervous systems. In *Drosophila*, a variety of neurons in both the central nervous system (CNS) and peripheral nervous system (PNS) remodel their dendritic branches during metamorphosis from the larva to pupa and adult (Consoulas 2002; Kuo et al. 2005; Scott et al. 2011; Watts et al. 2003; Williams and Truman 2005). Taking advantage of the genetic tools available in *Drosophila*, a decade of extensive research has opened up new avenues for addressing the molecular and cellular basis of dendritic remodeling.

In this review, we summarize recent progress in understanding the cellular and molecular mechanisms of dendritic remodeling by focusing on layer IV neurons in the rodent somatosensory cortex and *Drosophila* PNS neurons, where major advances in knowledge have been made. The article begins with the cellular and molecular control of dendrite remodeling of developing neurons in the rodent somatosensory cortex. This is followed by a detailed account of genetic dissection of dendritic remodeling in the *Drosophila* PNS neurons. We also mention recent investigations using another invertebrate model system, *C. elegans* sensory neurons. We suggest that genetic studies using invertebrate model systems are better for gaining molecular insights into dendrite remodeling.

12.2 Dendritic Remodeling in the Mammalian Brain

12.2.1 Dendrite Remodeling of Developing Cortical Neurons in the Rodent Barrel Cortex

In the somatosensory cortex of rodents, the terminal arborization of thalamocortical afferents that represent each vibrissa or whisker is confined to separate barrels in layer IV. Layer IV spiny stellate cells initially develop symmetric trees, but they gradually orient their dendrites toward the center of their associated barrel, where the thalamic axons terminate. Interestingly, removal of one row of whiskers results in the formation of an expanded barrel. In contrast to the highly organized arborizations of normal animals, layer IV neurons in the center of the expanded barrel develop symmetric arborizations (Harris and Woolsey 1979; Steffen and Van der Loos 1980). It is thus likely that layer IV neurons reorganize their dendrites to preferentially contact active afferent inputs. Consistent with these notions, in the cortex-specific NR1 knockout, individual layer IV neurons lose oriented arborization and grow exuberant dendrites and spines (Datwani et al. 2002), suggesting that NMDA (*N*-methyl-D-aspartic acid) receptors mediate sensory stimulation-mediated dendrite orientation. Furthermore, by using the elegant single-cell knockout system MADM (mosaic analysis with double markers), Espinosa et al. (2009) demonstrated that NMDA activity is cell-autonomously required in cortical neurons for dendrite patterning but not for growth and branching. Indeed, a recent in vivo imaging study of layer IV dendrite dynamics during neonatal stages suggests that NMDA activity likely provides directional biases to dendrite dynamics during these stages (Mizuno et al. 2014).

12.2.2 Transcriptional Control of Dendrite Remodeling in the Barrel Cortex

A recent study has shown that the transcription factor BTBD3 is required for layer IV neurons to orient dendrites toward active axons (Matsui et al. 2013). BTBD3 is specifically expressed in the mouse somatosensory cortex during barrel formation, and its expression becomes detectable around postnatal day 2, at the time when thalamocortical axons innervate layer IV cortex. In cultured neural cell lines, BTBD3 is translocated from the cytosol into the nucleus in response to neural activity, implying its functions in activity-dependent neural processes. Indeed, short hairpin interference RNA (shRNAi)-mediated knockdown of *btbd3* in mouse layer IV neurons results in loss of dendrite orientation. More interestingly, ectopic *btbd3* expression in the visual cortex induces directed dendrite to active axon terminals. These data support a model in which BTBD3 mediates activity-dependent dendrite orientation in mammalian neurons. How the transcriptional factor may control dendrite orientation remains unclear. One possible scenario is that BTBD3 might promote trimming of dendritic branches that initially grow toward few active axon terminals. Alternatively, BTBD3 could promote biased extension of dendritic branches toward active axons. In either scenario, how the transcriptional factor locally regulates dendritic structures remains to be elucidated.

12.3 Dendritic Remodeling in Drosophila Neurons

12.3.1 Dendritic Changes During Transition from Larval to Adult Circuits in the Fly PNS

Peripheral neurons of *Drosophila* larvae are an excellent model to study dendrites (Emoto 2012; Parrish et al. 2007). In each hemisegment, 15 dendritic arborization (da) neurons elaborate stereotypic dendritic branches underneath the epidermal tissue. Da neurons are divided into four classes, I, II, III, and IV, according to the complexity of their dendritic morphology. Each class of da neurons, expressing different sets of genes including ion channels, can sense different external stimuli. For example, class III da (C3da) neurons express a transient receptor potential (TRP) channel, NompC, which confers the ability to sense gentle touch (Yan et al. 2013); class IV da (C4da) neurons sense noxious mechanical, heat, and light stimuli, which are converted into neural signals by the degenerin/epithelial sodium channel (DEG/ENaC) family Pickpocket (for mechanical stimuli), the newly discovered mechanosensory ion channel Piezo (also for mechanical stimuli), and the TRP channel TrpA1(for heat and light stimuli) (Kim et al. 2012; Zhong et al. 2010, 2012). Taking advantage of the class-specific gene expression, several transgenic reporter lines have been developed to label a specific class of da neurons with single-dendrite resolution in vivo.

Live imaging studies using the reporter lines have revealed that during metamorphosis from the larva to pupa and adult, dendrites of larval da neurons dynamically remodel to develop the adult PNS. The remodeling process of da neuron dendrites can be divided into three phases: destructive, latent, and regenerating (Fig. 12.1). The first destructive phase takes place within 16–20 h after puparium formation (APF). During this phase, ddaB class II da (C2da) and ddaA C3da neurons undergo programmed cell death within ~10 h APF. In contrast, ddaD/ddaE class I da (C1da) and ddaC C4da neurons remain alive and completely prune their dendritic arbors, while their axons and cell bodies remain intact (Williams and Truman 2005). Dendrite pruning of both classes of da neurons is mediated by local degeneration rather than retraction of branches (Williams and Truman 2004, 2005). After the completion of dendrite pruning, da neurons stay "dormant" without net regrowth of dendritic branches (the latent phase). The length of this phase differs between da neurons: up to ~42 and ~72 h APF for ddaE C1da and ddaC C4da neurons, respectively (Lyons et al. 2014; Williams and Truman 2004). Finally, during the regenerating phase, da neurons regenerate their dendritic branches

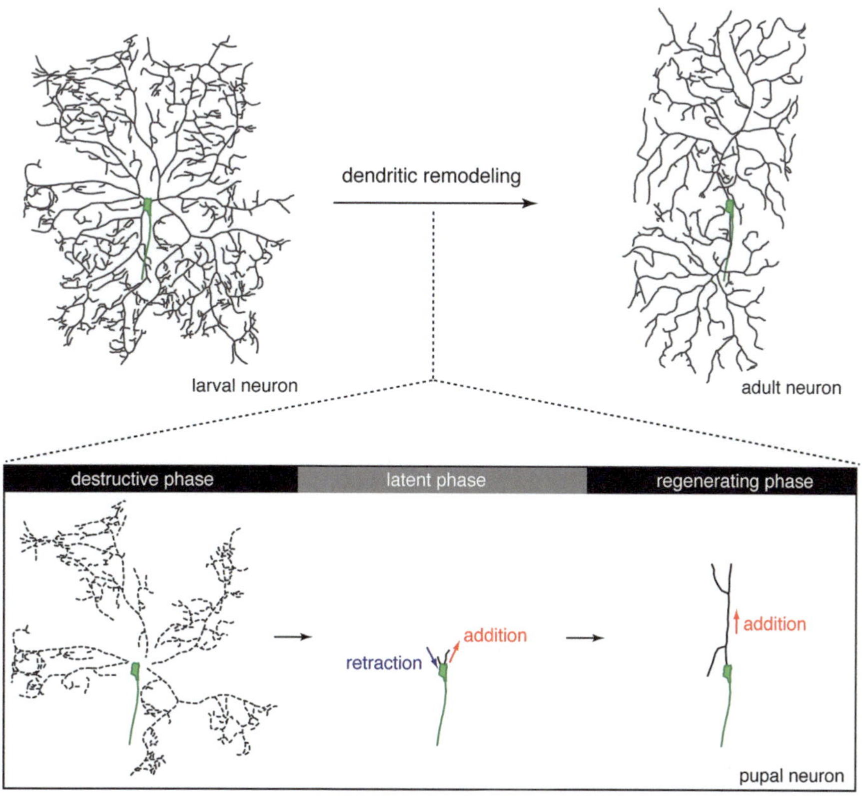

Fig. 12.1 Dendritic remodeling of *Drosophila* C4da sensory neurons. Dendritic branches of larval neurons (*upper left*) remodel during metamorphosis to form adult-specific dendrites (*upper right*). Dendrite remodeling is divided into three phases: destructive, latent, and regenerating phases (*bottom*). During the destructive phase, dendritic branches are pruned away by local degeneration, while axons and soma remain intact (*green*). After a latency of ~72 h, pupal neurons start to regenerate dendritic branches

through the rest of metamorphosis and are incorporated into the adult PNS circuits (Kuo et al. 2005; Lyons et al. 2014; Satoh et al. 2012; Shimono et al. 2009; Williams and Truman 2004). Hereafter, we will mainly focus on dendritic remodeling of ddaC C4da neurons, whose dendrites undergo pruning and regeneration during metamorphosis (Fig. 12.1), and review recently revealed molecular mechanisms underlying these processes.

12.3.2 Pruning of Larval Dendrites: The Molecular Mechanisms

12.3.2.1 Specification of Cell Responses

The molecular program of dendrite pruning is initiated by transcriptional control via the ecdysone-signaling cascade (Table 12.1). The steroid hormone ecdysone, which triggers a cascade of morphological changes during insect metamorphosis, acts through the ecdysone receptor (EcR), a member of the nuclear receptor superfamily, and its heterodimer partner ultraspiracle (Usp) (Fig. 12.2). On the prepupal pulse of ecdysone, the ecdysone receptor complex (EcR/Usp) is activated and induces the expression of a small set of genes, which triggers morphological changes of the larval tissue (Buszczak and Segraves 2000; Thummel 1996). Genetic inactivation of EcR/Usp functions inhibits dendrite pruning of C4da neurons in a cell-autonomous manner (Kuo et al. 2005; Williams and Truman 2005). It is thus plausible that dendrite pruning of C4da neurons is regulated by transcriptional induction of specific gene(s), which are vital for initiating and executing the pruning processes, through the ecdysone-signaling cascade.

Which are the genes induced by the ecdysone-signaling cascade to promote dendrite pruning of C4da neurons? A genetic study has identified a high-mobility group (HMG)-box transcription factor, Sox14, as a key gene that mediates dendrite pruning downstream of EcR/Usp (Kirilly et al. 2009). Sox14 expression is upregulated in the prepupal stage in an EcR/Usp-dependent manner. Interestingly, EcR forms a complex with a histone acetyltransferase CREB-binding protein (CBP) in response to ecdysone and is recruited to the *sox14* locus to facilitate H3K27 acetylation, the transcriptionally active chromatin mark (Kirilly et al. 2011). Since the upregulation of *sox14* also requires the function of CBP, it is likely that the EcR/Usp/CBP complex directly induces *sox14* expression. After being induced by ecdysone signaling, Sox14 upregulates the expression of at least two genes: molecule interacting with CasL (*mical*) and *cullin 1* (Kirilly et al. 2009; Wong et al. 2013). The expression level of Mical is increased in prepupal C4da neurons, similar to that of Sox14, and this increase is dependent not only on EcR/Usp but also on Sox14 (Kirilly et al. 2009). The mechanism by which Mical promotes dendrite pruning remains to be elucidated. However, given previous reports suggesting that Mical has a role in the regulation of actin cytoskeletons (Beuchle et al. 2007; Hung et al. 2010), this cytosolic protein might regulate the dynamics of actin filaments in pruning dendrites. The E3 ubiquitin ligase Cullin 1 (see below for details) is another possible downstream target of Sox14. Before initiation of axon pruning of the mushroom body (MB) γ neurons, an extensively studied model of developmental axon pruning, expression of Cullin 1 is upregulated in MB γ neurons in an EcR- and Sox14-dependent manner, although this upregulation in C4da neurons has not been directly examined (Wong et al. 2013). Given that this upregulation does not require Mical and that pruning defects of *mical* mutant C4da neurons are significantly enhanced by an additional mutation in

Table 12.1 Genes required for dendrite pruning of C4da neurons

Function in pruning	Fly gene name	Molecular function	References
Cell specification	*EcR*	Nuclear hormone receptor	Williams and Truman (2005); Kuo et al. (2005)
	ultraspiracle	Co-receptor of EcR	Williams and Truman (2005); Kuo et al. (2005)
	sox14	Transcriptional factor	Kirilly et al. (2009)
	CBP	Histone acetyltransferase	Kirilly et al. (2011)
	brahma	SWI/SNF chromatin remodeler	Kirilly et al. (2011)
	osa	SWI/SNF chromatin remodeler	Kirilly et al. (2011)
Nrg degradation	*rab5*	Rab family GTPase	Zhang et al. (2014)
	vps4	Disassembly of ESCRT-III complex	Zhang et al. (2014)
	vps28	ESCRT-I complex	Zhang et al. (2014)
	vps32	ESCRT-III complex	Zhang et al. (2014)
Thinning/severing of proximal regions	*mical*	Actin filament disassembly	Kirilly et al. (2009)
	headcase	Unknown	Loncle and Williams (2012)
	kat-60L1	Microtubule destabilization	Lee et al. (2009)
	ik2	Protein kinase	Lee et al. (2009)
	uba1	Ubiquitin activation enzyme	Kuo et al. (2005); Wong et al. (2013)
	mov34	19S proteasome particle	Kuo et al. (2005)
	ubcD1	E2 ubiquitin-conjugating enzyme	Kuo et al. (2006)
	vcp	Ubiquitin-selective AAA chaperone	Rumpf et al. (2011, 2014)
	pros26	20S proteasome core particle subunit	Rumpf et al. (2014)
	pros α7	20S proteasome core particle subunit	Rumpf et al. (2014)
	pros β5	20S proteasome core particle subunit	Rumpf et al. (2014)
	cullin 1	SCF complex, E3 ubiquitin ligase	Wong et al. (2013); Rumpf et al. (2014)
	roc1a	SCF complex, E3 ubiquitin ligase	Wong et al. (2013)
	skpA	SCF complex, E3 ubiquitin ligase	Wong et al. (2013)
	slimb	SCF complex, E3 ubiquitin ligase	Wong et al. (2013)
	nedd8	Neddylation of Cullin1	Wong et al. (2013)

(continued)

Table 12.1 (continued)

Function in pruning	Fly gene name	Molecular function	References
Compartmentalized Ca^{2+} transients	*ca-α1D*	Voltage-gated Ca^{2+} channel	Kanamori et al. (2013)
	cacophony	Voltage-gated Ca^{2+} channel	Kanamori et al. (2013)
	ca-beta	Voltage-gated Ca^{2+} channel	Kanamori et al. (2013)
	rab5	Rab family GTPase	Kanamori et al. (2015)
	shibire	Dynamin GTPase	Kanamori et al. (2015)
Distal degeneration	*dronc*	Caspase	Williams et al. (2006)
	drice	Caspase	Schoenmann et al. (2010)
	dcp-1	Caspase	Schoenmann et al. (2010)
	calpA	Calpain, Ca^{2+} dependent protease	Kanamori et al. (2013)
	calpB	Calpain, Ca^{2+} dependent protease	Kanamori et al. (2013)
Debris clearance	*draper*	Cell corpse engulfment receptor	Williams et al. (2006); Han et al. (2014)

a component of the Cullin 1 E3 ligase complex, *slimb*, it is likely that Mical and Cullin 1 act independently to promote dendrite pruning under the transcriptional control of Sox14 (Wong et al. 2013).

Another gene downstream of EcR/Usp is *headcase*. Screening genes that show genetic interactions with EcR mutants, Loncle and Williams identified *headcase* as a regulator of dendrite pruning (Loncle and Williams 2012). Similar to *sox14*, *headcase* expression follows the prepupal pulse of ecdysone in an EcR-dependent manner, although it remains to be elucidated whether EcR/Usp is recruited to the *headcase* locus to induce its transcription directly. Interestingly, the upregulation of *headcase* does not require *sox14* expression, and vice versa, suggesting that these two genes are independently regulated downstream of EcR/Usp. Consistent with this, pruning defects in a null allele of *headcase* are additively enhanced by RNAi knockdown of *mical*, a downstream gene of *sox14*. The molecular function of headcase in dendrite pruning remains unknown.

Sox14 and *headcase* are unlikely to be the only genes that C4da neurons express in response to the ecdysone-signaling cascade to initiate dendrite pruning. Although overexpression of *sox14*, but not that of *headcase* (Loncle and Williams 2012), accelerates dendrite pruning to occur earlier than in wild-type neurons, this precocious pruning is observed only after the prepupal pulse of ecdysone, suggesting that EcR/Usp regulates the expression of other genes in addition to these two. Identification of additional regulators that promote dendrite pruning downstream of EcR/Usp awaits further research.

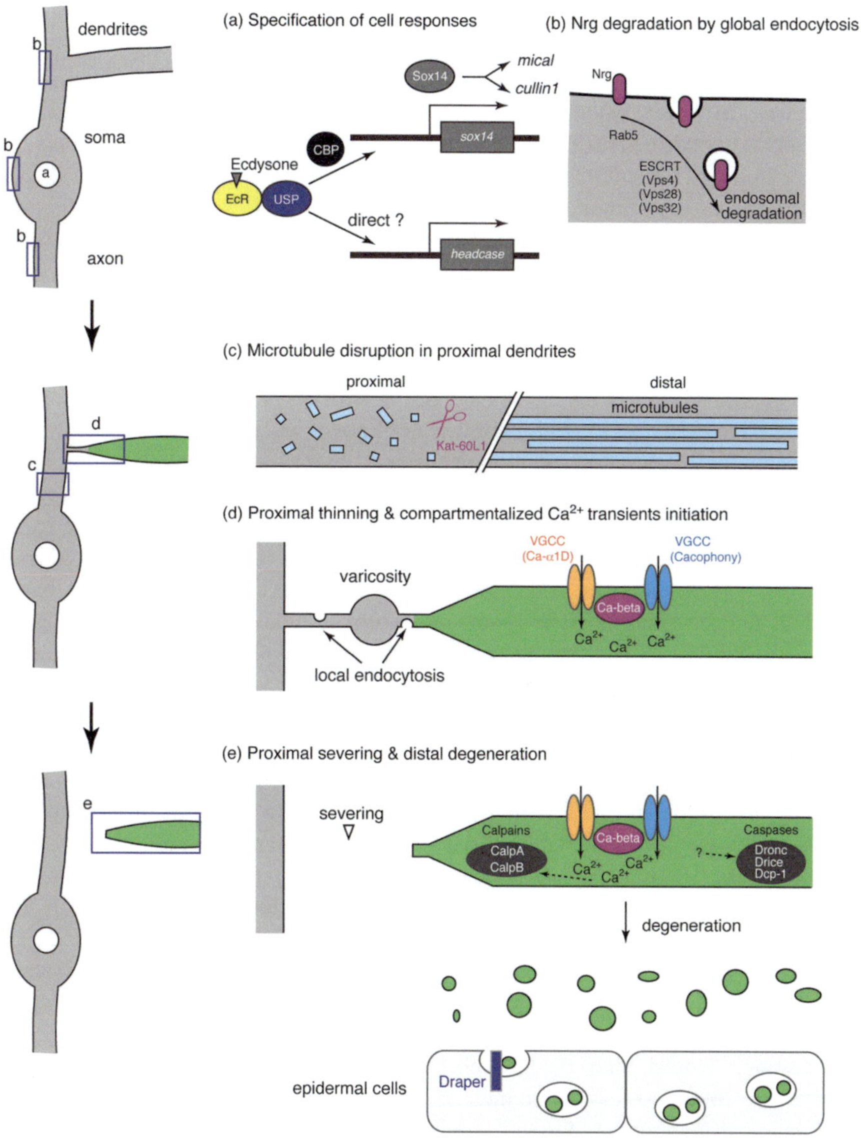

Fig. 12.2 Molecular mechanisms governing dendrite pruning of *Drosophila* C4da neurons. (**a**, **b**) Early events in the prepupal stage. Specification of cell responses is regulated by ecdysone signaling (**a**). The steroid hormone ecdysone binds to EcR/USP complex to induce expression of downstream genes, such as *sox14* and *headcase*. Induction of *sox14* expression also requires a histone acetyltransferase CBP. Sox14 induces expression of *mical* and *cullin1*. Downregulation of Nrg by endosomal degradation is a prerequisite for dendrite pruning (**b**). This involves Rab5 and the ESCRT complex components (Vps4, Vps28, and Vps32). (**c**, **d**) Local events in proximal dendrites. Microtubule disruption in proximal regions involves Kat-60L1 (**c**). Endocytic activity is locally elevated in proximal dendrites to induce branch thinning (**d**). Proximal thinning accompanies varicosity formation and tightly linked with the initiation of compartmentalized Ca^{2+} transients. Two types of VGCCs (L-type, Ca-α1D; and PQNR-type, cacophony), which require the beta subunit (Ca-beta) for their functions, are responsible for generating Ca^{2+} transients.

12.3.2.2 Thinning and Severing of Proximal Branches

After the initial specification driven by ecdysone signaling, the first morphological alterations are observable in the proximal regions of dendrites (Fig. 12.2): proximal dendrites actively form varicosities, around which dendritic branches progressively become thinner (Kanamori et al. 2015; Kirilly et al. 2009; Williams and Truman 2005). Some of the thinning proximal dendrites also form dynamic filopodia-like extensions (Williams and Truman 2005; our unpublished observations). The thinning dendritic branches are eventually severed (i.e., the distal regions become physically detached from the rest of the cell) (Kuo et al. 2005; Williams and Truman 2005). The initial severing of proximal dendrites occurs in a spatially correlated manner with the location of a distal edge of dendrites wrapped by the glial membrane, suggesting the existence of extrinsic mechanisms for dendrite severing (Han et al. 2011). Although it is not known whether each of these events is regulated by distinct molecular mechanisms, recent loss-of-function studies have begun to provide molecular insights into these early processes in dendrite pruning.

Regulators of the cytoskeleton are one group of genes involved in these early events. Cytoskeletal changes are prominent features of the thinning proximal dendrites: microtubules are fragmented locally in proximal dendrites in the early pupal stage (Lee et al. 2009; Williams and Truman 2005); and actin filaments are also disrupted in proximal regions of dendrites (Lee et al. 2009). Actin disruption is retarded compared with that of microtubules (Kanamori et al. 2015). *Katanin p60-like 1* (*kat-60L1*) is a key molecule for the microtubule destabilization in proximal dendrites. Given that Kat-60L1 is homologous to a p60 subunit of the microtubule-severing protein Katanin, this protein likely promotes dendrite pruning by directly severing microtubules in proximal dendrites. In addition, as mentioned above, Mical is likely to be involved in disrupting actin filaments to facilitate dendrite severing at the proximal regions. One important, but as yet unanswered, question is how these cytoskeletal regulators act locally in proximal dendrites during the pruning processes.

Another group of genes encodes regulators of the ubiquitin-proteasome system (UPS). Proper severing of proximal dendrites requires several UPS genes such as ubiquitin activation enzyme 1 (*Uba1*), a component of the 19S particle of the proteasome, *Mov34*, and the E2 ubiquitin-conjugating enzyme *UbcD1* (Kuo et al. 2005, 2006). At least three mechanisms are proposed for the function of the UPS in dendrite pruning. Firstly, it has been suggested that proximal severing requires UPS-mediated degradation of the E3-ubiquitin ligase *Drosophila* inhibitor of apoptosis protein 1 (DIAP1). This is mainly based on the following evidence:

Fig. 12.2 (continued) (**e**) Severing of proximal dendrites and degeneration of their distal regions. Proximal dendrites are severed during the repetition of Ca^{2+} transients, and the distal regions undergo degeneration via the local activation of calpains (CalpA and CalpB) and caspases (Dronc, Drice, and Dcp-1). Debris clearance is mediated by a cell surface protein Draper that is expressed in epidermal cells

(1) a gain-of-function mutation in *diap1*, which results in a DIAP1 protein insensitive to UPS-mediated degradation, compromises proximal severing (Kuo et al. 2006), and (2) inhibition of valosin-containing protein (VCP), a ubiquitin-selective AAA chaperone that is required for DIAP1 degradation in vitro, causes a similar severing defect, which is significantly suppressed by reducing DIAP1 expression (Rumpf et al. 2011). Supporting this idea, the noncanonical IκB kinase family member Ik2, which functions as a negative regulator of DIAP1 by promoting DIAP1 degradation (Kuranaga et al. 2006; Oshima et al. 2006), is also required for proper severing of proximal dendrites (Lee et al. 2009). Because severing of proximal dendrites apparently occurs in mutants of the cell death protease (caspase) Dronc, a well-known DIAP1 substrate, DIAP1 might degrade some substrates other than Dronc for proximal severing (Kanamori et al. 2013; Rumpf et al. 2014; Williams et al. 2006).

The second proposed mechanism by which the UPS controls dendrite severing involves downregulation of the insulin signaling pathway (Wong et al. 2013). Substrate specificity of the UPS-mediated protein degradation is achieved via E3 ligase substrate binding, which results in ubiquitination of the bound substrates. The Cullin1-based SCF (Skp1-Cullin-F-box) complex is an E3 ligase required for proximal severing. Genetic evidence suggests that a signaling pathway composed of insulin receptor (InR), phosphatidylinositol 3-kinase (PI3K), and target of rapamycin (TOR) (InR/PI3K/TOR pathway) is aberrantly activated in *cullin1* knockdown C4da neurons. Interestingly, the protein kinase Akt, a positive regulator of the InR/PI3K/TOR pathway, is more abundant in *cullin1* knockdown neurons compared with control neurons at the beginning of the pupal stage and physically interacts with the substrate-recognizing F-box Slimb. Based on these findings, it is proposed that the SCF E3 ligase complex targets Akt to downregulate the InR/PI3K/TOR pathway. Further investigation is needed to clarify how this downregulation is linked to the cell biological processes underlying proximal severing.

Regarding the third mechanism involving the UPS, Rumpf and colleagues recently reported that VCP also regulates proximal severing in a manner independent of protein degradation (Rumpf et al. 2014). Inhibition of VCP causes two kinds of defects in mRNA metabolism: missplicing of *mical* mRNA and aberrant localization and expression of RNA binding proteins, which could at least in part explain the pruning defects in *vcp* mutant neurons. Interestingly, these defects in mRNA metabolism are not observed in the mutants of the 20S proteasome component. Thus, the UPS likely promotes dendrite pruning both in a protein degradation-dependent and degradation-independent manner. Which cell biological process the UPS regulates to promote proximal severing remains to be elucidated.

12.3.2.3 Dendrite Degeneration and Its Clearance

After a proximal dendrite is severed, the dendrites distal to a site of severing degenerate rapidly after a latency of less than 2 h. Degeneration involves rapid blebbing and fragmentation of an entire dendritic branch into short segments (Williams and Truman 2005). These changes in cellular morphology are similar to Wallerian degeneration in live preparations of mouse axons, where the distal portion of axotomized peripheral nerves degenerates after about 1.5 days of latency without affecting the main cell bodies (Coleman and Freeman 2010). This argues for a potential conservation of degenerative mechanisms between these two different contexts (i.e., developmental dendrite pruning and Wallerian degeneration of mouse axons). This possibility was tested and supported by recent studies utilizing the causal gene product of the slow Wallerian degeneration mouse (C57BL/Wld^S) (Lunn et al. 1989), in which Wallerian degeneration is delayed tenfold. In flies overexpressing Wld^S proteins in C4da neurons, degeneration of the pupal dendrites during pruning was slightly but significantly retarded, while proximal dendritic branches seemed to be severed normally (Schoenmann et al. 2010; Tao and Rolls 2011), suggesting that developmental dendrite pruning may share molecular mechanisms with Wallerian degeneration (although this was not the case in developmental axon pruning of MB γ neurons (Hoopfer et al. 2006)). Because pruning defects seen in Wld^S overexpressing C4da neurons were subtle, it is likely that other mechanisms also contribute to the degeneration process.

Recently, local activation of caspases was suggested as an additional mechanism of dendritic degeneration in C4da neurons. Using the genetically encoded detector of effector caspases (such as Drice and Dcp-1), caspase activity was first readily detected within degenerating C4da dendrites, but not in the cell bodies or axons, during pupal stages (Williams et al. 2006). The authors also demonstrated that overexpression of the viral effector caspases inhibitor p35 had an inhibitory effect on dendrite pruning in C4da neurons (Kuo et al. 2006). Consistent with this are recent observations that flies lacking Drice or Dcp-1 were slightly but significantly defective in dendrite degeneration, while dendrite severing at proximal regions was apparently normal (Schoenmann et al. 2010). Therefore, caspase activation in dendritic branches could act as a degeneration mechanism in dendrite pruning. In addition, given that in C4da neurons overexpressing both Wld^S and p35 proteins dendrite pruning was inhibited slightly more strongly than in those expressing either of them (Schoenmann et al. 2010), it is possible that these two mechanisms contribute cooperatively to the degeneration process.

After dendritic branches are degenerated into short fragments, the neuronal debris is completely removed from the surrounding tissue, paving the way for subsequent dendritic regrowth during late metamorphosis. Which are the cell types responsible for clearing the degenerating dendrites? Hemocytes, highly motile macrophage-like cells, have been previously suggested as a possible candidate for this process (Williams and Truman 2005). Live imaging showed that some hemocytes were closely associated with degenerating sensory dendrites labeled with green fluorescent protein (GFP) and that they progressively exhibited green fluorescent cytoplasmic particles that are also labeled with the lysosome marker LysoTracker (Williams and Truman 2005). These results raise the possibility that hemocytes may phagocytose and clear the degenerating dendritic branches. Subsequently, a recent study from Jan and colleagues reexamined this possibility and showed that hemocytes play only a minor role (if any) in debris clearance although they appear to contribute partially to dendrite fragmentation during degeneration with an unknown mechanism (Han et al. 2014). Instead, the authors argue that epidermal cells adjacent to C4da dendrites are the major phagocytes clearing the neuronal debris (Fig. 12.2). They performed high-resolution 3D confocal imaging of degenerating dendrites and found that fragmented dendritic debris dispersed into the epidermal layer within several minutes after degeneration and eventually incorporated into the phagolysosomal compartments in the epidermal cells. This phagocytosis was inhibited in flies overexpressing mutant dynamin, which is known to regulate phagocytosis (Yu et al. 2006), specifically in epidermal cells; a significant amount of fragmented dendrites was not cleared and remained outside of the epidermal layer (Han et al. 2014). Taken together, these findings suggest that epidermal cells are likely the major phagocytes responsible for debris clearance.

The first clues to the molecular mechanisms of neuronal debris clearance came from studies of developmental pruning and injury-induced degeneration of *Drosophila* axons. These studies uncovered the role of the engulfment receptor *draper*, required for the clearance of apoptotic cells in *C. elegans* (Zhou et al. 2001) and phagocytosis in cultured *Drosophila* cells (Manaka et al. 2004), in the clearance of degenerating axons by glial cells (Awasaki et al. 2006; Hoopfer et al. 2006; MacDonald et al. 2006). Recently, astrocytes have been reported to be at least one of the glial cells that execute Draper-mediated phagocytosis during axon pruning of MB γ neurons (Hakim et al. 2014; Tasdemir-Yilmaz and Freeman 2014). Similarly, it was later shown that *draper* also has a role in the clearance of degenerating C4da dendrites during metamorphosis and its function is required, at least in part, in epidermal cells (Williams et al. 2006; Han et al. 2014). This Draper-mediated dendrite clearance by epidermal cells does not require the phagocytic receptor *six-microns-under* (*simu*), which functions upstream of *draper* in the context of apoptotic cell clearance in the developing CNS (Han et al. 2014; Kurant et al. 2008). This raises the intriguing possibility that different engulfment pathways are utilized depending on whether neuronal debris was generated through neurite pruning or apoptotic cell death. Evidence in support of this notion is also found in studies on developmental axon pruning of MB γ neurons (Kuraishi et al. 2009; Tasdemir-Yilmaz and Freeman 2014). The molecular and genetic dissection of the engulfment pathways in these different contexts should provide further insight into the molecular mechanisms underlying neuronal debris clearance.

Intriguingly, Draper-mediated phagocytic mechanisms might have an active role in the neural circuit remodeling. In other words, phagocytes (i.e., astrocytes and epidermal cells in the axon and dendrite pruning, respectively) are likely not just passive participants that scavenge the already degraded debris, but they could actively promote the degeneration of axonal and dendritic branches. This notion is based on the following observations: (1) before any visible signs of axon pruning, larval astrocytes transform into phagocytes in a cell-autonomous manner to infiltrate axon lobes destined for pruning and to increase the expression level of Draper and phagocytic activity (Hakim et al. 2014; Tasdemir-Yilmaz and Freeman 2014);

(2) genetic inactivation of glial infiltration into axon lobes significantly delays axon fragmentation (Awasaki and Ito 2004); (3) glial cells, presumably those other than astrocytes (Hakim et al. 2014), around MB γ axon branches destined to be pruned increase in number prior to branch fragmentation (Watts et al. 2004); (4) whole-body knockout or phagocyte-specific knockdown of *draper* causes defects not only in debris clearance but also in branch fragmentation (Han et al. 2014; Tasdemir-Yilmaz and Freeman 2014; but see also Hakim et al. 2014). In good agreement with this idea, Chung, Barres, and colleagues recently demonstrated that astrocytes utilize MEGF10 (mouse Draper homolog) and the engulfment receptor MERTK to eliminate excessive synapses formed between retinal ganglion cells (RGCs) and dorsal lateral geniculate nucleus (dLGN) neurons in the developing mouse retinogeniculate system. In addition, they showed that in mice lacking these two genes, the retinogeniculate connections were not refined normally and excessive weak but functional synapses did remain, suggesting an active role of Draper/MEGF10 in eliminating excessive connections (i.e., more than just scavenging neural debris) (Chung et al. 2013). It will be interesting to determine how phagocytes selectively eliminate unnecessary neurites and synapses among many others and whether or not there is a neuronal molecular machinery that could be activated by phagocytes to facilitate branch fragmentation.

12.3.2.4 An Emerging New Paradigm: Calcium Signaling and Endocytosis in Dendrite Pruning

The current view on dendrite pruning that proximal severing induces the distal degeneration of dendritic branches is based on the observation of dendritic shapes using membrane-targeted fluorescent markers. However, recent evidence suggests that this prevalent view is probably an oversimplification (Kanamori et al. 2013) (Fig. 12.2). Time-lapse imaging of pruning dendrites with simultaneous Ca^{2+} imaging has revealed that Ca^{2+} transients occur in a compartmentalized manner in dendritic branches destined to be pruned. Interestingly, compartmentalized Ca^{2+} transients are observed before dendrite severing and perfectly predict the location and timing of the dendrite pruning. Voltage-gated Ca^{2+} channels (VGCCs) are responsible for generating Ca^{2+} transients, and mutant C4da neurons lacking VGCC activity show significant defects in dendrite pruning. Subsequent calcium signaling activates the serine protease calpains, which promote the degeneration of dendrites cooperatively with the activity of caspases. Thus, compartmentalized Ca^{2+} transients act as spatial and temporal cues to trigger dendrite pruning before proximal severing occurs. How are dendrites compartmentalized into those with and without Ca^{2+} transients? A recent study has addressed this question and provided evidence suggesting that local endocytosis in proximal dendrites contributes to compartmentalization and initiation of Ca^{2+} transients by inducing branch thinning of proximal dendrites (Kanamori et al. 2015) (Fig. 12.2).

Involvement of endocytosis in dendrite pruning is also reported by Wong et al., showing that global endocytosis occurs in the entire neurons, including axons, dendrites, and cell bodies, and degrades the L1-type cell adhesion molecule Neuroglian (Nrg) through the endo-/lysosomal degradation pathway (Zhang et al. 2014) (Fig. 12.2). Nrg degradation starts from the onset of metamorphosis (i.e., the white pupal stage), and loss-of-function *nrg* mutant neurons show precocious dendrite pruning. Thus, removal of Nrg from the cell surface is a prerequisite for dendrite pruning. Interestingly, genetic evidence suggests that this global endocytosis for Nrg degradation promotes dendrite pruning cooperatively with local endocytosis for compartmentalized Ca^{2+} transients (Kanamori et al. 2015). These recent studies reveal that two distinct endocytic mechanisms act together in developmental dendrite pruning.

12.4 Regrowth and Reshaping of Adult Dendrites

After all larval dendrites are eliminated, C4da neurons regrow dendritic branches for the adult PNS circuit during the rest of the pupal stage. Dendrite regrowth in the late pupal stage can be divided into the latent and regenerating phases (Fig. 12.1). After the completion of dendrite pruning at 16 h APF, C4da neurons stay "dormant" until ~72 h APF without extensive dendritic regrowth (the latent phase), whereas numerous neurites extend and retract from the cell bodies. There is little net increase of dendritic length or branch point numbers during this phase (Lyons et al. 2014). Strikingly, however, dendritic branches grow in an exponential fashion between 72 and 96 h APF (the regenerating phase), eventually resulting in asymmetric dendritic fields that are characteristic of the adult ddaC C4da neurons (Lyons et al. 2014; Shimono et al. 2009) (Fig. 12.1). These observations suggest the existence of mechanisms that inhibit or promote dendritic regrowth during the latent or regenerating phase, respectively. Which molecular mechanisms contribute to dendritic regrowth? A recent study provided the first molecular insight into the regulation of dendritic regrowth during the late metamorphosis (Lyons et al. 2014). Kuo and colleagues showed that this process requires *Cysteine proteinase-1* (*Cp1*), the functions of which are not well characterized in the nervous system. Expression of Cp1 in C4da neurons increases during the latent phase in an EcR-dependent manner. Cp1 cleaves the homeodomain transcription factor cut to change its subnuclear localization from heterochromatin to the entire nucleus, suggesting that temporal Cp1 activation contributes to initiating the transcriptional program underlying the temporal switch between the two phases in dendrite regrowth. Interestingly, Cp1 is dispensable for dendritic regrowth of ddaE C1da neurons, which have a shorter latent phase for regrowth compared with that of C4da neurons (Lyons et al. 2014; Williams and Truman 2004). Thus, each class of da neurons might use distinct strategies for regrowing dendritic branches during late metamorphosis.

Which are the mechanisms that shape the regrowing dendrites for the adult PNS circuit? Several recent studies on v'ada neurons, another subclass of C4da neurons, have provided insights into this question. Firstly, actively regrowing v'ada dendrites extend in the same path with those of another cell type of da neurons (Satoh et al. 2012). This suggests a possible role of the physical interaction between dendrites of different neurons or between dendrites and extracellular substrates in the regulation of dendritic regrowth. Secondly, an HSP90 co-chaperone protein, CHORD, regulates the final size of dendritic arbors together with the insulin/IGF signaling pathway (Shimono et al. 2014). Finally, and most strikingly, v'ada C4da neurons reshape their newly regrown dendrites from radial shape to lattice shape in the early adult stage (Shimono et al. 2009; Yasunaga et al. 2010). This radial-to-lattice transformation involves remodeling of the extracellular matrix (ECM) surrounding dendrites, which is controlled by temporal upregulation of matrix metalloproteinase Mmp2 in the epidermal tissue (Yasunaga et al. 2010). How do dendritic branches sense the extracellular environment? Which molecular cues direct dendritic branches toward their final shapes? These are the questions to be addressed in future research.

12.5 Dendritic Remodeling in Other Invertebrate Model Systems

C. elegans sensory neurons are an emerging model system for studying dendrite development and remodeling. Recent genetic screens have identified several genes involved in dendrite branching, extension, and self-avoidance of the PVD multidendritic neurons (Liu and Shen 2011; Smith et al. 2012, 2013). For example, the type I membrane protein EFF-1, which was originally identified as a cell fusion-promoting factor, regulates the complexity of dendritic arbors by pruning excessive dendritic branches (Oren-Suissa et al. 2010). The pruning process involves not only dendrite severing and retraction but also dendrite-dendrite autofusion, which is never observed in C4da dendrite pruning, thus providing additional mechanistic insights into dendrite pruning.

With regard to dendrite remodeling, a recent study has reported that in response to adverse environmental conditions, the IL2 bipolar sensory neurons undergo dendrite arborization and shift from bipolar to multipolar neurons (Schroeder et al. 2013). This process is reversible: the arborized dendrites are pruned away after worms are returned back to the normal (non-stress) environment. The Furin proprotein convertase KPC-1, whose expression in IL2 neurons increases in stress conditions, is identified as a critical regulator of this reversible arborization of dendritic branches (Schroeder et al. 2013). Given that KPC-1/Furin also controls the dendritic development of PVD neurons in the same genetic pathway as the leucine-rich transmembrane receptor (LRR) DMA-1 (Liu and Shen 2011; Salzberg et al. 2014), KPC-1/Furin might regulate the IL2 dendrite remodeling by proteolytically processing DMA-1/LRR, which indeed contains a predicted Furin cleavage site. Further genetic and biochemical studies are required to elucidate this point.

Thus, studies of the worm sensory neurons have started to uncover a new path that will pave the way for a deeper understanding of dendrite remodeling.

12.6 Conclusions and Future Directions

What information have invertebrate model systems provided about dendrite remodeling? Cellular mechanisms underlying dendrite pruning are extensively described in *Drosophila* C4da neurons, which are amenable to live imaging during the whole pruning processes. Early events in dendrite pruning are observed in the proximal dendrites: cytoskeletal remodeling and dendritic thinning. Compartmentalized Ca^{2+} transients, spatiotemporal cues to trigger dendrite pruning, occur in a correlated manner with dendritic thinning (Kanamori et al. 2013, 2015). Proximal dendrites are eventually severed and their distal regions undergo rapid degeneration after a latency period. Genetic screens also uncovered key factors in dendrite pruning of C4da neurons. Interestingly, many of these factors are evolutionarily conserved from fruit flies to vertebrates. We end this review by outlining the key questions remaining to be addressed.

First, despite extensive research, our understanding of the molecular mechanisms downstream of the ecdysone-signaling cascade is rudimentary. Elucidating this point will require unbiased identification of downstream genes not only by performing conventional genetic screens, which are not yet saturated, but also by using microarray expression analysis. The latter approach was successfully taken to identify the key regulators of MB γ axon pruning (Hoopfer et al. 2008). Applying current state-of-the-art sequencing technologies, such as the single-cell RNA-seq technique, will facilitate identification of additional EcR-dependent regulators.

Second, although it is clear that in response to ecdysone-signaling C1da and C4da neurons prune their dendrites, while C2da and C3da neurons undergo programmed cell death, it is unclear how such distinct responses are induced in response to the same steroid hormone. Are there any genes expressed specifically in distinct classes of neurons to induce their cellular responses? Epigenetic mechanisms might be involved in the regulation of distinct cell responses. Although both *mical* and *headcase* are required for C4da dendrite pruning but not for apoptotic cell death of C3da neurons, the expression of these genes is induced in both classes of da neurons in an EcR-dependent manner (Kirilly et al. 2009; Loncle and Williams 2012). Therefore, expression analysis from single da neurons would be a more straightforward approach than loss-of-function screening for genes required specifically for dendrite pruning or apoptotic cell death.

Finally, it is important to consider whether molecular mechanisms elucidated from studies using invertebrate models are evolutionarily conserved from invertebrates to vertebrates. Establishing vertebrate model systems, which are amenable not only to live imaging but also to genetic manipulation, is the first step to answer this question. The recent availability of genetic tools in mice, genetically engineered viruses, and multiphoton microscopy makes it easier to establish such systems.

References

Awasaki T, Ito K (2004) Engulfing action of glial cells is required for programmed axon pruning during Drosophila metamorphosis. Curr Biol 14:668–677. doi:10.1016/j

Awasaki T, Tatsumi R, Takahashi K, Arai K, Nakanishi Y, Ueda R, Ito K (2006) Essential role of the apoptotic cell engulfment genes draper and ced-6 in programmed axon pruning during Drosophila metamorphosis. Neuron 50:855–867. doi:10.1016/j.neuron.2006.04.027

Beuchle D, Schwarz H, Langegger M, Koch I, Aberle H (2007) Drosophila MICAL regulates myofilament organization and synaptic structure. Mech Dev 124:390–406. doi:10.1016/j.mod.2007.01.006

Buszczak M, Segraves WA (2000) Insect metamorphosis: out with the old, in with the new. Curr Biol 10:R830–R833. doi:10.1016/S0960-9822(00)00792-2

Cajal R (1911) Histology of the nervous system of man and vertebrates translated from the French. Oxford University Press, Oxford

Chen JL, Lin WC, Cha JW, So PT, Kubota Y, Nedivi E (2011) Structural basis for the role of inhibition in facilitating adult brain plasticity. Nat Neurosci 14:587–594. doi:10.1038/nn.2799

Chung WS, Clarke LE, Wang GX, Stafford BK, Sher A, Chakraborty C, Joung J, Foo LC, Thompson A, Chen C, Smith SJ, Barres BA (2013) Astrocytes mediate synapse elimination through MEGF10 and MERTK pathways. Nature 504:394–400. doi:10.1038/nature12776

Coleman MP, Freeman MR (2010) Wallerian degeneration, Wld(S), and nmnat. Annu Rev Neurosci 33:245–267. doi:10.1146/annurev-neuro-060909-153248

Consoulas C (2002) Dendritic remodeling and growth of motoneurons during metamorphosis of Drosophila melanogaster. J Neurosci 22:4906–4917

Datwani A, Iwasato T, Itohara S, Erzurumlu RS (2002) Lesion-induced thalamocortical axonal plasticity in the S1 cortex is independent of NMDA receptor function in excitatory cortical neurons. J Neurosci 22:9171–9175

Ehlinger DG, Bergstrom HC, McDonald CG, Smith RF (2012) Nicotine-induced dendritic remodeling in the insular cortex. Neurosci Lett 516:89–93. doi:10.1016/j.neulet.2012.03.064

Emoto K (2012) Signaling mechanisms that coordinate the development and maintenance of dendritic fields. Curr Opin Neurobiol 22:805–811. doi:10.1016/j.conb.2012.04.005

Espinosa JS, Wheeler DG, Tsien RW, Luo L (2009) Uncoupling dendrite growth and patterning: single-cell knockout analysis of NMDA receptor 2B. Neuron 62:205–217. doi:10.1016/j.neuron

Gourley SL, Swanson AM, Koleske AJ (2013) Corticosteroid-induced neural remodeling predicts behavioral vulnerability and resilience. J Neurosci 33:3107–3112. doi:10.1523/JNEUROSCI.2138-12.2013

Hakim Y, Yaniv SP, Schuldiner O (2014) Astrocytes play a key role in Drosophila mushroom body axon pruning. PLoS One 9:e86178. doi:10.1371/journal.pone.0086178

Han C, Jan LY, Jan Y-N (2011) Enhancer-driven membrane markers for analysis of nonautonomous mechanisms reveal neuron-glia interactions in Drosophila. Proc Natl Acad Sci U S A 108:9673–9678. doi:10.1073/pnas.1106386108

Han C, Song Y, Xiao H, Wang D, Franc NC, Jan LY, Jan Y-N (2014) Epidermal cells are the primary phagocytes in the fragmentation and clearance of degenerating dendrites in Drosophila. Neuron 81:544–560. doi:10.1016/j.neuron.2013.11.021

Harris RM, Woolsey TA (1979) Morphology of Golgi-impregnated neurons in mouse cortical barrels following vibrissae damage at different postnatal ages. Brain Res 161:1177–1184

Hoopfer ED, McLaughlin T, Watts RJ, Schuldiner O, O'Leary DDM, Luo L (2006) Wlds protection distinguishes axon degeneration following injury from naturally occurring developmental pruning. Neuron 50:883–895. doi:10.1016/j.neuron.2006.05.013

Hoopfer ED, Penton A, Watts RJ, Luo L (2008) Genomic analysis of Drosophila neuronal remodeling: a role for the RNA-binding protein Boule as a negative regulator of axon pruning. J Neurosci 28:6092–6103. doi:10.1523/JNEUROSCI.0677-08.2008

Hung R-J, Yazdani U, Yoon J, Wu H, Yang T, Gupta N, Huang Z, van Berkel WJH, Terman JR (2010) Mical links semaphorins to F-actin disassembly. Nature 463:823–827. doi:10.1038/nature08724

Kanamori T, Kanai MI, Dairyo Y, Yasunaga K, Morikawa RK, Emoto K (2013) Compartmentalized calcium transients trigger dendrite pruning in Drosophila sensory neurons. Science 340: 1475–1478. doi:10.1126/science.1234879

Kanamori T, Yoshino J, Yasunaga K, Dairyo Y, Emoto K (2015) Local endocytosis triggers dendrite thinning and pruning in Drosophila sensory neurons. Nat Commun (in press)

Kaneko M, Yamaguchi K, Eiraku M, Sato M, Takata N, Kiyohara Y, Mishina M, Hirase H, Hashikawa T, Kengaku M (2011) Remodeling of monoplanar Purkinje cell dendrites during cerebellar circuit formation. PLoS One 6:e20108. doi:10.1371/journal.pone.0020108

Kim SE, Coste B, Chadha A, Cook B, Patapoutian A (2012) The role of Drosophila Piezo in mechanical nociception. Nature 483:209–212. doi:10.1038/nature10801

Kirilly D, Gu Y, Huang Y, Wu Z, Bashirullah A, Low BC, Kolodkin AL, Wang H, Yu F (2009) A genetic pathway composed of Sox14 and Mical governs severing of dendrites during pruning. Nat Neurosci 12:1497–1505. doi:10.1038/nn.2415

Kirilly D, Wong JJL, Lim EKH, Wang Y, Zhang H, Wang C, Liao Q, Wang H, Liou Y-C, Wang H, Yu F (2011) Intrinsic epigenetic factors cooperate with the steroid hormone ecdysone to govern dendrite pruning in Drosophila. Neuron 72:86–100. doi:10.1016/j.neuron.2011.08.003

Kuo CT, Jan LY, Jan YN (2005) Dendrite-specific remodeling of Drosophila sensory neurons requires matrix metalloproteases, ubiquitin-proteasome, and ecdysone signaling. Proc Natl Acad Sci U S A 102:15230–15235. doi:10.1073/pnas.0507393102

Kuo CT, Zhu S, Younger S, Jan LY, Jan YN (2006) Identification of E2/E3 ubiquitinating enzymes and caspase activity regulating Drosophila sensory neuron dendrite pruning. Neuron 51:283–290. doi:10.1016/j.neuron.2006.07.014

Kuraishi T, Nakagawa Y, Nagaosa K, Hashimoto Y, Ishimoto T, Moki T, Fujita Y, Nakayama H, Dohmae N, Shiratsuchi A, Yamamoto N, Ueda K, Yamaguchi M, Awasaki T, Nakanishi Y (2009) Pretaporter, a Drosophila protein serving as a ligand for Draper in the phagocytosis of apoptotic cells. EMBO J 28:3868–3878. doi:10.1038/emboj.2009.343

Kuranaga E, Kanuka H, Tonoki A, Takemoto K, Tomioka T, Kobayashi M, Hayashi S, Miura M (2006) Drosophila IKK-related kinase regulates nonapoptotic function of caspases via degradation of IAPs. Cell 126:583–596. doi:10.1016/j.cell.2006.05.048

Kurant E, Axelrod S, Leaman D, Gaul U (2008) Six-microns-under acts upstream of Draper in the glial phagocytosis of apoptotic neurons. Cell 133:498–509. doi:10.1016/j.cell.2008.02.052

Lee W-CA, Huang H, Feng G, Sanes JR, Brown EN, So PT, Nedivi E (2006) Dynamic remodeling of dendritic arbors in GABAergic interneurons of adult visual cortex. PLoS Biol 4:e29. doi:10.1371/journal.pbio.0040029

Lee H, Jan LY, Jan Y (2009) p60-like 1 regulate dendrite pruning of sensory neuron during metamorphosis. Proc Natl Acad Sci U S A 106:6363–6368

Li J, Liu N, Lu K, Zhang L, Gu J, Guo F, An S, Zhang L, Zhang L (2012) Cocaine-induced dendritic remodeling occurs in both D1 and D2 dopamine receptor-expressing neurons in the nucleus accumbens. Neurosci Lett 517:118–122. doi:10.1016/j.neulet.2012.04.040

Liston C, Miller MM, Goldwater DS, Radley JJ, Rocher AB, Hof PR, Morrison JH, McEwen BS (2006) Stress-induced alterations in prefrontal cortical dendritic morphology predict selective impairments in perceptual attentional set-shifting. J Neurosci 26:7870–7874. doi:10.1523/JNEUROSCI.1184-06.2006

Liu OW, Shen K (2011) The transmembrane LRR protein DMA-1 promotes dendrite branching and growth in C. elegans. Nat Neurosci 15:57–63. doi:10.1038/nn.2978

Loncle N, Williams DW (2012) An interaction screen identifies headcase as a regulator of large-scale pruning. J Neurosci 32:17086–17096. doi:10.1523/JNEUROSCI.1391-12.2012

Lunn ER, Perry VH, Brown MC, Rosen H, Gordon S (1989) Absence of Wallerian degeneration does not hinder regeneration in peripheral nerve. Eur J Neurosci 1:27–33

Lyons GR, Andersen RO, Abdi K, Song W-S, Kuo CT (2014) Cysteine proteinase-1 and cut protein isoform control dendritic innervation of two distinct sensory fields by a single neuron. Cell Rep 6:783–791. doi:10.1016/j.celrep.2014.02.003

MacDonald JM, Beach MG, Porpiglia E, Sheehan AE, Watts RJ, Freeman MR (2006) The Drosophila cell corpse engulfment receptor Draper mediates glial clearance of severed axons. Neuron 50: 869–881. doi:10.1016/j.neuron.2006.04.028

Malun D, Brunjes PC (1996) Development of olfactory glomeruli: temporal and spatial interactions between olfactory receptor axons and mitral cells in opossums and rats. J Comp Neurol 368:1–16

Manaka J, Kuraishi T, Shiratsuchi A, Nakai Y, Higashida H, Henson P, Nakanishi Y (2004) Draper-mediated and phosphatidylserine-independent phagocytosis of apoptotic cells by Drosophila hemocytes/macrophages. J Biol Chem 279:48466–48476. doi:10.1074/jbc.M408597200

Matsui A, Tran M, Yoshida AC, Kikuchi SS, U M, Ogawa M, Shimogori T (2013) BTBD3 controls dendrite orientation toward active axons in mammalian neocortex. Science 342: 114–118. doi:10.1126/science.1244505

Mizuno H, Luo W, Tarusawa E, Saito YM, Sato T, Yoshimura Y, Itohara S, Iwasato T (2014) NMDAR-regulated dynamics of layer 4 neuronal dendrites during thalamocortical reorganization in neonates. Neuron 82:365–379. doi:10.1016/j.neuron.2014.02.026

Oren-Suissa M, Hall DH, Treinin M, Shemer G, Podbilewicz B (2010) The fusogen EFF-1 controls sculpting of mechanosensory dendrites. Science 328:1285–1288. doi:10.1126/science.1189095

Oshima K, Takeda M, Kuranaga E, Ueda R, Aigaki T, Miura M, Hayashi S (2006) IKK epsilon regulates F actin assembly and interacts with Drosophila IAP1 in cellular morphogenesis. Curr Biol 16:1531–1537. doi:10.1016/j.cub.2006.06.032

Parrish JZ, Emoto K, Kim MD, Jan YN (2007) Mechanisms that regulate establishment, maintenance, and remodeling of dendritic fields. Annu Rev Neurosci 30:399–423. doi:10.1146/annurev.neuro.29.051605.112907

Rumpf S, Lee SB, Jan LY, Jan YN (2011) Neuronal remodeling and apoptosis require VCP-dependent degradation of the apoptosis inhibitor DIAP1. Development 138:1153–1160. doi:10.1242/dev.062703

Rumpf S, Bagley JA, Thompson-Peer KL, Zhu S, Gorczyca D, Beckstead RB, Jan LY, Jan YN (2014) Drosophila valosin-containing protein is required for dendrite pruning through a regulatory role in mRNA metabolism. Proc Natl Acad Sci U S A 111:7331–7336. doi:10.1073/pnas.1406898111

Salzberg Y, Ramirez-Suarez NJ, Bülow HE (2014) The proprotein convertase KPC-1/furin controls branching and self-avoidance of sensory dendrites in Caenorhabditis elegans. PLoS Genet 10:e1004657. doi:10.1371/journal.pgen.1004657

Satoh D, Suyama R, Kimura K, Uemura T (2012) High-resolution in vivo imaging of regenerating dendrites of Drosophila sensory neurons during metamorphosis: local filopodial degeneration and heterotypic dendrite-dendrite contacts. Genes Cells 17:939–951. doi:10.1111/gtc.12008

Schoenmann Z, Assa-Kunik E, Tiomny S, Minis A, Haklai-Topper L, Arama E, Yaron A (2010) Axonal degeneration is regulated by the apoptotic machinery or a NAD+-sensitive pathway in insects and mammals. J Neurosci 30:6375–6386. doi:10.1523/JNEUROSCI.0922-10.2010

Schroeder NE, Androwski RJ, Rashid A, Lee H, Lee J, Barr MM (2013) Dauer-specific dendrite arborization in C. elegans is regulated by KPC-1/Furin. Curr Biol 23:1527–1535. doi:10.1016/j.cub.2013.06.058

Scott JA, Williams DW, Truman JW (2011) The BTB/POZ zinc finger protein Broad-Z3 promotes dendritic outgrowth during metamorphic remodeling of the peripheral stretch receptor dbd. Neural Dev 6:39. doi:10.1186/1749-8104-6-39

Shimono K, Fujimoto A, Tsuyama T, Yamamoto-Kochi M, Sato M, Hattori Y, Sugimura K, Usui T, Kimura K, Uemura T (2009) Multidendritic sensory neurons in the adult Drosophila abdomen: origins, dendritic morphology, and segment- and age-dependent programmed cell death. Neural Dev 4:37. doi:10.1186/1749-8104-4-37

Shimono K, Fujishima K, Nomura T, Ohashi M, Usui T, Kengaku M, Toyoda A, Uemura T (2014) An evolutionarily conserved protein CHORD regulates scaling of dendritic arbors with body size. Sci Rep 4:4415. doi:10.1038/srep04415

Smith CJ, Watson JD, VanHoven MK, Colón-Ramos DA, Miller DM 3rd (2012) Netrin (UNC-6) mediates dendritic self-avoidance. Nat Neurosci 15:731–737. doi:10.1038/nn.3065

Smith CJ, O'Brien T, Chatzigeorgiou M, Spencer WG, Feingold-Link E, Husson SJ, Hori S, Mitani S, Gottschalk A, Schafer WR, Miller DM 3rd (2013) Sensory neuron fates are distinguished by a transcriptional switch that regulates dendrite branch stabilization. Neuron 79: 266–280. doi:10.1016/j.neuron.2013.05.009

Steffen H, Van der Loos H (1980) Early lesions of mouse vibrissal follicles: their influence on dendritic orientation in the cortical barrelfield. Exp Brain Res 40:410–431

Tao J, Rolls MM (2011) Dendrites have a rapid program of injury-induced degeneration that is molecularly distinct from developmental pruning. J Neurosci 31:5398–5405. doi:10.1523/JNEUROSCI.3826-10.2011

Tasdemir-Yilmaz OE, Freeman MR (2014) Astrocytes engage unique molecular programs to engulf pruned neuronal debris from distinct subsets of neurons. Genes Dev 28:20–33. doi:10.1101/gad.229518.113

Thummel C (1996) Flies on steroids—Drosophila metamorphosis and the mechanisms of steroid hormone action. Trends Genet 12:306–310

Wang G, Liets L, Chalupa L (2001) Unique functional properties of on and off pathways in the developing mammalian retina. J Neurosci 21:4310–4317

Watts R, Hoopfer E, Luo L (2003) Axon pruning during Drosophila metamorphosis: evidence for local degeneration and requirement of the ubiquitin-proteasome system. Neuron 38:871–885

Watts RJ, Schuldiner O, Perrino J, Larsen C, Luo L (2004) Glia engulf degenerating axons during developmental axon pruning. Curr Biol 14:678–684. doi:10.1016/j.cub.2004.03.035

Williams DW, Truman JW (2004) Mechanisms of dendritic elaboration of sensory neurons in Drosophila: insights from in vivo time lapse. J Neurosci 24:1541–1550. doi:10.1523/JNEUROSCI.4521-03.2004

Williams DW, Truman JW (2005) Cellular mechanisms of dendrite pruning in Drosophila: insights from in vivo time-lapse of remodeling dendritic arborizing sensory neurons. Development 132:3631–3642. doi:10.1242/dev.01928

Williams DW, Kondo S, Krzyzanowska A, Hiromi Y, Truman JW (2006) Local caspase activity directs engulfment of dendrites during pruning. Nat Neurosci 9:1234–1236. doi:10.1038/nn1774

Wong JJL, Li S, Lim EKH, Wang Y, Wang C, Zhang H, Kirilly D, Wu C, Liou Y-C, Wang H, Yu F (2013) A Cullin1-based SCF E3 ubiquitin ligase targets the InR/PI3K/TOR pathway to regulate neuronal pruning. PLoS Biol 11:e1001657. doi:10.1371/journal.pbio.1001657

Yan Z, Zhang W, He Y, Gorczyca D, Xiang Y, Cheng LE, Meltzer S, Jan LY, Jan YN (2013) Drosophila NOMPC is a mechanotransduction channel subunit for gentle-touch sensation. Nature 493:221–225. doi:10.1038/nature11685

Yasunaga K, Kanamori T, Morikawa R, Suzuki E, Emoto K (2010) Dendrite reshaping of adult Drosophila sensory neurons requires matrix metalloproteinase-mediated modification of the basement membranes. Dev Cell 18:621–632. doi:10.1016/j.devcel.2010.02.010

Yu X, Odera S, Chuang C-H, Lu N, Zhou Z (2006) C. elegans Dynamin mediates the signaling of phagocytic receptor CED-1 for the engulfment and degradation of apoptotic cells. Dev. Cell 10:743–757. doi:10.1016/j.devcel.2006.04.007

Zhang H, Wang Y, Wong JJL, Lim K-L, Liou Y-C, Wang H, Yu F (2014) Endocytic pathways downregulate the L1-type cell adhesion molecule neuroglian to promote dendrite pruning in Drosophila. Dev Cell 30:463–478. doi:10.1016/j.devcel.2014.06.014

Zhong L, Hwang RY, Tracey WD (2010) Pickpocket is a DEG/ENaC protein required for mechanical nociception in Drosophila larvae. Curr Biol 20:429–434. doi:10.1016/j.cub.2009.12.057

Zhong L, Bellemer A, Yan H, Honjo K, Robertson J, Hwang RY, Pitt GS, Tracey WD (2012) Thermosensory and non-thermosensory isoforms of Drosophila melanogaster TRPA1 reveal heat sensor domains of a thermoTRP channel. Cell Rep 1:43–55. doi:10.1016/j.celrep.2011.11.002

Zhou Z, Hartwieg E, Horvitz HR (2001) CED-1 is a transmembrane receptor that mediates cell corpse engulfment in C. elegans. Cell 104:43–56. doi:10.1016/S0092-8674(01)00190-8

Chapter 13
Experience-Dependent Dendritic Arbor Development

Hollis T. Cline

Abstract The architecture of dendritic arbors determines afferent inputs and dendritic integration. Aberrant dendrite development is predicted to lead to aberrant information processing and circuit output. This chapter will review experience-dependent mechanisms regulating dendritic arbor development and plasticity, focusing on work done in vivo. Dendrite arbor development is strongly regulated by activity-dependent mechanisms, which in many cases can be traced to mechanisms originating in synaptic development or function. As proteomic analysis provides a more complete inventory of synaptic proteins and genomic analysis identifies candidate proteins associated with neurodevelopmental disorders, it is becoming clear that synaptic proteins which regulate synaptogenesis, synaptic function, and synapse stability also regulate dendritic arbor development and brain circuit function.

Keywords Dendrite • Synaptogenesis • Synaptotropic hypothesis • Structural plasticity

13.1 Introduction

Dendrite arbor structure determines the spatial extent over which inputs are received by a neuron. In sensory systems, such as the visual system, sensory inputs enter central circuits through major axon trajectories and form organized connections onto postsynaptic dendritic arbors. In these circumstances, dendritic arbors are a major gatekeeper for information transfer from the sensory periphery into central circuits. Dendrite arbor structure also regulates the biophysical and integrative properties of dendrites and by extension, neuronal circuits. Aberrant dendritic arbor structure would also be expected to interfere with dendritic integration of synaptic inputs and information processing within central circuits. Consequently, dendritic arbor structure has a direct impact on three critical features of neuronal circuits: governing afferent input, information processing, and circuit output.

H.T. Cline (✉)
The Scripps Research Institute, La Jolla, CA, USA
e-mail: cline@scripps.edu

K. Emoto et al. (eds.), *Dendrites*, DOI 10.1007/978-4-431-56050-0_13

The essential functions of dendritic arbors in neural circuitry have motivated extensive research into the mechanisms controlling dendritic arbor development and plasticity. Sensory systems are favorites among the brain circuits to study dendritic arbor development because sensory input can be manipulated in a relatively controlled manner and because neurons which receive and process sensory inputs can be readily identified and studied in intact animals. Indeed, many studies on dendritic arbor development are done in vivo which has the clear advantage that neuronal development is studied in the context of intact developing circuits where synaptic inputs, extrinsic signals, and other parameters may change over time.

This chapter will review experience-dependent or activity-dependent cellular and molecular mechanisms of dendritic arbor development and plasticity, focusing on work done in vivo. Because of the extensive work done in this area, this will not be a comprehensive review; rather, work is selected to highlight major points of interest.

13.2 General Overview of Dendrite Development

The combination of single-cell labeling and in vivo time-lapse imaging has revealed the cellular mechanisms underlying dendritic development (Bestman and Cline 2008; Cline 2001; Hua and Smith 2004; Wong and Ghosh 2002). Advances in microscopy, cell biology, and molecular genetic methods have also been essential for significant advances in our understanding of the mechanisms behind the molecular and activity-dependent regulation of dendrite development. The retinotectal circuit of *Xenopus* (Fig. 13.1), in which excitatory and inhibitory tectal neurons receive direct visual input from the retinal ganglion cells (RGCs) in the eye (Holt 1989), is a powerful experimental system to study both structural (Bestman and Cline 2008; Haas et al. 2006; Shen et al. 2009; Sin et al. 2002) and functional plasticity (Aizenman and Cline 2007; Bestman and Cline 2008; Engert et al. 2002; Pratt and Aizenman 2007; Pratt et al. 2008; Shen et al. 2009; Tao and Poo 2005; Zhang et al. 2000) in vivo.

In vivo time-lapse imaging of *Xenopus* optic tectal neurons showed that dendritic arbor elaboration goes through distinct phases (Fig. 13.2) (Cline 2001; Wu et al. 1999). About 1 day after evidence of morphological differentiation, dendritic arbors enter a rapid growth phase for a few days, after which dendritic arbor growth rates stabilize. During the rapid dendritic arbor growth period, one might think that the growth of the dendritic arbor can be easily achieved by continuously lengthening preexisting dendrites and sprouting new dendritic branches; however, time-lapse imaging at intervals of minutes to hours reveals that dendritic growth is highly dynamic, consisting of not only branch addition and extension but also retraction and loss of dendritic branches (Fig. 13.2b; (Cline and Haas 2008; Dailey and Smith 1996; Ewald et al. 2008; Haas et al. 2006; Rajan and Cline 1998)). The branch dynamics that occur during dendritic morphogenesis persist in mature neurons when their overall structure is stable although at a slower rate (Cline 2001; Lee

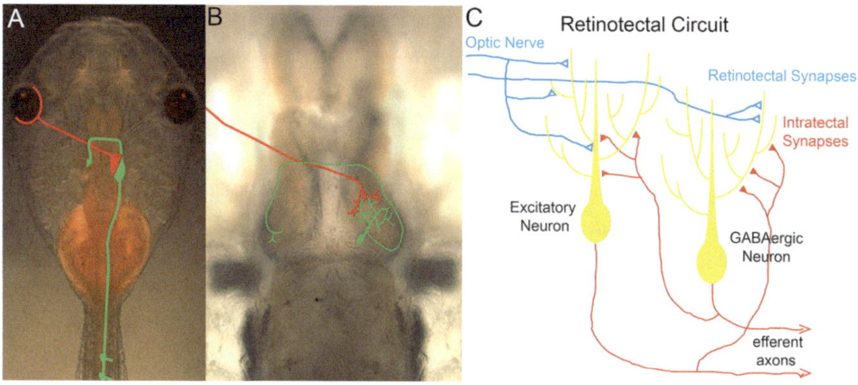

Fig. 13.1 The visual system of *Xenopus laevis* tadpoles. (**a**) Image of an albino *Xenopus* tadpole. The retinal ganglion cells extend axons, *red*, through the optic nerve into the contralateral optic tectum, *green*. Optic tectal neurons extend axons to several targets including the spinal cord, through which they govern visually guided swimming behaviors. (**b**) Image of the brain with a retinal axon, *red*, and tectal neuron, *green*, shown in this case extending an axon to the contralateral tegmentum. (**c**) Cartoon of the retinotectal circuit shows retinal axons, *blue*, synapsing on glutamatergic and GABAergic tectal neurons. Tectal neurons form intratectal connections within a recurrent circuit and extend efferent axons. Rostral is up in (**a**, **b**)

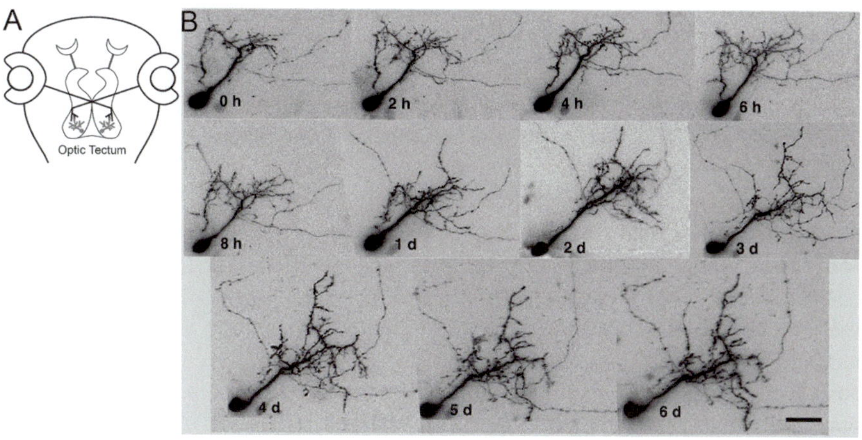

Fig. 13.2 Growth pattern of dendritic arbors. (**a**) Cartoon of tadpole visual system, showing retinal inputs and optic tectal neurons. (**b**) Images time-lapse laser scanning two-photon images of an optic tectal neuron collected at the indicated timepoints. Rapid increased arbor elaboration occurs over the first few days of imaging, after which dendritic arbor structure is more stable. Branch dynamics can be readily detected in images collected at 2 h intervals by comparing neighboring timepoints. Scale bar = 20 μm

et al. 2006, 2008; Wong and Ghosh 2002). Therefore, mechanisms that regulate developmental dendritic dynamics likely regulate dendritic structural plasticity in mature neurons.

13.3 Branch Dynamics Are the Heart of Dendritic Arbor Development

In vivo time-lapse imaging from a variety of experimental systems has demonstrated that dendritic arbors grow by a process in which branches are constantly added and retracted. A minority of newly added branches is stabilized and serves as a scaffold for subsequent branch additions and retractions (Fig. 13.2). Key to generating these conclusions was collecting the images at a range of intervals and over a range of total observation times. Time-lapse images of optic tectal neurons from *Xenopus* tadpoles and zebra fish collected at daily intervals show that tectal cell dendritic arbors elaborate from simple structures to complex arbors over a period of several days after which the arbor structure appears stable (Cline 2001; Cline et al. 1997; Niell et al. 2004). Images of developing neurons collected at intervals of 30 min over 2 h or at 2 h intervals over 6–8 h show little or no net arbor growth, but they do show significant branch dynamics in which about 80 % of branches observed are added and/or retracted over the observation period (Haas et al. 2006; Rajan and Cline 1998; Rajan et al. 1999). Shorter observation intervals reveal greater dynamics than longer intervals and can provide quantitative data on the average lifetime of branches in vivo (Niell et al. 2004; Sin et al. 2002; Witte et al. 1996).

Branch dynamics also serve as a mechanism for growing dendrites to extend into a specific target region. This has been beautifully demonstrated in the retina, where imaging studies show that retinal ganglion cell (RGC) dendritic lamination and dendritic elaboration in on and off laminae are concurrent. Prior to laminar targeting of dendrites, short interval imaging of RGC dendrites showed that all branches were dynamic and had short lifetimes, but branch lifetimes increased when they extended into targeted laminae (Mumm et al. 2006). This series of time-lapse observations indicated that RGC dendrites extend into specific target laminae and branches stabilize once they contact presynaptic axons. These conclusions contrast with those from previous studies based on static images in which dendrites were thought to elaborate profusely without laminar targeting and then establish laminar specificity by pruning. Furthermore, glutamatergic synaptic transmission and calcium transients in RGC dendrites are required for branch dynamics and branch stabilization (Lohmann et al. 2002; Wong et al. 2000). Together, these data provide strong support for the synaptotropic hypothesis, which proposes that axons and dendrites extend processes toward synaptic targets and that the formation and stabilization of synapses both stabilize growing processes against retraction and promote further elaboration of the neuronal arbor (Cline and Haas 2008; Niell et al. 2004; Vaughn et al. 1988).

13.4 Growing Signals

Synaptic Activity: Time-lapse imaging experiments of optic tectal neurons in *Xenopus* tadpoles demonstrated that individual neurons showed a wide range of growth rates in developing dendritic arbors (Hume and Purves 1981; Rajan and Cline 1998; Rajan et al. 1999; Wu and Cline 1998). Optic tectal neurons receive glutamatergic and GABAergic synaptic input even as they elaborate their dendritic arbors (Akerman and Cline 2006; Wu et al. 1996), so it seemed possible that the heterogeneity in growth rates might arise from differences in synaptic input activity. Furthermore, electrophysiological recordings show that glutamatergic inputs to immature neurons were predominantly mediated by NMDA-type glutamate receptors and that AMPA receptors gradually contributed more synaptic current as neurons mature (Rajan and Cline 1998; Wu et al. 1996). These electrophysiological data suggest that properties of developing synapses driven by sensory input might affect arbor development (Fig. 13.3). Indeed, blocking NMDA receptors at early stages of dendritic arbor development significantly decreased dendritic arbor growth rates and decreased the variance in growth rates, whereas blocking AMPA receptors or action potential activity in these newly differentiating neurons had no significant effect (Rajan and Cline 1998). By contrast, both AMPA and NMDA receptors were required for dendritic arbor elaboration in more mature neurons. These data correspond well with electrophysiological recordings showing that immature neurons have relatively little AMPA receptors-mediated current compared to more mature neurons (Wu et al. 1996). They also indicate that the variance in growth rates in neurons detected by time-lapse in vivo imaging is attributable to variation in synaptic inputs to the neurons. This further suggests that synaptic inputs targeted to subregions of dendritic arbors may result in local elaboration of the dendritic arbor and contribute to characteristic dendritic arbor morphologies in specific neuronal types.

Glutamatergic synaptic plasticity and AMPA receptor trafficking are important events in regulating dendritic arbor development, for instance, as shown in studies in the *Xenopus* optic tectum and rodent spinal cord (Cline and Haas 2008; Haas et al. 2006; Inglis et al. 2000, 2002; Kalb 1994). In *Xenopus*, visual experience increases the strength of glutamatergic synaptic transmission and increases dendritic arbor elaboration (Engert et al. 2002; Haas et al. 2006; Shen et al. 2011). Furthermore, blocking AMPA receptor trafficking into synaptic sites by expressing peptides corresponding to the C-terminal peptides (CTP) of AMPA receptor subunits decreases synaptic strength and stunts dendritic arbor elaboration (Haas et al. 2006). Together, these studies show that AMPA receptor-mediated synaptic transmission is necessary and sufficient for dendritic arbor elaboration in vivo (Fig. 13.3). Finally, CTP expression in *Xenopus* optic tectal neurons blocks visual avoidance behavior, suggesting that decreasing glutamatergic transmission ramifies to not only impair dendritic arbor growth and visual circuit function but may have behavioral consequences as well. AMPA receptor subunit composition and the splice variants of different subunits determine the biophysical properties of the

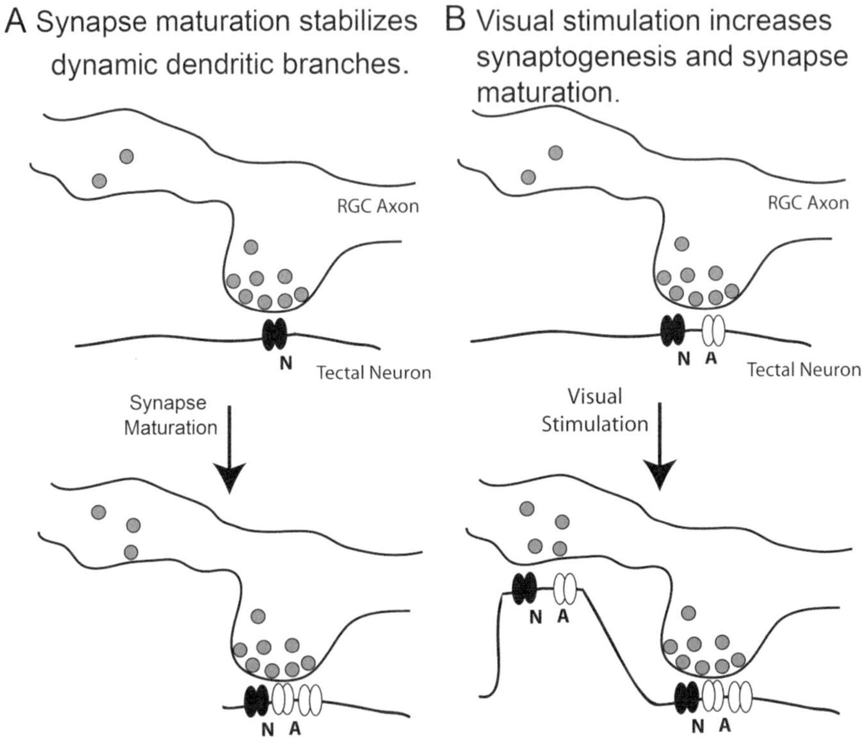

Fig. 13.3 Synaptic development and experience-dependent plasticity regulate synaptic stability. (**a**) Cartoon of a segment of RGC axon and tectal cell dendrite. NMDA receptors (N) are the predominant glutamate receptor at nascent retinotectal synapses. As synapses mature, AMPA receptors are trafficked into the postsynaptic membrane. This renders the synapses functional at resting potentials and stabilizes the synaptic contact. (**b**) Visual stimulation increases synaptogenesis and synapse maturation. Retinotectal synapses that were present before visual stimulation become stronger by the addition of postsynaptic AMPA receptors. In addition, new synapses form after visual experience

receptors and their ionic conductance. Furthermore, synaptic AMPA receptor composition changes over development (Monyer et al. 1991). Analysis of the subunit composition of different AMPA receptors relating to calcium permeability and channel kinetics suggests that AMPAR with prolonged conductances enhance dendritic arbor growth and that calcium-permeable AMPAR, in particular, promote dendritic arbor growth in rodent neocortical interneurons (Hamad et al. 2011). It appears that calcium influx through AMPA receptors, NMDA receptors, and voltage-dependent calcium channels may all contribute to dendritic arbor growth; however, experiments to demonstrate necessity and sufficiency for different sources of calcium influx have not been done systematically.

The studies summarized above suggested that the molecular mechanisms that regulate synaptogenesis and synapse maturation also regulate dendritic arbor development and by extension circuit development and function. Many studies have contributed to our rich understanding of pre- and postsynaptic density proteins that govern glutamatergic synaptic function (Huganir and Nicoll 2013; Kim and Sheng 2004; Siddiqui and Craig 2011). Some studies of synaptic proteins have borne out the prediction that proteins that play a role in trafficking and stability of AMPA receptors at synaptic sites, or more generally that culminate in changes in AMPA receptor signaling, also affect dendritic arbor development, neuronal integration into circuits, and circuit function (Sudhof 2008). For example, the transmembrane AMPA receptor regulatory proteins (TARPS), which are auxiliary subunits for AMPA receptors, regulate AMPA receptor trafficking to the cell surface and modulate AMPA receptor function (Chen et al. 2000). TARPs are differentially expressed during development and in different brain regions. The subset of type I TARPs which enhance trafficking of AMPA receptor to postsynaptic sites also regulate activity-dependent dendritic arbor development of pyramidal neurons in cortical slice cultures (Hamad et al. 2014). Different type I TARP proteins regulated dendritic arbor elaboration in different cellular compartments, such as apical or basal dendrites, and during particular developmental windows. Furthermore, the cytoplasmic tail of dendritogenic TARPs is required for the dendritic arbor growth-promoting activity, likely through its association with the PDZ protein-interaction domains of PSD95 (Chen et al. 2000). It is curious that the dendritogenic effect was specific for pyramidal neurons even though inhibitory neurons express the same growth-promoting TARPs, suggesting that other proteins regulate dendrite arbor development in inhibitory neurons.

Postsynaptic Density Proteins: The postsynaptic density (PSD) is a highly structured postsynaptic compartment that includes transmembrane proteins, including neurotransmitter receptors, ion channels, cell adhesion molecules, cytosolic signaling proteins, cytoskeletal proteins, and scaffolding proteins which are enriched in protein-interaction domains (Fig. 13.4) (Kim and Sheng 2004; Sheng and Hoogenraad 2007). Proteomic analysis of synaptoneurosomes identified PSD proteins and provided estimates of protein stoichiometry (Sheng and Hoogenraad 2007). It is particularly interesting that a significant number of postsynaptic density proteins are candidate disease genes for neurodevelopmental disorders such as autism and schizophrenia (Ebrahimi-Fakhari and Sahin 2015). PSD95 is one of the most abundant scaffold proteins in the postsynaptic density and has been a valuable entre for many studies of postsynaptic density structure and function. PSD95 regulates the numbers and distributions of AMPA receptors in the postsynaptic density. PSD95 itself is tightly regulated by posttranslational modifications (Feng and Zhang 2009; Kim and Sheng 2004).

Expression of PSD95-GFP has been a useful reagent to study synaptic dynamics (Fortin et al. 2014; Gray et al. 2006; Marrs et al. 2001; Marshak et al. 2012; Niell et al. 2004). PSD95-GFP expression demonstrated that synapses are extremely dynamic on the timecourse of hours (Marrs et al. 2001; Marshak et al. 2012; Niell et al. 2004) and also showed that PSD95 itself moves in and out of synapses (El-Husseini Ael et al. 2002; Gray et al. 2006). More recent work has taken advantage of gephyrin, a postsynaptic scaffold protein in inhibitory synapses, to examine dynamics of inhibitory synapses (Chen and Nedivi 2013; Chen

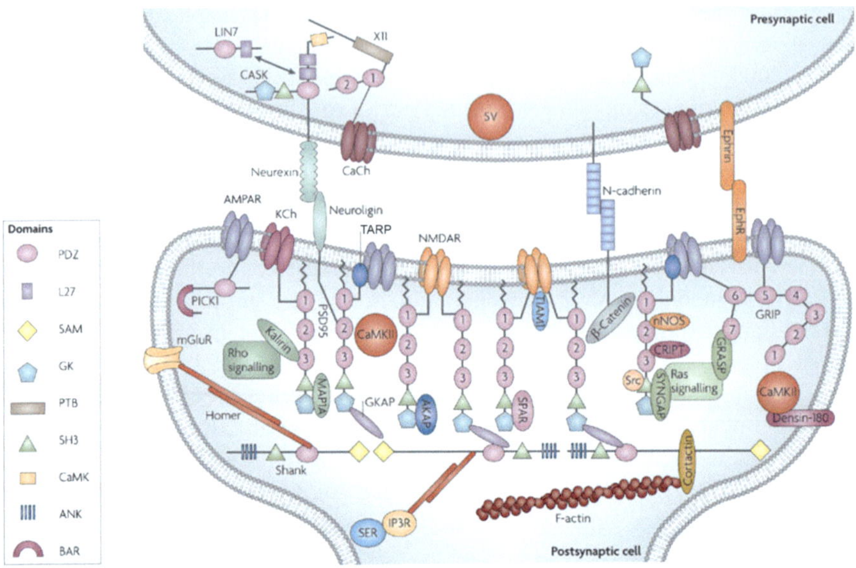

Fig. 13.4 Protein interactions in the synapse. Pre- and postsynaptic compartments are shown with representative proteins highlighting their protein-interaction domains. Transsynaptic proteins include N-cadherin, neurexin-neuroligin, and Ephrin-EphR. Glutamate receptors and their scaffold proteins, TARP, PSD95, and GRIP, are shown. Cytoskeletal proteins and regulators, as well as signaling proteins, are also shown (Adapted from Feng and Zhang 2009)

et al. 2012). Many details of synaptic protein dynamics have recently come to light as a result of super-resolution microscopy (Hosy et al. 2014). Of particular interest, here, PSD95 overexpression enhances synaptogenesis and synaptic strength, increases dendritic arbor complexity, and impairs synaptic plasticity (Ehrlich and Malinow 2004; El-Husseini et al. 2000; Futai et al. 2007; Niell et al. 2004; Stein et al. 2003). PSD95 is now recognized as a key player in the assembly of the postsynaptic density and control of synaptic AMPA receptor numbers (El-Husseini et al. 2000, 2002; Stein et al. 2003). Concurrent studies demonstrated that many proteins located in spines or in the postsynaptic density regulate synaptic function, including Shank, Homer, SynGAP, PTEN, kalarin, and CPG2, and in many cases, these proteins regulate dendritic arbor structure and circuit function (Loebrich and Nedivi 2009; Sheng and Hoogenraad 2007; Ting et al. 2012; Van Keuren-Jensen and Cline 2008). Mutations in several PSD proteins which regulate synaptic development and function are implicated in the pathogenesis of neurodevelopmental disorders including autism spectrum disorders and schizophrenia (Arons et al. 2012; Ting et al. 2012).

13.4.1 Cell Adhesion Proteins

Vaughn's original description of the synaptotropic hypothesis proposed that cell adhesion mechanisms initiate synaptogenesis and that synapses are stabilized by transsynaptic anterograde and retrograde communication. Once evidence accumulated to indicate that synaptogenic molecules can also play a role in dendritic arbor development, cell adhesion molecules immediately became likely candidates (Hua and Smith 2004). Cell adhesion molecules are thought to guide axonal growth cones to targets and organize projections, for instance, into laminae (Sanes and Zipursky 2010), but early studies showed that cell adhesion molecules such as NCAM were regulated by neuronal activity, suggesting that they might play a more interesting role in synaptic development (Dahm and Landmesser 1988, 1991; Rutishauser and Landmesser 1991; Tang and Landmesser 1993). Subsequent work demonstrated that synaptogenesis is initiated by contact and adhesion of processes from candidate pre- and postsynaptic neurons (Fujino et al. 2008; Ziv and Fisher-Lavie 2014). Many synaptic cell adhesion molecules have been identified and shown to play a role in early events in synaptogenesis, consistent with a role in cell adhesion. Surprisingly, synaptic cell adhesion molecules also play more complex roles in neurotransmitter receptor mobility/stability in the postsynaptic membrane, thereby regulating synaptic strength and synapse dynamics.

The presynaptic neurexins bind to postsynaptic neuroligins or leucine-rich repeat transmembrane (LRRTM) to form pre- and postsynaptic transsynaptic cell adhesion complexes across the synaptic cleft that can coordinate formation and maintenance of pre- and postsynaptic compartments (Aoto et al. 2013; Chanda et al. 2016; Craig and Kang 2007; Siddiqui and Craig 2011). Neuroligins, LRRTMs, and neurexins have been associated with autism and schizophrenia (Sudhof 2008). Several studies indicate that neuroligins affect synaptic function by regulating neurotransmitter receptor trafficking at both excitatory and inhibitory synapses (Chanda et al. 2016; Chubykin et al. 2007; Gibson et al. 2009). Loss of neuroligin-neurexin transsynaptic association may lead to synaptic pruning and microcircuit remodeling. Signaling downstream of NMDAR and CaMKII activity results in cleavage of neuroligin 1 by the extracellular metalloprotesase 9 (MMP9), and this in turn decreases synaptic strength and may lead to microcircuit remodeling by selective activity-dependent synaptic pruning (Peixoto et al. 2012). Similarly, contactin-associated protein 2, CASPR2, a cell addition protein that is highly homologous to neurexin, regulates the development of excitatory synapses and dendritic arbor development (Anderson et al. 2012). CASPR2 is encoded by *CNTNAP2* and is one of many synaptic cell adhesion proteins that are associated with autism (Anderson et al. 2012). Recent work in the *Xenopus* visual system using in vivo time-lapse imaging of dendritic arbor dynamics and electrophysiology tested whether the synaptogenesis and synapse stabilization functions of transsynaptic neuroligin and neurexin associations directly affect dendritic arbor development (Chen et al. 2010). The data support a model in which neurexin and neuroligin mediate initial cell adhesion between pre- and postsynaptic processes and that activity-dependent mechanisms recruit NMDA and AMPA receptors, as well as other pre- and postsynaptic proteins to strengthen and stabilize glutamatergic synapses. The stable synapses in turn increase the lifetimes of dendritic branches and over time lead to increased dendritic arbor growth and complexity. Furthermore, this study is a prime example of how direct visualization and quantitative analysis of dendritic arbor dynamics can identify novel functions for proteins of interest. It is likely that other transsynaptic cell adhesion molecules, such as SynCAM, play similar synaptotropic functions in dendritic arbor development (Missler et al. 2012).

13.4.2 Growth Factors

The effects of the growth factors on neuronal development have been extensively reviewed (Greenberg et al. 2009) and will be briefly mentioned here to highlight their functions in coordinating activity-dependent synaptic, dendritic, and circuit development. BDNF is widely recognized as an essential brain growth factor. Inspiring studies using cultured slices of the mammalian visual cortex showed dramatic effects of neurotrophins on dendritic arbor development (McAllister et al. 1995, 1997). Early studies in *Xenopus* tadpole visual system suggested that BDNF played a variety of role in CNS development (Cohen-Cory et al. 1996; Cohen-Cory and Fraser 1994, 1995). Subsequent work in *Xenopus* has been instrumental in identifying the broad range of cellular effects BDNF has on developing circuits. BDNF synthesis is increased by visual activity in *Xenopus* (Schwartz et al. 2011), as in other systems (Greenberg et al. 2009). In vivo imaging of retinal inputs and optic tectal neurons in *Xenopus* has shown that BDNF is synaptogenic and increases retinotectal synaptic stability (Alsina et al. 2001; Hu et al. 2005; Sanchez et al. 2006) and regulates experience-dependent synaptic strength (Du and Poo 2004; Schwartz et al. 2011). Consistent with the synaptotropic hypothesis, BDNF also increases optic tectal dendritic arbor elaboration as well as reciprocal elaboration of retinotectal axons (Alsina et al. 2001; Sanchez et al. 2006). Loss or impaired BDNF signaling has devastating effects on visual system function in *Xenopus* (Du and Poo 2004). Elegant studies in the *Xenopus* visual system further demonstrated that brief visual training protocols increased BDNF in the tectum and increased visual experience-dependent synaptic plasticity as well as the acquisition of visual response properties (Schwartz et al. 2011).

CPG15, aka neuritin, is an activity-induced growth factor that is induced by neural activity downstream of calcium influx through NMDA receptors and CREB activation (Fujino et al. 2003; Harwell et al. 2005; Javaherian and Cline 2005; Nedivi et al. 1996). CPG15 is a GPI-linked protein that is targeted to the axon surface (Nedivi et al. 2001). CPG15 must be attached to the plasma membrane by its GPI anchor in order to have its neurotrophic effects (Cantallops et al. 2000; Nedivi et al. 1998). In vivo imaging CPG15 tagged with the pH-sensitive ecliptic pHluorin (Miesenbock et al. 1998) showed that CPG15 is exocytosed to the surface of retinotectal axons by a calcium-dependent mechanism in response to retinal

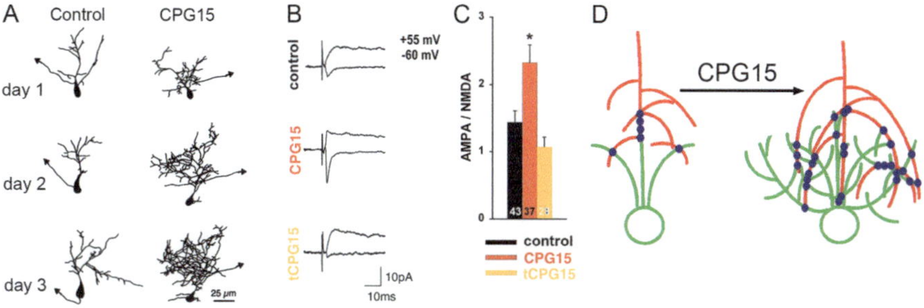

Fig. 13.5 The synaptogenic growth factor CGP15 is required for neuronal and circuit development. (**a**) Reconstructions of in vivo images of *Xenopus* optic tectal neurons collected at daily intervals show that CPG15 expression dramatically increases dendritic arbor elaboration. (**b**, **c**) CPG15 regulates synaptic maturation. (**b**) Whole cell recordings from optic tectal neurons from controls, or animals expressing CPG15 or dominant-negative truncated CPG15 (tCPG15) show that CPG15 regulates AMPA receptor-mediated excitatory synaptic transmission. Neurons were held at −60 mV or +55 mV holding potentials to record AMPA- or NMDA- and AMPA-mediated synaptic currents, respectively. (**c**) AMPA/NMDA ratios in tectal neurons. CPG15 increases AMPA/NMDA ratio. (**d**) Cartoon of CPG15 effects on tectal neuron dendritic arbors, presynaptic retinotectal axons, and glutamatergic retinotectal synapses

activity in *Xenopus* (Cantallops and Cline 2008; Nedivi et al. 1998). CPG15 increases excitatory synapse strength by increasing AMPA receptor trafficking into retinotectal synapses and dramatically increases dendritic arbor elaboration (Cantallops et al. 2000; Nedivi et al. 1998) and coordinately increases elaboration of presynaptic axon arbors (Fig. 13.5). In vivo time-lapse imaging of *Xenopus* spinal cord motor neurons showed that CPG15 expression is synaptogenic and increases axon arbor elaboration at muscle end plates (Javaherian and Cline 2005). Together, these studies indicate that afferent activity increases CPG15 on the surface of presynaptic axons, where it promotes synapse formation and maturation, which subsequently increases dendritic arbor elaboration and retrograde elaboration of the presynaptic axon arbor. CPG15 knockout mice have impaired development and plasticity of visual cortical circuits (Picard et al. 2014) consistent with the idea that CPG15-mediated effects on synaptic and neuronal developmental plasticity ramify to affect circuit development and function. Remaining open questions concerning CPG15 function include the identification of its postsynaptic binding partners and the mechanisms that coordinate pre- and postsynaptic neuronal development.

A third example of growth factor signaling in experience-dependent dendrite and circuit development is insulin receptor signaling. Insulin receptor signaling plays diverse roles in neuronal survival (Valenciano et al. 2006), synaptic plasticity (Beattie et al. 2000; Man et al. 2000; Passafaro et al. 2001; Skeberdis et al. 2001; Wan et al. 1997), and learning and memory (Dou et al. 2005; Zhao et al. 1999). The insulin receptor is a receptor tyrosine kinase that is enriched in the postsynaptic density (Abbott et al. 1999). Insulin is released from neurons in response to depolarization (Clarke et al. 1986), and the insulin receptor substrate, IRSp53, translocates to synapses in response to activity (Hori et al. 2005) suggesting that insulin receptor signaling is activity dependent. Studies in cell culture suggest that insulin receptor signaling regulates spine density and neurite growth (Choi et al. 2005; Govind et al. 2001). Several reports suggest that insulin receptor signaling regulates AMPA receptor trafficking; however, no consensus has been reached regarding the mechanisms or physiological significance of insulin receptor signaling on glutamatergic transmission (Beattie et al. 2000; Lin et al. 2000; Man et al. 2000; Passafaro et al. 2001; Zhou et al. 2001). Insulin has also been reported to recruit $GABA_A$ receptors to the postsynaptic membrane in cultured hippocampal neurons (Wan et al. 1997).

Several molecules downstream of the insulin receptor may play a role in excitatory synaptic connectivity and dendritic structure. IRSp53, an insulin receptor substrate, is localized to the PSD (Abbott et al. 1999), where it binds PSD scaffold proteins, Shank and PSD95 (Bockmann et al. 2002; Choi et al. 2005; Soltau et al. 2002, 2004), and multiple cytoskeletal regulators, such as Rac, Cdc42, WAVE2, and Mena (Govind et al. 2001; Krugmann et al. 2001; Miki et al. 2000; Soltau et al. 2002). These data suggest that insulin receptor signaling may affect the structural stabilization of excitatory synaptic contacts through the association of synaptic scaffolding proteins and the cytoskeleton. Indeed, overexpression of IRSp53 increased spine density in cultured hippocampal neurons, whereas IRSp53 knockdown decreased spine density (Choi et al. 2005). The insulin receptor can form heterodimers with the insulin-like growth factor 1 (IGF-1) receptor, which signal in response to IGF-1. Pyramidal neurons from the IGF-1 null mice have significantly reduced dendritic arbor length and complexity and decreased spine density (Cheng et al. 2003). Other studies have demonstrated that IGF-1 increased dendritic arbor complexity in the rodent somatosensory cortex (Niblock et al. 2000), but neither study tested whether these effects were accompanied by changes in synaptic function or plasticity or changes in circuit function.

Using the *Xenopus* visual system, studies showed that insulin receptor signaling is required for optic tectal neuronal dendritic arbor development, visual experience-dependent structural plasticity, and in vivo visually evoked responses. Furthermore, electron microscopy and electrophysiology indicated that insulin receptor signaling is required to maintain excitatory synapses. Decreased insulin receptor signaling, either by expression of dominant-negative receptor or by morpholino-mediated knockdown, results in synapse and dendritic branch retraction and reduced glutamatergic synaptic transmission, without altering GABAergic synaptic transmission (Chiu et al. 2008). Using electron microscopy to estimate synapse density on tectal neurons provides both definite identification of synaptic contacts onto transfected neurons and ultrastructural information about both pre- and post-synaptic profiles (Li et al. 2011). dnIR-expressing dendrites had less than half of the synapse density of GFP-labeled neuron controls, although no changes in other ultrastructural features or synapse maturation were observed. This study clearly demonstrated that insulin receptor signaling plays crucial roles in circuit function by regulating synapse density, synaptic transmission, experience-dependent dendritic structural plasticity, and response to afferent inputs within a circuit (Fig. 13.6). These diverse outcomes of reduced insulin receptor function may all

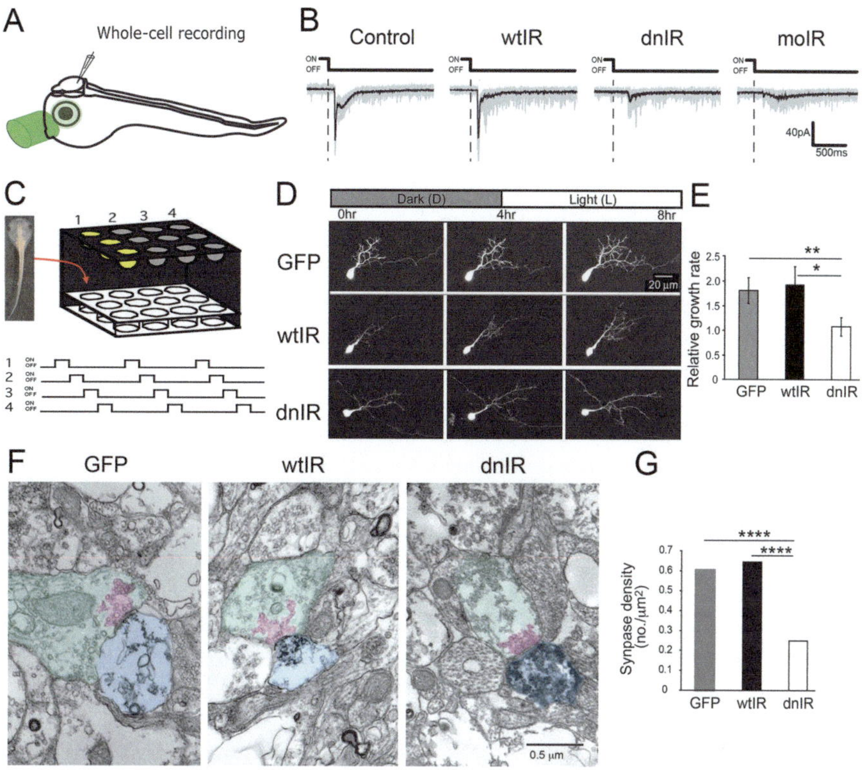

Fig. 13.6 Insulin receptor signaling governs multiple aspects of circuit formation. (**a**) Cartoon of experimental setup for in vivo electrophysiological recordings of visually evoked responses in optic tectal neurons. (**b**) Insulin receptor signaling is required for visual responses in vivo. Light-off-evoked responses recorded from optic tectal neurons expressing GFP (control), GFP + wild-type insulin receptor (wtIR), and GFP + dominant-negative insulin receptor (dnIR) or transfected with morpholinos against the insulin receptor (moIR). (**c–e**) Insulin receptor signaling is required for visual experience-dependent increase in dendritic arbor elaboration. (**c**) Cartoon of visual stimulation protocol. Animals are exposed to a simulated motion stimulus. (**d**) Time-lapse images collected from optic tectal neurons expressing GFP (control), GFP + wild-type insulin receptor (wtIR), and GFP + dominant-negative insulin receptor (dnIR). Images were collected before and after 4h in the dark and 4h after exposure to simulated motion stimulus (light). (**e**) Dominant-negative insulin receptor blocks experience-dependent increased dendritic arbor growth seen in control neurons. (**f**, **g**). Insulin receptor signaling regulates synaptic density. (**f**) Electron micrographs show ultrastructural morphology of synaptic terminals that contact dendrites of optic tectal neurons expressing GFP (control), GFP + wild-type insulin receptor (wtIR), and GFP + dominant-negative insulin receptor (dnIR). Postsynaptic areas, presynaptic area, and clustered synaptic vesicles were highlighted in *light blue*, *green*, and *pink*, respectively. (**g**) Dendrites from dnIR-expressing neurons receive significantly fewer synaptic contacts compared to dendrites from GFP- and wtIR-expressing neurons (Adapted from Chiu et al. 2008)

originate from a vital role for insulin receptor signaling in the maintenance of excitatory synapses.

Our observations are consistent with the synaptotropic hypothesis which states that the formation and maintenance of synapses promote the stabilization of dendritic branches and that dendritic arbor growth correlates positively with the number and strength of synapses (Cline and Haas 2008). In the visual systems of zebra fish and *Xenopus*, visual experience increases dendritic arbor growth rate, retinotectal synaptogenesis, and retinotectal synaptic strength (Aizenman and Cline 2007; Haas et al. 2006; Mu and Poo 2006; Niell et al. 2004; Sin et al. 2002; Zhang et al. 2000). The visual stimulation-induced increase in synapse number and strength (Aizenman and Cline 2007) then appears to stabilize newly extended dendritic branches. The failure of neurons with decreased insulin receptor signaling to increase their growth rate in response to visual stimulation could be a result of their low synapse density. One potential mechanism by which a lower synapse density could affect experience-dependent structural plasticity is that these neurons do not form and maintain synapses on newly added branches, and they are consequently retracted. The alternate, but not mutually exclusive mechanism, is based on the fact that, in these experiments, single tectal neurons were transfected within an otherwise normal optic tectum. Therefore, while surrounding tectal cells, which have twice the synapse density of the dnIR-expressing neurons, respond to visual stimulation normally and can increase their synapses and promote dendritic growth, the single dnIR-expressing neuron, which responds to visual inputs very weakly, may not be able to compete with normal neighboring tectal neurons for retinal inputs. Consequently, this might lead to dendritic branch retraction in the insulin receptor signaling deficient neurons.

Unexpectedly, dnIR-expressing neurons can still elaborate their dendritic over a period of several days even when synapse density is low during early development. A similar observation was reported with manipulation of levels of the neurotrophin BDNF which significantly changed synapse number but not dendritic arbor morphology (Sanchez et al. 2006). In the case of insulin receptor signaling where experience-dependent structural plasticity is decreased when assayed over a period of 4 h, these daily imaging data suggest that under conditions of decreased synaptic input, alternative mechanisms participate in dendritic arbor growth control.

13.5 Stop-Growing Signals

Just as it is critically important for dendritic arbors to grow and extend into specific target regions, it is equally important that arbors stop growing and are restricted from extending aberrantly into neighboring territory. Studies of peripheral sensory neurons in *Drosophila* and leech and in rodent retina show that dendrites of some sensory neurons tile across a target area, likely by mechanisms involving cell surface molecules, such as DSCAM (Jan and Jan 2010). Long-term in vivo imaging in adult mouse olfactory neurons showed that dendrite arbor structure is very stable (Mizrahi and Katz 2003), consistent with the apparent stability inferred from static images. The dendritic arbor elaboration in *Xenopus* optic tectal neurons reaches a plateau after which overall arbor structure in stable (Cline 2001). The work mentioned above indicates that mechanisms, such as insulin receptor signaling, promote synaptic maturation, which in turn stabilizes individual branches against retraction. It is interesting that activity-dependent mechanisms both enhance dendritic arbor development, as described above, and restrict dendritic arbor growth. For instance, electrophysiological experiments of synaptic maturation in *Xenopus* tectal optic neurons indicate that increased αCaMKII activity increases glutamatergic synaptic strength (Wu et al. 1996), likely by increasing AMPA receptor trafficking into developing synapses (Haas et al. 2006). In complementary in vivo imaging experiments of dendritic arbor structure, the same increase in αCaMKII activity stabilized dendritic arbor structure by reducing rates of branch additions and retractions (Wu and Cline 1998). αCaMKII, a multifunctional calcium and calmodulin-dependent kinase, acts downstream of synaptic activity-dependent increases in calcium to regulate synaptic strength (Lisman et al. 2002) and cytoskeletal dynamics (McVicker et al. 2014), for instance, via GTPases (Ghiretti et al. 2014; Sin et al. 2002). Mechanisms governing neuronal arbor size are an area of active investigation as reviewed in Koleske (2013) and in Chap. 15.

13.6 Summary

The architecture of the dendritic arbor contributes to the precise patterning of synaptic connections required for normal circuit function. Dendritic structure not only determines which axons are potential presynaptic partners but also determines how the inputs are integrated. The marriage of single-cell labeling and in vivo time-lapse imaging has made it possible to explore the cellular mechanisms underlying dendritic development. Advances in microscopy, cell biology, and molecular genetic methods have paved the way for significant advances in our understanding of the mechanisms behind the molecular and activity-dependent regulation of dendrite development. Numerous studies support a model of activity-dependent dendrite arbor development in which synaptogenesis and synapse maturation stabilize newly added dendrite branches and promote increased arbor growth. Activity-dependent stop-growing signals limit dendrite growth. Based on this model, proteins that affect synaptogenesis and synaptic function are predicted to have broad effects on dendritic arbor development and circuit function.

References

Abbott MA, Wells DG, Fallon JR (1999) The insulin receptor tyrosine kinase substrate p58/53 and the insulin receptor are components of CNS synapses. J Neurosci 19:7300–7308

Aizenman CD, Cline HT (2007) Enhanced visual activity in vivo forms nascent synapses in the developing retinotectal projection. J Neurophysiol 97:2949–2957

Akerman CJ, Cline HT (2006) Depolarizing GABAergic conductances regulate the balance of excitation to inhibition in the developing retinotectal circuit in vivo. J Neurosci 26:5117–5130

Alsina B, Vu T, Cohen-Cory S (2001) Visualizing synapse formation in arborizing optic axons in vivo: dynamics and modulation by BDNF. Nat Neurosci 4:1093–1101

Anderson GR, Galfin T, Xu W, Aoto J, Malenka RC, Sudhof TC (2012) Candidate autism gene screen identifies critical role for cell-adhesion molecule CASPR2 in dendritic arborization and spine development. Proc Natl Acad Sci U S A 109:18120–18125

Aoto J, Martinelli DC, Malenka RC, Tabuchi K, Sudhof TC (2013) Presynaptic neurexin-3 alternative splicing trans-synaptically controls postsynaptic AMPA receptor trafficking. Cell 154:75–88

Arons MH, Thynne CJ, Grabrucker AM, Li D, Schoen M, Cheyne JE, Boeckers TM, Montgomery JM, Garner CC (2012) Autism-associated mutations in ProSAP2/Shank3 impair synaptic transmission and neurexin-neuroligin-mediated transsynaptic signaling. J Neurosci 32:14966–14978

Beattie EC, Carroll RC, Yu X, Morishita W, Yasuda H, von Zastrow M, Malenka RC (2000) Regulation of AMPA receptor endocytosis by a signaling mechanism shared with LTD. Nat Neurosci 3:1291–1300

Bestman JE, Cline HT (2008) The RNA binding protein CPEB regulates dendrite morphogenesis and neuronal circuit assembly in vivo. Proc Natl Acad Sci U S A 105:20494–20499

Bockmann J, Kreutz MR, Gundelfinger ED, Bockers TM (2002) ProSAP/Shank postsynaptic density proteins interact with insulin receptor tyrosine kinase substrate IRSp53. J Neurochem 83:1013–1017

Cantallops I, Cline HT (2008) Rapid activity-dependent delivery of the neurotrophic protein CPG15 to the axon surface of neurons in intact Xenopus tadpoles. Dev Neurobiol 68:744–759

Cantallops I, Haas K, Cline HT (2000) Postsynaptic CPG15 promotes synaptic maturation and presynaptic axon arbor elaboration in vivo. Nat Neurosci 3:1004–1011

Chanda S, Aoto J, Lee SJ, Wernig M, Sudhof TC (2016) Pathogenic mechanism of an autism-associated neuroligin mutation involves altered AMPA-receptor trafficking. Mol Psychiatry 21:169–177

Chen JL, Nedivi E (2013) Highly specific structural plasticity of inhibitory circuits in the adult neocortex. Neuroscientist 19:384–393

Chen L, Chetkovich DM, Petralia RS, Sweeney NT, Kawasaki Y, Wenthold RJ, Bredt DS, Nicoll RA (2000) Stargazin regulates synaptic targeting of AMPA receptors by two distinct mechanisms. Nature 408:936–943

Chen SX, Tari PK, She K, Haas K (2010) Neurexin-neuroligin cell adhesion complexes contribute to synaptotropic dendritogenesis via growth stabilization mechanisms in vivo. Neuron 67: 967–983

Chen JL, Villa KL, Cha JW, So PT, Kubota Y, Nedivi E (2012) Clustered dynamics of inhibitory synapses and dendritic spines in the adult neocortex. Neuron 74:361–373

Cheng CM, Mervis RF, Niu SL, Salem N Jr, Witters LA, Tseng V, Reinhardt R, Bondy CA (2003) Insulin-like growth factor 1 is essential for normal dendritic growth. J Neurosci Res 73:1–9

Chiu SL, Chen CM, Cline HT (2008) Insulin receptor signaling regulates synapse number, dendritic plasticity, and circuit function in vivo. Neuron 58:708–719

Choi J, Ko J, Racz B, Burette A, Lee JR, Kim S, Na M, Lee HW, Kim K, Weinberg RJ et al (2005) Regulation of dendritic spine morphogenesis by insulin receptor substrate 53, a downstream effector of Rac1 and Cdc42 small GTPases. J Neurosci 25:869–879

Chubykin AA, Atasoy D, Etherton MR, Brose N, Kavalali ET, Gibson JR, Sudhof TC (2007) Activity-dependent validation of excitatory versus inhibitory synapses by neuroligin-1 versus neuroligin-2. Neuron 54:919–931

Clarke DW, Mudd L, Boyd FT Jr, Fields M, Raizada MK (1986) Insulin is released from rat brain neuronal cells in culture. J Neurochem 47:831–836

Cline HT (2001) Dendritic arbor development and synaptogenesis. Curr Opin Neurobiol 11: 118–126

Cline H, Haas K (2008) The regulation of dendritic arbor development and plasticity by glutamatergic synaptic input: a review of the synaptotrophic hypothesis. J Physiol 586: 1509–1517

Cline HT, Wu G-Y, Malinow R (1997) In vivo development of neuronal structure and function. Cold Spring Harbor Symp Quant Biol LXI:95–104

Cohen-Cory S, Fraser SE (1994) BDNF in the development of the visual system of Xenopus. Neuron 12:747–761

Cohen-Cory S, Fraser SE (1995) Effects of brain-derived neurotrophic factor on optic axon branching and remodelling in vivo. Nature 378:192–196

Cohen-Cory S, Escandon E, Fraser SE (1996) The cellular patterns of BDNF and trkB expression suggest multiple roles for BDNF during Xenopus visual system development. Dev Biol 179: 102–115

Craig AM, Kang Y (2007) Neurexin-neuroligin signaling in synapse development. Curr Opin Neurobiol 17:43–52

Dahm LM, Landmesser LT (1988) The regulation of intramuscular nerve branching during normal development and following activity blockade. Dev Biol 130:621–644

Dahm LM, Landmesser LT (1991) The regulation of synaptogenesis during normal development and following activity blockade. J Neurosci 11:238–255

Dailey ME, Smith SJ (1996) The dynamics of dendritic structure in developing hippocampal slices. J Neurosci 16:2983–2994

Dou JT, Chen M, Dufour F, Alkon DL, Zhao WQ (2005) Insulin receptor signaling in long-term memory consolidation following spatial learning. Learn Mem 12:646–655

Du JL, Poo MM (2004) Rapid BDNF-induced retrograde synaptic modification in a developing retinotectal system. Nature 429:878–883

Ebrahimi-Fakhari D, Sahin M (2015) Autism and the synapse: emerging mechanisms and mechanism-based therapies. Curr Opin Neurol 28:91–102

Ehrlich I, Malinow R (2004) Postsynaptic density 95 controls AMPA receptor incorporation during long-term potentiation and experience-driven synaptic plasticity. J Neurosci 24: 916–927

El-Husseini Ael D, Schnell E, Dakoji S, Sweeney N, Zhou Q, Prange O, Gauthier-Campbell C, Aguilera-Moreno A, Nicoll RA, Bredt DS (2002) Synaptic strength regulated by palmitate cycling on PSD-95. Cell 108:849–863

El-Husseini AE, Schnell E, Chetkovich DM, Nicoll RA, Bredt DS (2000) PSD-95 involvement in maturation of excitatory synapses. Science 290:1364–1368

Engert F, Tao HW, Zhang LI, Poo MM (2002) Moving visual stimuli rapidly induce direction sensitivity of developing tectal neurons. Nature 419:470–475

Ewald RC, Van Keuren-Jensen KR, Aizenman CD, Cline HT (2008) Roles of NR2A and NR2B in the development of dendritic arbor morphology in vivo. J Neurosci 28:850–861

Feng W, Zhang M (2009) Organization and dynamics of PDZ-domain-related supramodules in the postsynaptic density. Nat Rev Neurosci 10:87–99

Fortin DA, Tillo SE, Yang G, Rah JC, Melander JB, Bai S, Soler-Cedeno O, Qin M, Zemelman BV, Guo C et al (2014) Live imaging of endogenous PSD-95 using ENABLED: a conditional strategy to fluorescently label endogenous proteins. J Neurosci 34:16698–16712

Fujino T, Lee WC, Nedivi E (2003) Regulation of cpg15 by signaling pathways that mediate synaptic plasticity. Mol Cell Neurosci 24:538–554

Fujino T, Wu Z, Lin WC, Phillips MA, Nedivi E (2008) cpg15 and cpg15-2 constitute a family of activity-regulated ligands expressed differentially in the nervous system to promote neurite growth and neuronal survival. J Comp Neurol 507:1831–1845
Futai K, Kim MJ, Hashikawa T, Scheiffele P, Sheng M, Hayashi Y (2007) Retrograde modulation of presynaptic release probability through signaling mediated by PSD-95-neuroligin. Nat Neurosci 10:186–195
Ghiretti AE, Moore AR, Brenner RG, Chen LF, West AE, Lau NC, Van Hooser SD, Paradis S (2014) Rem2 is an activity-dependent negative regulator of dendritic complexity in vivo. J Neurosci 34:392–407
Gibson JR, Huber KM, Sudhof TC (2009) Neuroligin-2 deletion selectively decreases inhibitory synaptic transmission originating from fast-spiking but not from somatostatin-positive interneurons. J Neurosci 29:13883–13897
Govind S, Kozma R, Monfries C, Lim L, Ahmed S (2001) Cdc42Hs facilitates cytoskeletal reorganization and neurite outgrowth by localizing the 58-kD insulin receptor substrate to filamentous actin. J Cell Biol 152:579–594
Gray NW, Weimer RM, Bureau I, Svoboda K (2006) Rapid redistribution of synaptic PSD-95 in the neocortex in vivo. PLoS Biol 4:e370
Greenberg ME, Xu B, Lu B, Hempstead BL (2009) New insights in the biology of BDNF synthesis and release: implications in CNS function. J Neurosci 29:12764–12767
Haas K, Li J, Cline HT (2006) AMPA receptors regulate experience-dependent dendritic arbor growth in vivo. Proc Natl Acad Sci U S A 103:12127–12131
Hamad MI, Ma-Hogemeier ZL, Riedel C, Conrads C, Veitinger T, Habijan T, Schulz JN, Krause M, Wirth MJ, Hollmann M et al (2011) Cell class-specific regulation of neocortical dendrite and spine growth by AMPA receptor splice and editing variants. Development 138: 4301–4313
Hamad MI, Jack A, Klatt O, Lorkowski M, Strasdeit T, Kott S, Sager C, Hollmann M, Wahle P (2014) Type I TARPs promote dendritic growth of early postnatal neocortical pyramidal cells in organotypic cultures. Development 141:1737–1748
Harwell C, Burbach B, Svoboda K, Nedivi E (2005) Regulation of cpg15 expression during single whisker experience in the barrel cortex of adult mice. J Neurobiol 65:85–96
Holt CE (1989) A single-cell analysis of early retinal ganglion cell differentiation in Xenopus: from soma to axon tip. J Neurosci 9:3123–3145
Hori K, Yasuda H, Konno D, Maruoka H, Tsumoto T, Sobue K (2005) NMDA receptor-dependent synaptic translocation of insulin receptor substrate p53 via protein kinase C signaling. J Neurosci 25:2670–2681
Hosy E, Butler C, Sibarita JB (2014) Organization and dynamics of AMPA receptors inside synapses-nano-organization of AMPA receptors and main synaptic scaffolding proteins revealed by super-resolution imaging. Curr Opin Chem Biol 20:120–126
Hu B, Nikolakopoulou AM, Cohen-Cory S (2005) BDNF stabilizes synapses and maintains the structural complexity of optic axons in vivo. Development 132:4285–4298
Hua JY, Smith SJ (2004) Neural activity and the dynamics of central nervous system development. Nat Neurosci 7:327–332
Huganir RL, Nicoll RA (2013) AMPARs and synaptic plasticity: the last 25 years. Neuron 80: 704–717
Hume RI, Purves D (1981) Geometry of neonatal neurones and the regulation of synapse elimination. Nature 293:469–471
Inglis FM, Zuckerman KE, Kalb RG (2000) Experience-dependent development of spinal motor neurons. Neuron 26:299–305
Inglis FM, Crockett R, Korada S, Abraham WC, Hollmann M, Kalb RG (2002) The AMPA receptor subunit GluR1 regulates dendritic architecture of motor neurons. J Neurosci 22: 8042–8051
Jan YN, Jan LY (2010) Branching out: mechanisms of dendritic arborization. Nat Rev Neurosci 11:316–328

Javaherian A, Cline HT (2005) Coordinated motor neuron axon growth and neuromuscular synaptogenesis are promoted by CPG15 in vivo. Neuron 45:505–512

Kalb RG (1994) Regulation of motor neuron dendrite growth by NMDA receptor activation. Development 120:3063–3071

Kim E, Sheng M (2004) PDZ domain proteins of synapses. Nat Rev Neurosci 5:771–781

Koleske AJ (2013) Molecular mechanisms of dendrite stability. Nat Rev Neurosci 14:536–550

Krugmann S, Jordens I, Gevaert K, Driessens M, Vandekerckhove J, Hall A (2001) Cdc42 induces filopodia by promoting the formation of an IRSp53:Mena complex. Curr Biol 11:1645–1655

Lee WC, Huang H, Feng G, Sanes JR, Brown EN, So PT, Nedivi E (2006) Dynamic remodeling of dendritic arbors in GABAergic interneurons of adult visual cortex. PLoS Biol 4:e29

Lee WC, Chen JL, Huang H, Leslie JH, Amitai Y, So PT, Nedivi E (2008) A dynamic zone defines interneuron remodeling in the adult neocortex. Proc Natl Acad Sci U S A 105:19968–19973

Li J, Erisir A, Cline H (2011) In vivo time-lapse imaging and serial section electron microscopy reveal developmental synaptic rearrangements. Neuron 69:273–286

Lin JW, Ju W, Foster K, Lee SH, Ahmadian G, Wyszynski M, Wang YT, Sheng M (2000) Distinct molecular mechanisms and divergent endocytotic pathways of AMPA receptor internalization. Nat Neurosci 3:1282–1290

Lisman J, Schulman H, Cline H (2002) The molecular basis of CaMKII function in synaptic and behavioural memory. Nat Rev Neurosci 3:175–190

Loebrich S, Nedivi E (2009) The function of activity-regulated genes in the nervous system. Physiol Rev 89:1079–1103

Lohmann C, Myhr KL, Wong RO (2002) Transmitter-evoked local calcium release stabilizes developing dendrites. Nature 418:177–181

Man HY, Lin JW, Ju WH, Ahmadian G, Liu L, Becker LE, Sheng M, Wang YT (2000) Regulation of AMPA receptor-mediated synaptic transmission by clathrin-dependent receptor internalization. Neuron 25:649–662

Marrs GS, Green SH, Dailey ME (2001) Rapid formation and remodeling of postsynaptic densities in developing dendrites. Nat Neurosci 4:1006–1013

Marshak S, Meynard MM, De Vries YA, Kidane AH, Cohen-Cory S (2012) Cell-autonomous alterations in dendritic arbor morphology and connectivity induced by overexpression of MeCP2 in Xenopus central neurons in vivo. PLoS One 7:e33153

McAllister AK, Lo DC, Katz LC (1995) Neurotrophins regulate dendritic growth in developing visual cortex. Neuron 15:791–803

McAllister AK, Katz LC, Lo DC (1997) Opposing roles for endogenous BDNF and NT-3 in regulating cortical dendritic growth. Neuron 18:767–778

McVicker DP, Millette MM, Dent EW (2015) Signaling to the microtubule cytoskeleton: an unconventional role for CaMKII. Dev Neurobiol 75:423-434

Miesenbock G, De Angelis DA, Rothman JE (1998) Visualizing secretion and synaptic transmission with pH-sensitive green fluorescent proteins. Nature 394:192–195

Miki H, Yamaguchi H, Suetsugu S, Takenawa T (2000) IRSp53 is an essential intermediate between Rac and WAVE in the regulation of membrane ruffling. Nature 408:732–735

Missler M, Sudhof TC, Biederer T (2012) Synaptic cell adhesion. Cold Spring Harb Perspect Biol 4:a005694

Mizrahi A, Katz LC (2003) Dendritic stability in the adult olfactory bulb. Nat Neurosci 6: 1201–1207

Monyer H, Seeburg PH, Wisden W (1991) Glutamate-operated channels: developmentally early and mature forms arise by alternative splicing. Neuron 6:799–810

Mu Y, Poo MM (2006) Spike timing-dependent LTP/LTD mediates visual experience-dependent plasticity in a developing retinotectal system. Neuron 50:115–125

Mumm JS, Williams PR, Godinho L, Koerber A, Pittman AJ, Roeser T, Chien CB, Baier H, Wong RO (2006) In vivo imaging reveals dendritic targeting of laminated afferents by zebrafish retinal ganglion cells. Neuron 52:609–621

Nedivi E, Fieldust S, Theill LE, Hevron D (1996) A set of genes expressed in response to light in the adult cerebral cortex and regulated during development. Proc Natl Acad Sci U S A 93: 2048–2053

Nedivi E, Wu GY, Cline HT (1998) Promotion of dendritic growth by CPG15, an activity-induced signaling molecule. Science 281:1863–1866

Nedivi E, Javaherian A, Cantallops I, Cline HT (2001) Developmental regulation of CPG15 expression in Xenopus. J Comp Neurol 435:464–473

Niblock MM, Brunso-Bechtold JK, Riddle DR (2000) Insulin-like growth factor I stimulates dendritic growth in primary somatosensory cortex. J Neurosci 20:4165–4176

Niell CM, Meyer MP, Smith SJ (2004) In vivo imaging of synapse formation on a growing dendritic arbor. Nat Neurosci 7:254–260

Passafaro M, Piech V, Sheng M (2001) Subunit-specific temporal and spatial patterns of AMPA receptor exocytosis in hippocampal neurons. Nat Neurosci 4:917–926

Peixoto RT, Kunz PA, Kwon H, Mabb AM, Sabatini BL, Philpot BD, Ehlers MD (2012) Transsynaptic signaling by activity-dependent cleavage of neuroligin-1. Neuron 76:396–409

Picard N, Leslie JH, Trowbridge SK, Subramanian J, Nedivi E, Fagiolini M (2014) Aberrant development and plasticity of excitatory visual cortical networks in the absence of cpg15. J Neurosci 34:3517–3522

Pratt KG, Aizenman CD (2007) Homeostatic regulation of intrinsic excitability and synaptic transmission in a developing visual circuit. J Neurosci 27:8268–8277

Pratt KG, Dong W, Aizenman CD (2008) Development and spike timing-dependent plasticity of recurrent excitation in the Xenopus optic tectum. Nat Neurosci 11:467–475

Rajan I, Cline HT (1998) Glutamate receptor activity is required for normal development of tectal cell dendrites in vivo. J Neurosci 18:7836–7846

Rajan I, Witte S, Cline HT (1999) NMDA receptor activity stabilizes presynaptic retinotectal axons and postsynaptic optic tectal cell dendrites in vivo. J Neurobiol 38:357–368

Rutishauser U, Landmesser L (1991) Polysialic acid on the surface of axons regulates patterns of normal and activity-dependent innervation. Trends Neurosci 14:528–532

Sanchez AL, Matthews BJ, Meynard MM, Hu B, Javed S, Cohen Cory S (2006) BDNF increases synapse density in dendrites of developing tectal neurons in vivo. Development 133: 2477–2486

Sanes JR, Zipursky SL (2010) Design principles of insect and vertebrate visual systems. Neuron 66:15–36

Schwartz N, Schohl A, Ruthazer ES (2011) Activity-dependent transcription of BDNF enhances visual acuity during development. Neuron 70:455–467

Shen W, Da Silva JS, He H, Cline HT (2009) Type A GABA-receptor-dependent synaptic transmission sculpts dendritic arbor structure in Xenopus tadpoles in vivo. J Neurosci 29: 5032–5043

Shen W, McKeown CR, Demas JA, Cline HT (2011) Inhibition to excitation ratio regulates visual system responses and behavior in vivo. J Neurophysiol 106:2285–2302

Sheng M, Hoogenraad CC (2007) The postsynaptic architecture of excitatory synapses: a more quantitative view. Annu Rev Biochem 76:823–847

Siddiqui TJ, Craig AM (2011) Synaptic organizing complexes. Curr Opin Neurobiol 21:132–143

Sin WC, Haas K, Ruthazer ES, Cline HT (2002) Dendrite growth increased by visual activity requires NMDA receptor and Rho GTPases. Nature 419:475–480

Skeberdis VA, Lan J, Zheng X, Zukin RS, Bennett MV (2001) Insulin promotes rapid delivery of N-methyl-D- aspartate receptors to the cell surface by exocytosis. Proc Natl Acad Sci U S A 98:3561–3566

Soltau M, Richter D, Kreienkamp HJ (2002) The insulin receptor substrate IRSp53 links post-synaptic shank1 to the small G-protein cdc42. Mol Cell Neurosci 21:575–583

Soltau M, Berhorster K, Kindler S, Buck F, Richter D, Kreienkamp HJ (2004) Insulin receptor substrate of 53 kDa links postsynaptic shank to PSD-95. J Neurochem 90:659–665

Stein V, House DR, Bredt DS, Nicoll RA (2003) Postsynaptic density-95 mimics and occludes hippocampal long-term potentiation and enhances long-term depression. J Neurosci 23: 5503–5506
Sudhof TC (2008) Neuroligins and neurexins link synaptic function to cognitive disease. Nature 455:903–911
Tang J, Landmesser L (1993) Reduction of intramuscular nerve branching and synaptogenesis is correlated with decreased motoneuron survival. J Neurosci 13:3095–3103
Tao HW, Poo MM (2005) Activity-dependent matching of excitatory and inhibitory inputs during refinement of visual receptive fields. Neuron 45:829–836
Ting JT, Peca J, Feng G (2012) Functional consequences of mutations in postsynaptic scaffolding proteins and relevance to psychiatric disorders. Annu Rev Neurosci 35:49–71
Valenciano AI, Corrochano S, de Pablo F, de la Villa P, de la Rosa EJ (2006) Proinsulin/insulin is synthesized locally and prevents caspase- and cathepsin-mediated cell death in the embryonic mouse retina. J Neurochem 99:524–536
Van Keuren-Jensen KR, Cline HT (2008) Homer proteins shape Xenopus optic tectal cell dendritic arbor development in vivo. Dev Neurobiol 68:1315–1324
Vaughn J, Barber R, Sims T (1988) Dendritic development and preferential growth into synaptogenic fields: a quantitative study of Golgi-impregnated spinal motor neurons. Synapse 2:69–78
Wan Q, Xiong ZG, Man HY, Ackerley CA, Braunton J, Lu WY, Becker LE, MacDonald JF, Wang YT (1997) Recruitment of functional GABA(A) receptors to postsynaptic domains by insulin. Nature 388:686–690
Witte S, Stier H, Cline HT (1996) In vivo observations of timecourse and distribution of morphological dynamics in Xenopus retinotectal axon arbors. J Neurobiol 31:219–234
Wong RO, Ghosh A (2002) Activity-dependent regulation of dendritic growth and patterning. Nat Rev Neurosci 3:803–812
Wong WT, Faulkner-Jones BE, Sanes JR, Wong RO (2000) Rapid dendritic remodeling in the developing retina: dependence on neurotransmission and reciprocal regulation by Rac and Rho. J Neurosci 20:5024–5036
Wu GY, Cline HT (1998) Stabilization of dendritic arbor structure in vivo by CaMKII. Science 279:222–226
Wu G, Malinow R, Cline HT (1996) Maturation of a central glutamatergic synapse. Science 274:972–976
Wu GY, Zou DJ, Rajan I, Cline H (1999) Dendritic dynamics in vivo change during neuronal maturation. J Neurosci 19:4472–4483
Zhang LI, Tao HW, Poo MM (2000) Visual input induces long-term potentiation of developing retinotectal synapses. Nat Neurosci 3:708–715
Zhao W, Chen H, Xu H, Moore E, Meiri N, Quon MJ, Alkon DL (1999) Brain insulin receptors and spatial memory. Correlated changes in gene expression, tyrosine phosphorylation, and signaling molecules in the hippocampus of water maze trained rats. J Biol Chem 274: 34893–34902
Zhou Q, Xiao M, Nicoll RA (2001) Contribution of cytoskeleton to the internalization of AMPA receptors. Proc Natl Acad Sci U S A 98:1261–1266
Ziv NE, Fisher-Lavie A (2014) Presynaptic and postsynaptic scaffolds: dynamics fast and slow. Neuroscientist 20:439–452

Chapter 14
Dendrite Maintenance

Sara Marie Katrancha and Anthony J. Koleske

Abstract The structural maintenance of dendrite arbors and dendritic spines is crucial for proper brain function. Dendrite arbors and dendritic spines transition from dynamic plasticity to greater stability during maturation. Dendrite branches stabilize before dendritic spines to retain overall dendrite arbor integrity while preserving the ability to fine-tune synaptic connections. Nonetheless, shared signaling mechanisms exist to coordinate shaping and stabilization of both dendritic spines and dendrite arbors. Long-term dendrite arbor and dendritic spine stability is maintained by dynamic cytoskeletal elements that turn over quickly. This process is exquisitely controlled by small GTPases, kinases, and proteins that regulate actin and microtubule polymerization, stability, and turnover. These proteins collectively ensure the precise formation and balanced turnover of the actin and microtubule cytoskeletons to maintain dendritic spine and dendrite structure. In addition, cell surface adhesion receptors, including Eph receptors, immunoglobulin superfamily receptors, cadherins, and integrins, play important roles in dendritic spine formation and stability by signaling to cytoplasmic nonreceptor tyrosine kinases that mediate changes in cytoskeletal structure. The loss of dendritic spine and/or dendrite arbor stability in humans is a major contributing factor to normal aging and the pathology of psychiatric illnesses, neurodegenerative diseases, and stroke. In this review, we discuss the cellular and molecular mechanisms that differentially regulate dendritic spine and dendrite arbor stability, examine how these mechanisms interact functionally, and highlight how they become disrupted in different pathological states.

Keywords Maintenance • Dendritic spines • Small GTPases • Cytoskeleton • Actin stability • Alzheimer's disease • Rac1 • BDNF • CaMKII

S.M. Katrancha • A.J. Koleske (✉)
Interdepartmental Neuroscience Program, Yale University, New Haven, CT, USA

Department of Molecular Biophysics and Biochemistry, Yale University, New Haven, CT, USA

Department of Neuroscience, Yale University, New Haven, CT, USA
e-mail: sara.katrancha@yale.edu; anthony.koleske@yale.edu

K. Emoto et al. (eds.), *Dendrites*, DOI 10.1007/978-4-431-56050-0_14

14.1 Introduction

14.1.1 Dendrite Arbors and Dendritic Spines Transition from Dynamic Plasticity to Greater Stability During Development

The proper formation and long-term maintenance of neuronal connectivity is crucial for proper brain function. The size and shape of a neuron's dendrite arbor establish the blueprint for its connections by determining the number and distribution of receptive synaptic contacts it can make with afferent axons. Dendrites in developing neurons undergo continual dynamic changes in shape to facilitate proper wiring, synapse formation, and establishment of neural circuits. Growing dendrite arbors continually extend and retract branches as they mature, and this behavior is greatly influenced by synaptic activity. Importantly, only a subset of nascent dendrite branches become stabilized (Dailey and Smith 1996; Wu et al. 1999; Wong and Wong 2000; Wong et al. 2000; Niell et al. 2004). Many observations in diverse model systems have affirmed that afferent innervation or synapse formation on a nascent dendrite branch promotes its stabilization, whereas the loss or reduction of synaptic inputs destabilizes target dendrites (Clark 1957; Matthews and Powell 1962; Jones and Thomas 1962; Wiesel and Hubel 1963; Coleman and Riesen 1968; Vaughn 1989; Rajan et al. 1999; Wu et al. 1999; Niell et al. 2004; Cline and Haas 2008). To explain the observation that dendrite branches that receive synaptic inputs are preferentially stabilized, Vaughn proposed the "synaptotrophic hypothesis," offering a model by which afferents can influence dendrite arbor development (Vaughn 1989).

The structural plasticity of dendrites decreases greatly as circuits mature. Dendrite branches stabilize first, while dendritic spines continue to form, change shape, and turn over as circuits refine (Trachtenberg et al. 2002; Oray et al. 2004; Holtmaat et al. 2005, 2006; Majewska et al. 2006). Spine formation and pruning is especially dependent on experience and activity patterns during this early stage of development (Trachtenberg et al. 2002; Zuo et al. 2005a; Holtmaat et al. 2006; Yang et al. 2009). This period of spine dynamics is followed by a phase of extensive synapse and dendritic spine pruning, which extends through adolescence and early adulthood in some human brain regions (Huttenlocher 1979, 1990; Rakic et al. 1986; Markus and Petit 1987; Rakic et al. 1994; Holtmaat et al. 2005; Yang et al. 2009; Gourley et al. 2012). Dendritic spine and dendrite branch stability become mechanistically uncoupled during late refinement. Such uncoupling is crucial for long-term circuit stability, as it enables mature neurons to retain overall long-term dendrite arbor integrity and integration within networks while preserving the ability to fine-tune synaptic connections. The dynamic behavior of dendritic spines in the rodent cortex becomes greatly reduced by early adulthood. Transcranial two-photon imaging indicates that a significant fraction (50–85 %) of dendritic spines in adult rodent cortex is stable for extended time periods of several months and possibly years (Trachtenberg et al. 2002; Holtmaat et al. 2005, 2006; Zuo et al. 2005a; Majewska et al. 2006). Together, these findings suggest a scenario in which the majority of dendritic spines and dendrite arbors become stabilized for long periods within an organism's lifetime, perhaps decades in humans.

The loss of dendritic spine and/or dendrite arbor stability in humans is a major contributing factor to the pathology of psychiatric, neurological, and neurodegenerative disorders, as well as damage associated with stroke and normal aging. Importantly, different patterns of dendritic spine and dendrite branch loss are observed in different psychiatric and neurodegenerative disorders (reviewed in Kulkarni and Firestein 2012), suggesting that spine and branch stabilization mechanisms are differentially disrupted in different disease pathologies. The altered synaptic connectivity resulting from dendrite arbor and dendritic spine destabilization or loss is thought to contribute to the impaired perception, cognition, memory, mood, and decision-making that characterize these pathological conditions. Additionally, despite the evidence discussed for long-term dendritic spine stability during an organism's lifetime, a recent study has shown that cortical dendritic spines return to a more dynamic state in aged mice (>20 months) (Mostany et al. 2013). It is presently unclear how this increased dynamic spine behavior contributes to age-dependent decrements in cognition, learning, or memory.

Biochemical, imaging, and electrophysiological approaches have been used to dissect the mechanisms that mediate long-term dendritic spine and dendrite branch stability. We provide a perspective on the cellular and molecular mechanisms that differentially regulate dendritic spine versus dendrite branch stability and highlight how these mechanisms are disrupted in different pathological states.

14.2 Cytoskeletal Dynamics Play a Key Role in the Regulation of Dendrite Arbor and Dendritic Spine Stability

14.2.1 A Dynamic Actin Framework Supports Dendritic Spine Structure and Function

Dendritic spines are small protrusions from the dendrite branch that serve as postsynaptic contacts at the majority of excitatory synapses (Harris 1999; Tashiro and Yuste 2003; Bourne and Harris 2007; Izeddin et al. 2011). Spines are morphologically heterogeneous (e.g., thin, mushroom, and stubby), yet commonly have a narrow neck and bulbous head (Harris and Kater 1994; Harris 1999; Star et al. 2002; Tashiro and Yuste 2003; Bourne and Harris 2007; Arellano et al. 2007; Tønnesen et al. 2014). Spine shape is supported by a filamentous actin network, which reinforces overall spine structure and stability, organizes the postsynaptic signaling machinery, and powers changes in spine shape (Fifková and Delay 1982; Matus et al. 1982; Fischer et al. 1998; Racz and Weinberg 2006; Hotulainen and Hoogenraad 2010; Izeddin et al. 2011). The actin network consists of both linear and branched filaments, which interact with microtubules at the base of the spine neck and extend to the postsynaptic density (PSD) (Korobova and Svitkina 2010; Izeddin et al. 2011). Additionally, the actin network closely associates with the plasma membrane, and activity-dependent remodeling of the actin network drives changes in spine structure (Izeddin et al. 2011).

Even in stable spines, the actin cytoskeleton undergoes dynamic remodeling (Okamoto et al. 2004). Actin filaments are polarized with their barbed ends located near the membrane; polymerization at the barbed ends contributes to a retrograde flow from the plasma membrane to the spine center with an approximate turnover time in tens of seconds, as measured by both fluorescence recovery after photobleaching and two-photon photoactivation of fluorescent protein-tagged actin (Star et al. 2002; Pollard and Borisy 2003; Okamoto et al. 2004; Honkura et al. 2008). Changes in the polymerization and depolymerization rates or the degree of actin filament branching are believed to mediate the long-lasting changes in spine shape that accompany changes in synaptic efficacy (i.e., the degree to which a presynaptic afferent affects a postsynaptic spine) (Okamoto et al. 2004; Bourne and Harris 2007). For example, long-term potentiation (LTP) induced by tetanic stimulation causes increased actin polymerization and a rapid and selective enlargement of stimulated spine heads that coincides with increased synaptic efficacy (Matsuzaki et al. 2004; Okamoto et al. 2004; Bourne and Harris 2007). In contrast, long-term depression (LTD) induced by prolonged low-frequency stimulation results in actin depolymerization, spine shrinkage, and reduced synaptic efficacy (Fig. 14.1) (Okamoto et al. 2004; Zhou et al. 2004; Bourne and Harris 2007).

Interestingly, the dynamic behavior of actin is heterogeneous within different regions of the spine (Izeddin et al. 2011). Elegant measurements of the dispersal of photoactivated actin within discrete subspine regions strongly suggest that spines contain at least two distinct pools of actin filaments: a dynamic "shell" and a stable "core" (Fig. 14.1) (Izeddin et al. 2011). The more dynamic shell, consisting of the tip and periphery of the spine, determines spine shape and provides the protrusive force to change shape (Honkura et al. 2008). Not surprisingly, actin in the shell region displays a very fast turnover time (~tens of seconds) (Honkura et al. 2008). In contrast, the spine core, consisting of the central zone of the spine head extending to the base of the spine neck, turns over >20-fold more slowly than the shell (Honkura et al. 2008). Moreover, single actin molecule tracking in dendritic spines using photoactivation localization microscopy has supported this model by revealing that half of the actin molecules are stationary and one-third of the actin molecules undergo a retrograde motion, consistent with spines having a stable actin core and a more dynamic actin shell, respectively (Tatavarty et al. 2009).

The concept of distinct functional pools of actin in spines is reinforced by the observation that specific actin-binding proteins preferentially localize to discrete

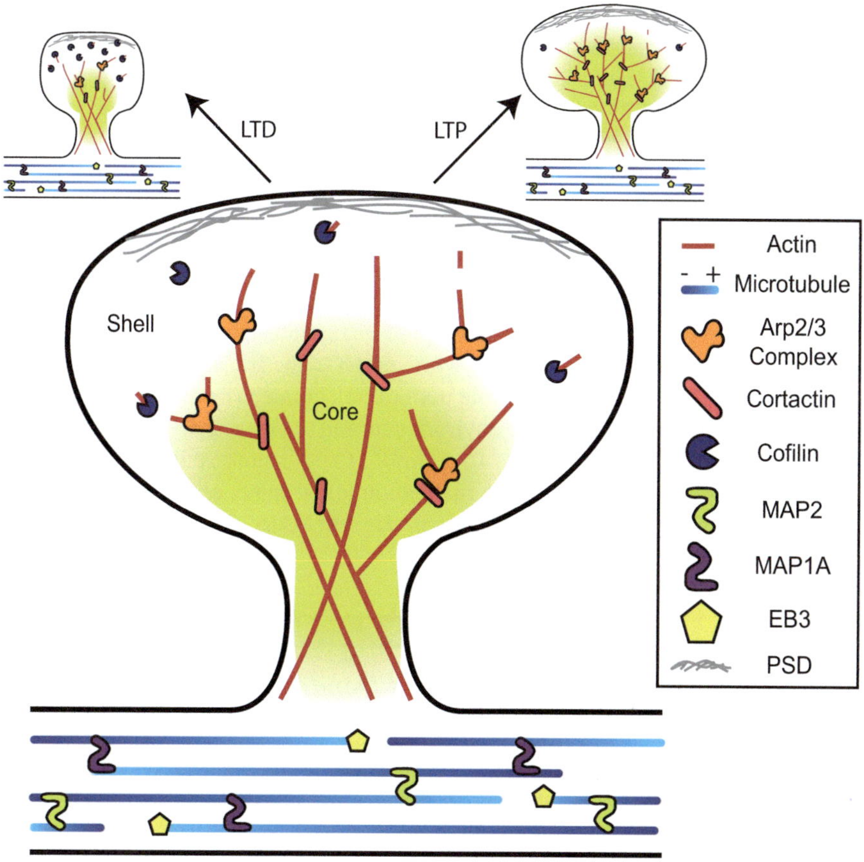

Fig. 14.1 *Key regulators of the actin and microtubule cytoskeletons.* The actin cytoskeleton in dendritic spines is composed of two distinct pools of actin: a dynamic "shell" at the tip and periphery of the spine and a more stable "core" extending from the center of the spine head to the base of the spine neck. Actin-binding proteins (ABPs) that increase actin dynamics, such as cofilin, which severs actin filaments, localize to the spine shell and postsynaptic density (PSD) to support actin remodeling. ABPs that stabilize actin, such as cortactin, localize to the spine core. Cortactin also activates the Arp2/3 complex, which localizes to a toroidal zone between the core and the shell where it nucleates new actin branches. The microtubule cytoskeleton is composed of mixed polarity microtubules that support dendrite arbor structure. Microtubule-associated proteins, including MAP1A and MAP2, form crossbridges between microtubules and mediate their assembly and stabilization. End-binding protein 3 (EB3) localizes to microtubule plus ends, which transiently enter spines. Finally, changes in the actin and microtubule cytoskeleton correlate with alterations in spine structure and synaptic efficacy. In particular, the spine shrinkage and reduced synaptic efficacy observed with long-term depression (LTD) correlate with actin depolymerization, increased actin severing by cofilin, and decreased actin stabilization by cortactin and the Arp2/3 complex. Accordingly, the spine enlargement and increased synaptic efficacy observed with long-term potentiation (LTP) correlate with increased actin polymerization, increased actin nucleation by the Arp2/3 complex, and increased cortactin-mediated actin stability

subdomains within the spine (Rácz and Weinberg 2013). For example, cortactin, which is implicated in actin filament nucleation, branching, and stabilization, localizes to the stable spine core where it helps to maintain actin filament stability (Rácz and Weinberg 2004; Hering and Sheng 2003; Lin et al. 2013). In contrast, cofilin, which severs actin filaments to initiate actin remodeling, localizes to the dynamic spine shell and PSD (Racz and Weinberg 2006). Additionally, the actin-related protein 2/3 (Arp2/3) complex is localized in a toroidal zone between the stable spine core and dynamic spine shell where it specializes in nucleating new actin branches that can extend into the shell from the stable core (Fig. 14.1) (Rácz and Weinberg 2008). Overall, these studies suggest that the distinct pools of actin may have discrete functions in dendritic spines. The stable actin pool allows spines to maintain a constant overall position with respect to the spine core and serves as a scaffold from which the more dynamic pool can elaborate. The dynamic actin pool allows spines to both probe their environment and make activity-dependent size adjustments at the spine periphery. In addition, the dynamic actin pool plays key roles in endocytosis and receptor internalization in the spine, as endocytic sites are enriched for actin regulatory proteins that are necessary for endocytosis (for a recent review, see Mooren et al. 2012).

14.2.2 Microtubule Arrays Support Dendrite Arbor Stability and Membrane Trafficking

Dendrite arbors in the mature nervous system are supported by dense, parallel arrays of regularly spaced microtubule bundles (Sasaki et al. 1983). These microtubules are required for dendrite arbor growth, dendrite arbor maintenance, and trafficking of cargoes, including organelles (Conde and Caceres 2009), PSD-associated proteins (El-Husseini et al. 2001), neurotransmitter receptors (Setou et al. 2000, 2002), specific mRNAs (Setou et al. 2002; Kanai et al. 2004), and proteins for local synthesis within the dendrite (Hirokawa and Takemura 2005). While axonal microtubules are uniformly polarized with their plus ends oriented away from the soma, dendritic microtubules are generally of mixed polarity to facilitate, in part, bidirectional trafficking of cargoes (Baas et al. 1988; Burton 1988). Nonetheless, microtubule polarization within dendrites is observed both close to the growth cone and in apical dendrites with large diameters, which may reflect the importance of transporting cargoes to the growth cone and away from the soma, respectively (Baas et al. 1988; Kwan et al. 2008).

14.2.3 Long-Term Dendrite Arbor and Dendritic Spine Stability Is Maintained by Dynamic Cytoskeletal Elements

While individual dendritic spines and dendrite branches can be stable for months to years, the individual cytoskeletal elements underlying these stable structures turn over in minutes to hours in both the developing and mature nervous system. In dendritic spines, more than 80 % of the actin turns over every minute, with the remaining portion undergoing turnover in tens of minutes (Star et al. 2002; Honkura et al. 2008; Tatavarty et al. 2009). Approximately 75 % of dendritic microtubules turn over within tens of minutes, while the rest turns over at a slower rate, likely to be on the order of a few hours, based on measurements for axonal microtubules (Okabe and Hirokawa 1990; Edson et al. 1993).

The structural and functional stability of dendrite branches and dendritic spines would appear to be in direct conflict with the high turnover rates observed for their underlying cytoskeletal elements. The resolution of this apparent paradox arises from the fact that formation and turnover of the actin and microtubule structures is under exquisite control. In particular, a huge collection of proteins is dedicated to the formation, organization, and stabilization of actin in dendritic spines and microtubules in dendrite branches. These regulatory proteins ensure carefully balanced turnover and maintenance of the actin and microtubule cytoskeletons in dendritic spines and dendrite branches, respectively, so that they can serve as scaffolds for their own replenishment and reorganization. A myriad of such proteins, some discussed below, ensure that only a fraction of the cytoskeleton is remodeled at any particular time.

Regulatory proteins that control the actin cytoskeleton in dendritic spines include the actin-stabilizing protein cortactin, the actin-severing protein cofilin, and the Arp2/3 complex. Of particular interest, the Arp2/3 complex localizes to dendritic spines (Rácz and Weinberg 2008) and mediates nucleation and initiation of new actin branches on preexisting filaments in vitro (Pollard and Borisy 2003). Conditional knockout of the Arp2/3 complex in excitatory neurons disrupts the ability of spines to refurbish actin, resulting in gradual spine and synapse loss in cortical and hippocampal neurons (Kim et al. 2013a), highlighting the critical role that the Arp2/3 complex plays in actin stability. In dendrite branches, the microtubule cytoskeleton is stabilized by microtubule-associated proteins (MAPs), such as MAP1A and MAP2, which localize in high concentrations to dendrite branches where they stabilize existing microtubules, form crossbridges between microtubules, and promote tubulin assembly into microtubules (Fig. 14.1) (Bloom et al. 1984; Huber and Matus 1984; De Camilli et al. 1984; Harada et al. 2002). Activity-dependent dendrite growth, branching, and stabilization depend critically on both MAP1A and MAP2 expression (Vaillant et al. 2002; Szebenyi et al. 2005). Correspondingly, upregulation of both MAPs in cultured neurons strongly correlates with dendrite branch stabilization, whereas knockout of both MAPs causes reduced microtubule density, disrupted microtubule spacing and bundling, and decreased overall dendrite length (Teng et al. 2001; Vaillant et al. 2002; Harada et al. 2002; Szebenyi et al. 2005). Additionally, MAP overexpression protects microtubules from microtubule-severing proteins, such as katanins, which can destabilize dendrite arbors by severing microtubules (Sudo and Baas 2010). In addition to the stabilizing effects of MAPs, microtubules can undergo covalent modifications that affect their stability. For example, detyrosination of the α-tubulin subunit stabilizes microtubules and inhibits their disassembly by decreasing their affinity for the kinesin-13 family of microtubule depolymerases (Peris et al. 2009). Overall, despite the dynamic nature of the individual cytoskeletal elements, the careful balance between stabilization and destabilization of microtubules and actin is essential for the long-term structural stability of dendrite branches and dendritic spines.

14.3 Actin Cytoskeletal Regulators are Essential for Dendritic Spine Maintenance

As the actin cytoskeleton determines spine structure, actin control machinery is critical for spine stability. Small GTPases, kinases, and actin-binding proteins regulate actin in spines to confer long-term structural stability.

14.3.1 Small GTPases Exert Differential Effects on Dendritic Spine Stability

Small GTPases, including Ras and the Rho family GTPases RhoA, Rac1, and Cdc42, are central regulators of cytoskeletal dynamics in neurons (Govek et al. 2005) and mediate both initial spine formation and activity-based changes in spine size. Even after initial spine formation, these GTPases continue to regulate spine shape and stability. For example, activation of single synapses by glutamate uncaging activates Ras, RhoA, Rac1, and Cdc42. Selective knockdown experiments reveal that each of these GTPases is critical for persistent spine head enlargement induced by repeated glutamate uncaging at single synapses (Murakoshi et al. 2011). Moreover, expression of dominant negative Rac1 in hippocampal neurons leads to progressive spine loss, indicating that Rac1 activity is required for long-term spine stability (Nakayama and Luo 2000). Rac1 activates the WAVE regulatory complex (WRC) to stimulate the Arp2/3 complex to nucleate new actin filament branches from the side of existing actin filaments (Machesky and Insall 1998; Machesky et al. 1999; Chen et al. 2014). WAVE1 and the Arp2/3 complex both localize to dendritic spines; knockout of the Arp2/3 complex and knockdown or knockout of *wave1* cause reductions in dendritic spine density (Kim et al. 2006; Soderling et al. 2007; Kim et al. 2013a). At the ultrastructural level, loss of WAVE1 function leads to abnormalities in the structure of both axon terminals

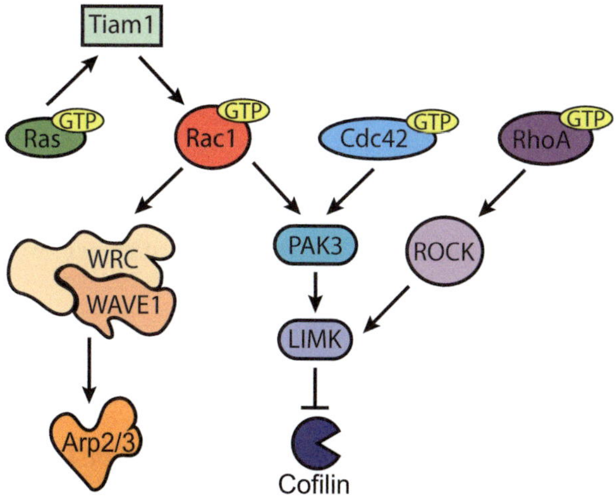

Fig. 14.2 *Small GTPases are central regulators of actin dynamics*. The Rho family GTPases, including Rac1, Cdc42, and RhoA, regulate changes in the actin cytoskeleton. In particular, GTP-bound Rac1 activates the WAVE regulatory complex (WRC) such that WAVE1 can stimulate the Arp2/3 complex and mediate actin nucleation. Additionally, GTP-bound Rac1 and Cdc42 stimulate PAK3 kinase to activate LIM kinase (LIMK), which phosphorylates cofilin and inhibits its actin-severing ability. GTP-bound RhoA activates ROCK, which also stimulates LIMK-based inhibition of cofilin. Finally, GTP-bound Ras GTPase influences actin dynamics through stimulation of T lymphoma invasion and metastasis-inducing protein 1 (Tiam1), which promotes activation of Rac1 by catalyzing the dissociation of GDP and allowing for the binding of GTP to Rac1

and dendritic spines and greatly alters the normal contacts between them (Hazai et al. 2013). These observations strongly suggest that Rac1 acts through WAVE1 and the Arp2/3 complex to refresh the spinoskeleton core and support long-term spine stability. Rac1 and Cdc42 also act through the PAK3 kinase to activate LIM kinase (LIMK) to phosphorylate and inhibit the actin-severing protein cofilin (Fig. 14.2). Knockdown or genetic inhibition of PAK3 activity converts hippocampal spines to a more immature thin morphology and reduces their stability (Boda et al. 2004; Kreis et al. 2007; Dubos et al. 2012).

Activated RhoA mutants or elevated RhoA activity causes reductions in dendritic spine density (Tashiro et al. 2000; Xing et al. 2012), while either RhoA inhibition or knockdown of the RhoA activator GEF1/Lfc increases spine density (Tashiro et al. 2000; Kang et al. 2009). Inhibition of the major RhoA target Rho-associated kinase (ROCK) can block spine loss resulting from elevated RhoA protein levels or activity (Xing et al. 2012; Lin et al. 2013). These studies suggest that RhoA-ROCK signaling antagonizes the stability of previously formed dendritic spines, although the target of RhoA signaling in regulating spine stability is not clear. Nonetheless, RhoA signaling through ROCK stimulates LIMK to phosphorylate cofilin and inhibit its actin-severing activity (Fig. 14.2) (Arber et al. 1998; Yang et al. 1998), but this might be expected to increase actin stability and promote overall spine stability. However, the effects of RhoA on spine stability may depend on the neuron type, developmental stage, or context. For example, in a number of cases, RhoA activation or loss of a RhoA inhibitory cascade in mature neurons leads to dramatic dendrite arbor shrinkage, but does not impact dendritic spine stability (Nakayama and Luo 2000; Sfakianos et al. 2007; Lin et al. 2013). In fact, repeated activation of individual spines by glutamate uncaging actually leads to increased RhoA activity, which is required for activity-based spine enlargement (Murakoshi et al. 2011). Another study found that localized ROCK inhibition decreased dendritic spine stability (Kang et al. 2009). Ongoing work in the field should resolve what factors determine these differential effects of RhoA-ROCK activation on overall spine stability.

14.3.2 Actin-Binding Proteins Play Key Roles in Long-Term Dendritic Spine Stabilization

Dendritic spines are enriched with several actin-binding proteins, including Abp1, cortactin, drebrin, and β-adducin, that play important roles in the long-term stability of dendritic spines (Shirao et al. 1988; Matsuoka et al. 1998; Yamazaki et al. 2001; Tanokashira et al. 2012). Abp1 plays a critical role in dendritic spine morphogenesis, particularly in the transition of thin immature dendritic spines to larger more stable mushroom-shaped spines in the hippocampus (Haeckel et al. 2008). Abp1 interacts with the Arp2/3 complex activator N-WASp and the synaptic scaffold protein ProSAP1/Shank2 during this process, presumably to promote formation of branched actin networks that support spine head enlargement and anchor them to the postsynaptic signaling machinery.

Cortactin is an actin-binding protein that stabilizes actin filaments, stimulates actin branch nucleation by the Arp2/3 complex, and stabilizes the resulting branches (MacGrath and Koleske 2012a; Courtemanche et al. 2015). Cortactin is enriched in the spinoskeleton core, where about 50 % of cortactin is largely immobile, as measured by fluorescence recovery after photobleaching (Rácz and Weinberg 2004; Iki et al. 2005). Cortactin knockdown leads to significant reductions in dendritic spine density (Hering and Sheng 2003). Interestingly, excessive NMDA receptor (NMDAR) stimulation or blockade of the tropomyosin receptor kinase B (TrkB) leads to cortactin redistribution from dendritic spines to dendrite shafts (Hering and Sheng 2003; Iki et al. 2005; Seese et al. 2012; Lin et al. 2013). This cortactin redistribution correlates with loss of spine actin content and spine destabilization, strongly suggesting that cortactin is a critical mediator of spine stability. Cortactin also interacts with several proteins with known or suspected roles in long-term spine stabilization, including cortactin-binding protein 2, the PSD scaffolding protein Shank3, p140Cap, and the Abl2/Arg nonreceptor tyrosine kinase (Du et al. 1998; Naisbitt et al. 1999; Jaworski et al. 2009; Chen and Hsueh 2012; Lin et al. 2013; Courtemanche et al. 2015).

DrebrinA, the main drebrin splice isoform expressed in neurons, has complex roles in dendritic spine development, function, and stability. DrebrinA promotes clustering of actin and the postsynaptic scaffolding protein 95 (PSD95) in dendritic filopodia and promotes pairing of these filopodia with presynaptic partners (Takahashi et al. 2003). Knocking down DrebrinA in immature cultured hippocampal neurons has been reported by one group to decrease the proportion of filopodia with PSD95 clusters, which may represent nascent synapses (Takahashi et al. 2003). However, another study demonstrated that DrebrinA knockdown in immature cultured hippocampal neurons at a similar age increased the proportion of mature spines (Biou et al. 2008). The basis for these different findings is at present unclear. Nonetheless, drebrin binding to actin filaments stabilizes them from depolymerization and protects them from cofilin-mediated severing (Mikati et al. 2013; Grintsevich and Reisler 2014), but drebrin also competes with several proteins, including tropomyosin, cofilin, and α-actinin, for binding to actin filaments (Ishikawa et al. 1994; Sasaki et al. 1996; Grintsevich and Reisler 2014). Interestingly, DrebrinA overexpression in established hippocampal neuron cultures selectively dislocates α-actinin from spines. This correlates with an elongation of spine shape and reduced spine stability (Hayashi and Shirao 1999; Biou et al. 2008), likely due to the loss of α-actinin as a key stabilizing link between the actin cytoskeleton and the N-cadherin adhesion receptor (Knudsen et al. 1995). Together, these data indicate that DrebrinA levels must be carefully balanced for it to contribute its critical roles to dendritic spine formation and stability. Interestingly, drebrin levels are significantly decreased not only in patients with Alzheimer's disease, but also in patients with mild cognitive impairment (Harigaya et al. 1996; Counts et al. 2012). This suggests that even slight reductions in drebrin levels can disrupt normal synaptic function.

β-adducin is an actin filament capping protein that serves as a bridge between actin and the membrane-associated cytoskeletal protein spectrin (Matsuoka et al. 2000). Spectrin/β-adducin complexes are key nodes in the very stable linkages between the red blood cell membrane and the underlying actin cytoskeleton. β-adducin likely plays a similar role in stabilizing membrane-actin cytoskeletal linkages in dendritic spines. Phosphorylation of β-adducin by protein kinase C, a key trigger for synaptic plasticity, can disrupt β-adducin's linkages to actin and spectrin, which may be critical to reshape spines during plasticity. Indeed, phosphorylated β-adducin is enriched in the dendritic spines of hippocampal neurons (Matsuoka et al. 1998). Although relatively normal in appearance, dendritic spines exhibit greatly increased rates of formation and turnover in *b-adducin* knockout mice (Bednarek and Caroni 2011). Stimuli can induce new spine formation in *b-adducin* knockout mice, but a significant fraction of these new spines fail to develop PSD95-positive postsynaptic densities (Bednarek and Caroni 2011; Jung et al. 2013). This increased spine lability and defective synapse formation likely contribute to the observed reductions in spine density in *b-adducin* knockout mice (Jung et al. 2013).

14.4 Activity-Dependent Mechanisms Regulate the Maintenance of Dendrite Arbors and Dendritic Spines

Neuronal activity controls the remodeling and stabilization of both dendritic spines and dendrite arbors. Specifically, the coordinated remodeling and stabilization of neuronal circuits requires crosstalk between dendritic spines and dendrite arbors through shared, activity-dependent signaling molecules that influence cytoskeletal shaping and stabilization. This crosstalk is, in part, mediated by brain-derived neurotrophic factor (BDNF) signaling through TrkB, which can influence both actin and microtubule dynamics. In support of this theory, BDNF release correlates with synaptic activity in cultured hippocampal neurons (Kohara et al. 2001; Kojima et al. 2001; Hartmann et al. 2001). In addition to BDNF-TrkB signaling, activation of calcium/calmodulin-dependent protein kinase II (CaMKII) in response to synaptic activity likely mediates dendritic spine and dendrite arbor stabilization, as premature dendrite stabilization is achieved by expression of constitutively active CaMKII (Wu and Cline 1998).

14.4.1 BDNF-TrkB Signaling Regulates Dendritic Spine Size and Stability

In response to synaptic activity, BDNF is released from both the pre- and postsynaptic terminals (Kohara et al. 2001; Kojima et al. 2001; Hartmann et al. 2001). BDNF acts on presynaptic TrkB to promote axon arbor maturation and to stabilize presynaptic release sites during development (Hu et al. 2005; Marshak et al. 2007). Intriguingly, blocking NMDAR activity disrupts the clustering of presynaptic release sites, but this effect can be overcome by treatment with BDNF, suggesting that NMDAR-mediated BDNF release stabilizes presynaptic boutons (Hu et al. 2005). While the mechanisms underlying these presynaptic changes mediated by BDNF-TrkB signaling remain unclear, TrkB may signal through cortactin or other actin regulatory proteins to mediate stabilization of the presynaptic actin network, which organizes synaptic vesicles and their release machinery.

The delivery of BDNF to postsynaptic TrkB modulates several pathways that mediate dendritic spine enlargement and stabilization. In particular, BDNF stimulation of TrkB activates the Rac1 GTPase, which stimulates the PAK3-LIMK cascade and inhibitory phosphorylation of cofilin, resulting in increased spine size and stability (Rex et al. 2007; Lai et al. 2012). Additionally, BDNF-induced Rac1 activation promotes Arp2/3 complex activation by WAVE1 and the simultaneous release of mRNAs encoding cytoskeletal elements from translational inhibition by the fragile X mental retardation protein complex (De Rubeis et al. 2013). BDNF signaling through TrkB also mediates activity-dependent activation of the Ras GTPase to promote spine enlargement (Fig. 14.3) (Atwal et al. 2000; Yasuda

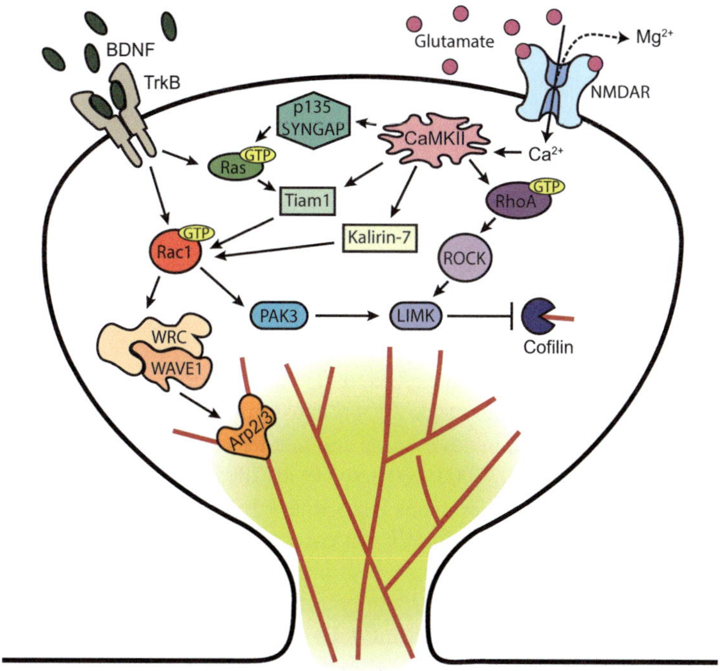

Fig. 14.3 *Activity-dependent dendritic spine stabilization*. During glutamatergic neurotransmission, glutamate is released from presynaptic terminals and binds to AMPA receptors and NMDA receptors (NMDARs) on the postsynaptic membrane. At the resting membrane potential, the NMDAR is blocked by Mg^{2+}; however, once sufficiently depolarized, the Mg^{2+} blockade is released and Ca^{2+} enters the dendritic spine. Ca^{2+} influx through NMDARs activates a variety of signaling mechanisms, including those controlled by calcium/calmodulin-dependent protein kinase II (CaMKII). CaMKII phosphorylates p135 SYNGAP to catalyze the activation of Ras GTPase, which in turn activates Tiam1, a Rac1 guanine nucleotide exchange factor (GEF). CaMKII can also directly and rapidly phosphorylate and activate Kalirin-7, another Rac1 GEF. Kalirin-7 and Tiam1 catalyze Rac1 activation, which signals through the PAK3-LIMK pathway to inhibit the actin-severing ability of cofilin and through the WAVE regulatory complex (WRC) and WAVE1 to stimulate the actin-nucleating ability of the Arp2/3 complex. In addition, CaMKII can activate RhoA, which inhibits the actin-severing ability of cofilin through the ROCK-LIMK pathway. Activity-dependent release of brain-derived neurotrophic factor (BDNF) and stimulation of the tropomyosin receptor kinase B (TrkB) also activate Ras and Rac1 to mediate their downstream effects on cofilin and the Arp2/3 complex

et al. 2006). Overall, BDNF-TrkB signaling employs these and likely additional mechanisms to mediate the activity-induced maintenance of dendritic spines. In addition, other activity-dependent mechanisms may potentiate the effects of BDNF-TrkB signaling, as research has shown that actin content increases following high-frequency stimulation of presynaptic afferents in hippocampal slices, yet this increase is only attenuated by TrkB blockade (Rex et al. 2007). Also, when combined with theta burst stimulation, BDNF treatment of hippocampal slices increases the actin content in spines more than either manipulation alone, suggesting that activity and BDNF act synergistically (Rex et al. 2007). Overall, BDNF-TrkB signaling regulates, at least in part, the activity-induced remodeling and stabilization of dendritic spines by controlling regulators of the actin cytoskeleton.

14.4.2 BDNF-TrkB Signaling Mediates Dendrite Arbor Stability Through the Regulation of Microtubule-Binding Proteins

Neuronal activity, which promotes BDNF-TrkB signaling (Kohara et al. 2001; Kojima et al. 2001; Hartmann et al. 2001), is essential for dendrite arbor growth, branching, and stabilization in developing and mature neurons (Wu and Cline 1998; Xu et al. 2000; Vaillant et al. 2002; Szebenyi et al. 2005). Specifically, knockout of TrkB from cortical pyramidal neurons leads to dendrite shrinkage (Xu et al. 2000). In addition, dendrites develop normally on cortical layer 2/3 neurons in forebrain-specific BDNF knockout mice, but they shrink by 3 weeks of age, indicating that cortical BDNF supports dendrite maintenance rather than initial development (Gorski et al. 2003). The mechanism(s) by which synaptic activity-dependent neurotrophin signaling influence microtubule stability are still being elucidated. Nonetheless, direct application of BDNF or nerve growth factor to developing neurons has been observed to promote increased expression of MAP2 and MAP1A as well as their association with microtubules to stabilize dendrite arbors (Bloom et al. 1984; Huber and Matus 1984; Vaillant et al. 2002; Szebenyi et al. 2005). Additionally, BDNF exposure increases mRNA and protein levels of cypin, a guanine deaminase that promotes microtubule polymerization to regulate dendrite branching and stabilization (Akum et al. 2004; Kwon et al. 2011). Overall, BDNF-TrkB signaling appears to influence dendrite arbor growth and stabilization by regulating proteins that mediate microtubule polymerization and stability.

14.4.3 Neuronal Activity Induces the Transient Invasion of Microtubules into Dendritic Spines and Mediates Crosstalk Between Dendrite Arbors and Dendritic Spines

In addition to the discrete effects of neuronal activity on the actin and microtubule cytoskeletons, activity-dependent mechanisms also mediate crosstalk between dendrite arbors and dendritic spines, influencing their maintenance. Even though spines are actin-rich and contain few microtubules, the dynamic plus ends of microtubules can make activity-dependent invasions into a subset of dendritic spines. In particular, dynamic or growing microtubule plus ends are associated with tip-binding proteins, such as end-binding protein 3 (EB3; also known as MAPRE3) (Fig. 14.1) (Jaworski et al. 2009). Live imaging studies have revealed that neuronal activity and BDNF application modulate the number and duration of transient invasions of EB3-labeled microtubule tips into dendritic spines (Gu et al. 2008; Hu et al. 2008, 2011; Jaworski et al. 2009). Invasion of microtubules into individual spines causes increased spine accumulation of PSD95 and the EB3-binding protein p140Cap, while no significant change in PSD95 or p140Cap is observed in spines that do not exhibit microtubule invasions (Jaworski et al. 2009; Hu et al. 2011). Both PSD95 and p140Cap are required for spine maintenance and stability, and their accumulation is associated with increased actin content and spine enlargement (Jaworski et al. 2009; Hu et al. 2011). In particular, p140Cap interacts with cortactin, which binds and stabilizes actin and recruits the Arp2/3 complex to facilitate actin branch formation, to promote spine stability (Weed et al. 2000; Uruno et al. 2001; Weaver et al. 2001; Jaworski et al. 2009; Courtemanche et al. 2015). Pharmacological inhibition of microtubule dynamics and EB3 knockdown impair PSD95 and p140Cap accumulation, respectively, leading to spine shrinkage or destabilization (Gu et al. 2008; Hu et al. 2008, 2011; Jaworski et al. 2009). In addition, under conditions that induce NMDAR-dependent LTD, EB3 dissociates from growing microtubules and associates with MAP2-positive microtubules in the dendritic shaft, resulting in the inhibition of dynamic microtubule elongation, loss of microtubule targeting to spines, and shrinkage and loss of dendritic spines (Kapitein et al. 2011). Finally, because microtubule invasions are associated with increased spine stability, a decrease in BDNF-induced microtubule invasions could, in part, explain the diminished spine density in animals with a reduced gene dosage of TrkB (von Bohlen und Halbach et al. 2003). Overall, the transient invasion of microtubules into dendritic spines occurs in response to synaptic activity and mediates crosstalk between the microtubule-rich dendrite arbor and the actin-rich dendritic spine.

14.4.4 CaMKII Mediates Synaptic Activity-Based Alterations in Dendrite Arbor and Dendritic Spine Structure

Reductions in synaptic input, such as experimental deafferentation and sensory input removal, result in fewer dendrites, shorter dendrite length, and decreased dendritic branching in many model systems (Clark 1957; Matthews and Powell 1962; Jones and Thomas 1962; Coleman and Riesen 1968; Trachtenberg et al. 2002). The maintenance of this input from presynaptic boutons to dendritic spines is crucial for the long-term stability of dendritic spines. Specifically, in vivo time-lapse two-photon imaging of dendritic spines has revealed that changes in sensory experience, such as whisker trimming, alter spine formation and elimination (Trachtenberg et al. 2002; Zuo et al. 2005b; Holtmaat et al. 2006). At the molecular level, activity-dependent mechanisms activate CaMKII and the MEK-

ERK pathway, both of which are required for activity-dependent stabilization of dendrites in cultured sympathetic neurons (Vaillant et al. 2002). CaMKII and ERK stabilize microtubules via the increased phosphorylation of MAP2, which promotes its binding to microtubules (Schulman 1984; Ray and Sturgill 1987; Sánchez et al. 2000; Vaillant et al. 2002). Precise temporal and activity-dependent control of CaMKII is important for dendrite maturation and stabilization. Expression of constitutively active CaMKII causes premature dendrite stabilization, while knocking down CaMKII expression correlates with reduced branch dynamics of developing dendrite arbors (Wu and Cline 1998). Specifically, CaMKIIα predominantly localizes to spines and dendritic subdomains near activated spines. Lemieux et al. have proposed that subspine CaMKIIα clusters may decode compartmental calcium transients to both support synapse remodeling through local changes in the actin cytoskeleton and interact with microtubules during activity-induced plasticity in cultured neurons (Takemoto-Kimura et al. 2010; Lemieux et al. 2012). The effect of CaMKIIα on dendrite arbor and dendritic spine stabilization may be mediated through its cytoskeleton-associated substrates, such as the Rac1 guanine nucleotide exchange factors Kalirin-7 (Xie et al. 2007) and T lymphoma invasion and metastasis-inducing protein 1 (Tiam1) (Fleming et al. 1999), as well as the Ras GTPase-activating protein p135 SYNGAP (Chen et al. 1998) (Fig. 14.3). Overall, CaMKIIα mediates activity-induced alterations in both the actin cytoskeleton of dendritic spines and the microtubule cytoskeleton of dendrite arbors. Further studies are needed to elucidate which specific targets of CaMKIIα mediate dendrite arbor and dendritic spine stabilization.

14.5 Adhesion Receptor and Neurotrophin Receptor Signaling to the Cytoskeleton Confer Long-Term Dendritic Spine Stability

Cell surface adhesion receptors, including ephrins and their Eph receptors, immunoglobulin superfamily receptors, cadherins, and integrins, play important roles in dendritic spine formation and stability (Biederer and Stagi 2008; Shi et al. 2009; Benson and Huntley 2012). Upon engagement with their extracellular binding partners, these receptors employ cytoplasmic nonreceptor tyrosine kinases of the focal adhesion kinase (FAK), Src, and Abl families to coordinate changes in cytoskeletal structure that impact spine stability.

14.5.1 EphB Signaling Through FAK Stabilizes Mature Dendritic Spines

The ephrin ligands for Eph receptors are composed of ephrin As, which activate EphA receptors and are anchored to the cell surface via covalent linkage to glycophosphoinositide lipids, and ephrin Bs, which activate EphB receptors and contain a transmembrane domain that can engage SH2 domain and PDZ domain-containing proteins to signal into the cell (Klein 2009). Importantly, just as ephrin binding activates the Eph receptor in so-called "forward" signaling, binding to an Eph receptor can also generate "reverse" signaling through the ephrin. Although originally identified as a kinase associated with integrin signaling, FAK can be activated by a variety of cell surface receptors, including EphB receptors, that regulate spine stability. Disruption of EphB2 signaling or knockdown of FAK in established hippocampal neuron cultures leads to significant loss of mature mushroom-shaped dendritic spines and an increase in immature thin "filopodia-like" spines (Shi et al. 2009). FAK may mediate spine stabilization by stimulating p190RhoGEF, which activates the RhoA-ROCK pathway and causes downstream phosphorylation and inactivation of cofilin by LIMK. Consistent with this model, an inactive phospho-mimetic cofilin mutant can block the spine destabilization induced by loss of EphB2-FAK signaling (Shi et al. 2009). In addition, activated FAK recruits and activates Src family kinases, such as Fyn, to phosphorylate p130CAS, a scaffolding molecule that promotes Rac1 activation (Tomar and Schlaepfer 2009). Interestingly, knockout of the Src family kinase Fyn leads to a progressive age-dependent loss of dendritic spines in mice (Babus et al. 2011), and knockdown of p130CAS leads to spine loss in established hippocampal neuron cultures (Bourgin et al. 2007). Both manipulations are expected to disrupt Rac1 activation in the spine, which suggests that FAK and Rac1 signaling are important mediators of dendritic spine stability. FAK can also phosphorylate the key spine protein cortactin and promote its binding to the Abl2/Arg kinase and the Nck1 adaptor, which activates the Arp2/3 complex to promote actin assembly (Fig. 14.4) (MacGrath and Koleske 2012b; Tomar et al. 2012). Overall, FAK signaling in response to EphB2 stimulation mediates stabilization of the actin cytoskeleton and dendritic spines through the Rac1 GTPase.

14.5.2 Astrocyte-Derived Ephrin A2 Regulates Dendritic Spine Stability by Controlling Glutamate Transmission in the Synaptic Cleft

Knockout of ephrin A2 in mice leads to dendritic spine loss in the mouse sensory cortex beginning at 1 month of age and leading to a significant reduction by 4 months of age. Interestingly, this spine loss requires activity, as trimming of whiskers to reduce sensory input or treatment with NMDAR antagonists greatly

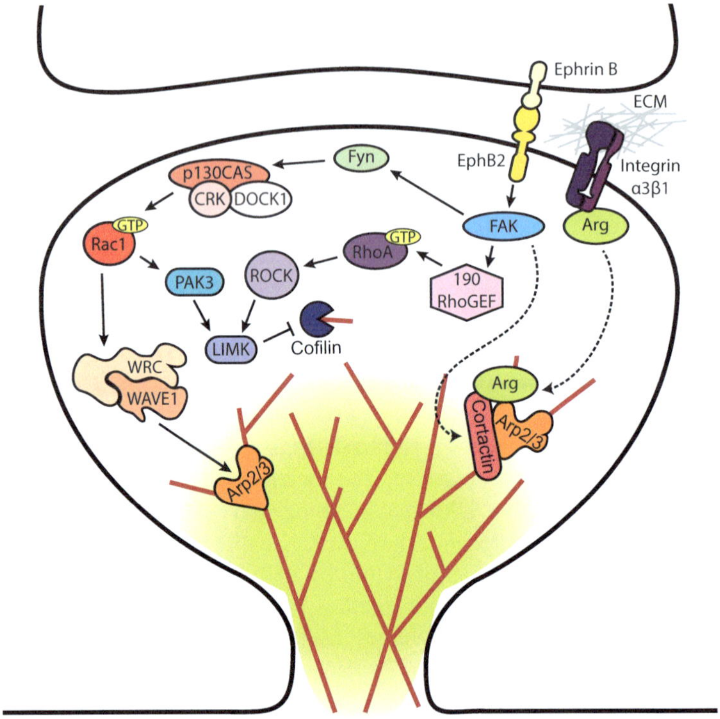

Fig. 14.4 *Adhesion-dependent dendritic spine stabilization.* Cell surface adhesion receptors, including ephrin B, EphB2 receptor, and integrin α3β1, coordinate changes in cytoskeletal structure and impact spine stability by employing cytoplasmic nonreceptor tyrosine kinases, such as the focal adhesion kinase (FAK) and Abl2/Arg kinase. Ephrin B on the presynaptic membrane stimulates the EphB2 receptor on the postsynaptic membrane to activate FAK. FAK activates Src family kinases, such as Fyn, to phosphorylate p130CAS, which, in complex with DOCK1 and the adaptor molecule CRK, mediates the activation of the Rac1 GTPase. Rac1 signals through the WAVE regulatory complex (WRC) and WAVE1 to stimulate actin nucleation by the Arp2/3 complex and through the PAK3-LIMK pathway to inhibit actin severing by cofilin. FAK also stimulates p190RhoGEF to activate RhoA, which signals through the ROCK-LIMK pathway to inhibit actin severing by cofilin. FAK also phosphorylates cortactin, which promotes its binding to Abl2/Arg kinase and activates the Arp2/3 complex to promote actin stabilization, branching, and nucleation. In addition, upon binding of its extracellular matrix (ECM) ligand, integrin α3β1 signals through Abl2/Arg kinase, which both cooperatively binds actin filaments to prevent their depolymerization and promotes cortactin binding to actin filaments to further stabilize them

reduces the enhanced spine loss observed in *ephrina2* knockouts. Ephrin A2 localizes to perisynaptic astrocytic processes where it regulates dendritic spine stability via a novel mechanism. Ephrin A2 stabilizes two distinct astrocytic glutamate transporters that help to clear glutamate from the synaptic cleft. Expression of these glutamate transporters is reduced in *ephrina2* knockout mice, leading to elevated levels of synaptic glutamate, which triggers spine elimination. It appears that one or more EphA receptors in the dendritic spine engage this astrocytic ephrin

A2-glutamate transporter system to prevent excessive glutamate buildup and maintain optimal synaptic function and spine stability (Yu et al. 2013).

14.5.3 Integrin α3β1 Signaling Through Abl2/Arg and Cortactin Mediates Dendritic Spine Stability

Integrins are heterodimeric receptors for the extracellular matrix that are composed of α and β subunits and are widely expressed in the nervous system (Pinkstaff et al. 1999). Interestingly, inhibition or loss of the integrin α3 or β1 subunits destabilizes spines and reduces synapse density (Bourgin et al. 2007; Warren et al. 2012; Kerrisk et al. 2013). Integrins signal through the Abl family kinases, Abl1/Abl and Abl2/Arg in vertebrates, to coordinate changes in cytoskeletal structure (Bradley and Koleske 2009). Arg is particularly abundant in the nervous system (Koleske et al. 1998), where it localizes to dendritic spines (Moresco et al. 2003; Warren et al. 2012; Lin et al. 2013). *arg* knockout mice exhibit normal spine and synapse development through postnatal day 21, followed by significant spine and synapse loss as the mice mature to adulthood (Sfakianos et al. 2007; Gourley et al. 2009b), phenotypes that closely resemble those resulting from neuron-specific ablation of integrin α3 or β1 (Warren et al. 2012; Kerrisk et al. 2013). Importantly, biochemical and genetic experiments indicate that the integrin β1 cytoplasmic tail binds directly to Arg and these interactions with Arg confer dendritic spine and dendrite stability (Warren et al. 2012; Simpson et al. 2015).

Dendritic spine stabilization by Arg in response to integrin α3β1 signaling occurs through two known mechanisms. First, Arg binds cooperatively to actin filaments and prevents their depolymerization (Fig. 14.4) (Wang et al. 2001; MacGrath and Koleske 2012a; Courtemanche et al. 2015), and second, Arg promotes increased cortactin binding to actin filaments to further stabilize them (MacGrath and Koleske 2012a). Knockdown of Arg in cultured neurons leads to cortactin loss from spines, a parallel decrease in spine actin content, and a 50 % loss of spines. NMDAR antagonists can prevent cortactin depletion from spines and rescue spine loss resulting from Arg knockdown. These data strongly suggest that activity-driven cortactin loss from spines is the cause of synapse loss in Arg knockdown neurons. Indeed, fusion of the Arg actin-binding domains to cortactin drives it into spines and rescues spine instability in Arg knockdown neurons (Lin et al. 2013). Whether and how Arg also interfaces functionally with NMDARs or other components of the glutamatergic signaling apparatus is a major unresolved question.

14.6 Dendritic Spines Are Pathologically Destabilized with Normal Aging and in Disease

14.6.1 Region-Specific Dendrite Arbor, Dendritic Spine, and Synaptic Alterations Occur with Normal Aging

Aging is the primary risk factor for most neurodegenerative diseases and is accompanied by cognitive decline even in individuals unaffected by neurodegeneration. Nonetheless, very little research has been done to understand whether and how aging is associated with alterations in dendrite and dendritic spine morphology and stability, and if so, what molecular mechanisms are involved. Despite early evidence that age-related cell loss occurred in all cortical layers (Brody et al. 1955) and the hippocampus (Ball 1977) of humans, stereological methods of neuronal quantification have since shown that cell death is not a characteristic of normal aging, but rather of neurodegeneration (West et al. 1994). More recent studies suggest that age-related impairments in cognitive performance may instead result from subtle dendritic and synaptic alterations in the hippocampus and prefrontal cortex (Morrison and Hof 1997; Burke and Barnes 2006; Morrison and Baxter 2012). In particular, region-specific alterations with advanced aging have been shown to occur in dendrite morphology (Flood et al. 1987b; Hanks and Flood 1991; De Brabander et al. 1998; Uylings and De Brabander 2002), dendritic spine density (Dumitriu et al. 2010; Bloss et al. 2011; Young et al. 2014), and dendritic spine stabilization (Mostany et al. 2013).

The functions of the human dorsolateral prefrontal cortex, such as working memory, decline with age (Gazzaley and D'Esposito 2007; Arnsten et al. 2012), and this is associated with changes in dendrite arbor and dendritic spine structure. In humans, decline of prefrontal function is accompanied by age-related decreases in cortical dendrite arbor branching (De Brabander et al. 1998; Uylings and De Brabander 2002). Studies of the prefrontal cortex in nonhuman primates have revealed a 30–33 % reduction of dendritic spines on pyramidal neurons with advancing age (Peters et al. 1998; Peters et al. 2008; Dumitriu et al. 2010; Young et al. 2014). These significant reductions occur primarily with long, thin spines in this region, while no decrease is observed in mushroom or stubby spine density (Dumitriu et al. 2010; Bloss et al. 2011; Young et al. 2014). Additionally, studies in the rodent cortex have shown that aged rats have decreased overall spine density but increased dendritic spine stability in a chronic stress paradigm in comparison to young rats, who exhibit dendritic spine loss during stress (Bloss et al. 2011). A recent study investigated these age-related changes in spine dynamics using transcranial in vivo two-photon imaging of mouse somatosensory cortex. Spine density was shown to increase through adulthood, stabilizing in mature and old mice. Interestingly, the authors found a decrease in the long-term stability and size of dendritic spines in aged mice and an increase in turnover spine dynamics in very old animals, suggesting a return of spines to a more plastic state later in life

(Mostany et al. 2013). These studies suggest that cortical network flexibility, spine remodeling, and spine stabilization are altered during normal aging.

In contrast to the prefrontal cortex, no changes in dendritic branching are observed with aging in hippocampal area CA1 (Hanks and Flood 1991) or CA3 (Flood et al. 1987b). Additionally, even though age-related CA3 spine loss is observed in the rat hippocampus (Smith et al. 2000; Adams et al. 2010), no CA1 spine loss is observed with normal aging (Geinisman et al. 2004). Instead, recent studies suggest that either the maintenance of complex synapses or the conversion from simple axospinous to complex synapses may be impaired with advancing age in the hippocampus, specifically in CA1. Complex synapses, such as perforated synapses or multisynaptic boutons, are enriched following LTP induction, and their density correlates with increased efficacy of synaptic transmission (Toni et al. 2001; Geinisman et al. 1993; Ganeshina et al. 2004a, b). Specifically, LTP induction increases the number of both multi-synapse boutons and perforated spines, which are characterized by a fenestrated PSD, with high AMPA receptor expression in CA1 (Toni et al. 2001). Additionally, cognitive decline in aged spatial learning-impaired rats correlates with decreased PSD size of CA1 perforated synapses (Nicholson et al. 2004), and aged rats display a reduction in perforated synapses in the sublayer of the dentate gyrus receiving input from entorhinal cortex (Geinisman et al. 1986). Aged nonhuman primates also exhibit age-related reductions in multiple-synapse boutons (Hara et al. 2011). Overall, these data suggest that alterations in the number and composition of complex synapses may underlie age-related changes in the hippocampus. Nonetheless, the molecular mechanisms underlying age-related alterations in both the cortex and hippocampus merit further investigation.

14.6.2 Dendritic Spines and Dendrite Arbors Are Pathologically Destabilized in Alzheimer's Disease

Alzheimer's disease (AD) is characterized by region-specific alterations in the hippocampus, including synapse loss, dendrite arbor atrophy, and neuronal death, that are not observed during normal aging (Flood et al. 1987a, b; Flood 1991; West et al. 1994; Anderton et al. 1998). The decreases in synaptic density, dendrite arborization, and neuronal density correlate with the degree of dementia and AD pathology, with reductions in synapse number being the strongest pathological correlate of cognitive and memory impairment (Terry et al. 1991; Falke et al. 2003). Accordingly, advances in the past decade have highlighted how disrupting molecular mechanisms mediating dendritic spine and dendrite arbor stability could lead to AD pathology. In particular, the observed synapse loss appears to be pathologically induced by the presence of soluble Aβ-derived diffusible ligands (ADDLs), which are small, soluble oligomers derived from Aβ that correlate with memory loss in AD patients, adversely impact synaptic plasticity, cause dendritic spine loss, and contribute to dendritic atrophy in AD (Gong et al. 2003; Haass and Selkoe 2007) (for a recent review, see Pozueta et al. 2013). The application of ADDLs to cultured hippocampal neurons results in rapid alterations in dendritic spine morphology accompanied by a significant decrease in spine density (Lacor et al. 2007; Calabrese et al. 2007), which is also observed in organotypic slice (Shankar et al. 2007). In particular, the effects of ADDLs on dendritic spines may be mediated, at least in part, by the selective binding of ADDLs to certain receptors, such as the α7 nicotinic acetylcholine receptor (Wang et al. 2000), leukocyte immunoglobulin-like receptor B2 (Kim et al. 2013b), and cellular prion protein (PrP^C) (Freir et al. 2011; Um et al. 2012). For example, ADDLs bind to PrP^C, which causes Fyn kinase activation and subsequent phosphorylation of the NMDAR NR2B subunit (Freir et al. 2011; Um et al. 2012). This ADDL-induced NMDAR phosphorylation causes an initial increase in NMDAR expression at the postsynaptic membrane of excitatory neurons, followed by a subsequent decrease in NMDAR expression mediated by receptor endocytosis (Lacor et al. 2007; Shankar et al. 2007; Um et al. 2012). The initial increase in NMDAR surface activity destabilizes dendritic spines by triggering loss of drebrin and cortactin from spines (Lacor et al. 2007; Um et al. 2012; Ishizuka et al. 2014) and causing the calcineurin-mediated dephosphorylation and activation of cofilin leading to actin disassembly (Fig. 14.4) (Wang et al. 2005; Wu et al. 2010). In support of this model, both calcineurin and cofilin are required for ADDL-mediated spine loss (Shankar et al. 2007). Additionally, the surface localization of other receptors implicated in dendritic spine stability and memory formation, such as EphB2, is disrupted by application of ADDLs (Lacor et al. 2007). Specifically, the application of ADDLs to neurons results in a 60 % decrease in EphB2 puncta within 6 h (Lacor et al. 2007), and disruption of EphB2 causes a decrease in mushroom-shaped dendritic spine density and an increase in "filopodia-like" spine formation in cultured hippocampal neurons (Shi et al. 2009).

In addition to their effects on dendritic spine structure, ADDLs have been shown to cause reductions in dendrite arborization through interactions with the microtubule-associated protein tau (Falke et al. 2003; Wu et al. 2010). Hyperphosphorylation of tau and its subsequent accumulation into neurofibrillary tangles is a hallmark of AD pathology that correlates with cognitive decline and the pathological loss of neurons and synapses (Giannakopoulos et al. 2003; Ingelsson et al. 2004). Hyperphosphorylated tau relocalizes from axons to somatodendritic compartments where it aggregates into neurofibrillary tangles, blocks trafficking on microtubules, and causes dystrophic patterning and simplification of dendrites (Wu et al. 2010; Kopeikina et al. 2012; Carlyle et al. 2014). These effects may be mediated by the ADDL-based activation of certain kinases, including MARK1, p70S6K, BRSK, and Fyn, that can phosphorylate tau, causing tau relocalization and microtubule depletion in response to elevated calcium levels (Fig. 14.5) (Zempel et al. 2010; Larson et al. 2012). In addition, these tau missorting events occur in proximity to areas of decreased spine density, suggesting that local elevations in calcium levels may mediate both spine loss and microtubule depletion (Zempel et al. 2010). Recently, tau has been suggested to play a physiologic role in the

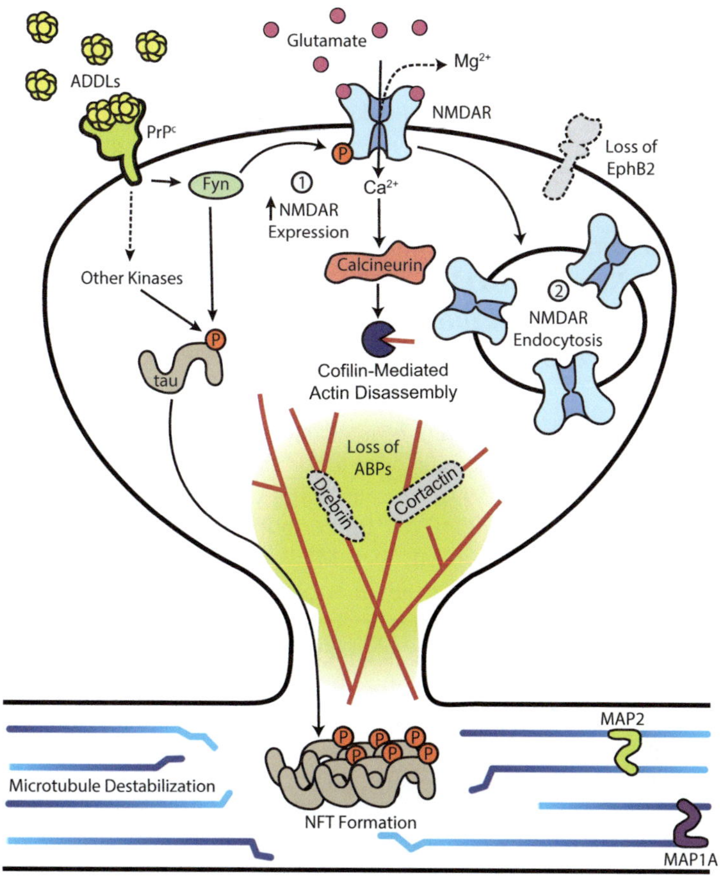

Fig. 14.5 *Dendrite arbor and dendritic spine destabilization in Alzheimer's disease*. Aβ-derived diffusible ligands (ADDLs) selectively bind to certain receptors, such as the cellular prion protein (PrP^C), which activate a variety of kinases, including Fyn kinase, that mediate the alterations in dendrite arbor and dendritic spine morphology observed in Alzheimer's disease. In particular, Fyn phosphorylates the NMDA receptor (NMDAR) GluN2B subunit, which causes an initial increase in NMDAR expression at the postsynaptic membrane (1). The increased surface activity of NMDARs destabilizes spines by triggering actin disassembly through the calcineurin-dependent activation of cofilin and the loss of actin-binding proteins (ABPs), such as drebrin and cortactin. Following this period of increased NMDAR expression, NMDAR expression decreases as the receptors undergo endocytosis (2). Fyn and other kinases, such as MARK1, BRSK, and p70S6K, also hyperphosphorylate tau in response to ADDL-based activation of PrP^C. Hyperphosphorylated tau accumulates into neurofibrillary tangles (NFTs) in the dendrite arbor, which blocks microtubule-based trafficking, triggers microtubule destabilization and depletion, and causes dystrophic patterning of dendrites. Additionally, ADDL application disrupts surface localization of receptors other than PrP^C, including EphB2

targeting of Fyn kinase to the PSD to regulate the NMDAR (Ittner et al. 2010). Thus, the ADDL-induced accumulation of hyperphosphorylated tau in spines could disrupt normal Fyn localization and thereby compromise NMDAR anchoring and activity at the synapse, leading to dendritic spine destabilization (Ittner et al. 2010; Hoover et al. 2010). Overall, pathologic ADDLs specifically target mechanisms that regulate dendritic spine and dendrite arbor stabilization and maintenance, resulting in loss of dendritic spines and dendrite arbors in AD.

14.6.3 Decreased Expression of Synaptic Proteins and Reduced BDNF Signaling Lead to Dendritic Spine Destabilization and Dendrite Atrophy in Depression

Major depressive disorder (MDD) is associated with reduced synapse density (Kang et al. 2012) and reduced dendrite arbor size in the prefrontal cortex and hippocampus (Drevets et al. 1997; Drevets 2000; Cotter et al. 2001, 2002). Interestingly, expression of the GATA1 transcriptional repressor is significantly increased in the brains of individuals with MDD (Kang et al. 2012). GATA1 is a master regulator of a cluster of genes that play critical roles in dendrite arbor maintenance and dendritic spine stability. Several GATA1 target genes are significantly downregulated in the MDD brain, including tubulin β4, the major β tubulin in neurons, and the trafficking regulators Rab3A and Rab4B (Kang et al. 2012). These reductions may decrease the size and stability of the microtubule network and perturb the flow of traffic within dendrites. This model is supported by the observation that GATA1 overexpression alone is sufficient to significantly reduce dendrite arbor size in cortical neurons (Kang et al. 2012). In addition, overexpression of GATA1 or the related GATA2 in the hippocampus is sufficient to reduce dendrite arbor size, decrease dendritic spine density, and lead to depression-like behaviors in mice (Choi et al. 2014). These observations support a model in which transcriptional repression by GATA1/GATA2 mediates the loss of connectivity and compromised neuronal function in MDD.

Chronic stress and elevated circulating corticosteroid levels are major risk factors for the development of MDD. Both conditions reduce BDNF and TrkB levels in the brain (Smith et al. 1995; Nibuya et al. 1999; Gourley et al. 2009a), which likely explains the reduction in BDNF in the brains of MDD patients (Schmidt and Duman 2007). Reduced BDNF signaling would be expected to compromise the BDNF-mediated support of dendritic spine and dendrite arbor stability, via mechanisms detailed in previous sections, thereby contributing to the neuronal destabilization found in this disease. Recent work shows that chronic corticosterone treatment also decreases expression of caldesmon, an actin-stabilizing molecule, increasing actin turnover in dendritic spines and resulting in smaller, more unstable spines (Tanokashira et al. 2012). Thus, reduced BDNF-TrkB signaling also appears to be a key factor in the altered connectivity and circuit dysfunction in MDD.

14.6.4 Dendritic Spine and Dendrite Arbor Stability Are Disrupted in Stroke

In addition to causing death of some neurons, stroke causes destabilization of spines and dendrites on neurons within and adjacent to the infarct area. For an individual neuron, this destabilization may clear dendrites away from areas where damage or death of afferents has disrupted connectivity and free up synaptic proteins and metabolic resources to remodel the dendrite arbor. In a photothrombotic murine ischemia model, light is used to occlude small vessels leading to lesions that are 1 mm in diameter. The reductions in dendrite arbor size in the peri-infarct region that follow this photothrombotic stroke are closely matched by growth of distal dendrites during the recovery period (Brown et al. 2010), suggesting that the neuron may attempt to reach some homeostatic "set point" in dendrite size and synaptic inputs as it reintegrates into networks. In contrast, occlusion of the middle cerebral artery leads to larger lesions averaging 7 mm^3; with this ischemia model, apical dendrites in the peri-infarct region become permanently reduced in size and lose higher order branches. Interestingly, intravital imaging reveals that dendrite shrinkage results both from tissue shrinkage in the peri-infarct region and also from retraction of the distal tips (Mostany and Portera-Cailliau 2011).

Ischemia also leads to alterations in dendritic spine stability and dynamics. Introduction of photothrombotic lesions leads to significantly increased dendritic spine turnover rates in the peri-infarct region for up to 4 weeks following the lesion. This increased dynamic behavior correlates with functional recovery and remapping of the sensory input from the affected area to adjacent cortical regions (Brown et al. 2009). In the middle cerebral artery occlusion model, significant spine loss is observed already at 4 days after stroke with 20 % of spines lost in the peri-infarct region by 12 days after stroke. These spine losses recover by about 4 weeks after stroke (Kalia et al. 2008). The observed reductions in spine number result both from an increase in spine loss and a significant decrease in new spine formation. Importantly, spines do not recover at similar rates on all neurons, and the best predictor of recovery is the amount of local blood flow at 1 h after stroke induction. In addition, neurons in areas distal to the stroke gain supranumerary spine densities, suggesting that these areas may be contributing to the functional remodeling of cortical circuits.

Elevated NMDAR activity is associated with neuronal damage and death following stroke (for a recent review, see Kalia et al. 2008). Application of NMDA to cultured neurons or focal application of NMDA to tissues in vivo has been widely used to model this glutamate-induced excitotoxicity. While prolonged, excessive stimulation leads to neuron death, sublethal NMDA stimulation of neurons, either in culture or focally in tissues, causes spine loss, dendrite branch swelling, and dendrite shrinkage (Hasbani et al. 1998, 2001; Halpain et al. 1998; Hering and Sheng 2003; Kitaoka et al. 2004; Iki et al. 2005; Tseng and Firestein 2011). These changes are closely associated with disruptions in actin and microtubule structure in spines and dendrites, respectively (Hering and Sheng 2003; Graber et al. 2004).

Excessive NMDA stimulation destabilizes spines by disrupting the function of actin regulatory proteins in the spine. For example, bath application of NMDA to cultured hippocampal neurons causes cortactin to relocalize from dendritic spines to the dendrite shaft, and this correlates with loss of spine stability (Hering and Sheng 2003; Iki et al. 2005). In addition, inhibitors of the cathepsin B family of proteases attenuate NMDA-induced spine loss, indicating that proteolysis of one or more key actin regulatory proteins may contribute to spine disruption (Graber et al. 2004). Indeed, NMDA stimulation leads to reduced levels of the spine stabilizing protein MARCKS, but this decrease can be prevented by cathepsin B inhibition. Elevated NMDA stimulation also disrupts the targeting of dynamic microtubules to spines, which prevents delivery of key proteins, including PSD95 and p140Cap (Kapitein et al. 2011), required for spine stability.

Exactly how elevated NMDA stimulation triggers dendrite destabilization is less clear. One potential candidate for mediating these effects is the RhoA GTPase. Focal synaptic stimulation can activate RhoA in the spine, and activated RhoA can then diffuse into the dendritic shaft (Murakoshi et al. 2011). The summed signaling from a group of spines in aggregate, such as that following excessive NMDA stimulation, could trigger RhoA-ROCK signaling at levels sufficient to disrupt MAP function (Yamamoto et al. 1983; Murthy and Flavin 1983) and cypin expression (Chen and Firestein 2007), thereby destabilizing dendrites. In support of this theory, the ROCK inhibitor fasudil can reduce the severity of damage to the dendrite-rich inner plexiform layer of the retina resulting from NMDA injection (Kitaoka et al. 2004). Moreover, RhoA activation and NMDA stimulation have both been shown to decrease cypin levels in cultured neurons (Tseng and Firestein 2011). Finally, it is possible that NMDA-mediated disruption of microtubule structure impacts dendrites by disrupting the microtubule motor-based delivery of key building blocks to the dendrite. Overall, the stabilization of dendrite arbors and dendritic spines is severely disrupted following stroke through a mechanism downstream of excessive NMDAR activation.

14.7 Summary

While dendrite arbors and dendritic spines are highly dynamic in early development, structural plasticity decreases and stability increases as circuits mature. Afferent innervation or synapse formation onto dendrite arbors or dendritic spines stabilizes them, while loss or decrease of synaptic inputs destabilizes dendrites and dendritic spines. During maturation, dendrite arbors stabilize before dendritic spines and help retain the overall dendrite arbor field integrity while preserving the ability to refine synaptic connections. A diverse collection of cellular and molecular mechanisms regulates this differential stabilization of dendritic spines and dendrite branches. Nonetheless, the coordinated stabilization of neuronal circuits requires the existence of shared signaling mechanisms, such as BDNF-TrkB signaling and CaMKII activation, that act on cytoskeletal elements and contribute to the stabilization of both dendritic spines and dendrite arbors. Accordingly, the long-term structural and functional stability of dendrite arbors and dendritic spines is maintained by these dynamic cytoskeletal elements that turn over in seconds to hours. To confer stability, the formation and turnover of actin and microtubule structures are exquisitely controlled by regulatory proteins, including Rho family GTPases, kinases, actin-binding proteins, and microtubule-binding proteins, that ensure the balanced turnover and maintenance of the actin and microtubule cytoskeletons so that they can serve as scaffolds for their own replenishment and reorganization. Cell surface adhesion receptors, including Eph receptors, immunoglobulin superfamily receptors, cadherins, and integrins, also play important roles in dendritic spine formation and stability through the employment of cytoplasmic nonreceptor tyrosine kinases, such as those of the FAK, Src, and Abl families, to mediate changes in cytoskeletal structure. Finally, the loss of dendritic spine and/or dendrite arbor stability in humans is a major contributing factor to normal aging and the pathology of psychiatric illnesses, such as major depressive disorder; neurodegenerative diseases, such as Alzheimer's disease; and stroke. Overall, the cellular and molecular mechanisms that mediate dendrite arbor and dendritic spine maintenance are crucial for proper brain function.

References

Adams MM, Donohue HS, Linville MC et al (2010) Age-related synapse loss in hippocampal CA3 is not reversed by caloric restriction. Neuroscience 171:373–382. doi:10.1016/j.neuroscience.2010.09.022

Akum BF, Chen M, Gunderson SI et al (2004) Cypin regulates dendrite patterning in hippocampal neurons by promoting microtubule assembly. Nat Neurosci 7:145–152. doi:10.1038/nn1179

Anderton BH, Callahan L, Coleman P et al (1998) Dendritic changes in Alzheimer's disease and factors that may underlie these changes. Prog Neurobiol 55:595–609

Arber S, Barbayannis FA, Hanser H et al (1998) Regulation of actin dynamics through phosphorylation of cofilin by LIM-kinase. Nature 393:805–809

Arellano JI, Benavides-Piccione R, Defelipe J, Yuste R (2007) Ultrastructure of dendritic spines: correlation between synaptic and spine morphologies. Front Neurosci 1:131–143. doi:10.3389/neuro.01.1.1.010.2007

Arnsten AFT, Wang MJ, Paspalas CD (2012) Neuromodulation of thought: flexibilities and vulnerabilities in prefrontal cortical network synapses. Neuron 76:223–239. doi:10.1016/j.neuron.2012.08.038

Atwal JK, Massie B, Miller FD, Kaplan DR (2000) The TrkB-Shc site signals neuronal survival and local axon growth via MEK and P13-kinase. Neuron 27:265–277

Baas PW, Deitch JS, Black MM, Banker GA (1988) Polarity orientation of microtubules in hippocampal neurons: uniformity in the axon and nonuniformity in the dendrite. Proc Natl Acad Sci U S A 85:8335–8339

Babus LW, Little EM, Keenoy KE et al (2011) Decreased dendritic spine density and abnormal spine morphology in Fyn knockout mice. Brain Res 1415:96–102. doi:10.1016/j.brainres.2011.07.059

Ball MJ (1977) Neuronal loss, neurofibrillary tangles and granulovacuolar degeneration in the hippocampus with ageing and dementia. A quantitative study. Acta Neuropathol 37:111–118

Bednarek E, Caroni P (2011) β-Adducin is required for stable assembly of new synapses and improved memory upon environmental enrichment. Neuron 69:1132–1146. doi:10.1016/j.neuron.2011.02.034

Benson DL, Huntley GW (2012) Building and remodeling synapses. Hippocampus 22:954–968. doi:10.1002/hipo.20872

Biederer T, Stagi M (2008) Signaling by synaptogenic molecules. Curr Opin Neurobiol 18:261–269. doi:10.1016/j.conb.2008.07.014

Biou V, Brinkhaus H, Malenka RC, Matus A (2008) Interactions between drebrin and Ras regulate dendritic spine plasticity. Eur J Neurosci 27:2847–2859. doi:10.1111/j.1460-9568.2008.06269.x

Bloom GS, Schoenfeld TA, Vallee RB (1984) Widespread distribution of the major polypeptide component of MAP 1 (microtubule-associated protein 1) in the nervous system. J Cell Biol 98:320–330

Bloss EB, Janssen WG, Ohm DT et al (2011) Evidence for reduced experience-dependent dendritic spine plasticity in the aging prefrontal cortex. J Neurosci 31:7831–7839. doi:10.1523/JNEUROSCI.0839-11.2011

Boda B, Alberi S, Nikonenko I et al (2004) The mental retardation protein PAK3 contributes to synapse formation and plasticity in hippocampus. J Neurosci 24:10816–10825. doi:10.1523/JNEUROSCI.2931-04.2004

Bourgin C, Murai KK, Richter M, Pasquale EB (2007) The EphA4 receptor regulates dendritic spine remodeling by affecting beta1-integrin signaling pathways. J Cell Biol 178:1295–1307. doi:10.1083/jcb.200610139

Bourne J, Harris KM (2007) Do thin spines learn to be mushroom spines that remember? Curr Opin Neurobiol 17:381–386. doi:10.1016/j.conb.2007.04.009

Bradley WD, Koleske AJ (2009) Regulation of cell migration and morphogenesis by Abl-family kinases: emerging mechanisms and physiological contexts. J Cell Sci 122:3441–3454. doi:10.1242/jcs.039859

Brody DA, Harris TR, Romans WE (1955) A simple method for obtaining esophageal electrocardiograms of good diagnostic quality. Am Heart J 50:923–926

Brown CE, Aminoltejari K, Erb H et al (2009) In vivo voltage-sensitive dye imaging in adult mice reveals that somatosensory maps lost to stroke are replaced over weeks by new structural and functional circuits with prolonged modes of activation within both the peri-infarct zone and distant sites. J Neurosci 29:1719–1734. doi:10.1523/JNEUROSCI.4249-08.2009

Brown CE, Boyd JD, Murphy TH (2010) Longitudinal in vivo imaging reveals balanced and branch-specific remodeling of mature cortical pyramidal dendritic arbors after stroke. J Cereb Blood Flow Metab 30:783–791. doi:10.1038/jcbfm.2009.241

Burke SN, Barnes CA (2006) Neural plasticity in the ageing brain. Nat Rev Neurosci 7:30–40. doi:10.1038/nrn1809

Burton PR (1988) Dendrites of mitral cell neurons contain microtubules of opposite polarity. Brain Res 473:107–115

Calabrese B, Shaked GM, Tabarean IV et al (2007) Rapid, concurrent alterations in pre- and postsynaptic structure induced by naturally-secreted amyloid-beta protein. Mol Cell Neurosci 35:183–193. doi:10.1016/j.mcn.2007.02.006

Carlyle BC, Nairn AC, Wang M et al (2014) cAMP-PKA phosphorylation of tau confers risk for degeneration in aging association cortex. Proc Natl Acad Sci U S A 111:5036–5041. doi:10.1073/pnas.1322360111

Chen H, Firestein BL (2007) RhoA regulates dendrite branching in hippocampal neurons by decreasing cypin protein levels. J Neurosci 27:8378–8386. doi:10.1523/JNEUROSCI.0872-07.2007

Chen Y-K, Hsueh Y-P (2012) Cortactin-binding protein 2 modulates the mobility of cortactin and regulates dendritic spine formation and maintenance. J Neurosci 32:1043–1055

Chen HJ, Rojas-Soto M, Oguni A, Kennedy MB (1998) A synaptic Ras-GTPase activating protein (p135 SynGAP) inhibited by CaM kinase II. Neuron 20:895–904

Chen B, Brinkmann K, Chen Z et al (2014) The WAVE regulatory complex links diverse receptors to the actin cytoskeleton. Cell 156:195–207. doi:10.1016/j.cell.2013.11.048

Choi M, Wang SE, Ko SY et al (2014) Overexpression of human GATA-1 and GATA-2 interferes with spine formation and produces depressive behavior in rats. PLoS One 9:e109253. doi:10.1371/journal.pone.0109253

Clark WL (1957) Inquiries into the anatomical basis of olfactory discrimination. Proc R Soc Lond B Biol Sci 146:299–319

Cline H, Haas K (2008) The regulation of dendritic arbor development and plasticity by glutamatergic synaptic input: a review of the synaptotrophic hypothesis. J Physiol 586:1509–1517. doi:10.1113/jphysiol.2007.150029

Coleman PD, Riesen AH (1968) Environmental effects on cortical dendritic fields. I. Rearing in the dark. J Anat 102:363–374

Conde C, Cáceres A (2009) Microtubule assembly, organization and dynamics in axons and dendrites. Nat Rev Neurosci 10:319–332

Cotter D, Mackay D, Landau S et al (2001) Reduced glial cell density and neuronal size in the anterior cingulate cortex in major depressive disorder. Arch Gen Psychiatry 58:545–553

Cotter D, Mackay D, Chana G et al (2002) Reduced neuronal size and glial cell density in area 9 of the dorsolateral prefrontal cortex in subjects with major depressive disorder. Cereb Cortex 12:386–394

Counts SE, He B, Nadeem M et al (2012) Hippocampal drebrin loss in mild cognitive impairment. Neurodegener Dis 10:216–219. doi:10.1159/000333122

Courtemanche N, Gifford SM, Simpson MA et al (2015) Abl2/Abl-related gene stabilizes actin filaments, stimulates actin branching by actin-related protein 2/3 complex and promotes actin filament severing by cofilin. J Biol Chem 290:4038–4046. doi:10.1074/jbc.M114.608117

Dailey ME, Smith SJ (1996) The dynamics of dendritic structure in developing hippocampal slices. J Neurosci 16:2983–2994. doi:10.1002/cne.901950106

De Brabander JM, Kramers RJ, Uylings HB (1998) Layer-specific dendritic regression of pyramidal cells with ageing in the human prefrontal cortex. Eur J Neurosci 10:1261–1269

De Camilli P, Miller PE, Navone F et al (1984) Distribution of microtubule-associated protein 2 in the nervous system of the rat studied by immunofluorescence. Neuroscience 11:817–846

De Rubeis S, Pasciuto E, Li KW et al (2013) CYFIP1 coordinates mRNA translation and cytoskeleton remodeling to ensure proper dendritic spine formation. Neuron 79:1169–1182. doi:10.1016/j.neuron.2013.06.039

Drevets WC (2000) Functional anatomical abnormalities in limbic and prefrontal cortical structures in major depression. Prog Brain Res 126:413–431. doi:10.1016/S0079-6123(00)26027-5

Drevets WC, Price JL, Simpson JRJ et al (1997) Subgenual prefrontal cortex abnormalities in mood disorders. Nature 386:824–827. doi:10.1038/386824a0

Du Y, Weed SA, Xiong WC et al (1998) Identification of a novel cortactin SH3 domain-binding protein and its localization to growth cones of cultured neurons. Mol Cell Biol 18:5838–5851

Dubos A, Combeau G, Bernardinelli Y et al (2012) Alteration of synaptic network dynamics by the intellectual disability protein PAK3. J Neurosci 32:519–527. doi:10.1523/JNEUROSCI.3252-11.2012

Dumitriu D, Hao J, Hara Y et al (2010) Selective changes in thin spine density and morphology in monkey prefrontal cortex correlate with aging-related cognitive impairment. J Neurosci 30:7507–7515. doi:10.1523/JNEUROSCI.6410-09.2010

Edson KJ, Lim SS, Borisy GG, Letourneau PC (1993) FRAP analysis of the stability of the microtubule population along the neurites of chick sensory neurons. Cell Motil Cytoskeleton 25:59–72. doi:10.1002/cm.970250108

El-Husseini AE-D, Craven SE, Brock SC, Bredt DS (2001) Polarized targeting of peripheral membrane proteins in neurons. J Biol Chem 276:44984–44992

Falke E, Nissanov J, Mitchell TW et al (2003) Subicular dendritic arborization in Alzheimer's disease correlates with neurofibrillary tangle density. Am J Pathol 163:1615–1621. doi:10.1016/S0002-9440(10)63518-3

Fifková E, Delay RJ (1982) Cytoplasmic actin in neuronal processes as a possible mediator of synaptic plasticity. J Cell Biol 95:345–350
Fischer M, Kaech S, Knutti D, Matus A (1998) Rapid actin-based plasticity in dendritic spines. Neuron 20:847–854
Fleming IN, Elliott CM, Buchanan FG et al (1999) Ca2+/calmodulin-dependent protein kinase II regulates Tiam1 by reversible protein phosphorylation. J Biol Chem 274:12753–12758
Flood DG (1991) Region-specific stability of dendritic extent in normal human aging and regression in Alzheimer's disease. II. Subiculum. Brain Res 540:83–95
Flood DG, Buell SJ, Horwitz GJ, Coleman PD (1987a) Dendritic extent in human dentate gyrus granule cells in normal aging and senile dementia. Brain Res 402:205–216
Flood DG, Guarnaccia M, Coleman PD (1987b) Dendritic extent in human CA2-3 hippocampal pyramidal neurons in normal aging and senile dementia. Brain Res 409:88–96
Freir DB, Nicoll AJ, Klyubin I et al (2011) Interaction between prion protein and toxic amyloid β assemblies can be therapeutically targeted at multiple sites. Nat Commun 2:336. doi:10.1038/ncomms1341
Ganeshina O, Berry RW, Petralia RS et al (2004a) Synapses with a segmented, completely partitioned postsynaptic density express more AMPA receptors than other axospinous synaptic junctions. Neuroscience 125:615–623. doi:10.1016/j.neuroscience.2004.02.025
Ganeshina O, Berry RW, Petralia RS et al (2004b) Differences in the expression of AMPA and NMDA receptors between axospinous perforated and nonperforated synapses are related to the configuration and size of postsynaptic densities. J Comp Neurol 468:86–95. doi:10.1002/cne.10950
Gazzaley A, D'Esposito M (2007) Top-down modulation and normal aging. Ann N Y Acad Sci 1097:67–83. doi:10.1196/annals.1379.010
Geinisman Y, de Toledo-Morrell L, Morrell F (1986) Loss of perforated synapses in the dentate gyrus: morphological substrate of memory deficit in aged rats. Proc Natl Acad Sci U S A 83:3027–3031
Geinisman Y, deToledo-Morrell L, Morrell F et al (1993) Structural synaptic correlate of long-term potentiation: formation of axospinous synapses with multiple, completely partitioned transmission zones. Hippocampus 3:435–445. doi:10.1002/hipo.450030405
Geinisman Y, Ganeshina O, Yoshida R et al (2004) Aging, spatial learning, and total synapse number in the rat CA1 stratum radiatum. Neurobiol Aging 25:407–416. doi:10.1016/j.neurobiolaging.2003.12.001
Giannakopoulos P, Herrmann FR, Bussière T et al (2003) Tangle and neuron numbers, but not amyloid load, predict cognitive status in Alzheimer's disease. Neurology 60:1495–1500
Gong Y, Chang L, Viola KL et al (2003) Alzheimer's disease-affected brain: presence of oligomeric A beta ligands (ADDLs) suggests a molecular basis for reversible memory loss. Proc Natl Acad Sci U S A 100:10417–10422. doi:10.1073/pnas.1834302100
Gorski JA, Zeiler SR, Tamowski S, Jones KR (2003) Brain-derived neurotrophic factor is required for the maintenance of cortical dendrites. J Neurosci 23:6856–6865
Gourley SL, Kedves AT, Olausson P, Taylor JR (2009a) A history of corticosterone exposure regulates fear extinction and cortical NR2B, GluR2/3, and BDNF. Neuropsychopharm: Off Publ Am Coll Neuropsychopharm 34:707–716. doi:10.1038/npp.2008.123
Gourley SL, Koleske AJ, Taylor JR (2009b) Loss of dendrite stabilization by the Abl-related gene (Arg) kinase regulates behavioral flexibility and sensitivity to cocaine. Proc Natl Acad Sci U S A 106:16859–16864. doi:10.1073/pnas.0902286106
Gourley SL, Olevska A, Warren MS et al (2012) Arg kinase regulates prefrontal dendritic spine refinement and cocaine-induced plasticity. J Neurosci 32:2314–2323. doi:10.1523/JNEUROSCI.2730-11.2012
Govek EE, Newey SE, Van Aelst L (2005) The role of the Rho GTPases in neuronal development. Genes Dev 19:1–49. doi:10.1101/gad.1256405

Graber S, Maiti S, Halpain S (2004) Cathepsin B-like proteolysis and MARCKS degradation in sub-lethal NMDA-induced collapse of dendritic spines. Neuropharmacology 47:706–713. doi:10.1016/j.neuropharm.2004.08.004

Grintsevich EE, Reisler E (2014) Drebrin inhibits cofilin-induced severing of F-actin. Cytoskeleton (Hoboken) 71:472–483. doi:10.1002/cm.21184

Gu J, Firestein BL, Zheng JQ (2008) Microtubules in dendritic spine development. J Neurosci 28:12120–12124. doi:10.1523/JNEUROSCI.2509-08.2008

Haass C, Selkoe DJ (2007) Soluble protein oligomers in neurodegeneration: lessons from the Alzheimer's amyloid beta-peptide. Nat Rev Mol Cell Biol 8:101–112. doi:10.1038/nrm2101

Haeckel A, Ahuja R, Gundelfinger ED et al (2008) The actin-binding protein Abp1 controls dendritic spine morphology and is important for spine head and synapse formation. J Neurosci 28:10031–10044. doi:10.1523/JNEUROSCI.0336-08.2008

Halpain S, Hipolito A, Saffer L (1998) Regulation of F-actin stability in dendritic spines by glutamate receptors and calcineurin. J Neurosci 18:9835–9844

Hanks SD, Flood DG (1991) Region-specific stability of dendritic extent in normal human aging and regression in Alzheimer's disease. I. CA1 of hippocampus. Brain Res 540:63–82

Hara Y, Park CS, Janssen WGM et al (2011) Synaptic characteristics of dentate gyrus axonal boutons and their relationships with aging, menopause, and memory in female rhesus monkeys. J Neurosci 31:7737–7744. doi:10.1523/JNEUROSCI.0822-11.2011

Harada A, Teng J, Takei Y et al (2002) MAP2 is required for dendrite elongation, PKA anchoring in dendrites, and proper PKA signal transduction. J Cell Biol 158:541–549. doi:10.1083/jcb.200110134

Harigaya Y, Shoji M, Shirao T, Hirai S (1996) Disappearance of actin-binding protein, drebrin, from hippocampal synapses in Alzheimer's disease. J Neurosci Res 43:87–92. doi:10.1002/jnr.490430111

Harris KM (1999) Structure, development, and plasticity of dendritic spines. Curr Opin Neurobiol 9:343–348

Harris KM, Kater SB (1994) Dendritic spines: cellular specializations imparting both stability and flexibility to synaptic function. Annu Rev Neurosci 17:341–371. doi:10.1146/annurev.ne.17.030194.002013

Hartmann M, Heumann R, Lessmann V (2001) Synaptic secretion of BDNF after high-frequency stimulation of glutamatergic synapses. EMBO J 20:5887–5897. doi:10.1093/emboj/20.21.5887

Hasbani MJ, Hyrc KL, Faddis BT et al (1998) Distinct roles for sodium, chloride, and calcium in excitotoxic dendritic injury and recovery. Exp Neurol 154:241–258. doi:10.1006/exnr.1998.6929

Hasbani MJ, Schlief ML, Fisher DA, Goldberg MP (2001) Dendritic spines lost during glutamate receptor activation reemerge at original sites of synaptic contact. J Neurosci 21:2393–2403

Hayashi K, Shirao T (1999) Change in the shape of dendritic spines caused by overexpression of drebrin in cultured cortical neurons. J Neurosci 19:3918–3925

Hazai D, Szudoczki R, Ding J et al (2013) Ultrastructural abnormalities in CA1 hippocampus caused by deletion of the actin regulator WAVE-1. PLoS One 8:e75248. doi:10.1371/journal.pone.0075248

Hering H, Sheng M (2003) Activity-dependent redistribution and essential role of cortactin in dendritic spine morphogenesis. J Neurosci 23:11759–11769

Hirokawa N, Takemura R (2005) Molecular motors and mechanisms of directional transport in neurons. Nat Rev Neurosci 6:201–214

Holtmaat AJGD, Trachtenberg JT, Wilbrecht L et al (2005) Transient and persistent dendritic spines in the neocortex in vivo. Neuron 45:279–291. doi:10.1016/j.neuron.2005.01.003

Holtmaat A, Wilbrecht L, Knott GW et al (2006) Experience-dependent and cell-type-specific spine growth in the neocortex. Nature 441:979–983. doi:10.1038/nature04783

Honkura N, Matsuzaki M, Noguchi J et al (2008) The subspine organization of actin fibers regulates the structure and plasticity of dendritic spines. Neuron 57:719–729. doi:10.1016/j.neuron.2008.01.013

Hoover BR, Reed MN, Su J et al (2010) Tau mislocalization to dendritic spines mediates synaptic dysfunction independently of neurodegeneration. Neuron 68:1067–1081. doi:10.1016/j.neuron.2010.11.030

Hotulainen P, Hoogenraad CC (2010) Actin in dendritic spines: connecting dynamics to function. J Cell Biol 189:619–629. doi:10.1083/jcb.201003008

Hu B, Nikolakopoulou AM, Cohen-Cory S (2005) BDNF stabilizes synapses and maintains the structural complexity of optic axons in vivo. Development 132:4285–4298. doi:10.1242/dev.02017

Hu X, Viesselmann C, Nam S et al (2008) Activity-dependent dynamic microtubule invasion of dendritic spines. J Neurosci 28:13094–13105. doi:10.1523/JNEUROSCI.3074-08.2008

Hu X, Ballo L, Pietila L et al (2011) BDNF-induced increase of PSD-95 in dendritic spines requires dynamic microtubule invasions. J Neurosci 31:15597–15603. doi:10.1523/JNEUROSCI.2445-11.2011

Huber G, Matus A (1984) Differences in the cellular distributions of two microtubule-associated proteins, MAP1 and MAP2, in rat brain. J Neurosci 4:151–160

Huttenlocher PR (1979) Synaptic density in human frontal cortex – developmental changes and effects of aging. Brain Res 163:195–205

Huttenlocher PR (1990) Morphometric study of human cerebral cortex development. Neuropsychologia 28:517–527

Iki J, Inoue A, Bito H, Okabe S (2005) Bi-directional regulation of postsynaptic cortactin distribution by BDNF and NMDA receptor activity. Eur J Neurosci 22:2985–2994. doi:10.1111/j.1460-9568.2005.04510.x

Ingelsson M, Fukumoto H, Newell KL et al (2004) Early Abeta accumulation and progressive synaptic loss, gliosis, and tangle formation in AD brain. Neurology 62:925–931

Ishikawa R, Hayashi K, Shirao T et al (1994) Drebrin, a development-associated brain protein from rat embryo, causes the dissociation of tropomyosin from actin filaments. J Biol Chem 269:29928–29933

Ishizuka Y, Shimizu H, Takagi E et al (2014) Histone deacetylase mediates the decrease in drebrin cluster density induced by amyloid beta oligomers. Neurochem Int 76:114–121. doi:10.1016/j.neuint.2014.07.005

Ittner LM, Ke YD, Delerue F et al (2010) Dendritic function of tau mediates amyloid-beta toxicity in Alzheimer's disease mouse models. Cell 142:387–397. doi:10.1016/j.cell.2010.06.036

Izeddin I, Specht CG, Lelek M et al (2011) Super-resolution dynamic imaging of dendritic spines using a low-affinity photoconvertible actin probe. PLoS One 6:e15611. doi:10.1371/journal.pone.0015611

Jaworski J, Kapitein LC, Gouveia SM et al (2009) Dynamic microtubules regulate dendritic spine morphology and synaptic plasticity. Neuron 61:85–100. doi:10.1016/j.neuron.2008.11.013

Jones WH, Thomas DB (1962) Changes in the dendritic organization of neurons in the cerebral cortex following deafferentation. J Anat 96:375–381

Jung Y, Mulholland PJ, Wiseman SL et al (2013) Constitutive knockout of the membrane cytoskeleton protein beta adducin decreases mushroom spine density in the nucleus accumbens but does not prevent spine remodeling in response to cocaine. Eur J Neurosci 37:1–9. doi:10.1111/ejn.12037

Kalia LV, Kalia SK, Salter MW (2008) NMDA receptors in clinical neurology: excitatory times ahead. Lancet Neurol 7:742–755. doi:10.1016/S1474-4422(08)70165-0

Kanai Y, Dohmae N, Hirokawa N (2004) Kinesin transports RNA: isolation and characterization of an RNA-transporting granule. Neuron 43:513–525

Kang MG, Guo Y, Huganir RL (2009) AMPA receptor and GEF-H1/Lfc complex regulates dendritic spine development through RhoA signaling cascade. Proc Natl Acad Sci U S A 106:3549–3554. doi:10.1073/pnas.0812861106

Kang HJ, Voleti B, Hajszan T et al (2012) Decreased expression of synapse-related genes and loss of synapses in major depressive disorder. Nat Med 18:1413–1417. doi:10.1038/nm.2886

Kapitein LC, Yau KW, Gouveia SM et al (2011) NMDA receptor activation suppresses microtubule growth and spine entry. J Neurosci 31:8194–8209. doi:10.1523/JNEUROSCI.6215-10.2011

Kerrisk ME, Greer CA, Koleske AJ (2013) Integrin alpha3 is required for late postnatal stability of dendrite arbors, dendritic spines and synapses, and mouse behavior. J Neurosci 33:6742–6752. doi:10.1523/JNEUROSCI.0528-13.2013

Kim Y, Sung JY, Ceglia I et al (2006) Phosphorylation of WAVE1 regulates actin polymerization and dendritic spine morphology. Nature 442:814–817. doi:10.1038/nature04976

Kim IH, Rácz B, Wang H et al (2013a) Disruption of Arp2/3 results in asymmetric structural plasticity of dendritic spines and progressive synaptic and behavioral abnormalities. J Neurosci 33:6081–6092. doi:10.1523/JNEUROSCI.0035-13.2013

Kim T, Vidal GS, Djurisic M et al (2013b) Human LilrB2 is a β-amyloid receptor and its murine homolog PirB regulates synaptic plasticity in an Alzheimer's model. Science 341:1399–1404. doi:10.1126/science.1242077

Kitaoka Y, Kitaoka Y, Kumai T et al (2004) Involvement of RhoA and possible neuroprotective effect of fasudil, a Rho kinase inhibitor, in NMDA-induced neurotoxicity in the rat retina. Brain Res 1018:111–118. doi:10.1016/j.brainres.2004.05.070

Klein R (2009) Bidirectional modulation of synaptic functions by Eph/ephrin signaling. Nat Neurosci 12:15–20

Knudsen KA, Soler AP, Johnson KR, Wheelock MJ (1995) Interaction of alpha-actinin with the cadherin/catenin cell-cell adhesion complex via alpha-catenin. J Cell Biol 130:67–77

Kohara K, Kitamura A, Morishima M, Tsumoto T (2001) Activity-dependent transfer of brain-derived neurotrophic factor to postsynaptic neurons. Science 291:2419–2423. doi:10.1126/science.1057415

Kojima M, Takei N, Numakawa T et al (2001) Biological characterization and optical imaging of brain-derived neurotrophic factor-green fluorescent protein suggest an activity-dependent local release of brain-derived neurotrophic factor in neurites of cultured hippocampal neurons. J Neurosci Res 64:1–10

Koleske AJ, Gifford AM, Scott ML et al (1998) Essential roles for the Abl and Arg tyrosine kinases in neurulation. Neuron 21:1259–1272

Kopeikina KJ, Hyman BT, Spires-Jones TL (2012) Soluble forms of tau are toxic in Alzheimer's disease. Transl Neurosci 3:223–233. doi:10.2478/s13380-012-0032-y

Korobova F, Svitkina T (2010) Molecular architecture of synaptic actin cytoskeleton in hippocampal neurons reveals a mechanism of dendritic spine morphogenesis. Mol Biol Cell 21:165–176. doi:10.1091/mbc.E09-07-0596

Kreis P, Thévenot E, Rousseau V et al (2007) The p21-activated kinase 3 implicated in mental retardation regulates spine morphogenesis through a Cdc42-dependent pathway. J Biol Chem 282:21497–21506. doi:10.1074/jbc.M703298200

Kulkarni VA, Firestein BL (2012) The dendritic tree and brain disorders. Mol Cell Neurosci 50:10–20. doi:10.1016/j.mcn.2012.03.005

Kwan AC, Dombeck DA, Webb WW (2008) Polarized microtubule arrays in apical dendrites and axons. Proc Natl Acad Sci U S A 105:11370–11375

Kwon M, Fernández JR, Zegarek GF et al (2011) BDNF-promoted increases in proximal dendrites occur via CREB-dependent transcriptional regulation of cypin. J Neurosci 31:9735–9745. doi:10.1523/JNEUROSCI.6785-10.2011

Lacor PN, Buniel MC, Furlow PW et al (2007) Abeta oligomer-induced aberrations in synapse composition, shape, and density provide a molecular basis for loss of connectivity in Alzheimer's disease. J Neurosci 27:796–807. doi:10.1523/JNEUROSCI.3501-06.2007

Lai K-O, Wong ASL, Cheung M-C et al (2012) TrkB phosphorylation by Cdk5 is required for activity-dependent structural plasticity and spatial memory. Nat Neurosci 15:1506–1515. doi:10.1038/nn.3237

Larson M, Sherman MA, Amar F et al (2012) The complex PrP(c)-Fyn couples human oligomeric Aβ with pathological tau changes in Alzheimer's disease. J Neurosci 32:16857–71a. doi:10.1523/JNEUROSCI.1858-12.2012

Lemieux M, Labrecque S, Tardif C et al (2012) Translocation of CaMKII to dendritic microtubules supports the plasticity of local synapses. J Cell Biol 198:1055–1073. doi:10.1083/jcb.201202058

Lin Y-C, Yeckel MF, Koleske AJ (2013) Abl2/Arg controls dendritic spine and dendrite arbor stability via distinct cytoskeletal control pathways. J Neurosci 33:1846–1857. doi:10.1523/JNEUROSCI.4284-12.2013

MacGrath SM, Koleske AJ (2012a) Arg/Abl2 modulates the affinity and stoichiometry of binding of cortactin to f-actin. Biochemistry 51:6644–6653. doi:10.1021/bi300722t

MacGrath SM, Koleske AJ (2012b) Cortactin in cell migration and cancer at a glance. J Cell Sci 125:1621–1626. doi:10.1242/jcs.093781

Machesky LM, Insall RH (1998) Scar1 and the related Wiskott-Aldrich syndrome protein, WASP, regulate the actin cytoskeleton through the Arp2/3 complex. Curr Biol 8:1347–1356

Machesky LM, Mullins RD, Higgs HN et al (1999) Scar, a WASp-related protein, activates nucleation of actin filaments by the Arp2/3 complex. Proc Natl Acad Sci U S A 96:3739–3744

Majewska AK, Newton JR, Sur M (2006) Remodeling of synaptic structure in sensory cortical areas in vivo. J Neurosci 26:3021–3029. doi:10.1523/JNEUROSCI.4454-05.2006

Markus EJ, Petit TL (1987) Neocortical synaptogenesis, aging, and behavior: lifespan development in the motor-sensory system of the rat. Exp Neurol 96:262–278

Marshak S, Nikolakopoulou AM, Dirks R et al (2007) Cell-autonomous TrkB signaling in presynaptic retinal ganglion cells mediates axon arbor growth and synapse maturation during the establishment of retinotectal synaptic connectivity. J Neurosci 27:2444–2456. doi:10.1523/JNEUROSCI.4434-06.2007

Matsuoka Y, Li X, Bennett V (1998) Adducin is an in vivo substrate for protein kinase C: phosphorylation in the MARCKS-related domain inhibits activity in promoting spectrin-actin complexes and occurs in many cells, including dendritic spines of neurons. J Cell Biol 142:485–497

Matsuoka Y, Li X, Bennett V (2000) Adducin: structure, function and regulation. Cell Mol Life Sci 57:884–895

Matsuzaki M, Honkura N, Ellis-Davies GCR, Kasai H (2004) Structural basis of long-term potentiation in single dendritic spines. Nature 429:761–766. doi:10.1038/nature02617

Matthews MR, Powell TP (1962) Some observations on transneuronal cell degeneration in the olfactory bulb of the rabbit. J Anat 96:89–102

Matus A, Ackermann M, Pehling G et al (1982) High actin concentrations in brain dendritic spines and postsynaptic densities. Proc Natl Acad Sci U S A 79:7590–7594

Mikati MA, Grintsevich EE, Reisler E (2013) Drebrin-induced stabilization of actin filaments. J Biol Chem 288:19926–19938. doi:10.1074/jbc.M113.472647

Mooren OL, Galletta BJ, Cooper JA (2012) Roles for actin assembly in endocytosis. Annu Rev Biochem 81:661–686. doi:10.1146/annurev-biochem-060910-094416

Moresco EM, Scheetz AJ, Bornmann WG et al (2003) Abl family nonreceptor tyrosine kinases modulate short-term synaptic plasticity. J Neurophysiol 89:1678–1687. doi:10.1152/jn.0892.2002

Morrison JH, Baxter MG (2012) The ageing cortical synapse: hallmarks and implications for cognitive decline. Nat Rev Neurosci 13:240–250. doi:10.1038/nrn3200

Morrison JH, Hof PR (1997) Life and death of neurons in the aging brain. Science 278:412–419

Mostany R, Portera-Cailliau C (2011) Absence of large-scale dendritic plasticity of layer 5 pyramidal neurons in peri-infarct cortex. J Neurosci 31:1734–1738. doi:10.1523/JNEUROSCI.4386-10.2011

Mostany R, Anstey JE, Crump KL et al (2013) Altered synaptic dynamics during normal brain aging. J Neurosci 33:4094–4104. doi:10.1523/JNEUROSCI.4825-12.2013

Murakoshi H, Wang H, Yasuda R (2011) Local, persistent activation of Rho GTPases during plasticity of single dendritic spines. Nature 472:100–104. doi:10.1038/nature09823

Murthy AS, Flavin M (1983) Microtubule assembly using the microtubule-associated protein MAP-2 prepared in defined states of phosphorylation with protein kinase and phosphatase. Eur J Biochem 137:37–46

Naisbitt S, Kim E, Tu JC et al (1999) Shank, a novel family of postsynaptic density proteins that binds to the NMDA receptor/PSD-95/GKAP complex and cortactin. Neuron 23:569–582

Nakayama AY, Luo L (2000) Intracellular signaling pathways that regulate dendritic spine morphogenesis. Hippocampus 10:582–586

Nibuya M, Takahashi M, Russell DS, Duman RS (1999) Repeated stress increases catalytic TrkB mRNA in rat hippocampus. Neurosci Lett 267:81–84

Nicholson DA, Yoshida R, Berry RW et al (2004) Reduction in size of perforated postsynaptic densities in hippocampal axospinous synapses and age-related spatial learning impairments. J Neurosci 24:7648–7653. doi:10.1523/JNEUROSCI.1725-04.2004

Niell CM, Meyer MP, Smith SJ (2004) In vivo imaging of synapse formation on a growing dendritic arbor. Nat Neurosci 7:254–260. doi:10.1038/nn1191

Okabe S, Hirokawa N (1990) Turnover of fluorescently labelled tubulin and actin in the axon. Nature 343:479–482. doi:10.1038/343479a0

Okamoto K-I, Nagai T, Miyawaki A, Hayashi Y (2004) Rapid and persistent modulation of actin dynamics regulates postsynaptic reorganization underlying bidirectional plasticity. Nat Neurosci 7:1104–1112. doi:10.1038/nn1311

Oray S, Majewska A, Sur M (2004) Dendritic spine dynamics are regulated by monocular deprivation and extracellular matrix degradation. Neuron 44:1021–1030. doi:10.1016/j.neuron.2004.12.001

Peris L, Wagenbach M, Lafanechère L et al (2009) Motor-dependent microtubule disassembly driven by tubulin tyrosination. J Cell Biol 185:1159–1166. doi:10.1083/jcb.200902142

Peters A, Sethares C, Moss MB (1998) The effects of aging on layer 1 in area 46 of prefrontal cortex in the rhesus monkey. Cereb Cortex 8:671–684

Peters A, Sethares C, Luebke JI (2008) Synapses are lost during aging in the primate prefrontal cortex. Neuroscience 152:970–981. doi:10.1016/j.neuroscience.2007.07.014

Pinkstaff JK, Detterich J, Lynch G, Gall C (1999) Integrin subunit gene expression is regionally differentiated in adult brain. J Neurosci 19:1541–1556

Pollard TD, Borisy GG (2003) Cellular motility driven by assembly and disassembly of actin filaments. Cell 112:453–465

Pozueta J, Lefort R, Shelanski ML (2013) Synaptic changes in Alzheimer's disease and its models. Neuroscience 251:51–65. doi:10.1016/j.neuroscience.2012.05.050

Rácz B, Weinberg RJ (2004) The subcellular organization of cortactin in hippocampus. J Neurosci 24:10310–10317. doi:10.1523/JNEUROSCI.2080-04.2004

Rácz B, Weinberg RJ (2006) Spatial organization of cofilin in dendritic spines. Neuroscience 138:447–456. doi:10.1016/j.neuroscience.2005.11.025

Rácz B, Weinberg RJ (2008) Organization of the Arp2/3 complex in hippocampal spines. J Neurosci 28:5654–5659. doi:10.1523/JNEUROSCI.0756-08.2008

Rácz B, Weinberg RJ (2013) Microdomains in forebrain spines: an ultrastructural perspective. Mol Neurobiol 47:77–89

Rajan I, Witte S, Cline HT (1999) NMDA receptor activity stabilizes presynaptic retinotectal axons and postsynaptic optic tectal cell dendrites in vivo. J Neurobiol 38:357–368. doi:10.1002/(SICI)1097-4695(19990215)38:3<357::AID-NEU5>3.0.CO;2-#

Rakic P, Bourgeois JP, Eckenhoff MF et al (1986) Concurrent overproduction of synapses in diverse regions of the primate cerebral cortex. Science 232:232–235

Rakic P, Bourgeois JP, Goldman-Rakic PS (1994) Synaptic development of the cerebral cortex: implications for learning, memory, and mental illness. Prog Brain Res 102:227–243. doi:10.1016/S0079-6123(08)60543-9

Ray LB, Sturgill TW (1987) Rapid stimulation by insulin of a serine/threonine kinase in 3T3-L1 adipocytes that phosphorylates microtubule-associated protein 2 in vitro. Proc Natl Acad Sci U S A 84:1502–1506
Rex CS, Lin C-Y, Kramár EA et al (2007) Brain-derived neurotrophic factor promotes long-term potentiation-related cytoskeletal changes in adult hippocampus. J Neurosci 27:3017–3029. doi:10.1523/JNEUROSCI.4037-06.2007
Sánchez C, Díaz-Nido J, Avila J (2000) Phosphorylation of microtubule-associated protein 2 (MAP2) and its relevance for the regulation of the neuronal cytoskeleton function. Prog Neurobiol 61:133–168
Sasaki S, Stevens JK, Bodick N (1983) Serial reconstruction of microtubular arrays within dendrites of the cat retinal ganglion cell: the cytoskeleton of a vertebrate dendrite. Brain Res 259:193–206
Sasaki Y, Hayashi K, Shirao T et al (1996) Inhibition by drebrin of the actin-bundling activity of brain fascin, a protein localized in filopodia of growth cones. J Neurochem 66:980–988
Schmidt HD, Duman RS (2007) The role of neurotrophic factors in adult hippocampal neurogenesis, antidepressant treatments and animal models of depressive-like behavior. Behav Pharmacol 18:391–418. doi:10.1097/FBP.0b013e3282ee2aa8
Schulman H (1984) Phosphorylation of microtubule-associated proteins by a Ca2+/calmodulin-dependent protein kinase. J Cell Biol 99:11–19
Seese RR, Babayan AH, Katz AM et al (2012) LTP induction translocates cortactin at distant synapses in wild-type but not Fmr1 knock-out mice. J Neurosci 32:7403–7413. doi:10.1523/JNEUROSCI.0968-12.2012
Setou M, Nakagawa T, Seog DH, Hirokawa N (2000) Kinesin superfamily motor protein KIF17 and mLin-10 in NMDA receptor-containing vesicle transport. Science 288:1796–1802
Setou M, Seog DH, Tanaka Y et al (2002) Glutamate-receptor-interacting protein GRIP1 directly steers kinesin to dendrites. Nature 417:83–87
Sfakianos MK, Eisman A, Gourley SL et al (2007) Inhibition of Rho via Arg and p190RhoGAP in the postnatal mouse hippocampus regulates dendritic spine maturation, synapse and dendrite stability, and behavior. J Neurosci 27:10982–10992. doi:10.1523/JNEUROSCI.0793-07.2007
Shankar GM, Bloodgood BL, Townsend M et al (2007) Natural oligomers of the Alzheimer amyloid-beta protein induce reversible synapse loss by modulating an NMDA-type glutamate receptor-dependent signaling pathway. J Neurosci 27:2866–2875. doi:10.1523/JNEUROSCI.4970-06.2007
Shi Y, Pontrello CG, DeFea KA et al (2009) Focal adhesion kinase acts downstream of EphB receptors to maintain mature dendritic spines by regulating cofilin activity. J Neurosci 29:8129–8142
Shirao T, Kojima N, Kato Y, Obata K (1988) Molecular cloning of a cDNA for the developmentally regulated brain protein, drebrin. Brain Res 464:71–74
Simpson MA, Bradley WD, Harburger D et al (2015) Direct interactions with the integrin β 1 cytoplasmic tail activate the Abl2/Arg kinase. J Biol Chem 290:8360–8372. doi:10.1074/jbc.M115.638874
Smith MA, Makino S, Kvetnansky R, Post RM (1995) Stress and glucocorticoids affect the expression of brain-derived neurotrophic factor and neurotrophin-3 mRNAs in the hippocampus. J Neurosci 15:1768–1777
Smith TD, Adams MM, Gallagher M et al (2000) Circuit-specific alterations in hippocampal synaptophysin immunoreactivity predict spatial learning impairment in aged rats. J Neurosci 20:6587–6593
Soderling SH, Guire ES, Kaech S et al (2007) A WAVE-1 and WRP signaling complex regulates spine density, synaptic plasticity, and memory. J Neurosci 27:355–365. doi:10.1523/JNEUROSCI.3209-06.2006
Star EN, Kwiatkowski DJ, Murthy VN (2002) Rapid turnover of actin in dendritic spines and its regulation by activity. Nat Neurosci 5:239–246. doi:10.1038/nn811

Sudo H, Baas PW (2010) Acetylation of microtubules influences their sensitivity to severing by katanin in neurons and fibroblasts. J Neurosci 30:7215–7226. doi:10.1523/JNEUROSCI.0048-10.2010

Szebenyi G, Bollati F, Bisbal M et al (2005) Activity-driven dendritic remodeling requires microtubule-associated protein 1A. Curr Biol 15:1820–1826. doi:10.1016/j.cub.2005.08.069

Takahashi H, Sekino Y, Tanaka S et al (2003) Drebrin-dependent actin clustering in dendritic filopodia governs synaptic targeting of postsynaptic density-95 and dendritic spine morphogenesis. J Neurosci 23:6586–6595

Takemoto-Kimura S, Suzuki K, Kamijo S et al (2010) Differential roles for CaM kinases in mediating excitation-morphogenesis coupling during formation and maturation of neuronal circuits. Eur J Neurosci 32:224–230. doi:10.1111/j.1460-9568.2010.07353.x

Tanokashira D, Morita T, Hayashi K et al (2012) Glucocorticoid suppresses dendritic spine development mediated by down-regulation of caldesmon expression. J Neurosci 32:14583–14591. doi:10.1523/JNEUROSCI.2380-12.2012

Tashiro A, Yuste R (2003) Structure and molecular organization of dendritic spines. Histol Histopathol 18:617–634

Tashiro A, Minden A, Yuste R (2000) Regulation of dendritic spine morphology by the rho family of small GTPases: antagonistic roles of Rac and Rho. Cereb Cortex 10:927–938

Tatavarty V, Kim E-J, Rodionov V, Yu J (2009) Investigating sub-spine actin dynamics in rat hippocampal neurons with super-resolution optical imaging. PLoS One 4:e7724. doi:10.1371/journal.pone.0007724

Teng J, Takei Y, Harada A et al (2001) Synergistic effects of MAP2 and MAP1B knockout in neuronal migration, dendritic outgrowth, and microtubule organization. J Cell Biol 155:65–76. doi:10.1083/jcb.200106025

Terry RD, Masliah E, Salmon DP et al (1991) Physical basis of cognitive alterations in Alzheimer's disease: synapse loss is the major correlate of cognitive impairment. Ann Neurol 30:572–580. doi:10.1002/ana.410300410

Tomar A, Schlaepfer DD (2009) Focal adhesion kinase: switching between GAPs and GEFs in the regulation of cell motility. Curr Opin Cell Biol 21:676–683. doi:10.1016/j.ceb.2009.05.006. PMID: 19525103

Tomar A, Lawson C, Ghassemian M, Schlaepfer DD (2012) Cortactin as a target for FAK in the regulation of focal adhesion dynamics. PLoS One 7:e44041. doi:10.1371/journal.pone.0044041. PMID: 22952866

Toni N, Buchs PA, Nikonenko I et al (2001) Remodeling of synaptic membranes after induction of long-term potentiation. J Neurosci 21:6245–6251

Tønnesen J, Katona G, Rózsa B, Nägerl UV (2014) Spine neck plasticity regulates compartmentalization of synapses. Nat Neurosci 17:678–685. doi:10.1038/nn.3682

Trachtenberg JT, Chen BE, Knott GW et al (2002) Long-term in vivo imaging of experience-dependent synaptic plasticity in adult cortex. Nature 420:788–794. doi:10.1038/nature01273

Tseng CY, Firestein BL (2011) The role of PSD-95 and cypin in morphological changes in dendrites following sublethal NMDA exposure. J Neurosci 31:15468–15480. doi:10.1523/JNEUROSCI.2442-11.2011

Um JW, Nygaard HB, Heiss JK et al (2012) Alzheimer amyloid-β oligomer bound to postsynaptic prion protein activates Fyn to impair neurons. Nat Neurosci 15:1227–1235. doi:10.1038/nn.3178

Uruno T, Liu J, Zhang P et al (2001) Activation of Arp2/3 complex-mediated actin polymerization by cortactin. Nat Cell Biol 3:259–266. doi:10.1038/35060051

Uylings HBM, De Brabander JM (2002) Neuronal changes in normal human aging and Alzheimer's disease. Brain Cogn 49:268–276

Vaillant AR, Zanassi P, Walsh GS et al (2002) Signaling mechanisms underlying reversible, activity-dependent dendrite formation. Neuron 34:985–998

Vaughn JE (1989) Fine structure of synaptogenesis in the vertebrate central nervous system. Synapse 3:255–285. doi:10.1002/syn.890030312

von Bohlen und Halbach O, Minichiello L, Unsicker K (2003) Haploinsufficiency in trkB and/or trkC neurotrophin receptors causes structural alterations in the aged hippocampus and amygdala. Eur J Neurosci 18:2319–2325
Wang HY, Lee DH, Davis CB, Shank RP (2000) Amyloid peptide Abeta(1–42) binds selectively and with picomolar affinity to alpha7 nicotinic acetylcholine receptors. J Neurochem 75:1155–1161
Wang Y, Miller AL, Mooseker MS, Koleske AJ (2001) The Abl-related gene (Arg) nonreceptor tyrosine kinase uses two F-actin-binding domains to bundle F-actin. Proc Natl Acad Sci U S A 98:14865–14870
Wang Y, Shibasaki F, Mizuno K (2005) Calcium signal-induced cofilin dephosphorylation is mediated by Slingshot via calcineurin. J Biol Chem 280:12683–12689. doi:10.1074/jbc.M411494200
Warren MS, Bradley WD, Gourley SL et al (2012) Integrin beta1 signals through Arg to regulate postnatal dendritic arborization, synapse density, and behavior. J Neurosci 32:2824–2834. doi:10.1523/JNEUROSCI.3942-11.2012
Weaver AM, Karginov AV, Kinley AW et al (2001) Cortactin promotes and stabilizes Arp2/3-induced actin filament network formation. Curr Biol 11:370–374
Weed SA, Karginov AV, Schafer DA et al (2000) Cortactin localization to sites of actin assembly in lamellipodia requires interactions with F-actin and the Arp2/3 complex. J Cell Biol 151:29–40
West MJ, Coleman PD, Flood DG, Troncoso JC (1994) Differences in the pattern of hippocampal neuronal loss in normal ageing and Alzheimer's disease. Lancet 344:769–772
Wiesel TN, Hubel DH (1963) Effects of visual deprivation on morphology and physiology of cells in the cats lateral geniculate body. J Neurophysiol 26:978–993
Wong WT, Wong RO (2000) Rapid dendritic movements during synapse formation and rearrangement. Curr Opin Neurobiol 10:118–124. doi:10.1016/S0959-4388(99)00059-8
Wong WT, Faulkner-Jones BE, Sanes JR, Wong RO (2000) Rapid dendritic remodeling in the developing retina: dependence on neurotransmission and reciprocal regulation by Rac and Rho. J Neurosci 20:5024–5036. doi:10.1016/0006-8993(76)90383-8
Wu GY, Cline HT (1998) Stabilization of dendritic arbor structure in vivo by CaMKII. Science 279:222–226
Wu GY, Zou DJ, Rajan I, Cline H (1999) Dendritic dynamics in vivo change during neuronal maturation. J Neurosci 19:4472–4483. doi:10.1016/S0896-6273(00)80471-7
Wu H-Y, Hudry E, Hashimoto T et al (2010) Amyloid beta induces the morphological neurodegenerative triad of spine loss, dendritic simplification, and neuritic dystrophies through calcineurin activation. J Neurosci 30:2636–2649. doi:10.1523/JNEUROSCI.4456-09.2010
Xie Z, Srivastava DP, Photowala H et al (2007) Kalirin-7 controls activity-dependent structural and functional plasticity of dendritic spines. Neuron 56:640–656. doi:10.1016/j.neuron.2007.10.005
Xing L, Yao X, Williams KR, Bassell GJ (2012) Negative regulation of RhoA translation and signaling by hnRNP-Q1 affects cellular morphogenesis. Mol Biol Cell 23:1500–1509. doi:10.1091/mbc.E11-10-0867
Xu B, Zang K, Ruff NL et al (2000) Cortical degeneration in the absence of neurotrophin signaling: dendritic retraction and neuronal loss after removal of the receptor TrkB. Neuron 26:233–245
Yamamoto H, Fukunaga K, Tanaka E, Miyamoto E (1983) Ca2+- and calmodulin-dependent phosphorylation of microtubule-associated protein 2 and tau factor, and inhibition of microtubule assembly. J Neurochem 41:1119–1125
Yamazaki H, Takahashi H, Aoki T, Shirao T (2001) Molecular cloning and dendritic localization of rat SH3P7. Eur J Neurosci 14:998–1008
Yang N, Higuchi O, Ohashi K et al (1998) Cofilin phosphorylation by LIM-kinase 1 and its role in Rac-mediated actin reorganization. Nature 393:809–812

Yang G, Pan F, Gan W-B (2009) Stably maintained dendritic spines are associated with lifelong memories. Nature 462:920–924. doi:10.1038/nature08577

Yasuda R, Harvey CD, Zhong H et al (2006) Supersensitive Ras activation in dendrites and spines revealed by two-photon fluorescence lifetime imaging. Nat Neurosci 9:283–291. doi:10.1038/nn1635

Young ME, Ohm DT, Dumitriu D et al (2014) Differential effects of aging on dendritic spines in visual cortex and prefrontal cortex of the rhesus monkey. Neuroscience 274:33–43. doi:10.1016/j.neuroscience.2014.05.008

Yu X, Wang G, Gilmore A et al (2013) Accelerated experience-dependent pruning of cortical synapses in ephrin-A2 knockout mice. Neuron 80:64–71. doi:10.1016/j.neuron.2013.07.014

Zempel H, Thies E, Mandelkow E, Mandelkow E-M (2010) Abeta oligomers cause localized Ca(2+) elevation, missorting of endogenous Tau into dendrites, Tau phosphorylation, and destruction of microtubules and spines. J Neurosci 30:11938–11950. doi:10.1523/JNEUROSCI.2357-10.2010

Zhou Q, Homma KJ, Poo M-M (2004) Shrinkage of dendritic spines associated with long-term depression of hippocampal synapses. Neuron 44:749–757. doi:10.1016/j.neuron.2004.11.011

Zuo Y, Lin A, Chang P, Gan W-B (2005a) Development of long-term dendritic spine stability in diverse regions of cerebral cortex. Neuron 46:181–189. doi:10.1016/j.neuron.2005.04.001

Zuo Y, Yang G, Kwon E, Gan WB (2005b) Long-term sensory deprivation prevents dendritic spine loss in primary somatosensory cortex. Nature 436:261–265. doi:10.1038/nature03715

Part V
Synapse Formation Onto Dendrites

Chapter 15
Molecular Assembly of Excitatory Synapses

Hirohide Iwasaki, Shinji Tanaka, and Shigeo Okabe

Abstract Excitatory synapses formed on dendrites are a key component of the functional neuronal network. Molecules present within excitatory synapses and their assembly mechanisms have been studied using multiple research strategies, including biochemistry, cell biology, imaging, and molecular genetics. These efforts have clarified the precise time courses and mechanisms of the synaptic molecular assembly, synaptic junction formation, and postsynaptic structure specialization. Using this knowledge and molecular manipulations, key molecules that regulate excitatory synapse formation have been identified. However, an integrated view of the molecular interactions that regulate excitatory synapse development has not yet been constructed. The difficulty in the integration of a wide range of experimental findings into a coherent model should be eliminated by the development of new imaging and computational approaches designed to examine excitatory synapses.

Keywords Glutamate receptors • Postsynaptic density • Dendritic spine • Actin cytoskeleton • Synapse maturation • Synapse organizer

15.1 The Structure of Excitatory Synapses

15.1.1 The Electron Microscopy of Excitatory Synapses

The ultrastructure of synapses in the central nervous system (CNS) was first described in the 1950s through electron microscopy (EM) studies (Palade and Palay 1954; Palay 1958). The chemical synapses revealed through electron microscopy comprised presynaptic nerve endings and postsynaptic elements separated by an extracellular space (synaptic cleft) with a width of 10–20 nm (Fig. 15.1).

H. Iwasaki • S. Tanaka • S. Okabe (✉)
Department of Cellular Neurobiology, Graduate School of Medicine, The University of Tokyo, 7-3-1 Hongo, Bunkyo-ku, Tokyo 113-0033, Japan

Core Research for Evolutional Science and Technology (CREST), Japan Science and Technology Corporation, Kawaguchi, Saitama 332-0012, Japan
e-mail: okabe@m.u-tokyo.ac.jp

K. Emoto et al. (eds.), *Dendrites*, DOI 10.1007/978-4-431-56050-0_15

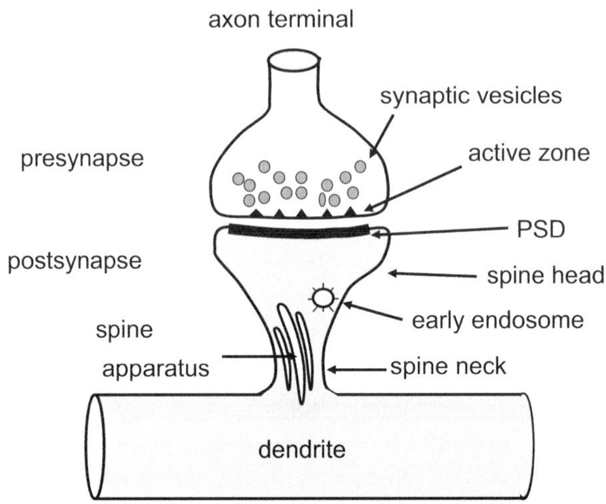

Fig. 15.1 The structure of excitatory spine synapses. Synapses are formed between axon terminals (presynaptic component) and dendrites (postsynaptic component). The axon terminal is filled with synaptic vesicles. The active zone and the PSD are both membrane thickenings located at either presynaptic or postsynaptic membranes. The cytoplasm of spines contains membrane organelles, such as early endosomes and the spine apparatus. Clathrin-coated pits and vesicles are major components of early endosomes and are important in endocytosis of membrane receptors in spines. The spine apparatus is a unique sER-related structure

The membranes on both the sides of the synaptic cleft exhibit densities on their cytoplasmic surfaces. The active zone corresponds to the presynaptic part of the membrane density and is important in the exocytosis of synaptic vesicles. Two types of membrane thickening have been described in the postsynaptic element (Gray 1959). The first type exhibits prominent postsynaptic membrane thickening (postsynaptic density, PSD) and is referred to as a type 1 synapse. In the synapses of the second type, the membrane density thicknesses on the presynaptic and postsynaptic sites are similar and this type of synapses was called type 2 synapses. Later studies revealed that type 1 synapses correspond to glutamatergic excitatory synapses (Petes et al. 1991). Type 2 synapses are inhibitory synapses with their neurotransmitters gamma-aminobutyric acid (GABA) and glycine. Therefore, the typical PSD is a component of glutamatergic excitatory synapses in the CNS.

Type 1 glutamatergic synapses often form on dendritic spines. The preferential formation of glutamatergic synapses on dendritic spines was confirmed in multiple types of neurons in the CNS [e.g., pyramidal neurons in the neocortex (Spacek and Hartmann 1983), pyramidal neurons in the hippocampus (Harris and Stevens 1989), and Purkinje neurons in the cerebellum (Harris and Stevens 1988)]. The PSD and spine can regulate signal processing in the postsynaptic cytoplasm, and the sizes of the two structures are correlated (Arellano et al. 2007; Harris and Stevens 1989). It should be emphasized that many neuron types receive glutamatergic synapses not on spines but rather on dendritic shafts. Together with motor neurons in the spinal cord, a majority of interneurons in the neocortex and hippocampus are classified as neuron types that form excitatory glutamatergic synapses on dendritic shafts. The basic structures of the PSDs in these neuron types are similar to those of neurons possessing dendritic spines, although excitatory synapse formation on the dendritic shafts of interneurons requires unique protrusive dendritic activity and active synaptic junction translocation (Kawabata et al. 2012).

15.1.2 The Structure of the PSD

The PSD, an essential component of the glutamatergic excitatory synapse, is a dense, submembranous, and filamentous structure with a diameter of 200–500 nm and thickness of 30–60 nm (Harris et al. 1992; Spacek and Harris 1998). The PSD comprises both the membrane and cytoplasmic proteins that play roles in molecular assembly and signal transduction in postsynaptic sites. Neurotransmitter receptors such as the AMPA-type and NMDA-type glutamate receptors are major protein components of the PSD (Cheng et al. 2006; Moon et al. 1994). Cell adhesion molecules can also be detected in the PSD (Jordan et al. 2004). Molecules important for intracellular signal transduction, such as protein kinases and phosphatases, are also present in the PSD (Pocklington et al. 2006). The recruitment and accumulation of neurotransmitter receptors, cell adhesion molecules, and intracellular signaling molecules at the PSD are mediated by interactions of these molecules with PSD scaffolding proteins. These interactions are often regulated by synaptic activity.

The PSD can be isolated biochemically, and EM analyses confirmed the morphology of an isolated PSD as a circular disk with a diameter of 300–400 nm (Petersen et al. 2003). Isolated PSDs can be visualized by making metal replicas after rapid freezing and freeze-drying (Fig. 15.2). This sample preparation technique revealed morphological features of the two surfaces of the PSDs. The cytoplasmic surface of the PSD exhibits irregular protrusions, whereas a dense layer of small particles was present on the cleft surface. These differences are assumed to result from the differential organization of molecules on the two surfaces. A more detailed nanostructure analysis of the PSD can be achieved via EM tomography of the PSD (Lucic et al. 2005). EM tomographic analysis of rapidly frozen and freeze-substituted PSD preparations revealed vertically oriented filaments in the core of the PSD structure; these filaments were labeled with an antibody against the predominant PSD scaffolding protein PSD-95 (Chen et al. 2008). Periodic filamentous structures linking the presynaptic and postsynaptic membranes were also identified during a cryo-EM analysis of vitreous sections (Zuber et al. 2005). Cryo-EM tomography of fully hydrated samples provides the ideal imaging conditions for detecting molecular organization within the PSD. Using cryo-EM, the architecture of native PSDs in cultured hippocampal neurons could be successfully visualized, and PSD structures reconstructed with this technique displayed filaments running parallel to the plasma membrane, together with shorter connecting molecules perpendicular to the filaments

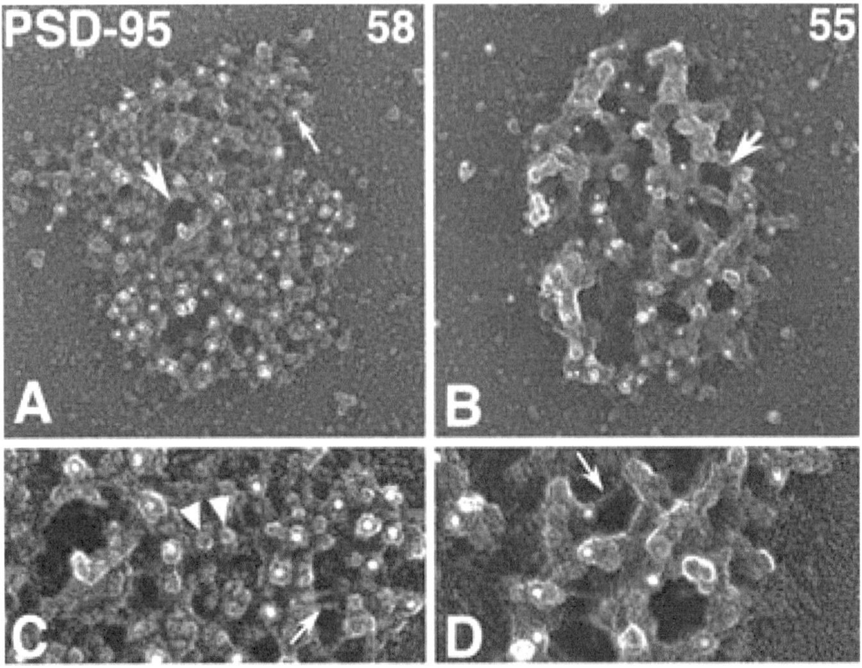

Fig. 15.2 Electron microscopic images of cleft and cytoplasmic surfaces of detergent-extracted PSDs labeled with a gold particle-conjugated anti-PSD-95 antibody (Petersen et al. 2003). **A** and **C** cleft surface of PSD at lower (**A**) and higher (**C**) magnifications. The large *arrow* in **A** is the opening in the central mesh structure. The small *arrow* in **A** indicates an immunogold particle for PSD-95. The *arrowheads* in **C** indicate granular particles. The *arrow* in **C** indicates underlying thin filaments within the central mesh. **B** and **D** cytoplasmic surface of PSD at lower (**B**) and higher (**D**) magnifications. The *arrow* in **B** indicates the opening in the central mesh structure. The *arrow* in **D** indicates underlying filaments of the central mesh. Numbers indicate the counts of gold particles

(Fernandez-Busnadiego et al. 2011; Lucic et al. 2007). The identification of the molecules corresponding to these parallel and perpendicular filaments may lead to an understanding of the general architecture of the PSD.

15.1.3 The Structure of Dendritic Spines

The dendritic spine is a tiny protrusion from dendritic shaft (Fig. 15.3) (Yuste 2010). Dendritic spine does not exhibit a uniform morphology and can be classified into three major categories: thin, mushroom, and stubby. The spine head is an enlarged structure at the end of the protrusion. The PSD is usually located within this spine head. The spine neck is a membranous tube with a width of 50–300 nm that connects the spine head and the dendritic shaft. The spine volume and surface area vary widely even within a single cell type. In the case of hippocampal

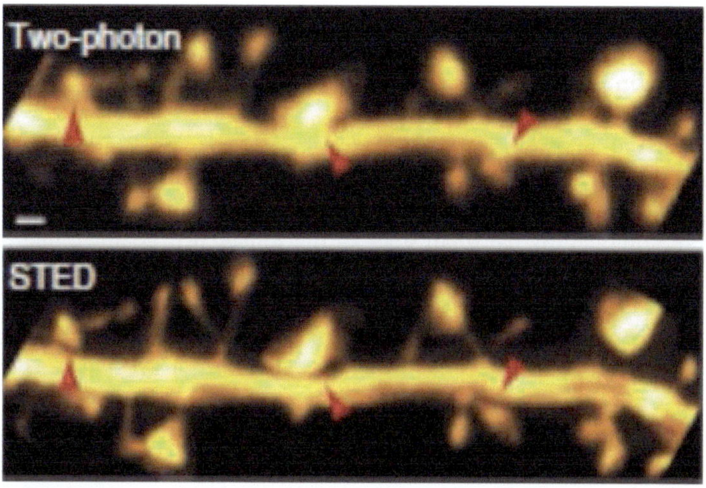

Fig. 15.3 The comparison of spine morphology imaged using two-photon excitation laser scanning and STED microscopy (Tonnesen et al. 2014). *Red arrows* indicate apparent stubby spines in two-photon images but resolved spine necks in STED images

pyramidal neurons in the stratum radiatum of area CA1, the spine volumes and surface areas range from 0.004 to 0.56 μm^3 and from 0.13 to 4.4 μm^2, respectively (Harris and Stevens 1989). Differences in the spine volumes and surface areas may reflect the previous histories of individual spines and their states of synaptic transmission. The spine volume correlates strongly with the area of the PSD (Arellano et al. 2007) and the number of functional AMPA-type glutamate receptors as estimated using the two-photon glutamate uncaging technique (Matsuzaki et al. 2001). This correlation indicates the possibility of parallel changes in spine volumes and postsynaptic function in response to synaptic activity.

Until recently, precise measurements of spine morphology were possible only through EM-based serial reconstruction. However, this situation was changed by the introduction of new super-resolution imaging technologies. Stimulated emission depletion (STED) microscopy, which can detect spine morphologies at a resolution of approximately 50 nm, has revealed the presence of thin spine necks in spines that had been previously classified as stubby using diffraction-limited imaging modalities (Fig. 15.3) (Tonnesen et al. 2014). Precise measurements of spine morphology based on super-resolution imaging represent a powerful approach for the efficient analysis of a large population of spines in live samples.

15.1.4 Spine Cytoplasm

15.1.4.1 Actin Cytoskeleton

The spine cytoplasm exhibits unique morphological features. Both the spine heads and necks are filled with a meshwork of actin filaments (Fischer et al. 1998; Hirokawa 1989). Detailed analyses of the orientations of this filamentous actin have been prohibited by the small scale of the spines. It is assumed that the actin filaments within spines exhibit a polarized orientation with their barbed ends located at the distal edge of spines, similar to the polarized orientation of filamentous actin in nonneuronal filopodia. Spines initiate from filopodia-like precursors, but detailed structural analyses of filamentous actin in spine precursors revealed a network-like organization of actin filaments distinct from the tight actin filament bundles present in nonneuronal filopodia (Korobova and Svitkina 2010; Okabe and Hirokawa 1989). In mature spines, actin filaments in the spine neck also exhibit a network-like appearance. From these observations, it has been postulated that tight actin filament bundling is not necessary for spine morphogenesis (Hotulainen and Hoogenraad 2010). The dynamics of actin filaments in spine synapses will be discussed in Sect. 15.3.4.

15.1.4.2 Smooth Endoplasmic Reticulum (sER)

sER is present in most of the dendritic spines on cerebellar Purkinje cells (Harris and Stevens 1988). This may be related to the fact that the release of calcium from IP3 receptors plays an essential role in synaptic plasticity in Purkinje cells (Miyata et al. 2000). Immuno-EM revealed the preferential enrichment of IP3 receptors in sER, suggesting that the sER within spines is a major source of calcium for IP3-dependent signaling (Walton et al. 1991). The fraction of dendritic spines that contains sER is much lower in hippocampal and cortical pyramidal neurons (Fig. 15.4) (Spacek and Harris 1997). The spine apparatus is a unique sER-related structure. This structure is present in a subset of spines in the hippocampus and is composed of sER-like membrane stacks connected by densely stained linkers (Deller et al. 2000). Synaptopodin is an actin-binding protein present in the spine apparatus, and synaptopodin knockout mice fail to form spine apparatuses (Deller et al. 2003). The knockout mice also exhibit deficits in LTP and impaired spatial learning. These results indicate that the spine apparatus plays important roles in the regulation of synaptic transmission. The spine apparatus is thought to be involved in either calcium release from internal stores or local protein synthesis. The latter possibility is supported by the fact that the spine apparatus has been shown to interact with polyribosomes (Steward and Reeves 1988). Direct evidence to support the roles of the spine apparatus in either calcium release or local protein synthesis has not yet been obtained.

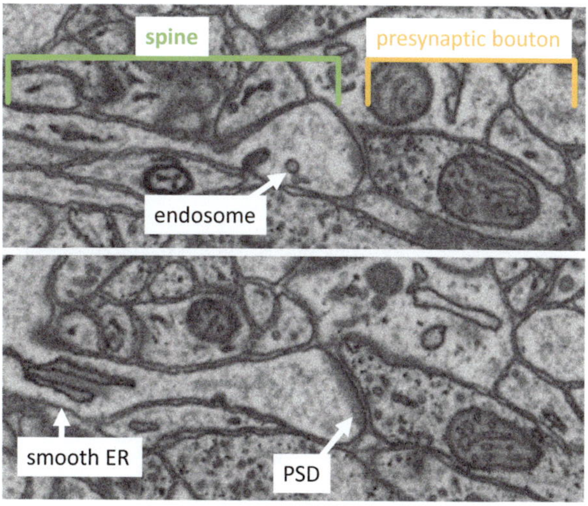

Fig. 15.4 The electron microscopic images of a mouse neocortical spine and its cytoplasmic specializations, such as PSDs, endosomes, and smooth ERs

15.1.4.3 Endocytotic Membranes

Typical clathrin-coated pit structures can be identified in a fraction of dendritic spines. Clathrin-coated pits are located in the vicinity of PSDs and are thought to be involved in AMPA receptor endocytosis (Racz et al. 2004). Active AMPA receptor endocytosis and its upregulation after the induction of synaptic plasticity have been reported (Petrini et al. 2009; Wang et al. 2008). Early endosomes and sorting endosomes have been reported to be present in a fraction of dendritic spines (Fig. 15.4) (Park et al. 2006). EM reconstruction analysis revealed that the number of endosomal-sorting complexes is much lower than the number of dendritic spines (only 10 % of the number of spines per unit dendrite length) (Cooney et al. 2002). This observation indicates that multiple spines in proximity share a common endosomal-sorting machinery for the delivery and recycling of endocytosed materials.

15.2 The Molecular Composition of the PSD

15.2.1 The Functional Roles of PSD Molecules

PSD-enriched fractions have been utilized for protein identification with a variety of biochemical methods. High-sensitivity mass spectrometry is a very powerful technique for protein species identification, and more than 400 proteins have been successfully identified (Husi et al. 2000; Jordan et al. 2004; Yoshimura et al. 2004). The identification of interaction partners using yeast two-hybrid screening also helped to increase the number of candidate proteins present within the PSD. Here

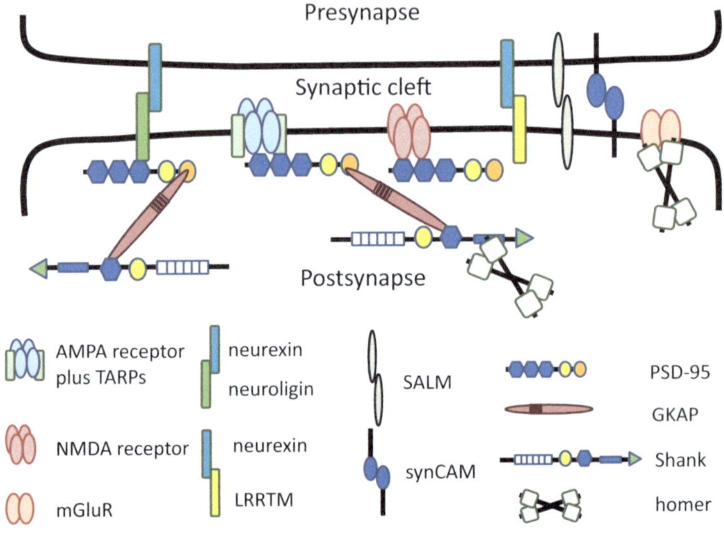

Fig. 15.5 Molecular assembly within spines. AMPA receptors and NMDA receptors are abundant membrane proteins within the PSD. Metabotropic glutamate receptors (mGluRs) are located at the periphery of the postsynaptic membrane. Postsynaptic neuroligin and LRRTM molecules form heterophilic binding with presynaptic neurexins. Homophilic interactions of SALM and synCAM cell adhesion molecules cross-bridge between presynaptic and postsynaptic membranes. PSD-95 interacts directly with NMDA receptors and indirectly with AMPA receptors via TARPs. Neuroligins also interact with PSD-95. GKAP interacts with both PSD-95 and Shank molecules. Shank and mGluRs are binding partners of another PSD scaffolding molecule, Homer, which forms homotetramers

we describe the properties of several major constituents of the PSD, including glutamate receptors, scaffolding molecules, and cell adhesion molecules (Fig. 15.5).

The purified PSD fraction is enriched for both AMPA-type and NMDA-type glutamate receptors (Cheng et al. 2006), and this enrichment has been confirmed by immunoelectron microscopy (Nusser 1999; Nusser et al. 1998; Petralia et al. 1994a, b; Tanaka et al. 2005). AMPA and NMDA receptors are essential functional elements of fast synaptic transmission. In the immature nervous system, a fraction of synapses exhibit relative AMPA receptor scarcity when compared with more mature synapses. The former synapses are “silent” in terms of normal synaptic transmission, but can be activated through NMDA receptor-dependent processes.

PSD scaffolding molecules accumulate within synapses and form molecular networks within PSDs. Predominant PSD scaffolding molecules include PSD-95, GKAP, Shank, and Homer. The guanylate kinase-like domain of PSD-95 directly binds to GKAP (Kim et al. 1997). The C-terminus of GKAP, in turn, interacts with the PDZ domain of Shank (Naisbitt et al. 1999). Shank also interacts with Homer via its proline-rich region (Tu et al. 1999). Accordingly, simple one-to-one interactions might exist between these four scaffolding proteins. PSD-95 belongs to the membrane-associated guanylate kinase (MAGUK) protein family (Cho et al. 1992) and binds both NMDA receptors and AMPA receptors. Interactions of PSD-95 with the NR2 subunits of NMDA receptors are direct (Kornau et al. 1995) and those with AMPA receptors are indirect and occur via transmembrane AMPA receptor regulatory proteins (TARPs), which are auxiliary components of the native AMPA receptor complex (Tomita et al. 2005). A variety of cell adhesion molecules, including neuroligins, can interact with PSD-95 (Irie et al. 1997; Meyer et al. 2004). Group I metabotropic glutamate receptors (mGluRs) can interact with another PSD scaffolding protein, Homer (Brakeman et al. 1997).

Synapses are specialized sites of cell-to-cell contact. Both the formation of synaptic contacts and maintenance of assembled synaptic structures are regulated by the cell adhesion molecules present in synapses. Neuroligins are synaptic cell adhesion molecules that mainly localize at postsynaptic membranes. The roles of neuroligins in both synapse formation and assembled synaptic junction regulation have been extensively studied. There are five isoforms of neuroligin in humans (NL1, NL2, NL3, NL4X, and NL4Y) and four isoforms in mice (NL1, NL2, NL3, and NL4) (Varoqueaux et al. 2006). In the rodent brain, NL1 and NL2 are present in the excitatory postsynaptic membrane and inhibitory postsynaptic membrane, respectively, whereas NL3 is present in both the excitatory and inhibitory synapses. Neuroligin binds to presynaptic receptor neurexins, and this interaction induces the differentiation of both presynaptic and postsynaptic structures (Sudhof 2008). Presynaptic neurexins have multiple binding partners, including neuroligin and leucine-rich repeat transmembrane neuronal (LRRTM) (Ko et al. 2009), and can also interact indirectly with delta2 receptors via Cbln1 (Ito-Ishida et al. 2012; Matsuda et al. 2010). The postsynaptic cell adhesion molecule LRRTM, which binds to neurexin, can induce presynaptic differentiation (Linhoff et al. 2009). Other synapse organizers include synaptic cell adhesion molecules (synCAMs) (Biederer et al. 2002), netrin-G ligand (NGL) (Kim et al. 2006; Woo et al. 2009b), synaptic adhesion-like molecules (SALMs) (Ko et al. 2006; Mah et al. 2010; Wang et al. 2006), TrkC (Takahashi et al. 2011), protein tyrosine phosphatases (including LAR, PTPδ, and PTPσ) (Takahashi et al. 2011; Woo et al. 2009a; Yoshida et al. 2011), slit and NTRK-like family member (Slitrk) (Takahashi et al. 2012), calsyntenin 3 (Pettem et al. 2013), and IgSF9b (Woo et al. 2013). Some of these synapse organizers can selectively promote either excitatory or inhibitory synapses.

15.2.2 The Molecular Contents of Scaffolding Proteins in the PSD

Quantification of proteins in biochemically purified PSD preparations can effectively estimate the relative abundances of multiple proteins. One problem associated with this method is the possibility that the purification steps induce the extraction of weakly associated proteins and/or nonspecific binding of proteins from other cellular compartments. To avoid this possibility, it is important to develop a method to quantify the local protein contents in single synapses. One possible method for obtaining measurements of local protein contents is the use of green fluorescent protein (GFP)-tagged PSD scaffolding proteins (Okabe 2013; Okabe et al. 1999, 2001). If one could estimate the number of GFP-tagged proteins in a single synapse, the number of native proteins could be deduced from the ratio of endogenous and GFP-tagged proteins as determined by immunolabeling (Sugiyama et al. 2005). GFP-based measurements of scaffolding protein contents in cultured hippocampal neurons revealed that a single postsynaptic site contained an average of 273 PSD-95 family proteins, 171 GKAP proteins, 310 Shank family proteins, and 343 Homer family proteins (Fig. 15.6). The estimated MAGUK protein content per synapse agrees well with the number of PSD-95 proteins determined from scanning transmission EM measurements of the average PSD mass and quantitative immunoblotting (Chen et al. 2005). This GFP-based quantification method also revealed similar concentrations of the four PSD scaffolding proteins per synapse, suggesting a relatively simple stoichiometry. The total mass of the four scaffolding proteins corresponds to 120 MDa, or 10 % of the total PSD mass, suggesting the importance of the four scaffolding proteins in the PSD structural framework (Okabe 2007).

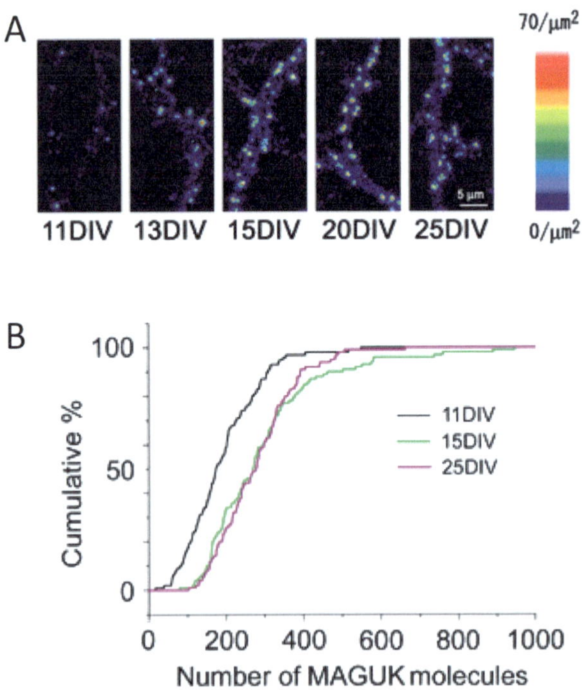

Fig. 15.6 Developmental changes in scaffolding protein numbers at single postsynaptic sites (Sugiyama et al. 2005). (**A**) Developmental changes in the Shank family protein. The *color* code indicates molecular density. (**B**) Developmental shift in the number of MAGUK proteins (PSD-95 family proteins) at single postsynaptic sites. Profiles range from 11 to 25 days in vitro (DIV)

15.3 Molecular Assembly and Synapse Development

15.3.1 Formation and Maturation of Dendritic Filopodia

Immature neurons express highly motile filopodia that protrude from dendritic shafts (Portera-Cailliau et al. 2003). Most dendritic filopodia are transient, and only a small proportion may stabilize and begin to form synaptic contacts (Okabe et al. 2001). There is an ongoing debate whether dendritic filopodia are a single entity or a mixture of protrusions with different properties and final fates (Portera-Cailliau et al. 2003). At minimum, a fraction of dendritic filopodia can serve as a spine synapse precursors. Actin filaments within dendritic filopodia are not organized into bundles but form meshworks of branched and linear actin filaments (Korobova and Svitkina 2010). In contrast, conventional filopodia in fibroblasts contain tightly bundled actin filaments (Okabe and Hirokawa 1989; Svitkina et al. 2003). This difference in actin organization may be of importance to specifying the sites of new actin polymerization within filopodia and could support more complex motility of these dendritic filopodia (Hotulainen et al. 2009).

Spine synapses may arise from interactions of motile filopodia with axons. Another possible route of excitatory synapse formation is direct contact between axons and dendritic shafts and the subsequent induction of dendritic spines at the sites of contact. The first model stresses the importance of active environmental scanning by filopodia (filopodial model) (Fiala et al. 1998), and the second model (Miller–Peters model) is based on EM observations and the categorization of nascent synapses in vivo (Miller and Peters 1981). When the order of appearance of the dendritic filopodia and postsynaptic molecular assembly was analyzed in a dissociated culture of neurons, filopodial formation generally preceded the acquisition of postsynaptic molecular assembly (Friedman et al. 2000; Okabe et al. 2001; Ziv and Smith 1996). However, imaging experiments of slice preparations revealed the presence of an alternative pathway in which protrusive dendrite activity occurs at the site of postsynaptic differentiation, with the subsequent stabilization of these structures as spine synapses (Marrs et al. 2001). These experiments involving dissociated neurons and slice preparations suggest the presence of two alternative spine differentiation developmental pathways. To detect the order of spine and synapse differentiation in a native tissue environment, an in vivo two-photon microscopic analysis of synapse development should be performed. Given the technical difficulties, reliable monitoring of synapse and spine formation in the early postnatal neocortex has not yet been accomplished. The adult mouse neocortex is a less challenging target for in vivo spine imaging, and in vivo imaging experiments combined with retrospective serial section EM have revealed that newly formed dendritic protrusions required maintenance without axonal interaction for 2 days before the gradual differentiation of synaptic junctions (Knott et al. 2006). This observation supports the filopodial model of neocortical synapse formation. It has not yet been demonstrated whether the same order of differentiation applies to synapses formed during early developmental stages.

15.3.2 The Recruitment of Cell Adhesion Molecules and Scaffolding Molecules to Synaptic Contacts

Without proper target recognition, appropriate synaptic contacts cannot be formed. Target recognition is mediated by molecular cues present on presynaptic and postsynaptic cell membranes. Cell adhesion molecules are considered the most important molecular cues for synaptic target recognition (Benson and Huntley 2012). Both homophilic and heterophilic interactions between cell adhesion molecules have been shown to be involved in synapse formation. Interactions between cell adhesion molecules are important to both the structural differentiation of synapses and determination of synapse subtypes and specificities.

Using assays that induced artificial synaptic structures between naïve neurons and nonneuronal cells expressing cell adhesion molecules, several cell adhesion molecules were identified as synapse organizers, molecules that can induce either presynaptic or postsynaptic structures in neurons (Krueger et al. 2012; Tallafuss et al. 2010). Neuroligins were the first identified synapse organizers (Scheiffele et al. 2000). These are heterophilic postsynaptic cell adhesion molecules that interact with presynaptic partner neurexins. Neurexins expressed on nonneuronal cells can induce postsynaptic differentiation in dendrites within a few days (Graf et al. 2004). The artificial postsynaptic sites contained a variety of postsynaptic molecules, including PSD-95, GKAP, and NMDA receptors (Graf et al. 2004; Nam and Chen 2005). Surface AMPA receptors are also recruited to the sites of neuroligin clusters via interactions with PSD-95 (Mondin et al. 2011). These results are consistent with the idea that postsynaptic neuroligins control molecular assembly at the postsynaptic sites via interactions with presynaptic neurexins. Interactions between neurexin and neuroligin at nascent synapses will trigger the simultaneous differentiation of both presynaptic and postsynaptic structures, thus synchronizing the differentiation process. Cultured immature cortical neurons exhibit the rapid recruitment of fluorescent protein-tagged NL1 clusters to sites of axodendritic contact (Barrow et al. 2009). Mobile NL1 clusters were also present in both dendritic shafts and filopodia in these immature neurons, suggesting vesicle-mediated NL1 transport. Rapid vesicle-mediated transport and local recruitment of neuroligins may represent a general strategy by which immature dendrites deliver sufficient amounts of synaptic cell adhesion molecules to local sites of contact with incoming axons.

The molecular interactions that enable neuroligin-induced postsynaptic differentiation have been studied using fluorescently tagged postsynaptic molecules (Giannone et al. 2013; Mondin et al. 2011). Neuroligins contain a C-terminal PDZ domain-binding motif that can bind to the PDZ domain of PSD-95. NL1 clusters that had been induced by antibodies against the extracellular epitope HA tag could not effectively induce subsequent intracellular PSD-95 clustering, in contrast to NL1 clusters induced by cross-linked neurexin 1β. This result indicates that neurexin binding facilitates interactions of clustered NL1 with PSD-95. Tyrosine phosphorylation of NL1 reduces its affinity for gephyrin, a scaffolding protein found in inhibitory synapses, and allows the preferential binding of NL1 to PSD-95. This tyrosine phosphorylation-regulated competition between PSD-95 and gephyrin may underlie the effective clustering of PSD-95 mediated by NL1–neurexin interactions.

15.3.3 The Recruitment of Glutamate Receptors to Synaptic Contacts

Imaging of cultured immature hippocampal neurons revealed the presence of two distinct populations of postsynaptic structures (Gerrow et al. 2006). One was mobile non-synaptic complex of multiple scaffolding proteins, including PSD-95, GKAP, and Shank, but not NL1. The other was stationary and also contained PSD-95, GKAP, and Shank as well as NL1. Several imaging studies reported the presence of mobile postsynaptic packets containing key receptor and scaffolding molecules and postulated their importance in terms of the supply of molecules needed for postsynaptic functions (Barrow et al. 2009; Washbourne et al. 2002, 2004). These studies also indicate key roles of neuroligin and similar synaptic cell adhesion molecules in the initial synapse differentiation.

The subsequent recruitment of glutamate receptors might be an important role played by neuroligins and associated scaffolding molecules within synapses. A clustered complex of NL1 and PSD-95 was shown to be effective in the recruitment of mobile AMPA receptors to dendritic surfaces (Mondin et al. 2011). It was also shown that AMPA receptor recruitment was mediated by the AMPA receptor auxiliary subunit transmembrane AMPA receptor regulatory proteins (TARPs) (Opazo et al. 2010). These experiments suggest that the accumulation of AMPA receptors at nascent synapses is mediated by intracellular interactions between NL1, PSD-95, and an AMPA receptor complex that includes TARPs. When NL1 is overexpressed in cultured neurons, the relative contents of AMPA and NMDA receptors, which are estimated from excitatory postsynaptic currents at different membrane potentials, shift toward a higher NMDA receptor content, suggesting a more direct impact of the NL1 abundance on NMDA receptor recruitment to excitatory synapses (Budreck et al. 2013). This observation can be explained by the direct interaction of the extracellular domain of NL1 with the GluN1 subunit of the NMDA receptor. Enhanced NMDA receptor clustering at synapses, which was mediated by the overexpression of NL1, was shown to be independent of the presence of PSD-95. This further supported the presence of interaction domains distinct from the C-terminal PDZ domain-binding motif of NL1. In summary, the intracellular and extracellular motifs of neuroligin molecules coordinate during nascent postsynaptic molecular assembly, which includes both AMPA receptors (Giannone et al. 2013; Mondin et al. 2011) and NMDA receptors (Bard et al. 2010; Budreck et al. 2013). The molecular assembly induced by other synapse organizers likely utilizes strategies similar to those identified in neuroligin-dependent postsynaptic organization. Further experimental evidence is required to confirm this point.

Glutamate receptor recruitment to synapses should be regulated by the transport of glutamate receptor-containing vesicles. In the case of NMDA receptors, NMDA receptor-containing vesicle transport was successfully visualized in cultured immature neurons (Washbourne et al. 2002). NMDA receptor recruitment to newly generated synapses is a rapid process, but the appearance of AMPA receptors on nascent synapses is delayed and may be initiated by the activation of NMDA receptors. AMPA receptor recruitment to synapses may be regulated by a process analogous to that of long-term potentiation (Ashby and Isaac 2011; Isaac et al. 1995). AMPA receptors may either be exposed to the surfaces of dendritic shafts and subsequently translocate along the plasma membrane into spines (Triller and Choquet 2008) or might be directly transported into spines via AMPA receptor-containing vesicles with subsequent exposure to the spine surface via local exocytosis (Patterson et al. 2010). In both surface receptor recruitment and local exocytosis, interactions of the C-termini of AMPA receptors with scaffolding molecules and regulation by posttranslational modifications play important roles (Bats et al. 2007; Kim et al. 2007; Opazo et al. 2010; Steiner et al. 2008; Xu et al. 2008).

15.3.4 Actin Organization and Dynamics in Spines

Dendritic spine morphology changes continuously (Majewska and Sur 2003; Majewska et al. 2006). Newly generated spines tend to be smaller and have less prominent heads. Older spines tend to be larger, with prominent heads (Yasumatsu et al. 2008). Spine morphology and life spans can be studied at better spatial and temporal resolutions in dissociated cultured neurons. Such studies have revealed rapid morphological changes in spines on the order of minutes, as well as the dependence of spine structural changes on the actin cytoskeleton (Fischer et al. 1998). Actin polymerization also drives structural changes in the PSDs (Blanpied et al. 2008). Partial PSD scaffold disassembly can be induced by actin depolymerization (Kuriu et al. 2006). Thus, both spine morphology and PSD molecular assembly are regulated by actin dynamics.

Spine maturation and stability may be closely related to the dynamic state of actin polymers. As described in Sect. 15.1.4.1, the orientations of actin filaments in dendritic protrusions are less organized than those in the filopodia of nonneuronal cells (Korobova and Svitkina 2010) (Fig. 15.7). To clarify the precise organization and dynamics of actin polymers, it is necessary to develop techniques that can monitor the state of actin polymers in the small volume of spine cytoplasm. Two-photon activation of photoactivatable (PA)-GFP-labeled actin provided evidence of retrograde actin flow in the spine head (Honkura et al. 2008). This finding indicates the addition of new actin monomers at the distal and peripheral domains of the spine head and subsequent filament treadmilling. Super-resolution imaging (PALM/STORM) of single actin molecules confirmed the presence of retrograde

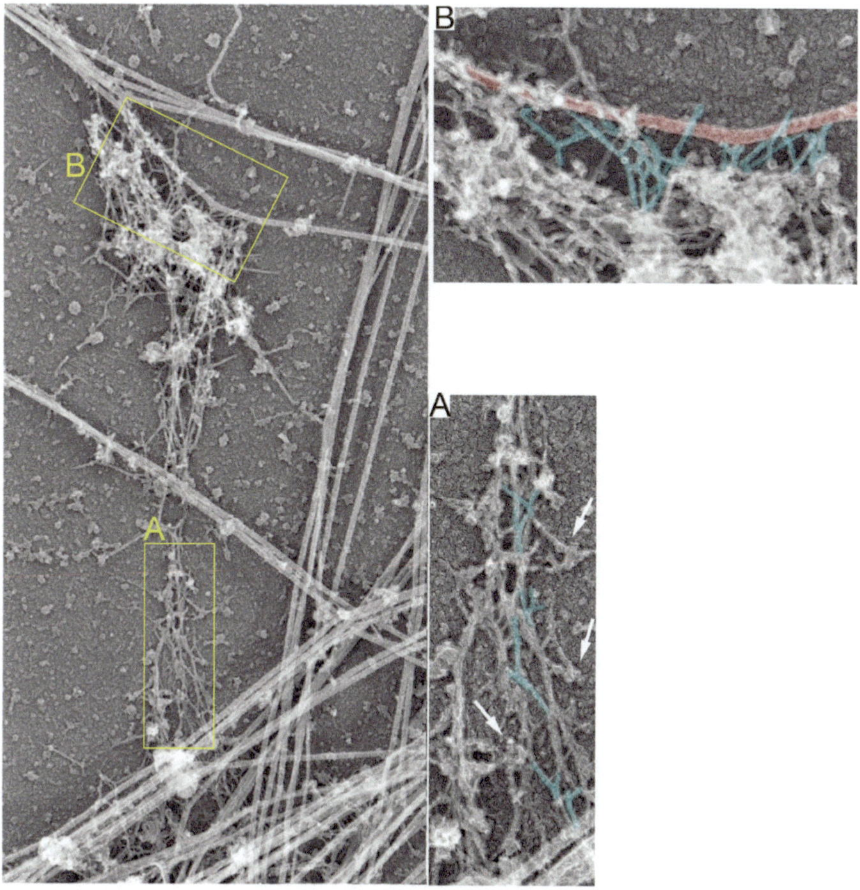

Fig. 15.7 Platinum replica electron microscopic images of a dendritic spine (Korobova and Svitkina 2010). Detergent extraction revealed cytoskeletal organization within the cytoplasm. *Yellow* boxes (**A**) (spine neck) and (**B**) (spine head) correspond to enlargements of images (**A**) and (**B**). Image B shows associations of actin filaments within a spine head (*cyan*) with an axonal microtubule (*red*). Image A illustrates branched actin filaments (*cyan*) in the spine neck. *Arrows* indicate putative filament ends

actin flow in the spine heads (Frost et al. 2010; Tatavarty et al. 2009). PALM imaging revealed that the velocities of individual actin molecules were heterogeneous and specifically enhanced in the vicinity of PSDs (Fig. 15.8). The heterogeneous actin movement and relatively short distance of the net actin flow are consistent with the idea that short actin filaments, with a less aligned orientation, form the main actin network within the spines.

Further subdomain-specific actin organization and dynamics were proposed following the super-resolution imaging of actin regulatory molecules (Chazeau et al. 2014). The formation of a branched actin network is driven by the Arp2/3

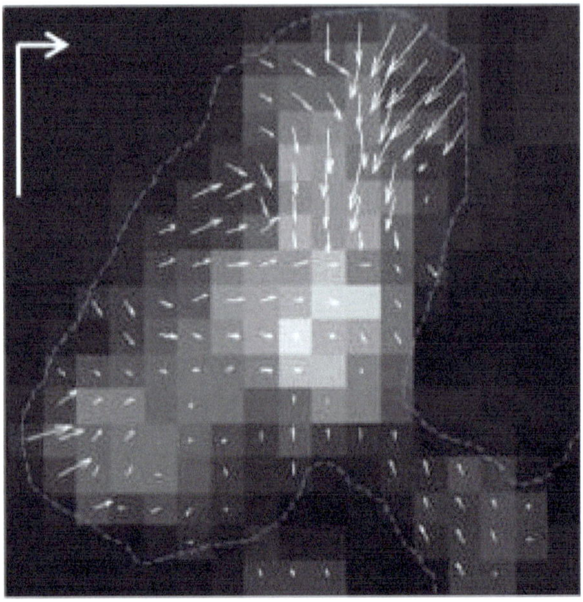

Fig. 15.8 PALM imaging of actin molecules in a spine (Frost et al. 2010). *Arrows* indicate the directions and relative velocities of single actin molecules tagged with photoactivatable fluorescent protein mEos2. The *gray* scale indicates the density of moving single particles. *Vertical scale bar*, 200 nm and *horizontal vector*, 100 nm/s

Fig. 15.9 Dual color super-resolution PALM images of mEos2-tagged Abi1 (a component of WAVE complex) and dSTORM images of endogenous PSD-95 labeled with Alexa647 (Chazeau et al. 2014). *Insets at left* are diffraction-limited fluorescence images. *Insets at right* are merged images of Abi1 and PSD-95. Higher magnification of the spines marked by *asterisks* are provided (*arrowhead*)

complex, following activation by WAVE (Takenawa and Suetsugu 2007). Conversely, the formation of aligned F-actin bundles is driven by the nucleation of single linear filaments through interactions with formins (Kovar et al. 2006). If the formation of an actin meshwork occurs uniformly at the distal submembranous domains of spines, WAVE complex proteins should be located at both the PSD and non-PSD subdomains of spines. However, PALM imaging revealed specific confinement of the WAVE complex in the vicinity of the PSD, indicating the formation of an actin meshwork specifically at the interface between the PSD and adjacent cytoplasm (Figs. 15.9 and 15.10). On the other hand, actin regulators that associate

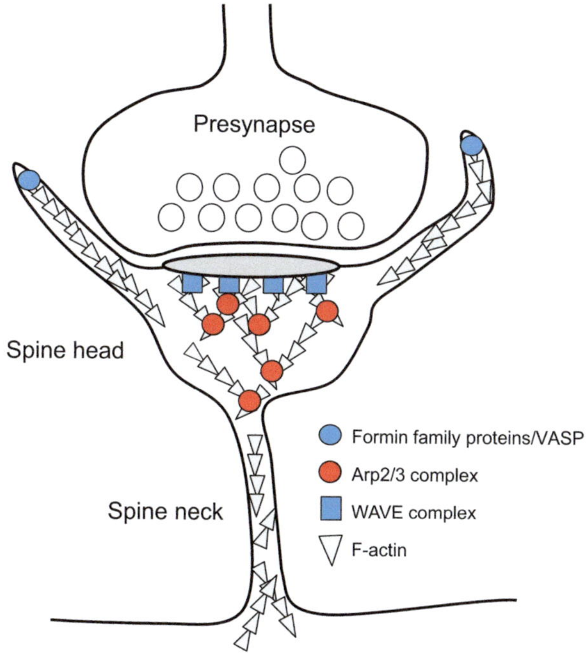

Fig. 15.10 A proposed model of actin filament organization and distribution of actin regulators in the spine cytoplasm. The WAVE complex is in the vicinity of the PSD and drives formation of an actin meshwork by activating the Arp2/3 complex. Actin regulators that associate with bundled actin filaments, such as formin family proteins and VASP, are preferentially associated with fingerlike protrusions from the spine head and regulate their dynamics. The barbed ends of F-actin correspond to fast-growing ends of these polymers

with bundled actin filaments, such as VASP and formin-like protein-2, were preferentially associated with fingerlike protrusions from the spine head. These observations indicate the nanoscale confinement of actin regulators within spines and the distinct roles of these regulators in spine morphology.

15.3.5 Spine Stability and Molecular Dynamics

Studies involving new optical techniques have revealed the precise regulation of actin dynamics within spines; however, the relationship between molecular dynamics and spine stability has not yet been clarified. The actin meshwork within spines is highly dynamic, with a half-life in minutes, at least in cultured neurons and organotypic slices (Honkura et al. 2008; Star et al. 2002). On the other hand, in the mouse neocortex, spines on pyramidal neurons can be maintained for more than several months (Grutzendler et al. 2002; Zuo et al. 2005). Thus, the system that regulates spine morphology should be designed to support long-term structural stabilization based on highly dynamic polymers. The finding that the PSD can function as a site of WAVE recruitment may provide information to fill the gap between spine stability and actin turnover, as the size of the PSD or, more specifically, the presence of specific WAVE-interacting partners within the PSD can regulate the speed of actin meshwork generation within spines (Chazeau et al. 2014). This model suggests that spines with prominent PSDs should have larger amounts of actin polymers within a meshwork-like architecture. Indeed, there is a gradual increase in the concentration of PSD scaffolding molecules during development (Sugiyama et al. 2005). This increase may underlie the transition in spine actin organization, which in turn regulates the overall shapes of spines. This model also explains how initial cell-to-cell contacts and PSD scaffold assembly can shift spine actin polymers from simple bundles to meshwork-like organizations (Okabe et al. 2001).

If actin organization in spines is regulated by the PSD, the next question asks how PSDs and spine stability are related. In vivo imaging of neocortical spine turnover revealed the life spans of spines. Most spines in the mature neocortex are stable, as only 5 % of spines are formed and eliminated (Grutzendler et al. 2002; Zuo et al. 2005). A comparison of the morphologies of stable and dynamic spines revealed that larger spines tend to be more stable (Holtmaat et al. 2005). In vivo imaging of spines, together with fluorescently tagged PSD-95, revealed that spines containing PSD-95 clusters were more stable (Fig. 15.11 (Cane et al. 2014; Isshiki et al. 2014). However, the relationship between spine stability and the amount of PSD-95 within spines is not straightforward (Cane et al. 2014). PSD-95-containing spines generally exhibited increased stability, but newly generated spines rarely

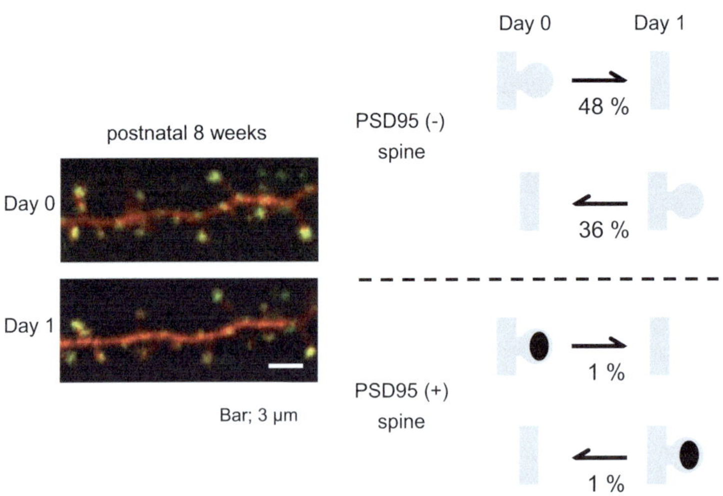

Fig. 15.11 In vivo imaging of mouse neocortical pyramidal neurons expressing DsRed and PSD-95–GFP at postnatal 8 weeks (*left*), at an interval of 24 h. Fractions of spine gain and loss in either PSD-95-negative or PSD-95-positive spines are shown

converted into persistent spines even if they had acquired PSD-95 assembly. This finding may indicate the presence of additional stabilization factors. On the other hand, a reduction in the PSD-95 content in spines could be detected well before the occurrence of spine pruning, indicating that a change in the PSD-95 content predicts the spine fate. These reports are consistent with the idea that the PSD scaffolds are important in spine stabilization.

Although in vivo imaging indicates the possibility that PSD scaffolds are important to spine stability, an in-depth understanding will require additional experimental evidence for a relationship between PSD scaffolds and actin dynamics. A recent analysis of WAVE-interacting partners identified several postsynaptic molecules, such as NL1 (Chen et al. 2014). It should be important to test whether actin dynamics and spine stability can be affected by mutations in the WAVE-interacting motifs of PSD molecules. Cortactin, which is enriched in PSDs, interacts with the major PSD scaffolding molecule, Shank, and can initiate branched actin polymer formation by recruiting the Arp2/3 complex (Hering and Sheng 2003; Iki et al. 2005). Although cortactin knockdown experiments revealed a strong spine phenotype, the detailed analyses of cortactin function within spine actin organization have not yet been performed. The regulation of actin meshwork generation by WAVE and cortactin at the interface between PSD scaffolds and the adjacent cytoplasm will be critical to our understanding of actin dynamics in spines.

15.4 Future Prospects

In the previous chapters, we described the quantitative properties of excitatory postsynaptic specialization and the developmental time course. An integrated view of the molecular interactions that regulate excitatory synapse development should be proposed using the accumulated data from synaptic molecules and their dynamics. However, this task remains difficult, and there are few proposed models of postsynaptic molecular assembly. Although biophysical models of molecular dynamics within spines have been constructed, their main focus was an explanation of glutamate receptor behaviors in the resting and activity-dependent states (Czondor et al. 2012; Earnshaw and Bressloff 2006). Theoretical models of actin polymer organization have been developed in several biological systems (Pollard et al. 2000). The relationship between local actin meshwork assembly and force generation has been extensively studied using the actin tail formation model system in the intracellular bacterial pathogen *Listeria monocytogenes* (Cameron et al. 2000). From these analyses, Mogilner and Oster proposed a "tethered ratchet" model in which the sequential events of actin branching, dissociation of new filaments from the load surface, and filament bending contribute to force generation (Mogilner and Oster 2003). At the interface between the PSD and adjacent spine cytoplasm, branched actin network formation may occur based on a similar molecular mechanism. In spines, the PSD structure is mechanically fixed by interactions with the presynaptic membrane. The elastic force created by actin polymerization may be transmitted to the spine plasma membrane outside of the PSD and contribute to changes in spine morphology. The overall organization of the actin meshwork should also depend on the rate of actin filament capping and cofilin-induced severing (Calabrese et al. 2014; Pontrello et al. 2012). To construct a realistic model, the application of a particulate-based model that simulates the behaviors of actin and actin-related proteins within spines may be required (Inoue et al. 2011). Quantitative optical measurements of multiple actin-related molecules will be required to achieve this, as comprehensive data regarding the dynamics of actin-related molecules in spines remain lacking. Modeling of actin dynamics in both extended space and time presents another challenge that will require new strategies for the integration of microscopic Brownian dynamic modeling (Yamaoka et al. 2012) and macroscopic modeling using ordinary differential equations, partial differential equations, or stochastic differential equations (Gardel et al. 2004). Imaging and computational technologies are developing rapidly, and we expect that the modeling of molecular dynamics within dendritic spines may be realized in the near future.

Acknowledgments This work was supported by Grants-in-Aid for Scientific Research (25117006 and 26250014 to S.O.) by the Ministry of Education, Culture, Sports, Science and Technology of Japan (S. O.) and by Core Research for Evolutional Science and Technology from the Japanese Science and Technology Agency (S.O.).

References

Arellano JI, Benavides-Piccione R, Defelipe J, Yuste R (2007) Ultrastructure of dendritic spines: correlation between synaptic and spine morphologies. Front Neurosci 1:131–143

Ashby MC, Isaac JT (2011) Maturation of a recurrent excitatory neocortical circuit by experience-dependent unsilencing of newly formed dendritic spines. Neuron 70:510–521

Bard L, Sainlos M, Bouchet D, Cousins S, Mikasova L, Breillat C, Stephenson FA, Imperiali B, Choquet D, Groc L (2010) Dynamic and specific interaction between synaptic NR2-NMDA receptor and PDZ proteins. Proc Natl Acad Sci U S A 107:19561–19566

Barrow SL, Constable JR, Clark E, El-Sabeawy F, McAllister AK, Washbourne P (2009) Neuroligin1: a cell adhesion molecule that recruits PSD-95 and NMDA receptors by distinct mechanisms during synaptogenesis. Neural Dev 4:17

Bats C, Groc L, Choquet D (2007) The interaction between Stargazin and PSD-95 regulates AMPA receptor surface trafficking. Neuron 53:719–734

Benson DL, Huntley GW (2012) Synapse adhesion: a dynamic equilibrium conferring stability and flexibility. Curr Opin Neurobiol 22:397–404

Biederer T, Sara Y, Mozhayeva M, Atasoy D, Liu X, Kavalali ET, Sudhof TC (2002) SynCAM, a synaptic adhesion molecule that drives synapse assembly. Science 297:1525–1531

Blanpied TA, Kerr JM, Ehlers MD (2008) Structural plasticity with preserved topology in the postsynaptic protein network. Proc Natl Acad Sci U S A 105:12587–12592

Brakeman PR, Lanahan AA, O'Brien R, Roche K, Barnes CA, Huganir RL, Worley PF (1997) Homer: a protein that selectively binds metabotropic glutamate receptors. Nature 386:284–288

Budreck EC, Kwon OB, Jung JH, Baudouin S, Thommen A, Kim HS, Fukazawa Y, Harada H, Tabuchi K, Shigemoto R et al (2013) Neuroligin-1 controls synaptic abundance of NMDA-type glutamate receptors through extracellular coupling. Proc Natl Acad Sci U S A 110:725–730

Calabrese B, Saffin JM, Halpain S (2014) Activity-dependent dendritic spine shrinkage and growth involve downregulation of cofilin via distinct mechanisms. PLoS ONE 9:e94787

Cameron LA, Giardini PA, Soo FS, Theriot JA (2000) Secrets of actin-based motility revealed by a bacterial pathogen. Nat Rev Mol Cell Biol 1:110–119

Cane M, Maco B, Knott G, Holtmaat A (2014) The relationship between PSD-95 clustering and spine stability in vivo. J Neurosci 34:2075–2086

Chazeau A, Mehidi A, Nair D, Gautier JJ, Leduc C, Chamma I, Kage F, Kechkar A, Thoumine O, Rottner K et al (2014) Nanoscale segregation of actin nucleation and elongation factors determines dendritic spine protrusion. EMBO J 33:2745–2764

Chen X, Vinade L, Leapman RD, Petersen JD, Nakagawa T, Phillips TM, Sheng M, Reese TS (2005) Mass of the postsynaptic density and enumeration of three key molecules. Proc Natl Acad Sci U S A 102:11551–11556

Chen X, Winters C, Azzam R, Li X, Galbraith JA, Leapman RD, Reese TS (2008) Organization of the core structure of the postsynaptic density. Proc Natl Acad Sci U S A 105:4453–4458

Chen B, Brinkmann K, Chen Z, Pak CW, Liao Y, Shi S, Henry L, Grishin NV, Bogdan S, Rosen MK (2014) The WAVE regulatory complex links diverse receptors to the actin cytoskeleton. Cell 156:195–207

Cheng D, Hoogenraad CC, Rush J, Ramm E, Schlager MA, Duong DM, Xu P, Wijayawardana SR, Hanfelt J, Nakagawa T et al (2006) Relative and absolute quantification of postsynaptic density proteome isolated from rat forebrain and cerebellum. Mol Cell Proteomics 5:1158–1170

Cho KO, Hunt CA, Kennedy MB (1992) The rat brain postsynaptic density fraction contains a homolog of the Drosophila discs-large tumor suppressor protein. Neuron 9:929–942

Cooney JR, Hurlburt JL, Selig DK, Harris KM, Fiala JC (2002) Endosomal compartments serve multiple hippocampal dendritic spines from a widespread rather than a local store of recycling membrane. J Neurosci 22:2215–2224

Czondor K, Mondin M, Garcia M, Heine M, Frischknecht R, Choquet D, Sibarita JB, Thoumine OR (2012) Unified quantitative model of AMPA receptor trafficking at synapses. Proc Natl Acad Sci U S A 109:3522–3527

Deller T, Merten T, Roth SU, Mundel P, Frotscher M (2000) Actin-associated protein synaptopodin in the rat hippocampal formation: localization in the spine neck and close association with the spine apparatus of principal neurons. J Comp Neurol 418:164–181

Deller T, Korte M, Chabanis S, Drakew A, Schwegler H, Stefani GG, Zuniga A, Schwarz K, Bonhoeffer T, Zeller R et al (2003) Synaptopodin-deficient mice lack a spine apparatus and show deficits in synaptic plasticity. Proc Natl Acad Sci U S A 100:10494–10499

Earnshaw BA, Bressloff PC (2006) Biophysical model of AMPA receptor trafficking and its regulation during long-term potentiation/long-term depression. J Neurosci 26:12362–12373

Fernandez-Busnadiego R, Schrod N, Kochovski Z, Asano S, Vanhecke D, Baumeister W, Lucic V (2011) Insights into the molecular organization of the neuron by cryo-electron tomography. J Electron Microsc (Tokyo) 60(Suppl 1):S137–S148

Fiala JC, Feinberg M, Popov V, Harris KM (1998) Synaptogenesis via dendritic filopodia in developing hippocampal area CA1. J Neurosci 18:8900–8911

Fischer M, Kaech S, Knutti D, Matus A (1998) Rapid actin-based plasticity in dendritic spines. Neuron 20:847–854

Friedman HV, Bresler T, Garner CC, Ziv NE (2000) Assembly of new individual excitatory synapses: time course and temporal order of synaptic molecule recruitment. Neuron 27:57–69

Frost NA, Shroff H, Kong H, Betzig E, Blanpied TA (2010) Single-molecule discrimination of discrete perisynaptic and distributed sites of actin filament assembly within dendritic spines. Neuron 67:86–99

Gardel ML, Shin JH, MacKintosh FC, Mahadevan L, Matsudaira P, Weitz DA (2004) Elastic behavior of cross-linked and bundled actin networks. Science 304:1301–1305

Gerrow K, Romorini S, Nabi SM, Colicos MA, Sala C, El-Husseini A (2006) A preformed complex of postsynaptic proteins is involved in excitatory synapse development. Neuron 49:547–562
Giannone G, Mondin M, Grillo-Bosch D, Tessier B, Saint-Michel E, Czondor K, Sainlos M, Choquet D, Thoumine O (2013) Neurexin-1beta binding to neuroligin-1 triggers the preferential recruitment of PSD-95 versus gephyrin through tyrosine phosphorylation of neuroligin-1. Cell Rep 3:1996–2007
Graf ER, Zhang X, Jin SX, Linhoff MW, Craig AM (2004) Neurexins induce differentiation of GABA and glutamate postsynaptic specializations via neuroligins. Cell 119:1013–1026
Gray EG (1959) Axo-somatic and axo-dendritic synapses of the cerebral cortex: an electron microscope study. J Anat 93:420–433
Grutzendler J, Kasthuri N, Gan WB (2002) Long-term dendritic spine stability in the adult cortex. Nature 420:812–816
Harris KM, Stevens JK (1988) Dendritic spines of rat cerebellar Purkinje cells: serial electron microscopy with reference to their biophysical characteristics. J Neurosci 8:4455–4469
Harris KM, Stevens JK (1989) Dendritic spines of CA 1 pyramidal cells in the rat hippocampus: serial electron microscopy with reference to their biophysical characteristics. J Neurosci 9:2982–2997
Harris KM, Jensen FE, Tsao B (1992) Three-dimensional structure of dendritic spines and synapses in rat hippocampus (CA1) at postnatal day 15 and adult ages: implications for the maturation of synaptic physiology and long-term potentiation. J Neurosci 12:2685–2705
Hering H, Sheng M (2003) Activity-dependent redistribution and essential role of cortactin in dendritic spine morphogenesis. J Neurosci 23:11759–11769
Hirokawa N (1989) The arrangement of actin filaments in the postsynaptic cytoplasm of the cerebellar cortex revealed by quick-freeze deep-etch electron microscopy. Neurosci Res 6:269–275
Holtmaat AJ, Trachtenberg JT, Wilbrecht L, Shepherd GM, Zhang X, Knott GW, Svoboda K (2005) Transient and persistent dendritic spines in the neocortex in vivo. Neuron 45:279–291
Honkura N, Matsuzaki M, Noguchi J, Ellis-Davies GC, Kasai H (2008) The subspine organization of actin fibers regulates the structure and plasticity of dendritic spines. Neuron 57:719–729
Hotulainen P, Hoogenraad CC (2010) Actin in dendritic spines: connecting dynamics to function. J Cell Biol 189:619–629
Hotulainen P, Llano O, Smirnov S, Tanhuanpaa K, Faix J, Rivera C, Lappalainen P (2009) Defining mechanisms of actin polymerization and depolymerization during dendritic spine morphogenesis. J Cell Biol 185:323–339
Husi H, Ward MA, Choudhary JS, Blackstock WP, Grant SG (2000) Proteomic analysis of NMDA receptor-adhesion protein signaling complexes. Nat Neurosci 3:661–669
Iki J, Inoue A, Bito H, Okabe S (2005) Bi-directional regulation of postsynaptic cortactin distribution by BDNF and NMDA receptor activity. Eur J Neurosci 22:2985–2994
Inoue Y, Tsuda S, Nakagawa K, Hojo M, Adachi T (2011) Modeling myosin-dependent rearrangement and force generation in an actomyosin network. J Theor Biol 281:65–73
Irie M, Hata Y, Takeuchi M, Ichtchenko K, Toyoda A, Hirao K, Takai Y, Rosahl TW, Sudhof TC (1997) Binding of neuroligins to PSD-95. Science 277:1511–1515
Isaac JT, Nicoll RA, Malenka RC (1995) Evidence for silent synapses: implications for the expression of LTP. Neuron 15:427–434
Isshiki M, Tanaka S, Kuriu T, Tabuchi K, Takumi T, Okabe S (2014) Enhanced synapse remodelling as a common phenotype in mouse models of autism. Nat Commun 5:4742
Ito-Ishida A, Miyazaki T, Miura E, Matsuda K, Watanabe M, Yuzaki M, Okabe S (2012) Presynaptically released Cbln1 induces dynamic axonal structural changes by interacting with GluD2 during cerebellar synapse formation. Neuron 76:549–564
Jordan BA, Fernholz BD, Boussac M, Xu C, Grigorean G, Ziff EB, Neubert TA (2004) Identification and verification of novel rodent postsynaptic density proteins. Mol Cell Proteomics 3:857–871

Kawabata I, Kashiwagi Y, Obashi K, Ohkura M, Nakai J, Wynshaw-Boris A, Yanagawa Y, Okabe S (2012) LIS1-dependent retrograde translocation of excitatory synapses in developing interneuron dendrites. Nat Commun 3:722

Kim E, Naisbitt S, Hsueh YP, Rao A, Rothschild A, Craig AM, Sheng M (1997) GKAP, a novel synaptic protein that interacts with the guanylate kinase- like domain of the PSD-95/SAP90 family of channel clustering molecules. J Cell Biol 136:669–678

Kim S, Burette A, Chung HS, Kwon SK, Woo J, Lee HW, Kim K, Kim H, Weinberg RJ, Kim E (2006) NGL family PSD-95-interacting adhesion molecules regulate excitatory synapse formation. Nat Neurosci 9:1294–1301

Kim MJ, Futai K, Jo J, Hayashi Y, Cho K, Sheng M (2007) Synaptic accumulation of PSD-95 and synaptic function regulated by phosphorylation of serine-295 of PSD-95. Neuron 56:488–502

Knott GW, Holtmaat A, Wilbrecht L, Welker E, Svoboda K (2006) Spine growth precedes synapse formation in the adult neocortex in vivo. Nat Neurosci 9:1117–1124

Ko J, Kim S, Chung HS, Kim K, Han K, Kim H, Jun H, Kaang BK, Kim E (2006) SALM synaptic cell adhesion-like molecules regulate the differentiation of excitatory synapses. Neuron 50:233–245

Ko J, Fuccillo MV, Malenka RC, Sudhof TC (2009) LRRTM2 functions as a neurexin ligand in promoting excitatory synapse formation. Neuron 64:791–798

Kornau HC, Schenker LT, Kennedy MB, Seeburg PH (1995) Domain interaction between NMDA receptor subunits and the postsynaptic density protein PSD-95. Science 269:1737–1740

Korobova F, Svitkina T (2010) Molecular architecture of synaptic actin cytoskeleton in hippocampal neurons reveals a mechanism of dendritic spine morphogenesis. Mol Biol Cell 21:165–176

Kovar DR, Harris ES, Mahaffy R, Higgs HN, Pollard TD (2006) Control of the assembly of ATP- and ADP-actin by formins and profilin. Cell 124:423–435

Krueger DD, Tuffy LP, Papadopoulos T, Brose N (2012) The role of neurexins and neuroligins in the formation, maturation, and function of vertebrate synapses. Curr Opin Neurobiol 22:412–422

Kuriu T, Inoue A, Bito H, Sobue K, Okabe S (2006) Differential control of postsynaptic density scaffolds via actin-dependent and -independent mechanisms. J Neurosci 26:7693–7706

Linhoff MW, Lauren J, Cassidy RM, Dobie FA, Takahashi H, Nygaard HB, Airaksinen MS, Strittmatter SM, Craig AM (2009) An unbiased expression screen for synaptogenic proteins identifies the LRRTM protein family as synaptic organizers. Neuron 61:734–749

Lucic V, Forster F, Baumeister W (2005) Structural studies by electron tomography: from cells to molecules. Annu Rev Biochem 74:833–865

Lucic V, Kossel AH, Yang T, Bonhoeffer T, Baumeister W, Sartori A (2007) Multiscale imaging of neurons grown in culture: from light microscopy to cryo-electron tomography. J Struct Biol 160:146–156

Mah W, Ko J, Nam J, Han K, Chung WS, Kim E (2010) Selected SALM (synaptic adhesion-like molecule) family proteins regulate synapse formation. J Neurosci 30:5559–5568

Majewska A, Sur M (2003) Motility of dendritic spines in visual cortex in vivo: changes during the critical period and effects of visual deprivation. Proc Natl Acad Sci U S A 100:16024–16029

Majewska AK, Newton JR, Sur M (2006) Remodeling of synaptic structure in sensory cortical areas in vivo. J Neurosci 26:3021–3029

Marrs GS, Green SH, Dailey ME (2001) Rapid formation and remodeling of postsynaptic densities in developing dendrites. Nat Neurosci 4:1006–1013

Matsuda K, Miura E, Miyazaki T, Kakegawa W, Emi K, Narumi S, Fukazawa Y, Ito-Ishida A, Kondo T, Shigemoto R et al (2010) Cbln1 is a ligand for an orphan glutamate receptor delta2, a bidirectional synapse organizer. Science 328:363–368

Matsuzaki M, Ellis-Davies GC, Nemoto T, Miyashita Y, Iino M, Kasai H (2001) Dendritic spine geometry is critical for AMPA receptor expression in hippocampal CA1 pyramidal neurons. Nat Neurosci 4:1086–1092

Meyer G, Varoqueaux F, Neeb A, Oschlies M, Brose N (2004) The complexity of PDZ domain-mediated interactions at glutamatergic synapses: a case study on neuroligin. Neuropharmacology 47:724–733

Miller M, Peters A (1981) Maturation of rat visual cortex. II. A combined Golgi-electron microscope study of pyramidal neurons. J Comp Neurol 203:555–573

Miyata M, Finch EA, Khiroug L, Hashimoto K, Hayasaka S, Oda SI, Inouye M, Takagishi Y, Augustine GJ, Kano M (2000) Local calcium release in dendritic spines required for long-term synaptic depression. Neuron 28:233–244

Mogilner A, Oster G (2003) Force generation by actin polymerization II: the elastic ratchet and tethered filaments. Biophys J 84:1591–1605

Mondin M, Labrousse V, Hosy E, Heine M, Tessier B, Levet F, Poujol C, Blanchet C, Choquet D, Thoumine O (2011) Neurexin-neuroligin adhesions capture surface-diffusing AMPA receptors through PSD-95 scaffolds. J Neurosci 31:13500–13515

Moon IS, Apperson ML, Kennedy MB (1994) The major tyrosine-phosphorylated protein in the postsynaptic density fraction is N-methyl-D-aspartate receptor subunit 2B. Proc Natl Acad Sci U S A 91:3954–3958

Naisbitt S, Kim E, Tu JC, Xiao B, Sala C, Valtschanoff J, Weinberg RJ, Worley PF, Sheng M (1999) Shank, a novel family of postsynaptic density proteins that binds to the NMDA receptor/PSD-95/GKAP complex and cortactin. Neuron 23:569–582

Nam CI, Chen L (2005) Postsynaptic assembly induced by neurexin-neuroligin interaction and neurotransmitter. Proc Natl Acad Sci U S A 102:6137–6142

Nusser Z (1999) A new approach to estimate the number, density and variability of receptors at central synapses. Eur J Neurosci 11:745–752

Nusser Z, Lujan R, Laube G, Roberts JD, Molnar E, Somogyi P (1998) Cell type and pathway dependence of synaptic AMPA receptor number and variability in the hippocampus. Neuron 21:545–559

Okabe S (2007) Molecular anatomy of the postsynaptic density. Mol Cell Neurosci 34:503–518

Okabe S (2013) Fluorescence imaging of synapse formation and remodeling. Microscopy (Oxf) 62:51–62

Okabe S, Hirokawa N (1989) Incorporation and turnover of biotin-labeled actin microinjected into fibroblastic cells: an immunoelectron microscopic study. J Cell Biol 109:1581–1595

Okabe S, Kim HD, Miwa A, Kuriu T, Okado H (1999) Continual remodeling of postsynaptic density and its regulation by synaptic activity. Nat Neurosci 2:804–811

Okabe S, Miwa A, Okado H (2001) Spine formation and correlated assembly of presynaptic and postsynaptic molecules. J Neurosci 21:6105–6114

Opazo P, Labrecque S, Tigaret CM, Frouin A, Wiseman PW, De Koninck P, Choquet D (2010) CaMKII triggers the diffusional trapping of surface AMPARs through phosphorylation of stargazin. Neuron 67:239–252

Palade GE, Palay SL (1954) Electron microscopic observations of interneuronal and neuromuscular synapses. Anat Rec 118:335–336

Palay SL (1958) The morphology of synapses in the central nervous system. Exp Cell Res 5:275–293

Park M, Salgado JM, Ostroff L, Helton TD, Robinson CG, Harris KM, Ehlers MD (2006) Plasticity-induced growth of dendritic spines by exocytic trafficking from recycling endosomes. Neuron 52:817–830

Patterson MA, Szatmari EM, Yasuda R (2010) AMPA receptors are exocytosed in stimulated spines and adjacent dendrites in a Ras-ERK-dependent manner during long-term potentiation. Proc Natl Acad Sci U S A 107:15951–15956

Petersen JD, Chen X, Vinade L, Dosemeci A, Lisman JE, Reese TS (2003) Distribution of postsynaptic density (PSD)-95 and Ca2+/calmodulin-dependent protein kinase II at the PSD. J Neurosci 23:11270–11278

Petes A, Palay SL, Webster Hd (1991) Fine structure of the nervous system: neurons and their supporting cells

Petralia RS, Wang YX, Wenthold RJ (1994a) The NMDA receptor subunits NR2A and NR2B show histological and ultrastructural localization patterns similar to those of NR1. J Neurosci 14:6102–6120

Petralia RS, Yokotani N, Wenthold RJ (1994b) Light and electron microscope distributions of the NMDA receptor subunit NMDAR1 in the rat nervous system using a selective anti-peptide antibody. J Neurosci 14:667–696

Petrini EM, Lu J, Cognet L, Lounis B, Ehlers MD, Choquet D (2009) Endocytic trafficking and recycling maintain a pool of mobile surface AMPA receptors required for synaptic potentiation. Neuron 63:92–105

Pettem KL, Yokomaku D, Luo L, Linhoff MW, Prasad T, Connor SA, Siddiqui TJ, Kawabe H, Chen F, Zhang L et al (2013) The specific alpha-neurexin interactor calsyntenin-3 promotes excitatory and inhibitory synapse development. Neuron 80:113–128

Pocklington AJ, Cumiskey M, Armstrong JD, Grant SG (2006) The proteomes of neurotransmitter receptor complexes form modular networks with distributed functionality underlying plasticity and behaviour. Mol Syst Biol 2(2006):0023

Pollard TD, Blanchoin L, Mullins RD (2000) Molecular mechanisms controlling actin filament dynamics in nonmuscle cells. Annu Rev Biophys Biomol Struct 29:545–576

Pontrello CG, Sun MY, Lin A, Fiacco TA, DeFea KA, Ethell IM (2012) Cofilin under control of beta-arrestin-2 in NMDA-dependent dendritic spine plasticity, long-term depression (LTD), and learning. Proc Natl Acad Sci U S A 109:E442–E451

Portera-Cailliau C, Pan DT, Yuste R (2003) Activity-regulated dynamic behavior of early dendritic protrusions: evidence for different types of dendritic filopodia. J Neurosci 23:7129–7142

Racz B, Blanpied TA, Ehlers MD, Weinberg RJ (2004) Lateral organization of endocytic machinery in dendritic spines. Nat Neurosci 7:917–918

Scheiffele P, Fan J, Choih J, Fetter R, Serafini T (2000) Neuroligin expressed in nonneuronal cells triggers presynaptic development in contacting axons. Cell 101:657–669

Spacek J, Harris KM (1997) Three-dimensional organization of smooth endoplasmic reticulum in hippocampal CA1 dendrites and dendritic spines of the immature and mature rat. J Neurosci 17:190–203

Spacek J, Harris KM (1998) Three-dimensional organization of cell adhesion junctions at synapses and dendritic spines in area CA1 of the rat hippocampus. J Comp Neurol 393:58–68

Spacek J, Hartmann M (1983) Three-dimensional analysis of dendritic spines. I. Quantitative observations related to dendritic spine and synaptic morphology in cerebral and cerebellar cortices. Anat Embryol (Berl) 167:289–310

Star EN, Kwiatkowski DJ, Murthy VN (2002) Rapid turnover of actin in dendritic spines and its regulation by activity. Nat Neurosci 5:239–246

Steiner P, Higley MJ, Xu W, Czervionke BL, Malenka RC, Sabatini BL (2008) Destabilization of the postsynaptic density by PSD-95 serine 73 phosphorylation inhibits spine growth and synaptic plasticity. Neuron 60:788–802

Steward O, Reeves TM (1988) Protein-synthetic machinery beneath postsynaptic sites on CNS neurons: association between polyribosomes and other organelles at the synaptic site. J Neurosci 8:176–184

Sudhof TC (2008) Neuroligins and neurexins link synaptic function to cognitive disease. Nature 455:903–911

Sugiyama Y, Kawabata I, Sobue K, Okabe S (2005) Determination of absolute protein numbers in single synapses by a GFP-based calibration technique. Nat Methods 2:677–684

Svitkina TM, Bulanova EA, Chaga OY, Vignjevic DM, Kojima S, Vasiliev JM, Borisy GG (2003) Mechanism of filopodia initiation by reorganization of a dendritic network. J Cell Biol 160:409–421

Takahashi H, Arstikaitis P, Prasad T, Bartlett TE, Wang YT, Murphy TH, Craig AM (2011) Postsynaptic TrkC and presynaptic PTPsigma function as a bidirectional excitatory synaptic organizing complex. Neuron 69:287–303

Takahashi H, Katayama K, Sohya K, Miyamoto H, Prasad T, Matsumoto Y, Ota M, Yasuda H, Tsumoto T, Aruga J et al (2012) Selective control of inhibitory synapse development by Slitrk3-PTPdelta trans-synaptic interaction. Nat Neurosci 15(389–398):S381–S382

Takenawa T, Suetsugu S (2007) The WASP-WAVE protein network: connecting the membrane to the cytoskeleton. Nat Rev Mol Cell Biol 8:37–48

Tallafuss A, Constable JR, Washbourne P (2010) Organization of central synapses by adhesion molecules. Eur J Neurosci 32:198–206

Tanaka J, Matsuzaki M, Tarusawa E, Momiyama A, Molnar E, Kasai H, Shigemoto R (2005) Number and density of AMPA receptors in single synapses in immature cerebellum. J Neurosci 25:799–807

Tatavarty V, Kim EJ, Rodionov V, Yu J (2009) Investigating sub-spine actin dynamics in rat hippocampal neurons with super-resolution optical imaging. PLoS ONE 4:e7724

Tomita S, Adesnik H, Sekiguchi M, Zhang W, Wada K, Howe JR, Nicoll RA, Bredt DS (2005) Stargazin modulates AMPA receptor gating and trafficking by distinct domains. Nature 435:1052–1058

Tonnesen J, Katona G, Rozsa B, Nagerl UV (2014) Spine neck plasticity regulates compartmentalization of synapses. Nat Neurosci 17:678–685

Triller A, Choquet D (2008) New concepts in synaptic biology derived from single-molecule imaging. Neuron 59:359–374

Tu JC, Xiao B, Naisbitt S, Yuan JP, Petralia RS, Brakeman P, Doan A, Aakalu VK, Lanahan AA, Sheng M et al (1999) Coupling of mGluR/Homer and PSD-95 complexes by the Shank family of postsynaptic density proteins. Neuron 23:583–592

Varoqueaux F, Aramuni G, Rawson RL, Mohrmann R, Missler M, Gottmann K, Zhang W, Sudhof TC, Brose N (2006) Neuroligins determine synapse maturation and function. Neuron 51:741–754

Walton PD, Airey JA, Sutko JL, Beck CF, Mignery GA, Sudhof TC, Deerinck TJ, Ellisman MH (1991) Ryanodine and inositol trisphosphate receptors coexist in avian cerebellar Purkinje neurons. J Cell Biol 113:1145–1157

Wang CY, Chang K, Petralia RS, Wang YX, Seabold GK, Wenthold RJ (2006) A novel family of adhesion-like molecules that interacts with the NMDA receptor. J Neurosci 26:2174–2183

Wang Z, Edwards JG, Riley N, Provance DW Jr, Karcher R, Li XD, Davison IG, Ikebe M, Mercer JA, Kauer JA et al (2008) Myosin Vb mobilizes recycling endosomes and AMPA receptors for postsynaptic plasticity. Cell 135:535–548

Washbourne P, Bennett JE, McAllister AK (2002) Rapid recruitment of NMDA receptor transport packets to nascent synapses. Nat Neurosci 5:751–759

Washbourne P, Liu XB, Jones EG, McAllister AK (2004) Cycling of NMDA receptors during trafficking in neurons before synapse formation. J Neurosci 24:8253–8264

Woo J, Kwon SK, Choi S, Kim S, Lee JR, Dunah AW, Sheng M, Kim E (2009a) Trans-synaptic adhesion between NGL-3 and LAR regulates the formation of excitatory synapses. Nat Neurosci 12:428–437

Woo J, Kwon SK, Kim E (2009b) The NGL family of leucine-rich repeat-containing synaptic adhesion molecules. Mol Cell Neurosci 42:1–10

Woo J, Kwon SK, Nam J, Choi S, Takahashi H, Krueger D, Park J, Lee Y, Bae JY, Lee D et al (2013) The adhesion protein IgSF9b is coupled to neuroligin 2 via S-SCAM to promote inhibitory synapse development. J Cell Biol 201:929–944

Xu W, Schluter OM, Steiner P, Czervionke BL, Sabatini B, Malenka RC (2008) Molecular dissociation of the role of PSD-95 in regulating synaptic strength and LTD. Neuron 57:248–262

Yamaoka H, Matsushita S, Shimada Y, Adachi T (2012) Multiscale modeling and mechanics of filamentous actin cytoskeleton. Biomech Model Mechanobiol 11:291–302

Yasumatsu N, Matsuzaki M, Miyazaki T, Noguchi J, Kasai H (2008) Principles of long-term dynamics of dendritic spines. J Neurosci 28:13592–13608

Yoshida T, Yasumura M, Uemura T, Lee SJ, Ra M, Taguchi R, Iwakura Y, Mishina M (2011) IL-1 receptor accessory protein-like 1 associated with mental retardation and autism mediates synapse formation by trans-synaptic interaction with protein tyrosine phosphatase delta. J Neurosci 31:13485–13499

Yoshimura Y, Yamauchi Y, Shinkawa T, Taoka M, Donai H, Takahashi N, Isobe T, Yamauchi T (2004) Molecular constituents of the postsynaptic density fraction revealed by proteomic analysis using multidimensional liquid chromatography-tandem mass spectrometry. J Neurochem 88:759–768

Yuste R (2010) Dendritic spines. The MIT Press, Cambridege, MA

Ziv NE, Smith SJ (1996) Evidence for a role of dendritic filopodia in synaptogenesis and spine formation. Neuron 17:91–102

Zuber B, Nikonenko I, Klauser P, Muller D, Dubochet J (2005) The mammalian central nervous synaptic cleft contains a high density of periodically organized complexes. Proc Natl Acad Sci U S A 102:19192–19197

Zuo Y, Lin A, Chang P, Gan WB (2005) Development of long-term dendritic spine stability in diverse regions of cerebral cortex. Neuron 46:181–189

Chapter 16
Localising Receptors and Channels Across the Dendritic Arbour

Arthur Bikbaev, Maël Duménieu, Jeffrey Lopez-Rojas, and Martin Heine

Abstract In central neurons, the strength and the number of excitatory and inhibitory synapses to a large extent determine the transfer of information from one neuron to another. The presence of multiple active conductances distributed throughout the highly arborised dendritic tree endows dendrites with the capacity to filter and modulate the evoked synaptic potentials and allows adequate representation of distal synaptic inputs in the soma for further integration and processing. Furthermore, active properties of the dendritic membrane enable dynamic modulation of neuronal excitability and thereby influence the firing output. In this chapter, we will review the distribution of dominant voltage- and ligand-gated ion channels and receptors in dendrites and their contribution to several aspects of dendritic computation.

Keywords Dendrites • Excitability • Ion channels • Receptors • Active dendrites • Lateral mobility

16.1 Introduction

The chemical synapse is the main interface for neuron-to-neuron communication: electrochemical conversion of presynaptic action potential (AP) into postsynaptic potential constitutes the directed transmission of information from one neuron to another. The vast majority of synaptic contacts reside on dendritic arbour that ensures further transmission of postsynaptic potentials along the dendritic membrane to somatic region. Historically, dendrites were considered as cable structures that merely perform electrotonic conduction of postsynaptic potentials for their algebraic summation in the soma/axon hillock (for retrospective, see Yuste and Tank 1996). However, it became evident in the last decades that the dendritic tree of

A. Bikbaev (✉) • M. Heine (✉)
Molecular Physiology Group, Leibniz Institute for Neurobiology, Magdeburg D-39118, Germany
e-mail: bikbaev@yahoo.com; martin.heine@lin-magdeburg.de

M. Duménieu • J. Lopez-Rojas
Neuroplasticity Group, Leibniz Institute for Neurobiology, Magdeburg D-39118, Germany

K. Emoto et al. (eds.), *Dendrites*, DOI 10.1007/978-4-431-56050-0_16

neurons in the CNS plays an active role in spatiotemporal integration of synaptic inputs. Furthermore, dendritic preprocessing of information is associated with dynamic modulation of neuronal excitability and has profound impact in firing behaviour of neurons (reviewed in Magee 2000; Williams and Stuart 2003; Spruston 2008). Combination of electrophysiological, immunohistochemical and imaging techniques has uncovered a very complex organisation of the dendritic membrane, including the presence of various ligand-gated and voltage-dependent ion channels and receptors, as well as their non-uniform distribution throughout the dendritic arbour (Fig. 16.1). Specific topology of the active conductances along the dendrite was proposed to reflect their different roles in dendritic segment- or branch-specific representation of distinct synaptic signals (Laudanski et al. 2014; Magee 1998). Non-uniform distribution of the channels and receptors throughout the dendritic tree allows a spatiotemporal segregation and selective amplification of the response of co-activated neighbouring inputs (reviewed in Spruston 2008; Migliore and Shepherd 2002; Hausser et al. 2000). Similarly to synaptic areas, the information in the dendritic arbour is encoded in changes of the membrane potential. Therefore, the localisation and density of ion channels and receptor complexes are key aspects to understand the integrative properties of dendrites that to a large extent determine the neuronal output, thereby shaping the contribution of an individual neuron to computational capacity of the network (reviewed in Migliore and Shepherd 2002; London and Häusser 2005; Reyes 2001) and allocation of memory traces in the brain (Rogerson et al. 2014). In this chapter, we will focus on pyramidal neurons of hippocampal CA1 area as a cell type most extensively explored in terms of structural and functional properties of their dendritic arbour.

16.2 Distribution of Ion Channels and Receptors in Dendritic Membrane

16.2.1 Ion Channels

16.2.1.1 Voltage-Gated Sodium Channels

Since pioneering works of Hodgkin and Huxley (Hodgkin and Huxley 1952a, b, c, d), the role of voltage-dependent activation of sodium and potassium currents in generation and propagation of AP remains one of the cornerstones of modern neuroscience. Inward sodium currents promote depolarisation of the cell and are mediated by voltage-gated sodium channels (VGSCs). Typical features of sodium channel-mediated currents include their fast voltage-dependent activation and inactivation, as well as slow (hundreds of milliseconds to seconds) inactivation for some VGSC isoforms (Rudy 1978; Chandler and Meves 1970; Adelman and Palti 1969). Functional VGSCs consist of pore-forming α-subunit associated with one or two auxiliary β-subunits that modulate the properties of the sodium channel.

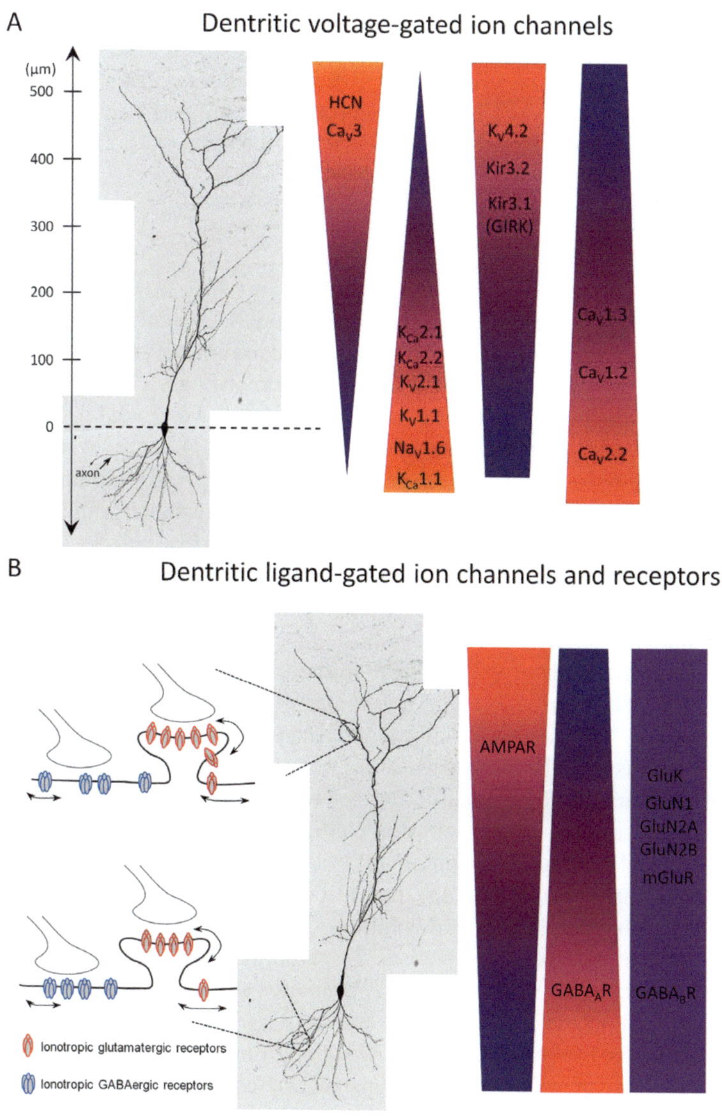

Fig. 16.1 Dendritic arbour of CA1 pyramidal neurons is characterised by non-uniform distribution of voltage- and ligand-gated ion channels and receptors. (**a**) Distal dendrites of CA1 pyramidal neuron (adult mouse, biocytin labelling) can be located hundreds of micrometres away from soma, with the density of several active conductances being dramatically changed along the proximo-distal axis of the dendritic tree. (**b**) Albeit the gradient of density is prominent for AMPARs and GABAARs, ligand-gated channels and receptors are distributed more homogeneously than voltage-gated ion channels

Ten isoforms of α-subunit (Nav1.1–Nav1.9 and voltage-independent channel-like Na_X) and four isoforms of β-subunit (β1–β4) were identified to date (Goldin 2001; Catterall 2012). In central neurons, Nav1.1–Nav1.3 and Nav1.6 are primary sodium channels, with Nav1.7–Nav1.9 being expressed predominantly in peripheral neurons (Wood and Baker 2001; Trimmer and Rhodes 2004).

Density and Distribution of VGSCs

In an early study, analysis of isoform-specific mRNA expression in rat central neurons revealed that Nav1.1 was expressed predominantly in somato-dendritic compartment, whereas Nav1.2 was localised primarily in axons (Westenbroek et al. 1989). Furthermore, in the human brain, Nav1.3 was found to be localised in the somato-dendritic compartment (Whitaker et al. 2001). More recently, *Lörincz* and *Nusser* employed highly sensitive electron microscopy and immunogold labelling to find that Nav1.1 in CA1 pyramidal cells was undetectable, while Nav1.2 was present in the proximal part of axon initial segment (AIS), but not in somata or proximal dendrites. Moreover, authors revealed the presence of Nav1.6 in dendrites of CA1 pyramidal neurons, although at 35–80-fold lower density than in the AIS (Lörincz and Nusser 2010).

Given the critical role of VGSCs in AP initiation and propagation, the analysis of their density and distribution was particularly detailed for axonal membrane. Predictably, the highest density of sodium channels found in axonal membrane (Lörincz and Nusser 2010; Kole et al. 2008) was associated with lowest threshold of AP initiation particularly in axon initial segment (Kole and Stuart 2008; Mainen et al. 1995, but see Colbert and Johnston 1996; Colbert and Pan 2002). In dendritic arbour, electrophysiological mapping showed relatively high and rather uniform density of sodium channels in the main apical dendrite in pyramidal neurons (Magee and Johnston 1995a) and inhibitory interneurons in hippocampal CA1 area (Martina et al. 2000), as well as in pyramidal neurons in layer 5 of the neocortex (Stuart and Sakmann 1994; Williams and Stuart 2000a). However, in vivo extracellular recording of CA1 and CA3 pyramidal neurons showed two- to eightfold faster backpropagation of APs along the somato-dendritic membrane and steep decrease of sodium channel density (Buzsaki et al. 1996), consistent with the gradual decrease of Nav1.6 density along the proximo-distal axis more recently reported in CA1 pyramidal neurons (Lörincz and Nusser 2010). A decrease of sodium channel density with the distance from soma was also found in thalamocortical neurons (Williams and Stuart 2000b) and Purkinje cells in the cerebellum (Callaway and Ross 1997), but not in mitral cells in olfactory bulb (Debarbieux et al. 2003; Bischofberger and Jonas 1997).

Modulation of VGSCs

Association of α-subunit with auxiliary β-subunits 1–4 strongly affects the trafficking and biophysical properties of the sodium channel. The β2- and β4-subunits are covalently linked to α-subunit via disulphide bridge, whereas β1 and β3 bind to α-subunit non-covalently. Association of β1-subunit is known to accelerate channel activation and inactivation and increases current density (Isom et al. 1992; Patton et al. 1994). Additionally, interaction of Navβ1 with pore-forming subunit of Kv4.2 was found to increase the density of Kv4.2-mediated A-type potassium currents (Marionneau et al. 2012). An increase of surface expression of β4-subunit resulted in larger fraction of persistent sodium currents that modulate the resting membrane potential (Aman et al. 2009). Sodium channels are also subject of modulation by phosphorylation and glycosylation (reviewed in Shao et al. 2009; Scheuer 1994), with the latter being essential for the expression of VGSCs in neurons (Zona et al. 1990). The α-subunit was found to be a good substrate for phosphorylation by cAMP-dependent protein kinase A (PKA) (Murphy et al. 1993; Smith and Goldin 1997) and C (PKC) (Costa and Catterall 1984a, b; Li et al. 1993; West et al. 1991; Numann et al. 1991), leading to isoform-specific reduction of maximal conductance and the changes of VGSC gating in neurons (Marban et al. 1998). Modulation of VGSCs by tyrosine phosphorylation in central neurons through association with receptor protein tyrosine phosphatase-β (RPTP-β) was associated with faster inactivation, negative shift of voltage dependence and reduction of sodium current (Ratcliffe et al. 2000).

16.2.1.2 Potassium Channels

Potassium channels are considered the main regulators of dendritic excitability. There is a very large diversity in this family, but they can in general be classified into two broad groups: voltage-dependent and calcium-dependent potassium channels.

Voltage-Dependent Potassium Channels (Kv)

Cell-attached patch-clamp recordings from apical dendrites of CA1 neurons show a voltage-dependent K^+ macroscopic current with a transient and a sustained component (Rudy 1988; Storm 1990). Interestingly, the magnitude of the transient component increases linearly with distance from soma. This transient component shows rapid kinetics of activation, deactivation and inactivation performing as an efficient "shock absorber". It starts to activate at relatively hyperpolarised potentials (around −50 mV), with around half of the channels being available at resting potential. It is quite remarkable that the channels present at more distal locations have an activation curve shifted towards hyperpolarised potentials (around 12 mV) in comparison to channels at soma or proximal dendrites (Hoffman et al. 1997).

The higher density of the channels and their easier activation (more hyperpolarised activation) boost the impact of the transient component in more distal dendrites, resulting in smaller amplitudes of backpropagating APs (bAPs), reduced excitatory postsynaptic potentials (EPSPs) and higher threshold for dendritic spike initiation. In contrast, the sustained component is homogeneously distributed along the apical dendrites. It has much slower activation kinetics and starts to activate at more depolarised potentials than the transient current. It shows very little inactivation (Hoffman et al. 1997).

The K^+ macroscopic currents show quite a variability that suggested the sampling of only a subpopulation of the total current, comprising different types of K^+ conductances. Indeed, single-channel recordings showed that there are four main voltage-dependent potassium conductances in dendrites of CA1 pyramidal cells: the transients A type and D type and the sustained delayed rectifier and M type (Chen and Johnston 2004). The single-channel recordings and the pharmacological profile of the currents indicate that the transient component of the macroscopic current is mostly due to A-type conductance, while the sustained component is mainly due to delayed rectifier channels (Hoffman et al. 1997; Chen and Johnston 2004).

A-Type Channels

Besides the above-mentioned peculiarities of A-type current in CA1 dendrites, it is also remarkable for the plastic features it has. In an elegant study, Frick et al. (2004) using dendritic patch-clamp recordings in combination with Ca^{2+} imaging found a local activity-dependent modulation of A-channels (Frick et al. 2004). Theta burst stimulation to apical dendritic branches of CA1 neurons induced a spatially restricted increase in the bAP and its associated Ca^{2+} influx. These changes were accompanied with a shift of the inactivation curve of A-type K^+ channels to hyperpolarised potentials, resulting in a reduction of the channels available for opening at resting potential. The study showed that changes in intrinsic excitability can also be local. Later work has suggested that the shift in the inactivation curve is just a beginning of a modulation process followed by the internalisation of the receptors from the activated synapses and a decrease in the total A-type current (Jung and Hoffman 2009; Kim et al. 2007).

From all the pore-forming subunits expressed in the hippocampus, the Kv1.4, Kv4.1, Kv4.2 and Kv4.3 have the kinetics and pharmacological profile of A-type currents in CA1 cells (Coetzee et al. 1999), but only the Kv4.2 shows a strong somato-dendritic expression in these neurons (Sheng et al. 1992; Maletic-Savatic et al. 1995; Serodio et al. 1996; Varga et al. 2000). Indeed, Kv4.2 is the major contributor to the A-type current in CA1 pyramidal cells (Kim et al. 2005), and it has been found enriched in dendritic spines (Kim et al. 2005; Alonso and Widmer 1997). Kv4.2 possesses phosphorylation sites for several kinases, including PKA, PKC, MAPK and CaMKII (Birnbaum et al. 2004; Schrader et al. 2002). The activation of these kinases affects the kinetics and amplitude of A-currents and channel expression (Hoffman and Johnston 1998; An et al. 2000; Anderson et al. 2000; Yuan et al. 2002; Varga et al. 2004). Kv4.2 has also numerous auxiliary subunits including the Kv channel-interacting proteins (KChIPs) and dipeptidyl peptidase-like proteins (DPPs) (Jerng and Pfaffinger 2014). Immunohistochemical studies suggest that hippocampal neurons express mainly KChIP2, KChIP4 and DPP6 (Rhodes et al. 2004; Zagha et al. 2005; Nadal et al. 2003). The auxiliary subunits can profoundly affect Kv4.2 channel properties and distribution (Wang et al. 2013; Sun et al. 2011). Different gradients of intracellular signalling molecules/auxiliary subunits have been proposed as a possible cause of the non-uniform distribution and properties of A-type channels along the apical dendrites. In fact, it was recently shown that DPP6 is one of the key factors in establishing not only the naturally occurring gradient of A-current density in CA1 apical dendrites but also the hyperpolarizing shift in the activation curve of the distally located channels (Sun et al. 2011). Since DPP6 has a large extracellular component, it could mediate the interaction with the extracellular matrix or glial cells (Lin et al. 2013, 2014). However, KChIPs and DPPs seem to be more than particular auxiliary subunits of A-type channels. Recent studies have shown that they can also modulate other ion channels/receptors or even act as intermediaries providing the necessary interface for the interaction of A-channels with other membrane proteins, including voltage-dependent Ca^{2+} channels and NMDA receptors (Anderson et al. 2010, 2013; Thomsen et al. 2009; Zhang et al. 2010). These cross-channel interactions widen even more our conception of the ion channels as components of large macromolecular complexes.

Calcium-Dependent Potassium Channels

In comparison to the voltage-dependent channels, little is known about calcium-dependent potassium channels in dendrites of CA1 pyramidal cells. At least three different channel types were shown to be involved in AP repolarisation, after hyperpolarisation (AHP) of different durations and spike frequency adaptation (Sah 1996; Sah and Faber 2002; Gu et al. 2007). The first dendritic recordings in CA1 neurons demonstrated that the duration of APs is greater in dendrites than in soma and that there is almost no AHP in dendrites after single or repetitive APs (Spruston et al. 1995). In later studies, it was found that the fast Ca^{2+}-activated K^+ current, Ic, mainly mediated by BK channels is present only at very proximal dendritic locations (Johnston et al. 2000). At least partially, this explains the widening of the AP and slowing repolarisation of the membrane as AP propagates back into the dendrites. Based on knockout mouse models, field potential recordings and modelling, BK channels were shown to promote high-frequency firing and to facilitate AP frequency adaptation (Gu et al. 2007). Small conductance calcium-activated potassium channels are mainly found in or close to the postsynaptic density along the dendrite (Stocker and Pedarzani 2000; Sailer et al. 2004), where they control postsynaptic input via sensing intracellular calcium changes induced by activation of NMDA receptors (Lin et al. 2008; Ngo-Anh et al. 2005). Channels distributed along the proximal dendrite and somatic membrane have been reported to control neuronal excitability in situations where other potassium channels (M-type) are not active (Adelman et al. 2012) or suppressed (Chen et al. 2014).

Clustering and density of potassium channels along the neuronal membrane only now start to be accessible. Using freeze fracture electron microscopy, several laboratories have shown that the functional diversity is partially reflected in the expression density and distribution (Kirizs et al. 2014; Kaufmann et al. 2010).

16.2.1.3 HCN Channels

Hyperpolarisation-activated cyclic-nucleotide-gated (HCN) channel subunits are responsible for h currents (Ih). Ih is a non-specific cation current activated by hyperpolarisation, a quite unusual feature, and deactivated by depolarisation. A portion of HCN channels is open at rest; therefore, Ih current significantly contributes to the resting membrane potential and input resistance of the cells (Wahl-Schott and Biel 2009; He et al. 2014; Pape 1996). Similarly to A-current, Ih shows nearly sevenfold increase in distal dendrites and a hyperpolarisation-shifted activation curve (Magee 1998, 2008; Notomi and Shigemoto 2004; Lorincz et al. 2002). Some of the main functional implications of the Ih current are a general reduction in dendritic excitability and modulation of temporal integration of inputs. Its non-homogeneous distribution along the apical dendrites of CA1 neurons contributes to the normalisation of the temporal integration, overcoming the filtering effects of the passive properties of dendrites (Magee 1998, 1999, 2000).

As observed with K^+ channels, HCN channels can be modified in an activity-dependent manner. Interestingly, both up-regulation (Fan et al. 2005) and down-regulation (Wang et al. 2003) of Ih currents have been reported following LTP-like stimulations. A possible explanation for such contradictory results was provided by *Campanac* and *Debanne* experiments (Campanac et al. 2008) showing that, depending on the stimulation strength used for the induction protocol, either increase or decrease of Ih current can occur. Therefore, Ih may undergo Hebbian or homeostatic plasticity according to the particular induction conditions.

HCN channels are highly modulated by the neurochemical environment. Small molecules, protein kinases and interacting protein can modify the kinetics, gating and expression of these channels (Wahl-Schott and Biel 2009; Lewis et al. 2010). This modulation is however subunit specific. For example, the classical modulator cAMP has a strong effect on HCN2 and HCN4 channels, but just a small one on HCN1 and HCN3 (Chen et al. 2001). Large variety of modulatory mechanisms opens as well the possibility for the regulation of HCN channels through the regulation of its modulators. In this sense, the accessory subunit tetratricopeptide-containing Rab8b-interacting protein (TRIP8b) is an interesting example (Bankston et al. 2012; Zolles et al. 2009). TRIP8b binds to HCN1 and HCN2, the main subunits expressed in the hippocampus, and strongly regulates their gating properties and trafficking. Animals with knocked out TRIP8b show reduced dendritic surface expression of HCN (Lewis et al. 2011). More than nine splice isoforms of

TRIP8b exist, with some up-regulating and others down-regulating HCN expression (Lewis et al. 2009; Santoro et al. 2009), adding further complexity to this regulatory network. Therefore, the possibility exists that plasticity changes not the net TRIP8b expression, but the relative expression of some isoforms of TRIP8b, thereby causing selective alterations in expression and gating of HCN channels (Shah et al. 2010). TRIP8b has multiple phosphorylation consensus sites for different protein kinases, such as CaMKII and PKC (Lewis et al. 2009; Santoro et al. 2009). Therefore, the question arises whether the changes in Ih seen upon increased activation of these kinases are a product of direct modification of HCN channels or rather an indirect modulatory effect through TRIP8b. A curious example occurs in prefrontal cortical neurons, where HCN1 channels colocalise with α2-adrenoreceptors (Wang et al. 2007). The activation of these receptors results in a decrease of cAMP and HCN1 activity. Since acute changes in cAMP have almost no effects on HCN1 receptors, regulation through regulatory subunits like TRIP8b seems like a plausible mechanism (Shah et al. 2010).

16.2.1.4 Voltage-Gated Calcium Channels

Voltage-gated calcium channels (VGCCs) are multiunit membrane complexes mediating transient calcium influx upon depolarisation and consisting of pore-forming α1-subunit and auxiliary α2δ-, β- and γ-subunits (reviewed in Catterall 2000; Eggermann et al. 2012). Calcium is one of the major signalling molecules in neurons; thus, fluxes of calcium not only change the membrane potential due to the electric charge Ca^{2+} carries but also trigger plethora of intracellular signalling pathways (Sjostrom et al. 2008). The crucial role of calcium channels in multiple phenomena, such as neurotransmitter release, membrane excitability and plasticity, has been shown for several types of VGCCs: L type (Grover and Teyler 1990; Magee and Johnston 1997; Bi and Poo 1998; Lee et al. 2009), P/Q type (Ito et al. 1995), T type (Magee and Johnston 1997; Ito et al. 1995; Isomura et al. 2002) and R type (Yasuda et al. 2003). However, the detailed analysis of subtype-specific density and localisation of VGCCs in dendrites and spines is hampered by the lack of selective immunochemical tools. Therefore, the mapping of dendritically expressed calcium channels was performed primarily using electrophysiological approaches and calcium imaging.

Localisation of VGCCs

High-Voltage-Activated (HVA) Calcium Channels

The data on distribution of P/Q channels in dendrites are the most controversial. In several studies, a uniform distribution of P/Q channels was found at least in dendritic shafts (Christie et al. 1995; Westenbroek et al. 1995; Kavalali et al. 1997). Other studies showed the absence of these channels both in shafts and in spines of CA1 neurons (Magee and Johnston 1995a; Sabatini and Svoboda 2000; Hoogland and Saggau 2004; Bloodgood and Sabatini 2007). Despite their lower density compared to other HVA channels (Christie et al. 1995), P/Q channels have been shown to be responsible, together with N type, for most of the HVA channel-mediated calcium fluxes in distal dendrites (Kavalali et al. 1997).

Several histological and electrophysiological studies have shown expression of N-type channels in a seemingly uniform fashion along the dendritic tree (Magee and Johnston 1995a; Kavalali et al. 1997; Westenbroek et al. 1992; Mills et al. 1994; Azimi-Zonooz et al. 2001; Losonczy and Magee 2006), although Christie et al. using calcium imaging found that these channels were present almost exclusively in the very proximal ($\leq$50 μm from the soma) dendrites (Christie et al. 1995). This discrepancy between channels' expression and calcium fluxes could be at least partially explained by the fact that activation of channels along the whole surface of neuronal membrane is not uniform. Indeed, subcellular localisation, structural alterations as well as interactions with other molecular partners may affect the sensitivity for activation of similar channels to an identical depolarisation (Wheeler et al. 2012). In addition, N-type channels were reported to be present also in spines, at least in proximal dendritic region (Sabatini and Svoboda 2000; Bloodgood and Sabatini 2007; Mills et al. 1994).

L-type channels are highly expressed in somatic and dendritic regions, where they form clusters (Christie et al. 1995; Westenbroek et al. 1990, 1992; Hell et al. 1993). They are also expressed along the dendrites with diminishing densities (Magee and Johnston 1995a; Christie et al. 1995; Sabatini and Svoboda 2000; Bloodgood and Sabatini 2007; Westenbroek et al. 1992; Hell et al. 1993), with a reduced or negligible role in distal dendrites (Christie et al. 1995; Kavalali et al. 1997). L-type channels were reported to be present in spines (Lee et al. 2009; Yasuda et al. 2003; Hoogland and Saggau 2004; Bloodgood and Sabatini 2007; Di Biase et al. 2011), although their contribution to synaptic calcium influx seems to be minor (Lee et al. 2009; Hoogland and Saggau 2004, but see Sabatini and Svoboda 2000). L-type participation to spine calcium fluxes has been shown to be strongly influenced by adrenergic stimulation through α2-receptors (Hoogland and Saggau 2004).

R-type channel-mediated currents and calcium fluxes have been described along dendrites in most of the studies (Magee and Johnston 1995a; Kavalali et al. 1997; Losonczy and Magee 2006) and seem to be distributed in a gradually increasing fashion with distance to the soma (Magee and Johnston 1995a; Christie et al. 1995; Kavalali et al. 1997). A spine-specific location of these channels renders them as the main source of VGCC-mediated influx in spines (Yasuda et al. 2003; Sabatini and Svoboda 2000; Hoogland and Saggau 2004; Bloodgood and Sabatini 2007).

T-type LVA channels are present in dendritic membrane with an apparent increase of density (or functional impact) in proximo-distal axis (Magee and Johnston 1995a; Christie et al. 1995; Kavalali et al. 1997; Losonczy and Magee 2006; Magee et al. 1995; McKay et al. 2006), leading to particularly strong impact of T-type channels in distal dendrites (Magee and Johnston 1997; Christie et al. 1995; Kavalali et al. 1997; Gillessen and Alzheimer 1997). The expression of T-type-mediated calcium fluxes/currents in spines is difficult to address, mainly due to pharmacological specificity issues. Some results argue in favour of a presence of T-type in spines (Sabatini and Svoboda 2000; Hoogland and Saggau 2004), but the possibility that they are present only in shafts cannot be ruled out (Yuste et al. 1999). Interestingly, alteration of the expression level of dendritic Cav3.2 subtype of the T-type family has been strongly linked with abnormal epileptic activity (Becker et al. 2008), confirming the dendritic distribution of T-type channels and highlighting their importance in regulating neuronal network activity. T-type channels exhibit a striking and subtype-specific ability to interact with diverse molecular partners, theoretically allowing them to serve different functions depending on their subcellular localisation (soma, axon and dendritic shafts or spines), binding properties and molecular environment, as well as lateral mobility and stability of membrane expression (reviewed in Turner and Zamponi 2014).

16.2.2 Ligand-Gated Ionotropic Receptors

Many neurotransmitter receptors are expressed in the dendritic membrane. Although most of the receptors are located primarily in the postsynaptic density, a large population has been described to be in the perisynaptic zone or along the dendritic shaft. The functional significance of such extrasynaptic receptors is well investigated for NMDA, AMPA, kainate and GABAA receptors and was partially addressed in other chapters of this book and several excellent reviews (Jaskolski et al. 2005; Brickley and Mody 2012; Carta et al. 2014; Traynelis et al. 2010; Vizi et al. 2010). Here, we will provide a brief overview.

16.2.2.1 Localisation of Ionotropic Receptors Along the Dendritic Membrane

Many ionotropic receptors are localised not exclusively within the postsynaptic zone but within a certain radius around the synapse. The specificity of targeting, local surface mobility and dynamic interactions with intra- and extracellular binding partners are responsible for gradually changing receptor density as a function of distance from the synapse. In addition to the mixed pool in and around the synapse, there are clear subsets of receptors along the dendritic membrane that are not part of the synaptic receptor population. This separation of mainly synaptic and mainly extrasynaptic receptors is best demonstrated for GABAA, kainate, AMPA and NMDA receptors. Their functional impact on the dendritic signal integration and propagation is driven mainly by the concentration of ambient glutamate or GABA being released by glial cells (Hassinger et al. 1995; Parpura et al. 1994) or spilled out of active synapses (Alger and Nicoll 1982; Kullmann et al. 1996). Due to differences in the molecular composition, these extrasynaptic ionotropic receptors often have different kinetic properties than those belonging to synaptic population. As a consequence, extrasynaptic receptors act as slow tonic conductances altering the excitability of the dendrite, in contrast to the fast activation and inactivation of synaptic receptors triggering transient changes of the membrane potential.

We have a very detailed picture about AMPA, NMDA, kainate and GABAA receptors due to their abundance and availability of experimental tools of detection. Much less is known about the impact of purinergic, serotonergic, acetylcholinergic and steroid hormone receptors. The multitude of ionotropic and metabotropic receptors and the space restriction limit this review only to some ionotropic receptor groups. To complete the picture, the reader will find further information in excellent recent reviews (Vizi et al. 2010; Niswender and Conn 2010; Bettler et al. 2004; Teles-Grilo Ruivo and Mellor 2013).

AMPA Receptors

Along the dendritic tree, there is a clear gradient in AMPA receptor density that increases severalfold in the distal regions of the dendrite (Andrasfalvy and Magee 2001, 2004; Shipman et al. 2013). The kinetic properties of AMPA receptors are not different along the dendrite of CA1 neurons (Andrasfalvy and Magee 2001, 2004). Endo- and exocytosis of AMPA receptors were found to take place in close neighbourhood to the postsynaptic density (PSD) (Racz et al. 2004; Anggono and Huganir 2012) or even in the PSD (Gerges et al. 2006). Here, AMPA receptors either diffuse randomly in the membrane or are trapped within endocytic pits. Depending on the position of these endocytic pits, AMPA receptors can participate in plasticity processes as a reserve pool of receptors to rapidly change the postsynaptic population of receptors (Petrini et al. 2006, 2009) in parallel to altered local exocytosis of AMPA receptors (Makino and Malinow 2009). The subunit composition of extrasynaptic AMPA receptors might transiently and activity-dependently change due to the increased exocytosis of GluA1-containing AMPA receptors (Makino and Malinow 2009; Kopec et al. 2006; Kessels et al. 2009), but has not been explored in detail. The molecular machinery involved in AMPA receptor recycling and de novo synthesis has been shown to be distinct from presynaptic mechanisms (Ehlers 2000; Kennedy et al. 2010). The exocytosis of AMPA receptor-containing vesicles depends on SNARE proteins and complexins and was reported to involve Syntaxin-3 (Jurado 2014; Jurado et al. 2013; Ahmad et al. 2012), instead of Syntaxin-4 as reported before (Kennedy et al. 2010). When AMPA receptors are inserted into the plasma membrane, they can explore the local environment by passive lateral diffusion agitated by lipids and other proteins due to thermal agitation (Groc et al. 2004; Heine et al. 2008; Czondor et al. 2012). Depending on their molecular composition, AMPA receptors have variable affinities to postsynaptic scaffold proteins within the PSD, as well as altered kinetic properties (Priel et al. 2005; Tomita et al. 2007; Bats et al. 2007; Zhang et al. 2014). Three groups of auxiliary AMPA receptor subunits, namely, TARPs, Shisa and cornichons, have been identified and are involved in receptor surface expression, localisation and anchoring to the PSD (Bats et al. 2007; von Engelhardt et al. 2010; Schwenk et al. 2009; Rouach et al. 2005; Constals et al. 2015). The localisation of AMPA receptor complexes equipped with different auxiliary subunits along the dendritic arbour has been not investigated in detail. Based on the differential tuning of kinetic properties, it might be relevant for integration of postsynaptic potentials from spatially distinct inputs and hence for shaping the contribution of AMPA receptors to the tuning of synaptic output and network activity. Whether a differential molecular composition of AMPA receptor complexes exists along the dendritic membrane of individual cells remains to be addressed. The dynamic combination of several different auxiliary subunits seems possible (Milstein et al. 2007; Khodosevich et al. 2014), since desensitised receptors can escape and diffuse along the dendrite (Constals et al. 2015). An ionotropic action of AMPA receptors along the dendritic membrane outside synapses has been suggested as a tonic depolarising current facilitating the activation of NMDA receptors (Schmidt-Salzmann et al. 2014). A transient activation seems to be rather unlikely in non-pathological conditions, since the glutamate affinity of AMPA receptors is very low and the concentration of ambient glutamate might force receptors rather to undergo desensitisation than activation (Featherstone and Shippy 2008). Therefore, the large difference in the diffusion coefficient between synaptic and extrasynaptic receptors might reflect the different activity status of these receptors, trapped in the desensitised status depending on the concentration of ambient glutamate (Constals et al. 2015).

NMDA Receptors

The synaptic function of NMDA receptors for synaptic activity, plasticity and structural changes is regularly reviewed in many aspects (Vizi et al. 2010; Shipton and Paulsen 2014; Paoletti et al. 2013; Sanz-Clemente et al. 2013). Different functions have been assigned to extrasynaptic NMDA receptors, including neurodegeneration, cell death, modulation of synapse to nucleus signalling and structural plasticity in immature and mature neurons (Iafrati et al. 2014; Hardingham and Bading 2010; Martel et al. 2012; van Zundert et al. 2004; Karpova et al. 2013; Espinosa et al. 2009). Extrasynaptic NMDA receptors mainly contain GluN2B subunits and have different kinetic properties (longer opening time) than the major population of synaptic NMDA receptors in mature synapses. Further details about composition, functional implications and signalling pathways were reported recently (Hardingham and Bading 2010; Papouin and Oliet 2014; Parsons and Raymond 2014; Miller et al. 2014). The function for dendritic signal integration of extrasynaptic NMDA receptors depends on the C-terminal domain of the GluN2 subunits and their interaction with PDZ-domain-containing scaffold proteins (Martel et al. 2012). NMDA receptors open only under the condition that the intracellular magnesium block is released (Nowak et al. 1984) and glutamate, as well as co-agonists, is bound to the receptor (Panatier et al. 2006). Thus, activation of extrasynaptic NMDA receptors depends on forward and backward propagation of synaptic potentials and backpropagating APs to remove the magnesium block, as well as on depolarising contribution of tonic depolarising conductances. Depending on the frequency and amplitude of bAPs, NMDA receptors can tune the activity of HCN channels in the dendrites of CA1 hippocampal neurons and thus influence, apart from their synaptic function, also the signal integration along the dendrite (Wu et al. 2012). The distribution of NMDA receptors containing the GluN2A or the GluN2B subunit in immature cells differs substantially from that in mature neurons. Nevertheless, the total amount of extrasynaptic NMDA receptors, for example, in granular cells of the dentate gyrus, does not change (Schmidt-Salzmann et al. 2014). The substantial alteration of subunit composition during neuronal development is caused by a stronger stabilisation of GluN2A-containing receptors in the synapse and looser coupling of GluN2B-containing receptors to the synaptic scaffold reflected in their altered surface diffusion (Groc et al. 2006). Additionally, the synaptic receptor composition is influenced during development by the extracellular matrix glycoprotein Reelin, which induces the switch in the postsynaptic NMDA receptor composition and affects the surface mobility of NMDA receptors (Groc et al. 2006).

Kainate Receptors

Kainate receptors are described to be located inside and outside the synapse, as well as on the presynaptic membrane (Carta et al. 2014). Due to the large number of subunit combinations and strong differences in their kinetic properties, different expression profiles along the dendrite have been postulated (Jaskolski et al. 2005; Contractor et al. 2011). Electrophysiological experiments on wild-type and knockout models confirm the specific expression pattern of different kainate receptor populations. However, the lack of antibodies, as well as the low density within the dendritic membrane, greatly complicates the direct localisation studies. Furthermore, the dynamics of endo- and exocytosis of kainate receptors makes the identification of a default state of receptor populations difficult. Interestingly, kainate receptors are as dynamic as AMPA receptors, but leave the synapse after CaMKII activation, in contrast to AMPA receptors that get stabilised in the synapse (Carta et al. 2013). This observation underlines the distinct function of kainate receptors as fast ionotropic modulators of synaptic activity in the dendritic and perisynaptic membrane, apart from their presynaptic function (Contractor et al. 2011). The receptors' sensitivity to kainate (glutamate) and their kinetic properties, but not the localisation pattern, are dependent on the expression of the auxiliary subunit NETO1 (Straub et al. 2011). Their specific location at several synapses and interneurons make kainate receptors particularly important for the temporal integration of network activity. Additionally to ionotropic signalling, kainate receptors have also metabotropic action and can shape, via modulation of potassium channels, the slow after hyperpolarisation of action potentials in CA1 neurons (Melyan et al. 2002).

GABAA Receptors

Depending on the developmental stage and the cell population, ionotropic GABA receptors (GABAARs) have very different functions in neuronal networks. GABAARs are divided into several subpopulations depending on their subunit stoichiometry and preferential distribution in the dendritic and synaptic membrane. As constitutive pentamers, GABAARs are very heterogeneous in their kinetic properties and pharmacology (Fritschy and Panzanelli 2014). The γ2-containing GABAARs are localised predominantly inside the postsynaptic membrane but are also present in the perisynaptic environment due to trafficking and lateral diffusion in and out of the synaptic zone (Bannai et al. 2009; Petrini and Barberis 2014; Petrini et al. 2014). These receptors are rapidly activated in response to GABA and control important neuronal functions, such as synchronisation of neuronal networks and timing of action potentials. The expression of different combinations of the pentamer out of different α1-, α2- and α3-subunits determine their inactivation and were suggested to tune the inhibitory impact of different interneuron populations (Thomson et al. 2000; Kasugai et al. 2010). Their affinity to GABAARs containing the δ-, α4- and 5-subunits to GABA is lower, and hence their impact on a tonic GABAergic conductance is under discussion. In the absence of γ2-subunits, mainly γ3-subunits overtake the clustering of the GABAARs in the synapse, which does not change activation but leads to a prolonged decay time (Kerti-Szigeti et al. 2014). Outside synapses, GABAARs without the γ2-subunit are more prominent, have a higher affinity to GABA and a smaller single-channel conductance and act as tonic modulators of the membrane excitability by increasing the leak conductance. Their impact in tuning dendritic signal propagation depends mainly on the spillover of GABA from inhibitory synapses, release of GABA from glial cells and the local density of GABA transporters in the membrane of neurons and adjacent glial cells (Scimemi 2014; Song et al. 2013; Beenhakker and Huguenard 2010). Thus, the glia-neuron interaction can modulate the impact of tonic inhibitory conductance in the dendritic membrane and very locally tune the integration properties of the neuronal membrane, independently from the synaptically localised GABAARs. To which extent the molecular composition of extrasynaptic GABAARs is altered along the dendrite is difficult to determine due to low density of the receptors and the limits to identify specific subpopulations of GABAARs (Kasugai et al. 2010).

Glycine Receptors

Glycine receptors are mainly reported to be expressed in the spinal cord and brainstem (Dutertre et al. 2012), but their expression in the hippocampus was reported in several studies (Chattipakorn and McMahon 2002; Song et al. 2006; Keck et al. 2008; Chen et al. 2011). Clear effects of glycine independent from the effect of the amino acid on the NMDARs were identified by blocking glycine receptors with strychnine. Glycine receptors function as a shunting conductance that reduces the amplitude of EPSPs travelling along the dendrite (Keck et al. 2008) and influence neuronal plasticity in a frequency-dependent manner via progressive closure of glycine receptors at higher EPSC frequency. The action of the receptors has been interpreted differently as being caused by synaptic or extrasynaptic expressed receptors. However, clear indications of a glycinergic innervation of the hippocampus are missing.

16.3 Dendritic Distribution of Ion Channels and Receptors: Functional Consequences for Dendritic Computation

The main function of dendritic arbour is to connect synaptic inputs of a neuron to the soma. For a central neuron, this task is associated with significant challenges – in the rat CA1 pyramidal neuron, a substantial part of the dendritic tree is represented by distal dendrites that are located hundreds of micrometres away from soma and receive several thousands of predominantly excitatory synaptic inputs (Megias et al. 2001). First, such spatial extension of dendrites poses a problem of distance-dependent attenuation and delay of postsynaptic potentials due to passive cable properties (Golding et al. 2005; Rall 1962; Mainen et al. 1996; Jaffe and Carnevale 1999). Second, extensive arborisation of the dendritic tree and the large number of synaptic inputs from other neurons renders the possibility of activation of multiple synapses producing a complex time- and space-varying input (Destexhe and Pare 1999). Given the plasticity of synaptic transmission (Bliss and Collingridge 1993) and influence of the network oscillations (Buzsaki and Draguhn 2004), neurons employ a variety of complementary mechanisms for detection of individual inputs in the choir of coincidentally active synapses, as well as for discrimination of specific spatiotemporal sequences. The dendritic integration of synaptic inputs over time and space was a subject of several extensive reviews (e.g. Spruston 2008; Hausser et al. 2000). We will overview the mechanisms and phenomena described for neurons in the CNS to overcome passive conductance-based limitations of the input integration, as well as to efficiently extract spatiotemporal features of input patterns (Fig. 16.2).

16.3.1 Compartmentalisation of Dendritic Arbour and Gradient of Active Conductances

The shape of dendritic arbour and dendritic wiring are aimed to optimise connectivity in terms of cable length and conduction time costs (Cuntz et al. 2012; Wen et al. 2009; Wen and Chklovskii 2008). Importantly, a growing body of evidence demonstrates that the dendritic arbour of central neurons is highly

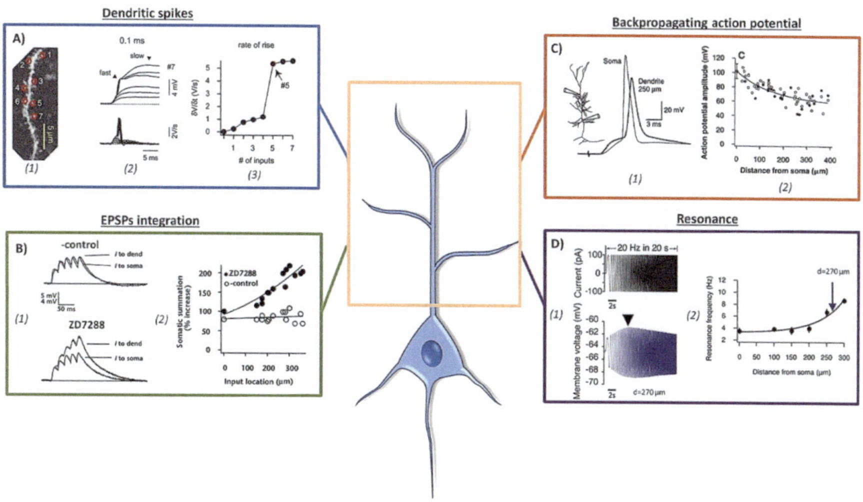

Fig. 16.2 Voltage-dependent ion channels on dendrites of CA1 pyramidal neurons are involved into a variety of physiological phenomena. (**a**) Dendritic spikes allow neurons to detect synchrony of inputs and overcome dendritic attenuation. These spikes can be supported by VGSCs, VGCCs or NMDA receptors. The voltage response to stimulation of multiple inputs (5 in this case) becomes non-linear. (**b**) Integration of subthreshold synaptic inputs is highly influenced by Ih conductance. The temporal summation of EPSPs at the soma of CA1 pyramidal cells is independent of input location ("normalised") under normal conditions, but is impaired by blockade of Ih currents. (**c**) Backpropagating action potentials actively invade dendrites of a CA1 pyramidal cell, following a somatic action potential. This phenomenon requires activation of dendritic VGSCs and, in turn, allows activation of dendritic VGCCs. (**d**) In CA1 pyramidal cells, the subthreshold resonance frequency is location dependent and significantly varies from soma to distal dendrites due to the non-uniform distribution of several active conductances along the dendrites
Panel **a**
(*1*) (Losonczy and Magee 2006) Neuron, Fig. 1b; (*2*) (Losonczy and Magee 2006) Neuron, Fig. 1f; (*3*) (Losonczy and Magee 2006) Neuron, Fig. 1h (Losonczy and Magee 2006)
Panel **b**
(*1*) *and* (*2*) (Magee 1999) Nature neurosciences Fig. 2b and d. (Magee 1999)
Panel **c**
(*1*)*and* (*2*) (Spruston et al. 1995) Science, Fig. 1a and c. (Spruston et al. 1995)
Panel **d**
(*1*) Narayanan and Johnston (2007) Neuron Supp, Fig. 1a and b (*blue* trace); (*2*) Narayanan and Johnston (2007) Neuron, Fig. 1c (−65 mV trace) (Narayanan and Johnston 2007)

compartmentalised, with several aspects of dendritic computation taking place in distinct compartments.

A first level of dendritic compartmentalisation is provided by functional segregation of the synaptic zone from the extrasynaptic area and dendritic shaft. In pyramidal neurons, most excitatory synapses are located at dendritic spines that are particularly prominent in distal dendritic segments (Bannister and Larkman 1995). Spines were shown to perform compartmentalisation of calcium fluxes and proposed as functional units of neuronal integration (Yuste and Denk 1995). In spiny neurons, the spine neck plays an important role in electrical isolation of synapses from the dendritic shafts (Araya et al. 2006, 2007; Tsay and Yuste 2002). Calcium transients restricted to dendritic spines after activation of single synapse (e.g. by glutamate uncaging) or during weak subthreshold EPSPs after excitation of very few afferent fibres are mediated mainly by NMDA receptors (Yuste et al. 1999; Kovalchuk et al. 2000), but can also result in activation of R-type VGCCs upon sufficiently strong depolarisation mediated by AMPA receptors (Bloodgood and Sabatini 2007). Similarly, functional isolation of synaptic zone from dendrites holds true also in aspiny neurons, where the spatial restriction of calcium fluxes is ensured, for example, by calcium-permeable AMPA receptors (Goldberg et al. 2003). This mechanism seems to be particularly important in mostly aspiny inhibitory interneurons that can operate at high spiking rates and therefore require higher temporal precision in processing of inputs.

A second level of compartmentalisation is associated with functional segregation of individual dendritic branches due to different distribution of active conductances (Fig. 16.1). Cooperative stimulation of several neighbouring synapses (e.g. by orthodromic stimulation of several afferent fibres) leads to the activation of VGCCs, in addition to NMDARs, in spines and adjacent dendritic shaft (Yuste and Denk 1995) within spatially restricted (~100 μm) area (Gasparini et al. 2004). These conductances comprise mainly T-type channels presumably located preferentially on dendritic shafts (Magee et al. 1995), as well as R-type channels in spines (Bloodgood and Sabatini 2007). Note that the threshold for activation of these LVA calcium channels renders them particularly suitable for weak subthreshold EPSPs. Upon strong synaptic stimulation, all types of dendritic VGGCs get activated and contribute to global calcium fluxes according to their density and distribution. Indeed, L- and N-type HVA channels determine most of the calcium fluxes in proximal dendrites (Christie et al. 1995; Sabatini and Svoboda 2000), whereas LVA T-type channels are predominant in distal parts (Isomura et al. 2002; Christie et al. 1995; Kavalali et al. 1997; Magee et al. 1995). In distal dendrites, SK-type calcium-activated potassium channels restrict the duration of calcium fluxes mediated by VGCCs, while A-type potassium channels confine these potentials to activated dendritic branches (Cai et al. 2004).

The variation of density of active conductances as a function of distance to the soma underlies further degree of compartmentalisation of the dendritic tree along the proximo-distal axis (Fig. 16.1). Specific subsets of active conductances amplify the distal inputs to overcome their attenuation along the dendritic shaft and thereby "normalise" the impact of distal and proximal dendritic branches on the firing behaviour of neurons. The HCN channels represent the most striking example of such variation, with the density in distal dendritic segments being 16 times higher than in proximal dendrites of similar diameter and 60 times higher than at the soma (Lorincz et al. 2002). Such peculiar distribution of HCN channels was demonstrated not only to compensate the location-dependent capacitive delay of dendritic inputs but also selectively and efficiently transfer rhythmic burst inputs in theta and gamma frequencies to the soma (Vaidya and Johnston 2013). Further, the increase of density with distance from the soma was shown for transient A-type K^+ channels that can serve for protection against membrane hyperexcitability due to excessive inward conductances mediated by sodium and calcium channels (Hoffman et al. 1997).

Some types of VGCCs are strikingly non-uniformly distributed in the dendrites. T-type channels are located predominantly in distal dendrites (Magee and Johnston 1997; Isomura et al. 2002; Christie et al. 1995; Kavalali et al. 1997; Gillessen and Alzheimer 1997), whereas density of L-type channels is higher in very proximal dendrites (<50 μm away from the soma) (Christie et al. 1995; Westenbroek et al. 1992; Hell et al. 1993). Higher density of T-type VGCCs in distal dendrites could have a role in opposing to the EPSP attenuation occurring during passive propagation along dendritic tree, allowing distal synapses to be still able to influence somatic membrane potential (Gillessen and Alzheimer 1997). Similarly to an increase of dendritic EPSP amplitude due to higher synaptic conductance (Magee and Cook 2000), amplification of the EPSP amplitude by T-type calcium channels particularly in distal dendrites (Isomura et al. 2002; Gillessen and Alzheimer 1997; Magee and Johnston 1995b) can serve as a mechanism for compensation of distance-dependent attenuation of impact of distal synapses in CA1 pyramidal neurons. On the other hand, L-type channels located in very proximal dendrites could serve to produce substantial calcium fluxes required for somatic signalling after integration of dendritic events (or inversely, provide calcium fluxes in dendrites following somatic events).

16.3.2 Dendritic Spikes

The multitude of excitatory synapses is distributed across the dendritic tree, providing basis for two modes of their activation, namely, asynchronous and synchronous. A logical assumption that synchronous activation of synaptic inputs must reflect stronger and more important signals leads to suggestion that their impact on the firing output must become more prominent and require additional amplifying mechanisms. Indeed, asynchronous activation of synaptic inputs was found to undergo linear summation (Losonczy and Magee 2006; Cash and Yuste 1999), whereas synchronous synaptic activation induces local membrane supra-linearities (non-linearities) termed dendritic spikes that enhance the impact of activated inputs on the firing output of the neuron (Losonczy and Magee 2006; Gasparini et al. 2004; Stuart et al. 1997a; Losonczy et al. 2008). In CA1 pyramidal neurons, dendritic spikes were found to be associated with high excitability of dendritic branches that, in turn, enabled precise AP output even in the presence of recurrent inhibition (Müller et al. 2012). Furthermore, dendritic spikes occurring in distal dendrites of CA1 pyramidal neurons were reported to favour Hebbian synaptic plasticity even in the absence of generation and backpropagation of axonal APs (Golding et al. 2002).

In line with the presence of active conductances in the dendritic membrane, several studies showed the contribution of Na^+ (Golding and Spruston 1998; Schwindt and Crill 1995; Makara et al. 2009) and Ca^{2+} (Golding et al. 2002) transients or combination of both (Kim and Connors 1993) into the genesis of dendritic spikes. Initiation of local dendritic spikes upon co-activation of clustered distal dendrites was shown to depend on NMDA receptor-mediated calcium currents (Golding et al. 2002; Schiller et al. 2000). Furthermore, synchronous activation of synaptic inputs can trigger a concerted action of calcium and potassium channels (Golding et al. 1999). The A-type K^+ currents suppressed the propagation of dendritic spikes (Losonczy et al. 2008).

Several approaches revealed spatial specificity of dendritic spikes to activated dendritic branches (Remy et al. 2009), providing additional mechanism of dendritic compartmentalisation and enhancing the capacity of dendritic arbour for selective processing of inputs. Furthermore, via detection and amplification of temporally correlated inputs, dendritic spikes pave the way for the integration of synaptic inputs. In CA1 pyramidal neuron, Katz et al. found that distribution of synapses throughout dendritic tree supports the two-stage model of integration, with individual synaptic inputs contributing to generation of branch-specific dendritic spikes followed by competition of branches for the control of the output (Katz et al. 2009).

Thus, the presence of active conductances in dendritic membrane has an important functional consequence: correlated (synchronous) activation of excitatory synaptic inputs can trigger dendritic spikes associated with sodium and calcium currents. Dendritically generated spikes act as coincidence detectors and translate synchronous activation of close synapses into supra-linear calcium signals (Yasuda et al. 2003; Losonczy and Magee 2006; Yuste and Denk 1995; Gasparini et al. 2004; Remy et al. 2009; Ariav et al. 2003). Furthermore, non-uniform distribution and involvement of active conductances result in spatially restricted dynamic modulation of excitability of activated dendritic branches.

16.3.3 Backpropagation of Action Potentials

In many neuronal types, initiation of fast sodium AP is followed by its active backpropagation into the dendritic tree, providing a retrograde signal of neuronal output and boosting local dendritic calcium signals (reviewed in Stuart et al. 1997b; Eilers and Konnerth 1997). Modelling of three-compartment dendrite (axial resistance, capacitance and conductance of dendrites (Rall 1962)) showed that same mechanisms responsible for compensation of passive cable properties might underlie the backpropagation of action potential (Cook and Johnston 1999).

Several lines of evidence demonstrate that activation of dendritic TTX-sensitive sodium channels is necessary and sufficient for backpropagation of somatically generated APs into dendrites (Stuart and Sakmann 1994; Stuart and Hausser 2001). Coincidence of synaptic activation (synaptic potentials) and bAPs was found to be associated with NMDA receptor-mediated Ca^{2+} transients restricted to postsynaptic spines and their adjacent dendrites (Rose and Konnerth 2001). Additionally, bAPs can be associated with calcium fluxes mediated mainly by R-type (Yasuda et al. 2003; Sabatini and Svoboda 2000; Hoogland and Saggau 2004; Bloodgood and Sabatini 2007) and, to a lesser extent, by L-type (Lee et al. 2009; Yasuda et al. 2003; Hoogland and Saggau 2004; Bloodgood and Sabatini 2007) VGCCS in spines. Participation of T-type channels in these calcium transients cannot be ruled out (Sabatini and Svoboda 2000; Hoogland and Saggau 2004; Bloodgood and Sabatini 2007; Yuste et al. 1999), especially in distal regions of the dendritic tree where somatic APs are usually unable to backpropagate (Golding et al. 2002). The amplitude of bAPs gradually decreases as a function of distance from soma (Kamondi et al. 1998; Buzsaki and Kandel 1998; Turner et al. 1991), although in spines no voltage attenuation or major electrical filtering of bAPs was found (Nuriya et al. 2006). Such decrease of the amplitude and duration of bAPs, as well as initiation of APs in dendrites, was dependent on activation of A-type K^+ channels (Hoffman et al. 1997; Colbert et al. 1997), at least in proximal dendritic segments (Stuart and Hausser 2001).

The backpropagation of APs into dendritic tree is as an essential retrograde signalling mechanism, because the temporal coordination of synaptic activity to bAPs can have particularly important physiological consequences in the brain (Buzsaki et al. 1996). Weak synaptic inputs correlated with bAP-triggered Ca^{2+} dendritic spike in distal dendrites that, in turn, evoked a burst of axonal APs (Larkum et al. 1999a). Strong stimulation leading to propagation of a spike into dendrites was shown to trigger pronounced activation of the broad spectrum of VGCCs expressed in spines and dendritic shafts both during backpropagating somatic APs (Yuste et al. 1999; Yuste and Denk 1995; Regehr et al. 1989; Jaffe et al. 1992; Miyakawa et al. 1992) or dendritically generated spikes (Golding and Spruston 1998; Golding et al. 1999; Kamondi et al. 1998). Indeed, bursting in CA1 pyramidal neurons was recently reported to require activation of NMDA receptors and depend on widespread large-amplitude calcium transients mediated by VGCCs in basal and apical dendrites (Grienberger et al. 2014).

16.3.4 Spontaneous Membrane Potential Oscillations (MPOs) and Frequency Resonance

Spontaneous membrane potential oscillations (MPOs) were found in various types of neurons in different brain regions, including pyramidal cells in CA1 (Ylinen et al. 1995; Soltesz and Deschenes 1993). Spontaneous MPOs are tightly related to the phenomenon of membrane resonance, which has been attributed to inductance in the membrane (Mauro et al. 1970; Puil et al. 1986) and can be considered as a global measure of its filtering capacity (London and Häusser 2005; Hutcheon et al. 1996; Hutcheon and Yarom 2000). For example, an injection of sinusoidal currents in hippocampal CA1 neurons upon slight depolarisation led to maximal response at theta frequency (3–10 Hz) (Leung and Yu 1998).

There is dissociation between mechanisms responsible for generation (emergence) of resonance and for its amplification (Hutcheon and Yarom 2000). Due to membrane potential-dependent fast activation and inactivation, many ion channels are ideally suited for frequency resonance. The contribution into resonant features of the membrane was shown for VGSCs (Hutcheon et al. 1996; Gutfreund et al. 1995), HCN channels (Hutcheon et al. 1996) and T-type Ca^{2+} channels (Llinas and Yarom 1986; Puil et al. 1994). Rhythmic activation of dendritic branches, in conjunction with gradient of active conductances along the dendritic tree and resulting propensity of dendritic membrane for subthreshold oscillations, results in the ability of neuron to amplify activation of specific branches.

In the context of frequency resonance, repetitive firing in bursts of action potentials becomes particularly important as compared to regular spiking in solitary APs. In layer 5 neocortical neurons, bursts of 4–5 bAPs at frequencies between 60 and 200 Hz elicited large regenerative VGCC-mediated potentials in the distal dendritic initiation zone, which in turn promoted further depolarisation at the soma (Larkum et al. 1999b). In hippocampal pyramidal neurons, this process is tightly controlled by A-type potassium channels that, through control of backpropagation of APs, can powerfully regulate the mode of firing in CA1 pyramidal neurons (single spiking or burst firing) (Johnston et al. 2000; Magee and Carruth 1999).

In conjunction with the aforementioned gradient of active conductances along the dendrites, frequency resonance can underlie distinct filtering properties of dendritic and somatic inputs (Zhuchkova et al. 2013) and temporal synchronisation of rhythmic inputs across wide regions of the dendritic arbour (Vaidya and Johnston 2013). Moreover, the resonant properties can vary throughout the dendritic tree of individual neurons: the results of modelling study suggest that dendritic resonance is a spatially distributed property and a neuron can pass through more than one frequency (Laudanski et al. 2014). Such uneven distribution of resonant features along the dendritic arbour represents yet another level of compartmentalisation and may serve a basis for selective impact of field oscillations on specific dendritic branches and for communication through resonance between weakly connected brain areas (Hahn et al. 2014).

16.4 Summary

Central neurons are embedded into tremendously complex brain network. They carry a multitude of synapses of varying strength that are distributed throughout dendritic arbour with complex shape and substantial spatial extension. In this chapter, we briefly reviewed data showing that the dendritic tree of neurons in the brain is characterised by highly compartmentalised structure and non-uniform distribution of numerous voltage- and ligand-gated ion channels and membrane receptors. Large body of experimental and computational data challenged the earlier view that dendritic membrane performs only passive linear summation of synaptic inputs. First, active conductances in dendritic membrane not only allow to overcome the attenuation of distal inputs and to retain temporal resolution but also enable neuron to distinguish and respond accordingly to ever-changing patterns of activation, with distinct impact of synaptic activation on the firing output depending on fine differences in spatiotemporal properties. Due to their active properties, dendrites perform non-linear/supra-linear integration of synaptic inputs and possess the capability for autonomous dendritic computation, i.e. input-specific local processing of incoming information, and greatly enhance the storage capacity of neurons (Poirazi and Mel 2001).

Further, we outlined several phenomena that are involved into dendritic computation and enable the detection of fine differences in spatiotemporal properties of synaptic activation. Dendrite-based phenomena rely on the structural complexity of dendritic arbour and strongly influence the somatic representation of synaptic inputs, thereby powerfully affecting their impact on neuronal output. Thus, highly complex structure of neuronal dendritic arbour is optimised for processing of information received through synaptic inputs, which constitutes the dendritic computation.

Interesting and potentially promising aspect emerging from this picture is analysis of crosstalk and interference between ion channels and membrane complexes organised into functional clusters. Dendritic VGCCs have been shown to be sensitive to neuromodulation (Hoogland and Saggau 2004), as well as to recent synaptic and membrane potential history (Yasuda et al. 2003; Golding and Spruston 1998). Such interaction would provide additional, higher level of modulation and allow dynamic modulation of local properties of dendritic membrane according to current network context (e.g. oscillatory phase), previous history of neuronal firing and additional input specificity.

Further, mobility of ion channels in general, and of their auxiliary subunits in particular, needs further investigation. Association with auxiliary subunits is characteristic for pore-forming subunits of many ion channels and is known to affect their trafficking, positioning and kinetic properties (Kaufmann et al. 2010). Given that the membrane is not static and lateral mobility is a typical feature of membrane-associated protein complexes, lateral diffusion and directed mobility can connect activity-dependent modulation of active conductances and intracellular signalling pathways.

References

Adelman WJ Jr, Palti Y (1969) The effects of external potassium and long duration voltage conditioning on the amplitude of sodium currents in the giant axon of the squid, Loligo pealei. J Gen Phys 54(5):589–606

Adelman JP, Maylie J, Sah P (2012) Small-conductance Ca2+ -activated K+ channels: form and function. Annu Rev Physiol 74:245–269. doi:10.1146/annurev-physiol-020911-153336

Ahmad M, Polepalli JS, Goswami D, Yang X, Kaeser-Woo YJ, Sudhof TC, Malenka RC (2012) Postsynaptic complexin controls AMPA receptor exocytosis during LTP. Neuron 73 (2):260–267. doi:10.1016/j.neuron.2011.11.020

Alger BE, Nicoll RA (1982) Feed-forward dendritic inhibition in rat hippocampal pyramidal cells studied in vitro. J Physiol 328:105–123

Alonso G, Widmer H (1997) Clustering of KV4.2 potassium channels in postsynaptic membrane of rat supraoptic neurons: an ultrastructural study. Neuroscience 77(3):617–621. doi: S0306452296005611 [pii]

Aman TK, Grieco-Calub TM, Chen C, Rusconi R, Slat EA, Isom LL, Raman IM (2009) Regulation of persistent Na current by interactions between beta subunits of voltage-gated Na channels. J Neurosci 29(7):2027–2042. doi:10.1523/JNEUROSCI.4531-08.2009

An WF, Bowlby MR, Betty M, Cao J, Ling HP, Mendoza G, Hinson JW, Mattsson KI, Strassle BW, Trimmer JS, Rhodes KJ (2000) Modulation of A-type potassium channels by a family of calcium sensors. Nature 403(6769):553–556. doi:10.1038/35000592

Anderson AE, Adams JP, Qian Y, Cook RG, Pfaffinger PJ, Sweatt JD (2000) Kv4.2 phosphorylation by cyclic AMP-dependent protein kinase. J Biol Chem 275(8):5337–5346

Anderson D, Mehaffey WH, Iftinca M, Rehak R, Engbers JD, Hameed S, Zamponi GW, Turner RW (2010) Regulation of neuronal activity by Cav3-Kv4 channel signaling complexes. Nat Neurosci 13(3):333–337. doi:nn.2493 [pii]; doi:10.1038/nn.2493

Anderson D, Engbers JD, Heath NC, Bartoletti TM, Mehaffey WH, Zamponi GW, Turner RW (2013) The Cav3-Kv4 complex acts as a calcium sensor to maintain inhibitory charge transfer during extracellular calcium fluctuations. J Neurosci 33(18):7811–7824. doi:33/18/7811 [pii];10.1523/JNEUROSCI.5384-12.2013 [doi]

Andrasfalvy BK, Magee JC (2001) Distance-dependent increase in AMPA receptor number in the dendrites of adult hippocampal CA1 pyramidal neurons. J Neurosci 21(23):9151–9159

Andrasfalvy BK, Magee JC (2004) Changes in AMPA receptor currents following LTP induction on rat CA1 pyramidal neurones. J Physiol 559(Pt 2):543–554. doi:10.1113/jphysiol.2004.065219 jphysiol.2004.065219 [pii]

Anggono V, Huganir RL (2012) Regulation of AMPA receptor trafficking and synaptic plasticity. Curr Opin Neurobiol 22(3):461–469. doi:10.1016/j.conb.2011.12.006

Araya R, Jiang J, Eisenthal KB, Yuste R (2006) The spine neck filters membrane potentials. Proc Natl Acad Sci U S A 103(47):17961–17966. doi:10.1073/pnas.0608755103

Araya R, Nikolenko V, Eisenthal KB, Yuste R (2007) Sodium channels amplify spine potentials. Proc Natl Acad Sci U S A 104(30):12347–12352. doi:10.1073/pnas.0705282104

Ariav G, Polsky A, Schiller J (2003) Submillisecond precision of the input-output transformation function mediated by fast sodium dendritic spikes in basal dendrites of CA1 pyramidal neurons. J Neurosci 23(21):7750–7758

Azimi-Zonooz A, Kawa CB, Dowell CD, Olivera BM (2001) Autoradiographic localization of N-type VGCCs in gerbil hippocampus and failure of omega-conotoxin MVIIA to attenuate neuronal injury after transient cerebral ischemia. Brain Res 907(1–2):61–70

Bankston JR, Camp SS, DiMaio F, Lewis AS, Chetkovich DM, Zagotta WN (2012) Structure and stoichiometry of an accessory subunit TRIP8b interaction with hyperpolarization-activated cyclic nucleotide-gated channels. Proc Natl Acad Sci U S A 109(20):7899–7904. doi:1201997109 [pii];10.1073/pnas.1201997109 [doi]

Bannai H, Levi S, Schweizer C, Inoue T, Launey T, Racine V, Sibarita JB, Mikoshiba K, Triller A (2009) Activity-dependent tuning of inhibitory neurotransmission based on GABAAR diffusion dynamics. Neuron 62(5):670–682. doi:10.1016/j.neuron.2009.04.023

Bannister NJ, Larkman AU (1995) Dendritic morphology of CA1 pyramidal neurones from the rat hippocampus: II. Spine distributions. J Comp Neurol 360(1):161–171. doi:10.1002/cne.903600112

Bats C, Groc L, Choquet D (2007) The interaction between Stargazin and PSD-95 regulates AMPA receptor surface trafficking. Neuron 53(5):719–734

Becker AJ, Pitsch J, Sochivko D, Opitz T, Staniek M, Chen CC, Campbell KP, Schoch S, Yaari Y, Beck H (2008) Transcriptional upregulation of Cav3.2 mediates epileptogenesis in the pilocarpine model of epilepsy. J Neurosci 28(49):13341–13353. doi:10.1523/JNEUROSCI.1421-08.2008 28/49/13341 [pii]

Beenhakker MP, Huguenard JR (2010) Astrocytes as gatekeepers of GABAB receptor function. J Neurosci 30(45):15262–15276. doi:10.1523/JNEUROSCI.3243-10.2010

Bettler B, Kaupmann K, Mosbacher J, Gassmann M (2004) Molecular structure and physiological functions of GABA(B) receptors. Physiol Rev 84(3):835–867. doi:10.1152/physrev.00036.2003

Bi GQ, Poo MM (1998) Synaptic modifications in cultured hippocampal neurons: dependence on spike timing, synaptic strength, and postsynaptic cell type. J Neurosci 18(24):10464–10472

Birnbaum SG, Varga AW, Yuan LL, Anderson AE, Sweatt JD, Schrader LA (2004) Structure and function of Kv4-family transient potassium channels. Physiol Rev 84(3):803–833. doi:10.1152/physrev.00039.2003 [doi];84/3/803 [pii]

Bischofberger J, Jonas P (1997) Action potential propagation into the presynaptic dendrites of rat mitral cells. J Physiol 504(Pt 2):359–365

Bliss TV, Collingridge GL (1993) A synaptic model of memory: long-term potentiation in the hippocampus. Nature 361(6407):31–39. doi:10.1038/361031a0

Bloodgood BL, Sabatini BL (2007) Nonlinear regulation of unitary synaptic signals by CaV(2.3) voltage-sensitive calcium channels located in dendritic spines. Neuron 53(2):249–260

Brickley SG, Mody I (2012) Extrasynaptic GABA(A) receptors: their function in the CNS and implications for disease. Neuron 73(1):23–34. doi:10.1016/j.neuron.2011.12.012 S0896-6273(11)01093-2 [pii]

Buzsaki G, Draguhn A (2004) Neuronal oscillations in cortical networks. Science 304(5679):1926–1929. doi:10.1126/science.1099745

Buzsaki G, Kandel A (1998) Somadendritic backpropagation of action potentials in cortical pyramidal cells of the awake rat. J Neurophysiol 79(3):1587–1591

Buzsaki G, Penttonen M, Nadasdy Z, Bragin A (1996) Pattern and inhibition-dependent invasion of pyramidal cell dendrites by fast spikes in the hippocampus in vivo. Proc Natl Acad Sci U S A 93(18):9921–9925

Cai X, Liang CW, Muralidharan S, Kao JP, Tang CM, Thompson SM (2004) Unique roles of SK and Kv4.2 potassium channels in dendritic integration. Neuron 44(2):351–364. doi:10.1016/j.neuron.2004.09.026

Callaway JC, Ross WN (1997) Spatial distribution of synaptically activated sodium concentration changes in cerebellar Purkinje neurons. J Neurophysiol 77(1):145–152

Campanac E, Daoudal G, Ankri N, Debanne D (2008) Downregulation of dendritic I(h) in CA1 pyramidal neurons after LTP. J Neurosci 28(34):8635–8643. doi:28/34/8635 [pii];10.1523/JNEUROSCI.1411-08.2008 [doi]

Carta M, Opazo P, Veran J, Athane A, Choquet D, Coussen F, Mulle C (2013) CaMKII-dependent phosphorylation of GluK5 mediates plasticity of kainate receptors. EMBO J 32(4):496–510. doi:10.1038/emboj.2012.334

Carta M, Fievre S, Gorlewicz A, Mulle C (2014) Kainate receptors in the hippocampus. Eur J Neurosci 39(11):1835–1844. doi:10.1111/ejn.12590

Cash S, Yuste R (1999) Linear summation of excitatory inputs by CA1 pyramidal neurons. Neuron 22(2):383–394

Catterall WA (2000) Structure and regulation of voltage-gated Ca2+ channels. Annu Rev Cell Dev Biol 16:521–555. doi:10.1146/annurev.cellbio.16.1.521

Catterall WA (2012) Voltage-gated sodium channels at 60: structure, function and pathophysiology. J Physiol 590(Pt 11):2577–2589. doi:10.1113/jphysiol.2011.224204

Chandler WK, Meves H (1970) Slow changes in membrane permeability and long-lasting action potentials in axons perfused with fluoride solutions. J Physiol 211(3):707–728

Chattipakorn SC, McMahon LL (2002) Pharmacological characterization of glycine-gated chloride currents recorded in rat hippocampal slices. J Neurophysiol 87(3):1515–1525

Chen X, Johnston D (2004) Properties of single voltage-dependent K+ channels in dendrites of CA1 pyramidal neurones of rat hippocampus. J Physiol 559(Pt 1):187–203. doi:10.1113/jphysiol.2004.068114 [doi];jphysiol.2004.068114 [pii]

Chen S, Wang J, Siegelbaum SA (2001) Properties of hyperpolarization-activated pacemaker current defined by coassembly of HCN1 and HCN2 subunits and basal modulation by cyclic nucleotide. J Gen Physiol 117(5):491–504

Chen RQ, Wang SH, Yao W, Wang JJ, Ji F, Yan JZ, Ren SQ, Chen Z, Liu SY, Lu W (2011) Role of glycine receptors in glycine-induced LTD in hippocampal CA1 pyramidal neurons. Neuropsychopharmacol Off Publ Am Coll Neuropsychopharmacol 36(9):1948–1958. doi:10.1038/npp.2011.86

Chen S, Benninger F, Yaari Y (2014) Role of small conductance Ca(2)(+)-activated K(+) channels in controlling CA1 pyramidal cell excitability. J Neurosci 34(24):8219–8230. doi:10.1523/JNEUROSCI.0936-14.2014

Christie BR, Eliot LS, Ito K, Miyakawa H, Johnston D (1995) Different Ca2+ channels in soma and dendrites of hippocampal pyramidal neurons mediate spike-induced Ca2+ influx. J Neurophysiol 73(6):2553–2557

Coetzee WA, Amarillo Y, Chiu J, Chow A, Lau D, McCormack T, Moreno H, Nadal MS, Ozaita A, Pountney D, Saganich M, de Vega-Saenz ME, Rudy B (1999) Molecular diversity of K+ channels. Ann N Y Acad Sci 868:233–285

Colbert CM, Johnston D (1996) Axonal action-potential initiation and Na+ channel densities in the soma and axon initial segment of subicular pyramidal neurons. J Neurosci 16(21):6676–6686

Colbert CM, Pan E (2002) Ion channel properties underlying axonal action potential initiation in pyramidal neurons. Nat Neurosci 5(6):533–538. doi:10.1038/nn857

Colbert CM, Magee JC, Hoffman DA, Johnston D (1997) Slow recovery from inactivation of Na+ channels underlies the activity-dependent attenuation of dendritic action potentials in hippocampal CA1 pyramidal neurons. J Neurosci 17(17):6512–6521

Constals A, Penn AC, Compans B, Toulme E, Phillipat A, Marais S, Retailleau N, Hafner AS, Coussen F, Hosy E, Choquet D (2015) Glutamate-induced AMPA receptor desensitization increases their mobility and modulates short-term plasticity through unbinding from Stargazin. Neuron 85(4):787–803. doi:10.1016/j.neuron.2015.01.012

Contractor A, Mulle C, Swanson GT (2011) Kainate receptors coming of age: milestones of two decades of research. Trends Neurosci 34(3):154–163. doi:10.1016/j.tins.2010.12.002

Cook EP, Johnston D (1999) Voltage-dependent properties of dendrites that eliminate location-dependent variability of synaptic input. J Neurophysiol 81(2):535–543

Costa MR, Catterall WA (1984a) Phosphorylation of the alpha subunit of the sodium channel by protein kinase C. Cell Mol Neurobiol 4(3):291–297

Costa MR, Catterall WA (1984b) Cyclic AMP-dependent phosphorylation of the alpha subunit of the sodium channel in synaptic nerve ending particles. J Biol Chem 259(13):8210–8218

Cuntz H, Mathy A, Hausser M (2012) A scaling law derived from optimal dendritic wiring. Proc Natl Acad Sci U S A 109(27):11014–11018. doi:10.1073/pnas.1200430109

Czondor K, Mondin M, Garcia M, Heine M, Frischknecht R, Choquet D, Sibarita JB, Thoumine OR (2012) Unified quantitative model of AMPA receptor trafficking at synapses. Proc Natl Acad Sci U S A 109(9):3522–3527. doi:10.1073/pnas.1109818109 1109818109 [pii]

Debarbieux F, Audinat E, Charpak S (2003) Action potential propagation in dendrites of rat mitral cells in vivo. J Neurosci 23(13):5553–5560

Destexhe A, Pare D (1999) Impact of network activity on the integrative properties of neocortical pyramidal neurons in vivo. J Neurophysiol 81(4):1531–1547

Di Biase V, Tuluc P, Campiglio M, Obermair GJ, Heine M, Flucher BE (2011) Surface traffic of dendritic CaV1.2 calcium channels in hippocampal neurons. J Neurosci 31(38):13682–13694. doi:31/38/13682 [pii] 10.1523/JNEUROSCI.2300-11.2011

Dutertre S, Becker CM, Betz H (2012) Inhibitory glycine receptors: an update. J Biol Chem 287(48):40216–40223. doi:10.1074/jbc.R112.408229 R112.408229 [pii]

Eggermann E, Bucurenciu I, Goswami SP, Jonas P (2012) Nanodomain coupling between Ca(2)(+) channels and sensors of exocytosis at fast mammalian synapses. Nat Rev Neurosci 13(1):7–21. doi:10.1038/nrn3125 nrn3125 [pii]

Ehlers MD (2000) Reinsertion or degradation of AMPA receptors determined by activity-dependent endocytic sorting. Neuron 28(2):511–525

Eilers J, Konnerth A (1997) Dendritic signal integration. Curr Opin Neurobiol 7(3):385–390

Espinosa JS, Wheeler DG, Tsien RW, Luo L (2009) Uncoupling dendrite growth and patterning: single-cell knockout analysis of NMDA receptor 2B. Neuron 62(2):205–217. doi:10.1016/j.neuron.2009.03.006

Fan Y, Fricker D, Brager DH, Chen X, Lu HC, Chitwood RA, Johnston D (2005) Activity-dependent decrease of excitability in rat hippocampal neurons through increases in I(h). Nat Neurosci 8(11):1542–1551. doi:nn1568 [pii];10.1038/nn1568 [doi]

Featherstone DE, Shippy SA (2008) Regulation of synaptic transmission by ambient extracellular glutamate. Neuroscientist 14(2):171–181. doi:1073858407308518 [pii] 10.1177/1073858407308518

Frick A, Magee J, Johnston D (2004) LTP is accompanied by an enhanced local excitability of pyramidal neuron dendrites. Nat Neurosci 7(2):126–135. doi:10.1038/nn1178 [doi];nn1178 [pii]

Fritschy JM, Panzanelli P (2014) GABAA receptors and plasticity of inhibitory neurotransmission in the central nervous system. Eur J Neurosci 39(11):1845–1865. doi:10.1111/ejn.12534

Gasparini S, Migliore M, Magee JC (2004) On the initiation and propagation of dendritic spikes in CA1 pyramidal neurons. J Neurosci Off J Soc Neurosci 24(49):11046–11056. doi:10.1523/JNEUROSCI.2520-04.2004

Gerges NZ, Backos DS, Rupasinghe CN, Spaller MR, Esteban JA (2006) Dual role of the exocyst in AMPA receptor targeting and insertion into the postsynaptic membrane. EMBO J 25 (8):1623–1634

Gillessen T, Alzheimer C (1997) Amplification of EPSPs by low Ni(2+)- and amiloride-sensitive Ca2+ channels in apical dendrites of rat CA1 pyramidal neurons. J Neurophysiol 77 (3):1639–1643

Goldberg JH, Tamas G, Aronov D, Yuste R (2003) Calcium microdomains in aspiny dendrites. Neuron 40(4):807–821

Goldin AL (2001) Resurgence of sodium channel research. Annu Rev Physiol 63:871–894. doi:10.1146/annurev.physiol.63.1.871

Golding NL, Spruston N (1998) Dendritic sodium spikes are variable triggers of axonal action potentials in hippocampal CA1 pyramidal neurons. Neuron 21(5):1189–1200

Golding NL, Jung HY, Mickus T, Spruston N (1999) Dendritic calcium spike initiation and repolarization are controlled by distinct potassium channel subtypes in CA1 pyramidal neurons. J Neurosci Off J Soc Neurosci 19(20):8789–8798

Golding NL, Staff NP, Spruston N (2002) Dendritic spikes as a mechanism for cooperative long-term potentiation. Nature 418(6895):326–331. doi:10.1038/nature00854

Golding NL, Mickus TJ, Katz Y, Kath WL, Spruston N (2005) Factors mediating powerful voltage attenuation along CA1 pyramidal neuron dendrites. J Physiol 568(Pt 1):69–82. doi:10.1113/jphysiol.2005.086793

Grienberger C, Chen X, Konnerth A (2014) NMDA receptor-dependent multidendrite Ca(2+) spikes required for hippocampal burst firing in vivo. Neuron 81(6):1274–1281. doi:10.1016/j.neuron.2014.01.014

Groc L, Heine M, Cognet L, Brickley K, Stephenson FA, Lounis B, Choquet D (2004) Differential activity-dependent regulation of the lateral mobilities of AMPA and NMDA receptors. Nat Neurosci 7(7):695–696

Groc L, Heine M, Cousins SL, Stephenson FA, Lounis B, Cognet L, Choquet D (2006) NMDA receptor surface mobility depends on NR2A-2B subunits. Proc Natl Acad Sci U S A 103 (49):18769–18774. doi:0605238103 [pii] 10.1073/pnas.0605238103

Grover LM, Teyler TJ (1990) Two components of long-term potentiation induced by different patterns of afferent activation. Nature 347(6292):477–479. doi:10.1038/347477a0

Gu N, Vervaeke K, Storm JF (2007) BK potassium channels facilitate high-frequency firing and cause early spike frequency adaptation in rat CA1 hippocampal pyramidal cells. J Physiol 580 (Pt.3):859–882. doi:10.1113/jphysiol.2006.126367

Gutfreund Y, yarom Y, Segev I (1995) Subthreshold oscillations and resonant frequency in guinea-pig cortical neurons: physiology and modelling. J Physiol 483(Pt 3):621–640

Hahn G, Bujan AF, Fregnac Y, Aertsen A, Kumar A (2014) Communication through resonance in spiking neuronal networks. PLoS Comput Biol 10(8), e1003811. doi:10.1371/journal.pcbi.1003811

Hardingham GE, Bading H (2010) Synaptic versus extrasynaptic NMDA receptor signalling: implications for neurodegenerative disorders. Nat Rev Neurosci 11(10):682–696. doi:10.1038/nrn2911

Hassinger TD, Atkinson PB, Strecker GJ, Whalen LR, Dudek FE, Kossel AH, Kater SB (1995) Evidence for glutamate-mediated activation of hippocampal neurons by glial calcium waves. J Neurobiol 28(2):159–170. doi:10.1002/neu.480280204

Hausser M, Spruston N, Stuart GJ (2000) Diversity and dynamics of dendritic signaling. Science 290(5492):739–744

He C, Chen F, Li B, Hu Z (2014) Neurophysiology of HCN channels: from cellular functions to multiple regulations. Prog Neurobiol 112:1–23. doi:S0301-0082(13)00101-9 [pii];10.1016/j.pneurobio.2013.10.001 [doi]

Heine M, Groc L, Frischknecht R, Beique JC, Lounis B, Rumbaugh G, Huganir RL, Cognet L, Choquet D (2008) Surface mobility of postsynaptic AMPARs tunes synaptic transmission. Science 320(5873):201–205. doi:320/5873/201 [pii] 10.1126/science.1152089

Hell JW, Westenbroek RE, Warner C, Ahlijanian MK, Prystay W, Gilbert MM, Snutch TP, Catterall WA (1993) Identification and differential subcellular localization of the neuronal class C and class D L-type calcium channel alpha 1 subunits. J Cell Biol 123(4):949–962

Hodgkin AL, Huxley AF (1952a) Currents carried by sodium and potassium ions through the membrane of the giant axon of Loligo. J Physiol 116(4):449–472

Hodgkin AL, Huxley AF (1952b) The components of membrane conductance in the giant axon of Loligo. J Physiol 116(4):473–496

Hodgkin AL, Huxley AF (1952c) The dual effect of membrane potential on sodium conductance in the giant axon of Loligo. J Physiol 116(4):497–506

Hodgkin AL, Huxley AF (1952d) A quantitative description of membrane current and its application to conduction and excitation in nerve. J Physiol 117(4):500–544

Hoffman DA, Johnston D (1998) Downregulation of transient K+ channels in dendrites of hippocampal CA1 pyramidal neurons by activation of PKA and PKC. J Neurosci 18 (10):3521–3528

Hoffman DA, Magee JC, Colbert CM, Johnston D (1997) K+ channel regulation of signal propagation in dendrites of hippocampal pyramidal neurons. Nature 387(6636):869–875. doi:10.1038/43119

Hoogland TM, Saggau P (2004) Facilitation of L-type Ca2+ channels in dendritic spines by activation of beta2 adrenergic receptors. J Neurosci 24(39):8416–8427. doi:10.1523/JNEUROSCI.1677-04.2004 24/39/8416 [pii]

Hutcheon B, Yarom Y (2000) Resonance, oscillation and the intrinsic frequency preferences of neurons. Trends Neurosci 23(5):216–222

Hutcheon B, Miura RM, Puil E (1996) Models of subthreshold membrane resonance in neocortical neurons. J Neurophysiol 76(2):698–714

Iafrati J, Orejarena MJ, Lassalle O, Bouamrane L, Gonzalez-Campo C, Chavis P (2014) Reelin, an extracellular matrix protein linked to early onset psychiatric diseases, drives postnatal development of the prefrontal cortex via GluN2B-NMDARs and the mTOR pathway. Mol Psychiatry 19(4):417–426. doi:10.1038/mp.2013.66

Isom LL, De Jongh KS, Patton DE, Reber BF, Offord J, Charbonneau H, Walsh K, Goldin AL, Catterall WA (1992) Primary structure and functional expression of the beta 1 subunit of the rat brain sodium channel. Science 256(5058):839–842

Isomura Y, Fujiwara-Tsukamoto Y, Imanishi M, Nambu A, Takada M (2002) Distance-dependent Ni(2+)-sensitivity of synaptic plasticity in apical dendrites of hippocampal CA1 pyramidal cells. J Neurophysiol 87(2):1169–1174

Ito K, Miura M, Furuse H, Zhixiong C, Kato H, Yasutomi D, Inoue T, Mikoshiba K, Kimura T, Sakakibara S et al (1995) Voltage-gated Ca2+ channel blockers, omega-AgaIVA and Ni2+, suppress the induction of theta-burst induced long-term potentiation in guinea-pig hippocampal CA1 neurons. Neurosci Lett 183(1–2):112–115

Jaffe DB, Carnevale NT (1999) Passive normalization of synaptic integration influenced by dendritic architecture. J Neurophysiol 82(6):3268–3285

Jaffe DB, Johnston D, Lasser-Ross N, Lisman JE, Miyakawa H, Ross WN (1992) The spread of Na + spikes determines the pattern of dendritic Ca2+ entry into hippocampal neurons. Nature 357 (6375):244–246. doi:10.1038/357244a0

Jaskolski F, Coussen F, Mulle C (2005) Subcellular localization and trafficking of kainate receptors. Trends Pharmacol Sci 26(1):20–26. doi:S0165-6147(04)00314-1 [pii] 10.1016/j.tips.2004.11.008

Jerng HH, Pfaffinger PJ (2014) Modulatory mechanisms and multiple functions of somatodendritic A-type K (+) channel auxiliary subunits. Front Cell Neurosci 8:82. doi:10.3389/fncel.2014.00082

Johnston D, Hoffman DA, Magee JC, Poolos NP, Watanabe S, Colbert CM, Migliore M (2000) Dendritic potassium channels in hippocampal pyramidal neurons. J Physiol 525(Pt 1):75–81

Jung SC, Hoffman DA (2009) Biphasic somatic A-type K channel downregulation mediates intrinsic plasticity in hippocampal CA1 pyramidal neurons. PLoS ONE 4(8):e6549. doi:10.1371/journal.pone.0006549

Jurado S (2014) The dendritic SNARE fusion machinery involved in AMPARs insertion during long-term potentiation. Front Cell Neurosci 8:407. doi:10.3389/fncel.2014.00407

Jurado S, Goswami D, Zhang Y, Molina AJ, Sudhof TC, Malenka RC (2013) LTP requires a unique postsynaptic SNARE fusion machinery. Neuron 77(3):542–558. doi:10.1016/j.neuron.2012.11.029

Kamondi A, Acsady L, Buzsaki G (1998) Dendritic spikes are enhanced by cooperative network activity in the intact hippocampus. J Neurosci Off J Soc Neurosci 18(10):3919–3928

Karpova A, Mikhaylova M, Bera S, Bar J, Reddy PP, Behnisch T, Rankovic V, Spilker C, Bethge P, Sahin J, Kaushik R, Zuschratter W, Kahne T, Naumann M, Gundelfinger ED, Kreutz MR (2013) Encoding and transducing the synaptic or extrasynaptic origin of NMDA receptor signals to the nucleus. Cell 152(5):1119–1133. doi:10.1016/j.cell.2013.02.002

Kasugai Y, Swinny JD, Roberts JD, Dalezios Y, Fukazawa Y, Sieghart W, Shigemoto R, Somogyi P (2010) Quantitative localisation of synaptic and extrasynaptic GABAA receptor subunits on hippocampal pyramidal cells by freeze-fracture replica immunolabelling. Eur J Neurosci 32(11):1868–1888. doi:10.1111/j.1460-9568.2010.07473.x

Katz Y, Menon V, Nicholson DA, Geinisman Y, Kath WL, Spruston N (2009) Synapse distribution suggests a two-stage model of dendritic integration in CA1 pyramidal neurons. Neuron 63 (2):171–177. doi:10.1016/j.neuron.2009.06.023

Kaufmann WA, Kasugai Y, Ferraguti F, Storm JF (2010) Two distinct pools of large-conductance calcium-activated potassium channels in the somatic plasma membrane of central principal neurons. Neuroscience 169(3):974–986. doi:10.1016/j.neuroscience.2010.05.070

Kavalali ET, Zhuo M, Bito H, Tsien RW (1997) Dendritic Ca2+ channels characterized by recordings from isolated hippocampal dendritic segments. Neuron 18(4):651–663

Keck T, Lillis KP, Zhou YD, White JA (2008) Frequency-dependent glycinergic inhibition modulates plasticity in hippocampus. J Neurosci 28(29):7359–7369. doi:10.1523/JNEUROSCI.5618-07.2008

Kennedy MJ, Davison IG, Robinson CG, Ehlers MD (2010) Syntaxin-4 defines a domain for activity-dependent exocytosis in dendritic spines. Cell 141(3):524–535. doi:10.1016/j.cell.2010.02.042

Kerti-Szigeti K, Nusser Z, Eyre MD (2014) Synaptic GABAA receptor clustering without the gamma2 subunit. J Neurosci 34(31):10219–10233. doi:10.1523/JNEUROSCI.1721-14.2014

Kessels HW, Kopec CD, Klein ME, Malinow R (2009) Roles of stargazin and phosphorylation in the control of AMPA receptor subcellular distribution. Nat Neurosci 12(7):888–896. doi:10.1038/nn.2340

Khodosevich K, Jacobi E, Farrow P, Schulmann A, Rusu A, Zhang L, Sprengel R, Monyer H, von Engelhardt J (2014) Coexpressed auxiliary subunits exhibit distinct modulatory profiles on AMPA receptor function. Neuron 83(3):601–615. doi:10.1016/j.neuron.2014.07.004

Kim HG, Connors BW (1993) Apical dendrites of the neocortex: correlation between sodium- and calcium-dependent spiking and pyramidal cell morphology. J Neurosci 13(12):5301–5311

Kim J, Wei DS, Hoffman DA (2005) Kv4 potassium channel subunits control action potential repolarization and frequency-dependent broadening in rat hippocampal CA1 pyramidal neurones. J Physiol 569(Pt 1):41–57. doi:jphysiol.2005.095042 [pii];10.1113/jphysiol.2005.095042 [doi]

Kim J, Jung SC, Clemens AM, Petralia RS, Hoffman DA (2007) Regulation of dendritic excitability by activity-dependent trafficking of the A-type K+ channel subunit Kv4.2 in hippocampal neurons. Neuron 54(6):933–947. doi:S0896-6273(07)00383-2 [pii];10.1016/j.neuron.2007.05.026 [doi]

Kirizs T, Kerti-Szigeti K, Lorincz A, Nusser Z (2014) Distinct axo-somato-dendritic distributions of three potassium channels in CA1 hippocampal pyramidal cells. Eur J Neurosci 39 (11):1771–1783. doi:10.1111/ejn.12526

Kole MH, Stuart GJ (2008) Is action potential threshold lowest in the axon? Nat Neurosci 11 (11):1253–1255. doi:10.1038/nn.2203

Kole MH, Ilschner SU, Kampa BM, Williams SR, Ruben PC, Stuart GJ (2008) Action potential generation requires a high sodium channel density in the axon initial segment. Nat Neurosci 11 (2):178–186. doi:10.1038/nn2040

Kopec CD, Li B, Wei W, Boehm J, Malinow R (2006) Glutamate receptor exocytosis and spine enlargement during chemically induced long-term potentiation. J Neurosci 26(7):2000–2009. doi:10.1523/JNEUROSCI.3918-05.2006

Kovalchuk Y, Eilers J, Lisman J, Konnerth A (2000) NMDA receptor-mediated subthreshold Ca(2 +) signals in spines of hippocampal neurons. J Neurosci 20(5):1791–1799

Kullmann DM, Erdemli G, Asztely F (1996) LTP of AMPA and NMDA receptor-mediated signals: evidence for presynaptic expression and extrasynaptic glutamate spill-over. Neuron 17(3):461–474. doi:S0896-6273(00)80178-6 [pii]

Larkum ME, Zhu JJ, Sakmann B (1999a) A new cellular mechanism for coupling inputs arriving at different cortical layers. Nature 398(6725):338–341. doi:10.1038/18686

Larkum ME, Kaiser KM, Sakmann B (1999b) Calcium electrogenesis in distal apical dendrites of layer 5 pyramidal cells at a critical frequency of back-propagating action potentials. Proc Natl Acad Sci U S A 96(25):14600–14604

Laudanski J, Torben-Nielsen B, Segev I, Shamma S (2014) Spatially distributed dendritic resonance selectively filters synaptic input. PLoS Comput Biol 10(8), e1003775. doi:10.1371/journal.pcbi.1003775

Lee SJ, Escobedo-Lozoya Y, Szatmari EM, Yasuda R (2009) Activation of CaMKII in single dendritic spines during long-term potentiation. Nature 458(7236):299–304. doi:10.1038/nature07842

Leung LS, Yu HW (1998) Theta-frequency resonance in hippocampal CA1 neurons in vitro demonstrated by sinusoidal current injection. J Neurophysiol 79(3):1592–1596

Lewis AS, Schwartz E, Chan CS, Noam Y, Shin M, Wadman WJ, Surmeier DJ, Baram TZ, Macdonald RL, Chetkovich DM (2009) Alternatively spliced isoforms of TRIP8b differentially control h channel trafficking and function. J Neurosci 29(19):6250–6265. doi:29/19/6250 [pii];10.1523/JNEUROSCI.0856-09.2009 [doi]

Lewis AS, Estep CM, Chetkovich DM (2010) The fast and slow ups and downs of HCN channel regulation. Channels (Austin) 4(3):215–231. doi:11630 [pii]

Lewis AS, Vaidya SP, Blaiss CA, Liu Z, Stoub TR, Brager DH, Chen X, Bender RA, Estep CM, Popov AB, Kang CE, Van Veldhoven PP, Bayliss DA, Nicholson DA, Powell CM, Johnston D, Chetkovich DM (2011) Deletion of the hyperpolarization-activated cyclic nucleotide-gated channel auxiliary subunit TRIP8b impairs hippocampal Ih localization and function and promotes antidepressant behavior in mice. J Neurosci 31(20):7424–7440. doi:31/20/7424 [pii];10.1523/JNEUROSCI.0936-11.2011 [doi]

Li M, West JW, Numann R, Murphy BJ, Scheuer T, Catterall WA (1993) Convergent regulation of sodium channels by protein kinase C and cAMP-dependent protein kinase. Science 261 (5127):1439–1442

Lin MT, Lujan R, Watanabe M, Adelman JP, Maylie J (2008) SK2 channel plasticity contributes to LTP at Schaffer collateral-CA1 synapses. Nat Neurosci 11(2):170–177. doi:10.1038/nn2041

Lin L, Sun W, Throesch B, Kung F, Decoster JT, Berner CJ, Cheney RE, Rudy B, Hoffman DA (2013) DPP6 regulation of dendritic morphogenesis impacts hippocampal synaptic development. Nat Commun 4:2270. doi:ncomms3270 [pii];10.1038/ncomms3270 [doi]

Lin L, Long LK, Hatch MM, Hoffman DA (2014) DPP6 domains responsible for its localization and function. J Biol Chem. doi:M114.578070 [pii];10.1074/jbc.M114.578070 [doi]

Llinas R, Yarom Y (1986) Oscillatory properties of guinea-pig inferior olivary neurones and their pharmacological modulation: an in vitro study. J Physiol 376:163–182

London M, Häusser M (2005) Dendritic computation. Annu Rev Neurosci 28:503–532. doi:10.1146/annurev.neuro.28.061604.135703

Lörincz A, Nusser Z (2010) Molecular identity of dendritic voltage-gated sodium channels. Science 328(5980):906–909. doi:10.1126/science.1187958

Lorincz A, Notomi T, Tamas G, Shigemoto R, Nusser Z (2002) Polarized and compartment-dependent distribution of HCN1 in pyramidal cell dendrites. Nat Neurosci 5(11):1185–1193. doi:10.1038/nn962

Losonczy A, Magee JC (2006) Integrative properties of radial oblique dendrites in hippocampal CA1 pyramidal neurons. Neuron 50(2):291–307. doi:10.1016/j.neuron.2006.03.016

Losonczy A, Makara JK, Magee JC (2008) Compartmentalized dendritic plasticity and input feature storage in neurons. Nature 452(7186):436–441. doi:10.1038/nature06725

Magee JC (1998) Dendritic hyperpolarization-activated currents modify the integrative properties of hippocampal CA1 pyramidal neurons. J Neurosci 18(19):7613–7624

Magee JC (1999) Dendritic Ih normalizes temporal summation in hippocampal CA1 neurons. Nat Neurosci 2(9):848. doi:10.1038/12229

Magee JC (2000) Dendritic integration of excitatory synaptic input. Nat Rev Neurosci 1 (3):181–190. doi:10.1038/35044552

Magee JC (2008) Dendritic voltage-gated ion channels. In: Stuart G, Spruston N, Hдusser M (eds) Dendrites, 2nd edn. Oxford University Press, New York, pp 225–250

Magee JC, Carruth M (1999) Dendritic voltage-gated ion channels regulate the action potential firing mode of hippocampal CA1 pyramidal neurons. J Neurophysiol 82(4):1895–1901

Magee JC, Cook EP (2000) Somatic EPSP amplitude is independent of synapse location in hippocampal pyramidal neurons. Nat Neurosci 3(9):895–903. doi:10.1038/78800

Magee JC, Johnston D (1995a) Characterization of single voltage-gated Na+ and Ca2+ channels in apical dendrites of rat CA1 pyramidal neurons. J Physiol 487(Pt 1):67–90

Magee JC, Johnston D (1995b) Synaptic activation of voltage-gated channels in the dendrites of hippocampal pyramidal neurons. Science 268(5208):301–304

Magee JC, Johnston D (1997) A synaptically controlled, associative signal for Hebbian plasticity in hippocampal neurons. Science 275(5297):209–213

Magee JC, Christofi G, Miyakawa H, Christie B, Lasser-Ross N, Johnston D (1995) Subthreshold synaptic activation of voltage-gated Ca2+ channels mediates a localized Ca2+ influx into the dendrites of hippocampal pyramidal neurons. J Neurophysiol 74(3):1335–1342

Mainen ZF, Joerges J, Huguenard JR, Sejnowski TJ (1995) A model of spike initiation in neocortical pyramidal neurons. Neuron 15(6):1427–1439

Mainen ZF, Carnevale NT, Zador AM, Claiborne BJ, Brown TH (1996) Electrotonic architecture of hippocampal CA1 pyramidal neurons based on three-dimensional reconstructions. J Neurophysiol 76(3):1904–1923

Makara JK, Losonczy A, Wen Q, Magee JC (2009) Experience-dependent compartmentalized dendritic plasticity in rat hippocampal CA1 pyramidal neurons. Nat Neurosci 12 (12):1485–1487. doi:10.1038/nn.2428

Makino H, Malinow R (2009) AMPA receptor incorporation into synapses during LTP: the role of lateral movement and exocytosis. Neuron 64(3):381–390. doi:10.1016/j.neuron.2009.08.035
Maletic-Savatic M, Lenn NJ, Trimmer JS (1995) Differential spatiotemporal expression of K+ channel polypeptides in rat hippocampal neurons developing in situ and in vitro. J Neurosci 15 (5 Pt 2):3840–3851
Marban E, Yamagishi T, Tomaselli GF (1998) Structure and function of voltage-gated sodium channels. J Physiol 508(Pt 3):647–657
Marionneau C, Carrasquillo Y, Norris AJ, Townsend RR, Isom LL, Link AJ, Nerbonne JM (2012) The sodium channel accessory subunit Navbeta1 regulates neuronal excitability through modulation of repolarizing voltage-gated K(+) channels. J Neurosci 32(17):5716–5727. doi:10.1523/JNEUROSCI.6450-11.2012
Martel MA, Ryan TJ, Bell KF, Fowler JH, McMahon A, Al-Mubarak B, Komiyama NH, Horsburgh K, Kind PC, Grant SG, Wyllie DJ, Hardingham GE (2012) The subtype of GluN2 C-terminal domain determines the response to excitotoxic insults. Neuron 74 (3):543–556. doi:10.1016/j.neuron.2012.03.021
Martina M, Vida I, Jonas P (2000) Distal initiation and active propagation of action potentials in interneuron dendrites. Science 287(5451):295–300
Mauro A, Conti F, Dodge F, Schor R (1970) Subthreshold behavior and phenomenological impedance of the squid giant axon. J Gen Phys 55(4):497–523
McKay BE, McRory JE, Molineux ML, Hamid J, Snutch TP, Zamponi GW, Turner RW (2006) Ca (V)3 T-type calcium channel isoforms differentially distribute to somatic and dendritic compartments in rat central neurons. Eur J Neurosci 24(9):2581–2594. doi:10.1111/j.1460-9568.2006.05136.x
Megias M, Emri Z, Freund TF, Gulyas AI (2001) Total number and distribution of inhibitory and excitatory synapses on hippocampal CA1 pyramidal cells. Neuroscience 102(3):527–540. doi: S0306-4522(00)00496-6 [pii]
Melyan Z, Wheal HV, Lancaster B (2002) Metabotropic-mediated kainate receptor regulation of IsAHP and excitability in pyramidal cells. Neuron 34(1):107–114
Migliore M, Shepherd GM (2002) Emerging rules for the distributions of active dendritic conductances. Nat Rev Neurosci 3(5):362–370. doi:10.1038/nrn810
Miller OH, Yang L, Wang CC, Hargroder EA, Zhang Y, Delpire E, Hall BJ (2014) GluN2B-containing NMDA receptors regulate depression-like behavior and are critical for the rapid antidepressant actions of ketamine. eLife 3, e03581. doi:10.7554/eLife.03581
Mills LR, Niesen CE, So AP, Carlen PL, Spigelman I, Jones OT (1994) N-type Ca2+ channels are located on somata, dendrites, and a subpopulation of dendritic spines on live hippocampal pyramidal neurons. J Neurosci 14(11 Pt 2):6815–6824
Milstein AD, Zhou W, Karimzadegan S, Bredt DS, Nicoll RA (2007) TARP subtypes differentially and dose-dependently control synaptic AMPA receptor gating. Neuron 55(6):905–918. doi:10.1016/j.neuron.2007.08.022
Miyakawa H, Ross WN, Jaffe D, Callaway JC, Lasser-Ross N, Lisman JE, Johnston D (1992) Synaptically activated increases in Ca2+ concentration in hippocampal CA1 pyramidal cells are primarily due to voltage-gated Ca2+ channels. Neuron 9(6):1163–1173
Müller C, Beck H, Coulter D, Remy S (2012) Inhibitory control of linear and supralinear dendritic excitation in CA1 pyramidal neurons. Neuron 75(5):851–864. doi:10.1016/j.neuron.2012.06.025
Murphy BJ, Rossie S, De Jongh KS, Catterall WA (1993) Identification of the sites of selective phosphorylation and dephosphorylation of the rat brain Na+ channel alpha subunit by cAMP-dependent protein kinase and phosphoprotein phosphatases. J Biol Chem 268 (36):27355–27362
Nadal MS, Ozaita A, Amarillo Y, de Vega-Saenz ME, Ma Y, Mo W, Goldberg EM, Misumi Y, Ikehara Y, Neubert TA, Rudy B (2003) The CD26-related dipeptidyl aminopeptidase-like protein DPPX is a critical component of neuronal A-type K+ channels. Neuron 37(3):449–461. doi:S0896627302011856 [pii]

Narayanan R, Johnston D (2007) Long-term potentiation in rat hippocampal neurons is accompanied by spatially widespread changes in intrinsic oscillatory dynamics and excitability. Neuron 56:1061–1075

Ngo-Anh TJ, Bloodgood BL, Lin M, Sabatini BL, Maylie J, Adelman JP (2005) SK channels and NMDA receptors form a Ca2+ -mediated feedback loop in dendritic spines. Nat Neurosci 8 (5):642–649

Niswender CM, Conn PJ (2010) Metabotropic glutamate receptors: physiology, pharmacology, and disease. Annu Rev Pharmacol Toxicol 50:295–322. doi:10.1146/annurev.pharmtox.011008.145533

Notomi T, Shigemoto R (2004) Immunohistochemical localization of Ih channel subunits, HCN1-4, in the rat brain. J Comp Neurol 471(3):241–276. doi:10.1002/cne.11039

Nowak L, Bregestovski P, Ascher P, Herbet A, Prochiantz A (1984) Magnesium gates glutamate-activated channels in mouse central neurones. Nature 307(5950):462–465

Numann R, Catterall WA, Scheuer T (1991) Functional modulation of brain sodium channels by protein kinase C phosphorylation. Science 254(5028):115–118

Nuriya M, Jiang J, Nemet B, Eisenthal KB, Yuste R (2006) Imaging membrane potential in dendritic spines. Proc Natl Acad Sci U S A 103(3):786–790. doi:10.1073/pnas.0510092103

Panatier A, Theodosis DT, Mothet JP, Touquet B, Pollegioni L, Poulain DA, Oliet SH (2006) Glia-derived D-serine controls NMDA receptor activity and synaptic memory. Cell 125 (4):775–784. doi:10.1016/j.cell.2006.02.051

Paoletti P, Bellone C, Zhou Q (2013) NMDA receptor subunit diversity: impact on receptor properties, synaptic plasticity and disease. Nat Rev Neurosci 14(6):383–400. doi:10.1038/nrn3504

Pape HC (1996) Queer current and pacemaker: the hyperpolarization-activated cation current in neurons. Annu Rev Physiol 58:299–327. doi:10.1146/annurev.ph.58.030196.001503

Papouin T, Oliet SH (2014) Organization, control and function of extrasynaptic NMDA receptors. Philos Trans R Soc Lond B Biol Sci 369(1654):20130601. doi:10.1098/rstb.2013.0601

Parpura V, Basarsky TA, Liu F, Jeftinija K, Jeftinija S, Haydon PG (1994) Glutamate-mediated astrocyte-neuron signalling. Nature 369(6483):744–747. doi:10.1038/369744a0

Parsons MP, Raymond LA (2014) Extrasynaptic NMDA receptor involvement in central nervous system disorders. Neuron 82(2):279–293. doi:10.1016/j.neuron.2014.03.030

Patton DE, Isom LL, Catterall WA, Goldin AL (1994) The adult rat brain beta 1 subunit modifies activation and inactivation gating of multiple sodium channel alpha subunits. J Biol Chem 269 (26):17649–17655

Petrini EM, Barberis A (2014) Diffusion dynamics of synaptic molecules during inhibitory postsynaptic plasticity. Front Cell Neurosci 8:300. doi:10.3389/fncel.2014.00300

Petrini EM, Lu J, Cognet L, Lounis B, Ehlers MD, Choquet D (2009) Endocytic trafficking and recycling maintain a pool of mobile surface AMPA receptors required for synaptic potentiation. Neuron 63(1):92–105. doi:S0896-6273(09)00400-0 [pii] 10.1016/j.neuron.2009.05.025

Petrini EM, Ravasenga T, Hausrat TJ, Iurilli G, Olcese U, Racine V, Sibarita JB, Jacob TC, Moss SJ, Benfenati F, Medini P, Kneussel M, Barberis A (2014) Synaptic recruitment of gephyrin regulates surface GABAA receptor dynamics for the expression of inhibitory LTP. Nat Commun 5:3921. doi:10.1038/ncomms4921

Poirazi P, Mel BW (2001) Impact of active dendrites and structural plasticity on the memory capacity of neural tissue. Neuron 29(3):779–796

Priel A, Kolleker A, Ayalon G, Gillor M, Osten P, Stern-Bach Y (2005) Stargazin reduces desensitization and slows deactivation of the AMPA-type glutamate receptors. J Neurosci 25 (10):2682–2686. doi:25/10/2682 [pii] 10.1523/JNEUROSCI.4834-04.2005

Puil E, Gimbarzevsky B, Miura RM (1986) Quantification of membrane properties of trigeminal root ganglion neurons in guinea pigs. J Neurophysiol 55(5):995–1016

Puil E, Meiri H, Yarom Y (1994) Resonant behavior and frequency preferences of thalamic neurons. J Neurophysiol 71(2):575–582

Racz B, Blanpied TA, Ehlers MD, Weinberg RJ (2004) Lateral organization of endocytic machinery in dendritic spines. Nat Neurosci 7(9):917–918. doi:10.1038/nn1303 nn1303 [pii]
Rall W (1962) Theory of physiological properties of dendrites. Ann N Y Acad Sci 96:1071–1092
Ratcliffe CF, Qu Y, McCormick KA, Tibbs VC, Dixon JE, Scheuer T, Catterall WA (2000) A sodium channel signaling complex: modulation by associated receptor protein tyrosine phosphatase beta. Nat Neurosci 3(5):437–444. doi:10.1038/74805
Regehr WG, Connor JA, Tank DW (1989) Optical imaging of calcium accumulation in hippocampal pyramidal cells during synaptic activation. Nature 341(6242):533–536. doi:10.1038/341533a0
Remy S, Csicsvari J, Beck H (2009) Activity-dependent control of neuronal output by local and global dendritic spike attenuation. Neuron 61(6):906–916. doi:10.1016/j.neuron.2009.01.032
Reyes A (2001) Influence of dendritic conductances on the input-output properties of neurons. Annu Rev Neurosci 24:653–675. doi:10.1146/annurev.neuro.24.1.653
Rhodes KJ, Carroll KI, Sung MA, Doliveira LC, Monaghan MM, Burke SL, Strassle BW, Buchwalder L, Menegola M, Cao J, An WF, Trimmer JS (2004) KChIPs and Kv4 alpha subunits as integral components of A-type potassium channels in mammalian brain. J Neurosci 24(36):7903–7915. doi:10.1523/JNEUROSCI.0776-04.2004 [doi];24/36/7903 [pii]
Rogerson T, Cai DJ, Frank A, Sano Y, Shobe J, Lopez-Aranda MF, Silva AJ (2014) Synaptic tagging during memory allocation. Nat Rev Neurosci 15(3):157–169. doi:10.1038/nrn3667
Rose CR, Konnerth A (2001) NMDA receptor-mediated Na+ signals in spines and dendrites. J Neurosci 21(12):4207–4214
Rouach N, Byrd K, Petralia RS, Elias GM, Adesnik H, Tomita S, Karimzadegan S, Kealey C, Bredt DS, Nicoll RA (2005) TARP gamma-8 controls hippocampal AMPA receptor number, distribution and synaptic plasticity. Nat Neurosci 8(11):1525–1533, Epub 2005 Oct 1529
Rudy B (1978) Slow inactivation of the sodium conductance in squid giant axons. Pronase resistance. J Physiol 283:1–21
Rudy B (1988) Diversity and ubiquity of K channels. Neuroscience 25(3):729–749
Sabatini BL, Svoboda K (2000) Analysis of calcium channels in single spines using optical fluctuation analysis. Nature 408(6812):589–593. doi:10.1038/35046076
Sah P (1996) Ca(2+)-activated K+ currents in neurones: types, physiological roles and modulation. Trends Neurosci 19(4):150–154. doi:S0166-2236(96)80026-9 [pii]
Sah P, Faber ES (2002) Channels underlying neuronal calcium-activated potassium currents. Prog Neurobiol 66(5):345–353. doi:S0301008202000047 [pii]
Sailer CA, Kaufmann WA, Marksteiner J, Knaus HG (2004) Comparative immunohistochemical distribution of three small-conductance Ca2+ -activated potassium channel subunits, SK1, SK2, and SK3 in mouse brain. Mol Cell Neurosci 26(3):458–469. doi:10.1016/j.mcn.2004.03.002
Santoro B, Piskorowski RA, Pian P, Hu L, Liu H, Siegelbaum SA (2009) TRIP8b splice variants form a family of auxiliary subunits that regulate gating and trafficking of HCN channels in the brain. Neuron 62(6):802–813. doi:S0896-6273(09)00360-2 [pii];10.1016/j.neuron.2009.05.009 [doi]
Sanz-Clemente A, Nicoll RA, Roche KW (2013) Diversity in NMDA receptor composition: many regulators, many consequences. Neuroscientist Rev J Bringing Neurobiol Neurol Psychiatry 19 (1):62–75. doi:10.1177/1073858411435129
Scheuer T (1994) Structure and function of voltage-gated sodium channels: regulation by phosphorylation. Biochem Soc Trans 22(2):479–482
Schiller J, Major G, Koester HJ, Schiller Y (2000) NMDA spikes in basal dendrites of cortical pyramidal neurons. Nature 404(6775):285–289. doi:10.1038/35005094
Schmidt-Salzmann C, Li L, Bischofberger J (2014) Functional properties of extrasynaptic AMPA and NMDA receptors during postnatal hippocampal neurogenesis. J Physiol 592 (Pt 1):125–140. doi:10.1113/jphysiol.2013.267203

Schrader LA, Anderson AE, Varga AW, Levy M, Sweatt JD (2002) The other half of Hebb: K+ channels and the regulation of neuronal excitability in the hippocampus. Mol Neurobiol 25 (1):51–66. doi:MN:25:1:051 [pii];10.1385/MN:25:1:051 [doi]

Schwenk J, Harmel N, Zolles G, Bildl W, Kulik A, Heimrich B, Chisaka O, Jonas P, Schulte U, Fakler B, Klocker N (2009) Functional proteomics identify cornichon proteins as auxiliary subunits of AMPA receptors. Science 323(5919):1313–1319. doi:323/5919/1313 [pii] 10.1126/science.1167852

Schwindt PC, Crill WE (1995) Amplification of synaptic current by persistent sodium conductance in apical dendrite of neocortical neurons. J Neurophysiol 74(5):2220–2224

Scimemi A (2014) Plasticity of GABA transporters: an unconventional route to shape inhibitory synaptic transmission. Front Cell Neurosci 8:128. doi:10.3389/fncel.2014.00128

Serodio P, de Vega-Saenz ME, Rudy B (1996) Cloning of a novel component of A-type K+ channels operating at subthreshold potentials with unique expression in heart and brain. J Neurophysiol 75(5):2174–2179

Shah MM, Hammond RS, Hoffman DA (2010) Dendritic ion channel trafficking and plasticity. Trends Neurosci 33(7):307–316. doi:S0166-2236(10)00036-6 [pii];10.1016/j.tins.2010.03.002 [doi]

Shao D, Okuse K, Djamgoz MB (2009) Protein-protein interactions involving voltage-gated sodium channels: post-translational regulation, intracellular trafficking and functional expression. Int J Biochem Cell Biol 41(7):1471–1481. doi:10.1016/j.biocel.2009.01.016

Sheng M, Tsaur ML, Jan YN, Jan LY (1992) Subcellular segregation of two A-type K+ channel proteins in rat central neurons. Neuron 9(2):271–284. doi:0896-6273(92)90166-B [pii]

Shipman SL, Herring BE, Suh YH, Roche KW, Nicoll RA (2013) Distance-dependent scaling of AMPARs is cell-autonomous and GluA2 dependent. J Neurosci 33(33):13312–13319. doi:10.1523/JNEUROSCI.0678-13.2013

Shipton OA, Paulsen O (2014) GluN2A and GluN2B subunit-containing NMDA receptors in hippocampal plasticity. Philos Trans R Soc Lond B Biol Sci 369(1633):20130163. doi:10.1098/rstb.2013.0163

Sjostrom PJ, Rancz EA, Roth A, Hausser M (2008) Dendritic excitability and synaptic plasticity. Physiol Rev 88(2):769–840. doi:10.1152/physrev.00016.2007

Smith RD, Goldin AL (1997) Phosphorylation at a single site in the rat brain sodium channel is necessary and sufficient for current reduction by protein kinase A. J Neurosci 17 (16):6086–6093

Soltesz I, Deschenes M (1993) Low- and high-frequency membrane potential oscillations during theta activity in CA1 and CA3 pyramidal neurons of the rat hippocampus under ketamine-xylazine anesthesia. J Neurophysiol 70(1):97–116

Song W, Chattipakorn SC, McMahon LL (2006) Glycine-gated chloride channels depress synaptic transmission in rat hippocampus. J Neurophysiol 95(4):2366–2379. doi:10.1152/jn.00386.2005

Song I, Volynski K, Brenner T, Ushkaryov Y, Walker M, Semyanov A (2013) Different transporter systems regulate extracellular GABA from vesicular and non-vesicular sources. Front Cell Neurosci 7:23. doi:10.3389/fncel.2013.00023

Spruston N (2008) Pyramidal neurons: dendritic structure and synaptic integration. Nat Rev Neurosci 9(3):206–221. doi:10.1038/nrn2286

Spruston N, Schiller Y, Stuart G, Sakmann B (1995) Activity-dependent action potential invasion and calcium influx into hippocampal CA1 dendrites. Science 268(5208):297–300

Stocker M, Pedarzani P (2000) Differential distribution of three Ca(2+)-activated K(+) channel subunits, SK1, SK2, and SK3, in the adult rat central nervous system. Mol Cell Neurosci 15 (5):476–493. doi:10.1006/mcne.2000.0842

Storm JF (1990) Potassium currents in hippocampal pyramidal cells. Prog Brain Res 83:161–187

Straub C, Hunt DL, Yamasaki M, Kim KS, Watanabe M, Castillo PE, Tomita S (2011) Distinct functions of kainate receptors in the brain are determined by the auxiliary subunit Neto1. Nat Neurosci 14(7):866–873. doi:10.1038/nn.2837

Stuart GJ, Hausser M (2001) Dendritic coincidence detection of EPSPs and action potentials. Nat Neurosci 4(1):63–71. doi:10.1038/82910
Stuart GJ, Sakmann B (1994) Active propagation of somatic action potentials into neocortical pyramidal cell dendrites. Nature 367(6458):69–72. doi:10.1038/367069a0
Stuart G, Schiller J, Sakmann B (1997a) Action potential initiation and propagation in rat neocortical pyramidal neurons. J Physiol 505(Pt 3):617–632
Stuart G, Spruston N, Sakmann B, Hausser M (1997b) Action potential initiation and backpropagation in neurons of the mammalian CNS. Trends Neurosci 20(3):125–131
Sun W, Maffie JK, Lin L, Petralia RS, Rudy B, Hoffman DA (2011) DPP6 establishes the A-type K(+) current gradient critical for the regulation of dendritic excitability in CA1 hippocampal neurons. Neuron 71(6):1102–1115. doi:S0896-6273(11)00722-7 [pii];10.1016/j.neuron.2011.08.008 [doi]
Teles-Grilo Ruivo LM, Mellor JR (2013) Cholinergic modulation of hippocampal network function. Front Synaptic Neurosci 5:2. doi:10.3389/fnsyn.2013.00002
Thomsen MB, Foster E, Nguyen KH, Sosunov EA (2009) Transcriptional and electrophysiological consequences of KChIP2-mediated regulation of CaV1.2. Channels (Austin) 3(5):308–310. doi:9560 [pii]
Thomson AM, Bannister AP, Hughes DI, Pawelzik H (2000) Differential sensitivity to Zolpidem of IPSPs activated by morphologically identified CA1 interneurons in slices of rat hippocampus. Eur J Neurosci 12(2):425–436
Tomita S, Shenoy A, Fukata Y, Nicoll RA, Bredt DS (2007) Stargazin interacts functionally with the AMPA receptor glutamate-binding module. Neuropharmacology 52(1):87–91. doi:10.1016/j.neuropharm.2006.07.012
Traynelis SF, Wollmuth LP, McBain CJ, Menniti FS, Vance KM, Ogden KK, Hansen KB, Yuan H, Myers SJ, Dingledine R (2010) Glutamate receptor ion channels: structure, regulation, and function. Pharmacol Rev 62(3):405–496. doi:10.1124/pr.109.002451 62/3/405 [pii]
Trimmer JS, Rhodes KJ (2004) Localization of voltage-gated ion channels in mammalian brain. Annu Rev Physiol 66:477–519. doi:10.1146/annurev.physiol.66.032102.113328
Tsay D, Yuste R (2002) Role of dendritic spines in action potential backpropagation: a numerical simulation study. J Neurophysiol 88(5):2834–2845. doi:10.1152/jn.00781.2001
Turner RW, Zamponi GW (2014) T-type channels buddy up. Pflugers Arch - Eur J Physiol 466 (4):661–675. doi:10.1007/s00424-013-1434-6
Turner RW, Meyers DE, Richardson TL, Barker JL (1991) The site for initiation of action potential discharge over the somatodendritic axis of rat hippocampal CA1 pyramidal neurons. J Neurosci 11(7):2270–2280
Vaidya SP, Johnston D (2013) Temporal synchrony and gamma-to-theta power conversion in the dendrites of CA1 pyramidal neurons. Nat Neurosci 16(12):1812–1820. doi:10.1038/nn.3562
van Zundert B, Yoshii A, Constantine-Paton M (2004) Receptor compartmentalization and trafficking at glutamate synapses: a developmental proposal. Trends Neurosci 27 (7):428–437. doi:10.1016/j.tins.2004.05.010
Varga AW, Anderson AE, Adams JP, Vogel H, Sweatt JD (2000) Input-specific immunolocalization of differentially phosphorylated Kv4.2 in the mouse brain. Learn Mem 7 (5):321–332
Varga AW, Yuan LL, Anderson AE, Schrader LA, Wu GY, Gatchel JR, Johnston D, Sweatt JD (2004) Calcium-calmodulin-dependent kinase II modulates Kv4.2 channel expression and upregulates neuronal A-type potassium currents. J Neurosci 24(14):3643–3654. doi:10.1523/JNEUROSCI.0154-04.2004 [doi];24/14/3643 [pii]
Vizi ES, Fekete A, Karoly R, Mike A (2010) Non-synaptic receptors and transporters involved in brain functions and targets of drug treatment. Br J Pharmacol 160(4):785–809. doi:10.1111/j.1476-5381.2009.00624.x
von Engelhardt J, Mack V, Sprengel R, Kavenstock N, Li KW, Stern-Bach Y, Smit AB, Seeburg PH, Monyer H (2010) CKAMP44: a brain-specific protein attenuating short-term synaptic plasticity in the dentate gyrus. Science 327(5972):1518–1522. doi:10.1126/science.1184178

Wahl-Schott C, Biel M (2009) HCN channels: structure, cellular regulation and physiological function. Cell Mol Life Sci 66(3):470–494. doi:10.1007/s00018-008-8525-0

Wang Z, Xu NL, Wu CP, Duan S, Poo MM (2003) Bidirectional changes in spatial dendritic integration accompanying long-term synaptic modifications. Neuron 37(3):463–472. doi: S0896627302011893 [pii]

Wang M, Ramos BP, Paspalas CD, Shu Y, Simen A, Duque A, Vijayraghavan S, Brennan A, Dudley A, Nou E, Mazer JA, McCormick DA, Arnsten AF (2007) Alpha2A-adrenoceptors strengthen working memory networks by inhibiting cAMP-HCN channel signaling in prefrontal cortex. Cell 129(2):397–410. doi:S0092-8674(07)00344-3 [pii];10.1016/j.cell.2007.03.015 [doi]

Wang HG, He XP, Li Q, Madison RD, Moore SD, McNamara JO, Pitt GS (2013) The auxiliary subunit KChIP2 is an essential regulator of homeostatic excitability. J Biol Chem 288 (19):13258–13268. doi:M112.434548 [pii];10.1074/jbc.M112.434548 [doi]

Wen Q, Chklovskii DB (2008) A cost-benefit analysis of neuronal morphology. J Neurophysiol 99 (5):2320–2328. doi:10.1152/jn.00280.2007

Wen Q, Stepanyants A, Elston GN, Grosberg AY, Chklovskii DB (2009) Maximization of the connectivity repertoire as a statistical principle governing the shapes of dendritic arbors. Proc Natl Acad Sci U S A 106(30):12536–12541. doi:10.1073/pnas.0901530106

West JW, Numann R, Murphy BJ, Scheuer T, Catterall WA (1991) A phosphorylation site in the Na+ channel required for modulation by protein kinase C. Science 254(5033):866–868

Westenbroek RE, Merrick DK, Catterall WA (1989) Differential subcellular localization of the RI and RII Na+ channel subtypes in central neurons. Neuron 3(6):695–704

Westenbroek RE, Ahlijanian MK, Catterall WA (1990) Clustering of L-type Ca2+ channels at the base of major dendrites in hippocampal pyramidal neurons. Nature 347(6290):281–284. doi:10.1038/347281a0

Westenbroek RE, Hell JW, Warner C, Dubel SJ, Snutch TP, Catterall WA (1992) Biochemical properties and subcellular distribution of an N-type calcium channel alpha 1 subunit. Neuron 9 (6):1099–1115

Westenbroek RE, Sakurai T, Elliott EM, Hell JW, Starr TV, Snutch TP, Catterall WA (1995) Immunochemical identification and subcellular distribution of the alpha 1A subunits of brain calcium channels. J Neurosci 15(10):6403–6418

Wheeler DG, Groth RD, Ma H, Barrett CF, Owen SF, Safa P, Tsien RW (2012) Ca(V)1 and Ca(V) 2 channels engage distinct modes of Ca(2+) signaling to control CREB-dependent gene expression. Cell 149(5):1112–1124. doi:10.1016/j.cell.2012.03.041

Whitaker WR, Faull RL, Waldvogel HJ, Plumpton CJ, Emson PC, Clare JJ (2001) Comparative distribution of voltage-gated sodium channel proteins in human brain. Brain Res Mol Brain Res 88(1–2):37–53

Williams SR, Stuart GJ (2000a) Backpropagation of physiological spike trains in neocortical pyramidal neurons: implications for temporal coding in dendrites. J Neurosci 20 (22):8238–8246

Williams SR, Stuart GJ (2000b) Action potential backpropagation and somato-dendritic distribution of ion channels in thalamocortical neurons. J Neurosci 20(4):1307–1317

Williams SR, Stuart GJ (2003) Role of dendritic synapse location in the control of action potential output. Trends Neurosci 26(3):147–154. doi:10.1016/S0166-2236(03)00035-3

Wood JN, Baker M (2001) Voltage-gated sodium channels. Curr Opin Pharmacol 1(1):17–21

Wu YW, Grebenyuk S, McHugh TJ, Rusakov DA, Semyanov A (2012) Backpropagating action potentials enable detection of extrasynaptic glutamate by NMDA receptors. Cell Rep 1 (5):495–505. doi:10.1016/j.celrep.2012.03.007

Yasuda R, Sabatini BL, Svoboda K (2003) Plasticity of calcium channels in dendritic spines. Nat Neurosci 6(9):948–955. doi:10.1038/nn1112 nn1112 [pii]

Ylinen A, Soltesz I, Bragin A, Penttonen M, Sik A, Buzsaki G (1995) Intracellular correlates of hippocampal theta rhythm in identified pyramidal cells, granule cells, and basket cells. Hippocampus 5(1):78–90. doi:10.1002/hipo.450050110

Yuan LL, Adams JP, Swank M, Sweatt JD, Johnston D (2002) Protein kinase modulation of dendritic K+ channels in hippocampus involves a mitogen-activated protein kinase pathway. J Neurosci 22(12):4860–4868. doi:22/12/4860 [pii]

Yuste R, Denk W (1995) Dendritic spines as basic functional units of neuronal integration. Nature 375(6533):682–684. doi:10.1038/375682a0

Yuste R, Tank DW (1996) Dendritic integration in mammalian neurons, a century after Cajal. Neuron 16(4):701–716

Yuste R, Majewska A, Cash SS, Denk W (1999) Mechanisms of calcium influx into hippocampal spines: heterogeneity among spines, coincidence detection by NMDA receptors, and optical quantal analysis. J Neurosci Off J Soc Neurosci 19(6):1976–1987

Zagha E, Ozaita A, Chang SY, Nadal MS, Lin U, Saganich MJ, McCormack T, Akinsanya KO, Qi SY, Rudy B (2005) DPP10 modulates Kv4-mediated A-type potassium channels. J Biol Chem 280(19):18853–18861. doi:M410613200 [pii];10.1074/jbc.M410613200 [doi]

Zhang Y, Su P, Liang P, Liu T, Liu X, Liu XY, Zhang B, Han T, Zhu YB, Yin DM, Li J, Zhou Z, Wang KW, Wang Y (2010) The DREAM protein negatively regulates the NMDA receptor through interaction with the NR1 subunit. J Neurosci 30(22):7575–7586. doi:30/22/7575 [pii];10.1523/JNEUROSCI.1312-10.2010 [doi]

Zhang W, Devi SP, Tomita S, Howe JR (2014) Auxiliary proteins promote modal gating of AMPA- and kainate-type glutamate receptors. Eur J Neurosci 39(7):1138–1147. doi:10.1111/ejn.12519

Zhuchkova E, Remme MW, Schreiber S (2013) Somatic versus dendritic resonance: differential filtering of inputs through non-uniform distributions of active conductances. PLoS ONE 8(11), e78908. doi:10.1371/journal.pone.0078908

Zolles G, Wenzel D, Bildl W, Schulte U, Hofmann A, Muller CS, Thumfart JO, Vlachos A, Deller T, Pfeifer A, Fleischmann BK, Roeper J, Fakler B, Klocker N (2009) Association with the auxiliary subunit PEX5R/Trip8b controls responsiveness of HCN channels to cAMP and adrenergic stimulation. Neuron 62(6):814–825. doi:S0896-6273(09)00359-6 [pii];10.1016/j.neuron.2009.05.008 [doi]

Zona C, Eusebi F, Miledi R (1990) Glycosylation is required for maintenance of functional voltage-activated channels in growing neocortical neurons of the rat. Proc R Soc Lond Ser B Containing Pap Biol Charact R Soc 239(1295):119–127

Chapter 17
Adhesion Molecules in Synapse Assembly and Function

Peter Chipman and Yukiko Goda

Abstract As the nervous system develops, cell adhesion molecules (CAMs) interact across the synaptic junction to govern synapse assembly and function. The CAM interactions are highly dependent on the molecular composition of the extracellular domains, of which there is a large variety, and on the intracellular domains, which provide a structural and signaling platform for the pre- and the postsynaptic terminals. The diversity of CAMs generates a myriad of potential extracellular interactions, yet CAM binding is highly specific and gives rise to unique adhesive codes that guide synapse assembly and tune the efficacy of synaptic communication. The aim of this chapter is to provide an overview of some of the major synaptic CAMs, highlighting their structural features that contribute to the specificity and promiscuity of CAM interactions, and to discuss how these mechanisms intersect to control synapse formation, organization, and plasticity.

Keywords Cell adhesion molecules • Synaptic plasticity • Synaptic strength • Homophilic adhesion • Heterophilic adhesion • Presynaptic terminal • Active zone • Postsynaptic terminal • Postsynaptic density • Scaffold proteins

17.1 Introduction

Synapses mediate spatially and temporally controlled cell-to-cell communication in the nervous system. The proper functioning of synapses requires a wide array of subcellular signaling components across both sides of the synaptic cleft. This transcellular organization at the synapse is highly dependent on cell adhesion molecules (CAMs). CAMs mechanically link the presynaptic and postsynaptic membranes to provide a *trans*-synaptic bridge. They also bind other CAMs or membrane-bound proteins along the same membrane surface in *cis* and contribute to synapse organization. Furthermore, they anchor to intracellular scaffolds to

P. Chipman (✉) • Y. Goda
Laboratory for Synaptic Plasticity and Connectivity, RIKEN Brain Science Institute, 2-1 Hirosawa, Wako-shi, Saitama 351-0198, Japan
e-mail: peter.chipman@riken.jp; goda@brain.riken.jp

K. Emoto et al. (eds.), *Dendrites*, DOI 10.1007/978-4-431-56050-0_17

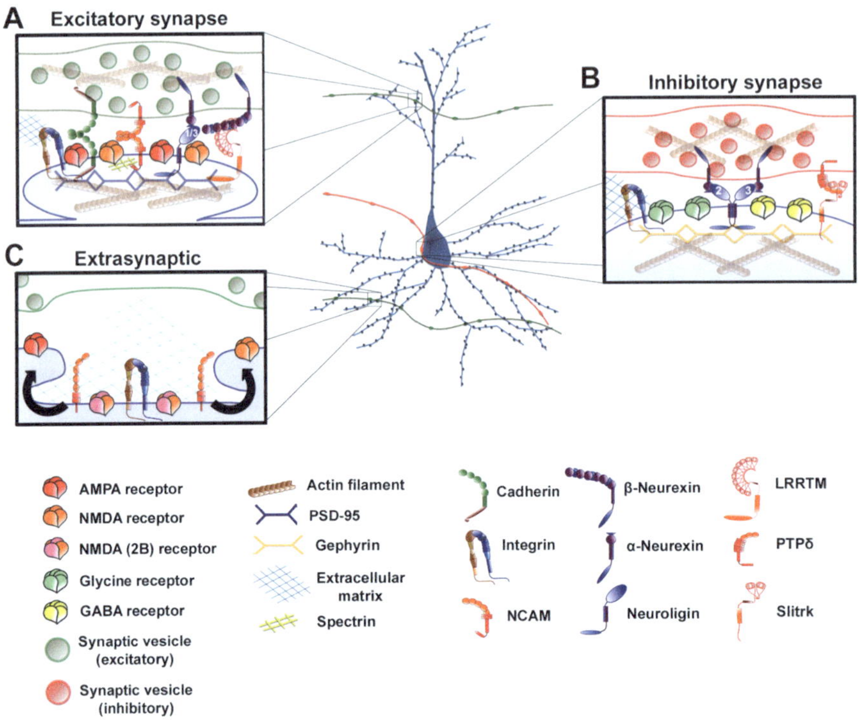

Fig. 17.1 Distribution profile of adhesion molecules across the dendritic tree. (**a**) Adhesion molecules that are localized to excitatory synapses (along the green axons) include the integrins, cadherins, NCAMs, neurexins, neuroligins (NLGN1 and NLGN3 isoforms), and LRRTMs. (**b**) Integrins, neurexins, and neuroligins (NLGN2 and NLGN3 isoforms) are also localized to inhibitory synapses (along the red axon), while a PTPδ/Slitrk constitute one of the only inhibitory synapse-specific adhesive complexes described to date. (**c**) Adhesion molecules such as NCAM and the integrins also distribute to extrasynaptic regions, as do NMDARs containing GluN2B subunits. Extrasynaptic NCAM may constitute a reserve pool that is shuttled into synaptic regions (*arrows*) to replenish and maintain the abundance of NCAM during ongoing turnover

promote synapse stability and initiate intracellular signaling events to regulate synapse structure and function.

There exists an incredible diversity of CAMs involved in distinct aspects of synapse function (Fig. 17.1), although in general, CAMs fall into two major classes: those that bind to the members of the same family (homophilic binding) or those that interact with different family members (heterophilic binding). Within each family, CAMs are further subdivided based on the structure and binding properties of their extracellular adhesion domains. The numerous classes of CAMs and the potential for heterophilic interactions between CAMs of different classes create an overwhelming abundance of CAM binding interactions. Such complexity of CAM binding interactions is underscored by observations that different classes of CAMs can cooperate or compete in synapse formation (Stan et al. 2010; Aiga et al. 2010). Nevertheless, CAM binding itself is highly specific and depends not only on the

Table 17.1 Binding affinities of common CAMs and their ligands

CAM	Ligand	K_d (nM)	Reference
N-cadherin	N-cadherin	22.6	Katsamba et al. (2009)
E-cadherin	E-cadherin	160	Katsamba et al. (2009)
N-cadherin	E-cadherin	22–160	Katsamba et al. (2009)
LRRTM2	NRXN1β (−SS4)	19.9	Siddiqui et al. (2010)
		7	de Wit et al. (2009)
		5.83	Ko et al. (2009)
	NRXN1α	16	de Wit et al. (2009)
LRRTM4	HSPG	24.3	Siddiqui et al. (2013)
NLGN1 (+SSB)	NRXN1β (−SS4)	9.6	Siddiqui et al. (2010)
	NRXN1β (+SS4)	26.9	Siddiqui et al. (2010)
NLGN1 (−SSA/−SSB)	NRXN1β (−SS4)	29	Comoletti et al. (2006)
NLGN2 (−SSA)	NRXN1β (−SS4)	8,840	Comoletti et al. (2006)
NLGN3 (−SSA)	NRXN1β (−SS4)	1,850	Comoletti et al. (2006)
NLGN4 (−SSA)	NRXN1β (−SS4)	132	Comoletti et al. (2006)
NGL3	LAR	37.4	Woo et al. (2009)
NCAM	NCAM	69.2	Moran and Bock (1988)
Integrin (α5 β1)	Fibronectin	1.3	Mould et al. (2003)
Integrin (αV β3)	RGD peptide (Fab-9)	13	Hu et al. (1996)
SynCAM1	SynCAM2	78	Fogel et al. (2010)

presence of specific ligands, but on their relative abundance and binding affinities (Table 17.1). Moreover, the binding affinities between CAMs and their ligands can be tuned by a variety of activity-dependent modifications (Ullrich et al. 1995; Iijima et al. 2011; Aoto et al. 2013) (Table 17.2 and Box 17.2).

In this chapter we will overview the key structural determinants that govern specific CAM interactions and outline how regulating such CAM interactions can influence the position, type, strength, and plasticity of particular types of synapses. While new synaptic CAMs and their ligands are continually being discovered (Biederer et al. 2002; Woo et al. 2009; Linhoff et al. 2009; Takahashi et al. 2012; de Wit et al. 2013) (Box 17.1), here we will highlight the following CAMs for which we have accumulated considerable knowledge: among the homophilic synaptic CAMs, the cadherins and Ig superfamily (IgSF) molecules, and among the heterophilic CAMs, the neuroligins (NLGNs), leucine-rich repeat transmembrane proteins (LRRTMs), netrin-G ligands (NGLs), and integrins. We end with a brief summary on the future outlook of how further knowledge on synapse adhesion proteins might contribute to the next advances in neuroscience research.

Table 17.2 Common modifications that influence CAM activity

CAM	Modification	Site	Function	Reference
N-cadherin	N-Glycosylation	EC1-5	Mutation of N-glycosylation sites promotes cell-to-cell adhesion by enhancing *cis* dimerization; modifies binding kinetics	Guo et al. (2009)
				Langer et al. (2012)
	Ectodomain shedding	Extracellular domain, 120 amino acids from transmembrane domain	Disrupts cell-to-cell adhesion, disperses β-catenin	Ito et al. (1999) and Monea et al. (2006)
NCAM	Polysialylation	Ig5	Disrupts cell–cell adhesion	Rutishauser et al. (1985)
	VASE	Ig4	May confer binding specificity	Chen et al. (1994)
	Ectodomain shedding	Juxtamembrane site between transmembrane domain and Ig5	Reduces adhesion	Hübschmann et al. (2005) and Hinkle et al. (2006)
SynCAM1	N-Glycosylation	Ig1 (N70, N104)	Enhances adhesion	Fogel et al. (2010)
	Polysialylation	Ig1	Reduces adhesion	Galuska et al. (2010) and Rollenhagen et al. (2012)
SynCAM2	N-Glycosylation	Ig1 (N60)	Reduces adhesion	Fogel et al. (2010)
NLGN1	Phosphorylation	Y782 (intracellular domain)	Controls intracellular interactions with gephyrin	Giannone et al. (2013)
		T739 (intracellular domain)	Influences activity-dependent NLGN surface expression and synapse formation	Bemben et al. (2013)
	Alternative splicing	SSB (proximal CLD)	Potently alters α- and β-NRXN binding affinity	Boucard et al. (2005)
		SSA (distal CLD)	Slightly alters β-NRXN binding affinity	Comoletti et al. (2006)
	N-glycosylation	A303 (within SSB)	Alters affinity for α- and β-NRXNs	Boucard et al. (2005)
NRXN	Alternative splicing	SS4 (LNS)	Alters affinity for NLGNs and LRRTMs	Boucard et al. (2005)
				Siddiqui et al. (2010)
LRRTM1-4	N-glycosylation	LRR3	Unknown	Lauren et al. (2003)
Integrin	Activation by intracellular ligands	β1 (Y788), β3 (Y747) cytoplasmic tails	Regulates extracellular domain conformation and ligand-binding affinity	O'Toole et al. (1994) and Calderwood et al. (1999)

Box 17.1: Common Methods Used to Assess Adhesion Molecule Function in Neural Circuits

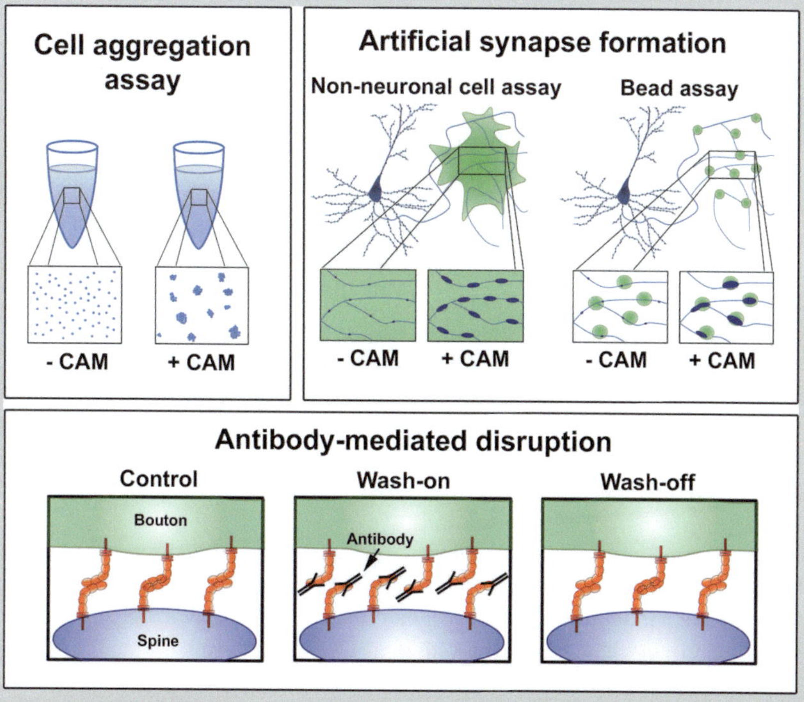

Assays

Cell Aggregation Assay

Some of the earliest investigations of adhesion molecules have made use of the cell aggregation assay (Rutishauser et al. 1982; Chuong et al. 1982; Miyatani et al. 1989). This assay involves assessing the degree to which suspended cells adhere to each other under different conditions. Prior transfection of cells with different adhesion molecules, their mutated isoforms, or treatment with antibodies or other agents can reveal the specifics of cell–cell adhesion.

(continued)

Box 17.1 (continued)

Artificial Synapse Formation Assay

Artificial synapse formation assays have proved to be a powerful tool in identifying the role of adhesion complexes in synapse assembly and function (Scheiffele et al. 2000; Graf et al. 2004; Linhoff et al. 2009). Two kinds of artificial synapse formation are used: one involves transfecting nonneuronal cells with DNA constructs encoding different adhesion molecule isoforms, while the other uses beads coated with particular ligands. Neurons are grown in the presence of transfected cells or coated beads, and synapse formation is assessed by monitoring the clustering of pre- or postsynaptic components to the sites of contact between neuron and the bead or the transfected cell. Other assays, such as styryl dye uptake into recycling vesicles, electron microscopy, and electrophysiology provide functional read-outs of synapses (Scheiffele et al. 2000; Graf et al. 2004; Sara et al. 2005).

Pharmacological or Antibody-Mediated Disruption

In some cases, adhesion molecule binding can be disrupted by introducing function-blocking antibodies or by naturally occurring toxins that target specific CAM interactions. These can be administered acutely, and the resulting effects can be monitored in real time. However, the results of these experiments should be interpreted carefully, as exogenous peptides mimicking ligand-binding domains could either inhibit or stimulate adhesion molecule signaling depending on the abundance of the endogenously bound ligand and/or the state of cell signaling. Ideally, these experiments are paired with gene manipulation approaches involving transgenic mice or acute overexpression or knockdown of CAM or its substrate expression, based on methods such as cell transfection, viral vectors, or electroporation.

Box 17.2: Common Activity-Dependent Modifications That Influence Adhesion Molecule Binding

Glycosylation

The addition or removal of oligosaccharides typically at N-linked (to an amide group of an asparagine residue) or O-linked (to a hydroxyl group of

(continued)

Box 17.2 (continued)

serine or threonine residues) glycosylation sites may influence protein folding to present or mask potential ligand-binding sites. Moreover, oligosaccharides may contain additional sugars such as polysialic acid (PSA) that contribute a strong negative charge and leads to membrane repulsion. Many adhesion molecules are regulated by glycosylation, with one prime example being the Ig superfamily molecules NCAM (Crossin et al. 1984) and SynCAM (Galuska et al. 2010; Fogel et al. 2010).

Alternative Splicing

Many adhesion molecules possess several alternative splice sites that include key residues that directly influence the three-dimensional structure of the protein and/or modify its binding properties. Consequently, proteins with distinct binding properties can be generated from single transcripts depending on the cellular context. For example, the neurexin/neuroligin complex is tightly regulated by alternative splicing (Ullrich et al. 1995; Boucard et al. 2005; Iijima et al. 2011).

Cleavage or Ectodomain Shedding

Several proteases are upregulated and secreted in an activity-dependent manner, and they target the extracellular domains of adhesion molecules. Cleavage normally leads to the generation of a soluble extracellular fragment and an intracellular fragment that may interact with additional components that would not be possible otherwise. N-cadherin and neuroligins are among those adhesion molecules whose specific susceptibility to proteinase-dependent cleavage has been reported (Reiss et al. 2005; Peixoto et al. 2012; Suzuki et al. 2012).

Phosphorylation

The addition of phosphate groups to proteins by protein kinases is one of the most common and potent signaling mechanisms in synapse cell biology. Adhesion molecules can be a substrate for phosphorylation (Giannone et al. 2013; Bemben et al. 2013) or can regulate the phosphorylation of other synaptic proteins (Bernard-Trifilo et al. 2005; Bodrikov et al. 2008). Given the proximity of synaptic adhesion molecules to neurotransmitter receptors and other synaptic machinery, these effects can be propagated quickly to tightly regulate synaptic strength and plasticity.

17.2 Homophilic Adhesion Molecules

17.2.1 The Cadherin Family

The cadherins are a superfamily of homophilic cell adhesion molecules that are classified into several subfamilies based on the organization of their protein domains and genomic structure. These include the classical (type I) cadherins, the atypical (type II) cadherins, the desmosomal cadherins, the flamingo cadherins, the protocadherins, and several independent cadherin isoforms (Nollet et al. 2000; Niessen et al. 2011). We will focus on the classical cadherins, which include the N-, E-, P-, and R-cadherins. Notably, the N- and E-cadherins are the most highly expressed in the nervous system.

17.2.1.1 Structure and Binding

All classical cadherins are single-pass type I transmembrane glycoproteins that possess a conserved extracellular domain consisting of five repeating modules (named EC1-5 from the most distal to the membrane proximal) (Fig. 17.2a) (Hatta et al. 1988; Niessen et al. 2011). In the cytoplasmic region, classical cadherins possess three major functional domains: a juxtamembrane domain that contributes to *cis* dimerization (Yap et al. 1998), a membrane proximal domain that participates in p120-catenin and δ-catenin binding (Arikkath and Reichardt 2008), and the highly conserved β-catenin-binding domain on the distal C-terminus (Niessen et al. 2011). β-Catenin binds to α-catenin, which in turn binds to

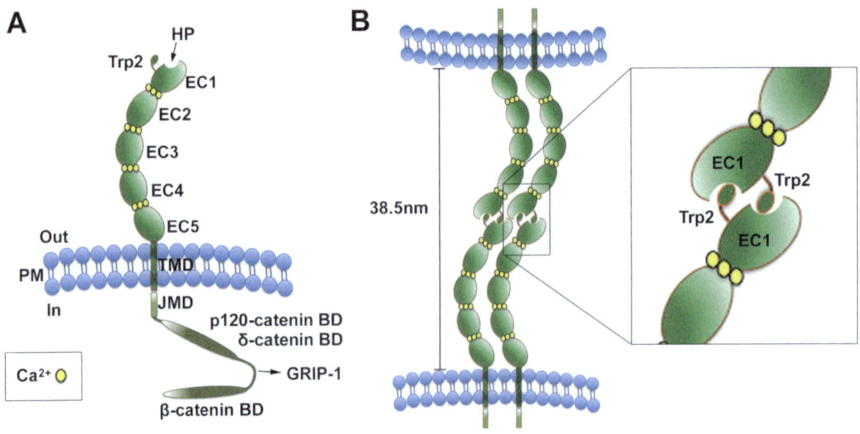

Fig. 17.2 The cadherins. (**a**) Cadherin domain organization. *EC* cadherin extracellular domain, *PM* plasma membrane, *TMD* transmembrane domain, *JMD* juxtamembrane domain, *HP* hydrophobic pocket, *BD* binding domain. (**b**) Illustration of the proposed "strand dimer"-binding mechanism for classical cadherins based on the crystal structure of C-cadherin described in Boggon et al. (2002)

filamentous actin (F-actin), thereby linking cadherin with the actin cytoskeleton. Interestingly, at an intermediate location between the p120-/δ-catenin and β-catenin-binding domains, an additional binding site for the *g*lutamate *r*eceptor-*i*nteracting *p*rotein-1 (GRIP1) has recently been identified (Heisler et al. 2014) and could play a role in postsynaptic regulation (see below).

In the extracellular domain, each EC region folds into a seven-stranded β-barrel containing three Ca^{2+}-binding sequences. Specific homophilic binding is thought to occur when a conserved tryptophan side chain (Trp2 or W2) in the EC1 domain of one cadherin molecule inserts into a hydrophobic pocket in the EC1 domain of a partner molecule (Fig. 17.2b) (Nose et al. 1990; Miyatani et al. 1989; Shapiro et al. 1995; Boggon et al. 2002). This "strand dimer" formation is essential for binding both in *cis* along the plane of the membrane (Shapiro et al. 1995) and in *trans* across the two membranes (Pertz et al. 1999; Boggon et al. 2002). While many structural studies have identified interactions between EC1 domains, additional binding mechanisms have been proposed that involve contributions from several other EC domains (Sivasankar et al. 2001; Chappuis-Flament et al. 2001; Zhu et al. 2003; Shan et al. 2004; Prakasam et al. 2006; Guo et al. 2009). Importantly, Ca^{2+} binds near the interface region between each EC domain (Nagar et al. 1996); Ca^{2+} binding facilitates *trans* cadherin interaction by stabilizing and extending the curved rod structure into the extracellular space and also by enhancing *cis* dimerization (Nagar et al. 1996; Brieher et al. 1996; Yap et al. 1998; Pertz et al. 1999).

Although homophilic cadherin interaction is highly specific (Miyatani et al. 1989), heterodimers composed of different cadherin isoforms have been observed (Niessen and Gumbiner 2002; Katsamba et al. 2009). N- and E-cadherin heterodimers bind with affinities that are intermediate between those observed for homodimers of either type, thus generating a hierarchy of cadherin-binding potency (N-N>N-E>E-E) (Katsamba et al. 2009).

17.2.1.2 Cadherins at Synapses

During neuronal development, N-cadherin and β-catenin are initially diffusely expressed along growing neurites and growth cones, and they subsequently become localized to synapses (Uchida et al. 1996; Fannon and Colman 1996; Benson and Tanaka 1998; Shan et al. 2000; Togashi et al. 2002). Ultrastructural analysis has revealed that N-cadherin and α- and β-catenin are localized to regions of both pre- and postsynaptic membranes adjacent to active zones and postsynaptic densities (Uchida et al. 1996; Fannon and Colman 1996), although a subsequent study has shown that cadherins are distributed evenly throughout the synaptic cleft in immature synapses and later become concentrated into hotspots within or adjacent to the synaptic cleft (Elste and Benson 2006). N-cadherin is initially present at both excitatory and inhibitory synapses in culture, but is progressively lost from inhibitory contacts as synapses mature (Benson and Tanaka 1998). Collectively these observations suggest that cadherin function in synapse organization is developmentally regulated and that N-cadherin plays an ongoing role only at excitatory synapses.

Interestingly, despite the early expression of N-cadherin in neurons prior to synapse formation, in artificial synapse formation assays N-cadherin does not induce synaptogenesis (Sara et al. 2005; Linhoff et al. 2009). Therefore, N-cadherin may contribute to the maturation of synapses rather than to their formation per se. Indeed, disrupting N-cadherin or β-catenin in cultured hippocampal neurons slows synapse formation and maturation (Togashi et al. 2002; Okuda et al. 2007); dendritic spines are longer with smaller spine heads, and exhibit reduced GluA2 surface expression and quantal amplitudes (Togashi et al. 2002; Okamura et al. 2004; Okuda et al. 2007; Vitureira et al. 2011). Moreover, in neuronal cultures prepared from αN-catenin knockout mice, dendritic spines remain immature and are less stable than wild-type spines (Togashi et al. 2002; Abe et al. 2004). Genetic deletion of p120-catenin also leads to reduced spine density, altered spine morphology, and reduced dendritic arbor complexity (Elia et al. 2006). Together, these observations demonstrate that synapse maturation, but not formation, is altered in the absence of N-cadherin and several of its intracellular partners. Whereas perturbing N-cadherin and its binding partners potently affect dendritic spine morphology, its effects on synapse function are more complex. For example, β-catenin null synapses show deficits in the postsynaptic receptor abundance, and this cellular phenotype could be rescued by exogenously expressing β-catenin mutants unable to bind to α-catenin but not by those harboring mutations in the PDZ-binding region (Okuda et al. 2007). Therefore, whereas β- and α-catenin binding is crucial for the link to actin and the spine morphology, it is dispensable for basal synaptic function.

The complexity of N-cadherin-dependent synapse regulation is further highlighted by a study in cultured hippocampal neurons that has addressed the extent of symmetric requirement across the synaptic cleft of this homophilic adhesion complex. Disrupting postsynaptic N-cadherin but not β-catenin retrogradely reduces basal release probability (p_r) of contacting presynaptic terminals (Vitureira et al. 2011). Interestingly, however, disrupting presynaptic N-cadherin does not replicate the effects on p_r, arguing against the role of N-cadherin homophilic binding for controlling p_r. Instead, postsynaptic N-cadherin may interact with the GluA2 AMPA receptor subunit to control presynaptic p_r *trans*-synaptically (Saglietti et al. 2007; Vitureira et al. 2011). Presynaptic N-cadherin/β-catenin complex however does play a role in presynaptic organization, specifically in recruiting the reserve and resting pool synaptic vesicles to sites of cadherin clustering (Bamji et al. 2003, 2006).

Synaptic activity strongly regulates the distribution pattern, internalization rate, and molecular composition of cadherin/catenin complexes (Tanaka et al. 2000; Bozdagi et al. 2000; Murase et al. 2002; Abe et al. 2004; Okamura et al. 2004; Tai et al. 2007; Brigidi et al. 2014). Following elevated network activity, N-cadherin redistributes along the surface of the membrane, forms *cis*-dimers, and becomes protease resistant (Tanaka et al. 2000). Activity also changes N-cadherin surface abundance in an NMDA receptor-dependent manner by slowing N-cadherin endocytosis via enhanced intracellular binding to β-catenin (Bozdagi et al. 2000;

Tai et al. 2007). These activity-dependent changes point to the role of cadherins as transducers of synaptic signaling.

The relationship between activity and cadherins is reciprocal in that cadherins also impact activity-dependent synaptic processes. Infusing function-blocking antibodies against N- and E-cadherin have been reported to attenuate LTP in hippocampal slices (Tang et al. 1998; Bozdagi et al. 2000), and introducing mutations into the W2 domain of N-cadherin alters depolarization-induced spine dynamics in culture (Okamura et al. 2004). Some of these effects may be due to disrupting a direct interaction between the extracellular domains of N-cadherin and the GluA2 subunit of the AMPA receptor (Saglietti et al. 2007) or compromising an intracellular interaction between N-cadherin and GRIP1 (Heisler et al. 2014) or other intracellular signaling molecules, including δ-catenin, which has been shown to be essential for spine head expansion and surface AMPA receptor insertion (Brigidi et al. 2014). N-cadherin has also been implicated in homeostatic modulation of postsynaptic strength in response to chronic activity alterations (Vitureira et al. 2011). However, contrary to its role in basal presynaptic strength regulation, postsynaptic N-cadherin does not retrogradely influence the homeostatic plasticity of p_r, although it is required for homeostatic modulation of postsynaptic AMPA receptors. Postsynaptic β-catenin, on the other hand, is required for homeostatically adjusting both pre- and postsynaptic strengths, the former occurring in a retrograde fashion independently of N-cadherin (Okuda et al. 2007; Vitureira et al. 2011).

Altogether, the cadherins and their accessory proteins play important roles in controlling synapse structural maturation, organization, and function. The highly complex nature of the regulation is reflected not only by the numerous interacting proteins, particularly in the intracellular domain of cadherin, but also by the developmental changes that are observed even at the same synapse type.

17.2.2 Ig Superfamily

The Ig superfamily (IgSF) CAMs participate in a wide range of cell-to-cell and cell-to-ECM interactions. All IgSF members are characterized by the number of extracellular Ig-like domains that participate in homophilic (Zhou et al. 1993, 1998; Rao et al. 1994) and heterophilic adhesion (Felding-Habermann et al. 1997; Blaess et al. 1998; Fogel et al. 2007). Most IgSF members possess a variable number of fibronectin type III (FNIII) domains, followed either by a GPI membrane anchor or a transmembrane domain (Walsh and Doherty 1997; Maness and Schachner 2007). Unlike cadherins, the intracellular domain of IgSF CAMs is highly variable, ranging from sequences that confer PDZ-like binding (Polo-Parada et al. 2005; Biederer 2006) and cytoskeletal interactions (Davis and Bennett 1994; Dahlin-Huppe et al. 1997; Leshchyns'ka et al. 2003) to phosphatase domains (Johnson and Van Vactor 2003).

Individual Ig domains are approximately 100 amino acids in length that fold into a tertiary structure composed of two β-sheets stabilized by disulfide bonds

(Wang 2013). Each β-sheet is itself composed of a variable number of β-strands that generate three distinct subgroups: a variable set (V-set), a conserved set (C-set; further subdivided into C1-set and C2-set), and an intermediate set (I-set; further subdivided into I1-set and I2-set) (Smith and Xue 1997; Wang 2013). Different IgSF CAMs are constructed of different combinations of these Ig domains and this variability contributes to their specific binding affinities (Zhou et al. 1993; Biederer 2006).

In this chapter, we will focus on the *n*eural *c*ell *a*dhesion *m*olecule (NCAM) and *syn*aptic *c*ell *a*dhesion *m*olecule (synCAM) subfamilies that have received considerable attention for their roles at central synapses.

17.2.2.1 NCAM

NCAM is a Ca^{2+}-independent CAM widely expressed throughout the developing and the adult nervous system. Although considered primarily a homophilic CAM, NCAM can also participate in a multitude of heterophilic interactions with other members of the Ig superfamily such as L1 (Horstkorte et al. 1993) and TAG-1/axonin (Milev et al. 1996) or with several growth factors and growth factor receptors such as the fibroblast growth factor receptor (FGFR) (Kiselyov et al. 2005), the glial cell-line-derived neurotrophic factor (GDNF) (Paratcha et al. 2003), and the brain-derived neurotrophic factor (BDNF) (Vutskits et al. 2001). Moreover, NCAM robustly interacts with several ECM-related molecules including the chondroitin sulfate proteoglycans, neurocan (Friedlander et al. 1994) and phosphacan (Milev et al. 1994), collagens (Probstmeier et al. 1992), heparin (Cole et al. 1986), and agrin (Storms and Rutishauser 1998).

NCAM Structure and Binding

Alternative splicing of a single NCAM gene generates a subfamily that is composed of at least three main isoforms expressed in neural tissue and named according to their molecular weights – NCAM120, NCAM140, and NCAM180 (Fig. 17.3a). All isoforms possess identical extracellular domains that mediate Ca^{2+}-independent adhesion, but differ in the presence or absence of the transmembrane domain and the intracellular sequence. NCAM120 is linked to the membrane via a glycophosphatidylinositol (GPI) anchor, while NCAM140 and NCAM180 both contain transmembrane domains followed by an intracellular domain that includes a 40 kDa insert containing a PDZ-like (KANESKA) motif on the distal C-terminus (Polo-Parada et al. 2005). An additional alternatively spliced sequence in the intracellular domain of NCAM180 contains dynein and cytoskeletal-binding domains (Polo-Parada et al. 2005; Perlson et al. 2013).

The extracellular domain of NCAM is composed of five I-type Ig domains linked by disulfide bonds followed by two FNIII domains (Jensen et al. 1999; Kiselyov et al. 2005). A long, negatively charged carbohydrate chain known as α2,8-linked

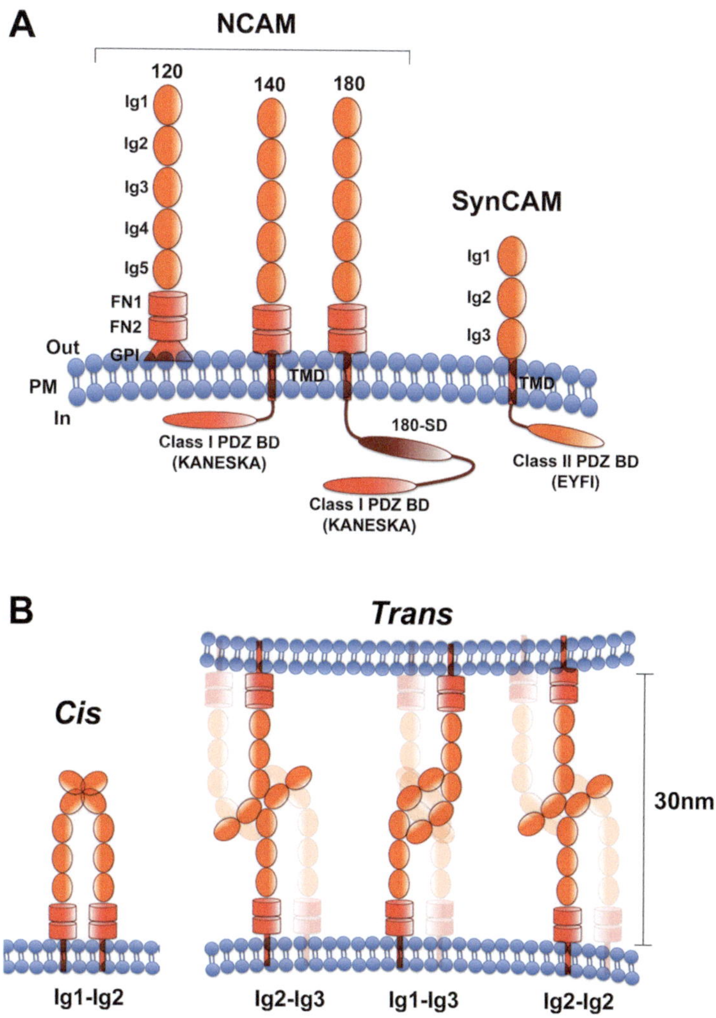

Fig. 17.3 Immunoglobulin superfamily members NCAM and SynCAM. (**a**) Domain organization of NCAM120, NCAM140, and NCAM180 isoforms and SynCAM. *Ig* *i*mmuno*g*lobulin domain; *FN* *f*ibro*n*ectin type III domain; *GPI* *g*lyco*p*hosphatidyl*i*nositol; *180-SD* NCAM180-specific domain; *PDZ BD* *p*ost*s*ynaptic *d*ensity 95, *d*isc large, *z*ona occludens 1 *b*inding *d*omain. Key amino acid sequences that participate in synaptic functions are shown in *brackets*. (**b**) Illustration of proposed homophilic binding mechanisms of NCAM in *cis* and *trans* based on crystal structures described in Soroka et al. (2003)

polysialic acid (PSA) can bind to the fifth Ig domain to inhibit NCAM-mediated adhesion (Rutishauser et al. 1985). The addition of 10-amino acid sequence, termed VASE (*v*ariable *a*lternatively *s*pliced *e*xon), to the fourth Ig domain of NCAM transforms it to resemble a V-type Ig domain (Small et al. 1988; Small and Akeson 1990) and may influence its binding affinities (Chen et al. 1994).

How NCAM Ig domains bind to each other has been disputed, and several different binding conformations have been proposed. These include a model where all five Ig domains participate in binding in an antiparallel manner (Rao et al. 1994; Ranheim et al. 1996), a model where only Ig domain 3 self-associates (Rao et al. 1993), and a model where Ig domains 1 and 2 bind in an antiparallel manner (Jensen et al. 1999; Kasper et al. 2000). More recent work has suggested that cis-binding occurs between Ig domains 1 and 2, while *trans*-binding occurs via three possible antiparallel conformations: between Ig domains 2 and 3, 1 and 3, or 2 and 2 (Fig. 17.2b) (Soroka et al. 2003). NCAM "zippers" would then be assembled as clusters of cis-dimers interacting in *trans* across the membrane.

NCAM at the Synapse

NCAM isoforms are shuttled to regions of neurite contact where they accumulate during the early stages of synaptogenesis (Sytnyk et al. 2002) and persist into maturity (Schuster et al. 2001). During initial stages of synapse formation, NCAM may facilitate the transition from neurite outgrowth (Leshchyns'ka et al. 2003; Chernyshova et al. 2011) to contact stabilization by regulating the delivery of trans-Golgi-derived cargo to the plasma membrane at specific sites in response to target-derived cues (Hata et al. 2007; Chernyshova et al. 2011). As such, several structural and functional deficits in both pre- and postsynaptic compartments have been observed at synapses from NCAM knockout animals (Rafuse et al. 2000; Polo-Parada et al. 2001; Dityatev et al. 2004). However, NCAM is not required for synapse formation. Like N-cadherin, NCAM is unable to induce synapse formation in artificial synapse formation assays (Sara et al. 2005), and instead, NCAM likely functions to promote synapse maturation and activity-dependent adjustments in synaptic strengths (Dityatev et al. 2000; Polo-Parada et al. 2001; Cambon et al. 2004; Puchkov et al. 2011; Chipman et al. 2014a).

NCAM140 and NCAM180 are most abundantly expressed at synapses. Presynaptically, they influence p_r, possibly by regulating a developmental switch in synaptic vesicle recycling and/or myosin light chain kinase activity involved in controlling synaptic vesicle pools (Polo-Parada et al. 2004; Hata et al. 2007; Shetty et al. 2013; Chipman et al. 2014a, b). NCAM180 is highly enriched in postsynaptic membranes where it is thought to contribute to postsynaptic organization and plasticity (Schuster et al. 2001; Sytnyk et al. 2006; Leshchyns'ka et al. 2011). Specifically, NCAM associates with $\alpha1\beta1$-spectrin that links NCAM to the NMDA receptor subunits, NR1 and NR2B, and CaMKIIα (Sytnyk et al. 2006). These interactions are critical for the appropriate clustering (Sytnyk et al. 2006), stability (Puchkov et al. 2011), and functionality (Kochlamazashvili et al. 2012) of these postsynaptic components. Moreover, perturbing NCAM function disrupts NMDAR-dependent LTP and LTD in the hippocampus (Lüthi et al. 1994; Muller et al. 1996; Eckhardt et al. 2000; Bukalo et al. 2004; Kochlamazashvili et al. 2012). However, some of these effects may represent the indirect consequences of losing NCAM as a carrier for PSA (Muller et al. 1996; Eckhardt et al. 2000).

NCAM adhesion and signaling are regulated by activity-dependent mechanisms. For instance, in hippocampal slices, PSA expression is modulated by activity in a spatially and temporally defined manner (Muller et al. 1996). Recent studies have demonstrated that ectodomain shedding of NCAM140 and NCAM180, a process that reduces their homophilic binding, can be induced by matrix metalloproteases such as ADAM (*a disintegrin and metalloprotease*)-8, ADAM-10, and ADAM-17 (Hübschmann et al. 2005; Hinkle et al. 2006; Kalus et al. 2006; Brennaman et al. 2013). Furthermore, once released, soluble NCAM ectodomain fragments have the potential to accumulate in the extracellular matrix and influence subsequent NCAM substrate adhesion and the synaptic activities that depend on it. Accordingly, transgenic mouse lines that overexpress NCAM ectodomain fragments in neurons show deficits in LTP and impaired inhibitory synapse formation in the cortex (Pillai-Nair et al. 2005; Brennaman et al. 2011). One study has reported increased amounts of extracellular NCAM following LTP induction in dentate gyrus (Fazeli et al. 1994), suggesting that activity may regulate NCAM ectodomain shedding. Notably, accumulation of NCAM ectodomain has been identified in patients with schizophrenia, and NCAM dysfunction is associated with cognitive disorders (Sandi 2004). These observations underscore the importance of NCAM-dependent mechanisms of synaptic circuit regulation.

17.2.2.2 SynCAM

The SynCAM subfamily is composed of four genes that code for four isoforms (SynCAM1–SynCAM4) that share significant sequence homology (Biederer 2006). In the brain, SynCAM1 and SynCAM4 are expressed embryonically, while SynCAM2 and SynCAM3 are detectable only postnatally (Fogel et al. 2007).

Structure and Binding

All SynCAM family members are Ca^{2+}-independent CAMs characterized by three extracellular Ig domains, a transmembrane domain, and a carboxyl tail that contains a PDZ-like type II-binding domain (Fig. 17.3a) (Biederer et al. 2002; Biederer 2006). According to the classification of Ig-like domain subclasses, the sequence information of SynCAM indicates that the first Ig domain of all SynCAM isoforms belongs to the so-called V-set, while the second belongs to the C1-set, and the third to the I-set (Biederer 2006). All Ig domains possess N-glycosylation sites (Biederer et al. 2002; Biederer 2006), but only the first Ig domains of SynCAM1 and 2 can be heavily glycosylated (Fogel et al. 2007, 2010) or polysialylated, both of which influence binding affinities (Galuska et al. 2010; Rollenhagen et al. 2012).

All SynCAM isoforms, with the exception of SynCAM4, participate in homophilic binding but preferentially assemble into heterodimers composed of either SynCAM1 and SynCAM2, or SynCAM3 and SynCAM4 (Fogel et al. 2007). The first Ig domain of SynCAM1 and SynCAM2 mediate their trans-synaptic adhesion (Fogel et al. 2010), while the second and third Ig domains of SynCAM1 engage in *cis* assembly in a manner that strengthens *trans* interactions (Fogel et al. 2011). N-glycosylation of each isoform modifies their binding affinities to inhibit SynCAM2 binding while promoting SynCAM1 binding (Fogel et al. 2010). Polysialylation also modifies SynCAM's adhesive properties (Galuska et al. 2010). These modifications occur in a developmental and region-specific manner and can potently modulate synapse formation.

SynCAM at the Synapse

All SynCAM isoforms have been detected at nascent and established excitatory and inhibitory synapses, although their distribution profiles differ somewhat between characterizations in vitro and in vivo (Biederer et al. 2002; Sara et al. 2005; Fogel et al. 2007; Stagi et al. 2010; Robbins et al. 2010). In vivo, SynCAM1 has been found to distribute primarily to excitatory synapses in the hippocampus (Robbins et al. 2010). SynCAM1 and SynCAM2 are potent synapse-organizing CAMs (Sara et al. 2005; Fogel et al. 2007; Stagi et al. 2010). When expressed in nonneuronal cells, they induce the clustering of functional presynaptic components, including synaptic vesicle proteins and SNARE machinery in contacting axons, and when overexpressed in neurons, they enhance excitatory but not inhibitory synapse formation (Biederer et al. 2002; Sara et al. 2005; Robbins et al. 2010). Sustained SynCAM expression into adulthood is required for the maintenance of excitatory synapses. Functionally, SynCAM constrains LTD but has little effect on LTP (Robbins et al. 2010).

Many of the synaptic effects of SynCAM disruption involve its intracellular domains: the juxtamembrane motif that is predicted to bind to members of the protein 4.1 family, and the PDZ type II-binding motif (Biederer et al. 2002; Sara et al. 2005). The protein 4.1 family consists of a cytoskeleton-interacting protein that stabilizes the spectrin/actin cytoskeleton (Hoover and Bryant 2000), while the PDZ type II-binding domain contains an EYFI sequence that can bind to a range of synapse-specific proteins implicated in synapse formation and plasticity (Sheng and Sala 2001). Altogether, among the homophilic synapse adhesion proteins, SynCAM displays the most potent synaptogenic activity. Future studies may further expand its repertoire in regulating aspects of synapse function.

17.3 Heterophilic Adhesion Molecules

17.3.1 Neuroligin/Neurexin Complex

The neuroligin (NLGN)/neurexin (NRXN) complex is one of the most widely studied synaptic adhesion complexes, in part because of its association with human cognitive disorders such as autism spectrum disorders, Tourette syndrome, and schizophrenia (Südhof 2008). As Ca^{2+}-dependent heterophilic CAMs, postsynaptic NLGNs bind to presynaptic NRXNs to promote the formation and maturation of stable synaptic contacts (Nguyen and Südhof 1997; Scheiffele et al. 2000; Missler et al. 2003; Graf et al. 2004; Varoqueaux et al. 2006; Chubykin et al. 2007; Berninghausen et al. 2007).

17.3.1.1 NRXN Structure

NRXNs are encoded by three genes (*NRXN1-3*), each of which generates a long isoform (α-NRXNs) and a short isoform (β-NRXNs) that are driven by distinct promoter regions (Ushkaryov et al. 1992). α-NRXNs possess six *l*aminin, *N*RXN, *s*ex-hormone-binding globulin (LNS) domains with three EGF (*e*pidermal *g*rowth *f*actor)-like domains that are interdigitated between the first and second, third and fourth, and fifth and sixth LNS domains to form three repeating LNS(A)-EGF-LNS (B) cassettes (Fig. 17.4a) (Ushkaryov et al. 1992; Missler et al. 1998). β-NRXNs, on the other hand, possess only one LNS(B) domain, which is identical to the sixth LNS domain of α-NRXN. In both α- and β-NRXN isoforms, the membrane proximal LNS(B) domain is followed by a linker region containing O-linked glycosylation sites, which, when occupied, may help extend the extracellular domains into the synaptic cleft. This region is followed by a transmembrane domain and a PDZ type II-binding domain, which can interact with N-type Ca^{2+} channels and the presynaptic actin network through CASK and protein 4.1 (Ushkaryov et al. 1992; Hata et al. 1996; Missler et al. 1998, 2003; Biederer and Südhof 2001; Mukherjee et al. 2008).

17.3.1.2 NLGN Structure

Neuroligins are single-pass transmembrane proteins and belong to the class of α/β hydrolase superfamily of proteins, which share significant sequence homology with acetylcholinesterases (Araç et al. 2007; Fabrichny et al. 2007). Four genes (*NLGN1-4*) code for NLGNs in mammals, and humans possess a fifth (*NLGN5*). Each isoform contains a highly conserved *c*holinesterase-*l*ike *d*omain (CLD) that participates in heterophilic binding with NRXNs in *trans* and homophilic dimerization in *cis* (Fig. 17.4b) (Dean et al. 2003; Araç et al. 2007; Fabrichny et al. 2007; Chen et al. 2007). The CLD domain, which contains several N-glycosylation sites as well as a splice site (Ichtchenko et al. 1995; Hoffman et al. 2004; Araç et al. 2007), adopts an ellipsoid shape composed of a twisted β-sheet surrounded by multiple helices (Araç et al. 2007; Fabrichny et al. 2007). A linker sequence bridges the CLD and the transmembrane region and contains several O-glycosylation sites (Ichtchenko et al. 1995; Hoffman et al. 2004) that may help

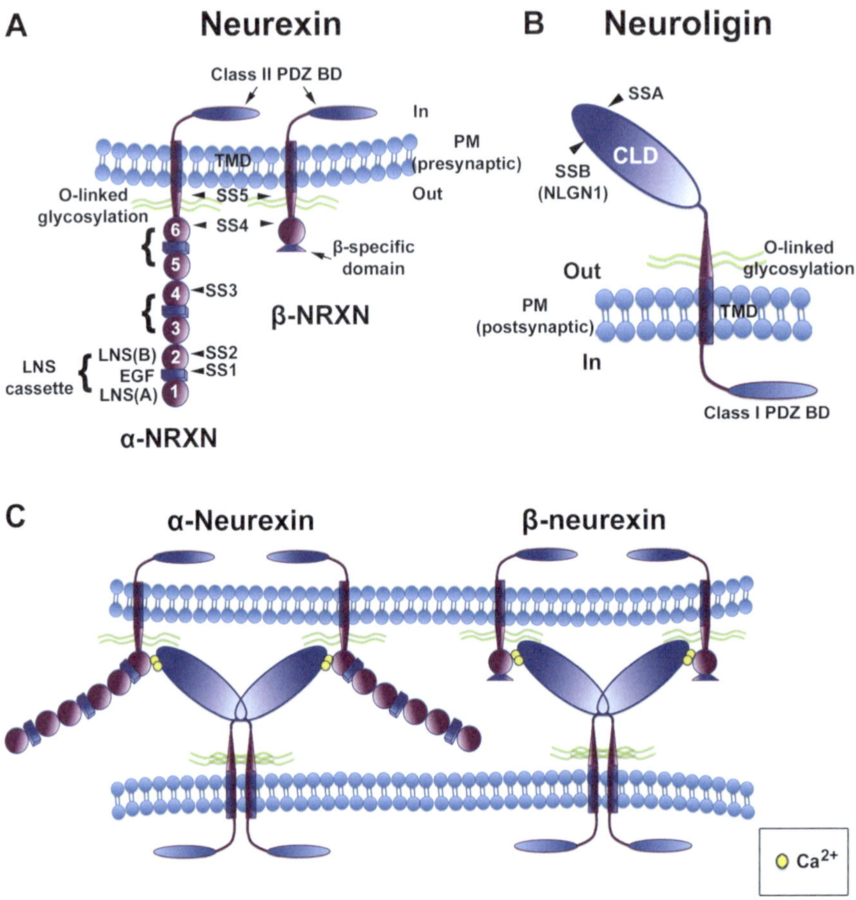

Fig. 17.4 Neurexins and neuroligins. (**a**) α- and β-neurexin domain organization. LNS, *l*aminin, *N*RXN, *s*ex-hormone-binding globulin domains; *EGF* epidermal growth factor domain; *SS1–5* *s*plice *s*ites 1–5. (**b**) Neuroligin domain organization. *CLD c*holinesterase-*l*ike *d*omain, *SSA* splice site A, *SSB* splice site B – only present in NLGN1. (**c**) Illustration of proposed binding mechanism between α- and β-NRXN and NLGN based on crystal structures derived by Araç et al. (2007) and presented in Sudhof (2008)

extend the globular CLD into the synaptic cleft in a manner similar to NRXN (Südhof 2008). The intracellular C-terminal end of NLGNs contains several domains that interact with postsynaptic signaling proteins and scaffolding molecules. For instance, NLGN1 contains a type I PDZ domain-binding motif and interacts with proteins such as PSD-95 (Irie et al. 1997).

NLGN constitutively forms dimers via key residues in the alpha helix region at the base of the CLD (Dean et al. 2003; Comoletti et al. 2007; Araç et al. 2007; Fabrichny et al. 2007; Chen et al. 2007; Shipman and Nicoll 2012a). Homo- and heterodimers, composed of NLGN1 and NLGN3 and NLGN2 and NLGN3, have been identified in the brain (Comoletti et al. 2006; Budreck and Scheiffele 2007; Shipman et al. 2011).

17.3.1.3 Binding

Both NLGNs and NRXNs are alternatively spliced to generate a wide array of potential binding interactions (Ullrich et al. 1995; Ichtchenko et al. 1995, 1996; Boucard et al. 2005) that may function to specify cell-type and circuit-specific connections with unique synaptic strength and plasticity properties (Südhof 2008). All NLGN isoforms possess a splice site within the CLD (called SSA) (Ichtchenko et al. 1995, 1996; Boucard et al. 2005). A second splice site (called SSB) is active only in NLGN1 and controls NLGN's binding affinity for α- and β-NRXNs (Boucard et al. 2005; Koehnke et al. 2010). α-NRXNs contain 5 splice sites in their extracellular domains, termed SS1–5; SS1 is after the first EGF-like domain; SS2, SS3, and SS4 are in the LNS(B) domains of cassettes 1, 2, and 3, respectively; and SS5 resides in the O-glycosylated proximal domain (Fig. 17.4a) (Missler et al. 1998). Due to their shorter extracellular domains β-NRXNs possess only SS4 and SS5. Altogether, alternative splicing of NRXN can generate thousands of individual isoforms (Ullrich et al. 1995), each of which may have different binding affinities for their postsynaptic partners and thus may vary in their potency of synaptic influence.

The heterophilic NLGN/NRXN adhesion is mediated by binding between the CLD of NLGNs and the sixth LNS domains of α- and β-NRXNs (Fig. 17.4b) (Boucard et al. 2005; Südhof 2008). However, the β-unique N-terminal domain must be present for β-NRXN binding (Nguyen and Südhof 1997) while the third EGF domain is required for α-NRXN binding (Boucard et al. 2005). Different NLGN isoforms exhibit different affinities for β-NRXNs (Comoletti et al. 2006), indicating that dimer composition could potently influence the binding affinity. Crystal structures of NLGN/β-NRXN1 complexes demonstrate that two β-NRXN1 monomers bind to each NLGN molecule on either side of the dimer interface (Araç et al. 2007; Fabrichny et al. 2007). Within the binding interface, Ca^{2+} ions occupy two distinct sites, and are coordinated by ligands from either molecule (Araç et al. 2007). The β-NRXN binding site on NLGN1 is located adjacent to the SSB site in the CLD (Boucard et al. 2005; Araç et al. 2007). Likewise, the NLGN binding site of β-NRXN is located next to the SS4 site in the LNS domain of β-NRXN. The proximity of these splice sites to the binding regions confers significant regulation of binding affinity when different splice variants are present at the synapse (Boucard et al. 2005; Comoletti et al. 2006; Siddiqui et al. 2010).

17.3.1.4 NLGN/NRXN at the Synapse

Both NLGNs and NRXNs are highly enriched throughout the developing and adult brain (Ullrich et al. 1995; Varoqueaux et al. 2006). Whereas NLGN1 distributes primarily to the PSD of excitatory synapses, NLGN2 distributes primarily to inhibitory synapses, and NLGN3 can heterodimerize with NLGN1 or NLGN2 and distributes to both types of synapses (Song et al. 1999; Graf et al. 2004; Chih et al. 2006; Chubykin et al. 2007; Budreck and Scheiffele 2007). Nevertheless, these associations are highly variable (see below) (Chih et al. 2006; Levinson et al. 2010).

The synaptogenic activity of NLGN (Scheiffele et al. 2000) and NRXN (Graf et al. 2004) is supported by artificial synapse formation assays. However, synapse formation proceeds rather normally in the complete absence of NLGNs, whereas synapses fail to develop mature properties of synaptic transmission (Varoqueaux et al. 2006). Interestingly, a recent study has shown that synapse density is significantly reduced only when NLGN1 is knocked-down in a subset of neurons rather than across all neurons (Kwon et al. 2012). Moreover, the same study suggests that relative NLGN abundance across neurons correlates with excitatory synapse density, although synapse-to-synapse variability has not been assessed. Therefore, competitive mechanisms may underlie NLGN/NRXN-dependent synapse formation.

Further experiments have focused on understanding the isoform and splice site variations in both NLGN and NRXN and how they contribute to specific heterophilic interactions. The SSB site of NLGN strongly regulates NRXN binding: the SSB-containing NLGN [NLGN(+SSB)] binds only to β-NRXNs that lack the SS4 site, whereas the SSB-lacking NLGNs [NLGN(−SSB)] bind to both α- and β-NRXNs regardless of the presence of SS4 (Ichtchenko et al. 1995; Boucard et al. 2005; Comoletti et al. 2006). Specific heterophilic interactions created by the isoform and splice site variations result in differences in synaptogenic activity. Neurons overexpressing full-length NLGN bear more numerous and larger synapses than those expressing NLGN(−SSA/−SSB) (Boucard et al. 2005). The subtle regulation of synapse assembly by the presence or absence of SSB in NLGN is further highlighted in the artificial synapse formation assay: postsynaptic cells expressing NLGN(+SSB) slow the accumulation of presynaptic components as compared to cells expressing NLGN(−SSB) (Lee et al. 2010). Impairing the N-linked glycosylation within the SSB that promotes binding to α-NRXN (Boucard et al. 2005) hastens synaptogenesis; this suggests a key role for N-glycosylation of SSB in slowing the rate of synaptogenesis by favoring NLGN binding to β-NRXNs (Lee et al. 2010).

Alternative splicing of NRXN isoforms also impacts synapse formation. In a well-studied case, β-NRXN(−SS4) isoforms potently cluster excitatory and inhibitory postsynaptic components in artificial synapse assembly assays. The presence of SS4 significantly reduces clustering of PSD-95 and NLGN1, NLGN3, and NLGN4 without an effect on gephyrin or NLGN2 (Graf et al. 2006).

Different isoforms and alternative splicing of NLGNs not only regulate the rate and extent of synaptogenesis, but they themselves are differentially localized to excitatory and inhibitory synapses. Such differential distribution of NLGNs has been proposed to contribute to the fine balance between excitation and inhibition that exists in neural circuits (Graf et al. 2004; Chih et al. 2005, 2006; Levinson and

El-Husseini 2005; Chubykin et al. 2007). For example, NLGN1(+SSA/+SSB) is preferentially localized to excitatory synapses, while NLGN1(−SSA/−SSB) is comparably distributed to both excitatory and inhibitory synapses; NLGN1 (+SSA/−SSB) is mostly present at inhibitory synapses, whereas NLGN1(−SSA/ +SSB) is targeted to excitatory synapses (Chih et al. 2006). Likewise, synapse targeting of NLGN2 is dependent on the SSA region in that NLGN2(+SSA) is preferentially localized to inhibitory synapses, while NLGN2(−SSA) is equally distributed between excitatory and inhibitory synapses (Chih et al. 2006). As one might expect, the ability of these isoforms to promote the assembly of a particular synapse type parallels their distribution patterns (Chih et al. 2006; Chubykin et al. 2007). The stoichiometry of PSD-95/gephyrin association with NLGNs is also important. Disrupting PSD-95 levels substantially alters the balance between excitation and inhibition in an NLGN-dependent manner (Prange et al. 2004). Moreover, knockdown of gephyrin shifts the association of NLGN2 from inhibitory to excitatory synapses, while a knockdown of PSD-95 shifts NLGN2 and NLGN3 from excitatory to inhibitory synapses (Levinson et al. 2010). However, several of NLGN's major synaptic effects have been shown to occur independently of its PDZ and gephyrin-binding domains (Shipman et al. 2011), suggesting that multiple intracellular domains participate in NLGN function at the synapse.

Interestingly, to add another layer of complexity, the synaptogenesis-promoting effect of NLGN isoforms is activity-dependent. NMDAR-mediated Ca^{2+}-influx and CaMKII activation are required for NLGN1-induced enhancement of excitatory synapse formation, while network activity is required for NLGN2-induced inhibitory synapse formation (Chubykin et al. 2007). Furthermore, a recent study has shown that activity shifts the interaction of NLGN1 with PSD-95 in favor of gephyrin by phosphorylation of NLGN1 at a tyrosine residue in its C-terminus (Giannone et al. 2013), which is homologous to a sequence in NLGN2 responsible for gephyrin interaction (Poulopoulos et al. 2009). Additionally, some recent studies have demonstrated activity-dependent cleavage of the extracellular domain of NLGNs by matrix metalloprotease 9 (MMP-9) and ADAM-10 (Peixoto et al. 2012; Suzuki et al. 2012). NLGN cleavage occurs within its juxtamembrane region and generates a secreted form of NLGN and a C-terminal stalk (Peixoto et al. 2012; Suzuki et al. 2012). Induction of NLGN cleavage destabilizes presynaptic β-NRXNs and reduces presynaptic p_r (Peixoto et al. 2012), whereas inhibiting NLGN cleavage increases dendritic spine density (Suzuki et al. 2012).

Altogether, NLGN/NRXN heterophilic adhesion controls many aspects of synapse assembly and function. Although not exclusively, NLGN's influence at excitatory synapses are executed via modifying postsynaptic AMPA (Heine et al. 2008; Etherton et al. 2011) and NMDA (Chubykin et al. 2007; Shipman and Nicoll 2012b; Kwon et al. 2012) receptor function at excitatory synapses or via gephyrin at inhibitory synapses (Poulopoulos et al. 2009; Levinson et al. 2010). NLGNs also influence presynaptic maturation in a retrograde manner (Futai et al. 2007; Wittenmayer et al. 2009; Lee et al. 2010; Peixoto et al. 2012).

17.3.2 Leucine-Rich Repeat-Containing Molecules (LRRTM/Slitrk/NGL/SALM)

Several known synaptic CAMs possess extracellular *l*eucine-*r*ich *r*epeat (LRR) domains and contribute to synapse formation, maturation, and plasticity. Similar to NLGNs, LRR-containing CAMs have recently received considerable attention because their mutations have been associated with autism spectrum disorders and schizophrenia (Francks et al. 2007; Sousa et al. 2010; Takashima et al. 2011; Ko 2012). The LRR-containing CAM families include the LRRTMs that bind to presynaptic NRXNs (Lauren et al. 2003; Linhoff et al. 2009; de Wit et al. 2009; Ko et al. 2009) and other ligands (de Wit et al. 2013; Siddiqui et al. 2013), the Slit and Trk-like family that bind to presynaptic *p*rotein *t*yrosine *p*hosphatases (PTPs) (Takahashi et al. 2011, 2012), *n*etrin-*G l*igands (NGLs) that bind to GPI-linked netrin-Gs and PTPs (Nakashiba et al. 2000; Lin et al. 2003b; Kim et al. 2006; Woo et al. 2009; DeNardo et al. 2012), and *s*ynapse *a*dhesion-*l*ike *m*olecules (SALMs) that are homophilic CAMs (Wang 2006; Ko et al. 2006; Seabold et al. 2008). Many members of these LRR-containing CAMs have been discovered using artificial synapse formation assays, suggesting that LRR-containing molecules are potent synaptic organizing CAMs. Moreover, these adhesion complexes can cooperate or compete with each other or other synaptic CAMs to profoundly influence the specifics of synapse formation and function (Siddiqui et al. 2010; Soler-Llavina et al. 2011).

17.3.2.1 The Structure of the LRR Domain

LRR domains are composed of repeating sequences of 20–29 amino acids with a conserved 11 amino acid sequence of LxxLxLxxN/CXL, where x can be any amino acid and L positions can be occupied also by valine, isoleucine, or phenylalanine (Kobe and Kajava 2001). The earlier crystal structures of the LRR domain have revealed that individual LRRs fold into a non-globular, horseshoe-shape (Fig. 17.5a) (Kobe and Deisenhofer 1993, 1995; Kobe and Kajava 2001); the parallel arrangement of α-helices and β-sheets forms an outer surface of α-helices that influence the curvature and an inner surface lined by β-sheets that provide a number of potential ligand-binding sites (Kobe and Kajava 2001; Ko 2012). The structure and binding specificity of the LRR domain can vary significantly in the number of repeats as well as in the composition of α-helix, β-sheets, and intermediate regions (Kobe and Kajava 2001).

LRRTM/Neurexin Complex

LRRTMs are type I transmembrane proteins, and in human and mice they are encoded by four genes (*LRRTM1-4*), three of which (*LRRTM1-3*) are located within

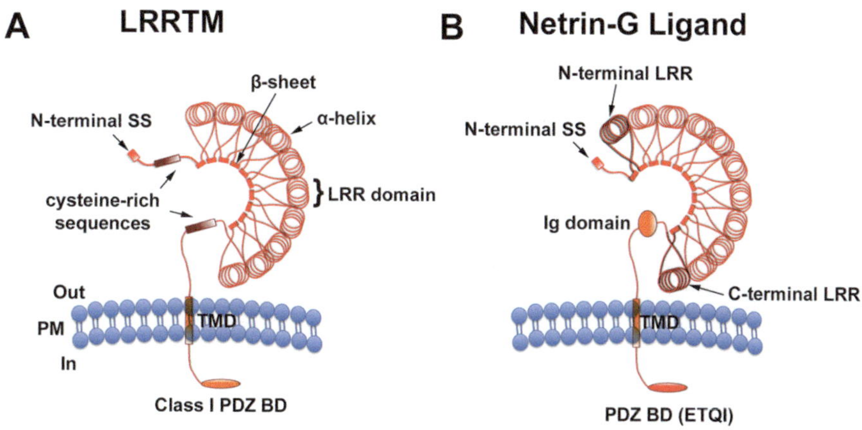

Fig. 17.5 Leucine-rich repeat-containing cell adhesion molecules, LRRTM, and netrin-G ligand. (**a**) Illustration of domain organization of LRRTM. The organization of LRR domains are based on crystal structures described by Kobe and Deisenhofer (1995). *LRR domain* leucine-rich repeat domain, *N-terminal SS* N-terminal signal sequence. (**b**) Illustration of the domain organization of netrin-G ligand

the gene coding for α-catenin (Lauren et al. 2003). All isoforms are highly expressed in the brain shortly after birth in a region-specific and non-overlapping manner (Lauren et al. 2003). During development, they become tightly localized to synapses (Linhoff et al. 2009; de Wit et al. 2009).

17.3.2.2 Binding

All LRRTM isoforms possess similar domain structures composed of an N-terminal signal sequence, ten LRRs flanked by cysteine-rich sequences, a transmembrane domain, and an intracellular tail that contains a highly conserved PDZ-like motif that binds to PSD-95 (Fig. 17.5a) (Lauren et al. 2003; de Wit et al. 2009). All isoforms have N-glycosylation sites in the third LRR domain, although additional sites are differentially distributed across isoforms (Lauren et al. 2003).

LRRTM1 and LRRTM2 can bind to all α- and β-NRXN isoforms lacking the SS4 insert in a Ca^{2+}-dependent manner (de Wit et al. 2009; Ko et al. 2009; Siddiqui et al. 2010). Specific extracellular domain interactions between NRXN and LRRTM have not been directly solved, but mutagenesis studies have shown that D260 and T262 residues within the concave β-sheet of the LRR domain of LRRTM2 are responsible for β-NRXN binding (Siddiqui et al. 2010). Moreover, the binding domain on NRXNs for LRRTMs is likely shared with that for NLGNs, leading to potential competing interactions between these three protein family members. Interestingly, however, LRRTMs and NLGNs cooperate in synapse development (Siddiqui et al. 2010), suggesting a complex interaction between NRXN/LRRTM and NRXN/NLGN complexes during synapse formation and maturation. The extracellular domain of LRRTM4 has recently been shown to bind in *trans* to the GPI-linked heparin sulfate proteoglycan, glypican, in an NRXN-dependent manner (de Wit et al. 2013; Siddiqui et al. 2013), thus expanding the molecular repertoire of LRRTM interactions.

17.3.2.3 LRRTMs at the Synapse

All LRRTM isoforms promote excitatory synapse formation in the artificial synapse formation assay: LRRTM1 and LRRTM2 are the most potent organizers, while LRRTM3 and LRRTM4 are less potent (Linhoff et al. 2009). Acute knockdown of LRRTM1 and LRRTM2 leads to a specific deficit in AMPA, but not NMDAR-mediated basal excitatory synaptic responses (de Wit et al. 2009; Soler-Llavina et al. 2011) along with deficits in LTP in young and mature hippocampal CA1 regions (Soler-Llavina et al. 2013). Knockdown of LRRTM4 in a subset of cortical neurons via in utero electroporation leads to a loss of synapses (de Wit et al. 2013). Constitutive loss of LRRTM1 (Linhoff et al. 2009) and LRRTM4 (Siddiqui et al. 2013) in mice show distinct phenotypes, which likely reflect their differential expression and binding partners. Whereas LRRTM1 knockout mice exhibit enhanced vesicle accumulation at synapses in CA1 *stratum radiatum* and *stratum oriens*, LRRTM4 knockout mice have normal synaptic vesicle accumulation in CA1 but reduced synaptic vesicle accumulation throughout the dentate gyrus. The latter phenotype is accompanied by deficits in DG granule cells of dendritic spine density, PSD-95 accumulation, and activity-dependent AMPA receptor surface expression (Soler-Llavina et al. 2013; Siddiqui et al. 2013). While these findings demonstrate distinct functions of LRRTM isoforms according to their binding affinity and distribution patterns, they also suggest that the shared actions of LRRTMs converge on the regulation of AMPAR abundance and surface expression.

Netrin-G Ligands

Netrin-G ligand-1 (NGL1) was originally identified as a growth-promoting ligand of netrin-G in thalamocortical axons (Lin et al. 2003b). Subsequent studies have identified two additional NGL isoforms, NGL2 and NGL3, in a yeast two-hybrid screen for PSD-95-interacting molecules (Kim et al. 2006). All isoforms are Ca^{2+}-independent heterophilic CAMs and are highly enriched in and distributed throughout the brain from the early postnatal period where they localize to central regions of the postsynaptic density (Kim et al. 2006).

17.3.2.4 NGL Structure and Binding

NGLs are composed of an N-terminal signal peptide, followed by nine LRR domains flanked by N and C-terminal LRR domains, an Ig domain, a type 1 transmembrane domain, and a C-terminal domain containing an ETQI PDZ-binding motif (Fig. 17.5b) (Lin et al. 2003b; Kim et al. 2006). NLG3 isoforms possess an additional N-terminal motif which is absent from NGL1 and NGL2, along with an extended linker sequence between the transmembrane and Ig domains.

NGL1 binds to netrin-G1 (also called laminet-1) but not netrin-G2 (also called laminet-2), whereas NGL2 show preferential binding to netrin-G2 over netrin-G1 (Lin et al. 2003b; Kim et al. 2006; Nishimura-Akiyoshi et al. 2007). NGL1 binding to netrin-G1 requires the first four LRR domains and the N-terminal domains, but does not require the Ig domain (Lin et al. 2003b). NLG3 binds to leucocyte common antigen-related (LAR) molecules (Woo et al. 2009), PTPσ, and PTPδ (Kwon et al. 2010), but does not bind to netrin-G1 or netrin-G2 (Kim et al. 2006). NGL3 binding is enabled by interactions between an NGL3 unique glutamine (Q96) in the first LRR domain and the first two FNIII domains of LAR, PTPσ, and PTPδ (Kwon et al. 2010).

17.3.2.5 NGLs at the Synapse

All NLG isoforms induce clustering of presynaptic proteins when expressed in nonneuronal cells in artificial synapse formation assays (Woo et al. 2009). Specifically, the NLG2/netrin-G2 complex contributes exclusively to excitatory, but not inhibitory synapse formation by inducing the clustering of functional presynaptic machinery and postsynaptic components such as NMDARs and scaffolding proteins, PSD-95 and Shank (Kim et al. 2006). Likewise, NLG3/LAR complex also contributes specifically to excitatory synapse formation. In addition to NMDARs and scaffolding proteins, NLG3 promotes the clustering of AMPARs, a property which may contribute to its greater apparent synaptogenic efficiency as compared to NLG1 and NLG2 (Woo et al. 2009). Notably, the NGL3/PTPσ complex induces the clustering of postsynaptic components (i.e., PSD-95), but the NGL3/PTPδ complex does not (Kwon et al. 2010). This suggests that upon interacting with PTPδ, NGL3 may alter its intracellular binding properties. Moreover, given that PTPδ and PTPσ also engage different postsynaptic ligands to induce inhibitory and excitatory synapse formation (Takahashi et al. 2011, 2012), combinatorial expression of these molecules during synapse formation or plasticity may tightly regulate the organization of postsynaptic density components to impact synaptic strength.

A feature of NGLs distinct from other synaptic CAMs discussed thus far is its distribution to specific sub-compartments of neuronal dendrites where they have been suggested to contribute to the functional compartmentalization of neural circuit signaling (Nishimura-Akiyoshi et al. 2007; DeNardo et al. 2012; Soto et al. 2013). For example, in the hippocampal CA1 region, NGL1 distributes to dendritic segments within the *stratum lacunosum moleculare* (SLM) where it encounters netrin-G1 positive axons originating from the entorhinal cortex. NGL2, on the other hand, distributes to dendritic segments in the *stratum radiatum* (SR) and *stratum oriens* (SO) of CA1 where it encounters netrin-G2-positive axons originating from the CA3 region (Nishimura-Akiyoshi et al. 2007). Both the extracellular LRR domain and the PDZ-binding domain are essential for this subcellular compartmentalization. Moreover, this differential distribution of NGLs has functional consequences: knocking out NGL2 expression selectively decreases the number of excitatory synapses within the SR but not the SLM, leading to altered neurotransmission and impaired dendritic integration of SR and SLM inputs (DeNardo et al. 2012).

17.3.3 Integrins

Integrins are allosteric, Ca^{2+}-dependent, heterodimeric cell adhesion molecules that participate primarily in cell-to-ECM but also in cell-to-cell adhesion (Hynes 2002). Major ligands include fibronectins, vitronectins, collagens, laminins, and IgSF members like L1 (Ruppert et al. 1995; Montgomery et al. 1996; Felding-Habermann et al. 1997; Yip et al. 1998; Hynes 2002). In vertebrates, integrin heterodimers are composed of 1 of 18 α and 1 of 8 β subunits, forming up to 24 distinct integrin αβ pairs (Luo et al. 2007). Two major integrin subgroupings are based on a common β1 subunit (binds αV, α1–11 subunits) and a common αV subunit (binds β1, 3, 5, 6, 8 subunits). Integrins indirectly link to the actin cytoskeleton and are capable of initiating diverse signaling events involving "outside-in" signaling (from outside of the cell to control the intracellular environment) and "inside-out" signaling (from the inside of the cell to adjust the extracellular ligand-binding state) (Luo et al. 2007). The ability to mediate such dynamic signaling is reflected in the modular and flexible structure of the α and β subunits that allows the molecule to adopt a wide range of conformational states depending on the cellular context. The complex signaling capacity makes the integrins an effective and highly versatile CAM at the synapse.

17.3.3.1 Integrin Structure and Binding

Both α- and β-integrins are type I transmembrane glycoproteins with a globular extracellular head domain whose α and β subunit interface in the heterodimer supports ligand binding. Also present in the extracellular domain is a "leg" or stalk region that is essential for integrin activation, which is followed by a single-pass transmembrane domain and a short cytoplasmic domain (Fig. 17.6a) (Nermut et al. 1988; Xiong et al. 2001; Luo et al. 2007).

The extracellular globular head domain of α-integrins consists of a seven-segment β-propeller structure containing multiple Ca^{2+}-binding sites (Xiong

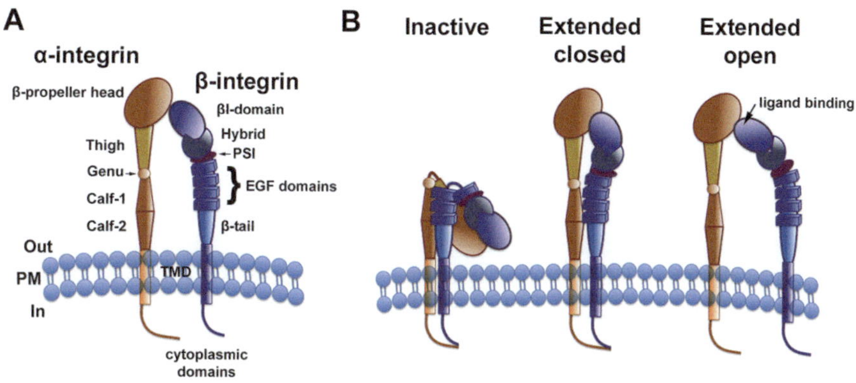

Fig. 17.6 The integrins. (**a**) Domain organization of α- and β-integrins. *PSI* *p*lexin/*s*emaphorin/*i*ntegrin domain. (**b**) Illustration of conformational changes in the structure and interaction of α- and β-integrins upon activation by extracellular or intracellular ligands based on crystal structures described in Kim et al. (2003). The extended open conformation presents a ligand-binding site near the interface region of α- and β-integrin

et al. 2001, 2002; Luo et al. 2007). The "leg" region contains three β-sandwich domains that are subdivided into a "thigh" domain and two "calf" domains (calf-1 and calf-2). An additional Ca^{2+}-binding loop separates the thigh region from calf-1 and calf-2 and is termed the "genu" because it is a key pivot point during integrin activation (Fig. 17.6b). The head domain of β-integrins is composed of a βI domain that participates in α subunit binding and is flanked by a hybrid domain that forms the upper part of the β "leg" region. The hybrid domain itself is positioned within a plexin/semaphorin/integrin (PSI) domain that contributes to the stability of the head region. The remainder of the leg region is composed of four EGF-like domains followed by a β-tail domain. When inactive, the leg of β subunit is thought to bend at a region between EGF domains 1 and 2.

The conformation of an inactive integrin αβ heterodimer is heavily bent with the head region drooping toward the membrane (Xiong et al. 2001). In this conformation, the extracellular domain forms an interface with the lower legs (Takagi et al. 2002), and the molecule is not effectively positioned for ligand binding. However, upon ligand binding (Takagi et al. 2002; Xiao et al. 2004) or cytoplasmic activation (Calderwood et al. 1999; Tsuboi 2002; Vinogradova et al. 2002), the αβ heterodimer adopts an active conformation where the head region is extended away from the membrane (Xiao et al. 2004; Luo et al. 2007). The extended conformation can exist in one of two orientations: an open headpiece conformation that presents the ligand-binding region and a closed headpiece conformation where the ligand-binding region is more obscured (Fig. 17.6b) (Luo et al. 2007). The difference between the two is the relative positions of the subunit legs, transmembrane domains, and intracellular domains that become separated in the open conformation (Kim et al. 2003) and promote the swing-out of the hybrid domain of the β subunit.

17.3.3.2 Integrins at the Synapse

Integrin subtypes are present throughout the brain in a region-specific manner (Pinkstaff et al. 1999). Different αβ pairs localize to specific domains within the hippocampus (McGeachie et al. 2011), where they are enriched at synapses (Shi and Ethell 2006) and participate in distinct functions in synapse maturation and plasticity (Chavis and Westbrook 2001; Kramár et al. 2003; Bernard-Trifilo et al. 2005; Chan et al. 2006; Cingolani et al. 2008; McGeachie et al. 2012). Two of the major integrin subtypes in the nervous system that participate in synaptic function are composed of either β1 or β3 subunits and recognize ligands containing the Arg-Gly-Asp (RGD) sequence present in ECM-related molecules such as fibronectin (Hynes 2002).

Many of the studies that have addressed the contributions of integrins to synaptic regulation have used the infusion of peptides containing an RGD sequence to interfere with the ECM interaction of integrins. Notably, these studies have revealed acute effects on AMPA (Cingolani et al. 2008) and NMDA (Lin et al. 2003a; Bernard-Trifilo et al. 2005) receptor currents, although the latter may have involved direct effect of peptides on the glycine-binding site of NMDAR (Cingolani et al. 2008). In cultured hippocampal neurons, RGD peptide application rapidly reduces the surface abundance of AMPARs, accompanied by a concomitant decrease in quantal amplitudes and evoked synaptic transmission (Cingolani et al. 2008). The specific interaction between β3-containing integrins with the GluA2 subunit (Pozo et al. 2012) is likely involved in controlling the abundance and composition of synaptic AMPARs. Whereas the β3-integrin modulation of synaptic AMPARs occurs independently of actin, RGD peptide application over a longer duration elicits changes in synaptic actin organization and dendrite morphology, where F-actin within spine heads is reduced and dendritic spines become elongated (Shi and Ethell 2006). Such effects on spine morphology could be mediated by β1-integrin (Shi and Ethell 2006), which plays a role in LTP stability (see below).

Chronic application of RGD peptides during synapse maturation in cultured hippocampal neurons arrests synaptic development in a state characterized by high p_r and the prevalence of NR2B-containing NMDARs. These changes can be reproduced if RGD peptides are replaced with function-blocking antibodies directed toward β3 but not β1-containing integrins (Chavis and Westbrook 2001). Likewise, glycinergic synapses of spinal cord neurons also show a differential effect of integrin subtypes on synaptic strength. While application of RGD peptides increases overall glycine receptor surface levels, this effect is the sum of very different influences of β1- and β3-containing integrins: function-blocking antibodies against β1-integrins decrease, while those against β3-integrins increase surface glycine receptor abundance (Charrier et al. 2010). These findings reveal a complex interplay between integrin subtypes in regulating basal synaptic activity.

Several integrins play a role in long-term synaptic plasticity (McGeachie et al. 2011). In particular, infusing RGD peptides or anti-integrin antibodies before or shortly after LTP induction disrupts the stability of LTP in the hippocampus (Staubli et al. 1990; Bahr et al. 1997; Stäubli et al. 1998). Studies using conditional β1-integrin knockout mice support the role for β1-integrins in the stabilization but not the induction of LTP (Chan et al. 2006; Huang et al. 2006; McGeachie et al. 2012). Interestingly, β3-containing integrins do not seem to influence LTP (McGeachie et al. 2012).

In contrast to LTP, homeostatic synaptic plasticity seems particularly dependent on β3- rather than β1-integrins (Cingolani et al. 2008). Eliciting compensatory, homeostatic changes in synaptic AMPAR levels by chronically altering network activity also induces a parallel change in surface β3-integrin levels without a change in β1-integrin. The requirement for β3-integrin in homeostatic plasticity is supported by observations that β3-integrin-deficient neurons do not show a compensatory increase in synaptic AMPARs upon chronic activity blockade. Together, these findings highlight thc differential roles for integrin subtypes in distinct forms of synaptic plasticity: whereas the β1-integrins regulate Hebbian synaptic plasticity, β3-integrins are involved in homeostatic synaptic plasticity. Much remains to be explored on the molecular mechanisms by which β1- and β3-integrins exert differential control over the two seemingly opposing synaptic processes that nonetheless target the AMPARs, as well as uncovering the roles for other integrin subtypes in synaptic regulation.

17.4 Outlook

We have discussed how synapse formation and synaptic strength can be dynamically controlled by the action of a variety of synaptic CAMs. Apart from their traditional role in conferring the stability of cell–cell junctions – i.e., at synapses – we have shown how different CAMs organize the pre- and the postsynaptic assemblies to affect the efficacy of neurotransmitter release and the availability of functional neurotransmitter receptors and participate in synaptic signaling events. The degree to which particular CAMs are able to control these functions depends on the binding interactions in which they are engaged, extracellularly and intracellularly. The binding activity of CAMs, in turn, is finely tuned by the presence or absence of additional ligands, their glycosylation state, activity of proteases that act on CAMs, and intracellular signaling events, just to name a few. Despite the enormous complexity of CAMs at the synapse, the rules that govern specific CAM interaction modes are slowly being revealed. The difficult task ahead will be to sort out how different synaptic CAMs cooperate with each other in time and space at individual synapses and across synapses. In addition, some very basic questions still remain unknown, such as the minimal number of CAM complexes needed at each synapse to effect a particular adhesive or regulatory function, or their spatiotemporal dynamics in relation to key pre- and postsynaptic components such as the readily releasable vesicles and the number of postsynaptic receptors. In this respect, recent developments in super-resolution imaging are highly promising in being able to tackle detailed arrangements of CAMs in synapses undergoing plastic changes. The strong implications for synaptic CAMs in neurological diseases should fuel further research efforts on this fascinating topic.

References

Abe K, Chisaka O, van Roy F, Takeichi M (2004) Stability of dendritic spines and synaptic contacts is controlled by αN-catenin. Nat Neurosci 7:357–363. doi:10.1038/nn1212

Aiga M, Levinson JN, Bamji SX (2010) N-cadherin and neuroligins cooperate to regulate synapse formation in hippocampal cultures. J Biol Chem 286:851–858. doi:10.1074/jbc.M110.176305

Aoto J, Martinelli DC, Malenka RC et al (2013) Presynaptic neurexin-3 alternative splicing trans-synaptically controls postsynaptic AMPA receptor trafficking. Cell 154:75–88. doi:10.1016/j.cell.2013.05.060

Araç D, Boucard AA, Ozkan E et al (2007) Structures of neuroligin-1 and the neuroligin-1/neurexin-1β complex reveal specific protein-protein and protein-Ca2+ interactions. Neuron 56:992–1003. doi:10.1016/j.neuron.2007.12.002

Arikkath J, Reichardt LF (2008) Cadherins and catenins at synapses: roles in synaptogenesis and synaptic plasticity. Trends Neurosci 31:487–494. doi:10.1016/j.tins.2008.07.001

Bahr BA, Stäubli U, Xiao P et al (1997) Arg-Gly-Asp-Ser-selective adhesion and the stabilization of long-term potentiation: pharmacological studies and the characterization of a candidate matrix receptor. J Neurosci 17:1320–1329

Bamji SX, Shimazu K, Kimes N et al (2003) Role of beta-catenin in synaptic vesicle localization and presynaptic assembly. Neuron 40:719–731

Bamji SX, Rico B, Kimes N, Reichardt LF (2006) BDNF mobilizes synaptic vesicles and enhances synapse formation by disrupting cadherin- -catenin interactions. J Cell Biol 174:289–299. doi:10.1083/jcb.200601087

Bemben MA, Shipman SL, Hirai T et al (2013) CaMKII phosphorylation of neuroligin-1 regulates excitatory synapses. Nat Publ Group 17:56–64. doi:10.1038/nn.3601

Benson DL, Tanaka H (1998) N-cadherin redistribution during synaptogenesis in hippocampal neurons. J Neurosci 18:6892–6904

Bernard-Trifilo JA, Kramár EA, Torp R et al (2005) Integrin signaling cascades are operational in adult hippocampal synapses and modulate NMDA receptor physiology. J Neurochem 93:834–849. doi:10.1111/j.1471-4159.2005.03062.x

Berninghausen O, Rahman MA, Silva J-P et al (2007) Neurexin Iβ and neuroligin are localized on opposite membranes in mature central synapses. J Neurochem 103:1855–1863. doi:10.1111/j.1471-4159.2007.04918.x

Biederer T (2006) Bioinformatic characterization of the SynCAM family of immunoglobulin-like domain-containing adhesion molecules. Genomics 87:139–150. doi:10.1016/j.ygeno.2005.08.017

Biederer T, Südhof TC (2001) CASK and protein 4.1 support F-actin nucleation on neurexins. J Biol Chem 276:47869–47876. doi:10.1074/jbc.M105287200

Biederer T, Sara Y, Mozhayeva M et al (2002) SynCAM, a synaptic adhesion molecule that drives synapse assembly. Science (New York, N Y) 297:1525–1531. doi:10.1126/science.1072356

Blaess S, Kammerer RA, Hall H (1998) Structural analysis of the sixth immunoglobulin-like domain of mouse neural cell adhesion molecule L1 and its interactions with αvβ3, αIIbβ3, and α5β1 integrins. J Neurochem 71:2615–2625

Bodrikov V, Sytnyk V, Leshchyns'ka I et al (2008) NCAM induces CaMKII -mediated RPTP phosphorylation to enhance its catalytic activity and neurite outgrowth. J Cell Biol 182:1185–1200. doi:10.1083/jcb.200803045

Boggon TJ, Murray J, Chappuis-Flament S et al (2002) C-cadherin ectodomain structure and implications for cell adhesion mechanisms. Science (New York, N Y) 296:1308–1313. doi:10.1126/science.1071559

Boucard AA, Chubykin AA, Comoletti D et al (2005) A splice code for trans-synaptic cell adhesion mediated by binding of neuroligin 1 to α- and β-neurexins. Neuron 48:229–236. doi:10.1016/j.neuron.2005.08.026

Bozdagi O, Shan W, Tanaka H et al (2000) Increasing numbers of synaptic puncta during late-phase LTP: N-cadherin is synthesized, recruited to synaptic sites, and required for potentiation. Neuron 28:245–259

Brennaman LH, Kochlamazashvili G, Stoenica L et al (2011) Transgenic mice overexpressing the extracellular domain of NCAM are impaired in working memory and cortical plasticity. Neurobiol Dis Neurobiol Dis 43:372–378. doi:10.1016/j.nbd.2011.04.008

Brennaman LH, Moss ML, Maness PF (2013) EphrinA/EphA-induced ectodomain shedding of neural cell adhesion molecule regulates growth cone repulsion through ADAM10 metalloprotease. J Neurochem 128:267–279. doi:10.1111/jnc.12468

Brieher WM, Yap AS, Gumbiner BM (1996) Lateral dimerization is required for the homophilic binding activity of C-cadherin. J Cell Biol 135:487–496

Brigidi GS, Sun Y, Beccano-Kelly D et al (2014) Palmitoylation of δ-catenin by DHHC5 mediates activity-induced synapse plasticity. Nat Publ Group 17:522–532. doi:10.1038/nn.3657

Budreck EC, Scheiffele P (2007) Neuroligin-3 is a neuronal adhesion protein at GABAergic and glutamatergic synapses. Eur J Neurosci 26:1738–1748. doi:10.1111/j.1460-9568.2007.05842.x

Bukalo O, Fentrop N, Lee AYW et al (2004) Conditional ablation of the neural cell adhesion molecule reduces precision of spatial learning, long-term potentiation, and depression in the CA1 subfield of mouse hippocampus. J Neurosci 24:1565–1577. doi:10.1523/JNEUROSCI.3298-03.2004

Calderwood DA, Zent R, Grant R et al (1999) The Talin head domain binds to integrin subunit cytoplasmic tails and regulates integrin activation. J Biol Chem 274:28071–28074. doi:10.1074/jbc.274.40.28071

Cambon K, Hansen SM, Venero C et al (2004) A synthetic neural cell adhesion molecule mimetic peptide promotes synaptogenesis, enhances presynaptic function, and facilitates memory consolidation. J Neurosci 24:4197–4204. doi:10.1523/JNEUROSCI.0436-04.2004

Chan C-S, Weeber EJ, Zong L et al (2006) Beta 1-integrins are required for hippocampal AMPA receptor-dependent synaptic transmission, synaptic plasticity, and working memory. J Neurosci 26:223–232. doi:10.1523/JNEUROSCI.4110-05.2006

Chappuis-Flament S, Wong E, Hicks LD et al (2001) Multiple cadherin extracellular repeats mediate homophilic binding and adhesion. J Cell Biol 154:231–243

Charrier C, Machado P, Tweedie-Cullen RY et al (2010) A crosstalk between β1 and β3 integrins controls glycine receptor and gephyrin trafficking at synapses. Nat Neurosci 13:1388–1395. doi:10.1038/nn.2645

Chavis P, Westbrook G (2001) Integrins mediate functional pre- and postsynaptic maturation at a hippocampal synapse. Nature 411:317–321. doi:10.1038/35077101

Chen A, Haines SL, Maxson K, Akeson RA (1994) VASE exon expression alters NCAM-mediated cell-cell interactions. J Neurosci Res 38:483–492

Chen X, Liu H, Shim AHR et al (2007) Structural basis for synaptic adhesion mediated by neuroligin-neurexin interactions. Nat Struct Mol Biol 15:50–56. doi:10.1038/nsmb1350

Chernyshova Y, Leshchyns'ka I, Hsu S-C et al (2011) The neural cell adhesion molecule promotes FGFR-dependent phosphorylation and membrane targeting of the exocyst complex to induce exocytosis in growth cones. J Neurosci 31:3522–3535. doi:10.1523/JNEUROSCI.3109-10.2011

Chih B, Engelman H, Scheiffele P (2005) Control of excitatory and inhibitory synapse formation by neuroligins. Science (New York, N Y) 307:1324–1328. doi:10.1126/science.1107470

Chih B, Gollan L, Scheiffele P (2006) Alternative splicing controls selective trans-synaptic interactions of the neuroligin-neurexin complex. Neuron 51:171–178. doi:10.1016/j.neuron.2006.06.005

Chipman PH, Schachner M, Rafuse VF (2014a) Presynaptic NCAM is required for motor neurons to functionally expand their peripheral field of innervation in partially denervated muscles. J Neurosci 34:10497–10510. doi:10.1523/JNEUROSCI.0697-14.2014

Chipman PH, Zhang Y, Rafuse VF (2014b) A stem-cell based bioassay to critically assess the pathology of dysfunctional neuromuscular junctions. PLoS One 9:e91643. doi:10.1371/journal.pone.0091643.t002

Chubykin AA, Atasoy D, Etherton MR et al (2007) Activity-dependent validation of excitatory versus inhibitory synapses by neuroligin-1 versus neuroligin-2. Neuron 54:919–931. doi:10.1016/j.neuron.2007.05.029

Chuong CM, McClain DA, Streit P, Edelman GM (1982) Neural cell adhesion molecules in rodent brains isolated by monoclonal antibodies with cross-species reactivity. Proc Natl Acad Sci 79:4234–4238

Cingolani LA, Thalhammer A, Yu LMY et al (2008) Activity-dependent regulation of synaptic AMPA receptor composition and abundance by β3 integrins. Neuron 58:749–762. doi:10.1016/j.neuron.2008.04.011

Cole GJ, Loewy A, Glaser L (1986) Neuronal cell-cell adhesion depends on interactions of N-CAM with heparin-like molecules. Nature 320:445–447

Comoletti D, Flynn RE, Boucard AA et al (2006) Gene selection, alternative splicing, and post-translational processing regulate neuroligin selectivity for β-neurexins †. Biochemistry 45:12816–12827. doi:10.1021/bi0614131

Comoletti D, Grishaev A, Whitten AE et al (2007) Synaptic arrangement of the neuroligin/β-neurexin complex revealed by X-ray and neutron scattering. Structure 15:693–705. doi:10.1016/j.str.2007.04.010

Crossin KL, Edelman GM, Cunningham BA (1984) Mapping of three carbohydrate attachment sites in embryonic and adult forms of the neural cell adhesion molecule. J Cell Biol 99:1848–1855

Dahlin-Huppe K, Berglund EO, Ranscht B, Stallcup WB (1997) Mutational analysis of the L1 neuronal cell adhesion molecule identifies membrane-proximal amino acids of the cytoplasmic domain that are required for cytoskeletal anchorage. Mol Cell Neurosci 9:144–156. doi:10.1006/mcne.1997.0608

Davis JQ, Bennett V (1994) Ankyrin binding activity shared by the neurofascin/L1/NrCAM family of nervous system cell adhesion molecules. J Biol Chem 269:27163–27166

de Wit J, Sylwestrak E, Sullivan MLO et al (2009) LRRTM2 interacts with neurexin 1 and regulates excitatory synapse formation. Neuron 64:799–806. doi:10.1016/j.neuron.2009.12.019

de Wit J, O'Sullivan ML, Savas JN et al (2013) Unbiased discovery of glypican as a receptor for LRRTM4 in regulating excitatory synapse development. Neuron 79:696–711. doi:10.1016/j.neuron.2013.06.049

Dean C, Scholl FG, Choih J et al (2003) Neurexin mediates the assembly of presynaptic terminals. Nat Neurosci 6:708–716. doi:10.1038/nn1074

DeNardo LA, de Wit J, Otto-Hitt S, Ghosh A (2012) NGL-2 regulates input-specific synapse development in CA1 pyramidal neurons. Neuron 76:762–775. doi:10.1016/j.neuron.2012.10.013

Dityatev A, Dityateva G, Schachner M (2000) Synaptic strength as a function of post- versus presynaptic expression of the neural cell adhesion molecule NCAM. Neuron 26:207–217

Dityatev A, Dityateva G, Sytnyk V et al (2004) Polysialylated neural cell adhesion molecule promotes remodeling and formation of hippocampal synapses. J Neurosci 24:9372–9382. doi:10.1523/JNEUROSCI.1702-04.2004

Eckhardt M, Bukalo O, Chazal G et al (2000) Mice deficient in the polysialyltransferase ST8SiaIV/PST-1 allow discrimination of the roles of neural cell adhesion molecule protein and polysialic acid in neural development and synaptic plasticity. J Neurosci 20:5234–5244
Elia LP, Yamamoto M, Zang K, Reichardt LF (2006) p120 catenin regulates dendritic spine and synapse development through Rho-family GTPases and cadherins. Neuron 51:43–56. doi:10.1016/j.neuron.2006.05.018
Elste AM, Benson DL (2006) Structural basis for developmentally regulated changes in cadherin function at synapses. J Comp Neurol 495:324–335. doi:10.1002/cne.20876
Etherton MR, Tabuchi K, Sharma M et al (2011) An autism-associated point mutation in the neuroligin cytoplasmic tail selectively impairs AMPA receptor-mediated synaptic transmission in hippocampus. EMBO J 30:2908–2919. doi:10.1038/emboj.2011.182
Fabrichny IP, Leone P, Sulzenbacher G et al (2007) Structural analysis of the synaptic protein neuroligin and its β-neurexin complex: determinants for folding and cell adhesion. Neuron 56:979–991. doi:10.1016/j.neuron.2007.11.013
Fannon AM, Colman DR (1996) A model for central synaptic junctional complex formation based on the differential adhesive specificities of the cadherins. Neuron 17:423–434
Fazeli MS, Breen K, Errington ML, Bliss TV (1994) Increase in extracellular NCAM and amyloid precursor protein following induction of long term potentiation in the dentate gyrus of anesthetized rats. Neurosci Lett 169:77–80
Felding-Habermann B, Silletti S, Mei F et al (1997) A single immunoglobulin-like domain of the human neural cell adhesion molecule L1 supports adhesion by multiple vascular and platelet integrins. J Cell Biol 139:1567–1581
Fogel AI, Akins MR, Krupp AJ et al (2007) SynCAMs organize synapses through heterophilic adhesion. J Neurosci 27:12516–12530. doi:10.1523/JNEUROSCI.2739-07.2007
Fogel AI, Li Y, Giza J et al (2010) N-glycosylation at the SynCAM (Synaptic Cell Adhesion Molecule) immunoglobulin interface modulates synaptic adhesion. J Biol Chem 285:34864–34874. doi:10.1074/jbc.M110.120865
Fogel AI, Stagi M, de Arce KP, Biederer T (2011) Lateral assembly of the immunoglobulin protein SynCAM 1 controls its adhesive function and instructs synapse formation. EMBO J 30:4728–4738. doi:10.1038/emboj.2011.336
Francks C, Maegawa S, Lauren J et al (2007) LRRTM1 on chromosome 2p12 is a maternally suppressed gene that is associated paternally with handedness and schizophrenia. Mol Psychiatry 12:1129–1139. doi:10.1038/sj.mp.4002053
Friedlander DR, Milev P, Karthikeyan L et al (1994) The neuronal chondroitin sulfate proteoglycan neurocan binds to the neural cell adhesion molecules Ng-CAM/L1/NILE and N-CAM, and inhibits neuronal adhesion and neurite outgrowth. J Cell Biol 125:669–680
Futai K, Kim MJ, Hashikawa T et al (2007) Retrograde modulation of presynaptic release probability through signaling mediated by PSD-95–neuroligin. Nat Neurosci 10:186–195. doi:10.1038/nn1837
Galuska SP, Rollenhagen M, Kaup M et al (2010) Synaptic cell adhesion molecule SynCAM 1 is a target for polysialylation in postnatal mouse brain. Proc Natl Acad Sci U S A 107:10250–10255. doi:10.1073/pnas.0912103107
Giannone G, Mondin M, Grillo-Bosch D et al (2013) Neurexin-1b binding to neuroligin-1 triggers the preferential recruitment of PSD-95 versus gephyrin through tyrosine phosphorylation of neuroligin-1. Cell Rep 3:1996–2007. doi:10.1016/j.celrep.2013.05.013
Graf ER, Zhang X, Jin S-X et al (2004) Neurexins induce differentiation of GABA and glutamate postsynaptic specializations via neuroligins. Cell 119:1013–1026. doi:10.1016/j.cell.2004.11.035
Graf ER, Kang Y, Hauner AM, Craig AM (2006) Structure function and splice site analysis of the synaptogenic activity of the neurexin-1β LNS domain. J Neurosci 26:4256–4265
Guo HB, Johnson H, Randolph M, Pierce M (2009) Regulation of homotypic cell-cell adhesion by branched N-glycosylation of N-cadherin extracellular EC2 and EC3 domains. J Biol Chem 284:34986–34997. doi:10.1074/jbc.M109.060806

Hata Y, Butz S, Südhof TC (1996) CASK: a novel dlg/PSD95 homolog with an N-terminal calmodulin-dependent protein kinase domain identified by interaction with neurexins. J Neurosci 16:2488–2494

Hata K, Polo-Parada L, Landmesser LT (2007) Selective targeting of different neural cell adhesion molecule isoforms during motoneuron myotube synapse formation in culture and the switch from an immature to mature form of synaptic vesicle cycling. J Neurosci 27:14481–14493. doi:10.1523/JNEUROSCI.3847-07.2007

Hatta K, Nose A, Nagafuchi A, Takeichi M (1988) Cloning and expression of cDNA encoding a neural calcium-dependent cell adhesion molecule: its identity in the cadherin gene family. J Cell Biol 106:873–881

Heine M, Thoumine O, Mondin M et al (2008) Activity-independent and subunit-specific recruitment of functional AMPA receptors at neurexin/neuroligin contacts. Proc Natl Acad Sci 105:20947–20952

Heisler FF, Lee HK, Gromova KV et al (2014) GRIP1 interlinks N-cadherin and AMPA receptors at vesicles to promote combined cargo transport into dendrites. Proc Natl Acad Sci U S A 111:5030–5035. doi:10.1073/pnas.1304301111

Hinkle CL, Diestel S, Lieberman J, Maness PF (2006) Metalloprotease-induced ectodomain shedding of neural cell adhesion molecule (NCAM). J Neurobiol 66:1378–1395. doi:10.1002/neu.20257

Hoffman RC, Jennings LL, Tsigelny I et al (2004) Structural characterization of recombinant soluble rat neuroligin 1: mapping of secondary structure and glycosylation by mass spectrometry †. Biochemistry 43:1496–1506. doi:10.1021/bi035278t

Hoover KB, Bryant PJ (2000) The genetics of the protein 4.1 family: organizers of the membrane and cytoskeleton. Curr Opin Cell Biol 12:229–234

Horstkorte R, Schachner M, Magyar JP et al (1993) The fourth immunoglobulin-like domain of NCAM contains a carbohydrate recognition domain for oligomannosidic glycans implicated in association with L1 and neurite outgrowth. J Cell Biol 121:1409–1421

Hu DD, Barbas CF, Smith JW (1996) An allosteric Ca2+ binding site on the beta3-integrins that regulates the dissociation rate for RGD ligands. J Biol Chem 271:21745–21751

Huang Z, Shimazu K, Woo NH et al (2006) Distinct roles of the beta1-class integrins at the developing and the mature hippocampal excitatory synapse. J Neurosci 26:11208–11219. doi:10.1523/JNEUROSCI.3526-06.2006

Hübschmann MV, Skladchikova G, Bock E, Berezin V (2005) Neural cell adhesion molecule function is regulated by metalloproteinase-mediated ectodomain release. J Neurosci Res 80:826–837. doi:10.1002/jnr.20530

Hynes RO (2002) Integrins: bidirectional, allosteric signaling machines. Cell 110:673–687

Ichtchenko K, Hata Y, Nguyen T et al (1995) Neuroligin 1: a splice site-specific ligand for beta-neurexins. Cell 81:435–443

Ichtchenko K, Nguyen T, Südhof TC (1996) Structures, alternative splicing, and neurexin binding of multiple neuroligins. J Biol Chem 271:2676–2682. doi:10.1074/jbc.271.5.2676

Iijima T, Wu K, Witte H et al (2011) SAM68 regulates neuronal activity-dependent alternative splicing of neurexin-1. Cell 147:1601–1614. doi:10.1016/j.cell.2011.11.028

Irie M, Hata Y, Takeuchi M et al (1997) Binding of neuroligins to PSD-95. Science 277:1511–1515

Ito K, Okamoto I, Araki N et al (1999) Calcium influx triggers the sequential proteolysis of extracellular and cytoplasmic domains of E-cadherin, leading to loss of beta-catenin from cell-cell contacts. Oncogene 18:7080–7090. doi:10.1038/sj.onc.1203191

Jensen PH, Soroka V, Thomsen NK et al (1999) Structure and interactions of NCAM modules 1 and 2, basic elements in neural cell adhesion. Nat Struct Biol 6:486–493. doi:10.1038/8292

Johnson KG, Van Vactor D (2003) Receptor protein tyrosine phosphatases in nervous system development. Physiol Rev 83:1–24. doi:10.1152/physrev.00016.2002

Kalus I, Bormann U, Mzoughi M et al (2006) Proteolytic cleavage of the neural cell adhesion molecule by ADAM17/TACE is involved in neurite outgrowth. J Neurochem 98:78–88. doi:10.1111/j.1471-4159.2006.03847.x

Kasper C, Rasmussen H, Kastrup JS et al (2000) Structural basis of cell-cell adhesion by NCAM. Nat Struct Biol 7:389–393. doi:10.1038/75165

Katsamba P, Carroll K, Ahlsen G et al (2009) Linking molecular affinity and cellular specificity in cadherin-mediated adhesion. Proc Natl Acad Sci U S A 106:11594–11599. doi:10.1073/pnas.0905349106

Kim M, Carman CV, Springer TA (2003) Bidirectional transmembrane signaling by cytoplasmic domain separation in integrins. Science (New York, N Y) 301:1720–1725. doi:10.1126/science.1084174

Kim S, Burette A, Chung HS et al (2006) NGL family PSD-95–interacting adhesion molecules regulate excitatory synapse formation. Nat Neurosci 9:1294–1301. doi:10.1038/nn1763

Kiselyov VV, Soroka V, Berezin V, Bock E (2005) Structural biology of NCAM homophilic binding and activation of FGFR. J Neurochem 94:1169–1179. doi:10.1111/j.1471-4159.2005.03284.x

Ko J (2012) The leucine-rich repeat superfamily of synaptic adhesion molecules: LRRTMs and Slitrks. Mol Cells 34:335–340. doi:10.1007/s10059-012-0113-3

Ko J, Kim S, Chung HS et al (2006) SALM synaptic cell adhesion-like molecules regulate the differentiation of excitatory synapses. Neuron 50:233–245. doi:10.1016/j.neuron.2006.04.005

Ko J, Fuccillo MV, Malenka RC, Südhof TC (2009) LRRTM2 functions as a neurexin ligandin promoting excitatory synapse formation. Neuron 64:791–798. doi:10.1016/j.neuron.2009.12.012

Kobe B, Deisenhofer J (1993) Crystal structure of porcine ribonuclease inhibitor, a protein with leucine-rich repeats. Nature 366:751–756. doi:10.1038/366751a0

Kobe B, Deisenhofer J (1995) A structural basis of the interactions between leucine-rich repeats and protein ligands. Nature 374:183–186. doi:10.1038/374183a0

Kobe B, Kajava AV (2001) The leucine-rich repeat as a protein recognition motif. Curr Opin Struct Biol 11:725–732

Kochlamazashvili G, Bukalo O, Senkov O et al (2012) Restoration of synaptic plasticity and learning in young and aged NCAM-deficient mice by enhancing neurotransmission mediated by GluN2A-containing NMDA receptors. J Neurosci 32:2263–2275. doi:10.1523/JNEUROSCI.5103-11.2012

Koehnke J, Katsamba PS, Ahlsen G et al (2010) Splice form dependence of b-neurexin/neuroligin binding interactions. Neuron 67:61–74. doi:10.1016/j.neuron.2010.06.001

Kramár EA, Bernard JA, Gall CM, Lynch G (2003) Integrins modulate fast excitatory transmission at hippocampal synapses. J Biol Chem 278:10722–10730. doi:10.1074/jbc.M210225200

Kwon SK, Woo J, Kim SY et al (2010) Trans-synaptic adhesions between Netrin-G Ligand-3 (NGL-3) and receptor tyrosine phosphatases LAR, Protein-tyrosine Phosphatase (PTP), and PTP via specific domains regulate excitatory synapse formation. J Biol Chem 285:13966–13978. doi:10.1074/jbc.M109.061127

Kwon H-B, Kozorovitskiy Y, Oh W-J et al (2012) Neuroligin-1–dependent competition regulates cortical synaptogenesis and synapse number. Nat Neurosci 15:1667–1674. doi:10.1038/nn.3256

Langer MD, Guo H, Shashikanth N et al (2012) N-glycosylation alters cadherin-mediated intercellular binding kinetics. J Cell Sci 125:2478–2485. doi:10.1242/jcs.101147

Lauren J, Airaksinen MS, Saarma M (2003) A novel gene family encoding leucine-rich repeat transmembrane proteins differentially expressed in the nervous system. Genomics 81:411–421

Lee H, Dean C, Isacoff E (2010) Alternative splicing of neuroligin regulates the rate of presynaptic differentiation. J Neurosci 30:11435–11446. doi:10.1523/JNEUROSCI.2946-10.2010

Leshchyns'ka I, Sytnyk V, Morrow JS, Schachner M (2003) Neural cell adhesion molecule (NCAM) association with PKCbeta2 via betaI spectrin is implicated in NCAM-mediated neurite outgrowth. J Cell Biol 161:625–639. doi:10.1083/jcb.200303020

Leshchyns'ka I, Tanaka MM, Schachner M, Sytnyk V (2011) Immobilized pool of NCAM180 in the postsynaptic membrane is homeostatically replenished by the flux of NCAM180 from extrasynaptic regions. J Biol Chem 286:23397–23406. doi:10.1074/jbc.M111.252098

Levinson JN, El-Husseini A (2005) Building excitatory and inhibitory synapses: balancing neuroligin partnerships. Neuron 48:171–174. doi:10.1016/j.neuron.2005.09.017

Levinson JN, Li R, Kang R et al (2010) Postsynaptic scaffolding molecules modulate the localization of neuroligins. Neuroscience 165:782–793. doi:10.1016/j.neuroscience.2009.11.016

Lin B, Arai AC, Lynch G, Gall CM (2003a) Integrins regulate NMDA receptor-mediated synaptic currents. J Neurophysiol 89:2874–2878. doi:10.1152/jn.00783.2002

Lin JC, Ho W-H, Gurney A, Rosenthal A (2003b) The netrin-G1 ligand NGL-1 promotes the outgrowth of thalamocortical axons. Nat Neurosci 6:1270–1276. doi:10.1038/nn1148

Linhoff MW, LaurEn J, Cassidy RM et al (2009) An unbiased expression screen for synaptogenic proteins identifies the LRRTM protein family as synaptic organizers. Neuron 61:734–749. doi:10.1016/j.neuron.2009.01.017

Luo B-H, Carman CV, Springer TA (2007) Structural basis of integrin regulation and signaling. Annu Rev Immunol 25:619–647. doi:10.1146/annurev.immunol.25.022106.141618

Lüthi A, Laurent JP, Figurov A et al (1994) Hippocampal long-term potentiation and neural cell adhesion molecules L1 and NCAM. Nature 372:777–779. doi:10.1038/372777a0

Maness PF, Schachner M (2007) Neural recognition molecules of the immunoglobulin superfamily: signaling transducers of axon guidance and neuronal migration. Nat Neurosci 10:19–26. doi:10.1038/nn1827

McGeachie AB, Cingolani LA, Goda Y (2011) Neurosci Res Neurosci Res 70:24–29. doi:10.1016/j.neures.2011.02.006

McGeachie AB, Skrzypiec AE, Cingolani LA et al (2012) β3 integrin is dispensable for conditioned fear and hebbian forms of plasticity in the hippocampus. Eur J Neurosci 36:2461–2469. doi:10.1111/j.1460-9568.2012.08163.x

Milev P, Friedlander DR, Sakurai T et al (1994) Interactions of the chondroitin sulfate proteoglycan phosphacan, the extracellular domain of a receptor-type protein tyrosine phosphatase, with neurons, glia, and neural cell adhesion molecules. J Cell Biol 127:1703–1715

Milev P, Maurel P, Häring M et al (1996) TAG-1/axonin-1 is a high-affinity ligand of neurocan, phosphacan/protein-tyrosine phosphatase-zeta/beta, and N-CAM. J Biol Chem 271:15716–15723

Missler M, Fernández-Chacón R, Südhof TC (1998) The making of neurexins. J Neurochem 71:1339–1347

Missler M, Zhang W, Rohlmann A et al (2003) α-Neurexins couple Ca2+ channels to synaptic vesicle exocytosis. Nature 423:939–948

Miyatani S, Shimamura K, Hatta M et al (1989) Neural cadherin: role in selective cell-cell adhesion. Science 245:631–635

Monea S, Jordan BA, Srivastava S et al (2006) Membrane localization of membrane type 5 matrix metalloproteinase by AMPA receptor binding protein and cleavage of cadherins. J Neurosci 26:2300–2312. doi:10.1523/JNEUROSCI.3521-05.2006

Montgomery AM, Becker JC, Siu C-H et al (1996) Human neural cell adhesion molecule L1 and rat homologue NILE are ligands for integrin alpha v beta 3. J Cell Biol 132:475–485

Moran N, Bock E (1988) Characterization of the kinetics of neural cell adhesion molecule homophilic binding. FEBS Lett 242:121–124

Mould AP, Barton SJ, Askari JA et al (2003) Role of ADMIDAS cation-binding site in ligand recognition by integrin 5 1. J Biol Chem 278:51622–51629. doi:10.1074/jbc.M306655200

Mukherjee K, Sharma M, Urlaub H et al (2008) CASK functions as a Mg2+-independent neurexin kinase. Cell 133:328–339. doi:10.1016/j.cell.2008.02.036

Muller D, Wang C, Skibo G et al (1996) PSA-NCAM is required for activity-induced synaptic plasticity. Neuron 17:413–422

Murase S, Mosser E, Schuman EM (2002) Depolarization drives beta-catenin into neuronal spines promoting changes in synaptic structure and function. Neuron 35:91–105
Nagar B, Overduin M, Ikura M, Rini JM (1996) Structural basis of calcium-induced E-cadherin rigidification and dimerization. Nature 380:360–364. doi:10.1038/380360a0
Nakashiba T, Ikeda T, Nishimura S et al (2000) Netrin-G1: a novel glycosyl phosphatidylinositol-linked mammalian netrin that is functionally divergent from classical netrins. J Neurosci 20:6540–6550
Nermut MV, Green NM, Eason P et al (1988) Electron microscopy and structural model of human fibronectin receptor. EMBO J 7:4093–4099
Nguyen T, Südhof TC (1997) Binding properties of neuroligin 1 and neurexin 1 reveal function as heterophilic cell adhesion molecules. J Biol Chem 272:26032–26039. doi:10.1074/jbc.272.41.26032
Niessen CM, Gumbiner BM (2002) Cadherin-mediated cell sorting not determined by binding or adhesion specificity. J Cell Biol 156:389–400. doi:10.1083/jcb.200108040
Niessen CM, Leckband D, Yap AS (2011) Tissue organization by Cadherin Adhesion molecules: dynamic molecular and cellular mechanisms of morphogenetic regulation. Physiol Rev 91:691–731. doi:10.1152/physrev.00004.2010
Nishimura-Akiyoshi S, Niimi K, Nakashiba T, Itohara S (2007) Axonal netrin-Gs transneuronally determine lamina-specific subdendritic segments. Proc Natl Acad Sci 104:14801–14806. doi:10.1073/pnas.0706919104
Nollet F, Kools P, van Roy F (2000) Phylogenetic analysis of the cadherin superfamily allows identification of six major subfamilies besides several solitary members. J Mol Biol 299:551–572. doi:10.1006/jmbi.2000.3777
Nose A, Tsuji T, Takeichi M (1990) Location in specificity determining sites in cadherin cell adhesion molecules. Cell 61:147–155
O'Toole TE, Katagiri Y, Faull RJ et al (1994) Integrin cytoplasmic domains mediate inside-out signal transduction. J Cell Biol 124:1047–1059
Okamura K, Tanaka H, Yagita Y et al (2004) Cadherin activity is required for activity-induced spine remodeling. J Cell Biol 167:961–972. doi:10.1083/jcb.200406030
Okuda T, Yu LMY, Cingolani LA et al (2007) Beta-catenin regulates excitatory postsynaptic strength at hippocampal synapses. Proc Natl Acad Sci 104:13479–13484. doi:10.1073/pnas.0702334104
Paratcha G, Ledda F, Ibáñez CF (2003) The neural cell adhesion molecule NCAM is an alternative signaling receptor for GDNF family ligands. Cell 113:867–879
Peixoto RT, Kunz PA, Kwon H et al (2012) Transsynaptic signaling by activity-dependent cleavage of neuroligin-1. Neuron 76:396–409. doi:10.1016/j.neuron.2012.07.006
Perlson E, Hendricks AG, Lazarus JE et al (2013) Dynein interacts with the Neural Cell Adhesion Molecule (NCAM180) to tether dynamic microtubules and maintain synaptic density in cortical neurons. J Biol Chem 288:27812–27824. doi:10.1074/jbc.M113.465088
Pertz O, Bozic D, Koch AW et al (1999) A new crystal structure, Ca2+ dependence and mutational analysis reveal molecular details of E-cadherin homoassociation. EMBO J 18:1738–1747
Pillai-Nair N, Panicker AK, Rodriguiz RM et al (2005) Neural cell adhesion molecule-secreting transgenic mice display abnormalities in GABAergic interneurons and alterations in behavior. J Neurosci 25:4659–4671. doi:10.1523/JNEUROSCI.0565-05.2005
Pinkstaff JK, Detterich J, Lynch G, Gall C (1999) Integrin subunit gene expression is regionally differentiated in adult brain. J Neurosci 19:1541–1556
Polo-Parada L, Bose CM, Landmesser LT (2001) Alterations in transmission, vesicle dynamics, and transmitter release machinery at NCAM-deficient neuromuscular junctions. Neuron 32:815–828
Polo-Parada L, Bose CM, Plattner F, Landmesser LT (2004) Distinct roles of different Neural Cell Adhesion Molecule (NCAM) isoforms in synaptic maturation revealed by analysis of NCAM 180 kDa isoform-deficient mice. J Neurosci 24:1852–1864. doi:10.1523/JNEUROSCI.4406-03.2004

Polo-Parada L, Plattner F, Bose CM, Landmesser LT (2005) NCAM 180 acting via a conserved C-terminal domain and MLCK is essential for effective transmission with repetitive stimulation. Neuron 46:917–931. doi:10.1016/j.neuron.2005.05.018

Poulopoulos A, Aramuni G, Meyer G et al (2009) Neuroligin 2 drives postsynaptic assembly at perisomatic inhibitory synapses through gephyrin and collybistin. Neuron 63:628–642. doi:10.1016/j.neuron.2009.08.023

Pozo K, Cingolani LA, Bassani S et al (2012) β3 integrin interacts directly with GluA2 AMPA receptor subunit and regulates AMPA receptor expression in hippocampal neurons. Proc Natl Acad Sci 109:1323–1328. doi:10.1073/pnas.1113736109/-/DCSupplemental

Prakasam AK, Maruthamuthu V, Leckband DE (2006) Similarities between heterophilic and homophilic cadherin adhesion. Proc Natl Acad Sci 103:15434–15439. doi:10.1073/pnas.0606701103

Prange O, Wong TP, Gerrow K et al (2004) A balance between excitatory and inhibitory synapses is controlled by PSD-95 and neuroligin. Proc Natl Acad Sci 101:13915–13920. doi:10.1073/pnas.0405939101

Probstmeier R, Fahrig T, Spiess E, Schachner M (1992) Interactions of the neural cell adhesion molecule and the myelin-associated glycoprotein with collagen type I: involvement in fibrillogenesis. J Cell Biol 116:1063–1070

Puchkov D, Leshchyns'ka I, Nikonenko AG et al (2011) NCAM/spectrin complex disassembly results in PSD perforation and postsynaptic endocytic zone formation. Cereb Cortex 21:2217–2232. doi:10.1093/cercor/bhq283

Rafuse VF, Polo-Parada L, Landmesser LT (2000) Structural and functional alterations of neuromuscular junctions in NCAM-deficient mice. J Neurosci 20:6529–6539

Ranheim TS, Edelman GM, Cunningham BA (1996) Homophilic adhesion mediated by the neural cell adhesion molecule involves multiple immunoglobulin domains. Proc Natl Acad Sci 93:4071–4075

Rao Y, Wu XF, Yip P et al (1993) Structural characterization of a homophilic binding site in the neural cell adhesion molecule. J Biol Chem 268:20630–20638

Rao Y, Zhao X, Siu CH (1994) Mechanism of homophilic binding mediated by the neural cell adhesion molecule NCAM. Evidence for isologous interaction. J Biol Chem 269:27540–27548

Reiss K, Maretzky T, Ludwig A et al (2005) ADAM10 cleavage of N-cadherin and regulation of cell–cell adhesion and β-catenin nuclear signalling. EMBO J 24:742–752

Robbins EM, Krupp AJ, de Arce KP et al (2010) SynCAM 1 adhesion dynamically regulates synapse number and impacts plasticity and learning. Neuron 68:894–906. doi:10.1016/j.neuron.2010.11.003

Rollenhagen M, Kuckuck S, Ulm C et al (2012) Polysialylation of the synaptic cell adhesion molecule 1 (SynCAM 1) depends exclusively on the polysialyltransferase ST8SiaII in vivo. J Biol Chem 287:35170–35180. doi:10.1074/jbc.M112.375642

Ruppert M, Aigner S, Hubbe M et al (1995) The L1 adhesion molecule is a cellular ligand for VLA-5. J Cell Biol 131:1881–1891

Rutishauser U, Hoffman S, Edelman GM (1982) Binding properties of a cell adhesion molecule from neural tissue. Proc Natl Acad Sci U S A 79:685–689

Rutishauser U, Watanabe M, Silver J et al (1985) Specific alteration of NCAM-mediated cell adhesion by an endoneuraminidase. J Cell Biol 101:1842–1849

Saglietti L, Dequidt C, Kamieniarz K et al (2007) Extracellular interactions between GluR2 and N-cadherin in spine regulation. Neuron 54:461–477. doi:10.1016/j.neuron.2007.04.012

Sandi C (2004) Stress, cognitive impairment and cell adhesion molecules. Nat Rev Neurosci 5:917–930. doi:10.1038/nrn1555

Sara Y, Biederer T, Atasoy D et al (2005) Selective capability of SynCAM and neuroligin for functional synapse assembly. J Neurosci 25:260–270. doi:10.1523/JNEUROSCI.3165-04.2005

Scheiffele P, Fan J, Choih J et al (2000) Neuroligin expressed in nonneuronal cells triggers presynaptic development in contacting axons. Cell 101:657–669

Schuster T, Krug M, Stalder M et al (2001) Immunoelectron microscopic localization of the neural recognition molecules L1, NCAM, and its isoform NCAM180, the NCAM-associated polysialic acid, beta1 integrin and the extracellular matrix molecule tenascin-R in synapses of the adult rat hippocampus. J Neurobiol 49:142–158

Seabold GK, Wang PY, Chang K et al (2008) The SALM family of adhesion-like molecules forms heteromeric and homomeric complexes. J Biol Chem 283:8395–8405. doi:10.1074/jbc.M709456200

Shan WS, Tanaka H, Phillips GR et al (2000) Functional cis-heterodimers of N- and R-cadherins. J Cell Biol 148:579–590

Shan W, Yagita Y, Wang Z et al (2004) The minimal essential unit for cadherin-mediated intercellular adhesion comprises extracellular domains 1 and 2. J Biol Chem 279:55914–55923. doi:10.1074/jbc.M407827200

Shapiro L, Fannon AM, Kwong PD et al (1995) Structural basis of cell-cell adhesion by cadherins. Nature 374:327–337. doi:10.1038/374327a0

Sheng M, Sala C (2001) PDZ domains and the organization of supramolecular complexes. Annu Rev Neurosci 24:1–29. doi:10.1146/annurev.neuro.24.1.1

Shetty A, Sytnyk V, Leshchyns'ka I et al (2013) The neural cell adhesion molecule promotes maturation of the presynaptic endocytotic machinery by switching synaptic vesicle recycling from adaptor protein 3 (AP-3)- to AP-2-dependent mechanisms. J Neurosci 33:16828–16845. doi:10.1523/JNEUROSCI.2192-13.2013

Shi Y, Ethell IM (2006) Integrins control dendritic spine plasticity in hippocampal neurons through NMDA receptor and Ca2+/calmodulin-dependent protein kinase II-mediated actin reorganization. J Neurosci 26:1813–1822. doi:10.1523/JNEUROSCI.4091-05.2006

Shipman SL, Nicoll RA (2012a) Dimerization of postsynaptic neuroligin drives synaptic assembly via transsynaptic clustering of neurexin. Proc Natl Acad Sci 109:19432–19437. doi:10.1073/pnas.1217633109/-/DCSupplemental

Shipman SL, Nicoll RA (2012b) A subtype-specific function for the extracellular domain of neuroligin 1 in hippocampal LTP. Neuron 76:309–316. doi:10.1016/j.neuron.2012.07.024

Shipman SL, Schnell E, Hirai T et al (2011) Functional dependence of neuroligin on a new non-PDZ intracellular domain. Nat Publ Group 14:718–726. doi:10.1038/nn.2825

Siddiqui TJ, Pancaroglu R, Kang Y et al (2010) LRRTMs and neuroligins bind neurexins with a differential code to cooperate in glutamate synapse development. J Neurosci 30:7495–7506. doi:10.1523/JNEUROSCI.0470-10.2010

Siddiqui TJ, Tari PK, Connor SA et al (2013) An LRRTM4-HSPG complex mediates excitatory synapse developmenton dentate gyrus granule cells. Neuron 79:680–695. doi:10.1016/j.neuron.2013.06.029

Sivasankar S, Gumbiner B, Leckband D (2001) Direct measurements of multiple adhesive alignments and unbinding trajectories between cadherin extracellular domains. Biophys J 80:1758–1768. doi:10.1016/S0006-3495(01)76146-2

Small SJ, Akeson R (1990) Expression of the unique NCAM VASE exon is independently regulated in distinct tissues during development. J Cell Biol 111:2089–2096

Small SJ, Haines SL, Akeson RA (1988) Polypeptide variation in an N-CAM extracellular immunoglobulin-like fold is developmentally regulated through alternative splicing. Neuron 1:1007–1017

Smith DK, Xue H (1997) Sequence profiles of immunoglobulin and immunoglobulin-like domains. J Mol Biol 274:530–545. doi:10.1006/jmbi.1997.1432

Soler-Llavina GJ, Fuccillo MV, Ko J et al (2011) The neurexin ligands, neuroligins and leucine-rich repeat transmembrane proteins, perform convergent and divergent synaptic functions in vivo. Proc Natl Acad Sci U S A 108:16502–16509. doi:10.1073/pnas.1114028108

Soler-Llavina GJ, Arstikaitis P, Morishita W et al (2013) Leucine-rich repeat transmembrane proteins Are essential for maintenance of long-term potentiation. Neuron 79:439–446. doi:10.1016/j.neuron.2013.06.007

Song JY, Ichtchenko K, Südhof TC, Brose N (1999) Neuroligin 1 is a postsynaptic cell-adhesion molecule of excitatory synapses. Proc Natl Acad Sci 96:1100–1105

Soroka V, Kolkova K, Kastrup JS et al (2003) Structure and interactions of NCAM Ig1-2-3 suggest a novel zipper mechanism for homophilic adhesion. Structure 11:1291–1301

Soto F, Watkins KL, Johnson RE et al (2013) NGL-2 regulates pathway-specific neurite growth and lamination, synapse formation, and signal transmission in the retina. J Neurosci 33:11949–11959. doi:10.1523/JNEUROSCI.1521-13.2013

Sousa I, Clark TG, Holt R et al (2010) Polymorphisms in leucine-rich repeat genes are associated with autism spectrum disorder susceptibility in populations of European ancestry. Mol Autism 1:7. doi:10.1093/brain/awh287

Stagi M, Fogel AI, Biederer T (2010) SynCAM 1 participates in axo-dendritic contact assembly and shapes neuronal growth cones. Proc Natl Acad Sci U S A 107:7568–7573. doi:10.1073/pnas.0911798107

Stan A, Pielarski KN, Brigadski T et al (2010) Essential cooperation of N-cadherin and neuroligin-1 in the transsynaptic control of vesicle accumulation. Proc Natl Acad Sci 107:11116–11121. doi:10.1073/pnas.0914233107

Staubli U, Vanderklish P, Lynch G (1990) An inhibitor of integrin receptors blocks long-term potentiation. Behav Neural Biol 53:1–5

Stäubli U, Chun D, Lynch G (1998) Time-dependent reversal of long-term potentiation by an integrin antagonist. J Neurosci 18:3460–3469

Storms SD, Rutishauser U (1998) A role for polysialic acid in neural cell adhesion molecule heterophilic binding to proteoglycans. J Biol Chem 273:27124–27129

Südhof TC (2008) Neuroligins and neurexins link synaptic function to cognitive disease. Nature 455:903–911. doi:10.1038/nature07456

Suzuki K, Hayashi Y, Nakahara S et al (2012) Activity-dependent proteolytic cleavage of neuroligin-1. Neuron 76:410–422. doi:10.1016/j.neuron.2012.10.003

Sytnyk V, Leshchyns'ka I, Delling M et al (2002) Neural cell adhesion molecule promotes accumulation of TGN organelles at sites of neuron-to-neuron contacts. J Cell Biol 159:649–661. doi:10.1083/jcb.200205098

Sytnyk V, Leshchyns'ka I, Nikonenko AG, Schachner M (2006) NCAM promotes assembly and activity-dependent remodeling of the postsynaptic signaling complex. J Cell Biol 174:1071–1085. doi:10.1083/jcb.200604145

Tai C-Y, Mysore SP, Chiu C, Schuman EM (2007) Activity-regulated N-cadherin endocytosis. Neuron 54:771–785. doi:10.1016/j.neuron.2007.05.013

Takagi J, Petre BM, Walz T, Springer TA (2002) Global conformational rearrangements in integrin extracellular domains in outside-in and inside-out signaling. Cell 110:599–611

Takahashi H, Arstikaitis P, Prasad T et al (2011) Postsynaptic TrkC and presynaptic PTPs function as a bidirectional excitatory synaptic organizing complex. Neuron 69:287–303. doi:10.1016/j.neuron.2010.12.024

Takahashi H, Katayama K-I, Sohya K et al (2012) Selective control of inhibitory synapse development by Slitrk3-PTPd trans-synaptic interaction. Nat Publ Group 15:389–398. doi:10.1038/nn.3040

Takashima N, Odaka YS, Sakoori K et al (2011) Impaired cognitive function and altered hippocampal synapse morphology in mice lacking Lrrtm1, a gene associated with schizophrenia. PLoS One 6:e22716. doi:10.1371/journal.pone.0022716.s001

Tanaka H, Shan W, Phillips GR et al (2000) Molecular modification of N-cadherin in response to synaptic activity. Neuron 25:93–107

Tang L, Hung CP, Schuman EM (1998) A role for the cadherin family of cell adhesion molecules in hippocampal long-term potentiation. Neuron 20:1165–1175

Togashi H, Abe K, Mizoguchi A et al (2002) Cadherin regulates dendritic spine morphogenesis. Neuron 35:77–89

Tsuboi S (2002) Calcium integrin-binding protein activates platelet integrin IIb 3. J Biol Chem 277:1919–1923. doi:10.1074/jbc.M110643200

Uchida N, Honjo Y, Johnson KR et al (1996) The catenin/cadherin adhesion system is localized in synaptic junctions bordering transmitter release zones. J Cell Biol 135:767–779
Ullrich B, Ushkaryov YA, Südhof TC (1995) Cartography of neurexins: more than 1000 isoforms generated by alternative splicing and expressed in distinct subsets of neurons. Neuron 14:497–507
Ushkaryov YA, Petrenko AG, Geppert M, Südhof TC (1992) Neurexins: synaptic cell surface proteins related to the alpha-latrotoxin receptor and laminin. Science 257:50–56
Varoqueaux F, Aramuni G, Rawson RL et al (2006) Neuroligins determine synapse maturation and function. Neuron 51:741–754. doi:10.1016/j.neuron.2006.09.003
Vinogradova O, Velyvis A, Velyviene A et al (2002) A structural mechanism of integrin alpha(IIb) beta(3) "inside-out" activation as regulated by its cytoplasmic face. Cell 110:587–597
Vitureira N, Letellier M, White IJ, Goda Y (2011) Differential control of presynaptic efficacy by postsynaptic N-cadherin and β-catenin. Nat Neurosci 15:81–89. doi:10.1038/nn.2995
Vutskits L, Djebbara-Hannas Z, Zhang H et al (2001) PSA-NCAM modulates BDNF-dependent survival and differentiation of cortical neurons. Eur J Neurosci 13:1391–1402
Walsh FS, Doherty P (1997) Neural cell adhesion molecules of the immunoglobulin superfamily: role in axon growth and guidance. Annu Rev Cell Dev Biol 13:425–456. doi:10.1146/annurev.cellbio.13.1.425
Wang CY (2006) A novel family of adhesion-like molecules that interacts with the NMDA receptor. J Neurosci 26:2174–2183. doi:10.1523/JNEUROSCI.3799-05.2006
Wang J-H (2013) The sequence signature of an Ig-fold. Protein Cell 4:569–572. doi:10.1007/s13238-013-3903-2
Wittenmayer N, Körber C, Liu H et al (2009) Postsynaptic neuroligin 1 regulates presynaptic maturation. Proc Natl Acad Sci U S A 106:13564–13569. doi:10.1073/pnas.0905819106
Woo J, Kwon S-K, Choi S et al (2009) Trans-synaptic adhesion between NGL-3 and LAR regulates the formation of excitatory synapses. Nat Neurosci 12:428–437. doi:10.1038/nn.2279
Xiao T, Takagi J, Coller BS et al (2004) Structural basis for allostery in integrins and binding to fibrinogen-mimetic therapeutics. Nature 432:59–67
Xiong J-P, Stehle T, Diefenbach B et al (2001) Crystal structure of the extracellular segment of integrin αVβ3. Science 294:339–345
Xiong J-P, Stehle T, Zhang R et al (2002) Crystal structure of the extracellular segment of integrin alpha Vbeta3 in complex with an Arg-Gly-Asp ligand. Science (New York, N Y) 296:151–155. doi:10.1126/science.1069040
Yap AS, Niessen CM, Gumbiner BM (1998) The juxtamembrane region of the cadherin cytoplasmic tail supports lateral clustering, adhesive strengthening, and interaction with p120ctn. J Cell Biol 141:779–789
Yip PM, Zhao X, Montgomery AM, Siu CH (1998) The Arg-Gly-Asp motif in the cell adhesion molecule L1 promotes neurite outgrowth via interaction with the alphavbeta3 integrin. Mol Biol Cell 9:277–290
Zhao X, Yip PM, Siu CH (1998) Identification of a homophilic binding site in immunoglobulin-like domain 2 of the cell adhesion molecule L1. J Neurochem 71:960–971
Zhou H, Fuks A, Alcaraz G et al (1993) Homophilic adhesion between Ig superfamily carcinoembryonic antigen molecules involves double reciprocal bonds. J Cell Biol 122:951–960
Zhu B, Chappuis-Flament S, Wong E et al (2003) Functional analysis of the structural basis of homophilic cadherin adhesion. Biophys J 84:4033–4042. doi:10.1016/S0006-3495(03)75129-7

Chapter 18
Excitatory and Inhibitory Synaptic Placement and Functional Implications

Katherine L. Villa and Elly Nedivi

Abstract Synaptic transmission between neurons is the basic unit of communication in neural circuits. The relative number and distribution of excitatory and inhibitory synaptic inputs across individual dendrites and neurons are the hardware of local dendritic and cellular computations. In this chapter, we discuss the structural and functional observations that have guided the understanding of excitatory and inhibitory synaptic organization across the neuronal arbor, the subcellular targeting properties of different neuronal subtypes, and the effects of synaptic placement on local integration within dendritic segments. We focus primarily on the adult mammalian cortex and hippocampus, where excitatory and inhibitory cell types, their connectivity, and its functional implications have been best characterized.

Keywords Synaptic placement • Excitatory-inhibitory balance • Dendritic integration

18.1 Introduction

The neurotransmitter synthesized and released at the synapse is the basis for the classification of neurons as excitatory or inhibitory. Glutamatergic neurons, releasing glutamate, excite postsynaptic cells, while GABAergic neurons, releasing GABA (γ-aminobutyric acid), inhibit them. The delicate balance between

K.L. Villa
Picower Institute for Learning and Memory, Massachusetts Institute of Technology, Cambridge, MA 02139, USA

Department of Biology, Massachusetts Institute of Technology, Cambridge, MA 02139, USA

E. Nedivi (✉)
Picower Institute for Learning and Memory, Massachusetts Institute of Technology, Cambridge, MA 02139, USA

Department of Biology, Massachusetts Institute of Technology, Cambridge, MA 02139, USA

Department of Brain and Cognitive Sciences, Massachusetts Institute of Technology, Cambridge, MA 02139, USA
e-mail: nedivi@mit.edu

K. Emoto et al. (eds.), *Dendrites*, DOI 10.1007/978-4-431-56050-0_18

excitation and inhibition is essential for precise nervous system function and plasticity, as its perturbation has been associated with disorders ranging from epilepsy (Mohler et al. 2004), to autism (Rubenstein and Merzenich 2003), and mental retardation (Dani et al. 2005; Kleschevnikov et al. 2004), as well as neuropsychiatric disorders such as schizophrenia (Lewis et al. 2005). A key to this balance is the precise coordination of excitatory and inhibitory synaptic activity at specific cellular locales. Yet, our knowledge regarding global distribution, as well as the local placement, of excitatory and inhibitory synaptic inputs across different neuronal subtypes is still quite poor. Below we summarize the current state of the field and the evolution of our understanding to date.

18.2 Organization of Synapses Across the Dendritic Arbor

18.2.1 First Views Through Golgi Staining and Electron Microscopy

18.2.1.1 Pyramidal Neurons

In the late nineteenth century, development of the Golgi stain, a silver stain that sparsely but intensely labels neurons, enabled the first clear discrimination of individual neuronal morphology (Fig. 18.1A–C). In his meticulously detailed drawings, Ramon y Cajal characterized the elaborate dendritic arbors and axonal

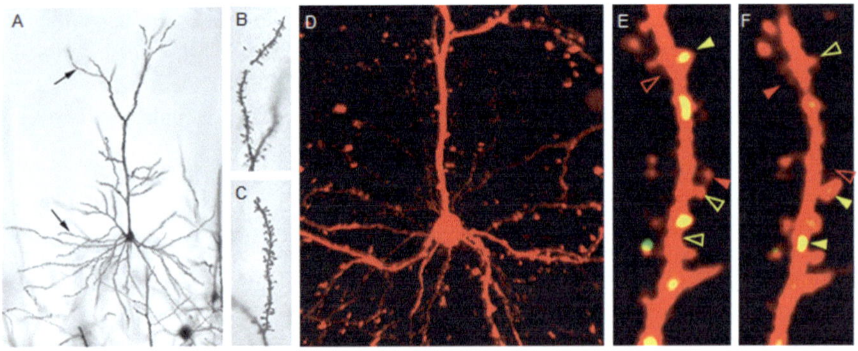

Fig. 18.1 Visualization of L2/3 pyramidal neurons. (**A**) Golgi-stained pyramidal neuron (Image kindly provided by Dr. Terry Robinson, University of Michigan). (**B**, **C**) High-magnification views of the two dendrites marked by *arrows* in panel **A**. Note dendritic spines. (**D**) 3D volume projection of a pseudo-colored pyramidal neuron imaged in vivo using two-photon microscopy. (**E**) An individual dendrite expressing YFP (*red*) and teal-gephyrin (*green*) to label inhibitory synapses. (**F**) The same dendrite imaged 8 days later. *Yellow arrows* indicate dynamic inhibitory synapses and *red arrows* indicate dynamic spines. *Filled arrows* show when structures are present and *open arrows* indicate their absence

processes of neurons with a wide range of morphologies. One obvious distinction was that some neurons have smooth dendrites, while others have many spiny protrusions. In general, most neurons with spiny dendrites were later revealed to be glutamatergic and excitatory, while neurons with smooth dendrites for the most part release GABA and are inhibitory (Gabbott and Somogyi 1986; Kubota 2014; Morishima et al. 2011; Thomson and Deuchars 1997). Elegant electron microscopy (EM) studies on Golgi-stained cells observed that each spine contains a synapse characterized by round presynaptic vesicles and a robust postsynaptic density (Hersch and White 1981; LeVay 1973; Parnavelas et al. 1977). These asymmetric synapses, classified as type 1 synapses, are innervated by axons of glutamatergic neurons (Baude et al. 1993). While there are a few reports of specific cell regions on particular types of neurons where type 1 synapses are localized directly on the dendritic shaft (Megias et al. 2001; Parnavelas et al. 1977), this is generally not the case. Given that the vast majority of spines, with the exception of some very thin spines (about 2–4 % of total cortical spines) (Arellano et al. 2007; Hersch and White 1981; White and Rock 1980), have a single type 1 excitatory synapse (Harris et al. 1992; LeVay 1973), it was reasonable to assume that each spine can serve as a readily identifiable structural surrogate for an excitatory synapse. Thus, for spiny pyramidal neurons, dendritic spine distributions as seen by Golgi stain or by filling cells with a fluorescent dye could inform as to the placement of excitatory synapses across the dendritic arbor (Elston and Rosa 1997; Larkman 1991; Trommald et al. 1995).

There is a great variety in the density of spines between cells and between different branches on the same cell. To give a general sense of spine distributions, spine density is low within 40 μm of the cell body, reaching a maximum density in a region 40–130 μm from the soma, and then gradually decreasing toward a dendrite's distal tips (Megias et al. 2001). On the distal branches of pyramidal cells, spine density can range from 0 to 70 spines per 10 μm (Megias et al. 2001) with the highest spine density often found on the thickest dendrites, usually the primary apical dendrite. The total number of spines on spiny excitatory neurons typically ranges from 5,000 to 35,000 spines per cell, ultimately depending on the total dendritic length, with L5 pyramidal cells generally having longer total dendrite length and therefore more synapses than L2/3 pyramidal cells (Larkman 1991).

Inhibitory synapses innervated by the axons of GABAergic neurons, classified by EM as type 2, or symmetric synapses, are typified by a symmetric synaptic cleft, due to a minimal postsynaptic density, as well as flattened presynaptic vesicles (Davis and Sterling 1979). Unfortunately, since type 2 inhibitory synapses lack a morphological surrogate, their distribution patterns cannot be discriminated by Golgi staining alone. Laborious EM reconstructions of relatively small neuronal regions have shown excitatory to inhibitory synaptic ratios along pyramidal dendrites ranging from 6.5–12.5 to 1 (Davis and Sterling 1979). As previously mentioned, the cell body and proximal dendrites of excitatory neurons typically lack spines. However, they are often densely innervated by type 2 synapses. In the aspiny region of the proximal dendrites, inhibitory synapse density can be as high as 17 synapses per 10 μm (Megias et al. 2001). EM studies also show that on excitatory dendrites, the majority of type 2 inhibitory synapses are located on the dendritic shaft at a density of about 3 per 10 μm (Hersch and White 1981). Inhibitory synapses can also be located on dendritic spines adjacent to excitatory synapses (Parnavelas et al. 1977), as well as on the axon initial segment (Hersch and White 1981; Westrum 1966). Inhibitory axons from different inhibitory neuron subtypes specifically target discrete regions of their postsynaptic pyramidal cell partners (we discuss this in greater detail further in the chapter).

18.2.1.2 Non-pyramidal Neurons

The dendrites of inhibitory neurons, in general, do not contain spines. However, a small subset of inhibitory neurons have dendritic spines (Azouz et al. 1997; Feldman and Peters 1978), with densities that range from 0.3 to 7 spines per 10 μm (Kawaguchi et al. 2006; Keck et al. 2011). Like spiny pyramidal cells, the cell body and most proximal dendrites of spiny interneurons within 30 μm of the soma lack spines (Kawaguchi et al. 2006). There are several distinct subtypes of spiny inhibitory neurons, but Martinotti cells are the ones with the highest spine density with about three to seven spines per 10 μm, approximately one fourth of the density on pyramidal cells (Gulyas et al. 1992; Kawaguchi et al. 2006). Martinotti cells also have longer spines than other spiny interneuron subtypes and more multiheaded spines (Kawaguchi et al. 2006). Immunohistochemistry experiments show that the majority of spines on these interneurons colocalize with Vglut1 but not VGAT, indicating that they mostly harbor excitatory synapses (Keck et al. 2011). Unlike dendritic spines on excitatory neurons, which typically contain only one excitatory synapse, the spines of some spiny inhibitory neurons in the hippocampus can contain up to six distinct excitatory synapses (Gulyas et al. 1992). The proportion of excitatory synapses located along the shaft of spiny inhibitory dendrites has not been established. We also know little about the distribution of inhibitory synapses on spiny interneurons (Gulyas et al. 1992).

Aspiny interneurons receive both excitatory and inhibitory inputs onto their soma and proximal dendrites (Davis and Sterling 1979), with a higher density along their distal dendrites (Parnavelas et al. 1977). Similarly to inhibitory synapses on pyramidal cells, both excitatory synapses and inhibitory synapses onto dendrites of aspiny inhibitory neurons cannot be visualized through a morphological surrogate and could initially be examined only by EM on relatively small arbor segments or by immunohistochemistry. EM reconstructions of isolated branches from different inhibitory neuron types provide anecdotal evidence that parvalbumin (PV)-positive cells have the highest density of synapses with a density of 30 excitatory and 2 inhibitory synapses per 10 μm of dendritic length for branches sampled (Gulyas et al. 1999). Inhibitory synapses are also relatively rare on calretinin (CR)-positive spiny interneurons (Gulyas et al. 1992) which have a lower density of seven excitatory synapses and two inhibitory synapses per 10 μm (Gulyas et al. 1999).

18.2.2 Synaptic Visualization by New In Vivo *Fluorescent Labeling Methods*

Golgi staining gave us a first view of dendritic spine distributions, and EM studies were first to shed light on the fundamental layout of excitatory and inhibitory synaptic distributions on different neuronal types. Yet, both methods have inherent limitations. As a cell fill, Golgi stain can at best identify spines on cells with spiney dendrites, while all inhibitory synapses as well as excitatory synapses on aspiny dendrites remain invisible. EM is limited by the difficulty of reconstructing large dendritic segments. EM reconstruction of an entire cell would require a heroic effort, one rarely attempted (see White and Rock 1980 for a reconstruction of an entire spiny stellate neuron). Both Golgi staining and EM reconstructions necessitate tissue fixation and cannot be used for visualizing structural dynamics. Thus, our initial view of synaptic distributions was constrained by the limitations of the Golgi and EM methods.

Recently, imaging of fluorescently labeled cells in vivo has provided a new view not only of dendritic spine distributions but also of excitatory and inhibitory synapses as well as the dynamics of these structures (Fig. 18.1D–F). First studies on neurons fluorescently labeled with green fluorescent protein (GFP) showed that in adult animals dendrites of pyramidal neurons are very stable (Grutzendler et al. 2002; Mizrahi and Katz 2003; Trachtenberg et al. 2002), but dendritic spines are highly dynamic (Trachtenberg et al. 2002), implying a capacity for synaptic removal and addition. Spine dynamics of excitatory as well as spiny inhibitory neurons can be further increased upon sensory deprivation (Hofer et al. 2009; Holtmaat et al. 2006; Keck et al. 2008; Trachtenberg et al. 2002). Dynamics differ dependent on deprivation protocol and cortical lamina, consistent with the view that they reflect specific circuit alterations.

In contrast to the stability of excitatory dendritic branches, inhibitory dendrites are capable of growth and retraction in vivo (Chen et al. 2011a, b; Lee et al. 2006, 2008), and their dynamics are influenced by sensory manipulations (Chen et al. 2011b). The boutons of inhibitory neurons are also capable of remodeling, and their dynamics also increase in response to sensory deprivation (Chen et al. 2011b; Keck et al. 2011; Pieraut et al. 2014; Schuemann et al. 2013; Wierenga et al. 2008). Thus, synaptic distributions are not necessarily rigid. Rather, both excitatory and inhibitory synapses are dynamic structures, and their remodeling potentially underlies functional plasticity.

More recently, the expression of fluorescent proteins fused to postsynaptic scaffolding molecules has enabled direct synaptic visualization in vivo. Expression of fluorescent proteins fused either to PSD95, as a postsynaptic marker of excitatory synapses, or gephyrin, as a postsynaptic marker of inhibitory synapses, together with a separate cell fill to outline their location on the arbor has allowed for the first time a direct assessment of synaptic distribution patterns on cortical pyramidal cells, as well as synapse dynamics at different cellular locales (Cane et al. 2014; Chen et al. 2012; Isshiki et al. 2014; van Versendaal et al. 2012; Villa et al. 2016).

A critical aspect of this approach is the sparse labeling of only a subset of neurons so that accurate counting of synaptic distributions is feasible, in a way that is not possible with antibody staining. PSD95-GFP labeling in vivo shows that most stable spines have large PSD95 puncta, but a small population are devoid of a PSD95 label (Cane et al. 2014; Villa et al. 2016). While these spines may contain other PSD95 scaffolding molecules, such as PSD93 or SAP 102 typical of immature synapses (Aoki et al. 2001; Elias et al. 2008; Sans et al. 2000), this finding highlights the fact that spines are an imperfect surrogate for excitatory synapses.

When labeling postsynaptic gephyrin as a marker for inhibitory synapses on pyramidal neurons, a surprising finding was the prevalence of inhibitory synapses on dendritic spines (Chen et al. 2012; van Versendaal et al. 2012). Dually innervated spines harboring both an excitatory and an inhibitory synapse had previously been reported by EM (Fifkova et al. 1992; Kisvarday et al. 1985; Knott et al. 2002; Kubota et al. 2007; Megias et al. 2001; Parnavelas et al. 1977). However, the fluorescent labeling provided an orders of magnitude larger sample size as compared to EM, leading to the realization that on average, a third of inhibitory synapses on L2/3 pyramidal neurons reside on spines. Large volume imaging of entire L2/3 pyramidal neurons, labeled with fluorescent gephyrin in addition to a cell fill, allowed a first quantitative look at inhibitory synapse distributions across the dendritic arbors of this cell type: on average five spines and two inhibitory synapses per 10 μm (Chen et al. 2012). About 15 % of spines contain an inhibitory synapse, with an overall density of one inhibitory synapse on a spine per 10 μm (Chen et al. 2012). Both spines and inhibitory shaft synapses are distributed relatively uniformly across the spiny region of the dendritic arbor. In contrast, inhibitory spine synapses were found to be denser on the apical tuft than on basal and proximal dendrites (Chen et al. 2012).

18.3 Segregation of Distinct Inputs onto Specific Cellular Locales

18.3.1 Interneuron Targeting of Pyramidal Neurons

While synaptic distributions and the valance of the input at each location are important aspects of an individual neuron's wiring diagram, identifying the precise source of input at each synaptic locale is also critical for understanding a neuron's computational properties. GABAergic cells are particularly heterogeneous with respect to their axonal and dendrite morphology, firing patterns, molecular markers, and postsynaptic targeting (Klausberger and Somogyi 2008) and can be classified into subtypes based on these features (DeFelipe et al. 2013; Petilla Interneuron Nomenclature et al. 2008) or by staining for molecular markers, such as parvalbumin (PV), somatostatin (SOM), the serotonin 5HT3a receptor (5HT3aR), vasoactive intestinal peptide (VIP), calbindin D (CB), and CR (Kawaguchi and Kubota 1997; Kubota et al. 1994, 2011). The PV, SOM, and 5HT3aR markers define nonoverlapping classes of neurons and together represent nearly 100 % of cortical interneurons (Lee et al. 2010; Rudy et al. 2011; Uematsu et al. 2008).

The majority of inhibitory neurons expressing the calcium-binding protein PV, classified as fast-spiking basket neurons, predominantly form synapses onto the soma and proximal dendrites of excitatory neurons (Miles et al. 1996) (Fig. 18.2a). In the G42 transgenic line, which labels a subset (~50 %) of PV-positive cells (Chattopadhyaya et al. 2004), 20 % of synapses onto pyramidal cells target the soma and 40 % the proximal dendrites (within 40 μm of the soma) (Di Cristo et al. 2004). The remaining 40 % target the more distal pyramidal neuron dendrites (Di Cristo et al. 2004). Thus, although PV cells are more likely to innervate the soma and proximal dendrites of pyramidal neurons, they are not strictly limited to these subcellular domains. PV cells strongly inhibit neighboring pyramidal neurons (60 % probability of connectivity) (Avermann et al. 2012; Lee et al. 2013). Because of the large number of PV cell inputs targeting the soma and proximal dendrites of pyramidal neurons, PV neurons are capable of regulating the timing and magnitude of pyramidal output and can limit the temporal window of action potential generation (Higley 2014; Pouille and Scanziani 2001).

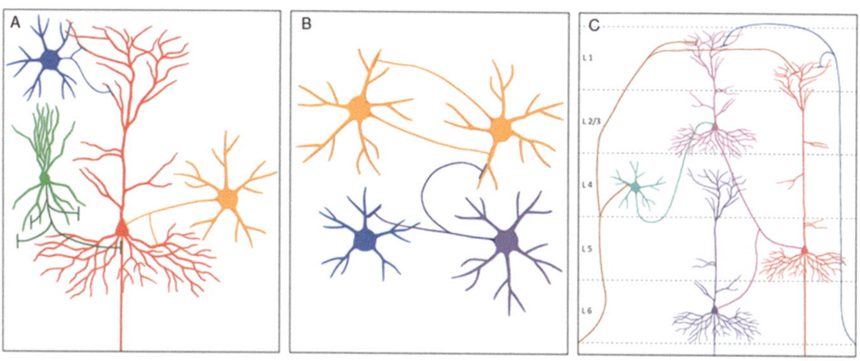

Fig. 18.2 Schematic circuit diagrams of cortical neurons illustrating general principles of connectivity. (**A**) Schematic of inhibitory innervation of a cortical pyramidal neuron (*red*) by cell types that target distinct cellular locations. Chandelier cells (*green*) target the axon initial segment. Parvalbumin (PV)-expressing interneurons (*orange*) mostly target the soma and proximal dendritic segments. Somatostatin (SOM)-expressing interneurons (*blue*) mostly target distal dendrites. (**B**) Schematic of how interneurons target other interneurons. In addition to pyramidal neurons, PV cells (*orange*) mostly innervate other PV cells. VIP-positive interneurons (*purple*) primarily target other interneurons and rarely innervate pyramidal neurons. In L2/3, SOM interneurons (*blue*) mostly target pyramidal neurons and are less likely to innervate other interneurons. (**C**) Schematic of excitatory cortical circuitry. Thalamic axons (*orange*) drive activity of L4 spiny stellate neurons (*light blue*) and innervate L1 apical tufts. Stellate neurons innervate the proximal dendrites of L2/3 neurons (*pink*) which in turn project distantly to other cortical regions and to the proximal dendrites of nearby L5 and L6 pyramidal neurons (*red* and *purple*), which project to subcortical and thalamic regions, respectively. Top-down cortical inputs (*blue*) primarily innervate L1 apical tufts

Chandelier cells, or axo-axonic cells, are a morphologically distinct GABAergic cell type recognizable by their axonal arborizations. These cells innervate the axon initial segment of neighboring pyramidal cells (Fig. 18.2a), giving their own axonal arbor a stereotypical chandelier-like appearance (Somogyi 1977; Somogyi et al. 1982; Woodruff et al. 2010, 2011). A widely held notion is that chandelier cells are a subset of PV-positive interneurons, but recent work shows that only a subset of chandelier cells stain positive for PV (Kawaguchi and Kubota 1998; Taniguchi et al. 2013). Through their innervation of the axon initial segment, chandelier cells can be seen as the last line of defense for preventing the post-synaptic cell from firing an action potential, although they can also have depolarizing effects, depending on the membrane potential of the postsynaptic neuron (Khirug et al. 2008; Szabadics et al. 2006; Woodruff et al. 2009).

SOM interneurons expressing the peptide hormone somatostatin typically target the dendrites of pyramidal neurons (Kawaguchi and Kubota 1998) (Fig. 18.2a). One well-known SOM cell subtype, the Martinotti cell, has a clearly identifiable morphology, with an axon that projects apically to layer 1 and there inhibits the apical tufts of deep-layer pyramidal cells. Although SOM-positive neurons are more inclined to innervate the most distal dendrites (Jiang et al. 2013), it is inaccurate to categorize their synapses as exclusive to that cellular locale. In the GIN transgenic line which labels a subset (~20 %) of SOM-positive cells (Oliva et al. 2000), pyramidal cell targeting by these neurons is such that 4 % of their synapses are directly onto the soma, 20 % are onto the proximal dendrites (within 40 μm of the soma), and the remaining 76 % are onto the more distal dendrites (Di Cristo et al. 2004). SOM interneurons are highly likely to innervate neighboring pyramidal cells (71 % probability of innervating cells within 200 μm) and also target other interneuron subtypes, but specifically avoid innervation of other SOM-positive neurons (Fig. 18.2b) (Fino and Yuste 2011; Pfeffer et al. 2013). By innervating the dendrites of pyramidal cells, SOM interneurons are likely to play an important role in controlling the generation of dendritic Ca^{2+} spikes (discussed later), and they can deliver more localized control over particular excitatory inputs. They are positioned to modulate the excitability of individual dendritic spines, segments, or branches.

18.3.2 Interneuron Targeting of Other Interneurons

The different subsets of interneurons also have specific targeting patterns onto other interneurons (Fig. 18.2b). PV cells have also been shown to strongly inhibit neighboring PV cells (with 55 % probability of connectivity), but are less likely to innervate other interneuron subtypes (24 % probability of connectivity with neighboring 5HT3aR neurons) (Avermann et al. 2012; Pfeffer et al. 2013). 5HT3aR-positive neurons have only been recently described and can be divided into two distinct subgroups. The first is VIP negative and typically reelin positive, which has a late-spiking firing pattern and a neurogliaform axonal arbor morphology restricted to L1 (Lee et al. 2010; Rudy et al. 2011). The second is VIP positive and represents about 15 % of all GABAergic cells. These neurons mostly target SOM and some PV interneurons, and thus their firing results in disinhibition of pyramidal cells (Hioki et al. 2013; Pfeffer et al. 2013; Pi et al. 2013). VIP-positive interneurons are highly unlikely to innervate neighboring pyramidal cells. Paired recordings show that only 12 % of VIP neurons innervate a pyramidal neighbor (Pfeffer et al. 2013). In a recently described mouse circuit, pyramidal neurons within vibrissal motor cortex (vM1) that project to somatosensory barrel cortex (S1) strongly innervate VIP neurons, which in turn innervate SOM neurons that target S1 pyramidal neurons (Lee et al. 2013). This circuit activates S1 VIP neurons during whisking, resulting in the inhibition of S1 SOM neurons and thus disinhibition of S1 pyramidal neurons, potentially influencing the coincident detection of sensory information and altering whisker movement (Lee et al. 2013). More generally, through their inhibitory control over other inhibitory circuits and their location in superficial cortical layers, VIP cells likely play important roles in regulating inter-areal cross-talk between different cortical regions.

Some SOM neurons in L4 of the somatosensory cortex have also been shown to target other GABAergic interneurons, so that their activation results in a disinhibition of pyramidal cells (Jiang et al. 2013; Xu et al. 2013). The SOM neurons in layer 4 differ in morphology, intrinsic electrophysiological properties, and output connectivity from L2/3 SOM interneurons. Specifically, L2/3 SOM interneurons make many strong contacts with pyramidal neurons and suppress pyramidal activity, while L4 SOM neurons make frequent strong contacts with PV-positive fast-spiking interneurons, which themselves inhibit L4 pyramidal neurons (Xu et al. 2013). Thus, increasing the activity of L4 SOM neurons results in increased activity of L4 pyramidal neurons (Xu et al. 2013). These L4 SOM neurons are not strongly activated by thalamic inputs, but are likely activated by acetylcholine (Fanselow et al. 2008; Kawaguchi 1997), which suggests that during arousal and attention, L4 SOM neurons can become activated, resulting in greater activation of L4 pyramidal neurons.

18.3.3 Pyramidal Cell Targeting of Other Pyramidal Neurons

Is there a subcellular bias to excitatory axon innervation of cortical excitatory targets? Historically, unlike inhibitory neurons, excitatory pyramidal cells have been considered a generally uniform population, with circuit specificity resulting from their laminar location. Recent molecular studies have identified genes that are expressed in specific cortical layers (Belgard et al. 2011; Molyneaux et al. 2007), as well as genetic markers resolving intra-layer subsets of excitatory neurons distinguished by their projections to specific target structures (Sorensen et al. 2015). When excitatory neurons are defined according to their axonal projection and target innervation, pyramidal neuron diversity may exceed that of inhibitory neurons (Huang 2014). Potentially, resolving excitatory neuron subclasses may also reveal distinct subcellular target specificity analogous to that of interneuron subclasses. In the hippocampus, considered a simplified cortical structure containing only one pyramidal cell layer (Thomson and Lamy 2007), genetically distinguishable pyramidal neuron subsets have not yet been identified. Cre recombinase driver lines created by the GENSAT project should prove useful for further resolution of pyramidal neuron subtypes (Gerfen et al. 2013; Gong et al. 2007).

Within the cortex, pyramidal cell dendrites can receive excitatory inputs from several sources: from local axons within the same cortical column, from adjacent cortical column axons, from other cortical regions in the same hemisphere, from commissural axons originating in the opposite hemisphere, as well as from direct feed-forward inputs that originate from the thalamus (Fig. 18.2c). The main target of feed-forward thalamic axons is L4, and indeed L4 cells are strongly activated by the thalamus despite the fact that cortical synapses onto L4 outnumber thalamic synapses by a factor of 10 (Benshalom and White 1986; Peters and Payne 1993). Thalamic synapses onto L4 are localized slightly more proximal to the soma than corticocortical synapses, but their synaptic strengths are approximately equal, suggesting that coincident activation of thalamic synapses is the main contributing factor to the strength of thalamocortical synapses (Schoonover et al. 2014). The classic thalamic information flow through the cortex is considered to pass from L4 to L2/3, then down to L5 and L6, and out to other cortical areas and back down to the thalamus and other subcortical regions. However, L2/3, L5, and L6 receive sparse thalamic inputs as well (for excellent reviews of cortical innervation patterns in the barrel cortex, see (Feldmeyer 2012; Lefort et al. 2009)), and recent in vivo studies show that thalamic inputs can drive L5/L6 independently of L4 (Constantinople and Bruno 2013).

Given the laminar nature of information flow within the cortex, the relative placement of a dendritic or other cellular locale within the cortical lamina is the strongest determinant of excitatory input specificity onto that locale. Apical dendrites of L2/3, L5, and L6 pyramidal cells arborizing in L1 receive feedback excitatory inputs from other cortical areas as well as from the posterior medial thalamic nucleus (POm) axons that travel in this lamina (Felleman and Van Essen 1991; Larkum 2013). Local, interlaminar cortical axons carrying feed-forward information are more likely to innervate the basal and proximal dendrites of their cortical targets (Feldmeyer 2012; Spruston 2008). For example, L4 axons mainly innervate the basal dendrites and proximal apical dendrites of L2/3 pyramidal cells within 70 μm of the soma (Feldmeyer et al. 2002; Silver et al. 2003), and L2/3 pyramids predominantly innervate the basal dendrites of L5 pyramidal neurons (Schubert et al. 2001). There is also a bias for thalamic axons to innervate the basal dendrites of L5 pyramidal neurons (Feldmeyer 2012). On the other hand, corticocallosally projecting L5 pyramidal neurons synapse mainly onto the distal region of the basal dendritic tree of their L5 partners at an average distance of about 130 μm (Larsen and Callaway 2006), suggesting that commissural axons are more likely to innervate the distal dendrites of their targets. There are also examples of subcellular innervation bias in the hippocampus, where the CA1 pyramidal cells are innervated by axons of CA3 and entorhinal cortex (EC); the EC axons preferentially target the distal dendrites of CA1 neurons, and the CA3 axons innervate the basal and proximal dendrites of the same CA1 neurons (Megias et al. 2001). Additionally, distant CA3 neurons are more likely to project to the apical CA1 dendrites, while CA3 neurons with cell bodies located close to CA1 are more likely to project to the basal dendrites of CA1 neurons (Li et al. 1994). Thus, a general theme seems to be that long-range feedback excitatory inputs are more likely to innervate distal dendrites, while feed-forward thalamic and interlaminar projections are more likely to target basal and proximal dendrites of their cortical targets. Although we have yet to delineate the full extent of cellular and subcellular targeting specificity for each cortical cell type, the precision we see thus far suggests that excitatory and inhibitory synaptic placement onto pyramidal cell dendrites is a critical aspect of their function within the circuit.

18.4 Functional Implications of Synaptic Placement Across the Dendritic Arbor

18.4.1 Spatial Clustering of Excitatory Inputs

Experiments suggest that neurons are capable of three different types of dendritic integration. Inputs can be summed sublinearly, when there is a decrease in driving force as the membrane potential nears the reversal potential; linearly, when inputs occur on different dendritic branches or are separated by sufficient distance to act independently; and supralinearly, when local conditions facilitate depolarization (Palmer 2014). These different modes of processing are possible within the same neuron, for example, if the dendrites operate linearly when activated by a non-preferred stimulus and nonlinearly when activated by a preferred stimulus (Grienberger et al. 2015). The type of integration performed is strongly influenced by the relative placement of excitatory and inhibitory synapses on the dendrite as well as the dendrite's location within the arbor.

Historically, synapses distant from the soma were thought to have less influence over action potential initiation in the axon, due to loss of charge as current flows from the dendrites to the soma and axon hillock in a model of passive dendrites (Golding et al. 2005). However, the discovery of dendritic spikes explained how distal excitatory synapses are able to propagate their signals effectively to the soma. Dendritic spikes are essentially action potentials that are generated in the dendrite, which acts as an active rather than passive signaling compartment. As dendritic spikes propagate toward the soma, they are capable of triggering action potentials (Golding and Spruston 1998), also recently shown in vivo (Grienberger et al. 2014). Dendritic Ca^{2+} spikes can be initiated by strong synchronous activation of several synapses (Gasparini and Magee 2006), and the probability of dendritic spike initiation in distal branches can be further increased by nearby initiation of NMDA spikes (Larkum et al. 2009). Theoretical models and in vitro experiments both support the view that spatially clustered, temporally coactivated excitatory inputs are more likely to initiate dendritic spikes than inputs that are distributed across multiple branches (Gasparini and Magee 2006; Larkum et al. 2009; Mel 1993; Winnubst and Lohmann 2012). Dendritic spikes can thus improve the computational properties of individual neurons (Larkum 2013; Major et al. 2013), by enabling the discrimination of patterns delivered to a single dendrite from those delivered randomly across the dendritic tree (Branco et al. 2010).

Another consequence of the fact that dendritic spikes elicited by nearby coactive synapses are capable of triggering action potential firing is that neighboring synapses would thus be more likely to undergo long-term potentiation (LTP) than synapses that are spatially distant. There are also examples of dendritic spike-induced LTP independent of action potential firing in vitro (Golding et al. 2002; Gordon et al. 2006) and recently in vivo (Gambino et al. 2014). Since dendritic spikes have a smaller spread than action potentials, these would preferentially produce LTP on locally clustered coactive synapses, especially in regions of the distal dendritic tree that are beyond the reach of backpropagating action potentials (Golding et al. 2001). The pervasiveness of functionally clustered inputs in vivo is under debate. Studies in the visual (Jia et al. 2010), auditory (Chen et al. 2011c), and barrel cortex (Varga et al. 2011) have suggested that functionally distinct excitatory inputs are randomly distributed along pyramidal dendrites. Other experiments provide evidence for clustering of functionally related excitatory inputs in the hippocampus (Druckmann et al. 2014; Kleindienst et al. 2011; Takahashi et al. 2012) and cortex (Chen et al. 2013; Makino and Malinow 2011; McBride et al. 2008; Takahashi et al. 2012) (reviewed in DeBello et al. 2014). These findings are not necessarily in conflict. Excitatory synapses that are functionally similar may have a higher probability of spatial clustering along the dendrite, but at the same time, dendrites are interspersed with a wide range of functional inputs. Spatial clustering of functionally related inputs may also be more prominent in certain subsets of neurons.

18.4.2 Effects of Inhibitory Synaptic Placement on Pyramidal Cell Dendritic Integration

Individual pyramidal cells receive thousands of excitatory synaptic inputs that can be functionally diverse. Yet, the action potential output of most pyramidal neurons is sparse and narrowly tuned. This restriction of action potential firing to a defined sensory range within a limited temporal window is brought about by the integration of excitatory synaptic events combined with the local influence of inhibition (reviewed in Chadderton et al. 2014). Inhibition has a major influence on the effective output of neurons at multiple levels. Individual inhibitory synapses on spines can veto individual spine conductances (Chiu et al. 2013). Inhibitory shaft synapses can influence excitatory integration within local dendritic segments (Kim et al. 1995; Perez-Garci et al. 2006; Willadt et al. 2013). On a more cellular scale, PV cell innervation of the soma (Pouille and Scanziani 2001) or chandelier cell innervation of the axon initial segment (Woodruff et al. 2011) can cancel axonal output. In the latter cases, the high density of inhibitory synapses on the cell body and axon initial segment (Di Cristo et al. 2004; Miles et al. 1996; Woodruff et al. 2010) suggests that synchronous firing of many synapses would be needed for effective inhibition in these locales, and a single inhibitory synapse would have a negligible effect on outcome. This is in contrast to dendritically targeted inhibition, where individual synapse changes could have a dramatic effect on the computation of dendritic integration for specific branches.

How are excitatory and inhibitory inputs integrated functionally at the level of individual dendritic branches and how does that influence overall neuronal firing? Clearly, local integration of excitatory and inhibitory synaptic activity is influenced by individual synaptic strength, timing, and location on the dendrite. Theoretical modeling predicts that activation of multiple distributed inhibitory synapses across the dendritic arbor can result in a global inhibition (Gidon and Segev 2012). However, experiments suggest that the influence of individual inhibitory synapses is locally restricted to the branch they target (Stokes et al. 2014), perhaps even to individual spines (Chiu et al. 2013). Thus, functional pairing or clustering of relevant excitatory synapses on particular dendritic segments with their inhibitory neighbors is highly relevant to dendritic integration. Moreover, the local constellation of such pairing or clustering is likely to have a different influence depending on their more global location along the arbor.

Dendritically targeted inhibition can limit initiation of dendritic Ca^{2+} spikes in both hippocampal and cortical pyramidal neurons (Buhl et al. 1994; Karube et al. 2004; Kim et al. 1995; Larkum et al. 1999; Perez-Garci et al. 2006; Willadt et al. 2013), so that they are initiated only in response to inputs encoding particular functional features. For example, in retinal ganglion cells, local integration of excitation and inhibition leads to initiation of dendritic spikes only in the preferred direction of motion, with inhibition preventing dendritic spike initiation in the non-preferred direction (Sivyer and Williams 2013). Mathematical models focusing on spike activity within dendrites have shown that "off path" inhibitory synapses, where the inhibitory synapse is located distally to the excitatory synapse, are more effective at preventing the initiation of dendritic spikes (Gidon and Segev 2012). Thus, the influence of inhibition on spike initiation can actually be more powerful when inhibitory synapses are located on distal dendrites rather than proximal dendrites and in a constellation where the inhibitory synapse is distal to its functionally paired excitatory synapse (Gidon and Segev 2012; Li et al. 2014). Conversely, in cases where a dendritic spike is successfully initiated, an "on path" inhibitory synapse, located between the excitatory synapse and the soma, is the most effective way to dampen the signal (Gidon and Segev 2012; Hao et al. 2009; Zhang et al. 2013).

Recent in vivo work has demonstrated that dynamic inhibitory synapses are likely to be located within 10 μm of dynamic spines, and this clustering of dynamic events increases upon sensory deprivation (Chen et al. 2012). It has been suggested that a loss of inhibition creates a permissive environment for excitatory synaptic changes (Chen et al. 2011b; Harauzov et al. 2010; Maya Vetencourt et al. 2008) because it broadens the window of spike-timing-dependent plasticity (Bi and Poo 2001; Higley and Contreras 2006; Song et al. 2000; Spruston 2008). The loss of an inhibitory synapse could create a permissive environment for neighboring excitatory synaptic gain or loss, while the addition of an inhibitory synapse could suppress plasticity in neighboring dendritic regions. Inhibitory inputs could also limit the ability to promote excitatory synaptic clustering through plasticity mechanisms. A recent study using calcium imaging in slice culture showed that GABA uncaging diminishes the Ca^{2+} transient resulting from a backpropagating action potential 20 μm in either direction of the uncaging site (Hayama et al. 2013). Thus, the specific placement of inhibitory synapses can regulate which excitatory synapses are able to undergo plasticity. Further, a recent modeling study suggests that the placement of inhibitory synapses can result in LTP, LTD, or a lack of plasticity in specific neighboring dendritic segments (Bar-Ilan et al. 2012), highlighting the importance on the placement of inhibitory synapses on the valance of plasticity at neighboring excitatory synapses.

18.5 Conclusion

How the excitatory and inhibitory inputs are integrated within a single neuron and how this integration supports computation in functioning networks are still critical questions in neuroscience today. Despite advances in our knowledge, we still have limited experimental evidence of the fine-scale synaptic architecture across the neuronal arbor of individual cells of different types, and we know even less of the functional interactions between synaptic excitation and inhibition (Chadderton et al. 2014). As we look to the future, combining structural two-photon imaging of synaptic locations with calcium functional imaging or electrophysiology provides a potential avenue to address these questions, in cell culture, in acute brain slices, and ideally in an intact in vivo system.

References

Aoki C, Miko I, Oviedo H, Mikeladze-Dvali T, Alexandre L, Sweeney N, Bredt DS (2001) Electron microscopic immunocytochemical detection of PSD-95, PSD-93, SAP-102, and SAP-97 at postsynaptic, presynaptic, and nonsynaptic sites of adult and neonatal rat visual cortex. Synapse 40:239–257

Arellano JI, Espinosa A, Fairen A, Yuste R, DeFelipe J (2007) Non-synaptic dendritic spines in neocortex. Neuroscience 145:464–469

Avermann M, Tomm C, Mateo C, Gerstner W, Petersen CC (2012) Microcircuits of excitatory and inhibitory neurons in layer 2/3 of mouse barrel cortex. J Neurophysiol 107:3116–3134
Azouz R, Gray CM, Nowak LG, McCormick DA (1997) Physiological properties of inhibitory interneurons in cat striate cortex. Cereb Cortex 7:534–545
Bar-Ilan L, Gidon A, Segev I (2012) The role of dendritic inhibition in shaping the plasticity of excitatory synapses. Front Neural Circ 6:118
Baude A, Nusser Z, Roberts JD, Mulvihill E, McIlhinney RA, Somogyi P (1993) The metabotropic glutamate receptor (mGluR1 alpha) is concentrated at perisynaptic membrane of neuronal subpopulations as detected by immunogold reaction. Neuron 11:771–787
Belgard TG, Marques AC, Oliver PL, Abaan HO, Sirey TM, Hoerder-Suabedissen A, Garcia-Moreno F, Molnar Z, Margulies EH, Ponting CP (2011) A transcriptomic atlas of mouse neocortical layers. Neuron 71:605–616
Benshalom G, White EL (1986) Quantification of thalamocortical synapses with spiny stellate neurons in layer IV of mouse somatosensory cortex. J Comp Neurol 253:303–314
Bi G, Poo M (2001) Synaptic modification by correlated activity: Hebb's postulate revisited. Ann Rev Neurosci 24:139–166
Branco T, Clark BA, Hausser M (2010) Dendritic discrimination of temporal input sequences in cortical neurons. Science 329:1671–1675
Buhl EH, Halasy K, Somogyi P (1994) Diverse sources of hippocampal unitary inhibitory postsynaptic potentials and the number of synaptic release sites. Nature 368:823–828
Cane M, Maco B, Knott G, Holtmaat A (2014) The relationship between PSD-95 clustering and spine stability in vivo. J Neurosci 34:2075–2086
Chadderton P, Schaefer AT, Williams SR, Margrie TW (2014) Sensory-evoked synaptic integration in cerebellar and cerebral cortical neurons. Nat Rev Neurosci 15:71–83
Chattopadhyaya B, Di Cristo G, Higashiyama H, Knott GW, Kuhlman SJ, Welker E, Huang ZJ (2004) Experience and activity-dependent maturation of perisomatic GABAergic innervation in primary visual cortex during a postnatal critical period. J Neurosci 24:9598–9611
Chen JL, Flanders GH, Lee WC, Lin WC, Nedivi E (2011a) Inhibitory dendrite dynamics as a general feature of the adult cortical microcircuit. J Neurosci 31:12437–12443
Chen JL, Lin WC, Cha JW, So PT, Kubota Y, Nedivi E (2011b) Structural basis for the role of inhibition in facilitating adult brain plasticity. Nat Neurosci 14:587–U573
Chen X, Leischner U, Rochefort NL, Nelken I, Konnerth A (2011c) Functional mapping of single spines in cortical neurons in vivo. Nature 475:501–505
Chen JL, Villa KL, Cha JW, So PT, Kubota Y, Nedivi E (2012) Clustered dynamics of inhibitory synapses and dendritic spines in the adult neocortex. Neuron 74:361–373
Chen TW, Wardill TJ, Sun Y, Pulver SR, Renninger SL, Baohan A, Schreiter ER, Kerr RA, Orger MB, Jayaraman V et al (2013) Ultrasensitive fluorescent proteins for imaging neuronal activity. Nature 499:295–300
Chiu CQ, Lur G, Morse TM, Carnevale NT, Ellis-Davies GC, Higley MJ (2013) Compartmentalization of GABAergic inhibition by dendritic spines. Science 340:759–762
Constantinople CM, Bruno RM (2013) Deep cortical layers are activated directly by thalamus. Science 340:1591–1594
Dani VS, Chang Q, Maffei A, Turrigiano GG, Jaenisch R, Nelson SB (2005) Reduced cortical activity due to a shift in the balance between excitation and inhibition in a mouse model of Rett syndrome. Proc Natl Acad Sci U S A 102:12560–12565
Davis TL, Sterling P (1979) Microcircuitry of cat visual cortex: classification of neurons in layer IV of area 17, and identification of the patterns of lateral geniculate input. J Comp Neurol 188:599–627
DeBello WM, McBride TJ, Nichols GS, Pannoni KE, Sanculi D, Totten DJ (2014) Input clustering and the microscale structure of local circuits. Front Neural Circ 8:112
DeFelipe J, Lopez-Cruz PL, Benavides-Piccione R, Bielza C, Larranaga P, Anderson S, Burkhalter A, Cauli B, Fairen A, Feldmeyer D et al (2013) New insights into the classification and nomenclature of cortical GABAergic interneurons. Nat Rev Neurosci 14:202–216

Di Cristo G, Wu C, Chattopadhyaya B, Ango F, Knott G, Welker E, Svoboda K, Huang ZJ (2004) Subcellular domain-restricted GABAergic innervation in primary visual cortex in the absence of sensory and thalamic inputs. Nat Neurosci 7:1184–1186
Druckmann S, Feng L, Lee B, Yook C, Zhao T, Magee JC, Kim J (2014) Structured synaptic connectivity between hippocampal regions. Neuron 81:629–640
Elias GM, Elias LA, Apostolides PF, Kriegstein AR, Nicoll RA (2008) Differential trafficking of AMPA and NMDA receptors by SAP102 and PSD-95 underlies synapse development. Proc Natl Acad Sci U S A 105:20953–20958
Elston GN, Rosa MG (1997) The occipitoparietal pathway of the macaque monkey: comparison of pyramidal cell morphology in layer III of functionally related cortical visual areas. Cereb Cortex 7:432–452
Fanselow EE, Richardson KA, Connors BW (2008) Selective, state-dependent activation of somatostatin-expressing inhibitory interneurons in mouse neocortex. J Neurophysiol 100: 2640–2652
Feldman ML, Peters A (1978) The forms of non-pyramidal neurons in the visual cortex of the rat. J Comp Neurol 179:761–793
Feldmeyer D (2012) Excitatory neuronal connectivity in the barrel cortex. Front Neuroanat 6:24
Feldmeyer D, Lubke J, Silver RA, Sakmann B (2002) Synaptic connections between layer 4 spiny neurone-layer 2/3 pyramidal cell pairs in juvenile rat barrel cortex: physiology and anatomy of interlaminar signalling within a cortical column. J Geophys Res 538:803–822
Felleman DJ, Van Essen DC (1991) Distributed hierarchical processing in the primate cerebral cortex. Cereb Cortex 1:1–47
Fifkova E, Eason H, Schaner P (1992) Inhibitory contacts on dendritic spines of the dentate fascia. Brain Res 577:331–336
Fino E, Yuste R (2011) Dense inhibitory connectivity in neocortex. Neuron 69:1188–1203
Gabbott PL, Somogyi P (1986) Quantitative distribution of GABA-immunoreactive neurons in the visual cortex (area 17) of the cat. Exp Brain Res 61:323–331
Gambino F, Pages S, Kehayas V, Baptista D, Tatti R, Carleton A, Holtmaat A (2014) Sensory-evoked LTP driven by dendritic plateau potentials in vivo. Nature 515:116–119
Gasparini S, Magee JC (2006) State-dependent dendritic computation in hippocampal CA1 pyramidal neurons. J Neurosci 26:2088–2100
Gerfen CR, Paletzki R, Heintz N (2013) GENSAT BAC cre-recombinase driver lines to study the functional organization of cerebral cortical and basal ganglia circuits. Neuron 80:1368–1383
Gidon A, Segev I (2012) Principles governing the operation of synaptic inhibition in dendrites. Neuron 75:330–341
Golding NL, Spruston N (1998) Dendritic sodium spikes are variable triggers of axonal action potentials in hippocampal CA1 pyramidal neurons. Neuron 21:1189–1200
Golding NL, Kath WL, Spruston N (2001) Dichotomy of action-potential backpropagation in CA1 pyramidal neuron dendrites. J Neurophysiol 86:2998–3010
Golding NL, Staff NP, Spruston N (2002) Dendritic spikes as a mechanism for cooperative long-term potentiation. Nature 418:326–331
Golding NL, Mickus TJ, Katz Y, Kath WL, Spruston N (2005) Factors mediating powerful voltage attenuation along CA1 pyramidal neuron dendrites. J Geophys Res 568:69–82
Gong S, Doughty M, Harbaugh CR, Cummins A, Hatten ME, Heintz N, Gerfen CR (2007) Targeting Cre recombinase to specific neuron populations with bacterial artificial chromosome constructs. J Neurosci 27:9817–9823
Gordon U, Polsky A, Schiller J (2006) Plasticity compartments in basal dendrites of neocortical pyramidal neurons. J Neurosci 26:12717–12726
Grienberger C, Chen X, Konnerth A (2014) NMDA receptor-dependent multidendrite Ca(2+) spikes required for hippocampal burst firing in vivo. Neuron 81:1274–1281
Grienberger C, Chen X, Konnerth A (2015) Dendritic function in vivo. Trends Neurosci 38:45–54
Grutzendler J, Kasthuri N, Gan WB (2002) Long-term dendritic spine stability in the adult cortex. Nature 420:812–816

Gulyas AI, Miettinen R, Jacobowitz DM, Freund TF (1992) Calretinin is present in non-pyramidal cells of the rat hippocampus--I. A new type of neuron specifically associated with the mossy fibre system. Neuroscience 48:1–27

Gulyas AI, Megias M, Emri Z, Freund TF (1999) Total number and ratio of excitatory and inhibitory synapses converging onto single interneurons of different types in the CA1 area of the rat hippocampus. J Neurosci 19:10082–10097

Hao J, Wang XD, Dan Y, Poo MM, Zhang XH (2009) An arithmetic rule for spatial summation of excitatory and inhibitory inputs in pyramidal neurons. Proc Natl Acad Sci U S A 106: 21906–21911

Harauzov A, Spolidoro M, DiCristo G, De Pasquale R, Cancedda L, Pizzorusso T, Viegi A, Berardi N, Maffei L (2010) Reducing intracortical inhibition in the adult visual cortex promotes ocular dominance plasticity. J Neurosci 30:361–371

Harris KM, Jensen FE, Tsao B (1992) Three-dimensional structure of dendritic spines and synapses in rat hippocampus (CA1) at postnatal day 15 and adult ages: implications for the maturation of synaptic physiology and long-term potentiation. J Neurosci 12:2685–2705

Hayama T, Noguchi J, Watanabe S, Takahashi N, Hayashi-Takagi A, Ellis-Davies GC, Matsuzaki M, Kasai H (2013) GABA promotes the competitive selection of dendritic spines by controlling local Ca2+ signaling. Nat Neurosci 16:1409–1416

Hersch SM, White EL (1981) Quantification of synapses formed with apical dendrites of Golgi-impregnated pyramidal cells: variability in thalamocortical inputs, but consistency in the ratios of asymmetrical to symmetrical synapses. Neuroscience 6:1043–1051

Higley MJ (2014) Localized GABAergic inhibition of dendritic Ca(2+) signalling. Nat Rev Neurosci 15:567–572

Higley MJ, Contreras D (2006) Balanced excitation and inhibition determine spike timing during frequency adaptation. J Neurosci 26:448–457

Hioki H, Okamoto S, Konno M, Kameda H, Sohn J, Kuramoto E, Fujiyama F, Kaneko T (2013) Cell type-specific inhibitory inputs to dendritic and somatic compartments of parvalbumin-expressing neocortical interneuron. J Neurosci 33:544–555

Hofer SB, Mrsic-Flogel TD, Bonhoeffer T, Hubener M (2009) Experience leaves a lasting structural trace in cortical circuits. Nature 457:313–317

Holtmaat A, Wilbrecht L, Knott GW, Welker E, Svoboda K (2006) Experience-dependent and cell-type-specific spine growth in the neocortex. Nature 441:979–983

Huang ZJ (2014) Toward a genetic dissection of cortical circuits in the mouse. Neuron 83: 1284–1302

Isshiki M, Tanaka S, Kuriu T, Tabuchi K, Takumi T, Okabe S (2014) Enhanced synapse remodelling as a common phenotype in mouse models of autism. Nat Commun 5:4742

Jia H, Rochefort NL, Chen X, Konnerth A (2010) Dendritic organization of sensory input to cortical neurons in vivo. Nature 464:1307–1312

Jiang X, Wang G, Lee AJ, Stornetta RL, Zhu JJ (2013) The organization of two new cortical interneuronal circuits. Nat Neurosci 16:210–218

Karube F, Kubota Y, Kawaguchi Y (2004) Axon branching and synaptic bouton phenotypes in GABAergic nonpyramidal cell subtypes. J Neurosci 24:2853–2865

Kawaguchi Y (1997) Selective cholinergic modulation of cortical GABAergic cell subtypes. J Neurophysiol 78:1743–1747

Kawaguchi Y, Kubota Y (1997) GABAergic cell subtypes and their synaptic connections in rat frontal cortex. Cereb Cortex 7:476–486

Kawaguchi Y, Kubota Y (1998) Neurochemical features and synaptic connections of large physiologically-identified GABAergic cells in the rat frontal cortex. Neuroscience 85:677–701

Kawaguchi Y, Karube F, Kubota Y (2006) Dendritic branch typing and spine expression patterns in cortical nonpyramidal cells. Cereb Cortex 16:696–711

Keck T, Mrsic-Flogel TD, Vaz Afonso M, Eysel UT, Bonhoeffer T, Hubener M (2008) Massive restructuring of neuronal circuits during functional reorganization of adult visual cortex. Nat Neurosci 11:1162–1167

Keck T, Scheuss V, Jacobsen RI, Wierenga CJ, Eysel UT, Bonhoeffer T, Hubener M (2011) Loss of sensory input causes rapid structural changes of inhibitory neurons in adult mouse visual cortex. Neuron 71:869–882

Khirug S, Yamada J, Afzalov R, Voipio J, Khiroug L, Kaila K (2008) GABAergic depolarization of the axon initial segment in cortical principal neurons is caused by the Na-K-2Cl cotransporter NKCC1. J Neurosci 28:4635–4639

Kim HG, Beierlein M, Connors BW (1995) Inhibitory control of excitable dendrites in neocortex. J Neurophysiol 74:1810–1814

Kisvarday ZF, Martin KA, Whitteridge D, Somogyi P (1985) Synaptic connections of intracellularly filled clutch cells: a type of small basket cell in the visual cortex of the cat. J Comp Neurol 241:111–137

Klausberger T, Somogyi P (2008) Neuronal diversity and temporal dynamics: the unity of hippocampal circuit operations. Science 321:53–57

Kleindienst T, Winnubst J, Roth-Alpermann C, Bonhoeffer T, Lohmann C (2011) Activity-dependent clustering of functional synaptic inputs on developing hippocampal dendrites. Neuron 72:1012–1024

Kleschevnikov AM, Belichenko PV, Villar AJ, Epstein CJ, Malenka RC, Mobley WC (2004) Hippocampal long-term potentiation suppressed by increased inhibition in the Ts65Dn mouse, a genetic model of down syndrome. J Neurosci 24:8153–8160

Knott GW, Quairiaux C, Genoud C, Welker E (2002) Formation of dendritic spines with GABAergic synapses induced by whisker stimulation in adult mice. Neuron 34:265–273

Kubota Y (2014) Untangling GABAergic wiring in the cortical microcircuit. Curr Opinion Neurobiol 26:7–14

Kubota Y, Hattori R, Yui Y (1994) Three distinct subpopulations of GABAergic neurons in rat frontal agranular cortex. Brain Res 649:159–173

Kubota Y, Hatada S, Kondo S, Karube F, Kawaguchi Y (2007) Neocortical inhibitory terminals innervate dendritic spines targeted by thalamocortical afferents. J Neurosci 27:1139–1150

Kubota Y, Shigematsu N, Karube F, Sekigawa A, Kato S, Yamaguchi N, Hirai Y, Morishima M, Kawaguchi Y (2011) Selective coexpression of multiple chemical markers defines discrete populations of neocortical GABAergic neurons. Cereb Cortex 21:1803–1817

Larkman AU (1991) Dendritic morphology of pyramidal neurones of the visual cortex of the rat: III. Spine distributions. J Comp Neurol 306:332–343

Larkum M (2013) A cellular mechanism for cortical associations: an organizing principle for the cerebral cortex. Trends Neurosci 36:141–151

Larkum ME, Zhu JJ, Sakmann B (1999) A new cellular mechanism for coupling inputs arriving at different cortical layers. Nature 398:338–341

Larkum ME, Nevian T, Sandler M, Polsky A, Schiller J (2009) Synaptic integration in tuft dendrites of layer 5 pyramidal neurons: a new unifying principle. Science 325:756–760

Larsen DD, Callaway EM (2006) Development of layer-specific axonal arborizations in mouse primary somatosensory cortex. J Comp Neurol 494:398–414

Lee WC, Huang H, Feng G, Sanes JR, Brown EN, So PT, Nedivi E (2006) Dynamic remodeling of dendritic arbors in GABAergic interneurons of adult visual cortex. PLoS Biol 4, e29

Lee WC, Chen JL, Huang H, Leslie JH, Amitai Y, So PT, Nedivi E (2008) A dynamic zone defines interneuron remodeling in the adult neocortex. Proc Natl Acad Sci U S A 105:19968–19973

Lee S, Hjerling-Leffler J, Zagha E, Fishell G, Rudy B (2010) The largest group of superficial neocortical GABAergic interneurons expresses ionotropic serotonin receptors. J Neurosci 30:16796–16808

Lee S, Kruglikov I, Huang ZJ, Fishell G, Rudy B (2013) A disinhibitory circuit mediates motor integration in the somatosensory cortex. Nat Neurosci 16:1662–1670

Lefort S, Tomm C, Floyd Sarria JC, Petersen CC (2009) The excitatory neuronal network of the C2 barrel column in mouse primary somatosensory cortex. Neuron 61:301–316

LeVay S (1973) Synaptic patterns in the visual cortex of the cat and monkey. Electron microscopy of Golgi preparations. J Comp Neurol 150:53–85

Lewis DA, Hashimoto T, Volk DW (2005) Cortical inhibitory neurons and schizophrenia. Nat Rev Neurosci 6:312–324
Li XG, Somogyi P, Ylinen A, Buzsaki G (1994) The hippocampal CA3 network: an in vivo intracellular labeling study. J Comp Neurol 339:181–208
Li S, Liu N, Zhang XH, Zhou D, Cai D (2014) Bilinearity in spatiotemporal integration of synaptic inputs. PLoS Comp Biol 10, e1004014
Major G, Larkum ME, Schiller J (2013) Active properties of neocortical pyramidal neuron dendrites. Ann Rev Neurosci 36:1–24
Makino H, Malinow R (2011) Compartmentalized versus global synaptic plasticity on dendrites controlled by experience. Neuron 72:1001–1011
Maya Vetencourt JF, Sale A, Viegi A, Baroncelli L, De Pasquale R, O'Leary OF, Castren E, Maffei L (2008) The antidepressant fluoxetine restores plasticity in the adult visual cortex. Science 320:385–388
McBride TJ, Rodriguez-Contreras A, Trinh A, Bailey R, Debello WM (2008) Learning drives differential clustering of axodendritic contacts in the barn owl auditory system. J Neurosci 28: 6960–6973
Megias M, Emri Z, Freund TF, Gulyas AI (2001) Total number and distribution of inhibitory and excitatory synapses on hippocampal CA1 pyramidal cells. Neuroscience 102:527–540
Mel BW (1993) Synaptic integration in an excitable dendritic tree. J Neurophysiol 70:1086–1101
Miles R, Toth K, Gulyas AI, Hajos N, Freund TF (1996) Differences between somatic and dendritic inhibition in the hippocampus. Neuron 16:815–823
Mizrahi A, Katz LC (2003) Dendritic stability in the adult olfactory bulb. Nat Neurosci 6: 1201–1207
Mohler H, Fritschy JM, Crestani F, Hensch T, Rudolph U (2004) Specific GABA(A) circuits in brain development and therapy. Biochem Pharmacol 68:1685–1690
Molyneaux BJ, Arlotta P, Menezes JR, Macklis JD (2007) Neuronal subtype specification in the cerebral cortex. Nat Rev Neurosci 8:427–437
Morishima M, Morita K, Kubota Y, Kawaguchi Y (2011) Highly differentiated projection-specific cortical subnetworks. J Neurosci 31:10380–10391
Oliva AA Jr, Jiang M, Lam T, Smith KL, Swann JW (2000) Novel hippocampal interneuronal subtypes identified using transgenic mice that express green fluorescent protein in GABAergic interneurons. J Neurosci 20:3354–3368
Palmer LM (2014) Dendritic integration in pyramidal neurons during network activity and disease. Brain Res Bull 103:2–10
Parnavelas JG, Sullivan K, Lieberman AR, Webster KE (1977) Neurons and their synaptic organization in the visual cortex of the rat. Electron microscopy of Golgi preparations. Cell Tissue Res 183:499–517
Perez-Garci E, Gassmann M, Bettler B, Larkum ME (2006) The GABAB1b isoform mediates long-lasting inhibition of dendritic Ca2+ spikes in layer 5 somatosensory pyramidal neurons. Neuron 50:603–616
Peters A, Payne BR (1993) Numerical relationships between geniculocortical afferents and pyramidal cell modules in cat primary visual cortex. Cereb Cortex 3:69–78
Petilla Interneuron Nomenclature G, Ascoli GA, Alonso-Nanclares L, Anderson SA, Barrionuevo G, Benavides-Piccione R, Burkhalter A, Buzsaki G, Cauli B, Defelipe J et al (2008) Petilla terminology: nomenclature of features of GABAergic interneurons of the cerebral cortex. Nat Rev Neurosci 9:557–568
Pfeffer CK, Xue M, He M, Huang ZJ, Scanziani M (2013) Inhibition of inhibition in visual cortex: the logic of connections between molecularly distinct interneurons. Nat Neurosci 16: 1068–1076
Pi HJ, Hangya B, Kvitsiani D, Sanders JI, Huang ZJ, Kepecs A (2013) Cortical interneurons that specialize in disinhibitory control. Nature 503:521–524
Pieraut S, Gounko N, Sando R 3rd, Dang W, Rebboah E, Panda S, Madisen L, Zeng H, Maximov A (2014) Experience-dependent remodeling of basket cell networks in the dentate gyrus. Neuron 84:107–122

Pouille F, Scanziani M (2001) Enforcement of temporal fidelity in pyramidal cells by somatic - feed-forward inhibition. Science 293:1159–1163
Rubenstein JL, Merzenich MM (2003) Model of autism: increased ratio of excitation/inhibition in key neural systems. Genes Brain Behav 2:255–267
Rudy B, Fishell G, Lee S, Hjerling-Leffler J (2011) Three groups of interneurons account for nearly 100% of neocortical GABAergic neurons. Dev Neurobiol 71:45–61
Sans N, Petralia RS, Wang YX, Blahos J 2nd, Hell JW, Wenthold RJ (2000) A developmental change in NMDA receptor-associated proteins at hippocampal synapses. J Neurosci 20: 1260–1271
Schoonover CE, Tapia JC, Schilling VC, Wimmer V, Blazeski R, Zhang W, Mason CA, Bruno RM (2014) Comparative strength and dendritic organization of thalamocortical and corticocortical synapses onto excitatory layer 4 neurons. J Neurosci 34:6746–6758
Schubert D, Staiger JF, Cho N, Kotter R, Zilles K, Luhmann HJ (2001) Layer-specific intracolumnar and transcolumnar functional connectivity of layer V pyramidal cells in rat barrel cortex. J Neurosci 21:3580–3592
Schuemann A, Klawiter A, Bonhoeffer T, Wierenga CJ (2013) Structural plasticity of GABAergic axons is regulated by network activity and GABAA receptor activation. Front Neural Circ 7:113
Silver RA, Lubke J, Sakmann B, Feldmeyer D (2003) High-probability uniquantal transmission at excitatory synapses in barrel cortex. Science 302:1981–1984
Sivyer B, Williams SR (2013) Direction selectivity is computed by active dendritic integration in retinal ganglion cells. Nat Neurosci 16:1848–1856
Somogyi P (1977) A specific 'axo-axonal' interneuron in the visual cortex of the rat. Brain Res 136:345–350
Somogyi P, Freund TF, Cowey A (1982) The axo-axonic interneuron in the cerebral cortex of the rat, cat and monkey. Neuroscience 7:2577–2607
Song S, Miller KD, Abbott LF (2000) Competitive Hebbian learning through spike-timing-dependent synaptic plasticity. Nat Neurosci 3:919–926
Sorensen SA, Bernard A, Menon V, Royall JJ, Glattfelder KJ, Desta T, Hirokawa K, Mortrud M, Miller JA, Zeng H et al (2015) Correlated gene expression and target specificity demonstrate excitatory projection neuron diversity. Cereb Cortex 25:433–449
Spruston N (2008) Pyramidal neurons: dendritic structure and synaptic integration. Nat Rev Neurosci 9:206–221
Stokes CC, Teeter CM, Isaacson JS (2014) Single dendrite-targeting interneurons generate branch-specific inhibition. Front Neural Circ 8:139
Szabadics J, Varga C, Molnar G, Olah S, Barzo P, Tamas G (2006) Excitatory effect of GABAergic axo-axonic cells in cortical microcircuits. Science 311:233–235
Takahashi N, Kitamura K, Matsuo N, Mayford M, Kano M, Matsuki N, Ikegaya Y (2012) Locally synchronized synaptic inputs. Science 335:353–356
Taniguchi H, Lu J, Huang ZJ (2013) The spatial and temporal origin of chandelier cells in mouse neocortex. Science 339:70–74
Thomson AM, Deuchars J (1997) Synaptic interactions in neocortical local circuits: dual intracellular recordings in vitro. Cereb Cortex 7:510–522
Thomson AM, Lamy C (2007) Functional maps of neocortical local circuitry. Front Neurosci 1: 19–42
Trachtenberg JT, Chen BE, Knott GW, Feng G, Sanes JR, Welker E, Svoboda K (2002) Long-term in vivo imaging of experience-dependent synaptic plasticity in adult cortex. Nature 420: 788–794
Trommald M, Jensen V, Andersen P (1995) Analysis of dendritic spines in rat CA1 pyramidal cells intracellularly filled with a fluorescent dye. J Comp Neurol 353:260–274
Uematsu M, Hirai Y, Karube F, Ebihara S, Kato M, Abe K, Obata K, Yoshida S, Hirabayashi M, Yanagawa Y et al (2008) Quantitative chemical composition of cortical GABAergic neurons revealed in transgenic venus-expressing rats. Cereb Cortex 18:315–330

van Versendaal D, Rajendran R, Saiepour MH, Klooster J, Smit-Rigter L, Sommeijer JP, De Zeeuw CI, Hofer SB, Heimel JA, Levelt CN (2012) Elimination of inhibitory synapses is a major component of adult ocular dominance plasticity. Neuron 74:374–383

Varga Z, Jia H, Sakmann B, Konnerth A (2011) Dendritic coding of multiple sensory inputs in single cortical neurons in vivo. Proc Natl Acad Sci U S A 108:15420–15425

Villa KL, Berry KP, Subramanian J, Cha JW, Oh WC, Kwon HB, Kubota Y, So PT, Nedivi E (2016) Inhibitory synapses are repeatedly assembled and removed at persistent sites in vivo. Neuron 89:756–769

Westrum LE (1966) Synaptic contacts on axons in the cerebral cortex. Nature 210:1289–1290

White EL, Rock MP (1980) Three-dimensional aspects and synaptic relationships of a Golgi-impregnated spiny stellate cell reconstructed from serial thin sections. J Neurocytol 9:615–636

Wierenga CJ, Becker N, Bonhoeffer T (2008) GABAergic synapses are formed without the involvement of dendritic protrusions. Nat Neurosci 11:1044–1052

Willadt S, Nenniger M, Vogt KE (2013) Hippocampal feedforward inhibition focuses excitatory synaptic signals into distinct dendritic compartments. PLoS One 8, e80984

Winnubst J, Lohmann C (2012) Synaptic clustering during development and learning: the why, when, and how. Front Mol Neurosci 5:70

Woodruff A, Xu Q, Anderson SA, Yuste R (2009) Depolarizing effect of neocortical chandelier neurons. Front Neural Circ 3:15

Woodruff AR, Anderson SA, Yuste R (2010) The enigmatic function of chandelier cells. Front Neurosci 4:201

Woodruff AR, McGarry LM, Vogels TP, Inan M, Anderson SA, Yuste R (2011) State-dependent function of neocortical chandelier cells. J Neurosci 31:17872–17886

Xu H, Jeong HY, Tremblay R, Rudy B (2013) Neocortical somatostatin-expressing GABAergic interneurons disinhibit the thalamorecipient layer 4. Neuron 77:155–167

Zhang D, Li Y, Rasch MJ, Wu S (2013) Nonlinear multiplicative dendritic integration in neuron and network models. Front Comp Neurosci 7:56

Chapter 19
Development of Synaptic Input Patterns on Dendrites of Retinal Neurons

Florence D. D'Orazi and Takeshi Yoshimatsu

Abstract Studies of the developing vertebrate retina have recently offered new insights into the molecular and cellular mechanisms that regulate the development and patterning of synapses onto dendrites. The dendrites and axons of retinal neurons are stratified, forming layers of functionally distinct connections. This architecture readily facilitates investigation of the developmental mechanisms that wire neural circuits. Like other central nervous system (CNS) neurons, each retinal neuron type exhibits a stereotypic wiring pattern that is critical for proper circuit function. To form its stereotyped connectivity, a neuron must select specific synaptic partners, make the appropriate number of contacts, and form synapses at proper sites on its dendrites. Both molecular cues and neurotransmission have been found to be responsible for shaping patterns of wiring. This chapter will review the most current knowledge of these developmental mechanisms, with a specific focus on the formation and patterning of synaptic connections onto dendrites in the inner and outer retina.

Keywords Synaptogenesis • Connectivity • Synapse • Development • Dendrite • Retina

19.1 Introduction

The retina is an especially tractable neural circuit for dissecting the relationship between dendritic structure, synaptic connections, and physiological function at the level of the individual neuron. The retina's layered organization facilitates relatively easy reconstruction of neuron morphology and identification of synapses, and as a sensory system, the retina offers access to neuronal output under physiological conditions. Diverse functions, from color vision to motion detection, arise in the retina. The structural features of dendritic arbors, including size and branching patterns, are found to greatly influence the physiological receptive field properties of the retinal output neurons, the ganglion cells (Schwartz et al. 2012). However,

F.D. D'Orazi • T. Yoshimatsu (✉)
Department of Biological Structure, University of Washington, Seattle, WA 98195, USA
e-mail: fddorazi@u.washington.edu; tyoshi@u.washington.edu

K. Emoto et al. (eds.), *Dendrites*, DOI 10.1007/978-4-431-56050-0_19

retinal neurons often have complex and diverse receptive field characteristics that cannot be explained by their dendritic structure alone. For example, neurons that respond maximally to a stimulus moving in one specific direction depend upon the differential strengths of inhibitory inputs across their dendritic arbors to determine their direction-selective responses (Wei and Feller 2011). Thus, stereotyped patterns of synaptic connections, in addition to dendritic arbor structure, are critical to generating specific neuronal functions in the retina.

Much work has aimed to tackle the question of how the dendrites of postsynaptic cells form complex yet precise connectivity in the retina during development. In this chapter, we briefly describe the organization of various neuronal circuits in the retina, which have recently been reviewed in great detail (Hoon et al. 2014). We then discuss how cellular interactions, molecular cues, and neurotransmission impact each of the four major steps necessary for neuronal dendrites to achieve stereotyped wiring during development: (1) assembling synapses, (2) selecting appropriate partners, (3) establishing the right number, and (4) distribution of synapses across the dendritic arbor. We will highlight important questions yet to be addressed and review evidence for hypotheses at the forefront of current research efforts.

19.2 Organization of Retinal Circuits

The retina is populated by five major neuronal cell classes, which form a variety of divergent and convergent patterns of connections with one another (Dunn and Wong 2014). While different neuron types are restricted to three nuclear layers, the processes and synaptic connections of these cells are further organized into synaptic laminae, the outer and inner plexiform layers (OPL and IPL), as schematized in Fig. 19.1a. Two major photoreceptor types at the outermost retina detect photons using light-sensitive opsins and transduce the light signal into an electrical signal through a cascade of molecular signaling steps (Arshavsky and Burns 2012). Rod photoreceptors (rods) are specialized to relay visual signals in dim light, whereas cone photoreceptors (cones) function at high light levels and encode light signals that contribute to color vision. Both photoreceptor types depolarize and release the excitatory neurotransmitter glutamate in response to decreases in light intensity and, conversely, decrease their glutamate release in response to increments in light. Photoreceptor axon terminals in the OPL synapse onto two interneuron types, inhibitory horizontal cells (HCs) and excitatory bipolar cells (BCs). In mammals, cones specifically synapse onto the dendrites of HCs, and rods synapse onto HC axons (Wässle and Boycott 1991). BCs receive photoreceptor inputs onto their dendrites and project their axons to the IPL. There, BCs form glutamatergic synapses and relay the light signal to the output neurons of the retina, retinal ganglion cells (RGCs), as well as to amacrine cell (AMC) interneurons. Neurons that synapse in the IPL (BCs, AMCs, and RGCs) can be categorized by the polarity of their response to light: ON cells depolarize whereas OFF cells

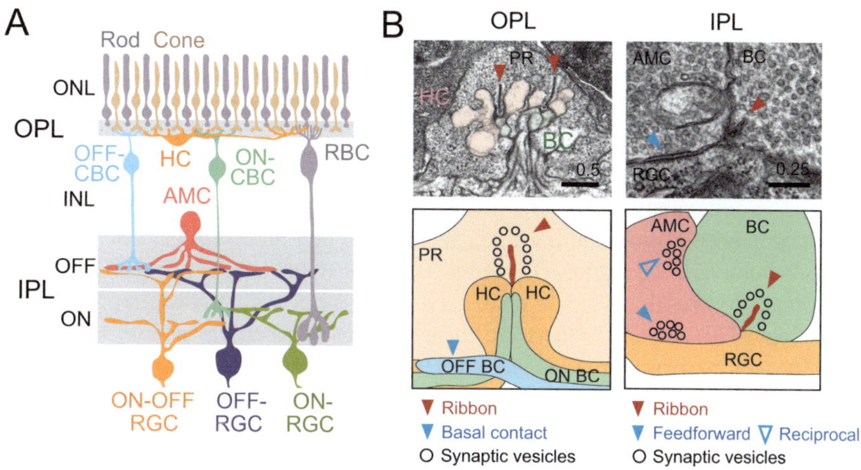

Fig. 19.1 Synaptic organization of the retinal circuit. (**a**) Schematic of the vertebrate retina. The major retinal neuron classes are cone and rod photoreceptors, cone bipolar cell (CBC), rod bipolar cell (RBC), horizontal cell (HC), amacrine cell (AMC), and retinal ganglion cell (RGC). These neurons are organized in laminae; cell bodies are located in the outer or inner nuclear layers (ONL or INL, respectively), and synapses are distributed to the outer or inner plexiform layers (OPL or IPL). The IPL is further divided into OFF and ON sublaminae. (**b**) Ultrastructure (*top*) and schematic (*bottom*) of the subcellular organization of ribbon synapses in the plexiform layers. (OPL) (*Top*) Electron micrograph of a triad synapse at a larval zebrafish cone terminal (from Yoshimatsu et al. 2014). HC processes and BC dendrites invaginate the terminal at sites apposed to ribbon structures (*red arrowhead*) in the photoreceptor (PR) terminal. Scale bar is 0.5 μm. (*Bottom*) OFF-BC dendrites form basal contacts onto the cone terminal (*blue arrowhead*). (IPL) (*Top*) Electron micrograph of a dyad synapse at an adult mouse BC terminal (from D'Orazi et al. 2014). AMCs form feed-forward conventional synapses onto RGC dendrites (*filled blue arrowhead*), as well as reciprocal conventional synapses back onto BC axons (*open blue arrowhead* in schematic). Scale bar is 0.25 μm

hyperpolarize in response to increments of light. The synapses of these functionally distinct neuronal classes are segregated in either the ON or OFF domain of the IPL (Fig. 19.1a). The parallel ON and OFF visual pathways first emerge in the OPL, at the photoreceptor-BC synapse, because BCs respond differently to light stimuli depending on their expression of either metabotropic or ionotropic glutamate receptors (mGluR or iGluR) (Euler et al. 2014). Mammalian BCs can be further distinguished by their dendritic connectivity. Cone BCs (CBCs) or rod BCs (RBCs) wire exclusively with either cones or rods (Fig. 19.1a), respectively, although recent evidence suggests that some OFF BCs may wire with a mix of cones and rods (Mataruga et al. 2007; Haverkamp et al. 2008; Tsukamoto and Omi 2014a).

Most, if not all, excitatory synapses in the vertebrate retina contain a specialized structure called a "ribbon" that can be identified by electron microscopy as an electron-dense structure in the presynaptic zone (Fig. 19.1b). Synaptic ribbons are specialized for fast and tonic neurotransmitter release in response to graded potentials, which optimizes rapid neural encoding of visual signals (Sterling and

Matthews 2005). Characteristic arrangements of postsynaptic processes form around the ribbon, completing the ribbon synapse. In the OPL, ribbon synapses are arranged as triads within the photoreceptor terminal. Triad synapses comprise a membrane-anchored ribbon in the photoreceptor terminal and two HC processes as well as one to three BC dendrites that appose the presynaptic ribbon (Fig. 19.1b). Both ON-BC dendrites and HC processes insert into the photoreceptor terminal, forming an invagination. The HC processes lie in closest proximity to the ribbon, which is hypothesized to allow them to regulate neurotransmitter flow from the photoreceptor to BC dendrites (Dowling 1970). In contrast, OFF-BC dendrites form superficial, or "basal," synaptic contacts at the base of the photoreceptor terminal (Dowling 1970). Basal contacts are located up to hundreds of nanometers away from the neurotransmitter release site, yet they still conduct glutamatergic signaling from the ribbon, albeit with slower responses compared to the deeply invaginating HCs (DeVries et al. 2006). It is suggested that the difference in proximity to photoreceptor ribbons between invaginating and basal dendritic contacts leads to different sensitivities and temporal frequency responses in ON and OFF BCs, respectively (Haverkamp et al. 2000).

Dyad ribbon synapses in the IPL comprise two postsynaptic dendrites from a combination of AMCs and RGCs, which are positioned opposite the ribbon in the BC axon terminal (Fig. 19.1b). Although ribbons in the IPL are not surrounded by invaginations as in the OPL, the postsynaptic processes are still located in close proximity to the ribbon. In addition to ribbon synapses, there is an abundance of conventional synapses in the IPL, which lack ribbons but feature other synaptic components, including presynaptic vesicles and postsynaptic densities. Most conventional synapses originate from inhibitory, GABAergic, or glycinergic AMCs synapsing onto BC axons, RGC dendrites, or other AMC processes in the IPL. Conventional synapses are often observed in the vicinity of ribbon synapses (Dowling 1970) and are thus well positioned to modulate these synapses. AMC processes that appose BC ribbons often make conventional synapses onto the same BC axon terminal and thus form a reciprocal synapse. Conventional synapses are also involved in a feed-forward circuit in the IPL; AMCs that are postsynaptic to a BC ribbon synapse make conventional synapses onto RGCs sharing the same ribbon. Consequently, whereas reciprocal synapses impact RGC responses by modulating their BC inputs, feed-forward synapses alter the electrical properties of the RGC itself (Chen et al. 2010; Grimes 2012).

19.3 Synapse Assembly

The assembly of retinal synapses is an ordered process during vertebrate development. Although the timing of the onset of visual information processing differs among vertebrate species, the overall order of developmental events is conserved across all species investigated thus far (Sernagor et al. 2006). This section describes the sequential steps of synapse formation on dendrites in the retina, focusing primarily on the retina of mice, a major model system in neuroscience research.

19.3.1 Synaptogenesis in the Inner and Outer Retina

19.3.1.1 Inner Retina

Synaptogenesis has been found to occur in the IPL before the OPL across vertebrate species (Olney 1968; Maslim and Stone 1986; Schmitt and Dowling 1999). In mice, IPL synapse assembly is already underway by the time of birth (postnatal day 0, P0) (Fig. 19.2a) and consequently in the absence of light-driven signaling, because mice do not open their eyes until P14. AMCs and RGCs have extended processes into the IPL, and a large fraction of the synaptic component proteins in these cell types are already expressed (von Kriegstein and Schmitz 2003; Wang et al. 2003; Sherry et al. 2003a, b). However, immunostaining reveals that the distributions of neurotransmitter receptors on postsynaptic membranes appear diffuse. In some cases, receptors are distributed first to the cell soma and later eliminated, as mature receptor clusters emerge on dendrites during the first postnatal week (summarized in Yoshimatsu et al. 2013). Electron microscopy reveals that electron-dense synaptic structures appear in the IPL during the period when receptor clusters form on the dendrites and axons of inner retinal cells (Fisher 1979) (Fig. 19.2b). However, at this early stage, only conventional synapses are observed. Because BC axons do not elaborate in the IPL until later in the first postnatal week (Morgan et al. 2006), it is assumed that these conventional synapses are formed among AMCs or between AMCs and RGCs. Visual information from the photoreceptors cannot be conveyed to AMCs and RGCs during the first postnatal week because photoreceptor-BC synapses have not yet formed, indicating that early-formed conventional synapses do not transmit light-driven information. Despite the absence of light-driven activity, early-formed conventional synapses are both functional and important for the development of visual pathways. Electrophysiological recordings showed that RGCs receive synaptic transmission in the first postnatal week (Sernagor et al. 2001). Pharmacological manipulations suggest that this early RGC activity is mediated by GABAergic and cholinergic inputs, likely from AMCs, because they are the only GABA or acetylcholine-releasing neurons in the IPL (Vaney 1990). Thus, early-formed conventional synapses in the IPL are functional. Indeed, cholinergic synapses participate in an excitatory synaptic network in the IPL that is responsible for generating waves of spontaneous bursting activity in RGCs and AMCs (Meister et al. 1991; Wong et al. 1995; Feller et al. 1996). Because retinal waves play a role in refining RGC axonal projections (Kirkby et al. 2013; Ackman and Crair 2014), the formation of conventional synapses in the IPL is essential for the proper development of visual areas beyond the retina.

The third synaptic element of IPL synapses, bipolar cell axon terminals, is observable around P5–7 (Morgan et al. 2006) (Fig. 19.2b). At this early stage of

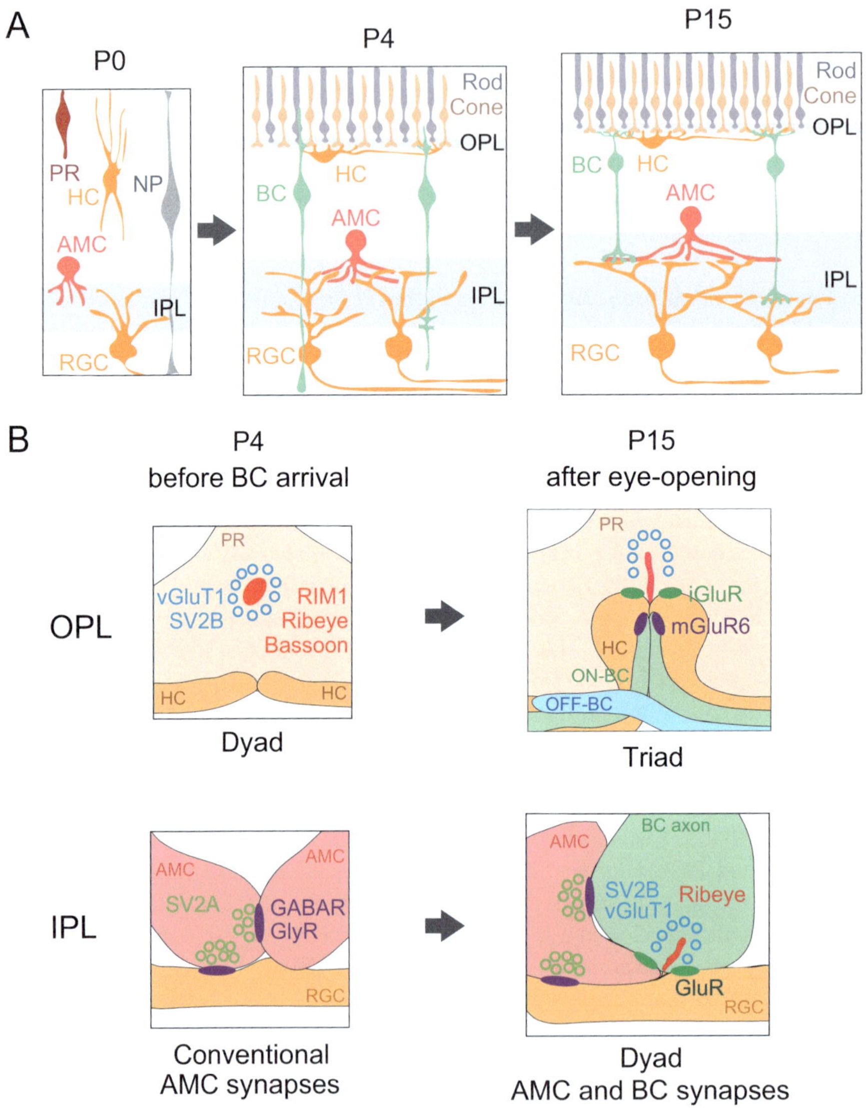

Fig. 19.2 Developmental time course of retinal synapse formation and lamination in the mouse retina. (**a**) Illustration of the sequence of events during outer and inner plexiform layer (OPL and IPL) development in mouse retina. Postnatal days (P) 0, 4, and 15. Amacrine cells (AMCs) and retinal ganglion cells (RGCs) elaborate first, while photoreceptors (PRs), horizontal cells (HCs), and neural precursors (NPs) that will give rise to bipolar cell (BC) processes are still maturing. Around P4, HCs extend lateral processes, BCs have differentiated and begin to retract their processes, and AMC and RGC dendrites have stratified in the IPL. The structure of the retinal circuit resembles its adult form by P15. (**b**) Sequence of synaptic assembly in the OPL (*top row*) and IPL (*bottom row*) of the mouse retina. (*Top*) At around P4, HC lateral processes form dyad synapses with PR terminals containing ribbon component proteins as well as synaptic vesicle glycoprotein 2B (SV2B) and vesicular glutamate transporter (vGlut1). By P15, ribbon synapses form the mature triad structure. HC and ON-BC processes express ionotropic glutamate receptor (iGluR) and metabotropic glutamate receptor 6 (mGluR6), respectively. (*Bottom*) AMCs form the

growth, BC axons already contain proteins essential to neurotransmitter release, such as synaptic vesicle glycoprotein 2B (SV2B) and vesicular glutamate transporter (vGluT1) (Wang et al. 2003; Sherry et al. 2003a). Further, punctate expression of postsynaptic density protein-95 (PSD-95), a postsynaptic scaffolding protein for glutamate receptors, is observed on the dendrites of RGCs at P5 (Morgan et al. 2008). Pharmacological and electrophysiological studies suggest that glutamatergic transmission from BCs onto RGC dendrites occurs as early as P7. Puff application of kainic acid evokes robust inward currents in RGCs at P5 (Rörig and Grantyn 1993). By P8, the majority of RGCs exhibit glutamatergic spontaneous excitatory postsynaptic currents (sEPSCs) (Morgan et al. 2008), and AMCs show glutamatergic sEPSCs slightly earlier, by P7 (Schubert et al. 2008). Because BCs are the major glutamatergic neurons in the inner retina, these observations suggest that BC axons form synapses onto RGCs and AMC dendrites by the end of the first week. However, ribbon structures are not apparent in the IPL until P10. The lack of ribbons at stages when glutamate-driven responses are apparent indicates that BCs may form conventional glutamatergic synapses prior to ribbon synapses, and further studies are required to explore this possibility.

Ribbon synapses mature during the second postnatal week and begin to provide glutamate signaling that will predominate excitatory transmission in the IPL. After P11, waves of spontaneous activity in the retina are driven by glutamate, rather than acetylcholine signaling in the IPL (Wong et al. 2000; Bansal et al. 2000). This glutamatergic signaling prior to eye-opening at P14 is functionally important, because glutamatergic coordination of synchronous wave activity contributes to additional refinement of RGC projection pathways. Further, BC glutamatergic input on RGC dendrites becomes robust by P12, as demonstrated by recordings of spontaneous and light-evoked activity in RGCs (Tian and Copenhagen 2001; Morgan et al. 2008). Because the time course of the development of RGC excitatory responses is matched by ribbon structure maturation, ribbon synapses likely provide glutamate transmission during this time. However, although ribbons are first observed in the IPL at P10, they are initially short in length. It is possible that immature ribbons may support non-light-evoked neural responses prior to eye-opening, but this remains to be determined. Nonetheless, ribbons become longer and more abundant, achieving adult morphology around P15 (Fisher 1979). Ribbon synapse maturation continues after eye-opening, including changes in neurotransmitter receptor subunit composition (Stafford et al. 2014), a classic developmental step observed in other brain regions (Yashiro and Philpot 2008).

Coincident with the emergence of early glutamatergic transmission from BC axons, AMCs form inhibitory synapses onto BC axon terminals as they elaborate in

Fig. 19.2 (continued) first IPL synapses and make conventional inhibitory synapses onto AMC or RGC dendrites, which express GABA receptors (GABAR) or glycine receptors (GlyR) at postsynaptic sites. By P15, the mature dyad structure has formed; the BC axon terminal expresses presynaptic proteins and contains ribbons anchored at sites apposed to glutamate receptor (GluR) puncta on AMC and RGC dendrites

the IPL (Schubert et al. 2008). There are differences among BC types in the timing of their inhibitory input development, but electrophysiological recording revealed that AMC inhibitory inputs are formed onto all RBCs, ON BCs, and OFF BCs by P10. Early-formed inhibitory synapses onto BC axons are critical to coordinating the activity of ON and OFF parallel pathways. AII AMCs stratify in ON and OFF IPL layers, allowing them to mediate neurotransmission between ON and OFF BCs at around P11–P13; AII AMCs receive input from ON BCs and provide crossover inhibition onto OFF BCs (Demb and Singer 2012). Maturation of crossover inhibition is responsible for the asynchronous spontaneous bursting activity of ON and OFF RGCs during early developmental stages (Akrouh and Kerschensteiner 2013). The asynchronous activity of ON and OFF RGCs is hypothesized to help segregate ON and OFF retinal projections to central visual targets (Wong and Oakley 1996). Thus, establishment of early inhibitory inputs in the retinal IPL is important for the development of proper retinal activity, as well as retinal projection patterns.

19.3.1.2 Outer Retina

Like the IPL, synapse assembly in the OPL follows a specific sequence of events, and for photoreceptor ribbon synapses, presynaptic maturation precedes that of postsynaptic elements. At the time of birth, mouse photoreceptors have been generated and localize to the outermost layer of the retina (Olney 1968) (Fig. 19.2a). By P2, photoreceptors express vGluT1 and SV2B (Wang et al. 2003; Sherry et al. 2003a) (Fig. 19.2b). However, precursor spheres, which contain ribbon structures surrounded by synaptic vesicles, as well as ribbon component proteins (such as Ribeye, Bassoon, Piccolo, and RIM1), are still being trafficked toward the photoreceptor axonal terminals at P2 (Regus-Leidig et al. 2009). Furthermore, HC processes have yet to target the OPL in part because HCs are still migrating from inner to outer layers of the INL. HCs also initially form vertical processes that allow them to tile with neighboring HCs, before retracting these processes at P4 (Huckfeldt et al. 2009). The newly formed HC lateral extensions first express ionotropic glutamate receptor, GluR4, at around P3, followed by GluR2/3 at P6 (Hack et al. 2002). HC neurites also begin to invaginate photoreceptor terminals at P4 (Blanks et al. 1974). At this point, a few ribbons have anchored to the membrane and are apposed by at least two postsynaptic processes, as seen by electron microscopy (Regus-Leidig et al. 2009). Thus, the assembly of pre- and postsynaptic arrangements in photoreceptor-HC ribbon synapses is initiated before that of BC dendrites.

BC dendrites begin to elaborate and insert into photoreceptor terminal invaginations at P5 (Morgan et al. 2006). As mature triad synapses appear and become abundant, the fraction of precursor spheres in photoreceptors decreases. BC postsynaptic assembly progresses slowly after dendritic invagination; the major glutamate receptor of ON BCs, mGluR6, appears diffuse by immunohistochemistry at P8 and does not assume a punctate pattern, indicative of clustering in mature synapses, until around P12 (Nomura et al. 1994). Despite an early start in presynaptic maturation, photoreceptor ribbon synapses are not complete until the second postnatal week. Mice open their eyes at P14, allowing photoreceptors to begin conducting light-driven neurotransmission. By this stage, mature ribbons are clearly evident in the OPL (Regus-Leidig et al. 2009) (Fig. 19.2b). However, OPL synapses undergo further maturation and refinement even after eye-opening. Localization of mGluR6 in the OPL becomes more distinct (Nomura et al. 1994), and the number of mGluR6 puncta associated with cone terminals increases, whereas non-cone-associated puncta are eliminated (Dunn et al. 2013). Altogether, the formation of synapses from photoreceptors onto BC dendrites in the second postnatal week marks the onset of visual information flow from photoreceptors to RGCs.

19.3.2 Molecular Control of Synapse Assembly

Synapse formation entails assembling and anchoring presynaptic proteins to the membrane, as well as targeting dendrites to presynaptic sites and allocating the appropriate neurotransmitter receptors on the postsynaptic dendrites. Much work has demonstrated that molecular cues regulate each of these steps in either excitatory or inhibitory synapse formation, as well as contribute to coordinating the assembly of pre- and postsynaptic specializations between different cell types.

19.3.2.1 Inner Retina

Little is known about the molecules and cellular interactions involved in assembling excitatory synapses in the IPL. Nonetheless, multiple molecules have been identified that regulate the step of receptor clustering at inhibitory conventional synapses. The scaffold protein gephyrin is essential to anchor, and thus cluster, inhibitory neurotransmitter receptors to the postsynaptic cytoskeleton during retinal development (Tyagarajan and Fritschy 2014). In the mammalian retina, gephyrin colocalizes with a large portion of conventional synapses, including all glycine receptors, and also colocalizes to different extents with a population of $GABA_A$ receptors, including α1–α3 subunit-containing receptors (Grünert and Wässle 1993; Sassoè-Pognetto et al. 1995). Gephyrin is required for glycine receptor clustering because glycine receptors are almost completely diminished in the mice lacking gephyrin (Fischer et al. 2000). In contrast, expression of α2, α3, and γ2 $GABA_A$ receptor subunits is reduced, but a substantial amount of punctate signal persists (Fischer et al. 2000). It remains to be explored why gephyrin is more important for clustering only a fraction of the inhibitory receptors it colocalizes with, but progress has been made in identifying other clustering factors. Neuroligin (NLG), a cell adhesion molecule, plays a slightly overlapping but also complementary role with gephyrin in clustering inhibitory receptors. NLGs determine postsynaptic differentiation (Graf et al. 2004; Nam and Chen 2005), and NLG-2 specifically colocalizes with $GABA_A$ receptors (α1, α3, and γ2 subunit containing) as well as a portion of glycine receptor synapses in mouse retina (Hoon et al. 2009). Knockout of NLG-2 leads to a reduction in $GABA_A$ receptor subunit α3 and γ2 punctate signal, without changing the number of glycine receptor and $GABA_A$ receptor α1 subunit puncta. Although both NLG-2 and gephyrin are involved in clustering a portion of the inhibitory neurotransmitter receptors at conventional synapses, further studies are required to elucidate how each receptor subunit is differentially regulated.

Conventional synapses among AMCs and RGCs are established well in advance of the elaboration of BC axons in the IPL and the onset of ribbon synapse formation (Fisher 1979). Thus, it is possible that AMC-RGC conventional synapses serve as a guide for BCs to form ribbon synapses. Although the route of ribbon trafficking and molecular mechanisms by which ribbons anchor in the IPL remain to be determined, insights into the requirements of BC ribbon formation come from observations in zebrafish. Deletion of *ptf1a* and *math1* transcription factors in zebrafish prevents the formation of HCs, AMCs, and RGCs, such that BCs are the only remaining neurons that project to the IPL (Randlett et al. 2013). Surprisingly, BC axons are still able to form a neuropil that resembles the IPL, despite the absence of their postsynaptic partners and early-formed conventional synapses. BCs also appear to form presynaptic structures in *ptf1a* and *math1* mutants; ribbons, surrounded by synaptic vesicles, appropriately anchor to the terminal membranes. Thus, in the IPL, the formation of AMC-RGC conventional synapses is unnecessary for BC ribbon assembly later in development. Further, BCs may not require the presence of RGC and AMC dendrites in order to form ribbons.

19.3.2.2 Outer Retina

Several molecular players have been identified that are involved in each step of ribbon synapse formation in the OPL (Fig. 19.3). Unsurprisingly, depleting proteins that are essential components of the presynaptic active zone results in severe synaptic phenotypes (Dick et al. 2003). Ribbons fail to anchor to the membrane when an essential active zone cytomatrix protein in photoreceptors, Bassoon, is knocked out in mouse. Disruption of presynaptic ribbon proteins also impacts postsynaptic structure, because Bassoon mutant mice have fewer invaginating processes in their cone and rod terminals. It is unclear whether HC processes fail to invaginate during development or are not maintained in the invagination in Bassoon mutants. Nonetheless, these results indicate that the organization of postsynaptic processes depends upon the normal formation of the presynaptic zone during synaptogenesis and that synaptic assembly may be coordinated between pre- and postsynaptic partners.

Experiments to perturb postsynaptic organization show disruptions in ribbon localization during synaptogenesis, suggesting that pre- and postsynaptic cells are mutually dependent in order to properly assemble their synaptic elements. Specifically, HC neurite targeting of photoreceptor invaginations may limit the localization of ribbons to the presynaptic terminal. Disruption of HC neurite targeting was first achieved by deletion of a transcription factor necessary to HC neurite

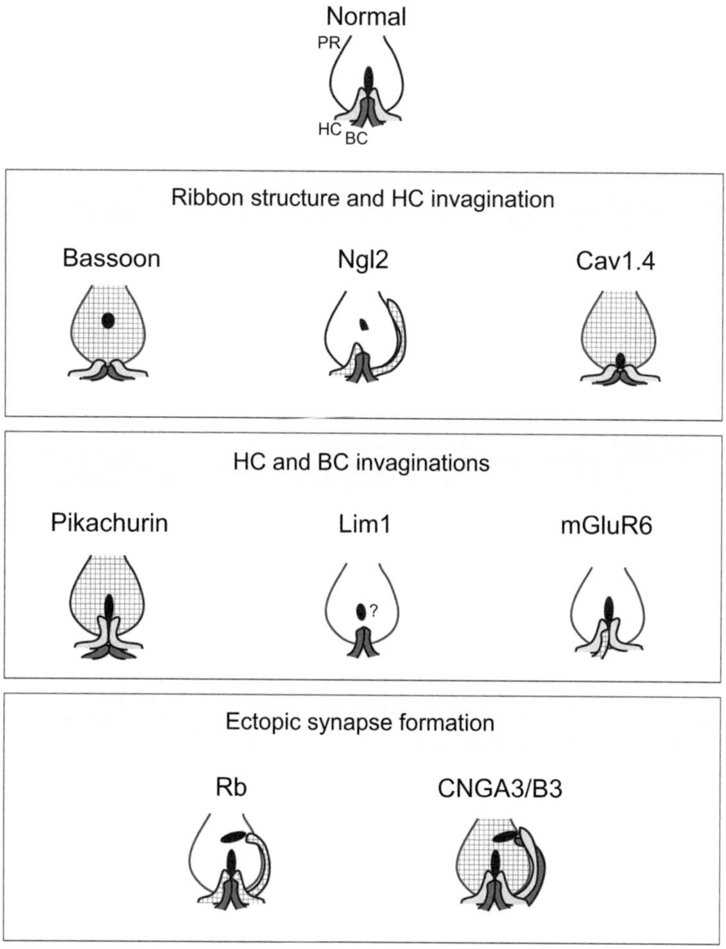

Fig. 19.3 Summary of structural changes to postsynaptic dendrites and ribbons at OPL synapses under various conditions. Photoreceptor (PR) terminals are normally invaginated by horizontal cell (HC) processes and bipolar cell (BC) dendrites that are apposed to presynaptic ribbons (e.g., *black oval*). Each schematic panel represents the ribbon synapse phenotype in animals with different mutated genes, indicated above each panel. Shading (*cross-hatch pattern*) indicates the cell type with a gene mutation. Note that the ribbon structure has not yet been studied in the *Lim1* conditional knockout mouse, as indicated by "?"

differentiation, *retinoblastoma* (*Rb*) (Martins et al. 2011). Deletion of *Rb* specifically in HCs causes aberrant vertical HC neurites to project to the ONL, where they form ectopic invaginations with photoreceptor soma (Fig. 19.3). Ribbon structures form at locations opposite sites of ectopic HC invaginations, suggesting that guidance of HC processes to the OPL is important to limit presynaptic ribbon formation to the OPL.

Recent research revealed a molecular mechanism to coordinate HC and photoreceptor synaptic maturation and further demonstrated that HC invagination is required for proper ribbon formation in the photoreceptor terminal. Numerous cell adhesion molecules localize to the synaptic site, making them primary candidates for mediating pre- and postsynaptic communication during synaptogenesis. In particular, the cell adhesion molecule netrin-G ligand-2 (NGL-2) interacts with postsynaptic scaffolding proteins and localizes specifically to HC axon tips (Soto et al. 2013). In NGL-2 knockout mice, HC axons sprout into the ONL and as a consequence fewer HC axons than normal invaginate rods (Fig. 19.3). Importantly, a presynaptic binding partner of NGL-2, netrin-G2 (Kim et al. 2006), is expressed in photoreceptors (Soto et al. 2013), priming NGL-2 to be a mediator of trans-synaptic interactions during synaptogenesis. Indeed, presynaptic maturation is disrupted in NGL-2 knockout mice; ribbon structures in rods are often misshapen or free-floating. Thus, HC invagination is likely required for proper ribbon formation in the OPL and is mediated by adhesive interactions between HCs and photoreceptors.

Whereas the presence of HC processes appears to be critical to the assembly of presynaptic ribbon structures, removing BC dendrites from photoreceptor terminals does not affect photoreceptor synaptogenesis. RBC dendrites stratify in the OPL but fail to invaginate photoreceptor terminals in mice lacking the extracellular matrix protein pikachurin (Sato et al. 2008) or its transmembrane receptor, dystroglycan, both of which are expressed in photoreceptors (Omori et al. 2012). Despite lacking invaginating BC processes, photoreceptor terminals still contain anchored ribbons in pikachurin or dystroglycan mutants (Fig. 19.3). In addition to ribbon formation, the expression of neurotransmitter receptors is independent of whether BCs invaginate; RBCs still express mGluR6 on their dendritic tips in pikachurin or dystroglycan mutant mice. Altogether, presynaptic ribbon localization, restriction of BC dendrites to the OPL, and allocation of synaptic receptors can proceed regardless of whether BC dendrites invaginate photoreceptor terminals.

Although HCs are major organizers of presynaptic maturation, the organization of HC processes and BC dendrites in the photoreceptor terminal appears to occur independently. The finding that HCs are not necessary for BC targeting of photoreceptors was first produced by experiments to eliminate HC processes from the OPL during retinal development. In mice with a conditional knockout of *Lim1* at early stages, HC migration is disrupted and results in HCs failing to extend processes into the OPL (Keeley et al. 2013) (Fig. 19.3). Even in the absence of HCs in the OPL, photoreceptor axon terminals and RBC and CBC dendrites stratify normally. BC synapses with photoreceptors appear largely normal, because invaginations of photoreceptor terminals and Ribeye protein are still clearly present in the OPL, although it is not clear whether ribbons are anchored at synaptic sites. These results are congruous with observations in NGL-2 knockout and *Rb*-deficient mice, in which RBC dendrites stratify normally and localize postsynaptic components in the OPL despite a reduction in the number of HC axons in the terminal (Martins et al. 2011; Soto et al. 2013). Thus, BC dendrites do not require the presence of HC processes in the OPL in order to target cone or rod terminal invaginations.

Likewise, HC processes invaginate independent of BCs; HC neurites still target photoreceptor invaginations opposite membrane-anchored ribbon structures in pikachurin or dystroglycan mutant mice (Sato et al. 2008; Omori et al. 2012). Thus, although BCs and HCs are intimately associated at photoreceptor ribbon synapses, their neurites localize independent of each other, and this localization is likely regulated by exclusive contingents of molecular cues.

19.3.3 Activity-Dependent Synapse Formation

Numerous studies have demonstrated that neurotransmission plays an important role in synapse formation throughout the brain (reviewed in Okawa et al. 2014b). Experiments manipulating neurotransmission in the retina, either by altering release from presynaptic cells or depleting receptor from postsynaptic cells, have provided insight into how neurotransmission regulates synapse formation onto dendrites.

19.3.3.1 Inner Retina

Examination of ON-BC connections with the dendrites of alpha-ON-RGCs (A-ON-RGCs) has provided strong evidence that neurotransmission directly impacts synaptogenesis in the IPL, to the extent that the level of transmission dictates the number of synapses formed. In retina in which ON BCs are silenced by expression of tetanus toxin, preventing synaptic vesicle release, A-ON-RGCs show reduced numbers of glutamatergic synapses on their dendrites (Kerschensteiner et al. 2009). Conversely, elevated glutamate release from BCs increases the number of synapses on A-ON-RGCs (Soto et al. 2012). Mutation of *crx*, a transcription factor necessary for photoreceptor maturation (Freund et al. 1997; Furukawa et al. 1997), leads to a lack of functional photoreceptor input onto BC dendrites and consequently induces hyperactive, spontaneous transmission from BCs.

Quantification of synapses on an RGC dendritic arbor over time provided important insight into how neurotransmission impacts synapse number during development, whether by influencing synapse formation, elimination, or a combination thereof. Time-lapse imaging of PSD-95 clusters on A-ON-RGC dendrites when BC transmission is reduced or increased showed that neurotransmission specifically regulates the rate of synapse formation, but not the rate of synapse elimination (Kerschensteiner et al. 2009; Soto et al. 2012). Synaptogenesis is increased in retinas with increased spontaneous activity and decreased in retina with silenced ON BCs, but altered neurotransmission yields no changes in the rate of synapse elimination on RGC dendrites. Observations in retina with silenced ON BCs further demonstrated that BC neurotransmission also affects the proper assembly of synaptic arrangements. A-ON-RGC dendrites retain about half their normal number of synapses in BC-silenced retina, but the structure of remaining BC ribbons is disrupted: multiple ribbons, rather than a single ribbon, cluster opposite the postsynaptic density (Kerschensteiner et al. 2009). Thus, in the IPL, neurotransmission from BCs regulates synapse formation onto RGC dendrites, as well as the proper localization of ribbons in BC axons.

19.3.3.2 Outer Retina

Functional transmission from the photoreceptors is critical for limiting HC synapse formation to the confines of the OPL. In mice lacking two essential components of the cone and rod phototransduction cascades, cyclic nucleotide-gated channel α3 and β1 subunits (CNGA3 and CNGB1), HCs initially target their processes to the OPL, but later sprout numerous neurites into the ONL by P12 (Michalakis et al. 2013) (Fig. 19.3). Because HC processes normally target the OPL early in retinal development, well before eye-opening at P14, it is likely that light-driven neurotransmission from photoreceptors is important for the maintenance of HC processes in the OPL, rather than their initial targeting. Reduced photoreceptor neurotransmission does not affect presynaptic assembly; some HC processes remain in the OPL and insert into photoreceptor invaginations, and ribbon structure and anchoring are unaffected in CNGA3/B1 knockout mice. However, electron microscopy revealed that by P36, the sprouted HC processes are associated with ectopic ribbon synapses, as seen in the *Rb* mutant mouse. Neurotransmission from photoreceptors is thus important for maintaining pre- and postsynaptic components of the photoreceptor-HC circuit at their proper locations in the OPL.

Neurotransmission further impacts the proper organization of dendritic synaptic components in the OPL by regulating BC synapses. RBC dendrites often fail to invaginate the rod axon terminal in mGluR6 knockout mice, which do not conduct glutamate signaling from photoreceptors (Tsukamoto and Omi 2014b). But, it is unclear whether RBC dendrites do not invaginate photoreceptor terminals as a result of initial mistargeting or as a result of failed maintenance in the terminal (Fig. 19.3). Sensory drive also affects postsynaptic assembly by regulating mGluR6 clustering on CBC dendrites. Although CBCs from dark-reared animals reduce receptor clusters on their dendrites at sites of contact with cone terminals, mGluR6 clustering at rod contacts is normal on RBC dendrites, despite light deprivation (Dunn et al. 2013). Exactly how dark-rearing alters neurotransmission from photoreceptors is unclear, and thus, the mechanisms by which sensory deprivation affects the synapses on BC dendrites remain unknown.

Glutamate transmission requires presynaptic calcium signaling, but calcium signaling alone is critical to synaptogenesis in the OPL. Mutation of the CaV1.4 voltage-gated calcium channel in mice perturbs calcium-dependent synaptic vesicle release (Mansergh et al. 2005) and ultimately leads to disruptions in multiple elements of proper synapse formation, including photoreceptor terminal stratification, ribbon morphology, ribbon protein allocation, and postsynaptic invagination (Bayley and Morgans 2007) (Fig. 19.3). Photoreceptor terminals initially stratify normally, but by P13, many retract into the ONL (Zabouri and Haverkamp 2013).

Similarly, ribbon-associated proteins, Bassoon and Ribeye, gradually accumulate at ectopic locations in the ONL. Calcium signaling does not only play a role in the maintenance of presynaptic elements at later stages, however, because electron microscopy revealed that photoreceptor terminals are malformed by P13 and contain shortened ribbon structures. The misshapen terminals also lacked invaginating HC processes and BC dendrites, revealing a more severe invagination phenotype than when glutamate transmission alone is altered in CNGA3/B1 and mGluR6 knockout mice. Thus, independent of its role in mediating glutamate release, the presence of the CaV1.4 channel is also required for the development of normal ribbons in the outer retina.

19.4 Synaptic Partner Selection

The specialized functions of different neurons in the retina are produced in large part by their connections with presynaptic partners. The stereotyped connectivity pattern of a retinal neuron is characterized by the identity of its presynaptic partners, as well as by the number and distribution of synapses it makes with each of these partner types. How do neurons in the retina establish their unique patterns of connections? The laminar organization of the retina provides the first limitation on synaptic partner choice. For example, RGCs with dendrites stratifying in the ON layer of the IPL only encounter axons from the functionally matched ON BCs, rather than the entire BC population. Nonetheless, even within a synaptic layer, retinal neurons wire exclusively with a subset of available partner types. Different cell types may generate their connection patterns by employing various strategies, from targeted partner selection to developmental refinement of inappropriate connections. The known molecular mechanisms to direct lamination and instruct target specificity, as well as the roles that afferents play in regulating the wiring of postsynaptic cells, are discussed below.

19.4.1 Molecular Control of Dendritic Lamination and Partner Matching

The role of molecular signaling in guiding dendritic arbor lamination is well established. Several membrane-bound molecules have been shown to regulate dendrite stratification, including repulsive guidance cues from the semaphorin/plexin family, as well as adhesion molecules, such as cadherins, and immunoglobulin superfamily (IgSF) members, including Sidekicks, Dscam, and Contactins (discussed in detail in Chap. 10, by Fuerst).

Like the dendrites of postsynaptic cells, the axons of presynaptic cells must be guided to the proper synaptic laminae in the retina in order to allow comingling of appropriate synaptic partners for synaptogenesis. Recent work in mouse retina demonstrated a role for cadherins in this process and, importantly, provided the first evidence of a molecular mechanism to match the dendrites of a postsynaptic cell with the terminals of the appropriate partner within a synaptic plexus in the retina (Duan et al. 2014). In the direction-selective circuit, type 5 and type 2 BCs synapse directly onto the ON or OFF dendritic arbors of the ON-OFF direction-selective ganglion cell (DSGC), respectively, as well as indirectly via connections with the dendrites of ON and OFF starburst amacrine cells (SACs) (Fig. 19.4a). Each BC type expresses a different cadherin, cadherin-8 (cdh8) in type 2 BCs and cadherin-9 (cdh9) in type 5 BCs. Evidently, both type 2 and 5 BCs employ these adhesion molecules to achieve their particular stratification levels in the IPL. Deletion of cdh8 or cdh9 misplaces many, but not all, type 2 or type 5 BC axon terminals to the other's approximate laminar location (Fig. 19.4a). Both the misplaced and normally stratifying BC axon terminals in cdh8 or cdh9 mutants are still positioned to make synaptic contact with either the ON or OFF dendritic arbors of the appropriate DSGC type. Yet, electrophysiological recording from

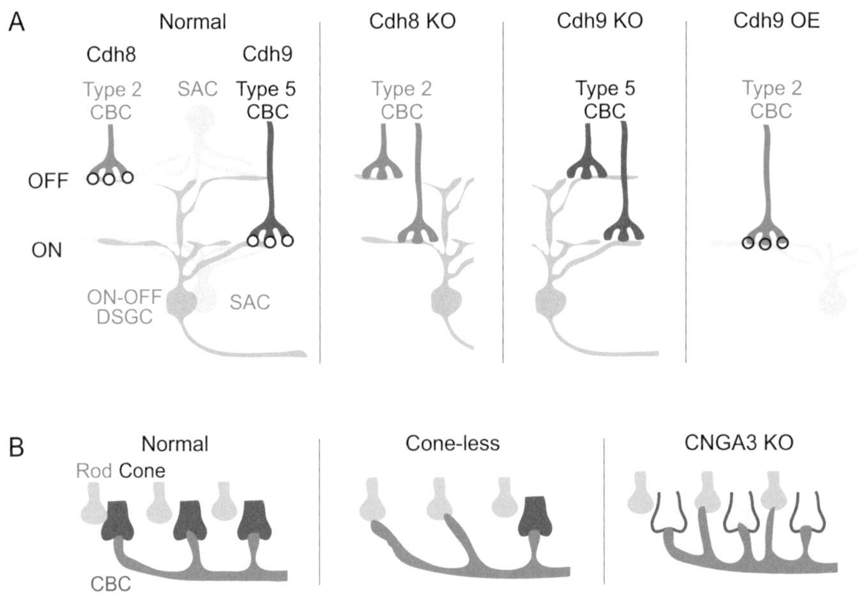

Fig. 19.4 Molecular and afferent regulation of synapse specificity onto retinal neuron dendrites. (**a**) Synaptic partner matching between cone bipolar cells (CBCs) and the dendrites of starburst amacrine cells (SACs) and direction-selective ganglion cells (DSGCs) in adult mouse retina. Cadherin-8 (Cdh8) and -9 (Cdh9) are normally expressed in type 2 and 5 cone bipolar cells (CBCs), respectively. Illustrations represent the differential effects of Cdh8 or Cdh9 knockout (KO) on type 2 or 5 BC axon stratification and connectivity with ON-OFF DSGC dendrites. Cdh9 overexpression (OE) redirects type 2 CBCs axon terminals to stratify and make ectopic synapses with ON-SAC dendrites. (**b**) Schematics of changes to CBC dendritic partner selection in response to alterations in cone composition (in "coneless" mutant mice) or neurotransmission (in CNGA3 KO mice). Cones with disrupted neurotransmission are outlined

DSGCs during optogenetic stimulation of single BCs showed that BC transmission onto DSGC dendrites is severely reduced, suggesting a lack of functional synapse formation from both normally stratifying and misplaced BCs. Thus, cadherins may not only facilitate contact between appropriate partners but directly impact synapse formation. This possibility is supported by experiments in which ectopic expression of cdh9 in type 2 BCs redirected their axons to the ON layer and induced ectopic synapse formation with ON SACs (Fig. 19.4a). Altogether, cadherins are clearly important for the proper axon lamination of two BC populations in the mouse retina and may be among the first molecules identified to "match" the dendrites of postsynaptic cells with their appropriate presynaptic partners.

19.4.2 Refinement of Dendritic Lamination

A combinatorial code of molecular cues can specify and confine neurite targeting to the plexiform layers, yet structural refinement may also occur to achieve appropriate stratification (reviewed in D'Orazi et al. 2014). In developing cat and ferret retina, the dendrites of ON or OFF alpha and beta RGCs initially span the entire IPL (Maslim and Stone 1986; Bodnarenko and Chalupa 1993). As the retina matures, the dendrites of these RGCs are eliminated from inappropriate depths, such that the final stratification level is sculpted from an initially diffuse arbor (Bodnarenko et al. 1995). However, work examining the lamination of RGCs over time in zebrafish and mouse retina has demonstrated that RGCs do not always rely on selective pruning from inappropriate depths to achieve lamination (Mumm et al. 2006; Kim et al. 2010). In particular, most RGCs examined during time-lapse imaging of transgenic zebrafish retina initially form an arbor that is biased toward the correct laminar zone. Further, bistratified zebrafish RGCs sequentially added multiple arbors at different levels, rather than refining an initially diffuse arbor.

Clearly, diverse strategies can be employed by retinal cells to achieve their final lamination, including different forms of laminar refinement as well as biased targeting and sequential addition of arbors. To date, it is unclear why a particular neuron type employs its chosen stratification strategy, and attempts to understand the relationship between neural activity and laminar refinement have provided only further evidence of diverse mechanisms. Hyperpolarizing ON BCs by chronic activation of mGluR6, effectively silencing input onto ON-RGCs, prevented restriction of ON alpha and beta RGCs' dendrites to the appropriate IPL (Bodnarenko and Chalupa 1993). Likewise, sensory deprivation experiments in mice found that an unusually high fraction of the RGC population had both ON and OFF responses, suggesting a failure to undergo refinement when neurotransmission is altered (Tian and Copenhagen 2003). By contrast, when BC transmission is disrupted by tetanus toxin expression or in mGluR6 knockout mice, RGCs arborize normally (Kerschensteiner et al. 2009). These results further highlight the need to study lamination and wiring with regard to specific cell types and by comparing different manipulations of neurotransmission (Bleckert and Wong 2011).

19.4.3 Regulation of Partner Choice Within a Synaptic Layer

Within a synaptic layer, the dendrites of a retinal neuron synapse with only some of the partners available to them, and their partner selection is cell-type specific and stereotyped. However, dendritic partner choice is not molecularly "hardwired," thus necessitating a mechanism to prevent the dendrites of neurons from wiring with inappropriate partners during normal development. This necessity was demonstrated by experiments in which rod or cone photoreceptors were eliminated during development in mouse retina. CBC dendrites form novel contacts with rods in "coneless" mice, in which cones are ablated by diphtheria toxin expression. Similarly, RBC dendrites, as well as the axons of HCs, make new synapses with cones in "rodless" mice, in which rods are respecified to become cones (Strettoi et al. 2004; Keeley and Reese 2010) (Fig. 19.4b). Thus, BC dendritic connectivity is plastic, but neurotransmission from the preferred afferents is likely the cue that prevents dendrites from selecting novel partner types during normal development. In mice with impaired rod function due to a mutation in rhodopsin, RBCs form and maintain connections with cones, before photoreceptors eventually degenerate (Peng et al. 2000; Cuenca et al. 2004). Likewise, in the CNGA3 mutant mouse, which lacks a cation channel subunit necessary for cone function, CBC dendrites form contacts with rods (Haverkamp et al. 2006) (Fig. 19.4b). Taken together with observations of rewiring in coneless and rodless mice, this indicates that neurotransmission from the appropriate afferents is necessary to prevent dendritic selection of inappropriate partner types in the photoreceptor-BC circuit in mice.

In the IPL, elimination of all the preferred partners of a postsynaptic cell results in severe effects on dendritic connections, including alterations to dendrite stratification as well as partner preference. Elimination of all ON BCs via expression of diphtheria toxin in mouse retina results in ectopic elaboration of the A-ON-RGC's dendrites into the OFF layer of the IPL (Okawa et al. 2014a). Remarkably, the OFF-stratifying dendrites selectively synapse with a specific OFF BC type, type 2, within the layer. Thus, in the A-ON-RGC circuit, the preferred BC partner specifies RGC connectivity by restricting the RGC dendrites to the appropriate lamina and by prohibiting synaptogenesis with functionally inappropriate cells. Unlike the photoreceptor-BC circuit, afferent control of partner selection is evidently exercised independently of neurotransmission from the RGCs' afferents; RGCs arborize appropriately even in retina in which all ON BCs are silenced (Kerschensteiner et al. 2009). Moreover, blocking neurotransmission alone does not affect dendritic partner specificity (Morgan et al. 2011), because the A-ON-RGC retains synapses with silenced type 6 and 7 ON BCs and still eliminates inappropriate contacts with silenced RBCs.

The dendrites of some neurons initially make "mismatched" connections with inappropriate partners, and thus, developmental refinement is critical to generate the mature connectivity patterns of these neurons. Evidence for refinement of partner choice comes from work in both mouse and zebrafish. Early on, the dendrites of A-ON-RGCs in mouse retina wire with both ON CBCs and RBCs in the ON layer of the IPL. However, at later stages, A-ON-RGCs eliminate their synapse with RBCs to establish connectivity specifically with ON CBCs (Morgan et al. 2011). Similarly, H3 HCs in zebrafish retina selectively synapse with UV and blue cones at maturity, but exhibit transient connections with red or green cones at immature stages (Yoshimatsu et al. 2014). In both cases, the developmental elimination of erroneous connections does not depend upon neurotransmission from the preferred partner types. A-ON-RGCs still lose their RBC connections when ON-BC neurotransmission is suppressed and H3 HCs do not maintain red/green cone connections despite a reduction in either or both UV and blue cone transmission. Furthermore, although the majority of H3 HC dendritic connections are made with UV cones, the presence of this major partner is not required for H3 HC synaptic specificity. Despite elimination of the UV cone population from the retina at early stages in the *lots-of-rods* mutant, H3 HC dendrites did not synapse with red or green cone types at maturity (Yoshimatsu et al. 2014). In the case of the H3 HCs, it is clear that the major partner type is unnecessary to prevent dendritic wiring with new partner types at maturity. Nonetheless, it remains to be seen whether transmission from the inappropriate afferents themselves induces elimination of their connections with H3 HCs during normal development.

19.5 Synapse Number and Density

In addition to choosing among presynaptic partner types, the dendrites of a cell must make a stereotyped number of connections with each of its afferents to establish the cell's connectivity pattern. Numerous presynaptic partners often lie within the territory of a postsynaptic cell's dendrites. How do dendrites regulate the number of partners with which they synapse, and how are synapses allocated per afferent? As in the process of partner selection, different cell types variably employ synapse elimination to achieve the mature number of synapses. This section summarizes the known mechanisms that regulate dendritic synapse number and synapse allocation per afferent.

19.5.1 Influence of Afferent Availability

Investigations of how dendrites acquire their final number of connections during development in mouse and zebrafish retina have demonstrated that different cell types may employ different strategies (Dunn and Wong 2012; Yoshimatsu et al. 2014). The H3 HC in zebrafish simply increases its synapses with UV cones as more become available due to the expansion of its dendritic arbor during development and eventually contacts all of the available UV cones. CBCs in mice engage diverse strategies to set their synapse number during development. Like H3 HCs, type 6 BCs continue to form synapses until they reach the mature number. In contrast, type 7 and 8 BCs exuberantly synapse with cones, only to reduce their total number of synapses after P13. Type 7 and 8 BCs also remodel the size of their dendritic fields over time. For all three CBC circuits described, the CBCs synapse with all of the afferents available to their dendrites, even in the case of type 7 and 8 CBCs, which exuberantly add and eliminate synapses over time. These findings suggest that the number of available afferents may determine the number of connections on a retinal cell's dendritic arbor.

Experiments to alter retinal cell densities in multiple species have shown that the number of afferents lying within a dendritic arbor's territory often dictates the number of cells it contacts (Fig. 19.5). For example, in mice, HC dendrites connect with a higher number of cones when the density of cones across the retina is increased by respecification of rods to the cone cell fate (Raven et al. 2007) (Fig. 19.5a). Similar results were observed for zebrafish HCs; a morpholino-induced increase in the density of the predominant partner, UV cones, causes H3 HC dendrites to contact more UV cones compared to wild type (Yoshimatsu et al. 2014) (Fig. 19.5b). Neurons may also increase the number of presynaptic partners available to them by increasing dendritic growth (Poché et al. 2008). In the *Lim1* conditional knockout mouse, HCs lacking neighboring HCs expand their dendritic territory and thus come in proximity to an increased number of afferents. Ultimately, the expanded HCs connect with more cones than in wild-type retina. Neurons make more dendritic connections than normal in all of the experiments discussed above, providing evidence that many retinal neurons are free of cell-intrinsic limitations on synapse number.

However, retinal neurons do not always synapse with all of the preferred afferents available to them at maturity, as seen in circuits with "biased" connectivity patterns. For example, H3 HCs in zebrafish only contact a fraction of their secondary partner, blue cones. Therefore, the availability of afferents alone cannot regulate the number of connections formed with secondary partners, and the strategies that dendrites employ to set their synapse number may depend upon the presynaptic partner type.

19.5.2 Homeostatic Mechanisms Controlling Synapse Density

Thus far, there are no clear examples of intrinsic mechanisms to set an upper limit on the number of afferents contacted by the dendrites of a retinal neuron. But, there is accumulating evidence of homeostatic mechanisms to compensate for a reduction

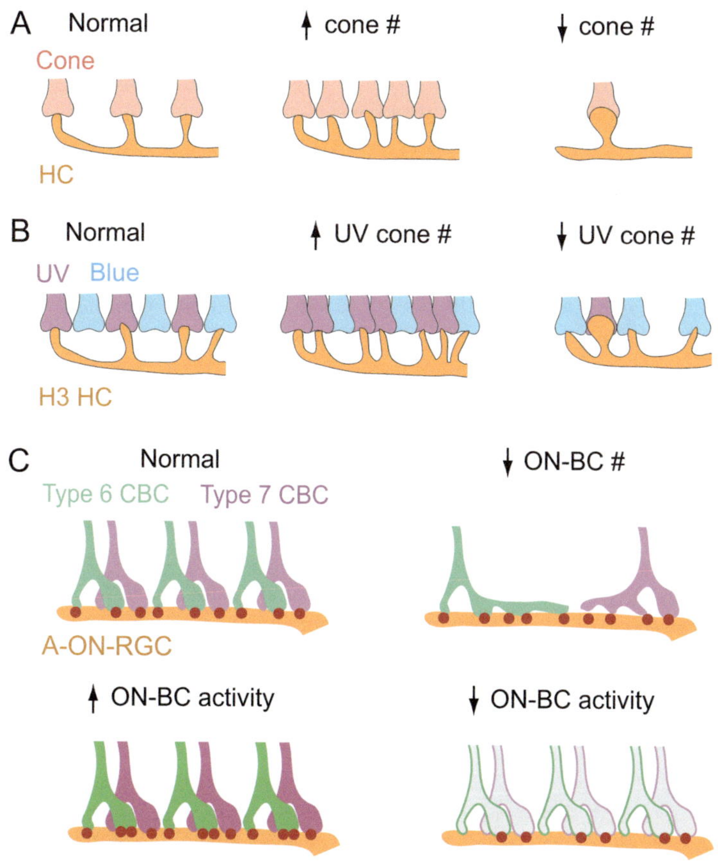

Fig. 19.5 Regulation of dendritic synapse number by afferent availability and activity. (**a**) Schematic of HC synaptic contact with cones after alteration of cone density in mouse retina (in "conefull" and "coneless" mutant mice). (**b**) H3 HCs in larval zebrafish retina connect with all the available UV cones (UV) within their dendritic fields, but only contact a fraction of the available blue cones (*Blue*). Schematics represent changes to H3 HC dendritic connections and morphology when UV cone numbers are altered by morpholino or in the *lots-of-rods* mutant. (**c**) (*Top*) Alpha-like ON-stratifying retinal ganglion cells (A-ON-RGCs) normally receive the majority of their inputs from type 6 cone bipolar cells (CBCs) and minor input from type 7 CBCs. (*Top right*) Illustration of the effects of a decreased number of ON bipolar cells (ON BCs) on BC presynaptic territory size and synaptic density on A-ON-RGC dendrites. (*Bottom*) Representation of the differential effects of modified presynaptic neurotransmission on the number of A-ON-RGC dendritic synapses with its major and minor input partners. Hyperactive neurotransmission from ON BCs in *crx* mutant mice and silencing of ON-BC neurotransmission by tetanus toxin expression

in afferent availability. For example, under circumstances in which the density of a cell's afferents is reduced, the cell's dendrites may expand their territory, thus increasing the number of available afferents in order to achieve a stereotypic number of synapses. This possibility was addressed by experiments to reduce, but not completely eliminate, the density of cones across the retina by genetic ablation in the coneless mouse and by respecification of cones to rods in the "rodfull" mouse. Examination of HC connectivity in these mice revealed that HCs do not expand their dendritic territories compared to normal, nor do they acquire the normal number of connections (Reese et al. 2005; Raven et al. 2007). Similar experiments in zebrafish demonstrated that retinal neurons may instead use a different homeostatic mechanism, increased synaptogenesis with a secondary partner, to compensate for a reduction in preferred partner availability (Yoshimatsu et al. 2014) (Fig. 19.5b). When the density of UV cones is decreased in the *lots-of-rods* mutant, H3 HC dendrites allocate more synapses to blue cones. However, this compensation is only partial, as the H3 HC's total number of dendritic synapses is still reduced compared to normal. Nonetheless, results from manipulations of HC afferents suggest another strategy whereby postsynaptic cells may compensate fully for lost connections. In both mice and zebrafish with reduced cone densities, the HC dendritic terminals that invaginate the remaining cone terminals are expanded in size (Reese et al. 2005; Yoshimatsu et al. 2014) (Fig. 19.5a, b). Although it remains to be determined, it is possible that this phenotype represents an increase in the number or strength of synapses per afferent. Thus, under circumstances in which it is not possible to increase the local availability of afferents, cells may achieve their total number of synapses, or level of synaptic input, by increasing the number and/or strength of synapses per afferent.

Direct evidence of a homeostatic increase in synapse number per presynaptic cell comes from the A-ON-RGC circuit. Even when a majority of ON BCs is ablated in mouse retina, A-ON-RGCs maintain the normal number of synapses on their dendrites (Okawa et al. 2014a). Analysis of the synapse number per individual BC axon terminal for a given RGC revealed two compensatory strategies (Fig. 19.5c). First, A-ON-RGC dendrites form more synapses per BC axon terminal. The remaining type 6 BCs expand their axons into the territories of their ablated neighbors, and the RGC forms more synapses with these expanded BCs. This increase in synaptogenesis matches axon territory expansion, such that the normal density of synapses onto the A-ON-RGC dendrites per type 6 BC terminal is maintained. Second, like the H3 HC, A-ON-RGCs increase synapses with their secondary partner, the type 7 BCs. The type 7 BCs also increase their axon terminal territories, but differ from type 6 BCs in that they increase their density of synapses with RGCs. Thus, A-ON-RGCs maintain their normal synapse number when afferent availability is reduced by increasing the number of synapses formed with both their major and minor afferents.

19.5.3 Activity-Dependent Mechanisms

The dendrites of retinal neurons are sensitive to the availability of their preferred presynaptic partners and may employ homeostatic mechanisms to compensate for a reduction in afferent availability from normal levels. How do presynaptic cells exert their control over the number of synapses on the postsynaptic cell's dendrites in the retina? As discussed in earlier sections, elimination of synapses with inappropriate partners does not appear to rely upon neurotransmission. But, the level of neurotransmission a dendrite receives directly impacts its rate of synaptogenesis and thus synapse number. Whether activity-dependent regulation of synapse number works at a local or global level on dendrites has been tested (Okawa et al. 2014a). Observations from retinas in which patches of BC populations express tetanus toxin showed that A-ON-RGCs decrease their synapse number in areas where BCs are silenced. Conversely, the normal number of synapses is maintained in areas in which BCs are not silenced. Thus, BC activity regulates A-ON-RGC dendritic synapse number locally, in a cell-autonomous manner.

Examinations of circuits in which connections are biased toward one input type provide further insights into how afferent neurotransmission influences the number of dendritic synapses formed per afferent type. In experiments to either increase or decrease ON-BC activity, only the number of RGC synapses specifically with type 6 BCs, but not type 7 BCs, is altered (Morgan et al. 2011; Soto et al. 2012) (Fig. 19.5c). A possible explanation for this difference is that presynaptic activity only impacts synapses with the major input type. This is unlikely to be true in the cone-H3 HC circuit, however, because H3 HCs connect with the normal number of UV cones even when the UV cones themselves are silenced (Yoshimatsu et al. 2014). However, in this circuit, neurotransmission specifically from the major input can control the number of synapses made with the minor input. In zebrafish retina in which UV cones are silenced, H3 HC dendrites increase the number of blue cones they synapse with, but silencing blue cones did not alter connections with UV cones. This unidirectional control of synapse number in a convergent circuit is unique, because investigations in other neural systems had previously found evidence for activity-dependent regulation that relied upon competition between converging afferents (Sanes and Lichtman 1999; Hashimoto and Kano 2013).

Lastly, recent work in mouse has demonstrated that the activity of the postsynaptic cell itself can "retrogradely" regulate its dendritic synapse number (Johnson and Kerschensteiner 2014). In experiments in which neurotransmitter release from the axon is abolished in a portion of the CBC population, the silenced CBCs have fewer cone connections on their dendrites, but only when the silenced cells are surrounded by active neighbors. Likewise, active CBCs increase their dendritic input number in regions where they are surrounded by silenced CBCs. Silenced CBC axons form fewer synapses regardless of their neighbors' status and therefore do not compete for synapses. These results show that, first, the dendrites of CBCs in mice compete for input from the same partners and, second, dendritic competition for synapses is surprisingly determined by the ability of CBCs to release neurotransmitter from their axons.

19.6 Synapse Distributions Across the Dendritic Arbor

Retinal neurons respond either to increments or decrements of light that fall within the center of their receptive fields. The spatial extent and structure of a retinal neuron's dendritic arbor dictates its sampling area and thus has a large impact on the size of its receptive field (Freed et al. 1992; Brown et al. 2000). Detailed light microscopy and electrophysiological studies of A-ON-RGCs revealed that the sparse pattern of the A-ON-RGC's dendritic arbor corresponds to the fine structure of its receptive field (Schwartz et al. 2012). This tight correlation between dendritic arbor pattern and receptive field structure is reflected by the uniform distribution of excitatory synapses across A-ON-RGCs' dendrites, and the development of this pattern is activity dependent. However, other examples exist in the retina in which the dendrite arborization pattern and connections of a neuron alone cannot account for its response properties. In the well-studied direction-selective circuit, a nonuniform distribution of inhibitory drive also plays a critical role in determining response characteristics (Fried et al. 2002; Wei and Feller 2011). In some convergent circuits, the locations of synapses from different input types are differentially distributed across the dendritic arbor. For example, in the H3 HC circuit, UV cone synapses are uniformly distributed across the dendritic arbor, but blue cone synapses are centrally located. The extent to which such unbalanced arrangements define response properties in different convergent circuits is still under investigation. Furthermore, the relative distributions of inhibitory and excitatory synapses on neuronal dendrites may impact local computation on the dendrite, as hypothesized for reciprocal or feed-forward synapses. This section describes the various patterns of dendritic synapse distributions found in the retina and discusses the known mechanisms to form such patterns during development.

19.6.1 Uniform Synapse Distributions

For several RGC types in mouse retina, excitatory synapses are distributed uniformly across the dendritic arbor, and this pattern is established early in the process of synaptogenesis (Fig. 19.6a). A recent study mapped the locations of glutamatergic inputs on the dendritic arbors of large-field RGCs by expressing fluorescently tagged PSD-95 via biolistic transfection (Morgan et al. 2008; Jakobs et al. 2008). At maturity, the spatial distributions of glutamatergic synapses on the dendrites of large-field RGCs show a slightly graded pattern, with a small increase in synaptic density in the central arbor, which is thought to match the spatial sensitivity of these RGCs' physiology (Morgan et al. 2008). Mapping at different stages revealed that the relatively uniform distribution of BC synapses on the dendrites of large-field RGCs is established from the onset of synaptogenesis at P5 and maintained through maturity at P30. However, the uniform synapse pattern of large-field RGCs is not a direct consequence of their dendritic morphology;

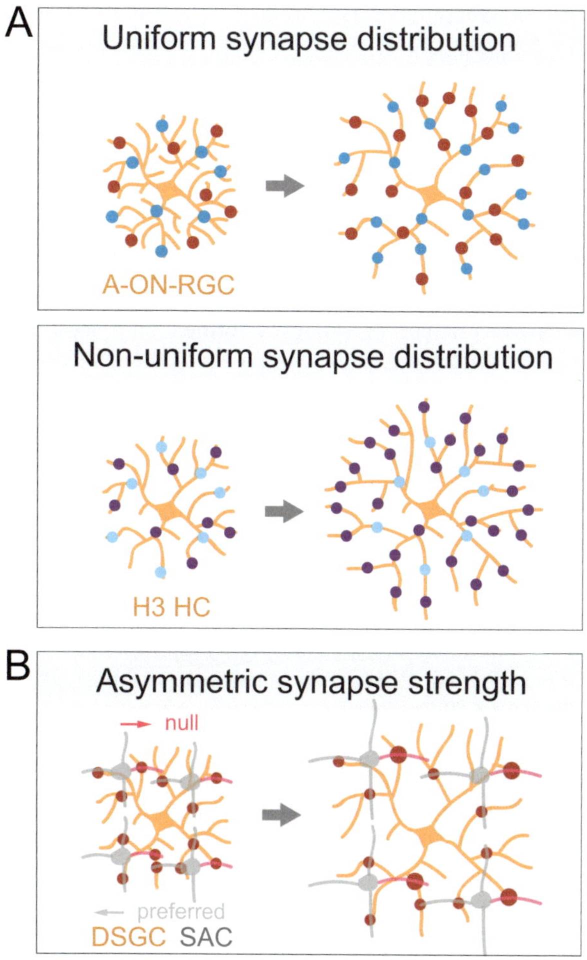

Fig. 19.6 Developmental mechanisms establishing distinct distributions of synaptic inputs onto dendrites of retinal neurons. The maturation of different adult synaptic distributions across dendritic arbors is illustrated from left to right. (**a**) (*Top panel*) From early to late stages, both excitatory (*red*) and inhibitory (*blue*) synapses onto alpha-like ON-stratifying retinal ganglion cell (A-ON-RGC) dendrites in mouse retina are uniformly distributed. (*Bottom panel*) H3 horizontal cells (HCs) in zebrafish retina undergo selective synaptogenesis with UV cones (*magenta*) in late development, resulting in unmatched synaptic distributions among UV and blue cone (*cyan*) inputs. (**b**) Direction-selective retinal ganglion cells (DSGCs) in mouse retina respond maximally to light moving in the preferred direction (*gray arrow*) and receive strong inhibition when light moves in the null direction (*pink arrow*). DSGCs first form synapses (*red circles*) with starburst amacrine cells (SACs) irrespective of the orientation of the SAC processes. Later in development, the synapses formed with SAC processes oriented in the null direction (*pink processes*) are selectively strengthened

during the period of synaptogenesis with BCs, the RGCs examined increase the territory of their dendritic arbors while undergoing extensive branch remodeling (Fig. 19.6a). Synapse formation onto and removal from the RGC dendrites are both ongoing during the period of RGC maturation, such that the global synapse density increases over time. Altogether, the dendrites of large-field RGCs maintain their early-established synaptic distributions despite both dendritic and synaptic remodeling during retinal development. The question of how these remodeling processes are regulated, such that a stable synaptic distribution is maintained, remains to be answered.

In the A-ON-RGC circuit, the level of afferent neurotransmission onto the RGC dendrites dictates the number of synapses formed with BCs (Kerschensteiner et al. 2009; Soto et al. 2012), and recent work has addressed how neurotransmission regulates the spatial distribution of these synapses. Previous investigations showed that the uniform distribution of type 6 BC synapses with A-ON-RGCs is a consequence of the RGCs forming a regular density of synapses with type 6 BCs over the entire territory of their dendritic arbors (Schwartz et al. 2012). Does afferent neurotransmission regulate the density of synapses that RGC dendrites form with BC axons globally? Or, is synaptic density controlled locally on an afferent-by-afferent basis? A recent study addressed these questions by quantifying the number of synapses a single A-ON-RGC forms per individual BC when only a fraction of the type 6 BC population is silenced during synaptogenesis, via expression of tetanus toxin (Okawa et al. 2014a). As previously established, reducing transmission from all type 6 BCs results in a decrease in synaptic density across an A-ON-RGC's dendritic arbor, thus maintaining an even synaptic distribution with type 6 BCs (Kerschensteiner et al. 2009). However, although reducing the transmission of an individual type 6 BC resulted in the A-ON-RGC dendrites forming fewer synapses with this BC, the density of synapses formed with active BCs converging onto the same RGC did not change. Consequently, the overall distribution of the A-ON-RGC's dendritic synapses with type 6 BCs is uneven when some of these BCs are silenced. This result demonstrated that neurotransmission from type 6 BCs during synaptogenesis regulates synaptic density on A-ON-RGC dendrites at the level of individual afferents, which in this circuit ultimately results in an even distribution of type 6 BC synapses on the dendritic arbors of A-ON-RGCs. Although it has yet to be demonstrated, this implies that the level of input provided across type 6 BCs is equivalent during synaptogenesis. In support of this assumption is the finding that ON BCs participate in patterned spontaneous activity during early stages of synaptogenesis (Wong et al. 2000; Blankenship et al. 2009), such that early on, type 6 BCs may provide comparable levels of excitatory neurotransmission onto A-ON-RGC dendrites.

It was recently shown that in the cases of some RGCs, the uniform pattern of excitatory synapses on RGC dendritic arbors is matched by the distributions of inhibitory synapses converging onto the same arbor. The local, spatial relationship between excitatory and inhibitory arrangements on retinal cell dendrites is likely functionally significant, because in other regions of the brain, individual dendrites can compute inputs at the level of single spines (Chen et al. 2011). Simultaneous labeling of excitatory and inhibitory synapses on RGC dendrites at different developmental stages in mice revealed that the precisely matched spatial organization of these synapse types in maturity is formed at early stages (Bleckert et al. 2013) (Fig. 19.6a). Mapping of GABAergic and glutamatergic synapses onto RGC dendrites was achieved by visualization of the GABA $\gamma 2$ subunit tagged with fluorescent protein, together with PSD-95 visualization. Like glutamatergic synapses, inhibitory synapse density increases with age, and inhibitory and excitatory densities coincide, such that a constant inhibitory/excitatory ratio is maintained throughout development (Soto et al. 2011; Bleckert et al. 2013). Computational comparisons of the measured densities allowed insight into how the relative distributions of excitatory and inhibitory synapses onto RGC dendrites might be functionally relevant; excitatory and inhibitory synapses are nonrandomly distributed from early through late stages, such that synapses of different types are localized closer together than synapses of the same type. Correlative light and electron microscopy reconstructions of excitatory and inhibitory synapses onto the RGC dendrites showed that the close proximity of these synapses corresponds to the structure of the feed-forward circuit (Fig. 19.1b). Given the proximity of excitatory and inhibitory synapses, as well as their overlapping time courses of synaptogenesis with RGC dendrites, it would be informative to explore whether excitatory neurotransmission influences inhibitory synapse formation and vice versa.

19.6.2 Nonuniform Synapse Distributions

Spatially nonuniform synapse distributions have also been found to shape the function of some retinal circuits. It has been proposed that the distribution of synapses from BCs onto SAC processes determines their sensitivity to centrifugal motion (Euler et al. 2002). SAC processes receive BC synapses along their length (Famiglietti 1991), however, type 2 BCs only synapse onto proximal dendrites, whereas type 3a BCs synapse on the distal dendrites (Kim et al. 2014). Computational modeling suggests that this spatially restricted distribution pattern, in combination with different input characteristics of type 2 and 3a BCs, may account for the SAC response to centrifugal motion. The question remains: how do cells confine synapses from different input types to specific regions of their dendrites?

Insights into the development of different synaptic distributions among input types come from the zebrafish HC circuit. As described in previous sections, mature H3 HCs synapse exclusively with UV and blue cones. UV cone synapses are distributed throughout the H3 HC arbor, whereas blue cone synapses are restricted to the proximal dendrites (Yoshimatsu et al. 2014) (Fig. 19.6a). Detailed connectivity analysis and time-lapse imaging during development revealed that H3 HC dendrites attain their bias for UV cones by becoming selective for UV cones as the distal dendrites grow (Fig. 19.6a). Genetic manipulation of neurotransmission from specific cone types demonstrated that UV cone transmission is required to restrict

H3 HC synaptogenesis with blue cones. Thus, H3 HC dendrites form and maintain different patterns of synaptic distributions with UV and blue cones by selective synaptogenesis, rather than by synapse elimination. However, due to technical limitations in recording from larval zebrafish retina, it remains unclear whether and how unmatched dendritic synapse patterns among different input types influence H3 HC responses.

In addition to the spatial patterning of synapses across the dendritic arbor, differences in synaptic strength among inputs can impact postsynaptic function. It is now evident that direction selectivity of DSGCs depends on an asymmetry of inhibitory input strength (Yoshida et al. 2001; Fried et al. 2002). There are four identified ON-OFF DSGC subtypes, each of which has a "preferred direction"; each DSGC responds maximally to light moving in one of the four cardinal directions and minimally to the anti-preferred direction ("null" direction). A single SAC provides inhibitory, GABAergic input to each of the four ON-OFF DSGC types. SACs respond to centrifugal directions of motion, such that a light stimulus moving away from the SAC soma only strongly depolarizes the dendrites that point in the direction that the light is moving (Euler et al. 2002). However, SACs synapses onto DSGCs are evenly distributed along the DSGC dendrites, on both the preferred and null sides of the DSGC arbor (Jeon et al. 2002; Chen and Chiao 2008) (Fig. 19.6b). Unlike the patterning of BC inputs onto SAC dendrites, asymmetric inhibition on DSGCs is generated by differences in synaptic strength (Fig. 19.6b). GABA-mediated inhibition evoked from null-direction SAC dendrites is stronger than the inhibition provided by SAC processes oriented in the preferred direction (Fried et al. 2002, 2005).

Ultimately, selective strengthening of a population of inhibitory synapses during maturation generates differences in synaptic strength among SACs converging onto the dendrites of a DSGC (Wei et al. 2011; Yonehara et al. 2011). Electrophysiological recording from pairs of SACs and DSGCs at different stages showed that, initially, GABAergic SAC synapses across the entire dendritic arbor are uniform in strength (Fig. 19.6b). Yet, only the synapses from the "null-direction" SAC dendrites are strengthened by later stages. The mechanisms that guide the eventual selective strengthening of specific synapses remain unknown. However, it is clear that light-evoked neurotransmission is not critical for this process, because direction selectivity is evident at stages when the light response has just begun to emerge (Chen et al. 2009). Neither modulation of GABAergic synapses, blockade of cholinergic synapses, nor blocking action potential activity by pharmacological treatment during development resulted in an alteration to direction selectivity in DSGCs (Sun et al. 2011; Wei et al. 2011).

19.7 Summary

Analysis of connectivity and physiology on a neuron-by-neuron basis has vastly expanded our understanding of the various factors that regulate dendritic wiring in the developing retina. Importantly, although a molecular code largely defines the organization of retinal neuron dendrites, the dendritic wiring of retinal neurons is also plastic. This demonstrates that in other brain circuits, molecular codes to organize dendrites may be overwritten if the neuronal environment changes. Insights from retinal development have also highlighted that many layers of control may influence each of the fundamental steps in normal synapse formation and patterning on neuron dendrites, including cellular interactions, molecular cues, and neurotransmission. Different retinal neurons depend variably upon each of these factors to form their stereotyped connections, and thus it is probable that no unifying rule can account for organizing the dendritic connections of the diverse circuits in the CNS.

Acknowledgments The authors acknowledge the support of NIH grant EY14358, as well as the support of NIH Training Grants EY07031 and HD07183 to F.D.D. The authors also thank Clare Gamlin, Mrinalini Hoon, Philip Mardoum, Haruhisa Okawa, and Rachel Wong for critical reading of the manuscript.

References

Ackman JB, Crair MC (2014) Role of emergent neural activity in visual map development. Curr Opin Neurobiol 24:166–175
Akrouh A, Kerschensteiner D (2013) Intersecting circuits generate precisely patterned retinal waves. Neuron 79:322–334
Arshavsky VY, Burns ME (2012) Photoreceptor signaling: supporting vision across a wide range of light intensities. J Biol Chem 287:1620–1626
Bansal A, Singer JH, Hwang BJ et al (2000) Mice lacking specific nicotinic acetylcholine receptor subunits exhibit dramatically altered spontaneous activity patterns and reveal a limited role for retinal waves in forming ON and OFF circuits in the inner retina. J Neurosci 20:7672–7681
Bayley PR, Morgans CW (2007) Rod bipolar cells and horizontal cells form displaced synaptic contacts with rods in the outer nuclear layer of the nob2 retina. J Comp Neurol 500:286–298
Blankenship AG, Ford KJ, Johnson J et al (2009) Synaptic and extrasynaptic factors governing glutamatergic retinal waves. Neuron 62:230–241
Blanks JC, Adinolfi AM, Lolley RN (1974) Synaptogenesis in the photoreceptor terminal of the mouse retina. J Comp Neurol 156:81–93
Bleckert A, Wong ROL (2011) Identifying roles for neurotransmission in circuit assembly: insights gained from multiple model systems and experimental approaches. BioEssays 33: 61–72
Bleckert A, Parker ED, Kang Y et al (2013) Spatial relationships between GABAergic and glutamatergic synapses on the dendrites of distinct types of mouse retinal ganglion cells across development. PLoS One 8:e69612
Bodnarenko SR, Chalupa LM (1993) Stratification of ON and OFF ganglion cell dendrites depends on glutamate-mediated afferent activity in the developing retina. Nature 364:144–146

Bodnarenko SR, Jeyarasasingam G, Chalupa LM (1995) Development and regulation of dendritic stratification in retinal ganglion cells by glutamate-mediated afferent activity. J Neurosci 15: 7037–7045

Brown SP, He S, Masland RH (2000) Receptive field microstructure and dendritic geometry of retinal ganglion cells. Neuron 27:371–383

Chen Y-C, Chiao C-C (2008) Symmetric synaptic patterns between starburst amacrine cells and direction selective ganglion cells in the rabbit retina. J Comp Neurol 508:175–183

Chen M, Weng S, Deng Q et al (2009) Physiological properties of direction-selective ganglion cells in early postnatal and adult mouse retina. J Physiol 587:819–828

Chen X, Hsueh H-A, Greenberg K, Werblin FS (2010) Three forms of spatial temporal feed-forward inhibition are common to different ganglion cell types in rabbit retina. J Neurophysiol 103:2618–2632

Chen X, Leischner U, Rochefort NL et al (2011) Functional mapping of single spines in cortical neurons in vivo. Nature 475:501–505

Cuenca N, Pinilla I, Sauvé Y et al (2004) Regressive and reactive changes in the connectivity patterns of rod and cone pathways of P23H transgenic rat retina. Neuroscience 127:301–317

Demb JB, Singer JH (2012) Intrinsic properties and functional circuitry of the aII amacrine cell. Vis Neurosci 29:51–60

DeVries SH, Li W, Saszik S (2006) Parallel processing in two transmitter microenvironments at the cone photoreceptor synapse. Neuron 50:735–748

Dick O, tom Dieck S, Altrock WD et al (2003) The presynaptic active zone protein bassoon is essential for photoreceptor ribbon synapse formation in the retina. Neuron 37:775–786

D'Orazi FD, Suzuki SC, Wong RO (2014) Neuronal remodeling in retinal circuit assembly, disassembly, and reassembly. Trends Neurosci 37:594–603

Dowling JE (1970) Organization of vertebrate retinas. Invest Ophthalmol 9:655–680

Duan X, Krishnaswamy A, De la Huerta I, Sanes JR (2014) Type II cadherins guide assembly of a direction-selective retinal circuit. Cell 158:793–807

Dunn FA, Wong ROL (2012) Diverse strategies engaged in establishing stereotypic wiring patterns among neurons sharing a common input at the visual system's first synapse. J Neurosci 32:10306–10317

Dunn FA, Wong ROL (2014) Wiring patterns in the mouse retina: collecting evidence across the connectome, physiology and light microscopy. J Physiol 592:4809–4823

Dunn FA, Della Santina L, Parker ED, Wong ROL (2013) Sensory experience shapes the development of the visual system's first synapse. Neuron 80:1159–1166

Euler T, Detwiler PB, Denk W (2002) Directionally selective calcium signals in dendrites of starburst amacrine cells. Nature 418:845–852

Euler T, Haverkamp S, Schubert T, Baden T (2014) Retinal bipolar cells: elementary building blocks of vision. Nat Rev Neurosci 15:507–519

Famiglietti EV (1991) Synaptic organization of starburst amacrine cells in rabbit retina: analysis of serial thin sections by electron microscopy and graphic reconstruction. J Comp Neurol 309: 40–70

Feller MB, Wellis DP, Stellwagen D et al (1996) Requirement for cholinergic synaptic transmission in the propagation of spontaneous retinal waves. Science 272:1182–1187

Fischer F, Kneussel M, Tintrup H et al (2000) Reduced synaptic clustering of GABA and glycine receptors in the retina of the gephyrin null mutant mouse. J Comp Neurol 427:634–648

Fisher LJ (1979) Development of synaptic arrays in the inner plexiform layer of neonatal mouse retina. J Comp Neurol 187:359–372

Freed MA, Smith RG, Sterling P (1992) Computational model of the on-alpha ganglion cell receptive field based on bipolar cell circuitry. Proc Natl Acad Sci U S A 89:236–240

Freund CL, Gregory-Evans CY, Furukawa T et al (1997) Cone-rod dystrophy due to mutations in a novel photoreceptor-specific homeobox gene (CRX) essential for maintenance of the photoreceptor. Cell 91:543–553

Fried SI, Münch TA, Werblin FS (2002) Mechanisms and circuitry underlying directional selectivity in the retina. Nature 420:411–414
Fried SI, Münch TA, Werblin FS (2005) Directional selectivity is formed at multiple levels by laterally offset inhibition in the rabbit retina. Neuron 46:117–127
Furukawa T, Morrow EM, Cepko CL (1997) Crx, a novel otx-like homeobox gene, shows photoreceptor-specific expression and regulates photoreceptor differentiation. Cell 91: 531–541
Graf ER, Zhang X, Jin S-X et al (2004) Neurexins induce differentiation of GABA and glutamate postsynaptic specializations via neuroligins. Cell 119:1013–1026
Grimes WN (2012) Amacrine cell-mediated input to bipolar cells: variations on a common mechanistic theme. Vis Neurosci 29:41–49
Grünert U, Wässle H (1993) Immunocytochemical localization of glycine receptors in the mammalian retina. J Comp Neurol 335:523–537
Hack I, Koulen P, Peichl L, Brandstätter JH (2002) Development of glutamatergic synapses in the rat retina: the postnatal expression of ionotropic glutamate receptor subunits. Vis Neurosci 19:1–13
Hashimoto K, Kano M (2013) Synapse elimination in the developing cerebellum. Cell Mol Life Sci 70:4667–4680
Haverkamp S, Grünert U, Wässle H (2000) The cone pedicle, a complex synapse in the retina. Neuron 27:85–95
Haverkamp S, Michalakis S, Claes E et al (2006) Synaptic plasticity in CNGA3(−/−) mice: cone bipolar cells react on the missing cone input and form ectopic synapses with rods. J Neurosci 26:5248–5255
Haverkamp S, Specht D, Majumdar S et al (2008) Type 4 OFF cone bipolar cells of the mouse retina express calsenilin and contact cones as well as rods. J Comp Neurol 507:1087–1101
Hoon M, Bauer G, Fritschy J-M et al (2009) Neuroligin 2 controls the maturation of GABAergic synapses and information processing in the retina. J Neurosci 29:8039–8050
Hoon M, Okawa H, Della Santina L, Wong ROL (2014) Functional architecture of the retina: development and disease. Prog Retin Eye Res 42:44–84
Huckfeldt RM, Schubert T, Morgan JL et al (2009) Transient neurites of retinal horizontal cells exhibit columnar tiling via homotypic interactions. Nat Neurosci 12:35–43
Jakobs TC, Koizumi A, Masland RH (2008) The spatial distribution of glutamatergic inputs to dendrites of retinal ganglion cells. J Comp Neurol 510:221–236
Jeon C-J, Kong J-H, Strettoi E et al (2002) Pattern of synaptic excitation and inhibition upon direction-selective retinal ganglion cells. J Comp Neurol 449:195–205
Johnson RE, Kerschensteiner D (2014) Retrograde plasticity and differential competition of bipolar cell dendrites and axons in the developing retina. Curr Biol 24:2301–2306
Keeley PW, Reese BE (2010) Role of afferents in the differentiation of bipolar cells in the mouse retina. J Neurosci 30:1677–1685
Keeley PW, Luna G, Fariss RN et al (2013) Development and plasticity of outer retinal circuitry following genetic removal of horizontal cells. J Neurosci 33:17847–17862
Kerschensteiner D, Morgan JL, Parker ED et al (2009) Neurotransmission selectively regulates synapse formation in parallel circuits in vivo. Nature 460:1016–1020
Kim S, Burette A, Chung HS et al (2006) NGL family PSD-95-interacting adhesion molecules regulate excitatory synapse formation. Nat Neurosci 9:1294–1301. doi:10.1038/nn1763
Kim I-J, Zhang Y, Meister M, Sanes JR (2010) Laminar restriction of retinal ganglion cell dendrites and axons: subtype-specific developmental patterns revealed with transgenic markers. J Neurosci 30:1452–1462
Kim JS, Greene MJ, Zlateski A et al (2014) Space-time wiring specificity supports direction selectivity in the retina. Nature 509:331–336
Kirkby LA, Sack GS, Firl A, Feller MB (2013) A role for correlated spontaneous activity in the assembly of neural circuits. Neuron 80:1129–1144

Mansergh F, Orton NC, Vessey JP et al (2005) Mutation of the calcium channel gene Cacna1f disrupts calcium signaling, synaptic transmission and cellular organization in mouse retina. Hum Mol Genet 14:3035–3046
Martins RAP, Davis D, Kerekes R et al (2011) Retinoblastoma (Rb) regulates laminar dendritic arbor reorganization in retinal horizontal neurons. Proc Natl Acad Sci U S A 108:21111–21116
Maslim J, Stone J (1986) Synaptogenesis in the retina of the cat. Brain Res 373:35–48
Mataruga A, Kremmer E, Müller F (2007) Type 3a and type 3b OFF cone bipolar cells provide for the alternative rod pathway in the mouse retina. J Comp Neurol 502:1123–1137
Meister M, Wong RO, Baylor DA, Shatz CJ (1991) Synchronous bursts of action potentials in ganglion cells of the developing mammalian retina. Science 252:939–943
Michalakis S, Schäferhoff K, Spiwoks-Becker I et al (2013) Characterization of neurite outgrowth and ectopic synaptogenesis in response to photoreceptor dysfunction. Cell Mol Life Sci 70: 1831–1847
Morgan JL, Dhingra A, Vardi N, Wong ROL (2006) Axons and dendrites originate from neuroepithelial-like processes of retinal bipolar cells. Nat Neurosci 9:85–92
Morgan JL, Schubert T, Wong ROL (2008) Developmental patterning of glutamatergic synapses onto retinal ganglion cells. Neural Dev 3:8
Morgan JL, Soto F, Wong ROL, Kerschensteiner D (2011) Development of cell type-specific connectivity patterns of converging excitatory axons in the retina. Neuron 71:1014–1021
Mumm JS, Williams PR, Godinho L et al (2006) In vivo imaging reveals dendritic targeting of laminated afferents by zebrafish retinal ganglion cells. Neuron 52:609–621
Nam CI, Chen L (2005) Postsynaptic assembly induced by neurexin-neuroligin interaction and neurotransmitter. Proc Natl Acad Sci U S A 102:6137–6142
Nomura A, Shigemoto R, Nakamura Y et al (1994) Developmentally regulated postsynaptic localization of a metabotropic glutamate receptor in rat rod bipolar cells. Cell 77:361–369
Okawa H, Della Santina L, Schwartz GW et al (2014a) Interplay of cell-autonomous and nonautonomous mechanisms tailors synaptic connectivity of converging axons in vivo. Neuron 82:125–137
Okawa H, Hoon M, Yoshimatsu T et al (2014b) Illuminating the multifaceted roles of neurotransmission in shaping neuronal circuitry. Neuron 83:1303–1318
Olney JW (1968) An electron microscopic study of synapse formation, receptor outer segment development, and other aspects of developing mouse retina. Invest Ophthalmol 7:250–268
Omori Y, Araki F, Chaya T et al (2012) Presynaptic dystroglycan-pikachurin complex regulates the proper synaptic connection between retinal photoreceptor and bipolar cells. J Neurosci 32: 6126–6137
Peng YW, Hao Y, Petters RM, Wong F (2000) Ectopic synaptogenesis in the mammalian retina caused by rod photoreceptor-specific mutations. Nat Neurosci 3:1121–1127
Poché RA, Raven MA, Kwan KM et al (2008) Somal positioning and dendritic growth of horizontal cells are regulated by interactions with homotypic neighbors. Eur J Neurosci 27: 1607–1614
Randlett O, MacDonald RB, Yoshimatsu T et al (2013) Cellular requirements for building a retinal neuropil. Cell Rep 3:282–290
Raven MA, Oh ECT, Swaroop A, Reese BE (2007) Afferent control of horizontal cell morphology revealed by genetic respecification of rods and cones. J Neurosci 27:3540–3547
Reese BE, Raven MA, Stagg SB (2005) Afferents and homotypic neighbors regulate horizontal cell morphology, connectivity, and retinal coverage. J Neurosci 25:2167–2175
Regus-Leidig H, Tom Dieck S, Specht D et al (2009) Early steps in the assembly of photoreceptor ribbon synapses in the mouse retina: the involvement of precursor spheres. J Comp Neurol 512: 814–824
Rörig B, Grantyn R (1993) Glutamatergic and GABAergic synaptic currents in ganglion cells from isolated retinae of pigmented rats during postnatal development. Brain Res Dev Brain Res 74: 98–110

Sanes JR, Lichtman JW (1999) Development of the vertebrate neuromuscular junction. Annu Rev Neurosci 22:389–442
Sassoè-Pognetto M, Kirsch J, Grünert U et al (1995) Colocalization of gephyrin and GABAA-receptor subunits in the rat retina. J Comp Neurol 357:1–14
Sato S, Omori Y, Katoh K et al (2008) Pikachurin, a dystroglycan ligand, is essential for photoreceptor ribbon synapse formation. Nat Neurosci 11:923–931
Schmitt EA, Dowling JE (1999) Early retinal development in the zebrafish, Danio rerio: light and electron microscopic analyses. J Comp Neurol 404:515–536
Schubert T, Kerschensteiner D, Eggers ED et al (2008) Development of presynaptic inhibition onto retinal bipolar cell axon terminals is subclass-specific. J Neurophysiol 100:304–316
Schwartz GW, Okawa H, Dunn FA et al (2012) The spatial structure of a nonlinear receptive field. Nat Neurosci 15:1572–1580
Sernagor E, Eglen SJ, Wong RO (2001) Development of retinal ganglion cell structure and function. Prog Retin Eye Res 20:139–174
Sernagor E, Eglen S, Harris B, Wong R (eds) (2006) Retinal development. Cambridge University Press, Cambridge
Sherry DM, Wang MM, Bates J, Frishman LJ (2003a) Expression of vesicular glutamate transporter 1 in the mouse retina reveals temporal ordering in development of rod vs. cone and ON vs. OFF circuits. J Comp Neurol 465:480–498
Sherry DM, Wang MM, Frishman LJ (2003b) Differential distribution of vesicle associated membrane protein isoforms in the mouse retina. Mol Vis 9:673–688
Soto F, Bleckert A, Lewis R et al (2011) Coordinated increase in inhibitory and excitatory synapses onto retinal ganglion cells during development. Neural Dev 6:31
Soto F, Ma X, Cecil JL et al (2012) Spontaneous activity promotes synapse formation in a cell-type-dependent manner in the developing retina. J Neurosci 32:5426–5439
Soto F, Watkins KL, Johnson RE et al (2013) NGL-2 regulates pathway-specific neurite growth and lamination, synapse formation, and signal transmission in the retina. J Neurosci 33: 11949–11959
Stafford BK, Park SJH, Wong KY, Demb JB (2014) Developmental changes in NMDA receptor subunit composition at ON and OFF bipolar cell synapses onto direction-selective retinal ganglion cells. J Neurosci 34:1942–1948
Sterling P, Matthews G (2005) Structure and function of ribbon synapses. Trends Neurosci 28: 20–29
Strettoi E, Mears AJ, Swaroop A (2004) Recruitment of the rod pathway by cones in the absence of rods. J Neurosci 24:7576–7582
Sun L, Han X, He S (2011) Direction-selective circuitry in rat retina develops independently of GABAergic, cholinergic and action potential activity. PLoS One 6:e19477
Tian N, Copenhagen DR (2001) Visual deprivation alters development of synaptic function in inner retina after eye opening. Neuron 32:439–449
Tian N, Copenhagen DR (2003) Visual stimulation is required for refinement of ON and OFF pathways in postnatal retina. Neuron 39:85–96
Tsukamoto Y, Omi N (2014a) Some OFF bipolar cell types make contact with both rods and cones in macaque and mouse retinas. Front Neuroanat 8:105
Tsukamoto Y, Omi N (2014b) Effects of mGluR6-deficiency on photoreceptor ribbon synapse formation: comparison of electron microscopic analysis of serial sections with random sections. Vis Neurosci 31:39–46
Tyagarajan SK, Fritschy J-M (2014) Gephyrin: a master regulator of neuronal function? Nat Rev Neurosci 15:141–156
Vaney DI (1990) Chapter 2 The mosaic of amacrine cells in the mammalian retina. Prog Retin Res 9:49–100
Von Kriegstein K, Schmitz F (2003) The expression pattern and assembly profile of synaptic membrane proteins in ribbon synapses of the developing mouse retina. Cell Tissue Res 311: 159–173

Wang MM, Janz R, Belizaire R et al (2003) Differential distribution and developmental expression of synaptic vesicle protein 2 isoforms in the mouse retina. J Comp Neurol 460:106–122
Wässle H, Boycott BB (1991) Functional architecture of the mammalian retina. Physiol Rev 71: 447–480
Wei W, Feller MB (2011) Organization and development of direction-selective circuits in the retina. Trends Neurosci 34:638–645
Wei W, Hamby AM, Zhou K, Feller MB (2011) Development of asymmetric inhibition underlying direction selectivity in the retina. Nature 469:402–406
Wong RO, Oakley DM (1996) Changing patterns of spontaneous bursting activity of on and off retinal ganglion cells during development. Neuron 16:1087–1095
Wong RO, Chernjavsky A, Smith SJ, Shatz CJ (1995) Early functional neural networks in the developing retina. Nature 374:716–718
Wong WT, Myhr KL, Miller ED, Wong RO (2000) Developmental changes in the neurotransmitter regulation of correlated spontaneous retinal activity. J Neurosci 20:351–360
Yashiro K, Philpot BD (2008) Regulation of NMDA receptor subunit expression and its implications for LTD, LTP, and metaplasticity. Neuropharmacology 55:1081–1094
Yonehara K, Balint K, Noda M et al (2011) Spatially asymmetric reorganization of inhibition establishes a motion-sensitive circuit. Nature 469:407–410
Yoshida K, Watanabe D, Ishikane H et al (2001) A key role of starburst amacrine cells in originating retinal directional selectivity and optokinetic eye movement. Neuron 30:771–780
Yoshimatsu T, Suzuki SC, Wong ROL (2013) Chapter 36 – circuit assembly in the developing vertebrate retina. In: Rakic JLRR (ed) Cellular migration and formation of neuronal connections. Academic, Oxford, pp 687–711
Yoshimatsu T, Williams PR, D'Orazi FD et al (2014) Transmission from the dominant input shapes the stereotypic ratio of photoreceptor inputs onto horizontal cells. Nat Commun 5:3699
Zabouri N, Haverkamp S (2013) Calcium channel-dependent molecular maturation of photoreceptor synapses. PLoS One 8:e63853

Part VI
Dendrites in Disease

Chapter 20
Dendrites in Autism Spectrum Disorders

Verónica Martínez-Cerdeño, Izumi Maezawa, and Lee-Way Jin

Abstract Autism spectrum disorders (ASDs), including classic autism and various syndromic ASDs, are a diverse group of complex developmental disability characterized by social impairments, communication difficulties, and restricted, repetitive, and stereotyped patterns of behavior. As many as 1 in 80 children in the United States are affected by some form of ASDs, an alarming statistics. The core symptoms of autism are considered due to under-connectivity among the brain regions participating in cortical networks, a consequence of complex pathological gene–environmental interactions. Recent investigations on ASD susceptible genes and environmental risk factors such as exposure to environmental toxins, perinatal or postnatal infections, obstetric factors, and medications have provided greater clarity to the clinical and biological complexity of ASDs. This chapter provides a comprehensive review to illustrate that most genetic and environmental risks for ASDs cause dendritic abnormalities underlying brain under-connectivity. Rodent models harboring mutated ASD susceptible genes or exposed to environmental risk factors have helped delineate molecular pathways leading to autism-like dendritic abnormalities, and the findings can be further validated by human neuropathological studies. It is intriguing to consider dendritic abnormalities as main pathological substrates of autism; targeting specific molecular pathways driving such abnormalities could lead to realization of the ideal of providing "precision medicine" to autistic children. This review also illustrates that the immune system with its myriad of cells and mediators has a great impact on the developing brain, and our "microglia hypothesis" proposes that microglia play a pivotal role in mediating environmental and genetic effects on dendritic pathology in autism.

V. Martínez-Cerdeño • I. Maezawa
Department of Pathology and Laboratory Medicine, M.I.N.D. (Medical Investigation of Neurodevelopmental Disorders) Institute, Intellectual and Developmental Disabilities Research Center, University of California Davis, Sacramento, CA 95817, USA

L.-W. Jin (✉)
Department of Pathology and Laboratory Medicine, M.I.N.D. (Medical Investigation of Neurodevelopmental Disorders) Institute, Intellectual and Developmental Disabilities Research Center, University of California Davis, Sacramento, CA 95817, USA

Department of Pathology and Laboratory Medicine, University of California Davis Medical Center, Sacramento, CA 95817, USA
e-mail: lwjin@ucdavis.edu

K. Emoto et al. (eds.), *Dendrites*, DOI 10.1007/978-4-431-56050-0_20

Keywords Autism spectrum disorders • Dendrite • Dendritic spine • Microglia • Rett syndrome • Fragile X syndrome • Susceptible genes • Gene–environmental interactions • Immune • Neurodevelopment • MeCP2 • FMRP

20.1 Introduction

Among several neurological disorders that pose serious public health threats, autism is particularly challenging because this group of disorders, generally categorized as autism spectrum disorders (ASDs), shows no characteristic "neuropathological hallmarks." Rather, ASDs are brain diseases of plasticity devoid of traditionally defined "lesions." In this context, it is likely profitable and translationally significant to study neuronal dendrites in ASDs, since dendrites and dendritic spines are essential anatomic and functional elements for the connectivity of the brain that constitutes our various mental faculties. Currently much remains to be learned about the characteristics of neuronal dendrites in subjects with autism. However, several new developments should make us feel confident that neuronal dendrites are one of the major anatomic substrates of the disease. Currently among the most informative studies on ASDs are those to identify genetic factors. As summarized below, the effects of most ASD-causing or ASD-modulating genetic factors converge on dendritic formation, dendritic complexity, and dendritic spine structure. Mice modeling exposure to environmental risk factors also show dendritic pathology. In addition, data obtained from humans are in agreement with those obtained from mouse models of autism. This general congruence will facilitate mechanistic studies on dendritic abnormalities in ASDs.

Autism, first described by Leo Kanner in 1943, is a complex developmental disability. Although initially considered a rare infantile disorder, recent surveys estimated that as many as 1 in 80 children in the United States are affected by some form of ASDs, an alarming statistics that could mount to an "autism epidemic" (Rice 2009; Williams 2012). Individuals with autism fail to properly share emotions or understand how others feel, have problems in verbal and nonverbal communications (such as eye contact or smiling), show repetitive behaviors, and obsessively follow certain routines. The term ASDs is used to include heterogeneous disorders grouped together based on similar features of atypical development in socialization, communication, and behavior. These conditions include autistic disorder (also called "classic" autism), Asperger syndrome, and pervasive developmental disorder-not otherwise specified (PDD-NOS). Children with Asperger syndrome or PDD-NOS have fewer diagnostic symptoms and milder impairment compared with classic autism. The clinical heterogeneity of ASDs is perhaps the result of etiological heterogeneity; it is now generally agreed upon that autism is the consequence of complex interplays between heritable genetic factors and environmental factors influencing individual's epigenome (Geschwind 2011; Hallmayer et al. 2011; LaSalle 2011). Symptoms of ASDs typically are present before 3 years of age and often are accompanied by abnormalities in cognitive functioning, learning, attention, and sensory processing (Yeargin-Allsopp et al. 2003).

There currently is no cure for autism. Although there are a growing number of proposed and actually administered treatments for patients with autism, behavioral intervention programs remain the treatment of choice (Vismara and Rogers 2010). Major challenges for finding a cure include the etiological heterogeneity and the lack of consistent and reliable genetic or biologic diagnostic markers for accurate classification and early diagnosis of ASDs. Functional imaging studies have established autism as a disorder of under-connectivity among the brain regions participating in cortical networks (Minshew and Keller 2010; Geschwind and Levitt 2007; Just et al. 2007), pointing to a role of dendritic abnormalities. Studies on the genetic etiology of autism have also uncovered genes that regulate synaptic functions (Geschwind 2008; Betancur et al. 2009). In summary, the neurobiological basis of autism seems to be a complex combination of common genetic variants, epigenetic regulation, environmental factors, glial cell abnormalities, aberrant neurogenesis, blood factors (e.g., autoantibodies), and other yet to be identified risks, which converge to produce a developmental disconnection of brain regions involved in higher-order associations.

One powerful approach to understand autism is to study "syndromic" ASDs; these are a group of genetically diverse neurodevelopmental disorders with high penetrance of ASD diagnosis (Levitt and Campbell 2009). They include fragile X syndrome (FXS), Rett syndrome (RTT), Angelman syndrome, Prader–Willi syndrome, 15q duplication syndrome, Timothy syndrome, and Smith–Lemli–Opitz syndrome as well as neurofibromatosis and tuberous sclerosis complex (TSC). Although they represent only a minority of children with autism, studying these disorders is likely to lead to important clues about pathogenetic pathways of autism because these disorders are relatively homogeneous compared to classic autism and have clear genetic and metabolic abnormalities that are amenable to animal modeling (Levitt and Campbell 2009). For example, autism occurs in approximately 30 % of FXS cases, and PDD-NOS occurs in an additional 30 % of FXS cases (Hagerman et al. 2010). Similarly, abnormal expression of methyl CpG binding protein 2 (MeCP2), the protein that is deficient in patients with RTT due to loss-of-function mutations of the X-linked *MECP2*, also significantly contributes to the development of autism, due to epigenetic dysregulation (LaSalle and Yasui 2009). Reduced MeCP2 expression was observed in 79 % of autism brain samples and correlated with aberrant methylation of the *MECP2* promoter in male autism samples (Nagarajan et al. 2006). As summarized below, dendritic abnormalities are also a significant pathology in syndromic ASDs, exemplified by FXS and RTT.

20.2 Dendritic Pathology in Autism: Human Neuropathology

Dendrite studies in the human brain are limited but have yielded consistent results indicating reduced dendritic number or complexity in autistic brains. A typical example is shown in Fig. 20.1. Raymond et al. reported that neurons in the region CA4 and CA1 of the hippocampus of autistic children have reduced dendritic branching compared to those in control hippocampus (Raymond et al. 1996). Mukaetova-Ladinska reported that dorsolateral prefrontal cortex in two adult individuals with autism presented reduced dendrite numbers (Mukaetova-Ladinska et al. 2004). Overall, data from human brains point to a generalized reduction in size and number and to a morphology alteration of dendrites in autism.

Interestingly, while in most of developmental disorders dendritic spine density is reduced, in autism an increased dendritic spine density has been shown in a study that analyzed dendritic spine densities on cortical pyramidal cells from ASD subjects (Larkman 1991). Hutsler and Zhang examined dendritic spines on Golgi-impregnated cortical pyramidal cells in the cortex of ASD subjects and age-matched control cases. Relative to controls, spine densities were increased in

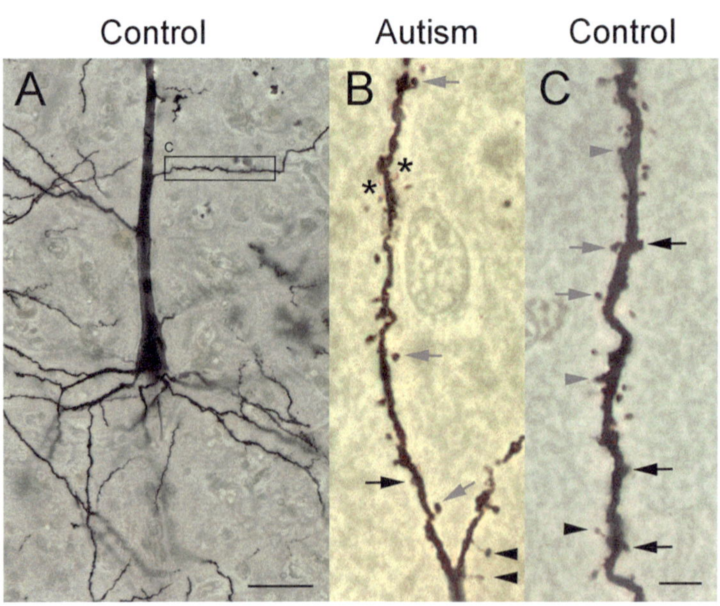

Fig. 20.1 Dendritic spines in autism and control tissue (Golgi silver impregnation). (**a**) A pyramidal neuron in the layer V of the prefrontal cortex in a control subject. (**b**) Layer V pyramidal dendrite in a subject with autism. (**c**) Dendrite in the control case outlined in **a**. The control case presents with numerous stubby and wide spines (**c**). However, the autism case presents with a higher number of thin and filiform spines (**b**). *Black arrows*: stubby spines; *Gray arrows*: mushroom spines; *Black arrowheads*: thin spines; *Gray arrowheads*: wide spines; *Asterisks*: filiform spines. Scale bar in **a** = 50 μm; scale bar in **c** (**b**) = 10 μm

ASD subjects. Greater spine densities were found predominantly within layer II of each cortical location and within layer V of the temporal lobe. High spine densities were associated with decreased brain weights and were most commonly found in ASD subjects with lower levels of cognitive functioning (Hutsler and Zhang 2010). In summary, spine densities seem to be increased in autism, with an increase of spines of an immature morphology and a decrease of more mature spines, indicating a general spine immaturity state. To better understand the effect of autism in spine morphology, number and function will require additional studies in humans.

20.3 Dendritic Pathology in Genetic Models of Autism

While the information obtained from human studies is nonspecific for mechanistic understanding of the disease, studies on animals harboring defects of ASD susceptible genes have revealed molecular pathways regulating formation of specific dendritic components. The gene encoding thousand-and-one amino acid kinase 2 (*TAOK2*), a member of the MAP kinase family that selectively activates mitogen-activated protein kinases, is located on chromosome 16p11.2 in human, a region that has been shown to carry substantial susceptibility to ASD (Weiss et al. 2008). Mice that received shRNA-mediated TAOK2 downregulation during embryonic development showed less complex basal dendritic arbors in cortical layer 2/3 neurons than control mice on postnatal day 7, but the apical dendrite complexity was not affected (de Anda et al. 2012). TAOK2 interacts with neuropilin 1 (Nrp1), a receptor protein that binds the secreted guidance cue semaphorin 3A (Sema3A). TAOK2 overexpression restores dendrite formation in cultured cortical neurons from Nrp1 (Sema) mice, which express Nrp1 receptors incapable of binding Sema3A. These data suggest that TAOK2 is involved in dendritic formation in autism (de Anda et al. 2012).

In addition to the TAOK2 pathway, Epac2, a guanine nucleotide exchange factor (GEF) for the Ras-like small GTPase Rap and highly enriched in dendrites, appears to have the same impact. Epac2 is also noted for its association with autism. An Epac2-coding variant that disrupts that interaction of Epac2 with Ras was found in human subjects diagnosed with autism, and Epac2-deficient mice exhibited autism-like specific impairments in sociability, but not in cognitive or motor behaviors (Srivastava et al. 2012a). In vivo overexpression of the autism-associated Epac2 variant or Epac2 knockdown via in utero electroporation induces a reduced basal dendritic architecture of cortical layer 2/3 neurons in mice. Moreover, Epac2 knockdown in mature cortical neurons in vitro mimics this effect. It is known that components of the Ras/Epac2/Rap pathway exhibit differential abundance in the basal versus apical dendritic compartments. Concomitantly, Epac2 knockdown robustly and selectively reduces basal dendrite complexity in cortical pyramidal neurons (Srivastava et al. 2012b).

Intriguingly, in a study using a similar in vivo approach, downregulation of the Angelman syndrome (a syndromic ASD) gene ubiquitin-protein ligase E3A

(Ube3a) in mice resulted in selective reduction of apical, rather than basal, dendritic outgrowth in cortical layer 2/3 neurons (Miao et al. 2013). Notwithstanding, the observed selective defects in basal or apical dendrite formation, induced by distinct autism-related molecular pathways, support the important notion that a primary defect in a dendritic component could result in under-connectivity associated with autism.

Additional studies of the dendritic arbor in animal models of autism have shown data that match previous observations in human. For example, siRNA-mediated KIAA2022 knockdown in mice, a model of the X-linked intellectual disability (XLID) syndrome related to autism, resulted in marked impairment in neurite outgrowth including both in dendrites and axons (Van Maldergem et al. 2013). Another example is Prickle2 (Pk2) that is a postsynaptic noncanonical Wnt signaling protein. Mice with disruption in Pk2 display behavioral abnormalities including altered social interaction, learning abnormalities, and behavioral inflexibility, symptoms comparable to those in patients with autism. Pk2 disruption in mouse hippocampal neurons leads to a reduction in dendrite branching and synapse formation (Sowers et al. 2013). Other proteins, such as the CNTN proteins (CNTN4, CNTN5, and CNTN6), a family of Ig cell adhesion molecules (IgCAMs), have been associated with ASD (Zuko et al. 2013) and are as well controlling neurite outgrowth (Mercati et al. 2013).

Further data in human suggest that MAP/microtubule affinity-regulating kinase 1 (MARK1) overexpression may be responsible for changes in dendritic functioning. High-resolution single-nucleotide polymorphisms (126 SNPs) genotyping across the chromosome 1q41–q42 region, followed by a MARK1-tagged-SNP association study in 276 families with autism, showed that several SNPs within the MARK1 gene are significantly associated with ASD. Both overexpression and silencing of MARK1 resulted in significantly shorter dendritic length and modified dendritic transport speed in mouse neocortical neurons (Maussion et al. 2008).

Proteins that repress the mechanistic target of rapamycin (mTOR) have been linked to autism and also play a role in dendritic and spine morphology. These include the phosphatase and tensin homolog (PTEN) and the tuberous sclerosis complex proteins (TSC) 1 and 2 (Weston et al. 2014). PTEN is a lipid and protein phosphatase that negatively regulates the PI3K/AKT/mTOR signaling pathway, ultimately modulating cell growth and protein translation (Orloff et al. 2013). The presence of mutations in the gene for the PTEN protein has been demonstrated in individuals with ASD. Mice with mutation or deletion of PTEN show abnormal social interaction and exaggerated responses to sensory stimuli. Deletion of PTEN in neurons in the cerebral cortex and hippocampus of these mice results in hypertrophic and ectopic dendrites and axonal tracts with increased synapses (Kwon et al. 2006). However, PTEN knockdown in basolateral amygdala leads to a significant decrease in total spine density in distal dendrites and increased mushroom spine density and size with correspondingly decreased thin protrusion density at more distal segments (Haws et al. 2014). The TSC1 gene also encodes a protein that represses mTOR signaling. Mutations in the TSC1 or TSC2 tumor suppressor genes lead to tuberous sclerosis (TS), a disorder that presents with mental retardation and autism. It is known that the TSC pathway regulates growth and synapse function in neurons, and the loss of Tsc1 or Tsc2 triggers enlargement of somas and dendritic spines in hippocampal pyramidal neurons of mice and rats (Tavazoie et al. 2005).

Other proteins that have been related with autism and have a role in spine morphology include Shank proteins and UBE3A. Mutations of the Shank family of scaffold proteins, in particular Shank3, are linked to a familial form of autism (Durand et al. 2007). When overexpressed in cultured hippocampal neurons, Shank proteins strongly promote the enlargement of dendritic spines, particularly the spine heads (Sala et al. 2001). As mentioned above, loss of function of the maternally inherited allele for the UBE3A ubiquitin ligase gene causes Angelman syndrome (AS), which is characterized by severe neurological impairment and motor dysfunction. In addition, UBE3A lies within chromosome 15q11–q13 region where duplications cause autism. In human AS dendritic spines have an inconsistent morphology, including high variability in spine neck length and head size. In Ube3a maternal-deficient mice, dendritic spine development, including spine morphology, number, and length, is affected on cerebellar Purkinje cells and on pyramidal neurons of the hippocampus and cortex (Dindot et al. 2008).

20.4 Dendritic Pathology in Models of Exposure to Autism Risk Factors

The above discussed mouse models, among many other models of autism, have been developed to mimic the genetics of specific autism spectrum and related disorders. However, this only represents one side of the story. Despite the strong genetic links of ASDs, a single gene that can account for the majority of ASD cases has yet to be identified and perhaps is nonexistent. The other side of etiology of autism is environmental insults inflicted from without, such as exposure to environmental toxins, perinatal or postnatal infections, obstetric factors, and medications (Hertz-Picciotto 2011). Models to mimic human exposures to environmental factors are emerging (Schwartzer et al. 2013a), and dendritic abnormalities appear to also underlie behavioral abnormalities shown in these models. For example, valproic acid (VPA) is a blocker of histone deacetylase widely used to treat epilepsy, bipolar disorders, and migraine, and its administration during pregnancy increases the risk of autism in the child. In the widely used VPA animal model of autism, there is a retracted neuronal arborization in the hippocampus (Bringas et al. 2013). However, in other brain areas, such as the prefrontal cortex, nucleus accumbens, and basolateral amygdala, an increase in neuronal arborization has been reported. VPA animals also present with a reduced number of spines in the prefrontal cortex, dorsal hippocampus, and basolateral amygdala, but an increase in the dendritic spine density in accumbens and ventral hippocampus (Bringas et al. 2013).

Several epidemiological studies have shown an association between infection and inflammation during pregnancy and increased risk of autism in the child (Zerbo et al. 2013; Brown 2012). The widely used maternal immune activation (MIA) model has supported the causal relationship between maternal inflammation during gestation and autism-relevant behaviors of the offspring in rodents (Malkova et al. 2012; Patterson 2012). In this model and its variations, the pregnant dams are exposed to lipopolysaccharides (LPS), a toll-like receptor(TLR)-4 agonist that mimics Gram-negative bacterial infection or polyinosinic–polycytidylic acid [poly (I:C)], a synthetic toll-like receptor (TLR)-3 agonist that mimics viral infection. A number of behavioral alterations observed in the offspring mimic features of autism and schizophrenia, including reduced exploratory behavior and impaired social interaction (Malkova et al. 2012; Schwartzer et al. 2013b; Shi et al. 2003). This experimental paradigm also induces dendritic reductions that mimic human findings (Burd et al. 2010; Balakrishnan et al. 2013; Abazyan et al. 2010). To test the gene–environmental interaction that would be most akin to pathogenesis of human autism, Abazyan et al. used a modification of the original MIA paradigm in which intraperitoneal injection of poly(I:C) was performed instead of intravenous injection to minimize possible ceiling effect of poly(I:C) itself. They discovered that although this mild form of MIA did not induce a reduction of dendritic spine density of hippocampal granule cells in wild-type mice, it did so in mice harboring the mutant human disrupted-in-schizophrenia 1 (DISC1), one of the genes that is associated with autism susceptibility (Kilpinen et al. 2008). In addition, the presence of mutant DISC1 also accentuated autism- or schizophrenia-like neurobehavioral and neuroanatomical phenotypes. This line of work illustrates the importance of using animal models to identify potential Gene–Environment interactions (Schwartzer et al. 2013a).

The MIA model illustrates the essential link between the immune system and the central nervous system; its pathogenesis is believed to involve LPS- or poly(I:C)-induced production of cytokines, particularly pro-inflammatory cytokines such as interleukin-1 (IL-2), IL-6, IL-12, and tumor necrosis factor-α (TNF-α). Among environmental risk factors for autism, prenatal or postnatal infection/inflammation and perhaps exposure to environmental toxins engender a complex interplay between the adaptive and innate immune systems (Filiano et al. 2014), with cytokine signaling possibly targeting myeloid cells in the brain to affect the neurodevelopmental trajectory. One key factor noted in the MIA model is the persistent cellular immune dysregulation. Offspring of immune-activated dams displayed altered immune profiles and function (Hsiao et al. 2012). Lymphocytes from such offspring developed into T helper (TH) 17 cells upon in vitro activation (Mandal et al. 2011). TH17 cells are known for its high propensity to induce autoimmune and inflammatory disorders (Korn et al. 2009). This preferential differentiation toward TH17 cell development could be a sign of an overall impaired immune development that manifests enhanced pro-inflammatory responses in the offspring, a phenotype that could persist into neonatal and adult life (Mandal et al. 2013). Indeed, systemic immune dysregulation is an inherent component of ASDs. Peripheral blood profiles often present irregularities including decreased numbers of B and T lymphocytes, reduced lymphocytic response to stimulation, increased numbers of monocytes, abnormal cytokine profiles, abnormal immunoglobulin levels, and increased myeloid dendritic cell frequencies (Breece et al. 2013; Goines et al. 2011). Numerous studies of serum cytokines demonstrated lower levels of transforming growth factor-β (TGF-β) and higher levels of macrophage inhibitory factor, leptin, IL-1β, IL-6, interferon-γ (IFN-γ), and IL-12 in various age groups of patients with ASDs (Goines et al. 2011; Goines and Ashwood 2013). Maternally generated cytokines can cross the placenta and regulate cell growth and immune development of the fetus. Placenta and fetus can also be a source of cytokines in the face of prenatal infection. These observations beg the question of how impaired immune development in the fetus or newborn interferes with neurodevelopment, especially dendritic and synaptic formation, and evidence is emerging from studying immune-brain interphase. One plausible path is that the peripheral pro-inflammatory phenotype, such as that seen in the MIA model, would, through TH17 cells or cytokines that pass through the immature blood–brain barrier, affect the brain myeloid cells, both resident microglia and monocytes/macrophages that infiltrate the CNS after injury (Walsh et al. 2014). Microglia or perhaps monocytes/macrophages can in turn release immune mediators or perform phagocytotic function to effect changes in neuronal dendrites and dendritic spines. It is now well established that microglia survey neuronal surface and periodically interact with synapses (Nimmerjahn et al. 2005; Davalos et al. 2005; Wake et al. 2009). During development, neuron–microglia interaction mediates synaptic pruning and activity-dependently sculpts postnatal neural circuits (Schafer et al. 2012; Zhan et al. 2014). It is well established that several cytokines (notably IL-1β, IL-6, and TNF-α) or microglia-released mediators (such as glutamate or reactive oxygen species) affect the development of dendritic arborization. A transient immune response to LPS induces long-lasting changes in microglia that could directly modulate dendritic spine dynamics (Kondo et al. 2011), although the mechanism of how maternal immune activation affects microglial function of the offspring remains to be studied. We have previously hypothesized that inherently abnormal phenotypes of microglia, either genetically determined or developmentally rendered, contribute to brain pathologies in ASD (Maezawa et al. 2011), and a pro-inflammatory phenotype of microglia that causes dendritic toxicity could be reasonably hypothesized for MIA to explain the dendritic reduction (the "microglia hypothesis").

In summary, we and others consider that cellular immune dysregulation, interacting with the effects of autism susceptible genes, could significantly alter microglial function into child and adulthood and therefore contribute to dendritic pathologies seen in ASDs (Maezawa et al. 2011; Monji et al. 2014; Derecki et al. 2014). A translationally significant question then is how this interference of normal neurodevelopment from the impaired peripheral immune system can be blocked. An enlightening finding is that behaviorally abnormal MIA offspring that had been irradiated and transplanted with immunologically normal bone marrow from either MIA or control offspring no longer exhibited deficits in stereotyped/repetitive and anxiety-like behaviors (Hsiao et al. 2012). Thus, dendritic defects in autistic brains could in principle be ameliorated by correcting the peripheral immune dysregulation.

20.5 Dendritic Pathology in FXS

Studies on dendrites in FXS, RTT, and related syndromes have shed light on the nature of dendritic arbors in autism. FXS is caused by the expansion of a CGG trinucleotide repeat on the fragile X mental retardation 1 (*Fmr1*) gene that results in a failure to express the fragile X mental retardation protein (FMRP). Repeat lengths less than 45 CGG are associated with typical development, while repeat lengths of more than 200 CGG result in FXS. Individuals with repeat lengths between 55 and 200 CGG are carriers of the fragile X premutation characterized by an increase in the levels of mRNA for FMRP and normal or slightly decreased FMRP protein. Premutation carriers are at risk for developing fragile X-associated tremor/ataxia syndrome (FXTAS), a neurodevelopmental disease characterized by attention deficit hyperactivity disorder (ADHD), social deficits, ASD, and occasionally intellectual disability. FXTAS is also a neurodegenerative disorder in late life, characterized by tremor, ataxia, and cognitive impairment (Hagerman 2013). FMRP is localized to neurons, specifically to dendrites (Devys et al. 1993; Feng et al. 1997), and it has been suggested that it plays a role in dendritic growth. In fact, patients with FXS, the most widespread single gene cause of autism, present with abnormal dendritic branching (Rudelli et al. 1985). Some authors reported that *Fmr1* knockout in mice leads to alterations in the distribution of dendritic arbor in motor neurons, consistent with slower rates of extension and abnormal pruning of intermediate dendritic segments (Thomas et al. 2008). However, others reported that *Fmr1* knockout mice do not demonstrate any significant differences from controls in dendritic tree complexity (Irwin et al. 2002). CGG repeat knock-in mice, a FXTAS model, show fewer dendritic branches proximal to the soma, reduced total dendritic length, and a higher frequency of longer dendritic spines in II/III pyramidal neurons in the primary visual cortex (Berman et al. 2012). The cytoplasmic FMRP-interacting protein 1 (CYFIP1), a functional partner of FMRP, is located within a hot spot for ASD (chr15q11.2), close to a region critical for ASD-related syndromes. *CYFIP1* mRNA is downregulated in a subgroup of FXS patients who present with Prader–Willi phenotype and show severe ASD and obsessive–compulsive behavior (Nowicki et al. 2007). Together with FMRP, CYFIP1 represses neuronal protein synthesis. In particular, in vivo and in vitro experiments demonstrated that CYFIP1 regulates actin in the cytoskeleton, controlling dendrite and spine formation (De Rubeis et al. 2013). Overexpression of CYFIP1 in vitro leads to increased dendritic complexity. On the other hand, neurons derived from a Cyfip1 haploinsufficience animal exhibit deficits in dendritic complexity as well as an altered ratio of immature to mature spines in hippocampal CA1 neurons. Both Cyfip1 overexpression and haploinsufficiency increase immature spine number (Pathania et al. 2014). Interestingly, a significant autism-associated gene related to FXS is *TAOK2*, a gene described above in Sect. 20.3. *TAOK2* mRNA is a direct target of FMRP (Darnell et al. 2011), further cementing the pathogenetic link between FXS and autism.

The other possible similarity between FXS and autism is that FXS patients also exhibit a higher density of dendritic spines than control subjects on distal segments of apical and basal dendrites in the cerebral cortex, with an increase of spines with an immature morphology together with a decrease of spines with a more mature morphology (Hinton et al. 1991). Layer V neurons in the barrel cortex of the *Fmr1* knockout mice show an increase in spine density and a decrease in spine length in the first postnatal days (Nimchinsky et al. 2001). These neurons further show a developmental delay in the downregulation of spine turnover and in the transition from immature to mature spine subtypes, demonstrating that FMRP delays spine stabilization (Cruz-Martin et al. 2010). In addition, Purkinje cell-specific knockout of *Fmr1* shows elongated spines (Koekkoek et al. 2005).

Several approaches in animal models and in vitro cell cultures have been able to reverse the effect of *Fmr1* knockdown on spine morphology. FMRP has been shown to interact in vitro with the p21-activated kinase (PAK), an enzyme known to play a critical role in actin polymerization and dendritic spine morphogenesis. Expression of a dominant-negative PAK transgene in the forebrain of *Fmr1* knockout mice leads to partially restored elevated spine density and elongated spine characteristic of the FXS cortex (Hayashi et al. 2007). The amyloid β-protein precursor (AβPP) is upregulated in the FXS mouse model (Napoli et al. 2008). Genetic reduction of AβPP in *Fmr1* knockout mice rescues the ratio of mature versus immature dendritic spines (Westmark et al. 2011). In addition, treatment of *Fmr1* knockout mice with the mGluR antagonist 2-methyl-6-(phenylethynyl)-pyridine (*MPEP*) also results in rescue of dendritic spine morphology in *Fmr1* knockout mice (Su et al. 2011).

20.6 Dendritic Pathology in RTT

Loss-of-function mutations of the X-linked gene for the methyl CpG binding protein 2 (MeCP2), a transcriptional repressor that also suppresses microRNA processing, lead to RTT (Chahrour and Zoghbi 2007; Amir et al. 1999; Chahrour et al. 2008; Yasui et al. 2007; Cheng et al. 2014a). RTT primarily affects young girls (*Mecp−/+*), who develop normally until 6–18 months of age, at which time they start showing progressive loss of neurodevelopmental milestones, a tragic process called "regression." Human neuropathology studies in early days of RTT research have shown reduced neuronal size, reduced dendritic arborization, thinned dendrites, and sparse and short dendritic spines in selected brain regions (Zoghbi 2003; Armstrong 2005; Jellinger 2003; Armstrong et al. 1995). Consistent with human findings, two independent lines of MECP2 knockout (MECP2-KO) mice showed reduced dendritic complexity, thin dendrites, decreased dendritic spine density, altered spine morphology, and subdued spine mobility in a variety of neurons (Fukuda et al. 2005; Belichenko et al. 2009; Tropea et al. 2009; Landi et al. 2011; Minh et al. 2012; Stuss et al. 2012; Kishi and Macklis 2004), although the extent and time of these changes vary between studies. Treatment with insulin-like growth factor 1 (IGF-1), currently under clinical trial for RTT, was able to restore dendritic spine density and dynamics (Tropea et al. 2009; Landi et al. 2011; Castro et al. 2014). Some groups reported that MeCP2 restricts dendritic growth and spine maturation (Cheng et al. 2014a) and that dendritic growth of hippocampal neurons is inhibited when MeCP2 is elevated (Zhou et al. 2006). This is in contrast to the report that MeCP2 overexpression induces dendritic overgrowth in mice; in particular apical dendritic arbors in layer V pyramidal neurons have more branches of higher order. These mice also suffer from higher spine gain and loss, with a net bias in favor of spine elimination reflecting the persistence of an immature state (Jiang et al. 2013). Of note, high genetic dosage of MECP2 also causes a childhood neurological disorder called MECP2 duplication syndrome, characterized by intellectual disability and autism (Ramocki et al. 2010). Dendritic pathology may result from any deviations of MeCP2 from the tightly controlled normal level, and the MeCP2 overexpression models could reflect the duplication syndrome. One possible mechanism via which MeCP2 affects dendrites is through regulating the expression of a subset of microRNAs or suppressing microRNA processing (Szulwach et al. 2010; Cheng et al. 2014b; Wu et al. 2010; Urdinguio et al. 2010). Some microRNAs regulated by MeCP2, such as miR-137, are known to regulate dendritic formation and spine morphogenesis (Smrt et al. 2010).

In line with the discussion in Sect. 20.4, both RTT and MECP2 duplication show immunological impairments that could contribute to neurological deficits (Ramocki et al. 2010; Yang et al. 2012; Alvarez-Saavedra et al. 2010; Collins et al. 2004; Derecki et al. 2010; Cortelazzo et al. 2014; Li et al. 2014; Theoharides et al. 2015). The immune system and CNS may share same MeCP2-controlled signaling pathways, such as the STAT-3 pathway (Jiang et al. 2014). Strong evidence suggests that microglia could be a common effector in RTT, supporting the microglia hypothesis discussed above (Maezawa and Jin 2010; Derecki et al. 2012; Jin et al. 2015). The Kipnis group showed that MECP2-KO mice, which usually die at 9–10 weeks, became almost normal and some lived to over 1 year after their brains were populated with wild-type myeloid cells/microglia by a bone marrow transplant approach. However, when cranial irradiation was blocked by lead shield in this procedure to prevent microglial engraftment, disease was not arrested, suggesting that the effect came from microglia. They also showed significant phenotypic reversal after genetic re-expression of WT *Mecp2* only in myeloid cells/microglia (Derecki et al. 2012). The dendritic abnormalities could in part attributed to inherent abnormalities of RTT microglia, which constituently release high levels of glutamate and reactive oxygen species that reduced dendritic extension in cultured neurons (Maezawa and Jin 2010; Jin et al. 2015). Thus, in principle, significant improvements in dendritic morphology and neurobehavioral outcomes in RTT children could be achieved by rectifying microglial abnormalities.

20.7 Conclusion

Figure 20.2 shows a synoptic summary of changes relevant to dendrites in ASDs. Neuronal dendrites are the major anatomic substrate of ASDs. Overall, data from human and most of the data obtained from animal models point to a generalized reduction in size and number and to a morphology alteration of dendrites in autism. Additional studies in human are needed to understand the properties of dendrites in the autistic brain. One critical question is that while dendritic pathology is significant, what constitutes changes specific to autism and related disorders? Animal models with autism-related molecular manipulations could one day accomplish a high-resolution mechanistic picture showing a multitude of signature pathways leading to ASDs. There is likely a myriad of dendritic pathologies that map to different components of the dendrite and correspond to different kinds of ASDs. Targeting specific cellular and molecular pathways driving such abnormalities could lead to realization of the ideal of providing "precision medicine" to autistic children. The models of syndromic ASDs, such as FXS and RTT highlighted here, are highly useful tools to advance our knowledge about molecular changes underlying dendritic pathology. We particularly emphasize that neural network in general and dendrites in particular are not an isolated system. Investigation into autism provides best examples that the immune system with its myriad of cells and mediators has a great impact, and our "microglia hypothesis" proposes that microglia play a pivotal role in mediating environmental and genetic effects on dendritic pathology in autism.

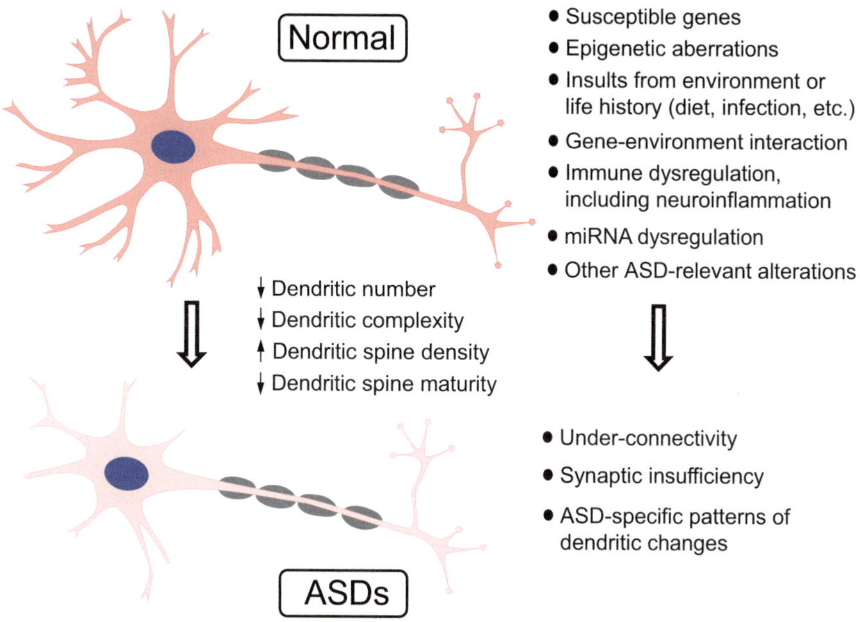

Fig. 20.2 Synoptic summary of dendritic changes in ASDs

References

Abazyan B et al (2010) Prenatal interaction of mutant DISC1 and immune activation produces adult psychopathology. Biol Psychiatry 68:1172–1181. doi:10.1016/j.biopsych.2010.09.022

Alvarez-Saavedra M et al (2010) Elevated expression of MeCP2 in cardiac and skeletal tissues is detrimental for normal development. Hum Mol Genet 19:2177–2190. doi:10.1093/Hmg/Ddq096

Amir RE et al (1999) Rett syndrome is caused by mutations in X-linked MECP2, encoding methyl-CpG-binding protein 2. Nat Genet 23:185–188. doi:10.1038/13810

Armstrong DD (2005) Neuropathology of Rett syndrome. J Child Neurol 20:747–753

Armstrong D, Dunn JK, Antalffy B, Trivedi R (1995) Selective dendritic alterations in the cortex of Rett syndrome. J Neuropathol Exp Neurol 54:195–201

Balakrishnan B, Dai H, Janisse J, Romero R, Kannan S (2013) Maternal endotoxin exposure results in abnormal neuronal architecture in the newborn rabbit. Dev Neurosci 35:396–405. doi:10.1159/000353156

Belichenko PV et al (2009) Widespread changes in dendritic and axonal morphology in Mecp2-mutant mouse models of Rett syndrome: evidence for disruption of neuronal networks. J Comp Neurol 514:240–258. doi:10.1002/cne.22009

Berman RF, Murray KD, Arque G, Hunsaker MR, Wenzel HJ (2012) Abnormal dendrite and spine morphology in primary visual cortex in the CGG knock-in mouse model of the fragile X premutation. Epilepsia 53(Suppl 1):150–160. doi:10.1111/j.1528-1167.2012.03486.x

Betancur C, Sakurai T, Buxbaum JD (2009) The emerging role of synaptic cell-adhesion pathways in the pathogenesis of autism spectrum disorders. Trends Neurosci 32:402–412. doi:S0166-2236(09)00092-7 [pii]10.1016/j.tins.2009.04.003

Breece E et al (2013) Myeloid dendritic cells frequencies are increased in children with autism spectrum disorder and associated with amygdala volume and repetitive behaviors. Brain Behav Immun 31:69–75. doi:10.1016/j.bbi.2012.10.006

Bringas ME, Carvajal-Flores FN, Lopez-Ramirez TA, Atzori M, Flores G (2013) Rearrangement of the dendritic morphology in limbic regions and altered exploratory behavior in a rat model of autism spectrum disorder. Neuroscience 241:170–187. doi:10.1016/j.neuroscience.2013.03.030

Brown AS (2012) Epidemiologic studies of exposure to prenatal infection and risk of schizophrenia and autism. Dev Neurobiol 72:1272–1276. doi:10.1002/Dneu.22024

Burd I et al (2010) Inflammation-induced preterm birth alters neuronal morphology in the mouse fetal brain. J Neurosci Res 88:1872–1881. doi:10.1002/jnr.22368

Castro J et al (2014) Functional recovery with recombinant human IGF1 treatment in a mouse model of Rett syndrome. Proc Natl Acad Sci U S A 111:9941–9946. doi:10.1073/pnas.1311685111

Chahrour M, Zoghbi HY (2007) The story of Rett syndrome: from clinic to neurobiology. Neuron 56:422–437. doi:S0896-6273(07)00756-8 [pii]10.1016/j.neuron.2007.10.001

Chahrour M et al (2008) MeCP2, a key contributor to neurological disease, activates and represses transcription. Science 320:1224–1229. doi:320/5880/1224 [pii] 10.1126/science.1153252

Cheng TL et al (2014) MeCP2 suppresses nuclear microRNA processing and dendritic growth by regulating the DGCR8/Drosha complex. Dev Cell 28:547–560. doi:10.1016/j.devcel.2014.01.032

Collins AL et al (2004) Mild overexpression of MeCP2 causes a progressive neurological disorder in mice. Hum Mol Genet 13:2679–2689. doi:10.1093/Hmg/Ddh282

Cortelazzo A et al (2014) Subclinical inflammatory status in Rett syndrome. Mediat Inflamm 2014:480980. doi:10.1155/2014/480980

Cruz-Martin A, Crespo M, Portera-Cailliau C (2010) Delayed stabilization of dendritic spines in fragile X mice. J Neurosci 30:7793–7803. doi:10.1523/JNEUROSCI.0577-10.2010

Darnell JC et al (2011) FMRP stalls ribosomal translocation on mRNAs linked to synaptic function and autism. Cell 146:247–261. doi:10.1016/j.cell.2011.06.013

Davalos D et al (2005) ATP mediates rapid microglial response to local brain injury in vivo. Nat Neurosci 8:752–758. doi:10.1038/nn1472
de Anda FC et al (2012) Autism spectrum disorder susceptibility gene TAOK2 affects basal dendrite formation in the neocortex. Nat Neurosci 15:1022–1031. doi:10.1038/Nn.3141
De Rubeis S et al (2013) CYFIP1 coordinates mRNA translation and cytoskeleton remodeling to ensure proper dendritic spine formation. Neuron 79:1169–1182. doi:10.1016/j.neuron.2013.06.039
Derecki NC, Privman E, Kipnis J (2010) Rett syndrome and other autism spectrum disorders – brain diseases of immune malfunction? Mol Psychiatry 15:355–363. doi:mp201021 [pii]10.1038/mp.2010.21
Derecki NC et al (2012) Wild-type microglia arrest pathology in a mouse model of Rett syndrome. Nature 484:105–109. doi:nature10907 [pii] 10.1038/nature10907
Derecki NC, Katzmarski N, Kipnis J, Meyer-Luehmann M (2014) Microglia as a critical player in both developmental and late-life CNS pathologies. Acta Neuropathol 128:333–345. doi:10.1007/s00401-014-1321-z
Devys D, Lutz Y, Rouyer N, Bellocq JP, Mandel JL (1993) The FMR-1 protein is cytoplasmic, most abundant in neurons and appears normal in carriers of a fragile X premutation. Nat Genet 4:335–340. doi:10.1038/ng0893-335
Dindot SV, Antalffy BA, Bhattacharjee MB, Beaudet AL (2008) The Angelman syndrome ubiquitin ligase localizes to the synapse and nucleus, and maternal deficiency results in abnormal dendritic spine morphology. Hum Mol Genet 17:111–118. doi:10.1093/hmg/ddm288
Durand CM et al (2007) Mutations in the gene encoding the synaptic scaffolding protein SHANK3 are associated with autism spectrum disorders. Nat Genet 39:25–27. doi:10.1038/ng1933
Feng Y et al (1997) Fragile X mental retardation protein: nucleocytoplasmic shuttling and association with somatodendritic ribosomes. J Neurosci 17:1539–1547
Filiano AJ, Gadani SP, Kipnis J (2014) Interactions of innate and adaptive immunity in brain development and function. Brain Res. doi:10.1016/j.brainres.2014.07.050
Fukuda T, Itoh M, Ichikawa T, Washiyama K, Goto Y (2005) Delayed maturation of neuronal architecture and synaptogenesis in cerebral cortex of Mecp2-deficient mice. J Neuropathol Exp Neurol 64:537–544
Geschwind DH (2008) Autism: many genes, common pathways? Cell 135:391–395. doi:S0092-8674(08)01307-X [pii] 10.1016/j.cell.2008.10.016
Geschwind DH (2011) Genetics of autism spectrum disorders. Trends Cogn Sci 15:409–416. doi:S1364-6613(11)00145-8 [pii] 10.1016/j.tics.2011.07.003
Geschwind DH, Levitt P (2007) Autism spectrum disorders: developmental disconnection syndromes. Curr Opin Neurobiol 17:103–111. doi:S0959-4388(07)00011-6 [pii] 10.1016/j.conb.2007.01.009
Goines PE, Ashwood P (2013) Cytokine dysregulation in autism spectrum disorders (ASD): possible role of the environment. Neurotoxicol Teratol 36:67–81. doi:10.1016/j.ntt.2012.07.006
Goines P, Zimmerman A, Ashwood P, Van de Water J (2011) In: Amaral DG, Dawson G, Geschwind DH (eds) Autism spectrum disorders. Oxford University Press, New York, pp 395–419. Ch. 24
Hagerman P (2013) Fragile X-associated tremor/ataxia syndrome (FXTAS): pathology and mechanisms. Acta Neuropathol 126:1–19. doi:10.1007/s00401-013-1138-1
Hagerman R, Hoem G, Hagerman P (2010) Fragile X and autism: intertwined at the molecular level leading to targeted treatments. Mol Autism 1:12. doi:2040-2392-1-12 [pii] 10.1186/2040-2392-1-12
Hallmayer J et al (2011) Genetic heritability and shared environmental factors among twin pairs with autism. Arch Gen Psychiatry. doi:archgenpsychiatry.2011.76 [pii]10.1001/archgenpsychiatry.2011.76

Haws ME et al (2014) PTEN knockdown alters dendritic spine/protrusion morphology, not density. J Comp Neurol 522:1171–1190. doi:10.1002/cne.23488
Hayashi ML et al (2007) Inhibition of p21-activated kinase rescues symptoms of fragile X syndrome in mice. Proc Natl Acad Sci U S A 104:11489–11494. doi:10.1073/pnas.0705003104
Hertz-Picciotto I (2011) In: Amaral DG, Dawson G, Gerschwind DH (eds) Autism spectrum disorders. Oxford University Press, New York, pp 827–826. Ch. 47
Hinton VJ, Brown WT, Wisniewski K, Rudelli RD (1991) Analysis of neocortex in three males with the fragile X syndrome. Am J Med Genet 41:289–294. doi:10.1002/ajmg.1320410306
Hsiao EY, McBride SW, Chow J, Mazmanian SK, Patterson PH (2012) Modeling an autism risk factor in mice leads to permanent immune dysregulation. Proc Natl Acad Sci U S A 109: 12776–12781. doi:10.1073/pnas.1202556109
Hutsler JJ, Zhang H (2010) Increased dendritic spine densities on cortical projection neurons in autism spectrum disorders. Brain Res 1309:83–94. doi:10.1016/j.brainres.2009.09.120
Irwin SA et al (2002) Dendritic spine and dendritic field characteristics of layer V pyramidal neurons in the visual cortex of fragile-X knockout mice. Am J Med Genet 111:140–146. doi:10.1002/ajmg.10500
Jellinger KA (2003) Rett syndrome – an update. J Neural Transm 110:681–701. doi:10.1007/s00702-003-0822-z
Jiang M et al (2013) Dendritic arborization and spine dynamics are abnormal in the mouse model of MECP2 duplication syndrome. J Neurosci 33:19518–19533. doi:10.1523/JNEUROSCI.1745-13.2013
Jiang S et al (2014) MeCP2 reinforces STAT3 signaling and the generation of effector CD4+ T cells by promoting miR-124-mediated suppression of SOCS5. Sci Signal 7:ra25. doi:10.1126/scisignal.2004824
Jin LW, Horiuchi M, Wulff H, Liu X-B, Cortopassi GA, Erickson JD, Maezawa I (2015) Dysregulation of glutamine transporter SNAT1 in Rett syndrome microglia: a mechanism for mitochondrial dysfunction and neurotoxicity. J Neurosci 35:2516–2529
Just MA, Cherkassky VL, Keller TA, Kana RK, Minshew NJ (2007) Functional and anatomical cortical underconnectivity in autism: evidence from an FMRI study of an executive function task and corpus callosum morphometry. Cereb Cortex 17:951–961. doi:10.1093/cercor/bhl006
Kilpinen H et al (2008) Association of DISC1 with autism and Asperger syndrome. Mol Psychiatry 13:187–196. doi:10.1038/sj.mp.4002031
Kishi N, Macklis JD (2004) MECP2 is progressively expressed in post-migratory neurons and is involved in neuronal maturation rather than cell fate decisions. Mol Cell Neurosci 27:306–321. doi:10.1016/j.mcn.2004.07.006
Koekkoek SK et al (2005) Deletion of FMR1 in Purkinje cells enhances parallel fiber LTD, enlarges spines, and attenuates cerebellar eyelid conditioning in fragile X syndrome. Neuron 47:339–352. doi:10.1016/j.neuron.2005.07.005
Kondo S, Kohsaka S, Okabe S (2011) Long-term changes of spine dynamics and microglia after transient peripheral immune response triggered by LPS in vivo. Mol Brain 4:27. doi:10.1186/1756-6606-4-27
Korn T, Bettelli E, Oukka M, Kuchroo VK (2009) IL-17 and Th17 cells. Annu Rev Immunol 27: 485–517. doi:10.1146/annurev.immunol.021908.132710
Kwon CH et al (2006) Pten regulates neuronal arborization and social interaction in mice. Neuron 50:377–388. doi:10.1016/j.neuron.2006.03.023
Landi S et al (2011) The short-time structural plasticity of dendritic spines is altered in a model of Rett syndrome. Sci Rep 1:45. doi:Artn 45 Doi 10.1038/Srep00045
Larkman AU (1991) Dendritic morphology of pyramidal neurones of the visual cortex of the rat: III. Spine distributions. J Comp Neurol 306:332–343. doi:10.1002/cne.903060209
LaSalle JM (2011) A genomic point-of-view on environmental factors influencing the human brain methylome. Epigenetics 6:862–869. doi:16353 [pii]
LaSalle JM, Yasui DH (2009) Evolving role of MeCP2 in Rett syndrome and autism. Epigenomics 1:119–130. doi:10.2217/epi.09.13

Levitt P, Campbell DB (2009) The genetic and neurobiologic compass points toward common signaling dysfunctions in autism spectrum disorders. J Clin Invest 119:747–754. doi:37934 [pii] 10.1172/JCI37934

Li C et al (2014) MeCP2 enforces Foxp3 expression to promote regulatory T cells' resilience to inflammation. Proc Natl Acad Sci U S A 111:E2807–E2816. doi:10.1073/pnas.1401505111

Maezawa I, Jin LW (2010) Rett syndrome microglia damage dendrites and synapses by the elevated release of glutamate. J Neurosci 30:5346–5356. doi:30/15/5346 [pii]10.1523/JNEUROSCI.5966-09.2010

Maezawa I, Calafiore M, Wulff H, Jin LW (2011) Does microglial dysfunction play a role in autism and Rett syndrome? Neuron Glia Biol 7:85–97. doi:10.1017/S1740925X1200004X

Malkova NV, Yu CZ, Hsiao EY, Moore MJ, Patterson PH (2012) Maternal immune activation yields offspring displaying mouse versions of the three core symptoms of autism. Brain Behav Immun 26:607–616. doi:10.1016/j.bbi.2012.01.011

Mandal M, Marzouk AC, Donnelly R, Ponzio NM (2011) Maternal immune stimulation during pregnancy affects adaptive immunity in offspring to promote development of TH17 cells. Brain Behav Immun 25:863–871. doi:10.1016/j.bbi.2010.09.011

Mandal M et al (2013) Maternal immune stimulation during pregnancy shapes the immunological phenotype of offspring. Brain Behav Immun 33:33–45. doi:10.1016/j.bbi.2013.04.012

Maussion G et al (2008) Convergent evidence identifying MAP/microtubule affinity-regulating kinase 1 (MARK1) as a susceptibility gene for autism. Hum Mol Genet 17:2541–2551. doi:10.1093/hmg/ddn154

Mercati O et al (2013) Contactin 4, -5 and -6 differentially regulate neuritogenesis while they display identical PTPRG binding sites. Biol Open 2:324–334. doi:10.1242/bio.20133343

Miao S et al (2013) The Angelman syndrome protein Ube3a is required for polarized dendrite morphogenesis in pyramidal neurons. J Neurosci 33:327–333. doi:10.1523/Jneurosci.2509-12.2013

Minh VCN et al (2012) MeCP2 is critical for maintaining mature neuronal networks and global brain anatomy during late stages of postnatal brain development and in the mature adult brain. J Neurosci 32:10021–10034. doi:10.1523/Jneurosci.1316-12.2012

Minshew NJ, Keller TA (2010) The nature of brain dysfunction in autism: functional brain imaging studies. Curr Opin Neurol 23:124–130. doi:10.1097/WCO.0b013e32833782d4

Monji A, Maezawa I, Mizoguchi Y, Kato TA, Jin L-W (2014) In: Tremblay M-E, Sierra A (eds) Microglia in health and disease. Springer, New York, pp 345–372. Ch. 14

Mukaetova-Ladinska EB, Arnold H, Jaros E, Perry R, Perry E (2004) Depletion of MAP2 expression and laminar cytoarchitectonic changes in dorsolateral prefrontal cortex in adult autistic individuals. Neuropathol Appl Neurobiol 30:615–623. doi:10.1111/j.1365-2990.2004.00574.x

Nagarajan RP, Hogart AR, Gwye Y, Martin MR, LaSalle JM (2006) Reduced MeCP2 expression is frequent in autism frontal cortex and correlates with aberrant MECP2 promoter methylation. Epigenetics 1:e1–e11

Napoli I et al (2008) The fragile X syndrome protein represses activity-dependent translation through CYFIP1, a new 4E-BP. Cell 134:1042–1054. doi:10.1016/j.cell.2008.07.031

Nimchinsky EA, Oberlander AM, Svoboda K (2001) Abnormal development of dendritic spines in FMR1 knock-out mice. J Neurosci 21:5139–5146

Nimmerjahn A, Kirchhoff F, Helmchen F (2005) Resting microglial cells are highly dynamic surveillants of brain parenchyma in vivo. Science 308:1314–1318. doi:1110647 [pii] 10.1126/science.1110647

Nowicki ST et al (2007) The Prader-Willi phenotype of fragile X syndrome. J Dev Behav Pediatr 28:133–138. doi:10.1097/01.DBP.0000267563.18952.c9

Orloff MS et al (2013) Germline PIK3CA and AKT1 mutations in Cowden and Cowden-like syndromes. Am J Hum Genet 92:76–80. doi:10.1016/j.ajhg.2012.10.021

Pathania M et al (2014) The autism and schizophrenia associated gene CYFIP1 is critical for the maintenance of dendritic complexity and the stabilization of mature spines. Transl Psychiatry 4:e374. doi:10.1038/tp.2014.16

Patterson PH (2012) Maternal infection and autism. Brain Behav Immun 26:393. doi:10.1016/j.bbi.2011.09.008

Ramocki MB, Tavyev YJ, Peters SU (2010) The MECP2 duplication syndrome. Am J Med Genet A 152A:1079–1088. doi:10.1002/ajmg.a.33184

Raymond GV, Bauman ML, Kemper TL (1996) Hippocampus in autism: a Golgi analysis. Acta Neuropathol 91:117–119

Rice C (2009) MMWR Surveill Summ 58:1–20

Rudelli RD et al (1985) Adult fragile X syndrome. Clinico-neuropathologic findings. Acta Neuropathol 67:289–295

Sala C et al (2001) Regulation of dendritic spine morphology and synaptic function by Shank and Homer. Neuron 31:115–130

Schafer DP et al (2012) Microglia sculpt postnatal neural circuits in an activity and complement-dependent manner. Neuron 74:691–705. doi:10.1016/j.neuron.2012.03.026

Schwartzer JJ, Koenig CM, Berman RF (2013a) Using mouse models of autism spectrum disorders to study the neurotoxicology of gene-environment interactions. Neurotoxicol Teratol 36: 17–35. doi:10.1016/j.ntt.2012.08.007

Schwartzer JJ et al (2013b) Maternal immune activation and strain specific interactions in the development of autism-like behaviors in mice. Transl Psychiatry 3:e240. doi:10.1038/tp.2013.16

Shi L, Fatemi SH, Sidwell RW, Patterson PH (2003) Maternal influenza infection causes marked behavioral and pharmacological changes in the offspring. J Neurosci 23:297–302

Smrt RD et al (2010) MicroRNA miR-137 regulates neuronal maturation by targeting ubiquitin ligase mind bomb-1. Stem Cells 28:1060–1070. doi:10.1002/stem.431

Sowers LP et al (2013) Disruption of the non-canonical Wnt gene PRICKLE2 leads to autism-like behaviors with evidence for hippocampal synaptic dysfunction. Mol Psychiatry 18:1077–1089. doi:10.1038/mp.2013.71

Srivastava DP et al (2012a) Social, communication, and cortical structural impairments in Epac2-deficient mice. J Neurosci 32:11864–11878. doi:10.1523/JNEUROSCI.1349-12.2012

Srivastava DP et al (2012b) An autism-associated variant of Epac2 reveals a role for Ras/Epac2 signaling in controlling basal dendrite maintenance in mice. PLoS Biol 10:e1001350. doi:10.1371/journal.pbio.1001350

Stuss DP, Boyd JD, Levin DB, Delaney KR (2012) MeCP2 mutation results in compartment-specific reductions in dendritic branching and spine density in layer 5 motor cortical neurons of YFP-H mice. PLoS One 7:e31896. doi:ARTN e31896 DOI 10.1371/journal.pone.0031896

Su T et al (2011) Early continuous inhibition of group 1 mGlu signaling partially rescues dendritic spine abnormalities in the Fmr1 knockout mouse model for fragile X syndrome. Psychopharmacology (Berl) 215:291–300. doi:10.1007/s00213-010-2130-2

Szulwach KE et al (2010) Cross talk between microRNA and epigenetic regulation in adult neurogenesis. J Cell Biol 189:127–U181. doi:10.1083/jcb.200908151

Tavazoie SF, Alvarez VA, Ridenour DA, Kwiatkowski DJ, Sabatini BL (2005) Regulation of neuronal morphology and function by the tumor suppressors Tsc1 and Tsc2. Nat Neurosci 8: 1727–1734. doi:10.1038/nn1566

Theoharides TC, Athanassiou M, Panagiotidou S, Doyle R (2015) Dysregulated brain immunity and neurotrophin signaling in Rett syndrome and autism spectrum disorders. J Neuroimmunol 279:33–38. doi:10.1016/j.jneuroim.2014.12.003

Thomas CC, Combe CL, Dyar KA, Inglis FM (2008) Modest alterations in patterns of motor neuron dendrite morphology in the Fmr1 knockout mouse model for fragile X. Int J Dev Neurosci 26:805–811. doi:10.1016/j.ijdevneu.2008.06.003

Tropea D et al (2009) Partial reversal of Rett Syndrome-like symptoms in MeCP2 mutant mice. Proc Natl Acad Sci U S A 106:2029–2034. doi:106/6/2029 [pii]10.1073/pnas.0812394106

Urdinguio RG et al (2010) Disrupted microRNA expression caused by Mecp2 loss in a mouse model of Rett syndrome. Epigenetics 5:656–663. doi:10.4161/epi.5.7.13055
Van Maldergem L et al (2013) Loss of function of KIAA2022 causes mild to severe intellectual disability with an autism spectrum disorder and impairs neurite outgrowth. Hum Mol Genet 22: 3306–3314. doi:10.1093/hmg/ddt187
Vismara LA, Rogers SJ (2010) Behavioral treatments in autism spectrum disorder: what do we know? Annu Rev Clin Psychol 6(6):447–468. doi:10.1146/annurev.clinpsy.121208.131151
Wake H, Moorhouse AJ, Jinno S, Kohsaka S, Nabekura J (2009) Resting microglia directly monitor the functional state of synapses in vivo and determine the fate of ischemic terminals. J Neurosci 29:3974–3980. doi:29/13/3974 [pii]10.1523/JNEUROSCI.4363-08.2009
Walsh JT, Watson N, Kipnis J (2014) T cells in the central nervous system: messengers of destruction or purveyors of protection? Immunology 141:340–344. doi:10.1111/imm.12187
Weiss LA et al (2008) Association between microdeletion and microduplication at 16p11.2 and autism. New Engl J Med 358:667–675. doi:10.1056/Nejmoa075974
Westmark CJ et al (2011) Reversal of fragile X phenotypes by manipulation of AbetaPP/Abeta levels in Fmr1KO mice. PLoS One 6:e26549. doi:10.1371/journal.pone.0026549
Weston MC, Chen H, Swann JW (2014) Loss of mTOR repressors Tsc1 or Pten has divergent effects on excitatory and inhibitory synaptic transmission in single hippocampal neuron cultures. Front Mol Neurosci 7:1
Williams SC (2012) Genetics: searching for answers. Nature 491:S4–S6
Wu H et al (2010) Genome-wide analysis reveals methyl-CpG-binding protein 2-dependent regulation of microRNAs in a mouse model of Rett syndrome. Proc Natl Acad Sci U S A 107:18161–18166. doi:10.1073/pnas.1005595107
Yang T et al (2012) Overexpression of methyl-CpG binding protein 2 impairs T(H)1 responses. Sci Transl Med 4:163ra158. doi:10.1126/scitranslmed.3004430
Yasui DH et al (2007) Integrated epigenomic analyses of neuronal MeCP2 reveal a role for long-range interaction with active genes. Proc Natl Acad Sci U S A 104:19416–19421. doi:0707442104 [pii] 10.1073/pnas.0707442104
Yeargin-Allsopp M et al (2003) Prevalence of autism in a US metropolitan area. JAMA 289: 49–55. doi:joc21908 [pii]
Zerbo O et al (2013) Is maternal influenza or fever during pregnancy associated with autism or developmental delays? Results from the CHARGE (CHildhood Autism Risks from Genetics and Environment) study. J Autism Dev Disord 43:25–33. doi:10.1007/s10803-012-1540-x
Zhan Y et al (2014) Deficient neuron-microglia signaling results in impaired functional brain connectivity and social behavior. Nat Neurosci 17:400–406. doi:10.1038/nn.3641
Zhou Z et al (2006) Brain-specific phosphorylation of MeCP2 regulates activity-dependent Bdnf transcription, dendritic growth, and spine maturation. Neuron 52:255–269. doi:S0896-6273(06)00775-6 [pii] 10.1016/j.neuron.2006.09.037
Zoghbi HY (2003) Postnatal neurodevelopmental disorders: meeting at the synapse? Science 302: 826–830. doi:10.1126/science.1089071 302/5646/826 [pii]
Zuko A et al (2013) Contactins in the neurobiology of autism. Eur J Pharmacol 719:63–74. doi:10.1016/j.ejphar.2013.07.016

Chapter 21
Mechanisms of Dendrite Degeneration in Amyotrophic Lateral Sclerosis

Sebum Lee, Yulei Shang, and Eric J. Huang

Abstract Amyotrophic lateral sclerosis (ALS) and frontotemporal dementia (FTD) are two ends of a disease spectrum. Among the many etiologies for ALS and FTD, the identification of mutations and proteinopathies in two RNA-binding proteins, TDP-43 (TARDBP or TAR DNA-binding protein 43) and its closely related RNA-/DNA-binding protein FUS (fused in sarcoma), raises the intriguing possibility that perturbations to the RNA homeostasis and metabolism in neurons may contribute to the pathogenesis of these diseases. Although the similarities between TDP-43 and FUS suggest that mutations and proteinopathy involving these two proteins may converge on the same mechanisms leading to neurodegeneration, there is emerging evidence that FUS mutations target distinct mechanisms to cause early disease onset and aggressive progression of disease. This chapter focuses on the recent advances on the molecular, cellular, and genetic approaches to uncover the mechanisms of endogenous FUS proteins in neural development and neurodegeneration. These findings provide important insights to understand how FUS mutations may perturb the maintenance of dendrites through fundamental processes in RNA splicing, RNA transport, and DNA damage response/repair. These results contribute to the understanding of phenotypic manifestations in neurodegeneration related to FUS mutations and to identify important directions for future investigations.

Grant Sponsors National Institute of Health; Grant number: OD011915; Veterans Administrations BLR&D Merit Review Award, Grant numbers: I01 BX001108; Pilot Award, Grant number: I21 BX1625; Muscular Dystrophy Association (MDA) Research Grant #217592; Consortium for Frontotemporal Dementia Research (CFR); and University of California MRPI Award #142675

S. Lee • Y. Shang
Department of Pathology, University of California San Francisco, 505 Parnassus Avenue, HSW450C, Box 0502, San Francisco, CA 94143-0502, USA

E.J. Huang (✉)
Department of Pathology, University of California San Francisco, 505 Parnassus Avenue, HSW450C, Box 0502, San Francisco, CA 94143-0502, USA

Pathology Service 113B, VA Medical Center, San Francisco, CA, USA
e-mail: eric.huang2@ucsf.edu

K. Emoto et al. (eds.), *Dendrites*, DOI 10.1007/978-4-431-56050-0_21

Keywords RNA-binding protein • Fused in sarcoma (FUS) • DNA damage repair • RNA splicing • Amyotrophic lateral sclerosis (ALS) • Frontotemporal dementia (FTD)

21.1 The Expanding Landscape of Neurodegenerative Diseases

21.1.1 Amyotrophic Lateral Sclerosis and Frontotemporal Dementia: Two Ends of a Disease Spectrum

Amyotrophic lateral sclerosis (ALS), also known as Lou Gehrig's disease, is an adult-onset neurodegenerative disease that affects upper and lower motor neurons. As initially described by Jean-Martin Charcot more than 140 years ago, the key clinical features in ALS patients include muscle wasting and progressive loss of spinal motor neurons and upper motor neurons and their axons in the lateral columns of the spinal cord. This century-old classic definition, however, has been under scrutiny by the emerging appreciations that some ALS patients who exhibit deficits in higher cognitive functions at the early stage of their clinical course eventually develop behavioral variant of frontotemporal dementia (bvFTD). It has been estimated that ~15 % of FTD patients develop features of ALS (Lomen-Hoerth et al. 2002; Ringholz et al. 2005), and up to 50 % of ALS patients show abnormal neuropsychological testing indicative of frontal lobe dysfunctions (Lillo et al. 2011; Lomen-Hoerth et al. 2003). Furthermore, patients with FTD-ALS usually have shortened life span compared to patients with pure FTD or ALS (Olney et al. 2005). Together, these studies provide a new framework to reevaluate and broaden the diagnostic criteria for ALS (Strong et al. 2009) and lead to a new clinical paradigm in which FTD and ALS are linked within a disease spectrum (Fig. 21.1).

21.1.2 Genetic and Pathological Correlates Connecting FTD and ALS

The idea that FTD and ALS are a disease spectrum is further supported by genetic evidence that patients with familial FTD-ALS and ALS often carry mutations in the same genes. In fact, of all the different subtypes of FTD, FTD-ALS cases have such a high propensity (~30 %) of familial inheritance that genetic counseling is now considered as the standard of care for these patients (Goldman et al. 2005). Among a growing number of genes involved in familial FTD-ALS and ALS (Fig. 21.1), mutations in three genes account for the majority of cases. These mutations include missense mutations in genes encoding two RNA-/DNA-binding proteins, *TDP-43* (*TARDBP* or TAR-DNA-binding protein-43) and *FUS/TLS* (*fused in sarcoma/*

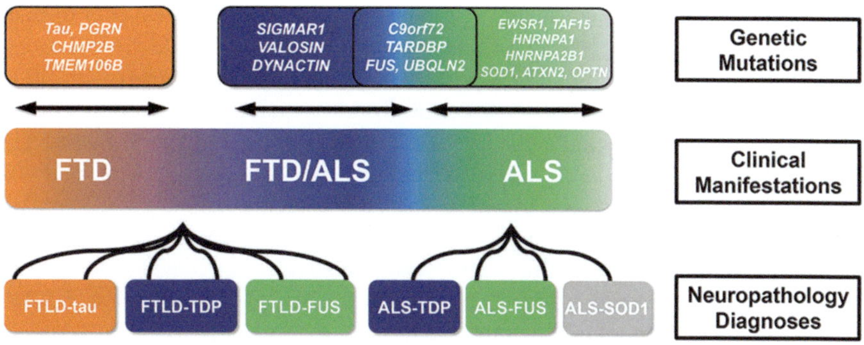

Fig. 21.1 A schematic diagram illustrating the disease spectrum, genetics, and proteinopathy of frontotemporal dementia (FTD) and amyotrophic lateral sclerosis (ALS). Clinical manifestations indicate that FTD and ALS are two ends of a disease spectrum. It is estimated that ~15 % of FTD patients develop ALS and up to 50 % of ALS patients show impairments in frontal executive functions. This new clinical paradigm is further supported by the findings that patients with FTD-ALS and ALS share similar genetic mutations, in particular mutations that involve the GGGGCC hexanucleotide expansion in the noncoding region of the *C9ORF72* gene and two RNA-/DNA-binding proteins TDP-43 and FUS. FTD-ALS and ALS patients with mutations in the same gene (*C9ORF72*, *TARDBP*, or *FUS*) often exhibit similar neuropathological features with neurons containing distinct protein aggregates of TDP-43 or FUS proteins. Therefore, the neuropathology of these patients can be categorized by the nature of the proteinopathy into FTLD (frontotemporal lobar degeneration)-tau, FTLD-TDP, FTLD-FUS, ALS-TDP, ALS-FUS, or ALS-SOD1

translocation in liposarcoma or *FUS*), and the GGGGCC hexanucleotide expansions in *C9ORF72* gene (Lee et al. 2012; Ling et al. 2013). The discovery of TDP-43 as a major component in the ubiquitin-positive, tau-negative insoluble protein aggregates in neurons and glia represents a major breakthrough in FTD research (Arai et al. 2006; Neumann et al. 2006). Moreover, the impact of this discovery goes beyond the identification of a single disease gene and essentially ushers in a new era of research that focuses on the potential contributions of transcription, RNA splicing, and RNA metabolism on neurodegenerative diseases.

TDP-43 is originally identified to bind to the TAR DNA sequence in HIV-1 genome to regulate viral gene expression (Ou et al. 1995). Under physiological conditions, TDP-43 is a ubiquitous nuclear protein; however, in FTD patients, TDP-43 aggregates are present predominantly in neuronal cytoplasm and dystrophic neuronal processes (Arai et al. 2006; Neumann et al. 2006). This distinct feature, defined as TDP-43 proteinopathy, constitutes a major neuropathological diagnosis entity in frontotemporal lobar degeneration (FTLD-TDP) and in sporadic ALS (ALS-TDP) (Fig. 21.1). Several subsequent studies show that dominant mutations in the *TARDBP* gene can also be identified in familial ALS and FTD patients (Lattante et al. 2013). The identification of autosomal dominant mutations in the *FUS* gene in a large kindred of familial ALS (FALS) further expanded the genetic and neuropathological landscape of ALS (Kwiatkowski et al. 2009; Vance et al. 2009). Similar to TDP-43, FUS proteins reside primarily in the neuronal nuclei, but in ALS-FUS patients, FUS proteins form large aggregates in the cytoplasm. The morphology of FUS proteinopathy in FALS ranges from diffuse and dense cytoplasmic aggregate present in late-onset cases to basophilic inclusions commonly found in juvenile FALS with FUS-P525L mutation. Finally, in 2011, two groups independently reported the GGGGCC hexanucleotide repeat expansions in the noncoding region of the *C9ORF72* gene as causal links to ALS and FTD (DeJesus-Hernandez et al. 2011; Renton et al. 2011). Although TDP-43 proteinopathy can be detected in FTLD and ALS patients with *C9ORF72* mutations, the neuropathological features in these cases are quite heterogeneous and also include prominent ubiquitin and p62-positive but TDP-43-negative intracytoplasmic and intranuclear inclusions (Bigio 2012; Mackenzie et al. 2014).

It is estimated that mutations in *TARDBP* and *FUS* each account for ~5 % of FALS, whereas the GGGGCC expansion mutations in *C9ORF72* account for 20–40 % of ALS and FTD-ALS cases, depending on the population studied. Given the large number of cases with *C9ORF72* mutations, there have been tremendous interests in understanding the underlying mechanisms. While several mechanisms, including RNA toxicity and dipeptide accumulation (Ash et al. 2013; Donnelly et al. 2013; Mori et al. 2013), have been proposed for *C9ORF72* mutations, the rapidly progressing research on this subject will definitively bring many more surprises in the future. One important feature noted in a recent study indicates that the age of disease onset for FALS caused by *FUS*, *TARDBP*, and *C9ORF72* mutations differs quite drastically in that mutations in *FUS* account for ~35 % of FALS in patients younger than 40 years old, whereas mutations in *C9ORF72* are much more common in patients older than 50 years of age (Millecamps et al. 2012). Indeed, meta-analyses of 154 ALS cases with *FUS* mutations (including FALS and SALS with de novo *FUS* mutations) show an average disease onset of 43.8 ± 17.4 years (Fig. 21.2) (Deng et al. 2014a; Lattante et al. 2013). More than 60 % of cases with FUS mutations show disease onset before 45 years of age, with many juvenile ALS cases presenting with disease onset in late teens and early 20s (Fig. 21.2) (Baumer et al. 2010; Huang et al. 2010). These findings are similar to those from another study using smaller sample size and show that the average disease onset for *FUS*, *SOD1*, or *TARDBP* mutations is 43.6 ± 15.8, 47.7 ± 13.0, and 54.7 ± 15.3, respectively (Yan et al. 2010). Kaplan-Meier survival analysis shows statistically significant differences in the trend of age of onset among these three mutations. This distinctive feature of *FUS* mutations raises the intriguing hypothesis that, despite the similarities between TDP-43 and FUS proteins, mutations in FUS may target divergent mechanisms that perturb the development, maintenance, and homeostasis of the nervous system in early postnatal life and in the aging process.

This chapter focuses on the recent progress on the molecular, cellular, and genetic approaches to uncover the mechanisms of endogenous FUS proteins in the organismal and neural development. These findings provide important insights to understand how FUS mutations may perturb the fundamental processes in DNA damage response/repair, RNA splicing, and RNA transport, to interpret the phenotypic manifestations in neurodegeneration related to FUS mutations, and to identify important directions for future investigations. It is important to point out that

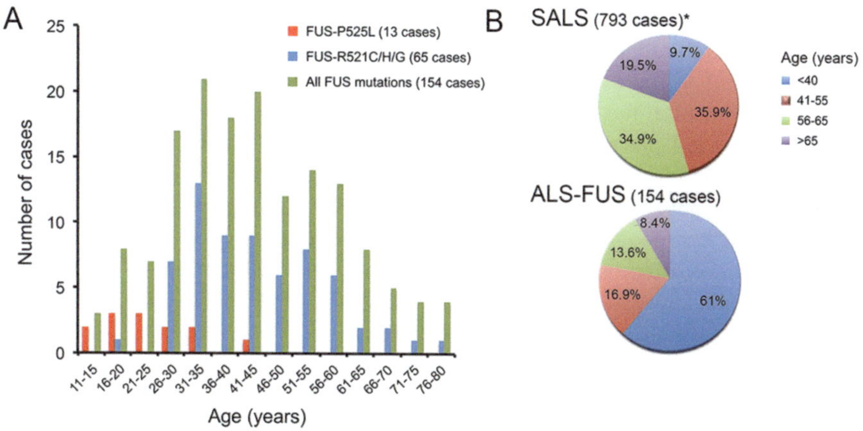

Fig. 21.2 The age of disease onset in ALS patients with FUS mutations. (**a**) Meta-analyses of 154 ALS patients (either familial ALS inherited mutations or sporadic ALS with de novo mutations) show a predilection early disease onset. Compared to all FUS mutations and the most common mutations that occur in amino acid 521 (FUS-R521C), FUS-P525L mutation tends to occur in late teens and early 20s and represents a much more aggressive form of disease. (**b**) For sporadic ALS (SALS), about 35.9 % and 34.9 % of patients show disease onset in the range of 41–55 and 56–65 years old. In contrast, more than 60 % of ALS patients with FUS mutations show disease onset before 40 years old. *Statistics for SALS have been adapted from the study by Testa and colleagues (2004)

patients with FTLD-FUS do not carry mutations in *FUS* gene. In addition, patients with FTLD-FUS pathology are quite rare, and the clinical manifestations and proteinopathy found in these patients tend to be heterogeneous (Mackenzie et al. 2011). Given the complexity and heterogeneity in the pathogenesis of FTLD-FUS, this chapter will not include further discussions on this entity.

21.2 Mechanisms of FUS in Neural Development

21.2.1 Biochemical Properties of FUS

FUS is identified as an oncogene that undergoes chromosomal translocation in myxoid liposarcoma, in which the N-terminal transcriptional activation domain of FUS is fused to CHOP (CAAT enhancer-binding homologous protein), a growth arrest and DNA damage inducible member of the C/EBP family of transcription factors (Crozat et al. 1993; Rabbitts et al. 1993). Subsequent studies further reveal that chromosomal translocations involving FUS can also be identified in several other human cancers, including acute myeloid leukemia, where the N-terminus of *FUS* gene is translocated to the *ERG* gene, a member of the ETS transcription factor family (Ichikawa et al. 1994; Prasad et al. 1994). Structurally, FUS belongs to a family of FET RNA-binding proteins, including FUS, Ewing's sarcoma RNA-binding protein 1 (EWSR1), and Tata-binding protein-associated factor 2N (TAF-15), that are known to interact with RNA polymerase II (RNAP II) and general transcription factor TFIID (Das et al. 2007; Schwartz et al. 2012; Tan and Manley 2009). Full-length human FUS protein contains 526 amino acids that can be divided into the N-terminal Q/G/S/Y domain (amino acids 1–165) and Gly-rich region (amino acids 166–267) (Fig. 21.3). The C-terminal half of FUS protein contains an RNA recognition motif (RRM) (amino acids 285–371); two Arg-Gly-Gly (RGG)-repeat regions (amino acids 371–422 and 453–501), interrupted by a Cys_2-Cys_2 zinc-finger motif (ZNF) (amino acids 422–453), and a nonconventional nuclear localization signal (NLS) (amino acids 510–526), which can be recognized by the nuclear transport receptor transportin 1 (Fig. 21.3) (Dormann et al. 2010, 2012; Iko et al. 2004). Most of the FALS-associated FUS mutations have been reported to cluster in the N-terminal Gly-rich region, the second RGG domain and NLS in the C-terminal (Fig. 21.3).

Following its discovery as an oncogene involved in chromosomal translocation in malignant tumors, several studies have elucidated the biochemical properties of FUS as an RNA-binding protein that regulates splicing. First, by UV cross-linking approaches, it has been shown that FUS can bind to RNA. The binding seems not to depend on the RRM in the C-terminus, but rather on the zinc-finger (ZnF) motif (Iko et al. 2004; Zinszner et al. 1997). Second, FUS is an abundant nuclear protein that can form stable complex with many members of the heterogeneous ribonuclear protein (hnRNP) family and can be co-purified from nuclear extracts by single-stranded DNA affinity chromatography (Calvio et al. 1995; Zinszner et al. 1994, 1997). The stability of the FUS-hnRNP complex is dependent on the integrity of its constituent RNA (Zinszner et al. 1994). While these results do not prove that FUS can directly interact with RNA, they suggest that at least some component(s) of the FUS-hnRNP complex has RNA-binding activity. Finally, the *Drosophila* homolog of FUS, SARFH (sarcoma-associated RNA-binding fly homolog), is localized to areas of active transcription in the polytene chromosomes of salivary glands (Immanuel et al. 1995).

To further determine whether FUS directly binds to specific RNA sequence, Lerga and colleagues used an in vitro selection assay and identified a common GGUG motif in RNA oligoribonucleotides that bind to recombinant FUS protein (Lerga et al. 2001). These results were verified using UV cross-linking combined with competition and immunoprecipitation in nuclear extracts. The ability of FUS to directly interact with RNA has been further examined using FUS antibody immunoprecipitates from mouse and human brain tissues, followed by CLIP (cross-linking immunoprecipitation)-RNA sequencing (CLIP-Seq) (Lagier-Tourenne et al. 2012). This approach shows that both mouse and human FUS proteins bind to RNAs that contain an enriched GUGGU motif, different from the GU-rich binding sequence reported for TDP-43 (Polymenidou et al. 2011; Tollervey et al. 2011). It is important to note, however, that the putative FUS-binding GUGGU motif has not been tested with biochemical approaches, such as electrophoretic mobility shift assays, and thus there may be significant differences in binding affinity and kinetics between the two reported FUS-binding sequences,

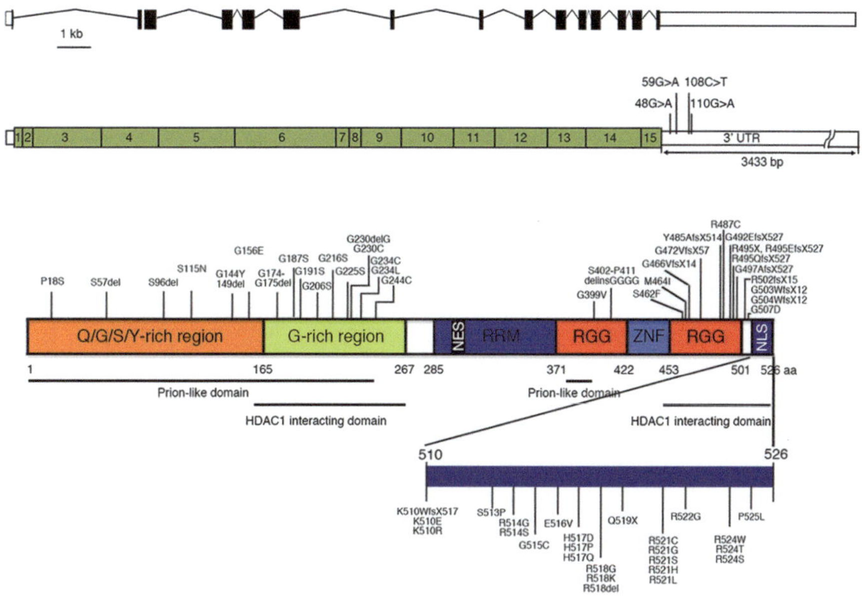

Fig. 21.3 Schematic diagrams of genomic organization of the human *FUS* gene, *FUS* mutations identified in ALS, and functional domains in FUS proteins. The human FUS gene consists of 15 exons, spanning ~14.9 Kb, and is located on chromosome 16p11.2. The *FUS* mRNA transcripts are predicted to contain a 3433 bp 3′UTR, which has been recently shown to contain four disease-related variants. The full-length human FUS protein contains 526 amino acids that can be further divided into several functional domains, including the Q/G/S/Y-rich region (amino acids 1–165); the G-rich region (amino acids 165–267); the arginine-rich motif (RRM, amino acids 285–371); two Arg-Gly-Gly (RGG)-repeat regions (amino acids 371–422 and 453–501), interrupted by a Cys_2-Cys_2 zinc-finger motif (ZNF) (amino acids 422–453); and a nonconventional nuclear localization signal (NLS) (amino acids 510–526). The majority of FALS-related mutations are more commonly found in (1) the G-rich region, (2) the 2nd RGG region, and (3) the NLS. Additional structural and functional domains in FUS include the prion-like domains and the HDAC1-interacting domains

GGUG and GUGGU (Lagier-Tourenne et al. 2012; Lerga et al. 2001). Moreover, several other studies using similar CLIP-seq technology do not find similar consensus RNA-binding sequences for FUS (Colombrita et al. 2012; Hoell et al. 2011; Ishigaki et al. 2012; Rogelj et al. 2012). Instead, the results from these studies favor the model wherein FUS-binding sites in RNA tend to form stable secondary structures, such as the stem and loop structure (Hoell et al. 2011; Ishigaki et al. 2012).

Most RNAs that bind to FUS contain intronic sequences. Perhaps the most unique feature is that in genes with long intron, FUS-RNA binding exhibits a distinct "sawtooth" CLIP pattern (Lagier-Tourenne et al. 2012; Rogelj et al. 2012), with substantially higher FUS cluster density at the beginning of introns and a gradual decrease toward the 3′ sequence. These results support the notion that FUS is co-transcriptionally deposited onto the nascent RNA transcripts. While TDP-43 and FUS are both involved in processing long RNAs that are critical for neurodevelopment and the maintenance of neuronal functions, their targets appear to be quite divergent (Colombrita et al. 2012; Lagier-Tourenne et al. 2012). In addition, FUS binding has also been identified around the alternatively spliced exons and in the promoter antisense strands in several genes implicated in neurodevelopmental and neurodegenerative diseases (Ishigaki et al. 2012), suggesting that FUS may involve in alternative splicing and transcription.

The support for FUS in RNA splicing is further underscored by the identification of FUS as an direct interacting partner with splicing factors, SC35 and SRSF10 (serine/arginine-rich splicing factor 10), and as one of the ~50 non-snRNP proteins in the pre-spliceosome (Behzadnia et al. 2007; Wahl et al. 2009; Yang et al. 1998). While these results provide physical and biochemical evidence of FUS in RNA splicing, the exact mechanisms by which FUS regulates the recognition of the 5′ splice junction, the formation and stability of complex A, and the efficiency of splicing remain unclear. In addition to the evidence supporting FUS in RNA splicing, Yamazaki and colleagues use GST-FUS to identify FUS interactome in nuclear extracts from HeLa cells and show that FUS can also interact with SMN and U1 snRNP (Yamazaki et al. 2012). The results that FUS can interact with SMN are intriguing because SMN is implicated in fatal childhood motor neuron disease spinal muscular atrophy (SMA) and the organization of Gemini of Cajal bodies (Gems). Consistent with these findings, knocking down FUS or expression of FALS-associated FUS mutant proteins severely compromises the formation of Gems in HeLa cells and fibroblasts from FALS patients, respectively. In primary neurons, FALS-associated FUS mutant proteins promote SMN protein aggregation in the cytoplasm and axons of primary neurons (Groen et al. 2013; Yamazaki et al. 2012). In a recent study, Sun and colleagues further show that the RGG domain in FUS and the Tudor domain in SMN are required for direct interaction. Surprisingly, FALS-associated FUS mutations enhance the interaction between FUS and SMN, thereby affecting the normal functions of SMN by reducing Gems bodies and changing the steady-state level of snRNA in transgenic mouse tissues and in fibroblasts from patients expressing mutant FUS proteins (Sun et al. 2015). Global analysis of RNA splicing reveals that mutant FUS-dependent splicing changes mimic partial FUS loss of activity, independent of cytosolic mislocalization. These results provide evidence for both gain and loss of function caused by ALS-linked mutations in FUS and the potential convergence in pathological pathways of ALS and SMA.

21.2.2 FUS Regulates Organismal and Neural Development

Several studies provide critical clues to the essential roles of FUS in the organismal and neural development in model organisms. For instance, two groups independently generated mice lacking FUS (*FUS*$^{-/-}$ mutants) and showed that FUS deficiency consistently leads to a marked increase in DNA damage that affect a wide range of cell types during perinatal development and in postnatal life (Table 21.1) (Hicks et al. 2000; Kuroda et al. 2000). When maintained in the C57Bl6

Table 21.1 Summary of FUS loss-of-function and ALS-associated FUS mutations in model organisms

Mutation	Species	Targeting methods	Phenotypes	References
LOF	Mouse	Knockout	Male sterility	Kuroda et al. (2000)
			↑Unpaired and mispaired chromosomal axes in spermatocytes	
		Knockout	Perinatal lethal	Hicks et al. (2000)
			Defects in B lymphocytes	
			↑ Genomic instability	
	Drosophila	Deletion in *caz* gene	Pupation and eclosion defects	Wang et al. (2011)
			Adult *caz* mutants show reduced survival and locomotor defects	
			TBPH cannot rescue *caz* phenotype	
	Xenopus	Antisense MO knockdown	Severe gastrulation defects	Dichmann and Hartland (2012)
			RNA-seq: splicing defects in genes in FGF, cell adhesion, and other major signaling pathways	
LOF/ GOF	Zebrafish	Antisense MO knockdown	Axon and behavior defects	Kabashi et al. (2011)
			Rescued by WT FUS, but not FUS-R521C or FUS-R521H	
			Toxic GOF phenotype in FUS-R521H	
		Antisense MO	Impaired motor activity	Armstrong and Drapeau (2013)
			Reduced NMJ synaptic transmission	
			FUS-R521H reduces synaptic fidelity	
GOF	Zebrafish	mRNA injection	Cytoplasmic mislocalization	Bosco et al. (2010)
		WT FUS, FUS-H517Q -R521G, -R495X or -G515X	Stress granule formation	

(continued)

Table 21.1 (continued)

Mutation	Species	Targeting methods	Phenotypes	References
GOF	Rat	Tet-inducible system	ALS-like phenotype	Huang et al. (2011)
		WT FUS and FUS-R521C	Early postnatal mortality	
		Camk2a-tTA inducible	Dendrite defects in neurons in the entorhinal cortex	Huang et al. (2012)
		WT FUS and FUS-R521C	Golgi and mitochondria defects	
			Protein ubiquitination defects	
GOF	*C. elegans*	Pan-neuronal promoter	Motor defects, shortened life span	Murakami et al. (2012)
		FUS-R514G, -R521G -R522G, R524S, P525L	Neuronal dysfunctions	
GOF	Mouse	Mouse PrP promoter	ALS-like phenotypes in homozygotes	Mitchell et al. (2013)
		WT human FUS	FUS+/ubiquitin– cytoplasmic inclusions	
GOF	Mouse	Thy1 promoter	Cytoplasmic FUS aggregates	Shelkovnikova et al. (2013, 2014)
		FUS 1–359	Motor neuron loss	
			Disruption of paraspeckle	
GOF	Mouse	FLAG-FUS-R521C transgenic expression by Syrian hamster PrP promoter	ALS-like phenotypes w/moderate loss of spinal motor neurons	Qiu et al. (2014)
			Severe dendrite/synapse defects	
			Neuroinflammation	
			RNA-seq: transcription and splicing defects in genes for neural development	
GOF	Mouse	FUS-WT or FUS-R521G transgenic expression by Cre-dependent expression via ubiquitous promoter	ALS-like phenotypes in FUS-R521G Tg	Sephton et al. (2014)
			Modest loss of spinal motor neurons	
			Severe dendritic defects	
			Neuroinflammation	
			Dysregulation of mGluR and synaptic proteins	
			No neurological phenotype in FUS-WT Tg mice	

(continued)

Table 21.1 (continued)

Mutation	Species	Targeting methods	Phenotypes	References
GOF	*Drosophila*	Global or neural expression of human FUS	Severe eye degeneration	Lanson et al. (2011)
			Pupal mortality and locomotor defects	Daigle et al. (2013)
			Cytoplasmic mislocalization	Shahidullah et al. (2013)
			Stress granule formation	
			Synapse defects precede neurodegeneration	

Abbreviations: *LOF* loss of function, *GOF* gain of function, *MO* morpholino, *PrP* prion

background, $FUS^{-/-}$ mice are perinatal lethal and exhibit severe deficiency in B lymphocyte development. In contrast, $FUS^{-/-}$ mice maintained in the mixed 129svev/CD1 genetic background survive into adulthood and show male sterility. Regardless of the genetic background, however, one consistent phenotype in both independent $FUS^{-/-}$ mouse lines is the presence of genomic instability and increased vulnerability to ionizing irradiation. Consequently, $FUS^{-/-}$ mice present with increased numbers of unpaired and mispaired chromosomal axes in premeiotic spermatocytes and in B lymphocytes (Hicks et al. 2000; Kuroda et al. 2000). Despite these intriguing results, the mechanism by which loss of FUS may affect neuronal development and maintenance in vivo remains poorly understood.

The robust genomic instability and increased DNA damage phenotype in $FUS^{-/-}$ mice raise the intriguing possibility that FUS might be involved in DNA damage response and repair machinery. Indeed, FUS has been suggested to be involved in the formation of D-loops, an essential step in homologous recombination, and normally presents in chromosome pairing, DNA repair, and telomeres (Baechtold et al. 1999; Bertrand et al. 1999). Interestingly, wild-type FUS, but not FUS-CHOP fusion protein, can be phosphorylated by ATM (ataxia telangiectasia mutated) in response to DNA double-strand breaks (DSBs) (Gardiner et al. 2008). In addition, FUS has been shown to promote homologous DNA pairing, a key step in homologous recombination (HR), whereas the oncogenic fusion protein FUS-CHOP, in which the C-terminal domain of FUS was replaced by the DNA-binding and leucine zipper dimerization domain of CHOP (Crozat et al. 1993), is unable to promote DNA pairing. Since the Gly-rich domain is retained in the FUS-CHOP protein, these data suggest that the N-terminal domain of FUS, through interaction with other proteins in the DNA repair machinery, may be involved to DNA repair. Consistent with this notion, FUS has been shown to directly interact with CBP/p300, an acetyltransferase, through its N-terminal domain, and leads to the inhibition of *CCND1* transcription following DNA damage (Wang et al. 2008), suggesting that FUS may play multiple roles in response to DNA damage. Surprisingly, the ability of FUS to respond to DNA damage depends on the allosteric interaction with single-stranded, low copy number long noncoding RNA transcripts (Wang et al. 2008). Together, these results further underscore the unique feature of FUS as an RNA- and DNA-binding protein in regulating the DNA damage response and repair process.

21.2.3 FUS and RNA Splicing During Embryonic Development

In addition to the analyses in *FUS*$^{-/-}$ mice, FUS loss-of-function phenotype has also been examined extensively using the developing *Xenopus laevis* embryos as a model system. The *Xenopus* FUS homologs are expressed abundantly in the animal and marginal zones of gastrulating embryos, and antisense morpholino (MO) knockdown of two FUS homologs in embryos results in severe developmental defects during gastrulation (Dichmann and Harland 2012). Interestingly, RNA-seq analyses of FUS MO-injected embryos reveal profound splicing defects in genes that have been implicated in FGF signaling pathways, cadherin-mediated cell adhesion, and several other major signaling pathways, including TGF-β, Hedgehog, Wnt, and Notch. In contrast, genes involved in tRNA, telomeres, and vesicle processes are not affected in FUS MO-injected embryos. Furthermore, this study uses a clever splice-blocking MO to prevent the inclusion of the C-terminus domain of FUS, which contains the last 12 amino acids frequently mutated in patients with FALS (Kwiatkowski et al. 2009; Vance et al. 2009). Curiously, *Xenopus* embryos lacking the C-terminal domain develop normally and do not show evidence of splicing defects in *cdh1*, *fgf8*, or *fgfr2* genes. Rather, these mutant morphants show strong phenotype in their inability to swim. While the underlying cause(s) for this phenotype remains to be determined, these results suggest that mutations in the C-terminus of FUS may uncover additional gain-of-function properties of FUS.

21.2.4 Roles of FUS in RNA Transport in Dendrites

The functional consequences of FUS binding to RNA appear to extend beyond transcription and RNA splicing. In an attempt to isolate protein complexes involved in conventional kinesin protein KIF5-dependent RNA transporting granules in neurons, Hirokawa and colleagues identify FUS as one of the 12 major proteins in the detergent-resistant and RNase-sensitive complexes (Kanai et al. 2004). Several proteins in this complex, including staufen, hnRNP-U, Pur α, and Pur β, have well-documented roles in RNA transport, whereas others, such as FMRPs, have been implicated in translational silencing in dendrites. Indeed, many proteins in this complex can be detected in the dendrites of hippocampal neurons. Intriguingly, in a proteomics analysis FUS has also been isolated as a component in the NMDA receptor-adhesion protein complex in adult mouse brain (Husi et al. 2000). Subsequent studies confirm the colocalization of FUS and PSD95 in the dendrites of cultured neurons and that the movement of FUS-positive granules seems to be dependent on synaptic activity (Fujii et al. 2005). Although $FUS^{-/-}$ neurons exhibit a marked reduction in dendritic spine density, the exact mechanisms contributing to this phenotype and the significance of this observation in vivo remain unclear. It will be important to design future studies that can distinguish the role of FUS in RNA splicing and transport during development and in the maintenance of mature neurons in postnatal life.

21.3 Mechanisms of FUS Mutations in Neurodegeneration

21.3.1 Dendritic Degeneration in Rodent Models of FUS Mutations

Several groups have used a wide range of strategies to characterize the neuropathological consequences of expressing mutant FUS proteins in the nervous system of mouse or rat. These studies have provided ample evidence that the neurodegenerative phenotypes caused by FUS mutations are quite consistent. For instance, both the transgenic mouse and rat models expressing mutant FUS-R521C proteins develop early-onset ALS-like symptoms, including hindlimb paralysis, muscle wasting, and reduced innervation at the neuromuscular junction (NMJ) (Huang et al. 2011; Qiu et al. 2014). The cardinal phenotypes are age-dependent reductions in dendritic arborization and synaptic density in the spinal motor neurons and cortical neurons in the sensorimotor cortex of the FUS-R521C transgenic mice (Fig. 21.4) (Qiu et al. 2014). Similar dendritic arborization defects have also been reported in neurons in the entorhinal cortex of *Camk2a-tTA* transgenic rats (Huang et al. 2012) and in the spinal motor neurons and cortical neurons of Cre-inducible transgenic mouse lines that express FUS-R521G in the nervous system (Sephton et al. 2014). The dendritic phenotype caused by FUS-R521C can be recapitulated in cultured cortical neurons and can be partially rescued by exogenous BDNF (Qiu et al. 2014). In side-by-side comparisons, FUS-R521C and FUS-P525L cause more severe dendritic growth defects compared to wild-type FUS. These results support the notion that wild-type and mutant FUS affect dendritic growth in gene dosage-dependent manner. In light of these findings, it is interesting to note that transgenic mice expressing higher level of wild-type FUS also show early-onset motor neuron degeneration in a dosage-dependent manner (Mitchell et al. 2013). Consistent with these results, mutations in the 3′ UTR of the FUS gene have been identified in several FALS patients. These mutations drastically increase the FUS protein expression in the patients' fibroblasts, at levels higher than that in FUS-R521C fibroblasts (Sabatelli et al. 2013), supporting the notion that wild-type FUS expressed at exceedingly high levels can be pathogenic.

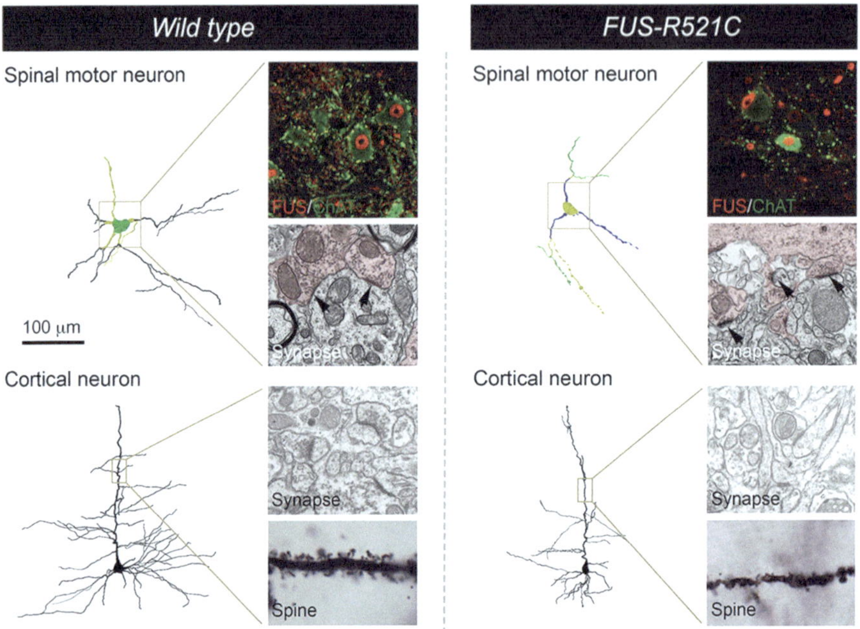

Fig. 21.4 Dendritic and synaptic phenotypes caused by FUS mutations. Neurolucida tracing shows that the dendritic arbors in control motor neurons, highlighted by Golgi staining techniques, had six to eight intersections per radial distance within 100 μm from the cell body, followed by a gradual reduction in the number of dendritic arbors from 100 to 250 μm. Compared to the control, the number of dendritic arbors in the FUS-R521C motor neurons shows no change within the first 50 μm from the cell body, but a significant reduction is noted from 50 to 250 μm, resulting in reduced cumulative dendritic area. To determine if the dendritic phenotype in FUS-R521C spinal motor neurons affected synaptic connectivity, we use ChAT (*green*) and FUS immunostaining to characterize the density of synapses surrounding motor neurons (Betley et al. 2009). Our results show that FUS proteins are present primarily in neuronal nuclei, but also show extensive colocalization with ChAT+ boutons and synaptophysin-immunoreactive presynaptic terminals. Remarkably, the density of ChAT+ boutons and SIPT showed significant reductions in the anterior horn of FUS-R521C spinal cord. To further characterize the synaptic defects, we use electron microscopy (EM) to ascertain the morphology and density of synapses within 100 μm radius of the cell body of spinal motor neurons and show that the cell bodies of control motor neurons are surrounded by synaptic terminals arranged as rosette-like structures (Betley et al. 2009). In contrast, the size of postsynaptic density and the number of synapse per unit area are reduced in FUS-R521C motor neurons. Similar dendrite and synaptic defects are also noted in the apical and basal dendrites of the pyramidal neurons in layer IV–V of the sensorimotor cortex (Qiu et al. 2014) and neurons in the entorhinal cortex (Huang et al. 2012)

Unlike the severe loss of spinal motor neurons in the $SOD1^{G93A}$ mice, the neuron loss phenotype in different FUS transgenic models appears to be more modest. At end stage, FUS-R521C transgenic mice and rats, and transgenic mice expressing FUS-R521G or a truncated FUS mutant protein (amino acids 1–359), show greater than 50 % preservation of spinal motor neurons (Table 21.1) (Huang et al. 2011; Qiu et al. 2014; Sephton et al. 2014; Shelkovnikova et al. 2013). The majority of mutant FUS-R521C proteins are located within the nuclei of spinal motor neurons in these transgenic animals, with few neurons showing evidence of FUS-R521C protein aggregates in the cytoplasm. The lack of prominent cytoplasmic FUS inclusion in the FUS-R521C transgenic models is quite different from the pathology observed in patients with FALS caused by FUS mutations (Huang et al. 2010; Kwiatkowski et al. 2009; Vance et al. 2009). Another interesting observation is that few neurons in FUS-R521C transgenic rats and wild-type FUS transgenic mice show accumulation of ubiquitin-positive inclusions in the cytoplasm. Curiously, however, most of the ubiquitin-positive cytoplasmic inclusions do not contain mutant or wild-type FUS proteins (Huang et al. 2011; Mitchell et al. 2013).

The discrepancy of neuropathology in the rodent models and human patients raises several intriguing questions regarding the cause and significance of FUS-positive cytoplasmic inclusions in FALS. One possible explanation for the lack of FUS+ cytoplasmic inclusions in transgenic rodent models is that the cytoplasmic aggregation of wild-type or mutant FUS proteins may be age and dosage dependent. Depending on the efficiency of nucleus-to-cytoplasm translocation for wild-type and mutant FUS proteins, the early postnatal lethality in most of the transgenic mice or rats may not have given FUS proteins sufficient time to accumulate in the neuronal cytoplasm. Alternatively, it is possible that neurons may develop inherent homeostatic mechanisms to maintain the certain level of FUS expression (Dini Modigliani et al. 2014). In this regard, only when expressed at exceedingly high level using viral vectors, such as AAV1, will the FUS proteins begin to accumulate in the neuronal cytoplasm (Verbeeck et al. 2012). Regardless of the mechanism, the observations that transgenic mice and rats develop severe neurodegenerative phenotypes even in the absence of prominent FUS proteinopathy in neuronal cytoplasm suggest that increase of wild-type FUS proteins or the presence of mutant FUS proteins in nucleus is sufficient to cause disease, most likely through the perturbations of DNA damage repair/response and RNA splicing machinery (Qiu et al. 2014; Wang et al. 2013). Interestingly, FALS-related mutation FUS-R521G exhibits a drastic shift in binding preference from the intronic sequences to sequences in the 3′UTR (Hoell et al. 2011). These results suggest that mutant FUS proteins may acquire gain-of-function properties to interact with new RNA targets, and hence it will be important to establish both in vitro and in vivo approaches to characterize how accumulation of mutant FUS proteins in neuronal cytoplasm affects ribosomal functions and RNA transport in dendrites and axons.

21.3.2 FUS-Mediated Neurodegeneration in Other Model Organisms

In addition to the rodent models of FUS mutations, results from other model organisms, including yeasts, *Xenopus*, and rodents, have revealed a wealth of information regarding the in vivo functions of FUS during organismal development and how FUS mutations may disrupt these functions and contribute to the neurodegenerative process. There are several additional studies performed in other model organisms, including *Drosophila melanogaster*, *C. elegans*, and zebrafish (*Danio rerio*), which provide novel insights on the genetic interactions between FUS and other FALS-related genes, such as *TARDBP* and *SOD1*.

Drosophila has a single homolog of FUS that shares 53 % amino acid identity to its mammalian counterpart and is encoded by the *cabeza* (*caz*) gene on X chromosome (Stolow and Haynes 1995). The *Drosophila* Caz protein contains 399 amino acids and is expressed in neurons, glial, and muscle cells. Loss-of-function analyses show that only 14 % of male *caz* mutant larvae successfully undergo pupation and eclose to become adults (Wang et al. 2011). The *caz* mutants that do survive into adulthood exhibit severe locomotion phenotype and a markedly reduced survival in postnatal life. The eclosion phenotype in *caz* mutants can be rescued by neuron-specific transgenic expression of Caz, wild-type human FUS ($hFUS^{WT}$), $hFUS^{R522H}$, or $hFUS^{P525L}$ at the same expression level, suggesting that $hFUS^{WT}$, $hFUS^{R522H}$, and $hFUS^{P525L}$ can functionally restore the role of Caz during eclosion. However, neuron-specific transgenic expression of Caz or $hFUS^{WT}$ only partially restores the locomotion and longevity phenotype, whereas neither $hFUS^{R522H}$ nor $hFUS^{P525L}$ is capable of restoring the locomotion or longevity phenotype. Interestingly, Drosophila *tbph* mutants lacking TDP-43 homolog TBPH also show similar phenotype in eclosion, locomotion, and longevity. Whereas expression of TBPH in *caz* mutants cannot rescue loss of Caz phenotype, overexpression of Caz in *tbph* mutants restores eclosion, locomotion, and longevity. These results support the model that *caz* and *tbph* genetically interact in *Drosophila* to regulate neuronal development and longevity (Wang et al. 2011). Several other studies using *Drosophila* as a model system also show that expressing mutant human FUS proteins, $hFUS^{R518K}$, $hFUS^{R521H}$, or $hFUS^{R521C}$, in the eye, motor neurons, or the nervous system leads to eye degeneration, defects in locomotion, and increase in mortality (Lanson et al. 2011). Detailed analyses of the locomotion defects indicate that mutant FUS proteins cause synaptic defects before the degeneration of motor neurons. The synaptic defects include disorganization of the presynaptic active zone protein bruchpilot, reduced quantal contents and miniature presynaptic currents, and reduced synaptic currents in the postsynaptic muscle cells (Shahidullah et al. 2013).

The effects of FUS in synaptic functions have also been investigated using the zebrafish larvae as a model system. Antisense morpholino (AMO) knockdown of FUS in zebrafish causes motor behavioral defects reflected as reduced touch-evoked escape response (TEER) and marked reductions in the branching and length of motor axons (Kabashi et al. 2011). Similar to the observations in *Drosophila*, AMO knockdown of FUS and expressing mutant human FUS proteins in zebrafish also cause defects in synaptic transmission at NMJ by reducing the presynaptic quantal contents (Armstrong and Drapeau 2013). These results are consistent with the RNA-seq results in FUS-R521C transgenic mice (Qiu et al. 2014) and support the notion that mutant FUS proteins can indeed affect the transcription and splicing of target genes critical for the maintenance of dendrite outgrowth and synapse formation.

21.4 Mechanisms of FUS Mutations in Dendritic Degeneration

21.4.1 FUS Mutations and RNA Transcription/Splicing Defects

The causal link between FUS mutations and DNA damage defects provides critical mechanistic insights to neurodegeneration. Interestingly, the process to repair DNA damage is tightly coupled to transcription through regulating the activity of RNA polymerase II and the subsequent RNA processing, including RNA splicing and transport (Cleaver et al. 2009; Kornblihtt et al. 2004; Munoz et al. 2009). This view is further supported by a plethora of evidence supporting that RNA transcription in the eukaryotic cells is a highly dynamic and tightly regulated process that involves multiple intricately connected steps, including splicing of pre-mRNA and transport of mature mRNA to its final destinations (Moore and Proudfoot 2009; Reed and Hurt 2002). In the nervous system, these regulatory mechanisms are known to generate a vast diversity of RNA transcripts that control cell fate determination, axon guidance, dendritic growth, and synaptic functions (Li et al. 2007; Martin and Ephrussi 2009). Perturbations to these critical mechanisms have been implicated in neuromuscular diseases, neurodevelopmental disorders, and neurodegenerative diseases (Cooper et al. 2009).

As discussed in the previous section, both human patients with FUS-R521C or FUS-P525L mutation and FUS-R521C transgenic mice exhibited evidence of DNA damage in cortical neurons and spinal motor neurons (Qiu et al. 2014; Wang et al. 2013). These results indicate that the FUS-R521C transgenic mice provide an invaluable system to identify neural genes implicated in DNA damage during neurodegeneration (Graff et al. 2012; Lu et al. 2004). Indeed, a PCR-based screening approach shows that the 5′ noncoding exons of the mouse *Bdnf* gene, which contain transcriptional start sites and required splicing of long intronic sequences to generate mature *Bdnf* mRNA, consistently exhibit evidence of DNA damage. These results lead to the identification of retentions of 5′ splice junctions in the *Bdnf* mRNA and defects in transporting *Bdnf* mRNA to distal dendrites. Using electrophoretic mobility shift assays (EMSA), it is further demonstrated that, compared to wild-type FUS, mutant FUS-R521C proteins form more stable protein-RNA complex to 5′ splice junction and the 3′UTR sequences of *Bdnf* pre-mRNA (Qiu et al. 2014) (Fig. 21.5b). These results support the model wherein FALS-associated FUS mutation FUS-R521C exhibits aberrant gain-of-function properties, including forming more stable protein-protein interactions with the endogenous wild-type FUS and more stable protein-RNA complex, which most likely alter the ability of FUS to recruit DNA damage repair machinery and the equilibrium of the interactions between FUS and RNA in the splicing machinery, respectively (Fig. 21.5) (Qiu et al. 2014; Wang et al. 2013). Similar phenotypes of FUS mutations in DNA damage repair and RNA splicing machinery have been reported in several other

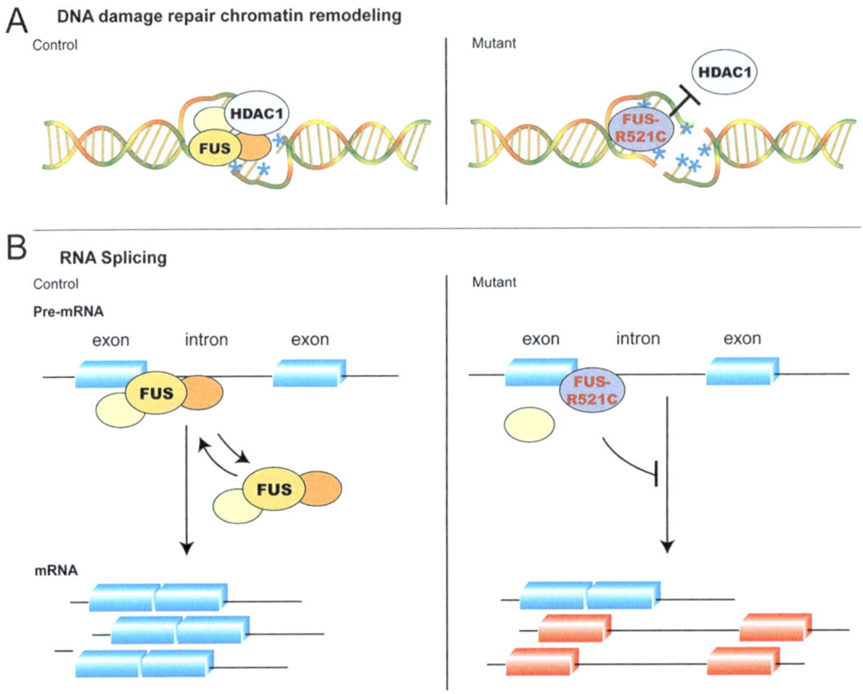

Fig. 21.5 Mechanisms of wild-type and mutant FUS in DNA damage repair/response and RNA splicing. (**a**) Wild-type FUS is rapidly recruited to DNA damage foci caused by double-stranded breaks, where it interacts with chromatin remodeling factor HDAC1. Although FUS-R521C can still be recruited to DNA damage foci, it fails to interact with HDAC1 and PARP1. Due to the defects in DNA repair/response machinery, neurons in FUS-R521C transgenic mice show increased DNA damage (indicated by *blue asterisks* and the presence of double-stranded breaks). (**b**) In addition to its role in DNA damage repair, several *lines* of evidence indicate that FUS can also regulate pre-mRNA splicing. Results from CLIP-RT-PCR and protein-RNA interactions in EMSA assays show that both wild-type FUS and FUS-R521C can interact with selective oligoribonucleotides from *Bdnf* exon-intron boundaries. Whereas the equilibrium of wild-type FUS-RNA interactions appears to be more dynamic, FUS-R521C tends to form more stable protein-RNA complexes that are more difficult to dissociate (Figure adapted from Qiu et al. 2014, with permissions from the *Journal of Clinical Investigation*)

studies using biochemical, cell biology, and bioinformatics analyses (Hoell et al. 2011; Mastrocola et al. 2013; Zhou et al. 2014).

Given the highly efficient process in RNA splicing, the observations that FUS-R521C can form more stable protein-RNA complex raise the intriguing question as to whether this gain-of-function property may have more widespread intron retention effects on the transcriptomes or only affect a selective subset of target genes. To distinguish these two possibilities, Qiu and colleagues perform genome-wide survey in the transcriptomes of FUS-R521C spinal cord using RNA-seq and show two primary defects involving the transcription and RNA splicing in selective genes that are critical for dendritic growth and synaptic functions (Qiu et al. 2014). For instance, RNA-seq results in the spinal cord of FUS-R521C mutants show perturbations in the expression or splicing of genes involved in the organization of extracellular matrix, including members of the *collagen* and *cadherin* gene families that regulate the specificity of axonal projection and target innervation (Robles and Baier 2012; Sanes and Yamagata 2009). Interestingly, similar targets have also been identified in the RNA-seq analyses of FUS MO-treated *Xenopus* morphants (Dichmann and Harland 2012), suggesting that FUS-R521C phenotype may recapitulate certain transcriptional and RNA splicing defects in FUS loss-of-function mutants. Another intriguing feature of the RNA-seq results in FUS-R521C spinal cord is that many target genes in the extracellular matrix assembly GO categories (GO:0005581, GO:0005201, GO:0005578 and GO:0031012) have also been shown to be transcriptional targets of DNA damage response gene Cockyane syndrome B (CSB) and HDAC1 (Newman et al. 2006) and are frequently misregulated and misspliced in the motor neurons of SALS patients (Rabin et al. 2010). While these results are correlative, they raise the interesting possibility that the recruitment of FUS, HDAC1, and CSB may constitute a critical step in the repair of damaged DNA in FALS caused by FUS mutations and in SALS.

Finally, one remarkable feature in the spinal cord of FUS-R521C transgenic mice is the upregulation of genes that are functionally related to immune response, complement activation, and chemotaxis (Qiu et al. 2014). Consistent with these findings, the FUS-R521C spinal cord shows pronounced microgliosis. Since neither wild-type FUS nor FUS-R521C proteins can be detected in the microglia, these results support the idea that non-cell autonomous mechanisms, such as damaged neurons or astrogliosis, may activate microglia and contribute to the neurodegeneration in ALS. Interestingly, similar non-cell autonomous mechanisms have been reported in the mutant SOD1 models (Boillee et al. 2006a, b). Alternatively, the defects in DNA damage repair and RNA splicing caused by mutant FUS-R521C may occur in glial cells, which promotes astroglial activation and/or degeneration of oligodendroglia, further contributing to the degeneration of spinal motor neurons in FUS-R521C mice.

21.4.2 Mechanisms of FUS Mutations in RNA Spicing Machinery

Both cell biology and biochemistry data suggest that FUS can physically interact with SMN and U1 snRNP and that loss of FUS or expressing FALS-associated FUS mutations disrupts the organization of Gemini of Cajal bodies (Gems), where the presence of TDP-43, FUS, and another fatal motor neuron disease gene product SMN is required to regulate the assembly of the Gems in several different cell types (Battle et al. 2006; Sun et al. 2015; Yamazaki et al. 2012). Consistent with these findings, $FUS^{-/-}$ hippocampal neurons show a near complete loss of Gems. Interestingly, the integrity of Gems and spliceosome is severely perturbed in the spinal motor neurons of patients with sporadic ALS, which most likely is due to the loss of nuclear TDP-43 and prominent upregulations of U snRNAs and snRNPs (Tsuiji et al. 2013). In addition to its roles in the organization of spliceosome, FUS has been implicated in the integrity of paraspeckles, which are subnuclear structures that regulate gene expression by nuclear retention of RNA (Bond and Fox 2009). The core paraspeckle proteins include DBHS (*Drosophila melanogaster* behavior, human splicing) proteins, PSF/SFPQ, P45NRB/NONO, and PSPC1. In addition, a long noncoding RNA (lncRNA) *NEAT1* is also required to maintain the integrity of paraspeckles. In a recent study, FUS and TDP-43 are shown to interact with *NEAT1* (Nishimoto et al. 2013), raising the possibility that perturbations to both proteins may disrupt the integrity or maintenance of paraspeckles. Consistent with this idea, spinal motor neurons in transgenic mice expressing truncated FUS mutant protein (amino acids 1–359) show cytoplasmic aggregates of P45NRB/NONO. Although confocal images from these transgenic neurons indicate that P45NRB/NONO proteins and mutant FUS proteins are in close proximity, it is unclear if the presumed complex between P45NRB/NONO and FUS is disrupted by the presence of mutant FUS proteins (Shelkovnikova et al. 2013). It is also unclear if the number and distribution of paraspeckles are disrupted in the spinal motor neurons of these transgenic mice and in patients with FALS or SALS. Another alternative mechanism for mutant FUS proteins to interfere with the integrity of paraspeckles is by altering the expression of lncRNA *NEAT1*. Indeed, RNA-seq analyses of spinal cord from FUS-R521C transgenic mice show that the *NEAT1* levels are two- to threefolds higher than that in the age-matched controls (Qiu et al. 2014). It will be interesting to characterize the mechanism by which FUS mutant proteins and the elevated *NEAT1* expression level cooperatively affect the structural integrity of paraspeckles.

21.4.3 FUS Autoregulation in the Pathogenesis of FALS

FUS employs several unique mechanisms to regulate a delicate balance of its own expression. For instance, Zhou and colleagues use CLIP-seq to show that FUS binds to a highly conserved region in exon 7 and the flanking intronic sequence of its own pre-mRNA (Zhou et al. 2013). Their results indicate that FUS is a repressor of exon 7 splicing and that the exon 7-skipped splice variant is subject to nonsense-mediated decay (NMD). Interestingly, several FUS mutations, including FUS-R521G, FUS-R522G, and FUS-ΔExon15, show deficiencies in both exon 7 repression and autoregulation of its own protein levels. These data suggest that compromised FUS autoregulation may contribute to the pathogenic accumulation of cytoplasmic FUS protein in ALS. In a recent study from Italy, several variants of FUS have been identified in the 3′UTR in FALS and SALS cases (48G>A, 59G>A, 108C>T, and 110G>A) (Fig. 21.3) (Sabatelli et al. 2013). Interestingly, skin fibroblasts from these patients show ~1.5- to 2.5-folds increase in *FUS* mRNA and three- to fourfolds increase in FUS protein levels. Biochemical data show that fibroblasts with FUS 3′UTR variants have a marked increase of FUS proteins in the cytosolic fractions, compared to those from fibroblasts of control and FUS-R521C patients. Using bioinformatic analysis, a subsequent study shows that mutation in one of the 3′UTR positions, 48G>A, alters the binding of two microRNA, miR-141 and miR-200a. Further in vitro assays show that FUS and these microRNAs show a mutual feed-forward regulatory mechanism that determines the level of FUS protein (Dini Modigliani et al. 2014). While it is still unclear how this novel mechanism may eventually affect the nucleus-cytoplasm distribution of FUS, these two studies reveal novel insights to the previously unrecognized mechanism of regulating FUS expression.

21.4.4 Roles of FUS in High-Order Assembly of Protein-RNA Complex

The ability of FUS to use its LC domain to form RNA-protein complexes has been further examined using a 48 nucleotide (nt)-long RNA (prD RNA) from the promoter region of DNMT3b gene, which has been confirmed to interact with FUS using cross-linking and sequencing (Schwartz et al. 2012, 2013). It is important to note that this prD RNA does not contain any of the previously identified FUS-binding motifs (Lagier-Tourenne et al. 2012; Lerga et al. 2001), yet exhibit robust binding to recombinant FUS proteins in electrophoretic mobility shift assays (Schwartz et al. 2013). This provides a convenient tool to characterize the essential role of the LC domain in FUS and its mutual interactions with RNA to form high-order protein-RNA complexes. Similar high-order assemblies have been reported using recombinant FUS proteins and synthetic RNA from the intron-exon boundary and 3′UTR of the *bdnf* gene (Qiu et al. 2014). Many of these *bdnf* RNA probes do not contain the reported FUS-binding motif, again supporting the notion that the RNA structure is likely to be more important for FUS assembly. Interestingly, the RNA-FUS assemblies appear to be critical for its interaction with the CTD of RNA polymerase II (Schwartz et al. 2013).

How might FALS-associated mutations in FUS alter the RNA-protein complex formation? Using the FUS-*bdnf* RNA interactions, Qiu and colleagues show that mutant FUS-R521C proteins form more stable and higher-order protein-RNA assemblies, which are more difficult to dissociate in competition assays (Qiu et al. 2014). Interestingly, in transgenic mice expressing FUS-R521C, the majority of mutant FUS proteins are in the nuclei of spinal motor neurons, suggesting that the presence of high-order mutant FUS-RNA assemblies may interfere with the transcription and/or RNA splicing. Consistent with these results, expression of mutant FUS proteins or siRNA knockdown of FUS in fibroblasts alters the distribution of RNA polymerase II within the nuclei. These results are further confirmed using fibroblasts derived from FALS patients with FUS mutations (Nomura et al. 2014; Schwartz et al. 2014).

21.4.5 Prion-Like Property of FUS and Its Implications in RNA Processing

The presence of TDP-43 and FUS proteinopathy in ALS and FTLD raises the intriguing possibility that TDP-43 and FUS may have a high propensity to form protein aggregates. Indeed, using a bioinformatics algorithm designed to identify proteins with "prion-like" domain (Alberti et al. 2009), it has been shown that the N-terminal Q/G/S/Y domain and part of the Gly-rich region in FUS (amino acids 1–239) and the C-terminal Gly-rich region of TDP-43 (amino acids 277–414) respectively rank 15th and 69th among 27,879 proteins in the human proteome for their prion-like property (Cushman et al. 2010; Gitler and Shorter 2011) (Fig. 21.3). Consistent with this notion, expression of TDP-43 and FUS in the baker's yeast *Saccharomyces cerevisiae* shows that both proteins are more prone to form protein aggregates that resemble proteinopathy in human diseases and that the protein aggregate formation is dependent on the prion-like domains (Fushimi et al. 2011; Sun et al. 2011). These findings underscore the value of yeasts as a model organism that can be used for genetic screens to identify modifiers that can alleviate FUS proteinopathy. Indeed, such screens reveal many potential candidates that are implicated in RNA metabolic process, ribosome biogenesis, response to cellular stress, etc. (Sun et al. 2011). Similar approach has also been exploited as an effective screen to identify the causal links between two other FET family members, EWSR1 and TAF15, to the pathogenesis of ALS (Couthouis et al. 2011, 2012). It is important to note that several other RNA-binding proteins with similar prion-like properties have been implicated in the organization of stress granules, and perturbations to this process can also contribute to ALS and other neurodegenerative diseases (Li et al. 2013).

Another line of evidence supporting the involvement of FUS in the formation of RNA granules comes from the observation of a small molecule chemical isoxazole, designated 5-aryl-isoxazole-3-carboxyamide, which is endowed with the unique property to aggregate and disaggregate RNA granules in a soluble, cell-free system (Han et al. 2012; Kato et al. 2012). This discovery leads to a highly efficient assay, the hydrogel formation, which identifies FUS and several other RNA-binding proteins in forming RNA-protein aggregates using their low complexity (LC) domain. Furthermore, transmission electron microscopy (TEM) and X-ray diffraction analyses of the FUS hydrogel droplets reveal the high propensity of FUS to form amyloid-like filamentous protein aggregates. Together, these results underscore the unique property of FUS in forming complex with RNA and provide additional insights that mutations in FUS and other RNA-binding proteins might alter this property and further increase the propensity to form less soluble and less reversible complex formation with RNA (Hoell et al. 2011; Qiu et al. 2014).

21.5 FUS Mutations and DNA Damage Repairs

21.5.1 DNA Damage and Neurodegeneration

Both prokaryotic and eukaryotic organisms use highly evolutionarily conserved mechanisms to repair DNA damages caused by radiation from the environment or by endogenous sources, such as free radicals produced within the cells. Mutations in DNA repair machinery have been linked to hereditary neurodegenerative diseases (Jackson and Bartek 2009; McKinnon 2009; Rass et al. 2007). For instance, ataxia telangiectasia (AT) is an autosomal recessive, early-onset neurodegenerative disease caused by mutations in the *ATM* gene, which encodes a protein kinase that regulates the cellular response to DNA double-strand breaks (DSBs). Patients with mutations in the components of the DNA damage sensor complex MRN (MRE11-RAD50-NBS1) also develop severe neurological symptoms, with hypersensitivity to ionizing radiation and genome instability. Finally, another DNA damage repair machinery involves base excision repair (BER), which is the primary mechanism that handles spontaneous DNA damage caused by free radicals and reactive oxygen species (ROS). Patients with mutations in critical components of the BER machinery, including CSA (also known as excision repair cross-complementing rodent repair deficiency, complementation group 6, or ERCC6), CSB (ERCC8), XPD (ERCC2), and XPG (ERCC5), develop Cockyane syndrome, characterized by retinal degeneration, microcephaly, deafness, and loss of Purkinje cells in the cerebellum (Cleaver et al. 2009). In addition to the inherited forms of neurodegenerative diseases, DNA damage and genomic instability have also been linked to late-onset neurodegenerative diseases, such as Alzheimer's disease and Parkinson's disease (Anderson et al. 1996; Bender et al. 2006; Lu et al. 2004).

Although loss-of-function in FUS has been implicated in DNA damage repair (Hicks et al. 2000; Kuroda et al. 2000), it is unclear how FUS regulates this process or whether FALS-associated mutations in FUS might perturb the physiological functions of DNA damage repair machinery. In a series of assays designed to determine the efficiency of DSB repair efficiency, Wang and colleagues show that siRNA knockdown of FUS resulted in impairments in homologous recombination and nonhomologous end joint (NHEJ)-mediated DSB repairs in both U2OS cells and primary neurons (Wang et al. 2013). Furthermore, using immunoreactivity for γH2AX as an early marker of DNA damage and a prerequisite marker for DSB repair (Fillingham et al. 2006; Pilch et al. 2003), this study further demonstrates that FUS is rapidly recruited to DNA damage sites, which precedes the accumulation of γH2AX. These results, independently confirmed by other groups (Mastrocola et al. 2013; Rulten et al. 2014), suggest that recruitment of FUS to damaged chromatin is required to elicit an effective DDR. In the study by Rulten and colleagues, it is reported that FUS recruitment to DNA damage foci is dependent on poly (ADP-ribose) polymerase 1 (PARP1). However, it is unclear if these results indicate the presence of hierarchy of protein complex formation involving PARP1 and FUS in the assembly of DNA response/repair machinery, or the requirement of

PARP1 is a cell type-specific phenomenon (Rulten et al. 2014; Wang et al. 2013). Finally, DSB can induce nucleus-to-cytoplasm translocation of FUS and causes phosphorylation of FUS in the C-terminus by DNA-dependent protein kinase (DNA-PK) (Deng et al. 2014b). Together, these studies provide the first molecular clue for the previous observations that *FUS*$^{-/-}$ mice exhibit enhanced radiation sensitivity, growth retardation, immunodeficiency, and increased genomic instability (Hicks et al. 2000; Kuroda et al. 2000).

21.5.2 FUS Mutations and HDAC-Dependent DNA Damage Response/Repair

Given the critical role of FUS in DNA damage response/repair, it is interesting to note that FALS-associated mutations in FUS do not affect the recruitment of mutant FUS proteins to DNA damage foci. Rather, the mechanism by which FUS regulates DNA damage repair machinery depends on its ability to directly interact with chromatin remodeling factor histone deacetylase 1 (HDAC1), which plays fundamental role in DNA repair and the maintenance of genomic stability (Fig. 21.5a). Deficiency in HDAC1 and the closely related HDAC2 causes severe hypersensitivity to DNA damaging agents and persistent phenotypes related to DNA repair defects, including dysregulation of histone acetylation upon DNA damage, abnormalities in heterochromatin formation, and aberrant expansion and recondensing of the chromatin structure in DNA repair process (Dinant et al. 2008; Lukas et al. 2011; Miller et al. 2010). Interestingly, loss of HDAC1 has been reported to sensitize neurons to DNA damage and induce aberrant cell cycle reentry, while the overexpression of HDAC1 protects neurons from genotoxic agents (Dobbin et al. 2013; Kim et al. 2008). Indeed, both neurons with FUS deficiency and *HDAC1*$^{-/-}$ neurons exhibit increased DNA damage following etoposide treatment (Dobbin et al. 2013; Wang et al. 2013), supporting the notion that the interaction between FUS and HDAC1 plays an important role in maintaining genome stability and integrity in neurons.

How might FALS mutations interfere with the functions of wild-type FUS proteins in HDAC-dependent DNA damage repair pathway? The fact that FALS mutations are transmitted in an autosomal dominant manner and FALS mutations do not affect FUS recruitment to DNA damage foci leads to the hypothesis that mutant FUS proteins may have dominant-negative effect that interferes with its interaction with HDAC1 and the subsequent assembly of DNA repair machinery (Fig. 21.5a). Indeed, structure-function analyses show that the glycine-rich domain (amino acids 156–262) and C-terminal domain (amino acids 450–526) of FUS are required for FUS-HDAC1 interaction (Wang et al. 2013). Remarkably, these two domains in FUS harbor most of the FALS mutations, and FUS mutations in these two domains, including FUS-R244C, FUS-R514S, and FUS-R521C, show impaired interaction with HDAC1 and reduced DSB repair efficiency when expressed in cells. Consistent with these results, wild-type FUS can be detected in a protein complex with HDAC1 in the control spinal cord tissues. In contrast, protein extracts from FUS-R521C transgenic mice show no detectable complex formation between FUS-R521C and HDAC1 (Qiu et al. 2014). Interestingly, the presence of FUS-R521C almost completely abolishes the protein-protein interaction between wild-type FUS and HDAC1 in FUS-R521C transgenic mice. This dominant effect of FUS-R521C is due to the abnormal gain-of-function property of the mutant FUS-R521C protein in forming more stable complex with wild-type FUS protein in both HEK293T cells and in FUS-R521C transgenic mice. Consistent with these results, spinal motor neurons and cortical neurons in FUS-R521C transgenic mice and in patients with FUS-R521C or FUS-P525L mutation show a robust increase of γH2AX staining (Qiu et al. 2014; Wang et al. 2013).

The demonstration that DNA damage repair defects contribute to the pathogenesis of neurodegeneration caused by FUS mutations further underscores the critical role of DNA damage repair in neurodegenerative conditions. Indeed, several previous studies have shown that increased levels of 8-hydroxydeoxyguanosine (8-OHdG) residues, a marker of oxidative DNA damage, can be identified in the spinal cord of both sporadic and familial ALS patients (Ferrante et al. 1997). Age-related motor neuron degeneration has been observed in mice lacking the DNA repair protein ERCC1, suggesting that the accumulation of DNA damage contributes to the motor neuron vulnerability (de Waard et al. 2010). While these results suggest that perturbations to multiple signaling pathways may converge on the DNA damage repair defects leading to neurodegeneration, it is important to note that DNA damage due to defects in oxidative stress and nucleotide excision repair is quite different from that caused by double-stranded DNA breaks or defects in the ATM pathways. Finally, it is unclear why motor neurons are more susceptible to FUS mutations despite the fact that almost all neurons express mutant FUS proteins and the evidence of DNA damage can also be detected in cortical neurons of FUS-R521C transgenic mice and human disease tissues (Qiu et al. 2014; Wang et al. 2013). Such "selective vulnerability" is a common theme in neurodegenerative diseases. One potential mechanism is that motor neurons may produce excessively higher amount of mitochondrial reactive oxygen species, which may create a vicious cycle that further promotes the accumulation of DNA damage repair defects (Cleaver et al. 2013; Rulten et al. 2014).

21.6 Future Directions

21.6.1 Mechanisms and Hierarchy of FUS Mutations and Proteinopathy

The discovery of dominant FUS mutations as one of the major causal links for FALS has opened up new windows to understanding the pathogenesis of ALS.

Judging from the aforementioned studies, it is evident that the mechanisms leading to FUS-mediated neurodegeneration are quite complex. However, the results from multiple model organisms indicate that both structural and functional defects in dendritic/axonal growth and synaptic transmission occur prior to neuronal loss and that these defects have major contributions to the neurodegeneration caused by FUS mutations. Mechanisms directly contribute to these phenotypes which include defects in DNA damage repair/response and in RNA transcription, splicing, and transport. The fundamental effects of mutant FUS proteins can be attributed to its aberrant gain-of-function properties that alter the homeostasis of the interactions of wild-type FUS and its interacting partners, including proteins, pre-mRNAs, and lncRNAs, in the DNA damage response/repair and RNA splicing machinery. In many aspects, these gain-of-function properties truly reflect the essential roles of wild-type FUS required for the embryonic and postnatal development and maintenance of the organisms and the nervous system.

Notwithstanding these new insights into FUS mutations, it remains a challenge to determine how these new mechanisms help in understanding the pathogenesis of FALS and SALS. With the availability of transgenic models for FUS and TDP-43, it will be important to determine whether mutations in FUS or TDP-43 target similar or divergent mechanisms that eventually lead to neurodegeneration. This is especially important given that TDP-43 proteinopathy is a major neuropathological feature not only in FALS but also in SALS. In addition to the insights from model organisms, the availability of patient-specific induced pluripotent stem cell (iPSC)-derived neurons also provides a new model system more closely related to human disease. Indeed, the combined strengths of using iPSC-derived neurons and the model organism *Drosophila* have led to the identification of novel functions of TDP-43 in mRNA transport in axons (Alami et al. 2014). Future studies using similar approaches can also provide more insights to unravel the mechanisms and their hierarchical interactions in the pathogenesis of human diseases. The iPSC approach will also provide important tools to identify potential therapeutic targets that are specific for different mutations in FUS-mediated FALS.

21.6.2 The Expanding Repertoire of RNA Machinery in Neurodegeneration

The identification of mutations in *FUS*, *TARDBP*, and *C9ORF72* in ALS and FTD-ALS has expanded the landscape of neurodegenerative diseases caused by defects in RNA metabolism machinery. These studies raise the intriguing questions as to whether *FUS*, *TARDBP*, and *C9ORF72* are just tips of an iceberg and whether dysregulations in RNA machinery may have broader roles in other neurodegenerative diseases. Results from several recent studies indicate that the answers to both questions are positive. For instance, mutations in the prion-like domains of hnRNPA2B1 and hnRNAPA1 accelerate filamentous protein aggregate formation, increase the propensity of stress granule formation in neuronal cytoplasm, and have been identified in patients with multisystem proteinopathy and ALS (Kim et al. 2013). Furthermore, chemical mutagenesis screens for recessive mutations (Neuroscience Mutagenesis Facility) have reported two mutations, *nmf291* and *nmf205*, that result in neurological phenotypes. The *nmf291* allele is caused by a 5-nucleotide deletion in the U2 snRNA, which leads to profound dysregulation in RNA splicing and an age-dependent, progressive degeneration of the cerebellum (An et al. 2008). In contrast, the *nmf205* mutation results in loss of GTPBP2 due to a point mutation in the consensus splice donor site of intron 6 of *Gtpbp2*, leading to missplicing of *Gtpbp2* mRNA and a premature stop codon. Since GTPBP2 is an essential binding partner for the ribosome recycling protein pelota, loss of GTPBP2 results in widespread ribosomal stalling and profound age-dependent neurodegeneration in the cerebellum of *nmf205* mutants (Ishimura et al. 2014). Finally, two recent studies show that mutations in the human *cleavage and polyadenylation factor subunit 1* (*CLP1*) gene, which encodes a multifunctional kinase implicated in the maturation of tRNA, mRNA and siRNA, cause severe neurodegeneration in the cerebellum (Karaca et al. 2014; Schaffer et al. 2014). Consistent with these findings, mouse mutants that express kinase-dead mutant CLP1 show neurodegeneration, characterized by a progressive loss of spinal motor neurons, axonal degeneration in the peripheral nerves, and denervation of neuromuscular junctions, leading to impaired motor function, muscle weakness, paralysis, and fatal respiratory failure (Hanada et al. 2013). Together, these findings further expand the critical role of RNA metabolism in maintaining the normal functions of the nervous system.

References

Alami NH, Smith RB, Carrasco MA, Williams LA, Winborn CS, Han SS, Kiskinis E, Winborn B, Freibaum BD, Kanagaraj A et al (2014) Axonal transport of TDP-43 mRNA granules is impaired by ALS-causing mutations. Neuron 81:536–543

Alberti S, Halfmann R, King O, Kapila A, Lindquist S (2009) A systematic survey identifies prions and illuminates sequence features of prionogenic proteins. Cell 137:146–158

An JJ, Gharami K, Liao GY, Woo NH, Lau AG, Vanevski F, Torre ER, Jones KR, Feng Y, Lu B et al (2008) Distinct role of long 3′ UTR BDNF mRNA in spine morphology and synaptic plasticity in hippocampal neurons. Cell 134:175–187

Anderson AJ, Su JH, Cotman CW (1996) DNA damage and apoptosis in Alzheimer's disease: colocalization with c-Jun immunoreactivity, relationship to brain area, and effect of postmortem delay. J Neurosci 16:1710–1719

Arai T, Hasegawa M, Akiyama H, Ikeda K, Nonaka T, Mori H, Mann D, Tsuchiya K, Yoshida M, Hashizume Y et al (2006) TDP-43 is a component of ubiquitin-positive tau-negative inclusions in frontotemporal lobar degeneration and amyotrophic lateral sclerosis. Biochem Biophys Res Commun 351:602–611

Armstrong GA, Drapeau P (2013) Loss and gain of FUS function impair neuromuscular synaptic transmission in a genetic model of ALS. Hum Mol Genet 22:4282–4292

Ash PE, Bieniek KF, Gendron TF, Caulfield T, Lin WL, Dejesus-Hernandez M, van Blitterswijk MM, Jansen-West K, Paul JW 3rd, Rademakers R et al (2013) Unconventional translation of

C9ORF72 GGGGCC expansion generates insoluble polypeptides specific to c9FTD/ALS. Neuron 77:639–646
Baechtold H, Kuroda M, Sok J, Ron D, Lopez BS, Akhmedov AT (1999) Human 75-kDa DNA-pairing protein is identical to the pro-oncoprotein TLS/FUS and is able to promote D-loop formation. J Biol Chem 274:34337–34342
Battle DJ, Kasim M, Yong J, Lotti F, Lau CK, Mouaikel J, Zhang Z, Han K, Wan L, Dreyfuss G (2006) The SMN complex: an assembly machine for RNPs. Cold Spring Harb Symp Quant Biol 71:313–320
Baumer D, Hilton D, Paine SM, Turner MR, Lowe J, Talbot K, Ansorge O (2010) Juvenile ALS with basophilic inclusions is a FUS proteinopathy with FUS mutations. Neurology 75:611–618
Behzadnia N, Golas MM, Hartmuth K, Sander B, Kastner B, Deckert J, Dube P, Will CL, Urlaub H, Stark H et al (2007) Composition and three-dimensional EM structure of double affinity-purified, human prespliceosomal A complexes. Embo J 26:1737–1748
Bender A, Krishnan KJ, Morris CM, Taylor GA, Reeve AK, Perry RH, Jaros E, Hersheson JS, Betts J, Klopstock T et al (2006) High levels of mitochondrial DNA deletions in substantia nigra neurons in aging and Parkinson disease. Nat Genet 38:515–517
Bertrand P, Akhmedov AT, Delacote F, Durrbach A, Lopez BS (1999) Human POMp75 is identified as the pro-oncoprotein TLS/FUS: both POMp75 and POMp100 DNA homologous pairing activities are associated to cell proliferation. Oncogene 18:4515–4521
Betley JN, Wright CV, Kawaguchi Y, Erdelyi F, Szabo G, Jessell TM, Kaltschmidt JA (2009) Stringent specificity in the construction of a GABAergic presynaptic inhibitory circuit. Cell 139:161–174
Bigio EH (2012) Motor neuron disease: the C9orf72 hexanucleotide repeat expansion in FTD and ALS. Nat Rev Neurol 8:249–250
Boillee S, Vande Velde C, Cleveland DW (2006a) ALS: a disease of motor neurons and their nonneuronal neighbors. Neuron 52:39–59
Boillee S, Yamanaka K, Lobsiger CS, Copeland NG, Jenkins NA, Kassiotis G, Kollias G, Cleveland DW (2006b) Onset and progression in inherited ALS determined by motor neurons and microglia. Science 312:1389–1392
Bond CS, Fox AH (2009) Paraspeckles: nuclear bodies built on long noncoding RNA. J Cell Biol 186:637–644
Bosco DA, Lemay N, Ko HK, Zhou H, Burke C, Kwiatkowski TJ Jr, Sapp P, McKenna-Yasek D, Brown RH Jr, Hayward LJ (2010) Mutant FUS proteins that cause amyotrophic lateral sclerosis incorporate into stress granules. Hum Mol Genet 19:4160–4175
Calvio C, Neubauer G, Mann M, Lamond AI (1995) Identification of hnRNP P2 as TLS/FUS using electrospray mass spectrometry. RNA 1:724–733
Cleaver JE, Lam ET, Revet I (2009) Disorders of nucleotide excision repair: the genetic and molecular basis of heterogeneity. Nat Rev Genet 10:756–768
Cleaver JE, Bezrookove V, Revet I, Huang EJ (2013) Conceptual developments in the causes of Cockayne syndrome. Mech Ageing Dev 134:284–290
Colombrita C, Onesto E, Megiorni F, Pizzuti A, Baralle FE, Buratti E, Silani V, Ratti A (2012) TDP-43 and FUS RNA-binding proteins bind distinct sets of cytoplasmic messenger RNAs and differently regulate their post-transcriptional fate in motoneuron-like cells. J Biol Chem 287:15635–15647
Cooper TA, Wan L, Dreyfuss G (2009) RNA and disease. Cell 136:777–793
Couthouis J, Hart MP, Shorter J, DeJesus-Hernandez M, Erion R, Oristano R, Liu AX, Ramos D, Jethava N, Hosangadi D et al (2011) A yeast functional screen predicts new candidate ALS disease genes. Proc Natl Acad Sci U S A 108:20881–20890
Couthouis J, Hart MP, Erion R, King OD, Diaz Z, Nakaya T, Ibrahim F, Kim HJ, Mojsilovic-Petrovic J, Panossian S et al (2012) Evaluating the role of the FUS/TLS-related gene EWSR1 in amyotrophic lateral sclerosis. Hum Mol Genet 21:2899–2911
Crozat A, Aman P, Mandahl N, Ron D (1993) Fusion of CHOP to a novel RNA-binding protein in human myxoid liposarcoma. Nature 363:640–644

Cushman M, Johnson BS, King OD, Gitler AD, Shorter J (2010) Prion-like disorders: blurring the divide between transmissibility and infectivity. J Cell Sci 123:1191–1201
Daigle JG, Lanson NA Jr, Smith RB, Casci I, Maltare A, Monaghan J, Nichols CD, Kryndushkin D, Shewmaker F, Pandey UB (2013) RNA-binding ability of FUS regulates neurodegeneration, cytoplasmic mislocalization and incorporation into stress granules associated with FUS carrying ALS-linked mutations. Hum Mol Genet 22:1193–1205
Das R, Yu J, Zhang Z, Gygi MP, Krainer AR, Gygi SP, Reed R (2007) SR proteins function in coupling RNAP II transcription to pre-mRNA splicing. Mol Cell 26:867–881
de Waard MC, van der Pluijm I, Zuiderveen Borgesius N, Comley LH, Haasdijk ED, Rijksen Y, Ridwan Y, Zondag G, Hoeijmakers JH, Elgersma Y et al (2010) Age-related motor neuron degeneration in DNA repair-deficient Ercc1 mice. Acta Neuropathol 120:461–475
DeJesus-Hernandez M, Mackenzie IR, Boeve BF, Boxer AL, Baker M, Rutherford NJ, Nicholson AM, Finch NA, Flynn H, Adamson J et al (2011) Expanded GGGGCC hexanucleotide repeat in noncoding region of C9ORF72 causes chromosome 9p-linked FTD and ALS. Neuron 72:245–256
Deng H, Gao K, Jankovic J (2014a) The role of FUS gene variants in neurodegenerative diseases. Nat Rev Neurol 10:337–348
Deng Q, Holler CJ, Taylor G, Hudson KF, Watkins W, Gearing M, Ito D, Murray ME, Dickson DW, Seyfried NT et al (2014b) FUS is phosphorylated by DNA-PK and accumulates in the cytoplasm after DNA damage. J Neurosci 34:7802–7813
Dichmann DS, Harland RM (2012) fus/TLS orchestrates splicing of developmental regulators during gastrulation. Genes Dev 26:1351–1363
Dinant C, Houtsmuller AB, Vermeulen W (2008) Chromatin structure and DNA damage repair. Epigenetics Chromatin 1:9
Dini Modigliani S, Morlando M, Errichelli L, Sabatelli M, Bozzoni I (2014) An ALS-associated mutation in the FUS 3′-UTR disrupts a microRNA-FUS regulatory circuitry. Nat Commun 5:4335
Dobbin MM, Madabhushi R, Pan L, Chen Y, Kim D, Gao J, Ahanonu B, Pao PC, Qiu Y, Zhao Y et al (2013) SIRT1 collaborates with ATM and HDAC1 to maintain genomic stability in neurons. Nat Neurosci 16:1008–1015
Donnelly CJ, Zhang PW, Pham JT, Haeusler AR, Mistry NA, Vidensky S, Daley EL, Poth EM, Hoover B, Fines DM et al (2013) RNA toxicity from the ALS/FTD C9ORF72 expansion is mitigated by antisense intervention. Neuron 80:415–428
Dormann D, Rodde R, Edbauer D, Bentmann E, Fischer I, Hruscha A, Than ME, Mackenzie IR, Capell A, Schmid B et al (2010) ALS-associated fused in sarcoma (FUS) mutations disrupt Transportin-mediated nuclear import. EMBO J 29:2841–2857
Dormann D, Madl T, Valori CF, Bentmann E, Tahirovic S, Abou-Ajram C, Kremmer E, Ansorge O, Mackenzie IR, Neumann M et al (2012) Arginine methylation next to the PY-NLS modulates Transportin binding and nuclear import of FUS. EMBO J 31:4258–4275
Ferrante RJ, Browne SE, Shinobu LA, Bowling AC, Baik MJ, MacGarvey U, Kowall NW, Brown RH Jr, Beal MF (1997) Evidence of increased oxidative damage in both sporadic and familial amyotrophic lateral sclerosis. J Neurochem 69:2064–2074
Fillingham J, Keogh MC, Krogan NJ (2006) GammaH2AX and its role in DNA double-strand break repair. Biochem Cell Biol 84:568–577
Fujii R, Okabe S, Urushido T, Inoue K, Yoshimura A, Tachibana T, Nishikawa T, Hicks GG, Takumi T (2005) The RNA binding protein TLS is translocated to dendritic spines by mGluR5 activation and regulates spine morphology. Curr Biol 15:587–593
Fushimi K, Long C, Jayaram N, Chen X, Li L, Wu JY (2011) Expression of human FUS/TLS in yeast leads to protein aggregation and cytotoxicity, recapitulating key features of FUS proteinopathy. Protein Cell 2:141–149
Gardiner M, Toth R, Vandermoere F, Morrice NA, Rouse J (2008) Identification and characterization of FUS/TLS as a new target of ATM. Biochem J 415:297–307

Gitler AD, Shorter J (2011) RNA-binding proteins with prion-like domains in ALS and FTLD-U. Prion 5:179–187

Goldman JS, Farmer JM, Wood EM, Johnson JK, Boxer A, Neuhaus J, Lomen-Hoerth C, Wilhelmsen KC, Lee VM, Grossman M et al (2005) Comparison of family histories in FTLD subtypes and related tauopathies. Neurology 65:1817–1819

Graff J, Rei D, Guan JS, Wang WY, Seo J, Hennig KM, Nieland TJ, Fass DM, Kao PF, Kahn M et al (2012) An epigenetic blockade of cognitive functions in the neurodegenerating brain. Nature 483:222–226

Groen EJ, Fumoto K, Blokhuis AM, Engelen-Lee J, Zhou Y, van den Heuvel DM, Koppers M, van Diggelen F, van Heest J, Demmers JA et al (2013) ALS-associated mutations in FUS disrupt the axonal distribution and function of SMN. Hum Mol Genet 22:3690–3704

Han TW, Kato M, Xie S, Wu LC, Mirzaei H, Pei J, Chen M, Xie Y, Allen J, Xiao G et al (2012) Cell-free formation of RNA granules: bound RNAs identify features and components of cellular assemblies. Cell 149:768–779

Hanada T, Weitzer S, Mair B, Bernreuther C, Wainger BJ, Ichida J, Hanada R, Orthofer M, Cronin SJ, Komnenovic V et al (2013) CLP1 links tRNA metabolism to progressive motor-neuron loss. Nature 495:474–480

Hicks GG, Singh N, Nashabi A, Mai S, Bozek G, Klewes L, Arapovic D, White EK, Koury MJ, Oltz EM et al (2000) Fus deficiency in mice results in defective B-lymphocyte development and activation, high levels of chromosomal instability and perinatal death. Nat Genet 24:175–179

Hoell JI, Larsson E, Runge S, Nusbaum JD, Duggimpudi S, Farazi TA, Hafner M, Borkhardt A, Sander C, Tuschl T (2011) RNA targets of wild-type and mutant FET family proteins. Nat Struct Mol Biol 18:1428–1431

Huang EJ, Zhang J, Geser F, Trojanowski JQ, Strober JB, Dickson DW, Brown RH Jr, Shapiro BE, Lomen-Hoerth C (2010) Extensive FUS-immunoreactive pathology in juvenile amyotrophic lateral sclerosis with basophilic inclusions. Brain Pathol 20:1069–1076

Huang C, Zhou H, Tong J, Chen H, Liu YJ, Wang D, Wei X, Xia XG (2011) FUS transgenic rats develop the phenotypes of amyotrophic lateral sclerosis and frontotemporal lobar degeneration. PLoS Genet 7:e1002011

Huang C, Tong J, Bi F, Wu Q, Huang B, Zhou H, Xia XG (2012) Entorhinal cortical neurons are the primary targets of FUS mislocalization and ubiquitin aggregation in FUS transgenic rats. Hum Mol Genet 21:4602–4614

Husi H, Ward MA, Choudhary JS, Blackstock WP, Grant SG (2000) Proteomic analysis of NMDA receptor-adhesion protein signaling complexes. Nat Neurosci 3:661–669

Ichikawa H, Shimizu K, Hayashi Y, Ohki M (1994) An RNA-binding protein gene, TLS/FUS, is fused to ERG in human myeloid leukemia with t(16;21) chromosomal translocation. Cancer Res 54:2865–2868

Iko Y, Kodama TS, Kasai N, Oyama T, Morita EH, Muto T, Okumura M, Fujii R, Takumi T, Tate S et al (2004) Domain architectures and characterization of an RNA-binding protein, TLS. J Biol Chem 279:44834–44840

Immanuel D, Zinszner H, Ron D (1995) Association of SARFH (sarcoma-associated RNA-binding fly homolog) with regions of chromatin transcribed by RNA polymerase II. Mol Cell Biol 15:4562–4571

Ishigaki S, Masuda A, Fujioka Y, Iguchi Y, Katsuno M, Shibata A, Urano F, Sobue G, Ohno K (2012) Position-dependent FUS-RNA interactions regulate alternative splicing events and transcriptions. Sci Rep 2:529

Ishimura R, Nagy G, Dotu I, Zhou H, Yang XL, Schimmel P, Senju S, Nishimura Y, Chuang JH, Ackerman SL (2014) RNA function. Ribosome stalling induced by mutation of a CNS-specific tRNA causes neurodegeneration. Science 345:455–459

Jackson SP, Bartek J (2009) The DNA-damage response in human biology and disease. Nature 461:1071–1078

Kabashi E, Bercier V, Lissouba A, Liao M, Brustein E, Rouleau GA, Drapeau P (2011) FUS and TARDBP but not SOD1 interact in genetic models of amyotrophic lateral sclerosis. PLoS Genet 7:e1002214

Kanai Y, Dohmae N, Hirokawa N (2004) Kinesin transports RNA: isolation and characterization of an RNA-transporting granule. Neuron 43:513–525

Karaca E, Weitzer S, Pehlivan D, Shiraishi H, Gogakos T, Hanada T, Jhangiani SN, Wiszniewski W, Withers M, Campbell IM et al (2014) Human CLP1 mutations alter tRNA biogenesis, affecting both peripheral and central nervous system function. Cell 157:636–650

Kato M, Han TW, Xie S, Shi K, Du X, Wu LC, Mirzaei H, Goldsmith EJ, Longgood J, Pei J et al (2012) Cell-free formation of RNA granules: low complexity sequence domains form dynamic fibers within hydrogels. Cell 149:753–767

Kim D, Frank CL, Dobbin MM, Tsunemoto RK, Tu W, Peng PL, Guan JS, Lee BH, Moy LY, Giusti P et al (2008) Deregulation of HDAC1 by p25/Cdk5 in neurotoxicity. Neuron 60:803–817

Kim HJ, Kim NC, Wang YD, Scarborough EA, Moore J, Diaz Z, MacLea KS, Freibaum B, Li S, Molliex A et al (2013) Mutations in prion-like domains in hnRNPA2B1 and hnRNPA1 cause multisystem proteinopathy and ALS. Nature 495:467–473

Kornblihtt AR, de la Mata M, Fededa JP, Munoz MJ, Nogues G (2004) Multiple links between transcription and splicing. RNA 10:1489–1498

Kuroda M, Sok J, Webb L, Baechtold H, Urano F, Yin Y, Chung P, de Rooij DG, Akhmedov A, Ashley T et al (2000) Male sterility and enhanced radiation sensitivity in TLS(–/–) mice. EMBO J 19:453–462

Kwiatkowski TJ Jr, Bosco DA, Leclerc AL, Tamrazian E, Vanderburg CR, Russ C, Davis A, Gilchrist J, Kasarskis EJ, Munsat T et al (2009) Mutations in the FUS/TLS gene on chromosome 16 cause familial amyotrophic lateral sclerosis. Science 323:1205–1208

Lagier-Tourenne C, Polymenidou M, Hutt KR, Vu AQ, Baughn M, Huelga SC, Clutario KM, Ling SC, Liang TY, Mazur C et al (2012) Divergent roles of ALS-linked proteins FUS/TLS and TDP-43 intersect in processing long pre-mRNAs. Nat Neurosci 15:1488–1497

Lanson NA Jr, Maltare A, King H, Smith R, Kim JH, Taylor JP, Lloyd TE, Pandey UB (2011) A Drosophila model of FUS-related neurodegeneration reveals genetic interaction between FUS and TDP-43. Hum Mol Genet 20:2510–2523

Lattante S, Rouleau GA, Kabashi E (2013) TARDBP and FUS mutations associated with amyotrophic lateral sclerosis: summary and update. Hum Mutat 34:812–826

Lee EB, Lee VM, Trojanowski JQ (2012) Gains or losses: molecular mechanisms of TDP43-mediated neurodegeneration. Nat Rev Neurosci 13:38–50

Lerga A, Hallier M, Delva L, Orvain C, Gallais I, Marie J, Moreau-Gachelin F (2001) Identification of an RNA binding specificity for the potential splicing factor TLS. J Biol Chem 276:6807–6816

Li Q, Lee JA, Black DL (2007) Neuronal regulation of alternative pre-mRNA splicing. Nat Rev Neurosci 8:819–831

Li YR, King OD, Shorter J, Gitler AD (2013) Stress granules as crucibles of ALS pathogenesis. J Cell Biol 201:361–372

Lillo P, Mioshi E, Zoing MC, Kiernan MC, Hodges JR (2011) How common are behavioural changes in amyotrophic lateral sclerosis? Amyotroph Lateral Scler 12:45–51

Ling SC, Polymenidou M, Cleveland DW (2013) Converging mechanisms in ALS and FTD: disrupted RNA and protein homeostasis. Neuron 79:416–438

Lomen-Hoerth C, Anderson T, Miller B (2002) The overlap of amyotrophic lateral sclerosis and frontotemporal dementia. Neurology 59:1077–1079

Lomen-Hoerth C, Murphy J, Langmore S, Kramer JH, Olney RK, Miller B (2003) Are amyotrophic lateral sclerosis patients cognitively normal? Neurology 60:1094–1097

Lu T, Pan Y, Kao SY, Li C, Kohane I, Chan J, Yankner BA (2004) Gene regulation and DNA damage in the ageing human brain. Nature 429:883–891

Lukas J, Lukas C, Bartek J (2011) More than just a focus: the chromatin response to DNA damage and its role in genome integrity maintenance. Nat Cell Biol 13:1161–1169
Mackenzie IR, Munoz DG, Kusaka H, Yokota O, Ishihara K, Roeber S, Kretzschmar HA, Cairns NJ, Neumann M (2011) Distinct pathological subtypes of FTLD-FUS. Acta Neuropathol 121:207–218
Mackenzie IR, Frick P, Neumann M (2014) The neuropathology associated with repeat expansions in the C9ORF72 gene. Acta Neuropathol 127:347–357
Martin KC, Ephrussi A (2009) mRNA localization: gene expression in the spatial dimension. Cell 136:719–730
Mastrocola AS, Kim SH, Trinh AT, Rodenkirch LA, Tibbetts RS (2013) The RNA-binding protein fused in sarcoma (FUS) functions downstream of poly(ADP-ribose) polymerase (PARP) in response to DNA damage. J Biol Chem 288:24731–24741
McKinnon PJ (2009) DNA repair deficiency and neurological disease. Nat Rev Neurosci 10:100–112
Millecamps S, Boillee S, Le Ber I, Seilhean D, Teyssou E, Giraudeau M, Moigneu C, Vandenberghe N, Danel-Brunaud V, Corcia P et al (2012) Phenotype difference between ALS patients with expanded repeats in C9ORF72 and patients with mutations in other ALS-related genes. J Med Genet 49:258–263
Miller KM, Tjeertes JV, Coates J, Legube G, Polo SE, Britton S, Jackson SP (2010) Human HDAC1 and HDAC2 function in the DNA-damage response to promote DNA nonhomologous end-joining. Nat Struct Mol Biol 17:1144–1151
Mitchell JC, McGoldrick P, Vance C, Hortobagyi T, Sreedharan J, Rogelj B, Tudor EL, Smith BN, Klasen C, Miller CC et al (2013) Overexpression of human wild-type FUS causes progressive motor neuron degeneration in an age- and dose-dependent fashion. Acta Neuropathol 125:273–288
Moore MJ, Proudfoot NJ (2009) Pre-mRNA processing reaches back to transcription and ahead to translation. Cell 136:688–700
Mori K, Weng SM, Arzberger T, May S, Rentzsch K, Kremmer E, Schmid B, Kretzschmar HA, Cruts M, Van Broeckhoven C et al (2013) The C9orf72 GGGGCC repeat is translated into aggregating dipeptide-repeat proteins in FTLD/ALS. Science 339:1335–1338
Munoz MJ, Perez Santangelo MS, Paronetto MP, de la Mata M, Pelisch F, Boireau S, Glover-Cutter K, Ben-Dov C, Blaustein M, Lozano JJ et al (2009) DNA damage regulates alternative splicing through inhibition of RNA polymerase II elongation. Cell 137:708–720
Murakami T, Yang SP, Xie L, Kawano T, Fu D, Mukai A, Bohm C, Chen F, Robertson J, Suzuki H et al (2012) ALS mutations in FUS cause neuronal dysfunction and death in Caenorhabditis elegans by a dominant gain-of-function mechanism. Hum Mol Genet 21:1–9
Neumann M, Sampathu DM, Kwong LK, Truax AC, Micsenyi MC, Chou TT, Bruce J, Schuck T, Grossman M, Clark CM et al (2006) Ubiquitinated TDP-43 in frontotemporal lobar degeneration and amyotrophic lateral sclerosis. Science 314:130–133
Newman JC, Bailey AD, Weiner AM (2006) Cockayne syndrome group B protein (CSB) plays a general role in chromatin maintenance and remodeling. Proc Natl Acad Sci U S A 103:9613–9618
Nishimoto Y, Nakagawa S, Hirose T, Okano HJ, Takao M, Shibata S, Suyama S, Kuwako K, Imai T, Murayama S et al (2013) The long non-coding RNA nuclear-enriched abundant transcript 1_2 induces paraspeckle formation in the motor neuron during the early phase of amyotrophic lateral sclerosis. Mol Brain 6:31
Nomura T, Watanabe S, Kaneko K, Yamanaka K, Nukina N, Furukawa Y (2014) Intranuclear aggregation of mutant FUS/TLS as a molecular pathomechanism of amyotrophic lateral sclerosis. J Biol Chem 289:1192–1202
Olney RK, Murphy J, Forshew D, Garwood E, Miller BL, Langmore S, Kohn MA, Lomen-Hoerth C (2005) The effects of executive and behavioral dysfunction on the course of ALS. Neurology 65:1774–1777

Ou SH, Wu F, Harrich D, Garcia-Martinez LF, Gaynor RB (1995) Cloning and characterization of a novel cellular protein, TDP-43, that binds to human immunodeficiency virus type 1 TAR DNA sequence motifs. J Virol 69:3584–3596

Pilch DR, Sedelnikova OA, Redon C, Celeste A, Nussenzweig A, Bonner WM (2003) Characteristics of gamma-H2AX foci at DNA double-strand breaks sites. Biochem Cell Biol 81:123–129

Polymenidou M, Lagier-Tourenne C, Hutt KR, Huelga SC, Moran J, Liang TY, Ling SC, Sun E, Wancewicz E, Mazur C et al (2011) Long pre-mRNA depletion and RNA missplicing contribute to neuronal vulnerability from loss of TDP-43. Nat Neurosci 14:459–468

Prasad DD, Ouchida M, Lee L, Rao VN, Reddy ES (1994) TLS/FUS fusion domain of TLS/FUS-erg chimeric protein resulting from the t(16;21) chromosomal translocation in human myeloid leukemia functions as a transcriptional activation domain. Oncogene 9:3717–3729

Qiu H, Lee S, Shang Y, Wang WY, Au KF, Kamiya S, Barmada S, Finkbeiner S, Lui H, Carlton CE et al (2014) ALS-associated mutation FUS-R521C causes DNA damage and RNA splicing defects. J Clin Invest 124:981–999

Rabbitts TH, Forster A, Larson R, Nathan P (1993) Fusion of the dominant negative transcription regulator CHOP with a novel gene FUS by translocation t(12;16) in malignant liposarcoma. Nat Genet 4:175–180

Rabin SJ, Kim JM, Baughn M, Libby RT, Kim YJ, Fan Y, La Spada A, Stone B, Ravits J (2010) Sporadic ALS has compartment-specific aberrant exon splicing and altered cell-matrix adhesion biology. Hum Mol Genet 19:313–328

Rass U, Ahel I, West SC (2007) Defective DNA repair and neurodegenerative disease. Cell 130:991–1004

Reed R, Hurt E (2002) A conserved mRNA export machinery coupled to pre-mRNA splicing. Cell 108:523–531

Renton AE, Majounie E, Waite A, Simon-Sanchez J, Rollinson S, Gibbs JR, Schymick JC, Laaksovirta H, van Swieten JC, Myllykangas L et al (2011) A hexanucleotide repeat expansion in C9ORF72 is the cause of chromosome 9p21-linked ALS-FTD. Neuron 72:257–268

Ringholz GM, Appel SH, Bradshaw M, Cooke NA, Mosnik DM, Schulz PE (2005) Prevalence and patterns of cognitive impairment in sporadic ALS. Neurology 65:586–590

Robles E, Baier H (2012) Assembly of synaptic laminae by axon guidance molecules. Curr Opin Neurobiol 22:799–804

Rogelj B, Easton LE, Bogu GK, Stanton LW, Rot G, Curk T, Zupan B, Sugimoto Y, Modic M, Haberman N et al (2012) Widespread binding of FUS along nascent RNA regulates alternative splicing in the brain. Sci Rep 2:603

Rulten SL, Rotheray A, Green RL, Grundy GJ, Moore DA, Gomez-Herreros F, Hafezparast M, Caldecott KW (2014) PARP-1 dependent recruitment of the amyotrophic lateral sclerosis-associated protein FUS/TLS to sites of oxidative DNA damage. Nucleic Acids Res 42:307–314

Sabatelli M, Moncada A, Conte A, Lattante S, Marangi G, Luigetti M, Lucchini M, Mirabella M, Romano A, Del Grande A et al (2013) Mutations in the 3′ untranslated region of FUS causing FUS overexpression are associated with amyotrophic lateral sclerosis. Hum Mol Genet 22:4748–4755

Sanes JR, Yamagata M (2009) Many paths to synaptic specificity. Annu Rev Cell Dev Biol 25:161–195

Schaffer AE, Eggens VR, Caglayan AO, Reuter MS, Scott E, Coufal NG, Silhavy JL, Xue Y, Kayserili H, Yasuno K et al (2014) CLP1 founder mutation links tRNA splicing and maturation to cerebellar development and neurodegeneration. Cell 157:651–663

Schwartz JC, Ebmeier CC, Podell ER, Heimiller J, Taatjes DJ, Cech TR (2012) FUS binds the CTD of RNA polymerase II and regulates its phosphorylation at Ser2. Genes Dev 26:2690–2695

Schwartz JC, Wang X, Podell ER, Cech TR (2013) RNA seeds higher-order assembly of FUS protein. Cell Rep 5:918–925

Schwartz JC, Podell ER, Han SS, Berry JD, Eggan KC, Cech TR (2014) FUS is sequestered in nuclear aggregates in ALS patient fibroblasts. Mol Biol Cell 25:2571–2578

Sephton CF, Tang AA, Kulkarni A, West J, Brooks M, Stubblefield JJ, Liu Y, Zhang MQ, Green CB, Huber KM et al (2014) Activity-dependent FUS dysregulation disrupts synaptic homeostasis. Proc Natl Acad Sci U S A 111:E4769–E4778

Shahidullah M, Le Marchand SJ, Fei H, Zhang J, Pandey UB, Dalva MB, Pasinelli P, Levitan IB (2013) Defects in synapse structure and function precede motor neuron degeneration in Drosophila models of FUS-related ALS. J Neurosci 33:19590–19598

Shelkovnikova TA, Peters OM, Deykin AV, Connor-Robson N, Robinson H, Ustyugov AA, Bachurin SO, Ermolkevich TG, Goldman IL, Sadchikova ER et al (2013) Fused in sarcoma (FUS) protein lacking nuclear localization signal (NLS) and major RNA binding motifs triggers proteinopathy and severe motor phenotype in transgenic mice. J Biol Chem 288:25266–25274

Shelkovnikova TA, Robinson HK, Troakes C, Ninkina N, Buchman VL (2014) Compromised paraspeckle formation as a pathogenic factor in FUSopathies. Hum Mol Genet 23:2298–2312

Stolow DT, Haynes SR (1995) Cabeza, a Drosophila gene encoding a novel RNA binding protein, shares homology with EWS and TLS, two genes involved in human sarcoma formation. Nucleic Acids Res 23:835–843

Strong MJ, Grace GM, Freedman M, Lomen-Hoerth C, Woolley S, Goldstein LH, Murphy J, Shoesmith C, Rosenfeld J, Leigh PN et al (2009) Consensus criteria for the diagnosis of frontotemporal cognitive and behavioural syndromes in amyotrophic lateral sclerosis. Amyotroph Lateral Scler 10:131–146

Sun Z, Diaz Z, Fang X, Hart MP, Chesi A, Shorter J, Gitler AD (2011) Molecular determinants and genetic modifiers of aggregation and toxicity for the ALS disease protein FUS/TLS. PLoS Biol 9:e1000614

Sun S, Ling SC, Qiu J, Albuquerque CP, Zhou Y, Tokunaga S, Li H, Qiu H, Bui A, Yeo GW et al (2015) ALS-causative mutations in FUS/TLS confer gain and loss of function by altered association with SMN and U1-snRNP. Nat Commun 6:6171

Tan AY, Manley JL (2009) The TET family of proteins: functions and roles in disease. J Mol Cell Biol 1:82–92

Testa D, Lovati R, Ferrarini M, Salmoiraghi F, Filippini G (2004) Survival of 793 patients with amyotrophic lateral sclerosis diagnosed over a 28-year period. Amyotroph Lateral Scler Other Mot Neuron Disord: Official Publ World Fed Neurol Res Group Mot Neuron Dis 5:208–212

Tollervey JR, Curk T, Rogelj B, Briese M, Cereda M, Kayikci M, Konig J, Hortobagyi T, Nishimura AL, Zupunski V et al (2011) Characterizing the RNA targets and position-dependent splicing regulation by TDP-43. Nat Neurosci 14:452–458

Tsuiji H, Iguchi Y, Furuya A, Kataoka A, Hatsuta H, Atsuta N, Tanaka F, Hashizume Y, Akatsu H, Murayama S et al (2013) Spliceosome integrity is defective in the motor neuron diseases ALS and SMA. EMBO Mol Med 5:221–234

Vance C, Rogelj B, Hortobagyi T, De Vos KJ, Nishimura AL, Sreedharan J, Hu X, Smith B, Ruddy D, Wright P et al (2009) Mutations in FUS, an RNA processing protein, cause familial amyotrophic lateral sclerosis type 6. Science 323:1208–1211

Verbeeck C, Deng Q, Dejesus-Hernandez M, Taylor G, Ceballos-Diaz C, Kocerha J, Golde T, Das P, Rademakers R, Dickson DW et al (2012) Expression of Fused in sarcoma mutations in mice recapitulates the neuropathology of FUS proteinopathies and provides insight into disease pathogenesis. Mol Neurodegener 7:53

Wahl MC, Will CL, Luhrmann R (2009) The spliceosome: design principles of a dynamic RNP machine. Cell 136:701–718

Wang X, Arai S, Song X, Reichart D, Du K, Pascual G, Tempst P, Rosenfeld MG, Glass CK, Kurokawa R (2008) Induced ncRNAs allosterically modify RNA-binding proteins in cis to inhibit transcription. Nature 454:126–130

Wang JW, Brent JR, Tomlinson A, Shneider NA, McCabe BD (2011) The ALS-associated proteins FUS and TDP-43 function together to affect Drosophila locomotion and life span. J Clin Invest 121:4118–4126

Wang WY, Pan L, Su SC, Quinn EJ, Sasaki M, Jimenez JC, Mackenzie IR, Huang EJ, Tsai LH (2013) Interaction of FUS and HDAC1 regulates DNA damage response and repair in neurons. Nat Neurosci 16:1383–1391

Yamazaki T, Chen S, Yu Y, Yan B, Haertlein TC, Carrasco MA, Tapia JC, Zhai B, Das R, Lalancette-Hebert M et al (2012) FUS-SMN protein interactions link the motor neuron diseases ALS and SMA. Cell Rep 2:799–806

Yan J, Deng HX, Siddique N, Fecto F, Chen W, Yang Y, Liu E, Donkervoort S, Zheng JG, Shi Y et al (2010) Frameshift and novel mutations in FUS in familial amyotrophic lateral sclerosis and ALS/dementia. Neurology 75:807–814

Yang L, Embree LJ, Tsai S, Hickstein DD (1998) Oncoprotein TLS interacts with serine-arginine proteins involved in RNA splicing. J Biol Chem 273:27761–27764

Zhou Y, Liu S, Liu G, Ozturk A, Hicks GG (2013) ALS-associated FUS mutations result in compromised FUS alternative splicing and autoregulation. PLoS Genet 9:e1003895

Zhou Y, Liu S, Ozturk A, Hicks GG (2014) FUS-regulated RNA metabolism and DNA damage repair: implications for amyotrophic lateral sclerosis and frontotemporal dementia pathogenesis. Rare Dis 2:e29515

Zinszner H, Albalat R, Ron D (1994) A novel effector domain from the RNA-binding protein TLS or EWS is required for oncogenic transformation by CHOP. Genes Dev 8:2513–2526

Zinszner H, Sok J, Immanuel D, Yin Y, Ron D (1997) TLS (FUS) binds RNA in vivo and engages in nucleo-cytoplasmic shuttling. J Cell Sci 110(Pt 15):1741–1750

Chapter 22
Dendrite Degeneration in Glaucoma

Luca Della Santina and Yvonne Ou

Abstract Glaucoma is a group of slowly progressive neurodegenerative optic neuropathies in which the final common pathway is death of the retinal ganglion cells (RGCs) that comprise the optic nerve. While many risk factors have been identified, including intraocular pressure, the mechanistic pathway between the initial insult and RGC loss is still not clearly delineated. One paradigm is that of compartmentalized neurodegeneration, in which the axons, dendrites, and synapses undergo independent degenerative programs prior to cell soma loss. This chapter will focus on the features of dendritic remodeling in glaucoma and the implications on RGC function. Mounting evidence from various animal models of glaucoma demonstrates that changes in dendritic architecture precede cell death, although the detailed mechanisms and consequences of these morphologic and functional alterations have yet to be fully dissected. Such early dendritic alterations and synaptic rearrangements influence the normal function and structure of RGCs, thus giving important insight into the mechanism of circuit disassembly. At the same time, imaging of early dendritic alterations and identification of specific features of visual function that are first impaired represent promising new approaches to earlier diagnosis of this disease and improved disease progression recognition in humans.

Keywords Glaucoma • Dendrite • Neurodegeneration • Retinal ganglion cell • Dendritic remodeling

---

Funding SupportUniversity of Pisa, intramural funding (LDS); NIH K08 Career Development Award KEY022676A (YO); Research to Prevent Blindness Career Development Award (YO).

L.D. Santina
Department of Pharmacy, University of Pisa, Via Bonanno 6, 56126 Pisa, Italy
e-mail: luca.dellasantina@unipi.it

Y. Ou (✉)
University of California, San Francisco, 10 Koret Way, Box 0730, San Francisco, CA 94143, USA
e-mail: yvonne.ou@ucsf.edu

K. Emoto et al. (eds.), *Dendrites*, DOI 10.1007/978-4-431-56050-0_22

22.1 The Neurodegenerative Process in Glaucoma

Glaucoma comprises a group of disorders characterized by progressive neurodegeneration and eventual loss of the retinal ganglion cells (RGCs) that compose the optic nerve and project to central visual targets in the brain. It is the leading cause of irreversible blindness worldwide, estimated to affect 70 million people (Quigley and Broman 2006). One of the major risk factors of glaucoma is intraocular pressure (IOP), although it is apparent that elevated IOP is neither necessary nor sufficient for either disease diagnosis or progression. However, reduction of IOP remains the mainstay of medical and surgical treatment of glaucoma, even for patients in whom the IOP is considered to be in the normal range (Collaborative Normal-Tension Glaucoma Study Group 1998). Much research has been focused on the pathologic sequence of degeneration in order to better understand the mechanism by which IOP or other factors results in loss of RGCs.

A major gap in our diagnostic and treatment paradigms for glaucoma is that they are often initiated after there is already evidence of RGC death and visual field loss. There is growing consensus that the site of initial injury is the RGC axon as it passes through the pores of the lamina cribrosa at the optic nerve head (Quigley 1999; Quigley et al. 2000; Howell et al. 2007). Therefore, focal axonal injury is thought to be a key component of RGC glaucomatous pathology. Eventually, the neuronal soma undergoes apoptosis and irreversible RGC death occurs. However, our understanding of the pathological cellular processes that occur between initial injury and eventual apoptotic cell death is incomplete. Emerging evidence supports the hypothesis that the pathological degeneration of a stressed RGC may be compartmentalized at the subcellular level (Whitmore et al. 2005; Morquette and Di Polo 2008; Nickells et al. 2012). There may be independent degenerative pathways in the soma, axon, dendrite, and synapse. Indeed, compartmentalized neurodegeneration in which the axons, dendrites, and synapses degenerate independently of the soma occurs in other brain neurodegenerative diseases and may develop prior to late clinical signs (Raff et al. 2002; Wishart et al. 2006). Thus, it is not surprising to find that recent attention has focused on the cellular events at distal synapses and dendritic morphologic changes that precede neuronal death. This chapter focuses on the dendritic architecture of RGCs in the context of glaucoma and the functional implication of dendritic remodeling in glaucoma pathogenesis.

22.2 Imaging Dendrite Degeneration

Before delving into a discussion of the dendritic changes in glaucoma, we will summarize the techniques used to assess RGC morphology. Imaging of RGCs undergoing degeneration in glaucoma has been performed in multiple animal species but still represents a technical challenge because of the detail needed to

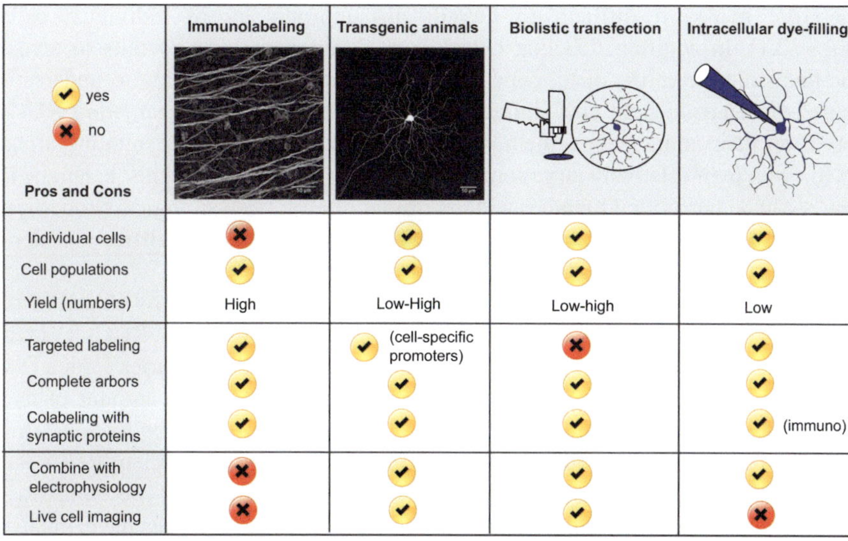

Fig. 22.1 Methods for labeling RGCs: advantages and limitations. Visualization of retinal ganglion cells (RGCs) in experimental glaucoma is critical to imaging and evaluating changes in morphology that could be related to functional alterations. Shown here are various techniques to label RGCs, together with their advantages and limitations. Immunolabeling can reveal RGC morphology with reasonable detail, but it is difficult to label individual cells with this method. Moreover, it cannot be combined with electrophysiology although it is possible to immunolabel cells after physiological recordings. Transgenic animals in which subpopulations of RGCs are labeled by fluorescent protein expression have many advantages, including labeling of specific RGC types, live cell imaging, and targeting RGCs for electrophysiological recording. Biolistic transfection to cause expression of fluorescent protein in RGCs also has similar advantages, but cells are randomly labeled. Finally, intracellular dye filling allows for simultaneous visualization of the cell and recording of the cell's responses, but time-lapse cell imaging is not possible

catch subtle or early changes in dendritic architecture. A summary of common methods used to label RGCs, along with their advantages and disadvantages, is illustrated in Fig. 22.1.

Most of the early imaging techniques were restricted to the use of Golgi staining, antibody labeling, and intracellular dye filling of RGCs (Weber et al. 1998), mainly in primates and cats. These techniques, which allow visualization of RGC dendrite morphology in fixed tissues, were usually applied to investigate late stages of the degenerative process. These approaches successfully identified key changes in dendritic architecture such as shrinkage of the dendritic arbor (Shou et al. 2003; Weber and Harman 2005). Concomitant labeling of the entire cell also revealed the relationship between dendritic alterations occurring in the axon and cell body. However, these labeling techniques exhibited several key limitations. For example, antibody labeling is limited by the relative low availability of antisera able to specifically reveal the dendritic morphology of RGCs. Frequently, investigators use antibodies against a generic component of the cytoskeleton, such as neurofilaments (Straznicky et al. 1992), which label multiple RGC types and therefore makes it difficult to identify the morphology of individual cells (Fig. 22.1). In addition, labeling cytoskeletal components usually fails to reveal the fine structure of the distal dendrites of the ganglion cell, leading to underestimates of dendritic field size. Intracellular dye injection enables targeting RGCs based on their soma size, a method useful, for example, in differentiating alpha RGCs with their relatively large somata. Another major strength of this technique is that it can be performed together with recording of electrophysiological properties, allowing a thorough characterization of both functional characteristics and dendritic morphology of the cell.

Effective labeling of RGC morphology can also be attained by injection of retrogradely transported dyes at the level of axonal terminals of RGCs (Vidal-Sanz et al. 1988; Dacey et al. 2003). The amount of labeled cells varies from a few cells to almost the entire population of RGCs depending on the amount of dye injection and the elapsed time between dye injection and tissue collection. This technique can be used to label RGCs in various species and all types of RGCs but relies on the presence of intact axons and their retrograde transport mechanisms.

Transgenic mice in which RGCs express fluorescent proteins have greatly facilitated imaging and comparison of RGC dendritic arbors in mouse models of glaucoma. For example, the dendritic arbor is labeled completely in *Thy1-YFP* mouse lines, and in some lines, sparse yellow fluorescent protein (YFP) expression enables the dendritic arbors of isolated RGCs to be distinguished from that of their neighbors (Feng et al. 2000). An alternative approach to obtain single-cell resolution for imaging RGC dendritic arbors is to use a biolistic method to acutely transfect retinas with plasmids encoding a fluorescent protein (Koizumi et al. 2011; Morgan et al. 2011; Della Santina et al. 2013). This "gene-gun" delivery method results in random labeling of RGCs with fluorescent proteins. Recently, the creation of transgenic mice expressing fluorescent protein in specific types of RGCs (Huberman et al. 2008, 2009; Dhande et al. 2013) permitted investigation of glaucoma-induced alterations in the dendrites of genetically identified RGC populations (El-Danaf and Huberman 2015).

A step beyond the analysis of the dendritic morphologic defects of RGCs in glaucoma is the investigation of the synapses between RGCs in the diseased retina and their presynaptic partners. Immunolabeling against presynaptic or postsynaptic proteins permits quantification of the density of excitatory synapses in the entire neuropil upon IOP elevation. However, in order to attain the density and distribution of synapses along individual RGC dendritic arbors, investigators have used biolistic methods to co-label dendrites with a fluorescent protein and excitatory postsynaptic sites with the postsynaptic protein, PSD95 (Morgan et al. 2008), tagged with a fluorescent protein of a different color.

Despite recent attempts to use *Thy1-FP* transgenic mice for imaging RGC death in vivo (Raymond et al. 2009) as well as their dendritic structure (Walsh and Quigley 2008), only a semiquantitative appreciation of the dendritic degeneration process in optic neuropathies has been possible thus far. Current limitations are dictated mainly by the difficulty in obtaining high-resolution in vivo images of RGC dendrites in the eyes of pigmented animals, where the imaging path must pass through the cornea and lens of the animal. Progress in longitudinal imaging of dendritic structure has been accomplished in rodents in optic nerve crush and retinal ischemia models, where the cornea remains clear (Leung et al. 2008a, b, 2011; Li et al. 2011). Using confocal scanning laser ophthalmoscopy (cSLO), Leung et al. demonstrated dendritic shrinkage as the earliest morphologic change after optic nerve crush using *Thy1-FP* mice. This technique does not permit evaluation of different RGC types since *en face* live imaging does not provide dendritic stratification information and is likely further limited by imaging brightly labeled dendritic arbors in the inner half of the IPL. Future improvement in cSLO optics such as the introduction of adaptive optics (Werkmeister et al. 2013) will hopefully bridge this technical gap. In order to extend the analysis of dendritic structure to humans, new methods need to be developed to label RGC dendrites. Current imaging techniques available in the clinic, such as cSLO or ocular coherence tomography (OCT), could then be used to follow the progression of dendritic alterations preceding cell death. One additional intriguing possibility is the use of spectral domain OCT (SD-OCT) to identify sublaminae within the IPL, which has recently been demonstrated using a commercially available SD-OCT instrument (Tanna et al. 2010).

22.3 Dendrite Remodeling in the Retina

Multiple lines of evidence in experimental glaucoma as well as human glaucoma suggest that dendrite remodeling occurs prior to cell death in response to intraocular pressure (IOP) elevation. One of the first studies to demonstrate that dendritic changes are an early marker of retinal ganglion cell degeneration was performed by intracellular labeling of individual RGCs in nonhuman primate retina (Weber et al. 1998). Using laser-induced scarification of the trabecular meshwork to chronically elevate IOP, Weber et al. examined 1298 intracellularly labeled RGCs and analyzed them both qualitatively and quantitatively in terms of soma size, dendritic field size, and axon diameter. This study demonstrated that the earliest measurable structural changes of RGC degeneration were dendritic in nature. The specific morphologic changes included thinning of proximal and distal dendrites, decreased dendritic process diameter at branch points, and reduced dendritic arbor complexity. Changes in axonal diameter occurred later, followed by smaller soma size. While animal models have provided many insights into the timing of RGC glaucomatous degeneration, human studies are much more rare given the difficulty in obtaining tissue and limitations in technical aspects of tissue preparation. One study using postmortem tissue of four patients with amaurotic (no light perception) glaucoma sheds some insight into dendritic changes in human retina (Pavlidis et al. 2003). Two of the four patients had a long-standing history of congenital glaucoma, while the remaining two had absolute secondary glaucoma with amaurosis and uncontrolled IOP greater than 40 mmHg. Although the RGC labeling technique using the fluorescent carbocyanine dye DiI has its limitations, the authors estimated that less than 1 % of the stainable RGCs remained in the glaucomatous retinas. Therefore, the labeled cells that were examined were considered resistant to glaucoma. Nevertheless, these cells exhibited abnormal axonal beading and less dendritic complexity, despite the fact that the cell soma size was not appreciably shrunken. This data, although representative of only a subpopulation of human RGCs, demonstrates that dendritic shrinkage is indeed one of the earlier morphologic changes in RGC degeneration in glaucoma patients, consistent with the nonhuman primate model.

While dendrite remodeling is one of the earlier structural changes in response to elevated IOP, the susceptibility and extent of these changes vary among RGC types. For example, of particular interest to the investigators using the nonhuman primate model discussed above (Weber et al. 1998) was whether there would be differences between the midget and parasol cells, as there had been reports in the literature that elevated IOP differentially affected RGCs based on cell size (Johnson et al. 1989; Glovinsky et al. 1991, 1993; Vickers et al. 1995; Quigley 1998). Indeed, in this study, parasol cells showed a significant reduction in dendritic field size and axon diameter, suggesting a possible cell type-specific response dependent on dendritic field size. Consistent with the finding that cells with larger dendritic fields may be more susceptible in experimental glaucoma, Shou et al. used a feline model of chronic IOP elevation (via injection of endogenous ghost red blood cells into the anterior chamber of the eye) (Shou et al. 2003) to demonstrate that RGC loss and dendritic changes were more pronounced in the Y-type (large-field alpha cells) than the X-type (small-field beta cells) cells. The authors retrogradely labeled RGCs by bilateral injection of horseradish peroxidase into layers A and A1 of the lateral geniculate nucleus (LGN) and carefully analyzed the morphology of the labeled cells, including maximum dendrite radius, total length of dendrites, and number of dendritic bifurcations. All of the Y-type cells showed significantly greater shrinkage in dendritic structure parameters compared to the X-type cells.

However, more recently, multiple laboratories using various rodent models of glaucoma have found that there are RGC type-specific responses to IOP elevation that are not consistent with the hypothesis that only large-field RGCs are more susceptible. One of the first such studies used *Thy1-YFP* mice to investigate whether specific types of RGCs or RGCs from a specific location within the retina were differentially affected by elevated IOP in a laser-induced model of chronic ocular hypertension (Feng et al. 2013). In this study, the authors classified RGCs based on their dendrite ramification depth and studied a subset of them. ON RGC dendrites stratify in the inner half of the IPL and OFF dendrites ramify in the outer half of the IPL. ON and OFF RGCs respond best to increases and decreases in light intensity, respectively (Nelson et al. 1978). ON-OFF RGCs respond to both increment and decrement of light intensity and have bistratified dendritic arbors. Interestingly, this study revealed that there are type-specific and location-dependent RGC morphologic changes. Specifically, dendritic shrinkage began on the vertical axis with a subgroup of ON RGCs exhibiting smaller dendritic areas and shorter dendrites in the superior quadrant of glaucomatous retinas, in contrast to the nonhuman primate and cat studies, which showed that RGCs with larger dendritic areas are more susceptible to IOP elevation (Weber et al. 1998; Shou et al. 2003). Furthermore, some RGCs had abnormal morphology, including dendrites that appeared to loop and fold back and dendritic trees pruned down to the primary dendrites. Consistent with other studies, dendritic degeneration preceded cell soma shrinkage. Although this data certainly supports the idea of RGC-type-specific dendritic morphologic changes in response to elevated IOP, only ON and ON-OFF RGCs could be analyzed in detail because there were few OFF RGCs labeled.

Not only does IOP elevation affect RGCs differentially depending on RGC type, but also distinct RGC types respond at different time scales and to varying degrees. Della Santina and colleagues investigated the morphologic and functional changes in various RGC types using the microbead injection technique to chronically elevate IOP in *Thy1-YFP* mice (Della Santina et al. 2013). This investigation focused its functional assessment on ON and OFF RGCs, specifically sustained or transient, depending on whether the RGC responds with a sustained or transient depolarization, and then focused the morphologic analysis on large-field alpha-like RGCs. OFF-transient RGCs were the most vulnerable to elevated IOP, exhibiting a more rapid decline in structural and functional parameters compared to OFF-sustained, ON-transient, and ON-sustained RGCs, with functional changes preceding the structural alterations. Indeed, when comparing these RGC types, only the OFF-transient RGCs showed a reduction in dendritic arbor size and complexity 30 days after IOP elevation. The distribution of excitatory postsynaptic sites along the dendritic arbor was also examined, and, as expected, OFF-transient RGCs with reduced dendritic arbor size and complexity had concomitant loss of excitatory postsynaptic sites. Interestingly, the OFF- and ON-sustained RGCs that had functional changes but no dendritic shrinkage also had loss of excitatory postsynaptic sites. Therefore, even before dendritic structure changes, the earliest biomarker of damage to RGCs is likely a loss of synapses.

Building on the hypothesis that dendritic alterations may be an early hallmark of RGC degeneration for specific RGC types, recent evidence suggests that the earliest dendritic remodeling events may occur in RGCs with dendrites that laminate in the OFF sublamina. El-Danaf and Huberman (2015) recently examined dendritic remodeling in genetically labeled RGCs using several different lines of mice after IOP elevation was induced by microbead injection. Furthermore, they examined these morphologic changes as early as 7 days after IOP elevation, highlighting the alterations that may be occurring in the earliest stages of disease. Consistent with the Della Santina study, El-Danaf and Huberman found that OFF-transient alpha RGCs (labeled in CB2-GFP mice) exhibited dendritic arbor changes, specifically decreased total dendritic length, albeit at a much earlier time point (7 days vs. 30 days after microbead injection). Furthermore, El-Danaf and Huberman noted that there was an asymmetric shrinkage of the dendritic arbors of these OFF-transient RGCs in a direction away from the temporal quadrant. Additional RGC types the authors examined using a genetically labeled line (Hoxd10-GFP) were the ON DSGCs (on direction-selective RGCs) and αON-OFF DSGCs. One week after IOP elevation, there was no apparent change in structural parameters in the ON DSGCs. In contrast, the αON-OFF DSGCs exhibited no change in total dendritic length but had laminar alterations; specifically, the dendritic length of ON dendrites increased while the dendritic length of OFF dendrites decreased. To explore whether these laminar specific changes are driven by the general location of dendritic ramification or are specific to the functional characteristics of the RGC, the authors also examined the intrinsically photosensitive RGCs, the M1 ipRGCs. M1 ipRGCs are functionally ON-type RGCs, but their dendrites stratify deep within the IPL in the OFF sublamina. Surprisingly, while there was no change in soma size or dendritic field area between control and microbead injected retinas, there was a decrease in dendritic complexity and length in the M1 ipRGCs of IOP-elevated retinas. Although examination of these 4 RGC types is an incomplete evaluation given the likely 30 plus RGC types in the mouse retina (Sanes and Masland 2015), this work suggests that RGCs that stratify primarily in the ON sublamina of the IPL are relatively resistant to IOP elevation, whereas RGCs with dendrites in the OFF sublamina exhibit significant remodeling and loss. A summary of what is currently known about the morphologic and functional alterations that occur in different mouse RGC types in response to IOP elevation is provided in Fig. 22.2.

While it is apparent that RGCs undergo dendrite remodeling in response to IOP elevation in a type-specific manner, it is also true that these differences vary depending on the experimental animal model used. For example, as discussed above, Shou et al. showed that dendritic perturbations and RGC loss were more dramatic in cells of the Y-type (large-field alpha cells) than the X-type (small-field beta cells) in a chronic IOP elevation model (Shou et al. 2003). However, previous work by the same group using a feline model of acute IOP elevation with higher IOPs than the chronic model demonstrated that Y-type cells were actually more

NA: Not available
NC: No change

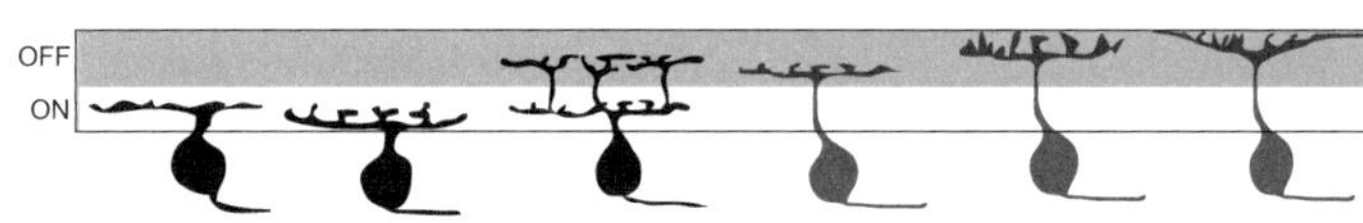

Measurements	DS-ON	sON	DS-ON/OFF	tOFF	sOFF	Melanopsin M1
Number of cells (1)	↓	NA	↓	↓↓	NA	NC
Dendritic arbor size (1,2,4)	NC (1)	NC (2) ↓ (4)	NC (4) ON ↑ OFF ↓ (1)	↓ (2)	NC (2)	NC (1) (branching ↓)
Excitatory synapses (2)	NA	↓	NA	↓	↓	NA
Light response (2,3)	NA	↓ (2,3)	NA	↓ (2)	↓ (2,3)	NA
Cation/Cl^- currents (3)	NA	↓	NA	NA	↓	NA
Receptive field size (2,4)	NA	NC (2) ↓ (4)	NA	↓ (2)	NC (2)	NA

Fig. 22.2 Structural and functional alterations of distinct types of RGCs. Summary of the structural and functional changes in distinct retinal ganglion cell (RGC) types across different models of experimental glaucoma. Please refer to the text for a detailed explanation of the similarities and differences observed across studies. Summarized from (*1*) El-Danaf and Huberman 2015; (*2*) Della Santina et al. 2013; (*3*) Pang et al. 2015; (*4*) Feng et al. 2013

resistant to short-term IOP elevation than X-type cells (Shou and Zhou 1989), suggesting that different models of injury can induce different outcomes depending on cell type. These differences may reflect acute versus chronic IOP elevation, as well as the magnitude of IOP elevation, with higher IOPs resulting in retinal ischemia. As another example, Kalesnykas et al. (2012) examined soma size, neurite outgrowth, and dendritic complexity in both an optic nerve crush and the microbead injection glaucoma model using *Thy1-YFP* mice. Interestingly, after optic nerve crush, there was decreased neurite outgrowth and dendritic complexity, but no change in RGC soma size. However, in the microbead injection glaucoma model, the soma and dendritic parameters examined (including soma size, dendritic length, and dendritic complexity) showed increases 3 weeks after IOP elevation, which is the opposite of what was seen in the optic nerve crush model. By 6 weeks after IOP elevation, these differences were no longer statistically significant, perhaps due to residual presence of only those RGCs more resilient to damage.

In addition to the animal model used, additional caveats to analyzing RGC-type-specific dendritic alterations across multiple studies include different techniques for imaging RGCs and the time points used for analysis. Thus far we have discussed mouse models in which glaucoma can be induced in the eyes of transgenic animals in which RGCs are already sparsely labeled by GFP expression. The *DBA/2J* mouse strain is a spontaneous mutant with elevated IOP that has been studied quite extensively (Savinova et al. 1998). It was previously shown using this mouse model of experimental glaucoma that axonal atrophy, dendritic loss and remodeling, and cell soma shrinkage preceded ganglion cell death (Jakobs et al. 2005). Also of importance is the fact that these investigators did not observe any type specificity in RGC death, although this may be due to the late stages of examination when only sectors of spared RGCs remained. In order to examine dendritic architecture in more detail, Williams et al. quantified dendritic field area, total dendritic length, and performed Sholl analysis to examine complexity (Williams et al. 2013). When dendritic parameters were measured in *DBA/2J.Thy1-YFP* mice at different stages of disease, specifically mice which have not yet shown signs of cell loss or signs of early damage (NOE), mice with moderate damage (MOD), and mice with severe damage (SEV) assessed from axon counts, there was no significant difference in dendritic field area, total dendritic length, or Sholl plots among these groups. However, characteristic dendritic shrinkage was observed when RGCs were labeled by carbocyanine dyes, thereby raising the concern that *Thy1* may be downregulated in rodent models of glaucoma and subsequently result in downregulation of YFP expression, as reported by Huang et al. (2006). Therefore, a second approach to assess RGC dendrite morphology involved diolistically labeling the RGCs using the carbocyanine dyes DiI and DiO. In contrast to the findings using the *DBA/2J.Thy1-YFP* mice, the labeled RGCs showed a significant decrease in dendritic parameters between DBA/2J retinas with no or early damage (NOE) and controls. Interestingly, there was no significant difference in dendritic parameters between the MOD and SEV groups and controls. The authors postulate several explanations, including sample bias, regional bias in diolistic RGC labeling, bias of labeling only in remaining healthy neurons, or recovery of dendritic architecture. Given the careful classification of the animal cohorts into NOE, MOD, and SEV groups based on optic nerve axon counts, this was the first study to definitively show that RGC dendritic shrinkage precedes significant axon degeneration.

To summarize, structural and functional changes of RGC dendrites appear to be influenced by the location and type of RGC (See Fig. 22.2). Feng et al. reported that a small subset of ON RGCs located in the dorsal retina of the *Thy1-YFP* mouse line are more likely to experience shrinkage of their dendritic arbor following IOP elevation (Feng et al. 2013). Recent progress in identification of RGC types more vulnerable to dendritic changes was made possible by the availability of transgenic mice labeling individual types of RGCs. El-Danaf and Huberman analyzed the dendritic complexity of multiple RGC types and discovered that RGCs with dendrites that laminate in the OFF sublamina display dendritic atrophy and rearrangements more readily than RGCs with dendrites that laminate in the ON sublamina (El-Danaf and Huberman 2015). Accordingly, multielectrode array recordings using the same mouse model of experimental glaucoma revealed that OFF-transient RGCs are the first type to experience functional receptive field size reduction (Della Santina et al. 2013). This "selective vulnerability" and differential involvement of RGC types in glaucoma may be critical in the early stages of the disease and thus important for the design of therapies before glaucoma reaches a late stage, when all types of RGCs are lost as observed in old DBA/2J mice (Jakobs et al. 2005).

22.4 Dendrite Remodeling in the Brain

While dendritic pruning in the retina appears to be a consistent feature across various animal models of experimental glaucoma, there is also evidence that dendritic atrophy is a hallmark of target neurons in the lateral geniculate nucleus (LGN) of the brain. Gupta et al. (2007) used MAP2 to label dendrites of LGN neurons in nonhuman primates with various degrees of optic nerve axon loss induced by argon laser scarification of the trabecular meshwork. Although the experimental design did not distinguish between LGN interneurons and relay neurons projecting to the visual cortex, both dendritic field area and dendritic complexity were significantly decreased in magnocellular layer 1 and parvocellular layer 6 compared to controls, even in cases where optic nerve axon loss was not detected. This study did not examine the retinas of these glaucomatous eyes, which would have been useful in determining the relationship between RGC dendritic degeneration in the retina with LGN dendritic atrophy. This group then went onto further detail the dendritic changes, specifically the LGN relay neurons that project to the primary visual cortex, using the same nonhuman primate model of glaucoma (Ly et al. 2011). Furthermore, they examined the effect of memantine, an NMDA glutamate open-channel blocker used as a neuroprotective agent to treat Alzheimer's disease, on dendrite morphologic parameters. As in the prior study, dendrite complexity and dendrite length were decreased in magnocellular layer 1 and parvocellular layer 6 compared to controls. In the memantine-treated group, reduction of dendrite complexity and length was significantly less than in the vehicle-treated group, suggesting that targeting the LGN neurons in glaucoma may be a worthwhile treatment goal.

More recently, the temporal relationship between RGC degeneration in the retina and in the brain has been examined using a rat model of ocular hypertension (Liu et al. 2014). In this study, dendritic morphologic parameters were examined in RGCs, superior colliculus (SC), and LGN cells over time in the same rat model of chronic glaucoma. As early as 1 week after IOP elevation, significant RGC dendritic degeneration had occurred. In some severely damaged RGCs, only the cell bodies and primary dendrites could be identified. The authors separated the RGCs based on morphologic types (RI, RII, and RIII) and concluded that RGC loss was not selective for larger cells since after 1 week the smaller RII cells had the most damage. Dendritic alterations in the SC and LGN were also examined in detail (five types of SC cells and two types of LGN cells). All types of SC and LGN cells exhibited dendritic degeneration, with more severe and earlier damage in the SC compared to the LGN. One explanation for this result may be that the majority of retinal projections in the rodent project to the SC rather than LGN. This study was unique in its examination of the temporal relationship of dendritic changes among these cells in different locations along the visual pathway, and showed that RGC dendritic shrinkage occurred earliest, around 1 week after IOP elevation, followed by neurons in the SC and then the LGN around 4 weeks.

22.5 Changes in Morphology vs. Function

The majority of functional data evaluating RGC status in glaucoma comes from mass recording of retinal activity, such as the pattern electroretinogram (PERG) (Porciatti 2015). Both in animal models and humans, this technique can be used to monitor disease progression reflected by progressively lower responses that parallel RGC death. Unfortunately, this approach does not provide information about function at the single-cell level, which is required for correlating changes in RGC morphology with alterations in function.

What is the relationship between the structural changes in dendrites and synapses and the functional changes observed in RGCs upon IOP elevation in experimental glaucoma models? Figure 22.3 delineates our current thinking with regard to the timing of morphologic and physiologic alterations that occur to RGCs in glaucoma. In early degeneration, one of the earliest morphologic changes is the loss of excitatory synapses, which precedes any structural changes to the dendrites (Fig. 22.3, second column). Functionally, there is decreased spontaneous firing of RGCs along with decreases in light responses and sensitivity. However, there is no change in receptive field size. Reduction of light sensitivity as well as cation currents for ON-sustained RGCs and chloride currents for OFF-sustained RGCs occur without alteration of bipolar cell light responses in multiple mouse models of experimental glaucoma (Pang et al. 2015). This recent evidence suggests that the

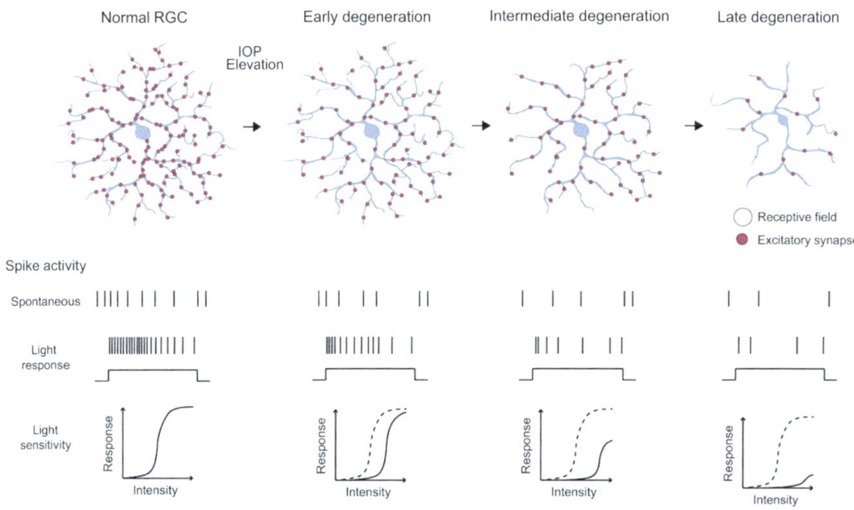

Fig. 22.3 Proposed sequence of morphologic and functional changes during RGC degeneration in experimental glaucoma. Early on in degeneration, excitatory synapses are eliminated, representing the first structural change of the retinal ganglion cell (RGC) in response to intraocular pressure (IOP) elevation. Coincident with the loss of synapses is diminished spontaneous activity and light responses. In intermediate stages of degeneration, the dendritic arbor shrinks, with further loss of function, most notably a decrease in the size of the functional receptive field. Finally, at late stages of degeneration, marked dendritic atrophy occurs, the cell soma shrinks, and the RGC eventually dies. Light sensitivity (*bottom row*) of degenerating RGCs (*solid line*) is progressively lower compared to control RGCs (*dashed line*). Similarly, maximal response amplitude decreases in parallel with the loss of excitatory synapses. The proposed sequence depicted here is summarized from Della Santina et al. (2013)

early impairment in RGC response sensitivity arises from reduced transmission between presynaptic partners and RGCs. Rod pathway-driven responses are reduced primarily due to a lack of input onto AII amacrine cells (Pang et al. 2015). Cone bipolar-driven responses may be altered due to a reduced number of excitatory synapses onto RGCs, as observed for ON- and OFF-sustained alpha RGCs early after IOP elevation, when their dendritic arbor morphology is unaltered (Della Santina et al. 2013).

At the intermediate stage of degeneration, there is further excitatory synapse loss, and the dendritic arbor begins to shrink, including decreased total dendritic length and dendritic complexity (Fig. 22.3, third column). Spontaneous activity and light responses further diminish. Indeed, Weber and Harman (Weber and Harman 2005) reported decreased responses in parasol RGCs to both spatial and temporal stimuli, in a nonhuman primate model of glaucoma where the same cells show dendritic arbor shrinkage. Although less responsive after glaucoma induction, parasol cells preserve normal phasic response to electrical stimulation (Weber and Harman 2005). Reduced response sensitivity has also been observed for alpha-like ganglion cells in the mouse by patch-clamp recording of their light responses (Pang et al. 2015).

Most notably, at this stage of degeneration, the functional receptive field has also shrunk. In the microbead injection mouse model, the reduction of dendritic arbor size coincides with a decrease in the size of the RGC's functional receptive field, a property that is evaluated by multielectrode array (MEA) recordings of RGC light responses (Feng et al. 2013; Della Santina et al. 2013).

Finally, at late stages of degeneration (Fig. 22.3, last column), there is considerable morphologic and functional loss of the RGC, including loss of entire dendritic branches and cell body shrinkage, along with loss of the light response and, finally, activation of apoptosis. While this is an overall framework of the morphologic and functional events in glaucomatous neurodegeneration, it remains to be determined how all of the different RGC types respond differentially to IOP elevation, both in terms of the tempo of perturbations and their specific characteristics.

While dendritic and synaptic changes impair RGC functionality, possibly up to the point where cell death mechanisms are initiated, a few recent reports suggest that mechanisms compensating for dendritic and synaptic degeneration may be engaged by RGCs. Interestingly, El-Danaf and Huberman found that for an individual ON-OFF RGC, dendrites stratifying in the OFF sublamina of the IPL degenerate, whereas dendrites in the ON sublamina expand their length and complexity (El-Danaf and Huberman 2015). This finding raises the possibility that a homeostatic mechanism in the RGC is evoked in response to the loss of OFF dendrites, possibly in an attempt by the cell to increase synaptic connectivity of the remaining dendrites and thus retain as much input as possible. Furthermore, Park et al. found that although there is a reduction in the total number of excitatory ribbon synapses in the IPL after IOP elevation, pre- and postsynaptic protein expression and the number of immature ribbon synapses were increased, suggesting that loss of dendrites may actually trigger synaptogenesis (Park et al. 2014).

22.6 Summary and Future Directions

Despite the increasing knowledge about dendritic degeneration in glaucoma, the cellular mechanisms underlying dendritic pruning and synapse loss are not yet well understood. It is becoming increasingly clear that not all RGCs degenerate with the same time course following ocular hypertension. In particular, OFF RGCs are reportedly more readily subjected to functional and morphologic alterations, whereas ON RGCs are more resilient. However, we recognize that until a complete RGC atlas and ability to identify individual types of RGCs is achieved, we will not fully understand which and why some RGC types are more vulnerable and others more resistant to damage in glaucoma. The cellular mechanisms driving this "selective vulnerability" remain to be elucidated. In parallel, a more in-depth characterization of the degeneration of each type of RGC will help reveal what RGC functional properties are rapidly perturbed and could be detected as specific vision defects.

Mechanisms driving the rearrangement of dendrites and synapses in glaucoma models are poorly understood. In vivo imaging of the degeneration process in action longitudinally should offer more detailed insight into how glaucoma leads to remodeling of synapses and dendritic architecture of the RGCs. Monitoring synaptic activity in various regions of the RGC dendritic tree could potentially reveal whether synapses that are preferentially eliminated belong to a particular input type, region of the dendritic tree, or display a characteristic level of activity. Unveiling the dynamic changes in degenerating RGC dendrites could also be an important step not only for identifying the initial effects of degeneration, but potentially also to reveal the critical steps needed for dendritic regeneration. Progress is already being made in the area of axonal regeneration, with certain RGC types demonstrated to have greater regenerative capacity (Duan et al. 2015).

From a therapy standpoint, it would also be of interest to identify compensatory mechanisms put in place by RGCs that are most resilient to degeneration. Such knowledge may provide key information to aid the design of strategies to slow down the disease. Therapeutic approaches will need to take into account the differential responses of RGC types to the insult, as well as capitalize on the differential timing of the steps leading to cell death in distinct RGC populations. Indeed, it will be necessary to tailor the intervention according to the disease state. Effective therapies administered at early stages of the disease could be aimed at preventing further RGC dysfunction and death, whereas late-stage interventions would require the repair of massively disrupted neural networks.

Dysfunction and death of neurons in both the outer and inner retina has been reported in experimental glaucoma models (Hernandez et al. 2009; Fernández-Sánchez et al. 2014), although these changes occur at later time points compared to damage and dysfunction of RGCs (Agudo-Barriuso et al. 2013). An in-depth analysis of the early changes in those neurons will reveal whether bipolar cells and amacrine cells that contact RGCs exhibit any compensatory behavior or remodel their connectivity over time. Moreover, current research and screening methods are particularly focused on the impact of degeneration on excitatory neurons, whereas alterations to inhibitory neurons remain unknown. Rearrangement in amacrine cell connections would suggest that inhibitory connections in the retina are also modified. Furthermore, changes in the outer retina involving photoreceptor synapses also need to be documented more thoroughly. Only when a comprehensive examination of how degeneration of the RGC, the last station of retinal processing, affects the entire retinal circuit will we attain a more complete understanding of how the CNS reacts and rearranges in the context of degenerative diseases in which the insult appears targeted to a single class of neuron.

References

Agudo-Barriuso M, Villegas-Pérez MP, de Imperial JM, Vidal-Sanz M (2013) Anatomical and functional damage in experimental glaucoma. Curr Opin Pharmacol 13:5–11. doi:10.1016/j.coph.2012.09.006

Collaborative Normal-Tension Glaucoma Study Group (1998) Comparison of glaucomatous progression between untreated patients with normal-tension glaucoma and patients with therapeutically reduced intraocular pressures. Am J Ophthalmol 126:487–497

Dacey DM, Peterson BB, Robinson FR, Gamlin PD (2003) Fireworks in the primate retina: in vitro photodynamics reveals diverse LGN-projecting ganglion cell types. Neuron 37:15–27

Della Santina L, Inman DM, Lupien CB et al (2013) Differential progression of structural and functional alterations in distinct retinal ganglion cell types in a mouse model of glaucoma. J Neurosci 33:17444–17457. doi:10.1523/JNEUROSCI.5461-12.2013

Dhande OS, Estevez ME, Quattrochi LE et al (2013) Genetic dissection of retinal inputs to brainstem nuclei controlling image stabilization. J Neurosci 33:17797–17813. doi:10.1523/JNEUROSCI.2778-13.2013

Duan X, Qiao M, Bei F et al (2015) Subtype-specific regeneration of retinal ganglion cells following axotomy: effects of osteopontin and mTOR signaling. Neuron 85:1244–1256. doi:10.1016/j.neuron.2015.02.017

El-Danaf RN, Huberman AD (2015) Characteristic patterns of dendritic remodeling in early-stage glaucoma: evidence from genetically identified retinal ganglion cell types. J Neurosci 35:2329–2343. doi:10.1523/JNEUROSCI.1419-14.2015

Feng G, Mellor RH, Bernstein M et al (2000) Imaging neuronal subsets in transgenic mice expressing multiple spectral variants of GFP. Neuron 28:41–51

Feng L, Zhao Y, Yoshida M et al (2013) Sustained ocular hypertension induces dendritic degeneration of mouse retinal ganglion cells that depends on cell type and location. Invest Ophthalmol Vis Sci 54:1106–1117. doi:10.1167/iovs.12-10791

Fernández-Sánchez L, De Sevilla Müller LP, Brecha NC, Cuenca N (2014) Loss of outer retinal neurons and circuitry alterations in the DBA/2J mouse. Invest Ophthalmol Vis Sci 55:6059–6072. doi:10.1167/iovs.14-14421

Glovinsky Y, Quigley HA, Dunkelberger GR (1991) Retinal ganglion cell loss is size dependent in experimental glaucoma. Invest Ophthalmol Vis Sci 32:484–491

Glovinsky Y, Quigley HA, Pease ME (1993) Foveal ganglion cell loss is size dependent in experimental glaucoma. Invest Ophthalmol Vis Sci 34:395–400

Gupta N, Ly T, Zhang Q et al (2007) Chronic ocular hypertension induces dendrite pathology in the lateral geniculate nucleus of the brain. Exp Eye Res 84:176–184. doi:10.1016/j.exer.2006.09.013

Hernandez M, Rodriguez FD, Sharma SC, Vecino E (2009) Immunohistochemical changes in rat retinas at various time periods of elevated intraocular pressure. Mol Vis 15:2696–2709. doi:10.1097/WNO.0b013e3182589589

Howell GR, Libby RT, Jakobs TC et al (2007) Axons of retinal ganglion cells are insulted in the optic nerve early in DBA/2J glaucoma. J Cell Biol 179:1523–1537. doi:10.1083/jcb.200706181

Huang W, Fileta J, Guo Y, Grosskreutz CL (2006) Downregulation of Thy1 in retinal ganglion cells in experimental glaucoma. Curr Eye Res 31:265–271. doi:10.1080/02713680500545671

Huberman AD, Manu M, Koch SM et al (2008) Architecture and activity-mediated refinement of axonal projections from a mosaic of genetically identified retinal ganglion cells. Neuron 59:425–438. doi:10.1016/j.neuron.2008.07.018

Huberman AD, Wei W, Elstrott J et al (2009) Genetic identification of an On-Off direction-selective retinal ganglion cell subtype reveals a layer-specific subcortical map of posterior motion. Neuron 62:327–334. doi:10.1016/j.neuron.2009.04.014

Jakobs TC, Libby RT, Ben Y et al (2005) Retinal ganglion cell degeneration is topological but not cell type specific in DBA/2J mice. J Cell Biol 171:313–325. doi:10.1083/jcb.200506099

Johnson MA, Drum BA, Quigley HA et al (1989) Pattern-evoked potentials and optic nerve fiber loss in monocular laser-induced glaucoma. Invest Ophthalmol Vis Sci 30:897–907

Kalesnykas G, Oglesby EN, Zack DJ et al (2012) Retinal ganglion cell morphology after optic nerve crush and experimental glaucoma. Invest Ophthalmol Vis Sci 53:3847–3857. doi:10.1167/iovs.12-9712

Koizumi A, Jakobs TC, Masland RH (2011) Regular mosaic of synaptic contacts among three retinal neurons. J Comp Neurol 519:341–357. doi:10.1002/cne.22522

Leung CK-S, Lindsey JD, Crowston JG et al (2008a) Longitudinal profile of retinal ganglion cell damage after optic nerve crush with blue-light confocal scanning laser ophthalmoscopy. Invest Ophthalmol Vis Sci 49:4898–4902. doi:10.1167/iovs.07-1447

Leung CKS, Lindsey JD, Crowston JG et al (2008b) In vivo imaging of murine retinal ganglion cells. J Neurosci Methods 168:475–478. doi:10.1016/j.jneumeth.2007.10.018

Leung CK-S, Weinreb RN, Li ZW et al (2011) Long-term in vivo imaging and measurement of dendritic shrinkage of retinal ganglion cells. Invest Ophthalmol Vis Sci 52:1539–1547. doi:10.1167/iovs.10-6012

Li Z-W, Liu S, Weinreb RN et al (2011) Tracking dendritic shrinkage of retinal ganglion cells after acute elevation of intraocular pressure. Invest Ophthalmol Vis Sci 52:7205–7212. doi:10.1167/iovs.10-6868

Liu M, Guo L, Salt TE, Cordeiro MF (2014) Dendritic changes in rat visual pathway associated with experimental ocular hypertension. Curr Eye Res 39:953–963. doi:10.3109/02713683.2014.884594

Ly T, Gupta N, Weinreb RN et al (2011) Dendrite plasticity in the lateral geniculate nucleus in primate glaucoma. Vision Res 51:243–250. doi:10.1016/j.visres.2010.08.003

Morgan JL, Schubert T, Wong ROL (2008) Developmental patterning of glutamatergic synapses onto retinal ganglion cells. Neural Dev 3:8. doi:10.1186/1749-8104-3-8

Morgan JL, Soto F, Wong ROL, Kerschensteiner D (2011) Development of cell type-specific connectivity patterns of converging excitatory axons in the retina. Neuron 71:1014–1021. doi:10.1016/j.neuron.2011.08.025

Morquette JB, Di Polo A (2008) Dendritic and synaptic protection: is it enough to save the retinal ganglion cell body and axon? J Neuroophthalmol 28:144–154. doi:10.1097/WNO.0b013e318177edf0

Nelson R, Famiglietti EV, Kolb H (1978) Intracellular staining reveals different levels of stratification for on- and off-center ganglion cells in cat retina. J Neurophysiol 41:472–483

Nickells RW, Howell GR, Soto I, John SWM (2012) Under pressure: cellular and molecular responses during glaucoma, a common neurodegeneration with axonopathy. Annu Rev Neurosci 35:153–179. doi:10.1146/annurev.neuro.051508.135728

Pang J-J, Frankfort BJ, Gross RL, Wu SM (2015) Elevated intraocular pressure decreases response sensitivity of inner retinal neurons in experimental glaucoma mice. Proc Natl Acad Sci U S A 112:2593–2598. doi:10.1073/pnas.1419921112

Park H-YL, Kim JH, Park CK (2014) Alterations of the synapse of the inner retinal layers after chronic intraocular pressure elevation in glaucoma animal model. Mol Brain 7:53. doi:10.1186/s13041-014-0053-2

Pavlidis M, Stupp T, Naskar R et al (2003) Retinal ganglion cells resistant to advanced glaucoma: a postmortem study of human retinas with the carbocyanine dye DiI. Invest Ophthalmol Vis Sci 44:5196–5205

Porciatti V (2015) Electrophysiological assessment of retinal ganglion cell function. Exp Eye Res. doi:10.1016/j.exer.2015.05.008

Quigley HA (1998) Selectivity in glaucoma injury. Arch Ophthalmol 116:396–398

Quigley HA (1999) Neuronal death in glaucoma. Prog Retin Eye Res 18:39–57

Quigley HA, Broman AT (2006) The number of people with glaucoma worldwide in 2010 and 2020. Br J Ophthalmol 90:262–267. doi:10.1136/bjo.2005.081224

Quigley HA, McKinnon SJ, Zack DJ et al (2000) Retrograde axonal transport of BDNF in retinal ganglion cells is blocked by acute IOP elevation in rats. Invest Ophthalmol Vis Sci 41:3460–3466

Raff MC, Whitmore AV, Finn JT (2002) Axonal self-destruction and neurodegeneration. Science 296:868–871. doi:10.1126/science.1068613

Raymond ID, Pool AL, Vila A, Brecha NC (2009) A Thy1-CFP DBA/2J mouse line with cyan fluorescent protein expression in retinal ganglion cells. Vis Neurosci 26:453–465. doi:10.1017/S095252380999023X

Sanes JR, Masland RH (2015) The types of retinal ganglion cells: current status and implications for neuronal classification. Annu Rev Neurosci 38:150421150146009. doi:10.1146/annurev-neuro-071714-034120

Savinova OV, Hawes NL, Chang B et al (1998) Essential iris atrophy, pigment dispersion, and glaucoma in DBA/2J mice. Invest Ophthalmol Vis Sci 39:951–962

Shou TD, Zhou YF (1989) Y cells in the cat retina are more tolerant than X cells to brief elevation of IOP. Invest Ophthalmol Vis Sci 30:2093–2098

Shou T, Liu J, Wang W et al (2003) Differential dendritic shrinkage of alpha and beta retinal ganglion cells in cats with chronic glaucoma. Invest Ophthalmol Vis Sci 44:3005–3010

Straznicky C, Vickers JC, Gábriel R, Costa M (1992) A neurofilament protein antibody selectively labels a large ganglion cell type in the human retina. Brain Res 582:123–128

Tanna H, Dubis AM, Ayub N et al (2010) Retinal imaging using commercial broadband optical coherence tomography. Br J Ophthalmol 94:372–376. doi:10.1136/bjo.2009.163501

Vickers JC, Schumer RA, Podos SM et al (1995) Differential vulnerability of neurochemically identified subpopulations of retinal neurons in a monkey model of glaucoma. Brain Res 680:23–35

Vidal-Sanz M, Villegas-Pérez MP, Bray GM, Aguayo AJ (1988) Persistent retrograde labeling of adult rat retinal ganglion cells with the carbocyanine dye diI. Exp Neurol 102:92–101

Walsh MK, Quigley HA (2008) In vivo time-lapse fluorescence imaging of individual retinal ganglion cells in mice. J Neurosci Methods 169:214–221. doi:10.1016/j.jneumeth.2007.11.029

Weber AJ, Harman CD (2005) Structure-function relations of parasol cells in the normal and glaucomatous primate retina. Invest Ophthalmol Vis Sci 46:3197–3207. doi:10.1167/iovs.04-0834

Weber AJ, Kaufman PL, Hubbard WC (1998) Morphology of single ganglion cells in the glaucomatous primate retina. Invest Ophthalmol Vis Sci 39:2304–2320

Werkmeister RM, Cherecheanu AP, Garhofer G et al (2013) Imaging of retinal ganglion cells in glaucoma: pitfalls and challenges. Cell Tissue Res 353:261–268. doi:10.1007/s00441-013-1600-3

Whitmore AV, Libby RT, John SWM (2005) Glaucoma: thinking in new ways-a rôle for autonomous axonal self-destruction and other compartmentalised processes? Prog Retin Eye Res 24:639–662. doi:10.1016/j.preteyeres.2005.04.004

Williams PA, Howell GR, Barbay JM et al (2013) Retinal ganglion cell dendritic atrophy in DBA/2J glaucoma. PLoS One 8:e72282. doi:10.1371/journal.pone.0072282

Wishart TM, Parson SH, Gillingwater TH (2006) Synaptic vulnerability in neurodegenerative disease. J Neuropathol Exp Neurol 65:733–739. doi:10.1097/01.jnen.0000228202.35163.c4

Index

K. Emoto et al. (eds.), *Dendrites*, DOI 10.1007/978-4-431-56050-0

T

MIX
Papier aus verantwortungsvollen Quellen
Paper from responsible sources
FSC® C105338

If you have any concerns about our products,
you can contact us on
ProductSafety@springernature.com

In case Publisher is established outside the EU,
the EU authorized representative is:
Springer Nature Customer Service Center GmbH
Europaplatz 3, 69115 Heidelberg, Germany

Printed by Libri Plureos GmbH
in Hamburg, Germany